DEFORMABLE BODIES
AND THEIR MATERIAL BEHAVIOR

DEFORMABLE BODIES AND THEIR MATERIAL BEHAVIOR

HENRY W. HASLACH, JR.
University of Maryland

RONALD W. ARMSTRONG
University of Maryland

WILEY

JOHN WILEY & SONS, INC.

Acquisitions Editor   *Joseph Hayton*
Marketing Manager   *Katherine Hepburn*
Senior Production Editor   *Valerie A. Vargas*
Senior Designer   *Madelyn Lesure*

This book was set in 10/12 Times Roman by Argosy and printed and bound by Hamilton Printing. The cover was printed by Lehigh Press.

This book is printed on acid-free paper.

Library of Congress Cataloging in Publication Data:
Haslach, Henry W.
    Deformable bodies and their material behavior / Henry W. Haslach, Jr., Ronald W. Armstrong.
        p. cm.
    ISBN 0-471-12578-4 (cloth)
        1. Deformations (Mechanics). 2. Materials-Mechanical properties. I. Armstrong, Ronald W. II. Title.
TA417.6.H36   2003
620.1'123–dc21
2003053845

PREFACE

Engineering design requires knowledge of the interaction between the mechanical loads on a device and the behavior of the materials in the components in order to control the response of the design. The integration of these topics in a mechanical engineering curriculum often awaits a capstone design course. The theme of this text is to describe and predict the mechanical behavior of a deformable body in terms of the relationships between its material properties and the forces and displacements applied to its boundary. The goal is to encourage more successful engineering design by joining techniques of mechanics and materials science.

This text introduces many of the tools that allow the designers of devices involving deformable bodies to avoid material functional failure. It includes stress and deformation analysis and also begins the study of material failure modes, such as permanent deformation, fracture, fatigue, and creep. These topics are discussed for structural metals, ceramics, composite materials, concrete, elastomers, polymers, wood, paper, and biological tissue in those cases that lend themselves to an explanation at the level of this text.

The micromechanical behavior of a material influences the mechanical response of the body. This interaction is often captured for analysis and mathematical modeling in a constitutive, or stress-strain, relation for the material, whether it be a metal, ceramic, polymer, or composite. Students are shown that not all materials have a linear stress-strain relation. Some experimental techniques are summarized, as well as some experimental results from materials science that influence design. Experiments usually must be represented by mathematical models to be useful in design. For example, in this text, the detailed steps in techniques such as heat treatments are left for a materials science course, but the mechanical consequences of some results of heat treatments, such as grain size, are modeled mathematically. Another aim of this text is to develop an appreciation for the approximations made in producing the mathematical models intended to predict mechanical response. Historical background indicates why definitions were made and shows that the models were invented by real people for practical reasons. Students then may have the courage to postulate their own models at some time in their engineering careers. The historically important research papers in the literature are often listed in the references at the end of the appropriate chapter, along with more recent sources providing additional details on the subject.

Stress, strain, and the stress-strain relations, with their material foundations, for most engineering materials are defined in the first three chapters to produce the 15 equations of linear elasticity in three dimensions and to lay the foundations for the study of nonlinear materials. The stress and strain tensors in Cartesian coordinates are defined as easily computable matrices; this description avoids the excessively complicated, geometric, three-dimensional Mohr's circle construction to obtain the states of stress or strain on arbitrary planes in a three-dimensional element. However, the two-dimensional Mohr's circle is employed in a number of places. Chapter 3 introduces stress-strain relations, the constitutive models, in order of complexity by first discussing linear elastic materials, such as metals, ceramics, and fibered composites. The idea of strain energy is introduced for later use in nonlinear elasticity and energy methods. Rubber and biological tissue are given as examples of nonlinear elastic materials. Simple time-dependent stress-strain relations for polymers are introduced as are the linearly orthotropic and other nonlinear models for the natural composites, wood and paper.

In the applications of linear elasticity to structures, the connection to the students' earlier work in stress and strain analysis is made by emphasizing that strength of materials is an approximation to linear elasticity, resulting from geometric assumptions such as plane sections remain planar. The Airy and Prandtl stress functions of elasticity are developed and applied. The polar equations of elasticity are used for thick-walled cylinders and for other structures such as curved beams, which are also analyzed by the classical advanced strength of materials technique, for comparison. Complex linear elastic structures and those made of nonlinear elastic materials are analyzed by energy methods including the classical Castigliano theorems, the principle of stationary potential energy, the Rayleigh-Ritz method, and the Euler-Lagrange equation.

Wherever possible, after Chapter 7, energy methods are used to determine the equilibrium configuration of a loaded deformable body rather than force equilibrium because this technique allows a study of the stability of the equilibrium state and is more common in current engineering research. Energy methods are used to develop the plate equations, lay the foundation for a study of the finite element method (FEM) and for the study of nonlinear elastic materials, derive the Griffith fracture theory, and determine elastic stability in buckling structures.

Today, the stress analysis of engineering systems often relies more on numerical techniques, such as the FEM, than on the simple stress estimates taught in traditional advanced strength of materials courses. But good reasons remain for learning the advanced strength of materials and the elasticity techniques of stress and deformation analysis. First, FEM is an approximation that can only be carried out if numerical values are chosen for all parameters of the problem. The elasticity and the advanced strength of materials techniques, in contrast, yield closed-form solutions in terms of the design parameters. Such solutions can be used to select the parameters to optimize the design by using techniques such as calculus or simple inequalities. However, optimization techniques are not discussed. Second, the engineer should develop a sense of the likely stress distribution throughout a body so that the engineer can judge when the FEM has given an invalid result, due to numerical analysis quirks or because the user incorrectly formulated the problem. An intuition about stress distributions developed from experience with the advanced strength of materials and the elasticity techniques also allows an engineer to predict the likely failure points in a design. But most importantly, if the FEM is used in design to decide the specified dimensions of a member, the engineer must proceed by trial and error, a time-consuming and therefore expensive process. Dimensions must first be guessed and a material chosen, then the numerical analysis carried out, and finally the results must be evaluated. The process is repeated until a satisfactory design is reached. This procedure does not guarantee that the design is in any way optimal. Alternatively, a closed-form expression for the stress or deformation available from elasticity, from advanced strength of materials, or from energy methods, and so forth, can be used to quickly compute dimensions that will ensure that the member will carry out its function, even if the resulting design is not optimal. The dimensions from the closed-form solutions can provide a starting point for an FEM analysis, say in conjunction with a computer-aided design (CAD) drawing program.

Such a starting point is likely to be close to the final design and thus cut down on the number of trial and error iterations in FEM and on guessing.

To use the finite element method as anything other than a "black box," however, an engineer must understand linear elasticity. A short description of a displacement-based FEM is given in this context. Both a change in the degree of the polynomial approximation to element displacements and an increase in mesh size are considered to improve the accuracy of the FEM approximation. This chapter is intended to support the use of a commercial FEM package to perform the stress and displacement analysis of an elastically loaded body.

The final segment of the text describes failure mechanisms, such as plasticity, fracture, fatigue, creep, and buckling, and builds on appropriate examples developed in earlier sections. In the first four cases, the material mechanisms contributing to the phenomenological behavior are summarized. A primary goal, in all five cases, is to present many of the mathematical models available to predict each type of failure in a design. Crystal dislocations in face-centered-cubic and body-centered-cubic metals are related to the material plastic stress-strain behavior and to hardness behavior. The measures making up the traditional hardness tests are developed. The classical biaxial failure theories lead to a discussion of three-dimensional plastic yielding based on the maximum shear stress and the distortional strain energy theories. The ideas of isotropic and kinematic hardening are explained. A short introduction is given to viscoplasticity and the use of internal state variables in modeling. Brittle fracture mechanics is fully developed for crack growth and fracture, and ductile fracture analysis for metals is introduced. Cyclic stress-strain behavior and resulting fatigue failure is modeled both from a classical life analysis and from a crack propagation analysis. Both stress- and strain-based models are presented. Creep is described in terms of material mechanisms as well as by mathematical models. Energy methods are emphasized in the description of buckling, which includes nonlinear models to permit the prediction of the buckling deformation and stability as well as the critical buckling load.

We made an effort to provide a well-organized presentation in order to give a clear outline of the interrelation of these topics. We believe that it is pedagogically important that students understand the reasons why the techniques presented work because this makes learning easier. For example, the solutions to particular differential equations arising in elasticity problems are developed in sufficient detail for the students to follow rather than merely giving a reference to a differential equations text.

The text aims to develop the students' sense of how stress and strain vary throughout a loaded body and to introduce techniques to compute the variation. Strength of materials and elasticity are just mathematical models for the behavior of a loaded body, and these models must be validated by experiment.

In the past, many curricula have isolated advanced stress analysis into courses such as advanced strength of materials and applied elasticity and so have restricted the scope to elastic loadings of linearly elastic materials. Such a division leaves the study of material failure modes, such as permanent deformation, fracture, fatigue, and creep to separate courses that not all students take. Therefore some students fail to make the connection of the stress and deformation analysis to material behavior for all types of materials.

The text lays the foundations for the design of mechanical devices. It can be used as a precursor to the capstone design course or can serve as an introduction to more advanced speciality courses in elasticity, the FEM, material microstructures, continuum mechanics, and detailed studies of plasticity, creep, fatigue, and fracture analysis. Several different types of courses can be based on this text. At the University of Maryland–College Park, this text is used in a two-semester sequence. The first semester focuses on stress and deformation analysis in Chapters 1–7. The second semester focuses on the failure modes of materials in Chapters 9–12. This sequence leads naturally into the capstone design course and for some students also lays the foundation for more advanced graduate level courses. A stress analysis course appropriate for aerospace engineering can be devised with more emphasis on Chapter 6 on thin-walled members. A course to establish the tools needed for a finite element methods course would include most of Chapters 1–8. A course to replace the traditional advanced strength of materials or applied elasticity one-semester courses on linearly elastic materials can be constructed using parts of Chapters 1–7 and 9 and omitting the nonlinear elastic parts of Chapter 3. The text offers many other opportunities for an instructor to tailor a one-semester course. Enough topics are presented for instructors to select those appropriate to the goals of their particular course and of interest to their students.

This course assumes material from the standard introductory courses on the calculus of several variables, vector analysis, matrix operations, elementary strength of materials, and elementary techniques for the solution of differential equations. The expected prerequisites are elementary strength of materials and introduction to materials science, as well as the full elementary calculus sequence with an introduction to differential equations.

The authors wish especially to thank Prof. Tony Farquhar of the University of Maryland Baltimore County (UMBC) who also taught from a manuscript version of this text and made valuable suggestions.

Henry W. Haslach, Jr.
Ronald W. Armstrong

CONTENTS

INTRODUCTION   **xiii**

CHAPTER 1   *THE STRESS TENSOR*   **1**

1.1   Definition of Stress   **2**
1.2   The Stress Tensor   **7**
1.3   The Stress Tensor is a Linear Transformation—The Cauchy Tetrahedron   **9**
1.4   Variation of the Stress Tensor from Point to Point in a Body in Equilibrium   **13**
1.5   Coordinate Changes and the Stress Tensor   **16**
1.6   Principal Stresses   **18**
1.7   Octahedral Stresses   **25**
1.8   Problems   **27**

CHAPTER 2   *STRAIN*   **31**

2.1   Definition of Strain   **31**
2.2   The Strain Tensor   **37**
2.3   Mohr's Circle for Strain   **39**
2.4   Lagrangian and Eulerian Coordinate Systems   **40**
2.5   Equations of Compatibility   **41**
2.6   Electronic Resistance Strain Gages   **42**
2.7   Other Strain Measurement Techniques   **45**
2.8   Problems   **46**

CHAPTER 3   *STRESS-STRAIN RELATIONS*   **49**

3.1   Uniaxial Stress-Strain Tests of Metals   **49**
    3.1.1   Plastic Strain   **52**
    3.1.2   Poisson's Ratio   **54**
3.2   Lattice Structure of Metals   **55**
    3.2.1   Bonding   **55**
    3.2.2   Formation and Structure of Crystals   **56**
    3.2.3   Atomic Bond Energy   **60**
    3.2.4   Dislocation Theory for Metals   **64**
    3.2.5   Ductile and Brittle Metals   **67**
3.3   Generalized Hooke's Law   **67**
3.4   Ceramics   **73**
    3.4.1   Typical Ceramic Lattice Structures   **74**
    3.4.2   Bending Tests for Strength   **77**
    3.4.3   Glass   **79**
3.5   Fibered Composites   **81**
3.6   Strain Energy   **83**
3.7   Concrete   **90**
3.8   Rubber   **93**
    3.8.1   Properties of Natural Rubber   **93**
    3.8.2   Kinetic Theory   **95**
    3.8.3   Vulcanization   **95**
    3.8.4   Synthetic Rubber   **96**
    3.8.5   Stretch and Models for Rubber   **97**
3.9   Incompressible Biological Tissues   **100**
3.10   Polymers and Viscoelastic Behavior   **101**
    3.10.1   Mechanics of Polymers   **103**
    3.10.2   Methods of Building Polymers   **105**
    3.10.3   Moisture Adsorption   **106**
    3.10.4   Modeling Polymer Behavior   **107**
    3.10.5   Dynamic Loading of Polymers   **111**
3.11   Wood   **112**
3.12   Paper   **114**
3.13   Problems   **117**

CHAPTER 4   *LINEAR ELASTICITY*   **122**

4.1   Boundary Conditions   **122**
4.2   Two-Dimensional Problems   **124**
4.3   Airy Stress Function   **125**
4.4   Techniques for Solving the Equations of Elasticity   **129**
4.5   Linear Thermoelasticity   **134**
4.6   Polar Coordinates   **137**
    4.6.1   Thick-walled Cylinders   **140**
    4.6.2   The Airy Stress Function in Polar Coordinates   **147**
4.7   Plate with a Hole   **149**
4.8   Stress Concentrations   **153**
    4.8.1   Stress Flow Model   **154**
    4.8.2   Stress and Strain Concentration Factors   **155**
    4.8.3   Experimental Determination of Stress Concentration Factors   **156**
4.9   Contact Stresses   **159**
    4.9.1   Geometric Considerations   **159**
    4.9.2   Displacement Analysis   **163**
4.10   Problems   **167**

CHAPTER 5   *APPLICATIONS OF LINEAR ELASTICITY AND ITS APPROXIMATIONS*   **173**

5.1   Torsion of Prismatic Rods   **173**
5.2   Bending of Curved Beams   **183**
    5.2.1   Circular Beam in Pure Bending   **183**
    5.2.2   Winkler-Bach Theory for Curved Beams   **187**
5.3   Beams on an Elastic Foundation   **193**
    5.3.1   A Beam Differential Equation   **194**
    5.3.2   Semi-infinite Beams   **199**
5.4   Problems   **199**

CHAPTER 6   *THIN-WALLED MEMBERS*   **203**

6.1   Principal Axes and Product Area Moments of Inertia in a Beam Cross Section   **203**
6.2   Unsymmetrical Bending of Beams   **205**
6.3   Bending of Thin-walled Beams   **209**
   6.3.1   Shear Center   **210**
6.4   Torsion of Thin-Walled Prismatic Members   **219**
   6.4.1   Open Section Rods   **222**
6.5   Problems   **225**

CHAPTER 7   *ENERGY METHODS*   **227**

7.1   Work, Strain Energy, and Complementary Energy   **228**
7.2   Castigliano's Theorems   **231**
   7.2.1   Method of Fictitious Loads   **236**
   7.2.2   Statically Indeterminate Problems   **238**
   7.2.3   Principle of Complementary Energy for Nonlinear Elastic Materials   **240**
7.3   Principle of Virtual Work   **242**
7.4   The Principle of Stationary Potential Energy   **244**
7.5   Internally Pressurized Spherical Membranes   **248**
7.6   Infinite Degree of Freedom Systems   **252**
7.7   The Rayleigh-Ritz Approximation Technique   **253**
7.8   Calculus of Variations and the Euler-Lagrange Equation for an Integral Function   **258**
7.9   Transversely Loaded Plates   **265**
   7.9.1   Small Bending Deflections of Rectangular Plates   **266**
   7.9.2   Bending of Circular Plates   **271**
   7.9.3   Orthotropic Plates   **274**
7.10   The Solution of Differential Equations Using Energy Methods   **277**
7.11   Problems   **278**

CHAPTER 8   *INTRODUCTION TO THE FINITE ELEMENT METHOD FOR NUMERICAL APPROXIMATIONS*   **283**

8.1   The Fundamental Equations for the FEM   **285**
8.2   Combined Equations for a Single Element   **287**
8.3   Assembly of the Finite Element Equations   **291**
8.4   A Beam Element   **294**
8.5   Constant Strain Triangle   **296**
   8.5.1   Area Coordinates   **301**
8.6   Isoparametric Elements   **302**
8.7   Error Estimation   **303**
8.8   Finite Element Method Projects   **306**
   8.8.1   Project 1: Thick-Walled Cylinders   **307**
   8.8.2   Project 2: Thick Members   **308**
   8.8.3   Project 3: Beams with and without a Notch   **310**
   8.8.4   Project 4: Beam with Fillet   **312**
8.9   Problems   **313**

CHAPTER 9   *PLASTICITY*   **316**

9.1   Introduction   **316**
   9.1.1   Product versus Processing   **316**
   9.1.2   Engineering Fundamentals   **318**
9.2   Plastic Behavior   **318**
   9.2.1   Constant Volume   **319**
   9.2.2   Strain Hardening   **319**
   9.2.3   Cyclic Phenomena   **321**
9.3   Material Models for the Flow of Metals   **323**
   9.3.1   Techniques for Strengthening a Metal   **331**
9.4   Modeling Uniaxial Plastic Behavior   **332**
   9.4.1   Perfect Plasticity   **333**
   9.4.2   Hardening Models   **337**
9.5   Modeling Multiaxial Plastic Behavior   **339**
   9.5.1   The Yield Surface   **340**
9.6   Threshold Criteria for Biaxial Loading   **341**
   9.6.1   Maximum Normal Stress Criterion   **342**
   9.6.2   Maximum Shear Stress Criterion   **343**
   9.6.3   Maximum Principal Normal Strain Criterion   **344**
   9.6.4   Maximum Distortion Energy Criterion   **345**
   9.6.5   Octahedral Shear Stress Criterion   **347**
   9.6.6   Comparison of the Various Failure Criteria   **347**
   9.6.7   Applications of the Biaxial Threshold Criteria in Design   **349**
9.7   Threshold Criteria for Triaxial Loading   **351**
   9.7.1   von Mises Criterion   **352**
   9.7.2   Tresca Criterion   **356**
   9.7.3   Isotropic and Kinematic Hardening Yield Surfaces   **357**
9.8   Three-Dimensional Plasticity Models   **358**
   9.8.1   Perfect Plasticity   **358**
   9.8.2   Internal State Variables and Dissipation   **361**
   9.8.3   Isotropic Hardening   **363**
   9.8.4   Kinematic Hardening   **366**
9.9   Hardness Tests   **367**
9.10   Viscoplasticity   **375**
9.11   Problems   **377**

CHAPTER 10   *FRACTURE*   **383**

10.1   Introduction   **383**
10.2   Energy and the Griffith Criterion for Brittle Fracture   **384**
   10.2.1   Stability   **390**
   10.2.2   Appendix: Plate with an Elliptical Hole   **391**
10.3   Stress Criteria   **394**
10.4   Non-brittle Fracture   **405**
   10.4.1   Plastic Zone at the Crack Tip   **406**
   10.4.2   The Crack Tip Opening Displacement   **409**
   10.4.3   The Parameter, J, for Nonlinear Elastic Materials   **412**
   10.4.4   Crack Extension Resistance   **413**

10.5 Dislocation Behavior Near a Crack Tip—Brittle and Ductile Materials 413
10.6 Crack Initiation 416
10.7 Design 417
10.8 Tearing of Rubber 419
10.9 Polymer Fracture 420
10.10 Paper Fracture 423
10.11 Flaw Detection 424
10.12 Problems 425

CHAPTER 11 FATIGUE 428
11.1 Introduction 428
11.2 Types of Fatigue Loading 428
11.3 Design Against Fatigue Failure 429
11.4 Failure Criteria for Uniaxial Metal Fatigue 432
    11.4.1 Helical Compression Springs 433
11.5 Multiaxial Failure Criteria for Metal Fatigue 435
11.6 Micromechanics of Fatigue 437
11.7 Fatigue Fracture 438
11.8 Fatigue Fracture Models 439
11.9 Fatigue of Ceramics 445
11.10 Fatigue of Polymers 447
    11.10.1 Fatigue of Rubber 449
    11.10.2 Fatigue of Paper 449
11.11 Strain Analysis of Fatigue 450
11.12 Problems 453

CHAPTER 12 CREEP 457
12.1 Introduction 457
12.2 Creep Tests 458
12.3 Creep Mechanisms 460
12.4 Mathematical Models for Uniaxial Creep 465
    12.4.1 Creep Life Estimates 466
    12.4.2 Secondary Region Strain Rate Models 467
    12.4.3 Strain Models 469
12.5 Multiaxial Creep 471
12.6 Relaxation 472
12.7 Mechanical Models for Polymer Creep and Relaxation 473
12.8 Moisture Accelerated Creep of Hydrophilic Materials 475
    12.8.1 Paper Creep In Variable Relative Humidity 475
12.9 Creep Fracture 476
12.10 Design for Creep 477
12.11 Problems 479

CHAPTER 13 STATIC STABILITY 481
13.1 Stability of Equilibrium States of Conservative Systems 482
13.2 Stability of Equilibrium States of Conservative Systems Subject to a Variable Design Parameter 483
13.3 Linear Elastic Columns 489
    13.3.1 The Critical Buckling Load for Various Boundary Conditions 491
    13.3.2 Design of Columns 493
    13.3.3 The Critical Buckling Load by the Rayleigh-Ritz Method 494
    13.3.4 The Exact Solution by Elliptical Integrals 497
13.4 Perturbation Methods 499
    13.4.1 Perturbation Technique 501
    13.4.2 The Energy Function 504
13.5 Buckling of Columns That Are Not Linear Elastic 504
13.6 Buckling of Thin-Walled Circular Rings and Tubes 507
13.7 Imperfections 510
    13.7.1 Initially Curved Columns 510
    13.7.2 The Eccentrically Loaded Column 511
    13.7.3 Imperfections Modeled in the Potential Energy 512
13.8 Compound Buckling 514
13.9 Problems 518

APPENDIX CURVATURE 521
A.1 The Curvature of a Curve 521
    A.1.1 Curves Lying in Two-Dimensional Space 521
    A.1.2 Curves Lying in Three-Dimensional Space 522
    A.1.3 Parametric Form 522
A.2 Surfaces 523
    A.2.1 The First Fundamental Form and the Speed Along a Path on a Surface 524
    A.2.2 The Curvature of a Surface 524

INDEX 527

INTRODUCTION

Design of useful devices requires knowledge of the stresses and/or strains in each component and understanding of the material response to loads and environmental conditions. The atoms which compose solids are arranged into complex substructures that influence the bulk material response. Such substructures include the lattice defect distribution and grains in metals or ceramics as well as the long-chain molecules formed into crystalline regions in polymers. Efficient design requires that the interaction between mechanics and materials science be carefully considered. While metals and wood are traditional engineering materials, today other materials such as ceramics, polymers, composites, and biological tissues are of concern to engineers. Not all of the various known responses of solid materials have been fully explained; much research remains to be done. For example, environmental conditions such as high temperatures for metals or high relative humidity for cellulosic materials and concrete can increase creep, which is time-dependent deformation occurring under a constant load or stress. Many loading cycles can cause fatigue failure of a solid member so that it fractures even though appearing to be only elastically loaded. Engineers need to describe the force-displacement relationships in deformable solids to take advantage of, or to modify, the properties of the solids in service.

Mechanics is the study of the effect of actions such as forces, heat, etc. on bodies. Each subfield either examines a different material phase, solid or fluid, or makes different simplifying assumptions about the nature of the bodies. Statics is the study of forces on stationary, rigid, solid bodies. The typical problem is to determine unknown forces on a rigid body in equilibrium. In elementary strength of materials, the rigid body restriction of statics is removed to study how solid bodies change shape under the action of forces. The deformation depends on the material of the body, as well as its original shape and the force applied. In statics, on the other hand, the body material is essentially irrelevant since the body is assumed rigid. Introductory courses on strength of materials are usually restricted to stationary bodies that change shape. Statics is applied to approximately determine all forces on the body. Once all the forces are known, strength of materials is used to predict the change in shape of the body and to determine whether or not a material part or the total structure will fail. Strength of materials considers three types of loading: axial, torsion, and bending, and their combinations.

Elasticity is the study of the change in shape and internal forces induced in an elastically loaded solid. A solid is elastically loaded if it returns to its original shape when the loads are removed. Every physical body changes shape under an applied force, but in some situations the change is very small. In this case, which includes elastically loaded metals and ceramics, the linear theory of elasticity may be used to describe the solid response. Strength of materials is an approximation to the solutions of linear elasticity theory because it makes simplifying geometric assumptions such as plane sections remain plane in beam bending. Unfortunately, the strength of materials results are not always consistent with equilibrium. To produce results that do satisfy equilibrium for linear elastic and static bodies, the theory of elasticity must be used. Furthermore, elasticity is applicable to an arbitrarily shaped body under any type of load.

Elastomeric materials, such as rubber and some biological tissues, undergo large deformations under loads. Nonlinear elasticity is required for these solids. Usually, such materials are described by a strain energy from which the response is deduced by a technique called energy methods. Polymer response is rate-dependent and so is described by viscoelasticity. Metals which are loaded by a sufficiently large load, that beyond their elastic limit, are said to behave plastically. Because plasticity involves permanent deformation, a description different from elasticity is needed. Nonequilibrium rate-dependent plastic loadings, such as metal creep, are described by viscoplasticity. In viscoelasticity, the equilibrium states to which a material relaxes under constant loads do not change with load or time; however in viscoplasticity, these equilibrium states do change with load.

A design engineer must know, for example, how the average grain size of a metal piece affects its elastic limit, perhaps to specify a heat treatment. The factors influencing fatigue crack formation and growth must be known to prevent fracture. To choose a material, the engineer must know the differences between the behavior of metals, polymers, and ceramics under loads. The elastic springback after plastically forming a metal must be determined to ensure the correct fit of parts. The original shape of an elastic self-expanding piston ring must be computed for internal combustion engine design. In many cases, even in a linear elastic body, the stress and strain distributions are so complex that they must be numerically approximated from a mathematical model to help decide whether or not the body is safely loaded. An engineer must have methods of predicting under what conditions a body will fail, or else engineering design reduces to an expensive and time-consuming trial-and-error process.

An engineer creates useful products, processes, and systems through the application of scientific theories, and so an engineer must fully understand the scientific ideas involved. In addition, engineers must develop practical experience and a sense of the safety, cost, and environmental consequences of their work. The process of planning a component, device, system, etc. for human use is called design. An engineer can be called on to design a completely new concept and put it into production, to modify an old design, or to replace an old product by a new one that performs the same function.

The design process requires the engineer to predict the success of the chosen solution to the problem. An engineer could test the design of a bridge, for example, by simply building the bridge. If the bridge collapses when the first truck drives onto it, the engineer learns that the design is a failure, at the expense of building the bridge and possibly of human injury or death. Then a new bridge would be built and so on. Such a trial-and-error technique is too dangerous and expensive. An engineer must be able to predict whether or not the bridge will be strong enough before it is built. The more a design task can be carried out before building prototypes, the more rapid and the less expensive the process. The prediction still needs to be tested in practice; the engineer might have made a mistake or overlooked an important part of the problem, or the scientific theory underlying the application may not be advanced enough to fully explain the behavior.

Prediction is based on mathematical models, such as elasticity, plasticity, viscoelasticity, and viscoplasticity, of the material and structural behavior. These models are the foundation of engineering theory and the core of engineering practice. An experienced engineer has learned the proper approximations to make in choosing a mathematical model to represent the physical situation faced.

A mathematical model is a mathematical expression, or set of expressions, which describes a physical situation. A well-known model is the ideal gas law from chemistry,

$$PV = NRT,$$

which describes the relationship between the pressure, P, and the volume, V, and the temperature, T, of N moles of an ideal gas. R is a universal constant. An ideal gas is, by definition, one whose properties obey this model. For example, if the volume V is known for a particular ideal gas, then the pressure can be predicted as the temperature changes. Without this model, an engineer would have to do experiments at each temperature to measure the corresponding pressure.

Design requires the choice of the materials from which the device is to be made. Robert Hooke (1635–1703) initiated the study of the elastic properties of materials. Hooke was a professor of geometry who had previously been curator of experiments for the Royal Society of London, a job he obtained through the efforts of Robert Boyle (of Boyle's law in chemistry) with whom he developed the air pump. He hung weights on the ends of wires made of various materials to see how the change in length of the rod depended on the weight and on the material from which the rod was made. He reported in 1678 that if the wires were elastically loaded, in the sense that they returned to their original length when the weight was removed, the stretching "will always bear the same proportions one to the other that the weights do that made them." He had to test quite long wires, from 20 to 40 feet long, in order to get elongations measurable with the instruments of the day. He similarly tested helical springs and found the same results. His name is now commonly attached to the deformation-force relation for elastic springs. He proposed that this result explains how a watch spring enables time to be kept, how to calculate the strength of a bow, and the uniform sound made by a vibrating string, among other applications. The proportionality between the force and deformation depends on the length and cross section, as well as the material, of the wire. Therefore it does not completely describe the behavior of the material irrespective of its shape. A modification of Hooke's ideas became the foundation for the study of linearly elastic bodies.

Hooke's work produced a mathematical model that relates the deformation of an elastic spring to the force acting on the spring. One end of the spring is mounted to a wall. A force, F, is applied at the other end and parallel to the spring. The amount the spring stretches from its undeformed length is denoted by x. Then Hooke's law, a mathematical model for an elastic spring, states

$$F = kx,$$

where k is a constant depending on the shape and material of the spring. The equation is very useful, for example, in designing spring scales by which the weight of an object is determined from the deformation of a spring.

A mathematical model is invented by seeking patterns in experimental data. Once the pattern is discovered, it is represented by an equation. The equation is tested by further experiment to verify that it makes correct predictions. Once the model is validated, the user no longer needs to reproduce lengthy and expensive experiments; the equation is used instead. The model saves both time and money, and it can help explain the nature of the physical processes involved. Research engineers devote their efforts to inventing such models.

A primary goal of engineers is to develop methods of predicting when structures subject to forces will fail, *i.e.* either break or change shape so much they cannot function. If a structure fails, most people would say that the forces on the structure were too large. An engineer must analyze exactly what this means and find a predictive tool to estimate the effect of the forces. This course develops mathematical models to predict the changes in shape of a solid body caused by the action of forces and environmental conditions. Other models for the relation between the material substructure and its behavior will also be described. To use these models, an engineer not only needs to know the equations, but also the assumptions used in deriving the model so that it is applied only in situations where it is valid.

EXAMPLE I.1

A valve shaft (cam shaft) on which cams are rotated to raise and lower the value stems for an internal combustion engine must be designed by choosing its dimensions and selecting the material from which it is to be made. An engineer has to predict how the shaft will behave under this load so that the shaft can be designed not to fail in its design lifetime. The engineer must consider, in the design process, all possible ways the shaft could fail by answering many questions. Will the shaft crack or fracture? Will it deflect too much so that it interferes with other parts? Will any deformation be elastic, so that when unloaded, the shaft can recover its shape, or will deformation be permanent? Could vibrations be induced which could cause fracture due to fatigue, much as a paper clip breaks after it is bent back and forth? Fatigue fracture might occur even if the shaft is repeatedly but elastically bent and unbent. If the shaft temperature is high enough, will the shaft continue to deform excessively even under a fixed load, due to creep? The dimensions and material of the shaft must be chosen to prevent all types of failure.

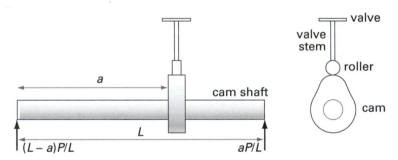

FIGURE I.1 A shaft with cam driving a valve.

Solution

Most engines have two valves for each cylinder and a corresponding cam for each valve. To illustrate a simple analytical process, only one cam is considered. The shaft is idealized as a rod with a constant circular cross section, and the cam near one end is replaced by a point load, P, transverse to the shaft so that elementary strength of materials results may be used. The load is assumed constant even though in an engine it would vary as the valve stem spring is compressed and released. A first step is to determine the internal forces induced by the load. The forces at a point are averaged over a plane, on which the forces act, through the point. The force per planar area is called the stress and exists with respect to any plane through any point. Fracture, for example, could occur on any of the infinite number of such planes, depending on the loads and the material.

As a first approximation, elementary strength of materials can give an idea of the stress distribution in the rod. For the purposes of this discussion, model the rod as simply supported at the bearings at each end. Then the reaction forces are $(L - a)P/L$ at the left and aP/L at the right, where a is the distance from the load to the left end and L is the length of the rod.

Put the origin at the center of the left end face. The x-axis is the longitudinal axis of the rod. The y-axis is parallel to the valve stem. Then the shear and moment functions are

$$V(x) = \frac{(L-a)P}{L} \quad \text{if} \quad 0 < x < a$$
$$\qquad - aP/L \quad \text{if} \quad a < x < L.$$

(I.1)

$$M(x) = \frac{(L-a)P}{L}x \quad \text{if} \quad 0 \le x \le a$$

$$\frac{(L-a)P}{L}x - P(x-a) = \frac{(L-x)aP}{L} \quad \text{if} \quad a \le x \le L,$$

(I.2)

where $V(x)$ is down and $M(x)$ is counterclockwise on the left face of a cut. Recall that, for bending, $dM(x)/dx = V(x)$. The bending stress at a point, (x, y, z) is $\sigma(x, y, z) = M(x)y/I(x)$, assuming that plane cross sections remain planar during bending. The cross-sectional area is a circle so that the area moment of inertia of the cross-sectional face at position x about the horizontal diameter is $I(x) = \pi R^4/4$, a constant.

The shear stress is

$$\tau(x, y, z) = \frac{V(x)Q}{I(x)t(x, y, z)}.$$

The thickness parallel to the horizontal diameter is

$$t(x, y, z) = 2z = 2\sqrt{R^2 - y^2}.$$

Let y' be a dummy variable.

$$\begin{aligned}
Q(x, y, z) &= \int_y^R y' dA \\
&= \int_y^R y' 2z dy' \\
&= 2 \int_y^R y' \sqrt{R^2 - (y')^2} dy' \\
&= -\frac{2}{3}(R^2 - (y')^2)^{3/2}\Big|_y^R \\
&= \frac{2}{3}(R^2 - y^2)^{3/2}.
\end{aligned}$$

(I.3)

Therefore, the shear stress in the y-direction on the cross-sectional face perpendicular to the x-axis is

$$\tau_{xy}(x, y, z) = V(x)Q/I(x)t(x, y, z) = \frac{4}{3}\frac{V(x)(R^2 - y^2)}{\pi R^4}.$$

(I.4)

At the horizontal diameter the shear stress reaches its maximum, $\tau_{xy}(x, 0, z) = \frac{4}{3}\frac{V(x)}{\pi R^2}$.

The stresses are, therefore,

$$\tau_{xy}(x, y, z) = \frac{4}{3}\frac{P(L-a)(R^2 - y^2)}{\pi R^4 L} \quad \text{if} \quad 0 < x < a$$

$$\frac{-4}{3}\frac{aP(R^2 - y^2)}{\pi R^4 L} \quad \text{if} \quad a < x < L.$$

(I.5)

$$\sigma_x(x, y, z) = \frac{4P(L-a)}{\pi R^4 L}xy \quad \text{if} \quad 0 \le x \le a$$

$$\frac{4(L-x)aPy}{\pi R^4 L} \quad \text{if} \quad a \le x \le L.$$

(I.6)

This equation says that the shear stress jumps at $x = a$ where the point load is applied. Such a jump is impossible if the material is continuous, although it might happen at a crack. The problem is that there is no such thing as a point load. In reality, the load is a force distributed over the surface area of the rod to which it is applied. Such a load is sometimes called a traction, a force per unit area.

The highest stresses might occur just as the rod begins to rotate or while the rod rotates with a constant angular velocity. The first case will be considered here and the latter in a subsequent chapter. At startup of the engine, a torque T is applied to the rod and a load P is applied to the cam. Determine the state of stress at startup.

This is a combined loading problem. The loads and dimensions will be left as letters for two reasons. First, all possible such problems are solved at one time. Second, the effect of each parameter on the result is then clear. The stresses due to bending and torsion are assumed to superpose. Since the chosen coordinate system is Cartesian, the stresses due to the torque must be broken into components. The magnitude of the stress at (x, y, z) is

$$\tau = \frac{T\rho}{J},$$

where

$$J = \frac{\pi R^4}{2}$$

is the polar area moment of inertia and

$$\rho = \sqrt{y^2 + z^2}$$

is the radial distance from the center to the point (x, y, z). The internal torque, T, is the same on all vertical cross sections. Therefore,

$$\tau = \frac{2T\sqrt{y^2 + z^2}}{\pi R^4}.$$

The position of (x, y, z) is determined by the angle, θ, the radial line makes with the positive z-axis, the horizontal diameter. The angle is $\theta = \arctan(y/z)$.

$$\tau_{xy}(x, y, z) = \tau \cos\theta = \frac{2T\sqrt{y^2 + z^2}}{\pi R^4} \frac{z}{\sqrt{y^2 + z^2}} = \frac{2Tz}{\pi R^4}; \qquad (I.7)$$

$$\tau_{xz}(x, y, z) = \tau \sin\theta = \frac{2T\sqrt{y^2 + z^2}}{\pi R^4} \frac{y}{\sqrt{y^2 + z^2}} = \frac{2Ty}{\pi R^4}. \qquad (I.8)$$

The sum of the bending and torsional stresses produces the full stress state at (x, y, z) on the coordinate planes. In strength of materials, stresses superpose.

$$\sigma_x(x, y, z) = \frac{4(L - a)P}{\pi R^4 L} xy \quad \text{if} \quad 0 \le x \le a$$

$$\frac{4(L - x)aPy}{\pi R^4 L} \quad \text{if} \quad a \le x \le L. \qquad (I.9)$$

$$\tau_{xy}(x, y, z) = \frac{4}{3} \frac{P(L - a)(R^2 - y^2)}{\pi R^4 L} + \frac{2Tz}{\pi R^4} \quad \text{if} \quad 0 < x < a$$

$$\frac{-4}{3} \frac{aP(R^2 - y^2)}{\pi R^4 L} + \frac{2Tz}{\pi R^4} \quad \text{if} \quad a < x < L. \qquad (I.10)$$

$$\tau_{xz}(x, y, z) = \frac{2Ty}{\pi R^4}. \qquad (I.11)$$

Thus far, the stress field, the stresses at each point on the planes parallel to the coordinate planes, has been determined. However, failure might occur due to stresses on some other plane through the point. For example, in elementary strength of materials, it is often suggested that a ductile metal rod under an axial load fails on a plane at a 45° angle with the coordinate planes because the shear stress is maximum on that plane. In more complicated loadings, the failure plane is not so easily estimated. A simple means must be developed to compute the stresses on arbitrary planes. Preferably, this method should relate the stresses on an arbitrary plane through the point of interest to the known stresses on planes parallel to the coordinate planes at that point.   ■

DEFORMABLE BODIES
AND THEIR MATERIAL BEHAVIOR

CHAPTER 1

THE STRESS TENSOR

From the time of Newton up to the early 1800s, the primary thrust of strength of materials was the analysis of beams and of elastic material behavior using Hooke's law. Most of the beam results were given in terms of the geometry of the beam and the external load. The idea that internal forces hold bodies together was common in physics, but had not been applied to strength of materials or the deformation of elastic bodies. In 1763, Boscovich developed the idea that attractive and repulsive forces act along the line between two particles and that equilibrium is achieved when these are in balance. Timoshenko [2] calls this the molecular theory of an elastic body. Each particle both tries to attract and repel neighboring particles. For example, in the classical view of an atom, the negatively charged electrons attract the positively charged nucleus of a nearby atom while repelling its electrons. The nucleus repels neighboring nuclei and attracts electrons. When two atoms are at the proper distance apart, all these forces are balanced. Poisson in 1812 tried with minimal success to apply this idea to the bending of plates. It was the engineer-mathematicians Navier and Cauchy who finally developed a coherent analysis of elastic behavior in the early 1820s, and our ideas of stress and strain are based on their work.

Navier first chose a Cartesian x-y-z coordinate system for the body and then defined the displacement of a point in the body by functions $u(x, y, z)$, $v(x, y, z)$, and $w(x, y, z)$. A point originally having coordinates (x, y, z) in the body has coordinates $(x + u(x, y, z), y + v(x, y, z), z + w(x, y, z))$ after the body is deformed. He obtained differential equations describing when the body is in equilibrium in terms of (u, v, w). Finally to relate the external forces on the body to the equilibrium displacements, he obtained an equation describing the interaction of the forces and displacements on the surface of the body. These equations involved the surface tractions (the distributed forces on the surface) and terms of the form

$$\partial u/\partial x, \partial v/\partial y, \partial w/\partial z, \partial u/\partial y + \partial v/\partial x, \partial u/\partial z + \partial w/\partial x, \quad \text{and} \quad \partial v/\partial z + \partial w/\partial y.$$

After carefully studying Navier's 1821 paper, Cauchy in the following few years wrote the papers that are the basis of the contemporary theory of linear elasticity. Even though Cauchy is most famous as a mathematician, he was educated as an engineer. Cauchy identified Navier's six expressions involving partial derivatives of the displacements as what we now call the strains. The physical interpretation followed the mathematics in this case. Cauchy's new idea was that Navier's equation for an outer surface also ought to work for any surface within the body if one pretends that a section of the body is cut out. This idea seems to have been suggested to him by his previous work on hydrodynamics. The force per unit area, similar to pressure, on an internal surface of a body, due to the actions of particles adjacent to the surface, is what we call stress. Cauchy simplified the conceptual basis and derivation of elasticity theory by dropping the idea of intermolecular forces and introducing those of stress and strain.

1

In elasticity, the stress on an internal face in the body, together with the displacements, is determined by the external loads and the solution to a set of differential equations. Such a solution can be found in closed form in only a few cases; usually numerical techniques are required. A typical elasticity problem is to determine the stress, strain, and displacement at each point inside a body given forces and displacements on different portions of the surface of the body.

1.1   DEFINITION OF STRESS

Stress is an internal force per unit area acting on a plane inside the body. The stress on the plane is the vector sum of the internal forces acting on that plane divided by the area of the plane. Since force is represented by a vector and area is a scalar, stress is the product of a scalar and a vector and is therefore itself a vector. The stress vector can be in any direction with respect to the plane. The stress vector can be written in components that are perpendicular and parallel to the plane or which are in some other fixed coordinate system.

The stress equations developed and emphasized in elementary strength of materials only give the magnitudes of the stresses acting on the cross-sectional plane of an axially loaded bar, a beam in bending, or a circular cross-section rod in torsion. Some failure theories assume that a material fails if the stress becomes too large on some plane through a point in the body; this maximum stress may occur on a plane whose stresses are not given by the elementary strength of materials equations. The problem then is to find a simple way to compute the stress vector on each and every plane through each point in the body.

To compute the stress on a particular plane, the plane first must be identified mathematically. A plane in three dimensional space is completely described if a unit vector, $\mathbf{n}$, normal to the plane and a point through which the plane passes are given. Recall from calculus that if the normal in some Cartesian coordinate system is given by $a\mathbf{i} + b\mathbf{j} + c\mathbf{k}$, and if the plane passes through the point (x_o, y_o, z_o), then the equation of the plane is

$$a(x - x_o) + b(y - y_o) + c(z - z_o) = 0.$$

Conversely, if the equation of the plane is given as $ax + by + cz = d$, where a, b, c, and d are numbers, then a normal to the plane is $\mathbf{N} = a\mathbf{i} + b\mathbf{j} + c\mathbf{k}$, which may or may not be a unit vector. If the plane has normal $\mathbf{N} = a\mathbf{i} + b\mathbf{j} + c\mathbf{k}$, then we say that the stress on this plane is the stress associated with $\mathbf{n} = \mathbf{N}/\|\mathbf{N}\| = (a\mathbf{i} + b\mathbf{j} + c\mathbf{k})/\sqrt{a^2 + b^2 + c^2}$.

EXAMPLE 1.1   A rod is axially loaded by a force on either end. The Cartesian coordinates are chosen so that the x-axis is along the longitudinal axis of the rod. For large enough force, such a rod made of a ductile metal is expected to yield due to shear stresses on a plane at an angle of $45°$ with the cross-section perpendicular to the longitudinal axis. There are infinitely many such planes. One of the two planes passing through the z-axis is defined by either the unit normal vector, $\mathbf{n} = \cos 45\mathbf{i} + \sin 45\mathbf{j}$ or by $-\mathbf{n} = -\cos 45\mathbf{i} - \sin 45\mathbf{j}$. The other is defined by $\mathbf{n} = \cos 45\mathbf{i} - \sin 45\mathbf{j}$.

Such a rod made of a brittle material, say a metal such as cast iron or a ceramic, is expected to fracture due to normal stresses on the cross section perpendicular to the longitudinal axis. This plane is defined by the unit normal $\mathbf{n} = \mathbf{i}$ or by $\mathbf{n} = -\mathbf{i}$.   ∎

Stress was defined in terms of the area of a plane as the average force per unit area of the plane. The definition needs to be modified to describe the stress at a point. A reasonable approach is to define the stress at a point in a direction, $\mathbf{n}$, as the limit of the stresses on smaller and smaller areas all perpendicular to $\mathbf{n}$.

Definition   Fix a point, P, in the body and a unit vector **n** at P. Consider all partial planes lying in the body which pass through P and have normal **n**. Let A represent the area of each of these planes. The stress vector at P acting in the direction **n** is the vector $\mathbf{t}(P, \mathbf{n}) = \lim_{A \to 0} \mathbf{F}/A$, where **F** is the resultant internal force on area A.

In general there is a different stress vector **t** for each direction **n** through P.

EXAMPLE 1.2   For axial loading there are stress components on any plane cut through the point except the y and z faces, those with normal vector either **j** or **k** or any combination of **j** or **k**. In strength of materials, attention was often restricted to the planes perpendicular to the x-y coordinate plane and a two-dimensional infinitesimal element was drawn.

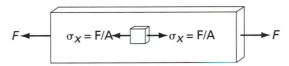

FIGURE 1.1   Two-dimensional stress element for uniaxial loading.

Looking at the face in the page, the edges of this element in Figure 1.1 represent a single horizontal plane and a single vertical plane through the point. These planes are perpendicular to the page. ∎

The Mohr's circle technique learned in strength of materials also gives the stress vector acting on a plane perpendicular to a given direction. The Mohr's circle technique produces the normal and shear components of this vector. Unfortunately, the Mohr's circle is easily applied only in two-dimensional problems. A new technique is required to quickly obtain the stress vector on a plane perpendicular to any direction in three space.

EXAMPLE 1.3   For the axially loaded member with uniform cross section A,

(a) Determine the stress vector on the plane with unit normal vector, $\mathbf{n} = \cos 30\mathbf{i} + \sin 30\mathbf{j} + 0\mathbf{k}$ through the point Q in Figure 1.2. Determine the magnitudes of the normal and shear components of the stress vector.

(b) Find the magnitudes of the normal and shear components of the stress vector using Mohr's circle and show that they are the same as the magnitudes of the components of the stress vector found in (a). *Hint*: first show that the plane makes a 30° angle with the vertical.

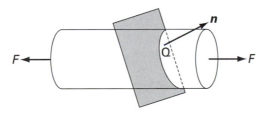

FIGURE 1.2   Plane defined by **n** cutting axially loaded rod.

Solution   **(a)** The average stress on the intersection of the plane with the rod can be found more easily from a two-dimensional figure, as in Figure 1.3.

Since the area of the slanted face is $A/\cos 30$, the stress vector on the slanted plane is

$$\mathbf{t}(P, \mathbf{n}) = \frac{F}{A} \cos 30\mathbf{i}.$$

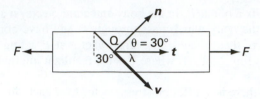

FIGURE 1.3 Two-dimensional schematic of the normal direction **n** and the shear direction **v** and the total stress **t**.

The magnitude of the normal component is the projection of $\mathbf{t}(P, \mathbf{n})$ onto **n** and is obtained from the dot product,

$$\sigma = \mathbf{t}(P, \mathbf{n}) \cdot \mathbf{n} = \frac{F}{A} \cos^2 30;$$

and the normal component of the stress vector is

$$\boldsymbol{\sigma} = \sigma\mathbf{n} = (\mathbf{t}(P, \mathbf{n}) \cdot \mathbf{n})\mathbf{n} = \frac{F}{A} \cos^2 30(\cos 30\mathbf{i} + \sin 30\mathbf{j}) = \frac{F}{A}\frac{3}{4}\left(\frac{\sqrt{3}}{2}\mathbf{i} + \frac{1}{2}\mathbf{j}\right).$$

The shear component, $\boldsymbol{\tau}$, lies in the plane and is perpendicular to the normal component, $\boldsymbol{\sigma}$. Since $\mathbf{t}(P, \mathbf{n}) = \boldsymbol{\sigma} + \boldsymbol{\tau}$, the magnitude of the shear component is, by the Pythagorean theorem, $\tau = [|\mathbf{t}(P, \mathbf{n})|^2 - \sigma^2]^{1/2} = (F/A) \cos 30 \sin 30$. Also,

$$\boldsymbol{\tau} = \frac{F}{A} \cos 30[(1 - \cos^2 30)\mathbf{i} - \cos 30 \sin 30\mathbf{j}] = \frac{F}{A}\frac{\sqrt{3}}{2}\left(\frac{1}{4}\mathbf{i} - \frac{\sqrt{3}}{4}\mathbf{j}\right).$$

The fact that the dot product of $\boldsymbol{\sigma}$ and $\boldsymbol{\tau}$ is zero verifies that they are perpendicular.

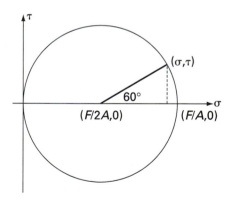

FIGURE 1.4 Mohr's circle for axial loading.

(b) The Mohr's circle, Figure 1.4, has radius $F/2A$ and center $(F/2A, 0)$ so that

$$\sigma = \frac{F}{2A} + \frac{F}{2A} \cos 60 = \frac{F}{A} \cos^2 30 = \frac{F}{A}\frac{3}{4};$$

$$\tau = \frac{F}{2A} \sin 60 = \frac{F}{A} \cos 30 \sin 30 = \frac{F}{A}\frac{\sqrt{3}}{4}.$$

The two methods give the same stress magnitudes. ∎

EXAMPLE 1.4   The study of metals shows that slip occurs under shear stress more easily on some planes than others and under smaller shear stress in certain directions on those planes. The stress parallel to the plane in this direction of easiest slip is called the critical resolved shear stress in materials science. The previous two-dimensional example determined the stress on an arbitrary plane that is perpendicular to the x-y plane. The stresses on other planes may also be important. For example, determine the normal and shear stress components on the plane making 45° angles with the coordinate axes, an octahedral plane (Fig. 1.5), which is particularly important for the study of certain metals. This plane is not perpendicular to any coordinate plane. Again assume a uniaxial load, F, parallel to the x-axis.

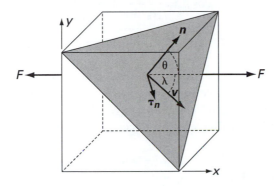

FIGURE 1.5   Stress components on the octahedral plane.

Solution   The cube is axially loaded in the x-direction with a stress, $\sigma\mathbf{i} = (F/A)\mathbf{i}$, where A is the area of the x-face. The total stress vector, $\mathbf{t}$, on a plane with normal, $\mathbf{n}$, is the force, $F\mathbf{i}$ divided by the area, A_n, which is the projection of A onto the plane with normal $\mathbf{n}$. Let θ be the angle between the normal to A and $\mathbf{n}$. A geometric argument shows that $A_n = A/\cos\theta$. Therefore the corresponding stress on the plane with normal $\mathbf{n}$ is $(F/A_n)\mathbf{i} = \sigma\cos\theta\mathbf{i}$. A normal to a particular plane is computed by taking the cross product of two vectors in the plane. In this case, $\mathbf{N} = (-\mathbf{i}+\mathbf{j}) \times (-\mathbf{i}+\mathbf{k}) = \mathbf{i}+\mathbf{j}+\mathbf{k}$. The unit normal vector to the plane is $\mathbf{n} = (\mathbf{i}+\mathbf{j}+\mathbf{k})/\sqrt{3}$ since the magnitude of $\mathbf{N}$ is $\sqrt{3}$. Because $\cos\theta = \mathbf{i} \cdot \mathbf{n} = 1/\sqrt{3}$, the total stress on the slanted plane is $(\sigma/\sqrt{3})\mathbf{i}$. The component of stress normal to the plane with normal $\mathbf{n}$ is the projection of $(\sigma/\sqrt{3})\mathbf{i}$ onto $\mathbf{n}$,

$$\sigma_n = \left(\frac{\sigma}{\sqrt{3}}\mathbf{i} \cdot \mathbf{n}\right)\mathbf{n} = \frac{\sigma}{3\sqrt{3}}(\mathbf{i}+\mathbf{j}+\mathbf{k}).$$

The shear component, τ_n, is obtained by

$$\tau_n = \frac{\sigma}{\sqrt{3}}\mathbf{i} - \sigma_n = \frac{2\sigma}{3\sqrt{3}}\mathbf{i} - \frac{\sigma}{3\sqrt{3}}\mathbf{j} - \frac{\sigma}{3\sqrt{3}}\mathbf{k}. \qquad \blacksquare$$

Exercise 1.1   Verify that σ_n and τ_n are perpendicular.

In materials science, the shear stress magnitude in a given direction lying in a particular plane is computed by "Schmid's law," which says that in a plane whose normal makes an angle θ with the force direction, the magnitude of shear stress in a direction, $\mathbf{v}$, which makes an angle λ with the force vector is $\tau = \sigma\cos\theta\cos\lambda$. Except for simple cases, it is usually difficult to determine the angles, θ and λ. This result can be recovered easily using vector operations. An example, which is easy to verify because the angles are unusually easy to determine from the triangles in three space, is to determine the magnitude of the shear stress in the $(\mathbf{i}-\mathbf{j})/\sqrt{2}$-direction on the octahedral plane.

The cosine of the angle, θ, between $\boldsymbol{\sigma}_n$ and $\mathbf{i}$ is, from the definition of the dot product,

$$\cos\theta = \frac{\boldsymbol{\sigma}_n \cdot \mathbf{i}}{|\boldsymbol{\sigma}_n||\mathbf{i}|} = \frac{\sigma}{3\sqrt{3}}\frac{3}{\sigma} = \frac{1}{\sqrt{3}};$$

and the cosine of the angle, β, between $\boldsymbol{\tau}_n$ and $\mathbf{i}$ is

$$\cos\beta = \frac{\boldsymbol{\tau}_n \cdot \mathbf{i}}{|\boldsymbol{\tau}_n||\mathbf{i}|} = \frac{2\sigma}{3}\frac{3}{\sqrt{6}\sigma} = \frac{\sqrt{2}}{\sqrt{3}}.$$

Exercise 1.2   Verify Schmid's Law for the magnitude of $\boldsymbol{\tau}_n$ of this example.

The magnitude of the shear stress component, τ, in the **v**-direction can be similarly computed. The angle between $\mathbf{v} = \mathbf{i} - \mathbf{j}$ and $\mathbf{i}$ is $\cos\lambda = 1/\sqrt{2}$. Schmid's law then says that $\tau = \sigma\cos\theta\cos\lambda = \sigma/\sqrt{6}$. This result can also be obtained directly from vector operations by simply projecting the stress vector, $\boldsymbol{\tau}_n$, onto the direction $(\mathbf{i} - \mathbf{j})/\sqrt{2}$.

$$\left(\frac{2\sigma}{3\sqrt{3}}\mathbf{i} - \frac{\sigma}{3\sqrt{3}}\mathbf{j} - \frac{\sigma}{3\sqrt{3}}\mathbf{k}\right) \cdot \frac{1}{\sqrt{2}}(\mathbf{i} - \mathbf{j}) = \frac{\sigma}{\sqrt{6}}.$$

The calculation of the stresses on the octahedral plane would be difficult to do using Mohr's circles. Recall that the generalization of the Mohr's circle diagram to three dimensional states of stress is comprised of three circles, each one of which has its horizontal maximum and minimum point specifying planes acted on only by normal stresses σ. The point corresponding to the state of stress on the plane of Example 1.4 is $(\sigma/3, \sqrt{2}\sigma/3)$. This point lies inside the largest circle and outside the inner circles (which are points at the origin for uniaxial loading). The points corresponding to the stress state on any plane which is not perpendicular to a coordinate plane lies in this region.

A method is needed to quickly compute the stress from some simple means of identifying the plane on which the stress is defined. The unit normal vector to the plane defines the plane uniquely once the point at which the stress is to be computed is given.

To describe the behavior of the solid, the stress at a point in each and every direction is required. This apparently requires an infinite number of stress vectors at each of an infinite number of points in the body. We need to develop tools to reduce this to a small finite number of vectors. Recall from strength of materials that a three-dimensional stress element showing the components of stress on the coordinate planes is given as in Figure 1.6.

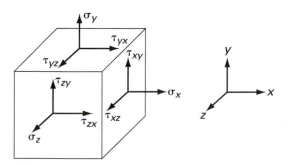

FIGURE 1.6   The three-dimensional stress element.

The index convention for the shear components of the stress is that the first index is the face on which the component acts and the second is the direction. For example, the x-face is that face perpendicular to the x-direction. The sign convention for shear is that the positive direction is the same as the positive direction of the axis given in the second index. The normal stresses are positive in tension (outward from the body) and negative in compression (inward). All the stresses on the element are shown in their positive direction. Figure 1.6 looks like a cube of material from the body, but it is only imagined to be a cube. It is actually three planes intersecting at the point. It is drawn as a cube so that the three planes are easily visualized.

The components of stress on the coordinate planes are enough to determine the stress components on any face through the point. To see this, the stress tensor must be defined and some linear algebra applied.

1.2   THE STRESS TENSOR

The stress tensor, T, at point P is a function that relates the direction $\mathbf{n}$ normal to a plane and the stress vector $\mathbf{t}(P, \mathbf{n})$ at a point P acting on that plane. The stress tensor changes from point to point in the body. To define this relationship and to find a simple way to compute $\mathbf{t}(P, \mathbf{n})$, some linear algebra preliminaries are necessary.

Definition   A vector space over the real numbers is a set of elements on which an addition, $+$, and a scalar multiplication are defined such that

1. The addition is commutative and associative. There is a zero vector, $\mathbf{O}$, and for each vector, $\mathbf{v}$, there is a vector $-\mathbf{v}$ such that $\mathbf{v} + (-\mathbf{v}) = \mathbf{O}$.
2. The product $\alpha\mathbf{v}$ of a real scalar α and a vector satisfies the following relations.

   (a) $(\mathbf{v} + \mathbf{u})\alpha = \mathbf{v}\alpha + \mathbf{u}\alpha$;
   (b) $\beta(\mathbf{v} + \mathbf{u}) = \beta\mathbf{v} + \beta\mathbf{u}$;
   (c) $(\alpha\beta)\mathbf{v} = \alpha(\beta\mathbf{v})$;
   (d) $(1)\mathbf{v} = \mathbf{v}$.

EXAMPLE 1.5   The set of all vectors

$$V^3 = \{a\mathbf{i} + b\mathbf{j} + c\mathbf{k}, \text{ such that } a, b, \text{ and } c \text{ are any real numbers}\}$$

is a vector space, as is the set of all vectors

$$V^2 = \{a\mathbf{i} + b\mathbf{j}, \text{ such that } a \text{ and } b \text{ are any real numbers}\}.$$ ∎

The dimension is determined by the number of basis vectors required. V^3, for example, is three dimensional because exactly three basis vectors $\mathbf{i}, \mathbf{j}$, and $\mathbf{k}$ are required to write all other vectors. The following definitions make this idea precise.

Definition   A set of vectors, $\mathbf{u}_1, \ldots, \mathbf{u}_n$, is said to span the vector space if every vector, $\mathbf{v}$, can be written as a linear combination of the $\mathbf{u}_i$, *i.e.* there are scalars α_i such that

$$\mathbf{v} = \alpha_1\mathbf{u}_1 + \alpha_2\mathbf{u}_2 + \cdots + \alpha_n\mathbf{u}_n.$$

Definition   A set of vectors, $\mathbf{u}_1, \ldots, \mathbf{u}_n$ is said to be linear independent if

$$\mathbf{O} = \alpha_1\mathbf{u}_1 + \alpha_2\mathbf{u}_2 + \cdots + \alpha_n\mathbf{u}_n$$

implies that each $\alpha_i = 0$.

Definition   A set of n vectors, $\mathbf{u}_1, \ldots, \mathbf{u}_n$, is a basis for a vector space if it is linearly independent and n is the smallest number such that the set spans the vector space.

The number, n, is called the dimension of the vector space. The dimension is well defined since any two different bases for the same vector space must contain the same number of vectors.

For example, the set of vectors, $\mathbf{i}, \mathbf{j}$, and $\mathbf{k}$, is linearly independent and spans the vectors in three-dimensional space, so that it is a basis. The set of vectors, $\mathbf{i}$ and $\mathbf{k}$, is also linearly independent but does not span the vectors in three-dimensional space, and so it is not a basis for three-dimensional space.

The set of vectors, $\mathbf{i}$ and $\mathbf{j}$, is linearly independent and spans the vectors in two-dimensional space, and so is a basis for two-dimensional space. Many bases for a vector space are possible. For example, the set $\mathbf{u}_1 = \mathbf{i} + \mathbf{j}$ and $\mathbf{u}_2 = \mathbf{i} - \mathbf{j}$ is also a basis for the two-dimensional vector space. The basis $\mathbf{u}_1, \mathbf{u}_2$ is not composed of unit vectors. Notice that $\mathbf{i} = (\mathbf{u}_1 + \mathbf{u}_2)/2$ and $\mathbf{j} = (\mathbf{u}_1 - \mathbf{u}_2)/2$. If coordinate axes are taken in the direction of the basis vectors, then the transformation from the basis $\mathbf{i}, \mathbf{j}$ to the basis of unit vectors $\mathbf{u}_1/\sqrt{2}, \mathbf{u}_2/\sqrt{2}$ corresponds to a $45°$ rotation of the axes in the plane. To see this, draw a sketch of the two sets of basis vectors in the plane.

A function from one vector space to another which preserves the vector space properties must be consistent with addition and scalar multiplication in the vector spaces. Such a function is called a linear transformation.

Definition   A linear transformation, T, from a vector space, V, to itself, denoted $T : V \rightarrow V$, is a function such that $T(\alpha\mathbf{u} + \beta\mathbf{v}) = \alpha T(\mathbf{u}) + \beta T(\mathbf{v})$, where $\mathbf{u}$ and $\mathbf{v}$ are vectors in V, and α and β are scalars.

The stress vector on a plane in a body at a point P and the unit vector $\mathbf{n}$ defining the plane are related by a function, T, from the three-dimensional vector space of unit normal vectors to planes through P to the three dimensional vector space of stress vectors at P,

$$T(\mathbf{n}) = \mathbf{t}(P, \mathbf{n}). \tag{1.1}$$

If it can be shown that T is a linear transformation, and if we know the value of T on just three vectors, $\mathbf{n}$, one in each of the directions of the basis vectors (*i.e.*, the stresses on the three coordinate planes), then any of the remaining infinite number of $\mathbf{t}(P, \mathbf{n})$ can be quickly computed by matrix multiplication.

Any unit vector normal to a plane through P can be written in the form, $\mathbf{n} = a\mathbf{i} + b\mathbf{j} + c\mathbf{k}$. If T is a linear transformation, then by definition

$$T(\mathbf{n}) = aT(\mathbf{i}) + bT(\mathbf{j}) + cT(\mathbf{k}). \tag{1.2}$$

Therefore to know $T(\mathbf{n})$, it is only necessary to determine $T(\mathbf{i})$, $T(\mathbf{j})$, and $T(\mathbf{k})$, the stress vectors on the planes respectively perpendicular to $\mathbf{i}, \mathbf{j}$, and $\mathbf{k}$, the coordinate planes. This explains why one draws a stress element at the point P with faces parallel to the coordinate planes. A free body diagram or some other technique is used to determine the stress vectors on the surface of the cubical element with such an orientation. Once this is done, equation (1.2) gives the stress vector on any other plane through P.

The next step, still assuming that T is a linear transformation, a fact yet to be proved, is to represent T as a matrix so that all calculations can be done by matrix multiplication. In linear algebra it is shown that any linear transformation from a vector space to itself has a unique matrix representation in a given coordinate system. Therefore to get the correct matrix representation for T, one need merely write the matrix that gives the correct value for T on the basis vectors. Represent the vector $\mathbf{i}$ by the column $(1, 0, 0)^t$, the vector $\mathbf{j}$ by the column $(0, 1, 0)^t$, and the vector $\mathbf{k}$ by the column $(0, 0, 1)^t$. One finds that by putting the coefficients of $\mathbf{t}(P, \mathbf{i})$ in the first column, of $\mathbf{t}(P, \mathbf{j})$ in the second column, and of $\mathbf{t}(P, \mathbf{k})$ in the third column, then matrix multiplication with the column vector on the right gives the correct stress resultant on the coordinate planes. For example, $\mathbf{t}(P, \mathbf{i}) = \sigma_x \mathbf{i} + \tau_{xy}\mathbf{j} + \tau_{xz}\mathbf{k}$ is recovered by

$$\begin{pmatrix} \sigma_x \\ \tau_{xy} \\ \tau_{xz} \end{pmatrix} = \begin{pmatrix} \sigma_x & \tau_{yx} & \tau_{zx} \\ \tau_{xy} & \sigma_y & \tau_{zy} \\ \tau_{xz} & \tau_{yz} & \sigma_z \end{pmatrix} \begin{pmatrix} 1 \\ 0 \\ 0 \end{pmatrix}.$$

Many times calculations are easier if the entries are given numerical indices in which x, y, and z correspond to 1, 2, and 3. To make the numerical indexing system match with the usual index notation for matrices, now let the first index correspond to the component and the second correspond to the face on which it acts. The stress vector on the $\mathbf{i}$ face at P is then denoted by $\mathbf{t}(P, \mathbf{i}) = T_{11}\mathbf{i} + T_{21}\mathbf{j} + T_{31}\mathbf{k}$. Likewise write the stress vector on the $\mathbf{j}$ face at P as $\mathbf{t}(P, \mathbf{j}) = T_{12}\mathbf{i} + T_{22}\mathbf{j} + T_{32}\mathbf{k}$, and the stress vector on the $\mathbf{k}$ face at P as $\mathbf{t}(P, \mathbf{k}) = T_{13}\mathbf{i} + T_{23}\mathbf{j} + T_{33}\mathbf{k}$. The two types of indices then have the following correspondence, T_{11} is σ_x, $T_{21} = \tau_{xy}$ and $T_{31} = \tau_{xz}$, etc. Then each entry of the matrix representation for T can be denoted T_{ij} where the indices vary from 1 to 3 and the first indicates the row and the second the column of the matrix. The matrix is often indicated by the shorthand, $T = (T_{ij})$. Therefore T is represented in the given x-y-z (1-2-3) coordinate system by

$$T = \begin{pmatrix} \sigma_x & \tau_{yx} & \tau_{zx} \\ \tau_{xy} & \sigma_y & \tau_{zy} \\ \tau_{xz} & \tau_{yz} & \sigma_z \end{pmatrix} = \begin{pmatrix} T_{11} & T_{12} & T_{13} \\ T_{21} & T_{22} & T_{23} \\ T_{31} & T_{32} & T_{33} \end{pmatrix}.$$

In other texts such as [1], the correspondence between x and 1, y and 2, z and 3 is preserved in the stress tensor denoted using numerical subscripts. In this convention, the key relation (1.1), $T(\mathbf{n}) = \mathbf{t}(P, \mathbf{n})$, then requires the use of the transpose defined below (see p. 12), and becomes $T^t(\mathbf{n}) = \mathbf{t}(P, \mathbf{n})$. Because of the symmetry properties of the stress tensor discussed below, the two conventions are equivalent in all applications considered in this text.

1.3 THE STRESS TENSOR IS A LINEAR TRANSFORMATION—THE CAUCHY TETRAHEDRON

In the previous section, under the assumption of linearity, the function relating the normal on a plane through P to the stress on the plane was represented by a matrix whose entries were the stress components on the coordinate planes. To show that this function is linear, and thus a tensor, the stress on an oblique plane must be related to the stresses on the coordinate planes. This proof was first constructed by Cauchy around 1822.

The point P is at the origin of Figure 1.7. Let $\mathbf{n} = a\mathbf{i} + b\mathbf{j} + c\mathbf{k}$ be the unit normal vector to the oblique plane BCD. Recall from statics or calculus that any vector, $\mathbf{F}$, can be written in terms of its direction cosines as

$$\mathbf{F} = F \cos(\theta_x)\mathbf{i} + F \cos(\theta_y)\mathbf{j} + F \cos(\theta_z)\mathbf{k},$$

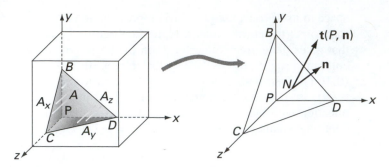

FIGURE 1.7 The Cauchy stress tetrahedron.

where θ_x is the angle from the positive x-axis to $\mathbf{F}$, and so on, and F is the magnitude of $\mathbf{F}$. Then the unit normal in the direction of $\mathbf{F}$ is $\mathbf{n}_F = \cos(\theta_x)\mathbf{i} + \cos(\theta_y)\mathbf{j} + \cos(\theta_z)\mathbf{k}$. Using this, the unit normal to the plane is written in terms of its direction cosines so that

$$a = \cos(\theta_x), \quad b = \cos(\theta_y), \quad \text{and} \quad c = \cos(\theta_z).$$

Pick the point N lying in plane BCD so that the line segment PN is perpendicular to the plane. Then by drawing right triangles, three different expressions can be written for the length of PN:

$$PN = PD\cos(\theta_x) = aPD;$$
$$PN = PB\cos(\theta_y) = bPB;$$
$$PN = PC\cos(\theta_z) = cPC.$$

These facts are used to relate the surface area, A, of the plane BCD to the surface areas of the coordinate planes, A_x, A_y, A_z, in the tetrahedron. The volume of a tetrahedron is one third the base times the height. This volume can be written in four different ways by considering PN, PD, PB, and PC as the heights in turn. PN is perpendicular to the base BCD, PD to PBC, PB to PCD, and PC to PBD so that

$$Vol = PN\frac{A}{3} = PC\frac{A_z}{3} = PB\frac{A_y}{3} = PD\frac{A_x}{3}.$$

Setting $PN \cdot A/3$ equal to each of the other three and using the above relations produces

$$A_x = aA, \quad A_y = bA, \quad \text{and} \quad A_z = cA. \tag{1.3}$$

The coordinate planes are the "backsides" of the standard stress element so that the stress components are shown in their positive directions in the stress sign convention, but are in the negative coordinate directions. The goal is to compute the stress components T_x, T_y, T_z of the stress vector $\mathbf{t}(P, \mathbf{n})$ acting on plane BCD. Notice that these are the components in the coordinate directions, not the normal and shear components, so that $\mathbf{t}(P, \mathbf{n}) = T_x\mathbf{i} + T_y\mathbf{j} + T_z\mathbf{k}$.

Now use Newton's law that $\mathbf{F} = m\mathbf{a}$, where $\mathbf{a} = a_x\mathbf{i} + a_y\mathbf{j} + a_z\mathbf{k}$ is the acceleration of the center of mass of the tetrahedron (Fig. 1.8). Mass is the density times the volume so that $m = \rho(PN)A/3$, where ρ is the density. Multiply the stress components by the area of the face on which they act to obtain the force components on each face.

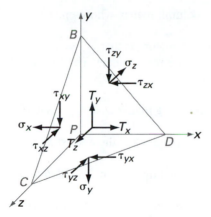

FIGURE 1.8   The stress components on the Cauchy stress tetrahedron.

Summing forces in the x-direction, in the y-direction, and in the z-direction gives

$$T_x A - \sigma_x A_x - \tau_{yx} A_y - \tau_{zx} A_z = \rho \frac{A}{3} P N a_x;$$

$$T_y A - \tau_{xy} A_x - \sigma_y A_y - \tau_{zy} A_z = \rho \frac{A}{3} P N a_y;$$

$$T_z A - \tau_{xz} A_x - \tau_{yz} A_y - \sigma_z A_z = \rho \frac{A}{3} P N a_z.$$

Use equations (1.3) to substitute for A_x, A_y, A_z. Then divide out A, and let PN approach zero so that BCD becomes a plane through P to obtain

$$\begin{aligned}
T_x &= a\sigma_x + b\tau_{yx} + c\tau_{zx}; \\
T_y &= a\tau_{xy} + b\sigma_y + c\tau_{zy}; \\
T_z &= a\tau_{xz} + b\tau_{yz} + c\sigma_z.
\end{aligned} \qquad (1.4)$$

Therefore the stress components in the coordinate directions on the oblique plane are linear functions of those on the coordinate planes. The relation, $T(\mathbf{n}) = \mathbf{t}(P, \mathbf{n})$, is linear. These three equations can be written in matrix form as

$$\begin{pmatrix} T_x \\ T_y \\ T_z \end{pmatrix} = \begin{pmatrix} \sigma_x & \tau_{yx} & \tau_{zx} \\ \tau_{xy} & \sigma_y & \tau_{zy} \\ \tau_{xz} & \tau_{yz} & \sigma_z \end{pmatrix} \begin{pmatrix} a \\ b \\ c \end{pmatrix}.$$

Recall that the first index is the face on which the stress acts. The tensor T is represented by the matrix

$$T = \begin{pmatrix} \sigma_x & \tau_{yx} & \tau_{zx} \\ \tau_{xy} & \sigma_y & \tau_{zy} \\ \tau_{xz} & \tau_{yz} & \sigma_z \end{pmatrix} \qquad (1.5)$$

in which the first column contains the components on the x-face and so forth. This is exactly the matrix that was found to give the correct resultants on the coordinate planes in the previous section. The function $T(\mathbf{n})$ is then obtained by multiplying the matrix for T on the right by the column matrix whose entries are the coefficients of the unit vector **n**.

Recall from strength of materials that $\tau_{xy} = \tau_{yx}$, $\tau_{xz} = \tau_{zx}$, and $\tau_{zy} = \tau_{yz}$ from the sum of the moments on the stress element. This implies that the entries in the matrix representing the stress tensor must satisfy $T_{ij} = T_{ji}$.

Definition The transpose of a matrix $\alpha = (\alpha_{ij})$ is $\alpha^t = (\alpha_{ji})$.

The entries of α are "flipped over" the main diagonal to obtain α^t.

Definition A matrix, α, such that $\alpha = \alpha^t$ is called a symmetric matrix.

The stress tensor is represented by a symmetric matrix, since $T = T^t$.

Exercise 1.3 Verify that this procedure gives the correct answers for the stresses on the coordinate planes.

EXAMPLE 1.6 The state of stress at a point P is given by the components of stress in MPa on the coordinate planes (Fig. 1.9).

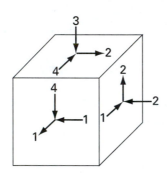

FIGURE 1.9 Stress element.

Compute the stress vector on the plane with unit normal vector, $\mathbf{n} = (2\mathbf{i} - \mathbf{j} + \mathbf{k})/\sqrt{6}$.

Solution The stress is

$$\mathbf{t}(P, \mathbf{n}) = \begin{pmatrix} -2 & 2 & -1 \\ 2 & -3 & -4 \\ -1 & -4 & 1 \end{pmatrix} \begin{pmatrix} \frac{2}{\sqrt{6}} \\ \frac{-1}{\sqrt{6}} \\ \frac{1}{\sqrt{6}} \end{pmatrix} = \begin{pmatrix} \frac{-7}{\sqrt{6}} \\ \frac{3}{\sqrt{6}} \\ \frac{3}{\sqrt{6}} \end{pmatrix} = (-7\mathbf{i} + 3\mathbf{j} + 3\mathbf{k})/\sqrt{6} \text{ (MPa).} \quad \blacksquare$$

On a free surface of a body, there can be no stresses in the direction normal to the surface. The state of stress at such a point is given a special name because the resultant stresses in the plane of the surface can be resolved into two mutually perpendicular directions. Assume for definiteness that the normal direction to the surface is the z-direction.

Definition A state of stress at a point is said to be plane stress if

$$\sigma_z = \tau_{xz} = \tau_{yz} = 0.$$

In general, the direction of zero stress need not coincide with a coordinate direction. The stress tensor at such a point is 2×2, and the analysis is much easier. A point in the interior of a body can also be in plane stress. For example, any point in an axially loaded rod is in plane stress.

1.4 VARIATION OF THE STRESS TENSOR FROM POINT TO POINT IN A BODY IN EQUILIBRIUM

In the previous sections, a stress tensor was constructed at each point in the body to relate the normal to a plane and the stress vector on that plane. When the body is not fractured (i.e., is continuous) and is in equilibrium, one would expect there to be a relation between the stress tensors at nearby points in the body. Since the stress tensor at a point is completely determined by its behavior on the three coordinate planes, it is only necessary to relate the nine components of stress on these planes (three components on each plane) to the nine components at a nearby point. It will be shown that the nine components must satisfy a set of three partial differential equations.

Assume that $\sigma_x, \sigma_y, \sigma_z, \tau_{yx}, \tau_{zx}, \tau_{xy}, \tau_{zy}, \tau_{xz}, \tau_{yz}$ are each functions of x, y, and z, or in the matrix notation for a stress tensor in Cartesian coordinates assume that each T_{ij} is a function of x, y, and z. The proof of the relation will be given in two dimensions; the generalization to three dimensions will then be clear. Near the point P, the stress diagram (Fig. 1.10) can be drawn with $P(x, y)$ in the lower left-hand corner, and a nearby point $Q(x + \Delta x, y + \Delta y)$ in the upper right-hand corner.

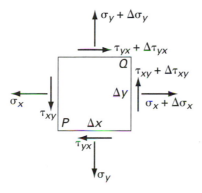

FIGURE 1.10   Stress element containing two nearby points.

View the rectangle as having sides Δx and Δy and thickness h in the z-direction. This rectangle is a piece of material in the body containing points P and Q. It is not an exploded view of the planes through P as in the case of the derivation of the stress tensor. The stress diagram shown must be converted to a free-body diagram to use the equilibrium condition. Multiplying each stress by the area of the face it acts on gives the total force on that face. Since the body is assumed to be in equilibrium, so is the element of the body on which we have drawn the stresses. The sum of the forces and moments must be zero on the free body diagram of the element.

Let $\mathbf{F} = F_x \mathbf{i} + F_y \mathbf{j} + F_z \mathbf{k}$ be the body force per unit volume, for example, the force due to gravity. Since the area of the x-face is $h \Delta y$ and the area of the y-face is $h \Delta x$, the sums of the forces in the x-direction and the y-directions are, respectively,

$$-\sigma_x \Delta y \, h + (\sigma_x + \Delta \sigma_x) \Delta y \, h - \tau_{yx} \Delta x \, h + (\tau_{yx} + \Delta \tau_{yx}) x \, h + F_x \Delta x \Delta y \, h = 0;$$
$$-\tau_{xy} \Delta y \, h + (\tau_{xy} + \Delta \tau_{xy}) \Delta y \, h - \sigma_y \Delta x \, h + (\sigma_y + \Delta \sigma_y) \Delta x \, h + F_y \Delta x \Delta y \, h = 0.$$

Dividing these equations by the volume, $\Delta x \Delta y \, h$ and taking the limits as Δx and Δy approach to zero independently yields the two-dimensional equilibrium equations,

$$\frac{\partial \sigma_x}{\partial x} + \frac{\partial \tau_{yx}}{\partial y} + F_x = 0;$$

$$\frac{\partial \tau_{xy}}{\partial x} + \frac{\partial \sigma_y}{\partial y} + F_y = 0.$$

The three-dimensional generalization follows the same pattern and is

$$\frac{\partial \sigma_x}{\partial x} + \frac{\partial \tau_{yx}}{\partial y} + \frac{\partial \tau_{zx}}{\partial z} + F_x = 0;$$

$$\frac{\partial \tau_{xy}}{\partial x} + \frac{\partial \sigma_y}{\partial y} + \frac{\partial \tau_{zy}}{\partial z} + F_y = 0; \qquad (1.6)$$

$$\frac{\partial \tau_{xz}}{\partial x} + \frac{\partial \tau_{yz}}{\partial y} + \frac{\partial \sigma_z}{\partial z} + F_z = 0.$$

EXAMPLE 1.7 A thin plate of width, w, of length, L, and of specific weight, ρ, is hung vertically from a ceiling (Fig. 1.11). Determine the state of stress at each point (x, y) in the plane of the strip. The plate is so thin that there is no significant stress variation through the thickness. Assume that $\tau_{xy}(x, y) = 0$ at each point.

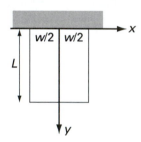

FIGURE 1.11 A thin plate hung vertically from a ceiling.

Solution Assume plane stress. The body forces are $F_x = 0$ and $F_y = \rho$. The equilibrium equation in the x-direction implies that $\partial \sigma_x / \partial x = 0$ and the equilibrium equation in the y-direction implies that $\partial \sigma_y / \partial y + \rho = 0$. Integration yields $\sigma_x(x, y) = c_1(y)$ and $\sigma_y(x, y) = -\rho y + c_2(x)$. The functions $c_1(y)$ and $c_2(x)$ are determined from the known stresses on the edges of the plate. On the sides, $\sigma_x(\pm w/2, y) = 0$ implies that $c_1(y) = 0$ and on the bottom, $\sigma_y(x, L) = 0$ implies that $c_2(x) = L\rho$. Therefore, $\sigma_x(x, y) = \tau_{xy}(x, y) = 0$ and $\sigma_y(x, y) = \rho(L - y)$. Note that the traction (force per area) exerted by the ceiling in the y-direction is $\sigma_y(x, 0) = \rho L$. ∎

EXAMPLE 1.8 The stresses given by strength of materials do not in some cases satisfy the stress equilibrium equations. In some cases, however, they do. An example is the cantilever beam with a point load at the free end (Fig. 1.12).

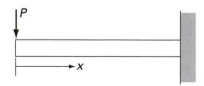

FIGURE 1.12 Side view of beam.

Solution   The internal moment is $M(x) = -Px$. The internal shear is $V = -P$. Let I be the area moment of inertia. The coordinates are taken with the origin at the centroid of the left face, the x-axis along the longitudinal axis of the beam and the y-axis downwards in the direction of the vertical edge of the beam. Then from strength of materials, the longitudinal normal stress is

$$\sigma_x = \frac{M(x)y}{I} = -\frac{Pxy}{I}.$$

Assume a rectangular cross section of thickness, t, and height, h.

The shear stress in the vertical direction equals the shear stress in the horizontal direction,

$$\tau = \frac{VQ}{It},$$

where $Q = y'A = \{[(h/2) + y]/2\}\{t[(h/2) - y]\}$, since the centroid of the shaded area is $y' = [(h/2) + y]/2$ and the area of the shaded region (Fig. 1.13) is $A = t[(h/2) - y]$.

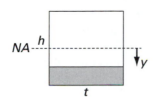

FIGURE 1.13   Beam cross-section with shaded area, A, for shear stress calculation.

Therefore

$$\tau = \frac{VQ}{It} = \frac{V[(h/2)^2 - y^2]}{2I}.$$

Then,

$$\frac{\partial \sigma_x}{\partial x} = -\frac{Py}{I}, \quad \text{and} \quad \frac{\partial \tau}{\partial y} = -\frac{2Vy}{2I} = \frac{Py}{I}.$$

Neglecting body forces,

$$\frac{\partial \sigma_x}{\partial x} + \frac{\partial \tau_{yx}}{\partial y} = -\frac{Py}{I} + \frac{Py}{I} = 0,$$

and the equilibrium equation is satisfied in the x direction.

Since $\tau = VQ/It$ is the only stress in the vertical direction ($\sigma_y = 0$ by the assumptions of strength of materials beam theory) and since $\partial V/\partial x = 0$, the equilibrium equation is satisfied in the vertical direction.

However, for any loading on a beam with rectangular cross-section in which $\partial V/\partial x \neq 0$, such as a constant distributed load, w, so that $V(x) = -wx$, the equilibrium equation is not satisfied in the vertical direction under the assumption of elementary strength of materials that plane sections remain planar. ∎

EXAMPLE 1.9   The example of the Introduction has a circular cross-section but is also a beam loaded by a transverse point load. Examine whether or not the strength of materials approximation to the stress given in the Introduction satisfies the stress equilibrium equations.

Solution   The equilibrium equation in the x-direction is

$$\frac{\partial \sigma_x}{\partial x} + \frac{\partial \tau_{yx}}{\partial y} + \frac{\partial \tau_{zx}}{\partial z} = \frac{4(L-a)P}{\pi R^4 L}y - \frac{4}{3}\frac{2y(L-a)P}{\pi R^4 L} \quad \text{if} \quad 0 < x < a$$

$$\frac{-4aPy}{\pi R^4 L} + \frac{4}{3}\frac{2aPy}{\pi R^4 L} \quad \text{if} \quad a < x < L$$

In neither portion of the rod is the sum zero for all points (x, y, z). Further, the derivatives are not defined at $x = a$ so that the stress equilibrium relation is also undefined at any point (a, y, z) in the rod. The strength of materials approximation to the stress field often fails to satisfy the stress equilibrium equation. However, for beams several times longer than their width, the strength of materials approximation is often satisfactory. ∎

Exercise 1.4   Show that the strength of materials equations for stress for a constant distributed load on a rectangular cross-section cantilever beam satisfy the equation of the equilibrium in the x-direction (*Hint*: $\partial M / \partial x = V$), but not always in the y-direction.

1.5   COORDINATE CHANGES AND THE STRESS TENSOR

If the matrix representation of the stress tensor is known, then the stress components on the coordinate planes can be read from the entries of the matrix. On the other hand, when the two-dimensional Mohr's circle is used to find the stress components on other planes, the normal and shear components are obtained, not the components in the coordinate directions. If we can rotate the coordinates in which the stress tensor is expressed so that they are in the normal and shear directions, then we could read the normal and shear components from the entries in the stress tensor represented as a matrix in the new coordinate system.

In strength of materials, when deriving the Mohr's circle from the equilibrium condition on a two-dimensional stress element, the following equations were obtained for the normal and shear components in the x'-y' coordinate system, which is rotated θ degrees counterclockwise from the x-y axes.

$$\sigma_{x'} = \sigma_x \cos^2 \theta + \sigma_y \sin^2 \theta + 2\tau_{xy} \cos \theta \sin \theta;$$
$$\tau_{x'y'} = \tau_{xy}(\cos^2 \theta - \sin^2 \theta) + (\sigma_y - \sigma_x) \cos \theta \sin \theta.$$

The x'-y' coordinate axes are normal and tangent to the plane of interest. This system is rewritten in terms of 2θ using trigonometric identities,

$$\sigma_{x'} = \frac{1}{2}(\sigma_x + \sigma_y) + \frac{1}{2}(\sigma_x - \sigma_y) \cos 2\theta + \tau_{xy} \sin 2\theta;$$

$$\tau_{x'y'} = \tau_{xy} \cos 2\theta + \frac{1}{2}(\sigma_y - \sigma_x) \sin 2\theta. \tag{1.7}$$

These values are interpreted geometrically as the coordinates of a point on Mohr's circle. The goal is to recover equations (1.7) by a matrix multiplication of the stress tensor in x-y coordinates at the point P. To do so, it will be much easier to use an index notation based on numbering the coordinate axes rather than naming them with letters. Let the x- and y-axes be called 1 and 2, respectively, and the x'- and y'-axes be called $1'$ and $2'$, respectively. Let $\mathbf{e}_i$ be the unit vector in the direction of the positive i^{th} axis, and likewise let $\mathbf{e}'_i$ be the unit vector in the direction of the positive i'^{th} axis. Define the matrix of direction cosines $\alpha = (\alpha_{ij})$ by defining α_{ij} to be the direction cosine of the "new" coordinate, $\mathbf{e}'_i$, with respect to the "old" coordinate, $\mathbf{e}_j$. By the definition of dot product of two vectors as the product of their magnitudes and the cosine of the angle between them, since the direction cosine is the cosine of the angle between the two vectors,

$$\alpha_{ij} = \mathbf{e}'_i \cdot \mathbf{e}_j.$$

In a two-dimensional rotation through θ radians counterclockwise,

$$\alpha_{11} = \cos\theta; \qquad \alpha_{12} = \cos(\pi/2 - \theta) = \sin\theta;$$
$$\alpha_{21} = \cos(\pi/2 + \theta) = -\sin\theta; \qquad \alpha_{22} = \cos\theta.$$

The matrix of direction cosines, for a counterclockwise rotation, is then

$$\alpha = \begin{pmatrix} \cos\theta & \sin\theta \\ -\sin\theta & \cos\theta \end{pmatrix}.$$

The matrix of direction cosines represents the linear transformation from the old coordinate system to the new coordinate system.

Definition An orthogonal matrix is one such that $\alpha^t = \alpha^{-1}$, that is, $\alpha^t\alpha = \alpha\alpha^t = I$. The identity matrix $I = (\delta_{ij})$, where δ_{ij} is the Kronecker delta defined by $\delta_{ij} = 1$ if $i = j$ and $\delta_{ij} = 0$ if $i \neq j$.

The matrix, α, of direction cosines is an orthogonal matrix; its determinant is 1. While the matrix α is a linear transformation from the old coordinate system to the new coordinate system, the transpose, α^t, takes the new coordinates back to the old since it is also the inverse transformation.

EXAMPLE 1.10 Transform a vector written in terms of the basis $\mathbf{i}$ and $\mathbf{j}$ to the same vector written in terms of the basis $\mathbf{u}_1 = (\mathbf{i} - \mathbf{j})/\sqrt{2}$ and $\mathbf{u}_2 = (\mathbf{i} + \mathbf{j})/\sqrt{2}$. The basis vectors act in the directions of the positive coordinate axes. Therefore this transformation is a clockwise rotation of $45°$. The matrix of direction cosines is

$$\alpha = \begin{pmatrix} \cos(-45) & \sin(-45) \\ -\sin(-45) & \cos(-45) \end{pmatrix} = \begin{pmatrix} \frac{1}{\sqrt{2}} & \frac{-1}{\sqrt{2}} \\ \frac{1}{\sqrt{2}} & \frac{1}{\sqrt{2}} \end{pmatrix}.$$

So, for example, the basis vector, $\mathbf{i}$, written as $1\mathbf{i} + 0\mathbf{j}$ is transformed to

$$\begin{pmatrix} \frac{1}{\sqrt{2}} & \frac{-1}{\sqrt{2}} \\ \frac{1}{\sqrt{2}} & \frac{1}{\sqrt{2}} \end{pmatrix} \begin{pmatrix} 1 \\ 0 \end{pmatrix} = \begin{pmatrix} \frac{1}{\sqrt{2}} \\ \frac{1}{\sqrt{2}} \end{pmatrix}$$

$= (\mathbf{u}_1 + \mathbf{u}_2)/\sqrt{2}$ in the new basis. This is the same vector as $\mathbf{i}$ but simply written in terms of a new set of basis vectors, as can be seen by substituting the expressions for $\mathbf{u}_1$ and $\mathbf{u}_2$ in terms of $\mathbf{i}$ and $\mathbf{j}$. ∎

The rotation matrix can be used to reach the goal of obtaining equations (1.7) by matrix multiplication. The tensor multiplication $T(\mathbf{n})$ gives the stress vector, $\mathbf{t}$, on the plane with normal, $\mathbf{n}$, written in components in the old coordinate system. We want the *same* vector, $\mathbf{t}$, but merely written in new coordinates whose directions are normal and tangent to the plane on which the desired stress vector acts. The scalar components of the stress vector, $\mathbf{t}$, in this new coordinate system are what we usually call the normal and shear stresses.

The tensor, T', in the new coordinates can be written in matrix multiplication as

$$T' = \alpha T \alpha^t. \tag{1.8}$$

Therefore

$$T' = \begin{pmatrix} \cos\theta & \sin\theta \\ -\sin\theta & \cos\theta \end{pmatrix} \begin{pmatrix} \sigma_x & \tau_{yx} \\ \tau_{xy} & \sigma_y \end{pmatrix} \begin{pmatrix} \cos\theta & -\sin\theta \\ \sin\theta & \cos\theta \end{pmatrix}$$

$$= \begin{pmatrix} \sigma_x\cos^2\theta + \sigma_y\sin^2\theta + 2\tau_{xy}\cos\theta\sin\theta & \tau_{xy}(\cos^2\theta - \sin^2\theta) + (\sigma_y - \sigma_x)\cos\theta\sin\theta \\ \tau_{xy}(\cos^2\theta - \sin^2\theta) + (\sigma_y - \sigma_x)\cos\theta\sin\theta & \sigma_y\cos^2\theta + \sigma_x\sin^2\theta - 2\tau_{xy}\cos\theta\sin\theta \end{pmatrix}.$$

The matrix for T in the new coordinates is

$$T' = \begin{pmatrix} \sigma_{x'} & \tau_{y'x'} \\ \tau_{x'y'} & \sigma_{y'} \end{pmatrix}. \tag{1.9}$$

Setting corresponding entries equal in the two expressions for T', the equilibrium equations (1.7) are recovered, showing that in fact $T' = \alpha T \alpha^t$. Notice that $\tau_{x'y'} = \tau_{y'x'}$, as required. Furthermore, the matrix multiplication technique also easily provides the expression for $\sigma_{y'}$. In three dimensions, the same relation holds, $T' = \alpha T \alpha^t$.

EXAMPLE 1.11   New three-dimensional coordinates are obtained from the old by rotating about the 3 axis θ degrees in the counterclockwise direction. Then the matrix of direction cosines is

$$\alpha = \begin{pmatrix} \cos\theta & \sin\theta & 0 \\ -\sin\theta & \cos\theta & 0 \\ 0 & 0 & 1 \end{pmatrix}. \qquad \blacksquare$$

This analysis means that to find the normal and shear stresses on a given plane, one needs only to multiply the stress tensor by the matrix of direction cosines as in equation (1.8) and then read off the stress components from the entries of the new matrix obtained. In two dimensions it is probably easier to use Mohr's circle. In three dimensions, however, the matrix multiplication technique is far superior to the somewhat mysterious geometric construction needed to extract the stress components from a three-dimensional Mohr's circle. Furthermore, the Mohr's circle geometric construction will only produce an approximation, while the matrix multiplication produces the exact answer quickly.

The behavior of a tensor T as above under changes of coordinates, that is, changes of basis vectors, is often included in the definition of tensor instead of saying explicitly that a tensor is a linear transformation. This definition is often used in continuum mechanics and other advanced solid mechanics courses. It says that a tensor is well defined independently of any chosen coordinate system. This is consistent with the requirement that any physical law must be independent of the coordinates in which it is described.

Definition   A second order tensor T_{ij} is a mathematical object that obeys the transformation rule $T' = \alpha T \alpha^t$.

The tensor transformation relation is often expressed in indicial notation as

$$T'_{ij} = \alpha_{ik} \alpha_{jr} T_{kr}.$$

The expression on the right is to be summed over each repeated pair of indices, k and r, from one to three so that T'_{ij} contains nine summands. This is called the Einstein convention. In the case of Cartesian coordinates, the index relation is the same as the component expression for the matrix multiplication, $T' = \alpha T \alpha^t$.

1.6   PRINCIPAL STRESSES

The study of strength of materials and Mohr's circles show that the normal stresses whose absolute values are largest of all normal stresses at a point occur on the planes with no shear stress components.

Definition   The principal stresses, **t**, at a point P in the body are those that are perpendicular to the plane on which they act; they have no shear components.

Even in three dimensions it is fairly easy to determine the magnitudes of the principal stresses directly from the three Mohr's circles. However, it is not simple in three dimensions to identify on which planes the principal stresses act by finding normals to the planes. The principal stress values and normal vectors to the planes on which they act can be computed directly from the matrix representing the stress tensor at the point.

If **t** is a principal stress acting on a plane with unit normal vector, **n**, then since both **t** and **n** are perpendicular to the same plane, they must be parallel. Two vectors that are parallel are scalar multiples of each other. Therefore, the definition of principal stress requires that there is some scalar, λ, which may be positive, negative, or zero such that

$$\mathbf{t} = T(\mathbf{n}) = \lambda\mathbf{n}.$$

Since **n** is a unit vector, the scalar λ is the magnitude of the principal stress, **t**. Therefore

$$(T - \lambda I)\mathbf{n} = \mathbf{O}. \tag{1.10}$$

Every linear transformation, T, takes **O** to **O**. If it takes any other vector to **O** as well, then it is not a one to one function. In this case it is called singular; it is shown in linear algebra that the determinant of any singular matrix is zero. The fact that $\mathbf{n} \neq \mathbf{O}$, or it would not be normal to a plane, implies that $T - \lambda I$ is singular and that its determinant is zero.

The scalars, λ, which make $T - \lambda I$ singular, are called the eigenvalues of T. The vectors, **n**, which correspond to the λ and satisfy equation (1.10), are called the eigenvectors of T. The eigenvalues are the magnitudes of the principal stresses, and the eigenvectors are in the directions of the principal stresses.

The condition to find the scalars, λ, has now been reduced to the scalar equation

$$\det(T - \lambda I) = 0. \tag{1.11}$$

Expanding the determinant in equation (1.11) produces the following cubic equation, called the characteristic equation of the matrix, T.

$$\lambda^3 - I_1\lambda^2 + I_2\lambda - I_3 = 0, \tag{1.12}$$

where

$$I_1 = \sigma_x + \sigma_y + \sigma_z \equiv \text{Trace}(T);$$
$$I_2 = \sigma_x\sigma_y + \sigma_x\sigma_z + \sigma_y\sigma_z - \tau_{xy}^2 - \tau_{xz}^2 - \tau_{yz}^2;$$
$$I_3 = \det(T).$$

Exercise 1.5   Verify equation (1.12) by expanding the determinant.

The numbers I_1, I_2, and I_3 are called invariants of T since they are the same no matter in what coordinate system the matrix for T is written. In other words, T and $T' = \alpha T \alpha^t$ have the same invariants. This is not a surprise since the magnitudes of the principal stresses are independent of the coordinate system. Therefore, the characteristic equation must be the same for any coordinate system, and its coefficients, I_1, I_2, I_3, must be independent of the coordinates. The scalars, λ, the eigenvalues, are the same in all coordinate systems for T.

A cubic equation has three roots, so equation (1.12) gives all three principal stress magnitudes. Furthermore, the three roots must be real; it is physically impossible for two of them to be complex. In linear algebra, it is proved that a real, symmetric matrix, such as T, has three real eigenvalues. This implies that the discriminant of the cubic equation cannot be positive. The only computational difficulty once the cubic equation is written is to determine its roots. If a factor can be guessed, then division by that factor gives a quadratic equation that can then be solved by the quadratic formula. Alternatively, the roots can be determined numerically on a computer or calculator. Otherwise, the roots can be found using the following procedure to obtain the roots of cubic equation of the form of equation (1.12). First compute the auxiliary terms,

$$a = \frac{1}{3}I_1^2 - I_2;$$

$$b = \frac{1}{3}I_1 I_2 - I_3 - \frac{2}{27}I_1^3;$$

$$c = \left(\frac{1}{27}a^3\right)^{1/2};$$

$$\alpha = \arccos\left(-\frac{b}{2c}\right);$$

$$\beta = \left(\frac{1}{3}a\right)^{1/2}.$$

Here α must be in degrees. Then the roots are

$$\sigma_1 = 2\beta \cos\left(\frac{1}{3}\alpha\right) + \frac{1}{3}I_1;$$

$$\sigma_2 = 2\beta \cos\left(\frac{1}{3}\alpha + 120\right) + \frac{1}{3}I_1;$$

$$\sigma_3 = 2\beta \cos\left(\frac{1}{3}\alpha + 240\right) + \frac{1}{3}I_1.$$

EXAMPLE 1.12   Determine the roots of $\lambda^3 + 11\lambda^2 - 144\lambda - 324 = 0$ using the above algorithm.

Solution   Note first that $I_1 = -11$, $I_2 = -144$, and $I_3 = 324$. Then $a = 184.33$, $b = 302.59$, $c = 481.64$, $\alpha = 108.31°$, and $\beta = 7.8386$. Therefore $\sigma_1 = 9$, $\sigma_2 = -18$, and $\sigma_3 = -2$.   ■

As shown in strength of materials by Mohr's circles, the three principal stresses act on three mutually perpendicular planes. Therefore, the normal vectors defining the planes on which the principal stresses act must also be mutually perpendicular. A coordinate system oriented in these directions is called the principal coordinate system. The representation of the stress tensor in principal coordinates is a diagonal matrix, T', whose entries along the diagonal are the magnitudes of the principal stresses and whose other entries are zero (since there are no shear stress components on the principal planes). The stress, T, in the given coordinate system is related to the diagonal stress matrix, T', in the principal coordinates by $T' = \alpha T \alpha^t$. It is shown in linear algebra that the rows of the transformation matrix α have entries which are the components of the three mutually perpendicular unit eigenvectors for T. In general, the principal coordinate systems at two different points in the body are not coincident; the principal directions change from point to point in the body.

EXAMPLE 1.13   At a point P in a body and in the given coordinate system, the stress element in Figure 1.14 has been determined.

(a) Write the matrix representing the stress tensor.

(b) Determine the magnitudes of the principal stresses and the normals to the planes on which they act.

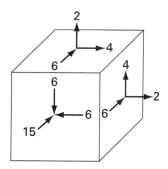

FIGURE 1.14   The stress element.

Solution   (a) The stress tensor is formed by reading the stress components from the element,

$$T = \begin{pmatrix} 2 & 4 & -6 \\ 4 & 2 & -6 \\ -6 & -6 & -15 \end{pmatrix}.$$

(b) The characteristic cubic equation is $\lambda^3 + 11\lambda^2 - 144\lambda - 324 = 0$. This factors into $(\lambda + 2)(\lambda + 18)(\lambda - 9) = 0$ so that the magnitudes of the principal stresses are $\lambda = 9, -2,$ and -18.

To find the normal vector, $\mathbf{n} = a\mathbf{i} + b\mathbf{j} + c\mathbf{k}$, to the plane on which the principal stress -2 acts, in other words the direction of the principal stress having magnitude -2, write the equations for the functional relation $T(\mathbf{n}) = -2\mathbf{n}$.

$$2a + 4b - 6c = -2a;$$
$$4a + 2b - 6c = -2b;$$
$$-6a - 6b - 15c = -2c.$$

Now collect terms to obtain

$$4a + 4b - 6c = 0;$$
$$4a + 4b - 6c = 0;$$
$$-6a - 6b - 13c = 0.$$

There is no unique solution to this set of three equations since the first two equations are the same. However, multiplying the first equation by 6 and the third by 4 and then adding the result shows that $c = 0$. Substituting $c = 0$ into all three equations produces the same equation in each case. The resulting equations are satisfied if $a = -b$. So the form of the set of solutions is $(a, -a, 0)$. But vectors of the form $a\mathbf{i} - a\mathbf{j}$ are all in the same direction as long as $a \neq 0$. So the required answer is the unit vector in that direction, $\mathbf{n}_1 = (1/\sqrt{2})\mathbf{i} - (1/\sqrt{2})\mathbf{j}$.   ∎

Exercise 1.6   Show in a similar manner that the other two directions of the principal stresses are $\mathbf{n}_2 = (1/\sqrt{18})\mathbf{i} + (1/\sqrt{18})\mathbf{j} + (4/\sqrt{18})\mathbf{k}$ and $\mathbf{n}_3 = (2/3)\mathbf{i} + (2/3)\mathbf{j} - (1/3)\mathbf{k}$. Then show, using the dot product, that the vectors, $\mathbf{n}_1, \mathbf{n}_2,$ and $\mathbf{n}_3$ are mutually perpendicular, as expected from strength of materials. Verify that these vectors form a right-handed coordinate system. If not, construct a right-handed system by taking the negatives of some of the unit vectors, $\mathbf{n}_i$.

The primary advantage of the matrix approach to finding stresses on noncoordinate planes is that the computation is simpler in three-dimensional problems. In two-dimensional problems, the Mohr's circle calculation is usually easier. However, to help clarify the matrix approach and to help relate it to what was learned in strength of materials, the following two-dimensional problem in which it is easy to see exactly what is going on is solved by both methods.

EXAMPLE 1.14

From the two-dimensional stress element at a point P in a body shown in Figure 1.15,

(a) Determine the stress normal and shear stress components on a face 30° counterclockwise from the x-face.

(b) Determine the magnitude and direction of the two principal stresses.

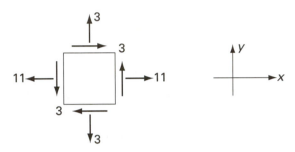

FIGURE 1.15 The two-dimensional stress element.

Solution

The solution will be found first from vectors and the stress tensor and then from Mohr's circle.

Tensor Approach

The matrix representation of the stress tensor in the indicated coordinate system is

$$T = \begin{pmatrix} 11 & 3 \\ 3 & 3 \end{pmatrix}.$$

(a) To determine the stress on the plane 30° counterclockwise from the x-face (perpendicular to the x-direction) in Figure 1.16, a unit normal to this plane must first be found.

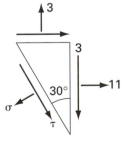

FIGURE 1.16 Two-dimensional stress element including the 30° face.

The unit normal vector is $\mathbf{n} = \cos 30\mathbf{i} + \cos 60\mathbf{j} = (\sqrt{3}/2)\mathbf{i} + (1/2)\mathbf{j}$. The stress vector in x-y coordinates is

$$\mathbf{t} = \begin{pmatrix} 11 & 3 \\ 3 & 3 \end{pmatrix} \begin{pmatrix} \frac{\sqrt{3}}{2} \\ \frac{1}{2} \end{pmatrix} = \begin{pmatrix} T_x \\ T_y \end{pmatrix} = \begin{pmatrix} 11.026 \\ 4.098 \end{pmatrix}.$$

Two methods are available to obtain the shear and normal components of this vector.

METHOD 1: The first method is to use the matrix of direction cosines to transform the x-y coordinates into an x'-y' system normal and tangent to the 30° plane,

$$\alpha = \begin{pmatrix} \cos 30 & \sin 30 \\ -\sin 30 & \cos 30 \end{pmatrix}.$$

Then using the fact that $T' = \alpha T \alpha'$, one obtains

$$T' = \begin{pmatrix} 11.60 & -1.96 \\ -1.96 & 2.402 \end{pmatrix}.$$

Therefore

$$\sigma_{x'} = 11.60, \quad \sigma_{y'} = 2.402, \quad \tau_{x'y'} = -1.96,$$

where both normal stresses are in tension and the shear stress on the x'-face points in the negative y'-direction.

METHOD 2: Project $\mathbf{t}$ onto $\mathbf{n}$ to obtain

$$\sigma_{x'} = \mathbf{t} \cdot \mathbf{n} = 11.026 \cos 30 + 4.098 \sin 30 = 11.60.$$

This is the component of $\mathbf{t}$ in the direction of $\mathbf{n}$. The shear component is $\tau = \mathbf{t} - \sigma_{x'}\mathbf{n}$. The Pythagorean theorem gives $\tau_{x'y'} = (|\mathbf{t}|^2 - \sigma_{x'}^2)^{1/2}$. See Figure 1.17.

FIGURE 1.17   The components of the stress $\mathbf{t}$.

(b) The principal stress magnitudes are found as the eigenvalues of the stress matrix. The characteristic equation,

$$0 = \begin{vmatrix} 11-\lambda & 3 \\ 3 & 3-\lambda \end{vmatrix} = \lambda^2 - 14\lambda + 24 = (\lambda - 12)(\lambda - 2),$$

so that $\lambda = 2$ and 12 are the principal stresses.

The direction of the principal stress with magnitude 12 is found from

$$\begin{pmatrix} 11 & 3 \\ 3 & 3 \end{pmatrix} \begin{pmatrix} a \\ b \end{pmatrix} = 12 \begin{pmatrix} a \\ b \end{pmatrix};$$

or

$$11a + 3b = 12a;$$
$$3a + 3b = 12b;$$

which reduce to

$$-a + 3b = 0;$$
$$3a - 9b = 0.$$

Both these equations are the same and imply that $a = 3b$. This says that all vectors perpendicular to this principal plane are of the form $\mathbf{N} = 3b\mathbf{i} + b\mathbf{j}$. The unit vector in this direction is $\mathbf{n}_1 = (3/\sqrt{10})\mathbf{i} + (1/\sqrt{10})\mathbf{j}$. The angle, θ, that this plane makes with the x-coordinate plane is given by $\cos \theta = 3/\sqrt{10}$ or $\sin \theta = 1/\sqrt{10}$, so that $\theta = 18.32°$ counterclockwise.

Exercise 1.7 Show that $\mathbf{n}_2 = (1/\sqrt{10})\mathbf{i} - (3/\sqrt{10})\mathbf{j}$ and that $\mathbf{n}_1$ and $\mathbf{n}_2$ are perpendicular.

Mohr's Circle Approach

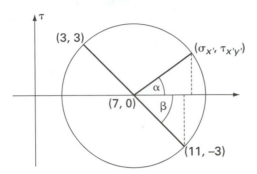

FIGURE 1.18 Mohr's circle for the two-dimensional state of stress.

(a) On a Mohr's circle (Figure 1.18), the stress on the y-face corresponds to the point $(3, 3)$ and the stress on the x-face corresponds to the point $(11, -3)$. Since these stresses act on faces which are at an angle of $90°$ with respect to each other, the points are located $180°$ from each other on the circle; in other words, the line between them is a diameter. Therefore the center is at the midpoint of this line segment and its coordinates are found by the midpoint formula to be $(7, 0)$. The radius, R, of the Mohr's circle is found by computing the distance from the center to one of the points, say that corresponding to the stress on the y-face.

$$R = [(7 - 3)^2 + (0 - 3)^2]^{1/2} = 5.$$

Now to find the point on the circle corresponding to the plane at a $30°$ counterclockwise angle from the x-face, compute the angles α and β shown on the circle.

$$\sin \beta = 3/5 \quad \text{implies} \quad \beta = 36.9;$$
$$\alpha = 60° - 36.9° = 23.1°.$$

Therefore

$$\sigma_{x'} = 7 + 5\cos(23.1) = 11.60;$$
$$\sigma_{y'} = 7 - 5\cos(23.1) = 2.401;$$
$$\tau_{x'y'} = 5\sin(23.1) = 1.96.$$

The sign of $\tau_{x'y'}$ from the tensor method is opposite the sign of $\tau_{x'y'}$ from the Mohr's circle method because of the different shear stress sign conventions used in the two methods. A drawing shows that the two coincide.

(b) The principal stresses from the Mohr's circle are

$$\sigma_1 = 7 + 5 = 12;$$
$$\sigma_2 = 7 - 5 = 2,$$

which agrees with the eigenvalues of the stress tensor matrix. The angle from the x-face to the point representing the principal stress 12 is $\beta = 36.9°/2 = 18.45°$ counterclockwise agreeing with the angle that the normal to that principal plane makes with the x-plane found in the tensor method. ∎

   The orientation of the planes on which the principal stresses act changes, in general, from point to point in the body. The location of an initial crack in a brittle member depends on the locations of flaws on the surface of the body as well as the magnitude of the principal stresses at the point. The crack will usually initiate on a principal plane, which may be oriented differently with respect to the body coordinates at different points in the body. Likewise, a ductile material may yield at the various points of the body along maximum shear planes oriented differently with respect to the body coordinates.

EXAMPLE 1.15

The variation in the orientation of the principal planes can be computed easily in the case of the strength of materials stress field for the cantilever beam with a downward point load, P, at the free end. The cross-section is rectangular. The coordinates are taken with the x-axis along the longitudinal axis of the beam so that $x = 0$ at the free end and the y-axis downward in the direction of the load and of the vertical edge of the beam. The internal moment at position x along the beam is $M(x) = -Px$ and the shear force is $V = P$. The stresses form a two-dimensional system as calculated in Example 1.8.

$$\sigma_x = -\frac{Pxy}{I} \quad \text{and} \quad \tau_{xy} = \frac{V[(h/2)^2 - y^2]}{2I}.$$

The angle, θ, that the x-face, the y-z plane, makes with the principal face on which the largest principal stress acts is computed from the Mohr's circle.

$$\sin(2\theta) = \frac{2\tau_{xy}}{\sqrt{\sigma_x^2 + 4\tau_{xy}^2}} = \frac{V[(h/2)^2 - y^2]}{\sqrt{P^2x^2y^2 + V^2[(h/2)^2 - y^2]^2}}.$$

At the top and bottom of the beam, the principal face is the y-z plane. At the middle of the beam ($y = 0$), the principal face is at a 45° angle with the y-z plane and perpendicular to the x-z plane. The continuous variation of principal plane orientation between these two positions depends on the position x along the length of the beam.

The direction that a crack propagates through the body cannot be predicted by these calculations because the state of stress is altered once the crack is initiated.

In many design calculations, the designer merely uses the points at which the radius of the Mohr's circle is largest as the most dangerous points. In this example, these points of largest principal stress lie at the wall at the top and bottom of the beam ($x = L$ and $y = \pm h/2$). However, this may be a risky strategy if flaws or other points of discontinuity occur in the material. ∎

1.7 OCTAHEDRAL STRESSES

The stresses on the octahedral planes are very useful in predicting failure of a material, especially ductile metals. A ductile metal in uniaxially loading yields along planes at a 45° angle with the plane normal to the load. Under a three-dimensional state of stress, yielding is likely to occur on the octahedral planes. It will be far easier to compute the stresses on these planes if the coordinate system is chosen in the direction of the three mutually perpendicular principal stresses. Recall the principal directions change from point to point in the body.

Definition The octahedral planes are those whose intersections with the principal coordinate planes make 45° angles with the principal coordinate axes.

To compute the stress vector on the octahedral plane in the first quadrant of the principal axes (Figure 1.19), one needs to find a unit normal vector to the plane and then multiply its column matrix representation with the diagonal matrix representing the stress tensor.

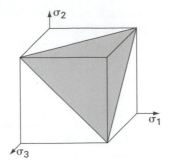

FIGURE 1.19   An octahedral plane in a principal coordinate system.

A normal vector $\mathbf{N}$ can be computed by taking the cross product of any two vectors in the octahedral plane. For example,

$$\mathbf{N} = (-\mathbf{i} + \mathbf{k}) \times (-\mathbf{j} + \mathbf{k}) = \mathbf{i} + \mathbf{j} + \mathbf{k}.$$

The unit normal vector is obtained by dividing $\mathbf{N}$ by its length,

$$\mathbf{n} = (1/\sqrt{3})\mathbf{i} + (1/\sqrt{3})\mathbf{j} + (1/\sqrt{3})\mathbf{k}.$$

Then $T(\mathbf{n}) = (\sigma_1/\sqrt{3})\mathbf{i} + (\sigma_2/\sqrt{3})\mathbf{j} + (\sigma_3/\sqrt{3})\mathbf{k}$. The components of this vector are in the principal directions. For failure theory, the normal and shear components are more useful. The normal component is the magnitude of the component, σ, perpendicular to the octahedral plane. This vector component is the projection of $T(\mathbf{n})$ onto the unit normal vector to the plane, $\mathbf{n} = (1/\sqrt{3})\mathbf{i} + (1/\sqrt{3})\mathbf{j} + (1/\sqrt{3})\mathbf{k}$. Its magnitude can be computed using the dot product of $T(\mathbf{n})$ and $\mathbf{n}$:

$$|\sigma| = T(\mathbf{n}) \cdot \mathbf{n} = \frac{1}{3}(\sigma_1 + \sigma_2 + \sigma_3).$$

The normal stress vector component is its magnitude times the unit vector in its direction,

$$\sigma = (T(\mathbf{n}) \cdot \mathbf{n})\mathbf{n} = \left[\frac{1}{3}(\sigma_1 + \sigma_2 + \sigma_3)\right]\left(\frac{1}{\sqrt{3}}\mathbf{i} + \frac{1}{\sqrt{3}}\mathbf{j} + \frac{1}{\sqrt{3}}\mathbf{k}\right).$$

The shear vector component is

$$\tau = T(\mathbf{n}) - \sigma.$$

Then the magnitude of the shear component can be found from the generalized Pythagorean theorem,

$$|\tau| = [|T(\mathbf{n})|^2 - |\sigma|^2]^{1/2} = \left\{\frac{1}{3}(\sigma_1^2 + \sigma_2^2 + \sigma_3^2) - \left[\frac{1}{3}(\sigma_1 + \sigma_2 + \sigma_3)\right]^2\right\}^{1/2}$$

$$= \frac{1}{3}[(\sigma_1 - \sigma_2)^2 + (\sigma_2 - \sigma_3)^2 + (\sigma_3 - \sigma_1)^2]^{1/2}. \tag{1.13}$$

This result is independent of any coordinate system because the principal stress magnitudes, σ_1, σ_2, and σ_3, are independent of coordinates. The technique used to find the shear and normal components by projections can be used to find those of any vector given $T(\mathbf{n})$ without transforming the coordinates.

Exercise 1.8 Verify the last equality in (1.13) by multiplying out the terms.

In two-dimensional states of stress, such as plane stress in which $\sigma_3 = 0$, the magnitude of the shear stress on the octahedral face is

$$|\tau| = \frac{\sqrt{2}}{3}\sqrt{\sigma_1^2 - \sigma_1\sigma_2 + \sigma_2^2}.$$

This stress is the key indicator for an important theory of yielding. Therefore, the expression

$$\sigma_m = \sqrt{\sigma_1^2 - \sigma_1\sigma_2 + \sigma_2^2} \tag{1.14}$$

is given a special name, the von Mises stress. This value is part of the output of many commercial finite element packages, which produce numerical approximations to the stress field. The shear stress magnitude can be calculated directly from the stress field in any coordinate system without computing the principal stresses. The octahedral shear stress can be written in an arbitrary Cartesian coordinate system as

$$|\tau| = \frac{1}{3}[(\sigma_x - \sigma_y)^2 + (\sigma_y - \sigma_z)^2 + (\sigma_z - \sigma_x)^2 + 6(\tau_{xy}^2 + \tau_{xz}^2 + \tau_{yz}^2)]^{1/2}. \tag{1.15}$$

Exercise 1.9 Verify this result in the case of a two-dimensional state of stress ($\sigma_z = \tau_{xz} = \tau_{yz} = 0$).

PROBLEMS

1.1 A cube of material with edges of length 1 is given a coordinate system parallel to its edges with the origin at the lower left front corner and the x-axis along the horizontal front edge.

(a) Determine the unit normal vector to the plane passing through the origin and the points $(1, 1/2, -1)$ and $(1, 1, 0)$.

(b) Does the point $(0, 0, -1)$ lie on this plane?

1.2 A cubical block of material is given a coordinate system parallel to its edges. Determine the shear stress component in the $-\mathbf{i} + \mathbf{j} + \mathbf{k}$ direction on the plane with normal $\mathbf{N} = \mathbf{i} + \mathbf{j}$ if the stress applied to the plane is

(a) $\mathbf{t} = \sigma\mathbf{k}$;

(b) $\mathbf{t} = \sigma(\mathbf{j} + \mathbf{k})$.

1.3 In what unit direction in the plane with normal $\mathbf{i} + \mathbf{j} + \mathbf{k}$ is the shear stress magnitude a maximum if the stress on this plane is that given in the two cases (a) and (b) of problem 1.2?

1.4 Use the state of stress defined in strength of materials to write the three-dimensional stress tensor at a point (x, y, z) in the following members.

(a) A square cross-sectional bar with edge 2 inches and

length 20 inches is mounted in a rigid wall and is loaded axially by a force, P. Put the origin at the wall and at the center of the cross section.

(b) A circular cross-sectional rod with radius 2 inches and length 20 inches is mounted in a rigid wall and is loaded at the free end by a torque, M_t. Put the origin at the wall and at the center of the cross section. Use cylindrical coordinates.

(c) A square cross-sectional beam with edge 2 inches and length 20 inches is mounted in a rigid wall and is loaded by a vertical force, P, applied at the free end. Put the origin at the wall and at the center of the cross section. Do not neglect shear.

1.5 Explain why the state of stress at a point on a free surface of a body is plane stress.

1.6 Determine whether or not the stress field

$$\sigma_x = y^2 + \nu(x^2 - y^2), \quad \sigma_y = x^2 + \nu(y^2 - x^2),$$
$$\tau_{xy} = -2\nu xy$$

satisfies equilibrium. The body force is $\mathbf{F} = 547y\mathbf{i} - 731xy\mathbf{j}$ (newtons).

1.7   Determine whether or not the stress field

$$\sigma_x = xy^2 + v(x^2 - y^2), \quad \sigma_y = x^2y + v(y^2 - x^2),$$
$$\tau_{xy} = -2vxy - y^3/3$$

satisfies equilibrium. The body force is $\mathbf{F} = -3xy\mathbf{j}$ (newtons).

1.8   An axially loaded rod is given a coordinate system with the x-axis along the longitudinal axis of the rod.

(a)   Is it possible for the rod to be in equilibrium if the axial normal stress is $\sigma_x = 55x$ (kPa)?

(b)   What form must σ_x take if the rod is to be in equilibrium under an axial load, assuming that body forces are neglected?

1.9   A thin plate of width, w, of length, L, and of specific weight, ρ, is somehow balanced vertically on the floor on one of its width edges. Determine the state of stress at each point (x, y) of the plate. Assume that $\tau_{xy}(x, y) = 0$ at each point.

1.10   Does the tensor, T, for a rectangular cross-section beam with a constant distributed load, w, and constant area moment of inertia, I, satisfy the stress equilibrium equations? The longitudinal direction is x, and y is the direction perpendicular to the bending axis. The constant h is half the height of the beam. Neglect body forces.

$$T = \begin{pmatrix} \frac{wx^2y}{2I} & \frac{wx(h^2-y^2)}{2I} \\ \frac{wx(h^2-y^2)}{2I} & 0 \end{pmatrix}.$$

1.11   The stress at a point P on a body is described by $\sigma_x, \sigma_y, \sigma_z, \tau_{xy}, \tau_{xz}, \tau_{yz}$, in a given coordinate system x-y-z.

(a)   Write the direction cosine matrix, α, for a 30° clockwise rotation of the coordinates about the x-axis.

(b)   Write the stress tensor in the new coordinate system in terms of the stress tensor in the original coordinate system and α.

1.12   An $n \times n$ matrix $A = (a_{ij})$ is called orthogonal if $\sum_{k=1}^{n} a_{ik}a_{jk} = \delta_{ij}$, where δ_{ij} is the Kronecker delta ($\delta_{ij} = 1$ if $i = j$ and $\delta_{ij} = 0$ if $i \neq j$).

(a)   Verify that this is the representation of $AA^t = A^tA = I$ in terms of the elements of A.

(b)   Show that, for any angle of rotation θ, the two-dimensional matrix of direction cosines, α, is orthogonal.

(c)   Prove that if A is orthogonal, then $A^t = A^{-1}$.

1.13   Two $n \times n$ matrices T and T' are said to be similar if there is a nonsingular matrix P such that $PTP^{-1} = T'$. Two square matrices are similar if and only if they represent the same linear transformation, but with respect to different choices of coordinates. Show that the matrix of direction cosines, α, can be used to prove the matrix similarity of stress tensors with respect to different coordinate systems.

1.14   A body is in plane stress at a point, where $\sigma_x = 10$ ksi, $\sigma_y = -12$ ksi, and $\tau_{xy} = -4$ ksi.

(a)   Use Mohr's circle to determine the magnitudes of the stresses, $\sigma_{x'}$ and $\tau_{x'y'}$ on a face rotated 30° clockwise from the x-face.

(b)   Use the stress tensor, T, and the rotation matrix, α, to determine the magnitudes of the stresses, $\sigma_{x'}$ and $\tau_{x'y'}$ on a face rotated 30° clockwise from the x-face.

(c)   Use the stress tensor, T, and the rotation matrix, α, to determine the magnitudes of the stresses, $\sigma_{x'}$ and $\tau_{x'y'}$ on a face rotated 30° counterclockwise from the x-face.

1.15   For the member of problem 1.4 (c), determine the eigenvalues of the stress tensor at a point on the bottom of the beam at the wall. Check your answer by a Mohr's circle calculation.

1.16   A square cross-sectional beam with edge s and length 20 inches is mounted in a rigid wall and is loaded by a downward vertical force, P, applied at the free end. Do not neglect shear. Determine the smallest edge dimension, s, in terms of P and σ_Y so that the maximum principal stress in the beam is less than the yield stress, σ_Y.

1.17   A rod of radius, R, is axially loaded by a tensile force, P, and also by a constant torque, T, about its longitudinal axis.

(a)   Determine the orientation of the principal face on which the largest principal stress acts, as a function of the radial position r of the point in the rod. Define the orientation by the angle the principal face makes with the cross section perpendicular to the longitudinal axis of the rod.

(b)   Assuming the rod is made of a ductile material which yields due to shear stress, on what plane will it first yield at a point on the surface of the rod?

1.18   The stress tensor, in ksi, at a point P of a body is, in x-y-z coordinates,

$$\begin{pmatrix} 2 & -1 & 1 \\ -1 & 0 & 1 \\ 1 & 1 & 2 \end{pmatrix}.$$

(a)   Draw a cube showing the stress components on each coordinate face.

(b)   From the matrix, compute the principal stresses. Do not use Mohr's circle. (*Hint*: One of the principal stresses is -1 ksi.)

(c)   Determine the stress on the face defined by $2\mathbf{i} + \mathbf{j} - 2\mathbf{k}$ in x-y-z components.

1.19   At a point in a machine part, the principal stresses are $\sigma_1 = 150$ MPa, $\sigma_2 = 60$ MPa, and $\sigma_3 = -80$ MPa.

(a)   Determine the maximum shear stress at the point.

(b)   Determine the normal stress in the direction making equal angles with the positive principal axes as a vector in principal coordinates. Then compute its magnitude.

1.20 A rod of length 14 inches and radius 2 inches is mounted in a rigid wall. It is loaded with a torque of $M_t = 32,000\pi$ in-lbs about the x-axis and a bending moment of $M = 24,000\pi$ in-lbs about the z-axis. A uniaxial test shows that the material of the rod yields at 10,600 psi.

(a) Write the stress tensor at the point $(6, 1, 0)$ in the rod as a matrix. (Use strength of materials to compute the stress components).

(b) Determine the resultant stress at point $(6, 1, 0)$ acting on the plane with normal $\mathbf{N} = 2\mathbf{i} - 5\mathbf{j} + 3\mathbf{k}$.

(c) Using determinants, compute the principal stresses at point $(6, 1, 0)$.

(d) Check your results for (c) using the Mohr's circle construction.

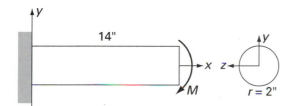

1.21 The principal stresses at a point P on a body are $\sigma_1 = 10$ MPa, $\sigma_2 = -30$ MPa, and $\sigma_3 = 20$ MPa. Label the principal axes by 1, 2, and 3.

(a) Write the direction cosine matrix, α, for a 40° clockwise rotation about the 3-axis.

(b) Determine the normal stress in the direction 40° clockwise from the 1-axis and lying in the 1-2 plane.

1.22 The stress tensor, in ksi, at a point P of a body is, in x-y-z coordinates,

$$\begin{pmatrix} 2 & 0 & 4 \\ 0 & 0 & 0 \\ 4 & 0 & -3 \end{pmatrix}.$$

(a) Draw a cube showing the stress components on each coordinate face.

(b) From the matrix compute the principal stresses. Do not use Mohr's circle.

(c) Determine, in x-y-z components, the stress on the face with normal $4\mathbf{i} + 2\mathbf{j} - 4\mathbf{k}$.

(d) Compute the magnitude of the normal component and of the shear component of the stress found in part (c). Do this without using the rotation matrix or Mohr's circle.

1.23 The stress tensor, in MPa, at a point P of a body is, in x-y-z coordinates,

$$\begin{pmatrix} 3 & -10 & 0 \\ -10 & 0 & 30 \\ 0 & 30 & -27 \end{pmatrix}.$$

(a) Draw a cube showing the stress components on each coordinate face.

(b) From the matrix, compute the principal stresses. Do not use Mohr's circle.

(c) Determine the stress, in x-y-z components, on the face perpendicular to $-2\mathbf{i} + 6\mathbf{j} - 3\mathbf{k}$.

(d) Compute the magnitude of the normal component and of the shear component of the stress found in part (c). Do this using vector projections, etc. Do not use the rotation matrix.

1.24 Draw the three-dimensional Mohr's circle for the state of stress in problem 1.23. Can this be done easily without first determining the principal stresses as in part (b) of problem 1.23?

1.25 Determine the octahedral stresses in (a) problem 1.20 and (b) problem 1.23.

1.26 Compare the magnitude of the octahedral shear stress to the maximum shear stress in problem 1.20 and in problem 1.23. Is there any general relation that holds between the octahedral shear stress and the maximum shear stress? Do these two shear stresses always occur on the same plane?

1.27 The cube has edges of length a, b, and c. The cube is uniaxially loaded by $\sigma\mathbf{i}$.

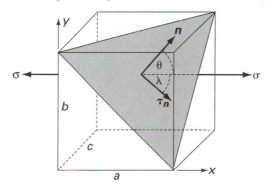

(a) Determine the shear and normal stress components, σ_n and τ_n, on the plane shown by vector methods.

(b) Verify that σ_n and τ_n are perpendicular.

(c) Does the point corresponding to the stress on this plane lie on any of the three circles in the three dimensional Mohr's circle diagram?

(d) Compute the angles, θ and λ.

(e) Set $a = b = c = 1$ and compare the results to those for the octahedral plane.

1.28 Use the Mohr's circle to write the principal stresses in terms of σ_x, σ_y, and τ_{xy} in the case of plane stress. Show that the Mises stress (1.14) can be written as $\sqrt{\sigma_x^2 - \sigma_x\sigma_y + \sigma_y^2 + 3\tau_{xy}^2}$. Compare this result to equation (1.15).

REFERENCES

[1] L. E. MALVERN, *Introduction to the Mechanics of a Continuous Media*, Prentice Hall, Englewood Cliffs, NJ, 1969.

[2] S. P. TIMOSHENKO, *History of Strength of Materials*, Dover, New York, 1983.

CHAPTER **2**

STRAIN

The mechanical response of a deformable body is, in part, described by the manner in which it changes shape. The deformation is described, roughly, by the deformation per unit length, strain. Navier initiated this idea in 1821 by examining the response of a body in terms of the displacements of its surface points. Cauchy extended this idea to all points of the body. Since then, different definitions of strain have been proposed. Some are with respect to the original length and some with respect to the current length of the body. Some are appropriate for very small deformations and others for large deformations. Several of these definitions will be used in this book; a full range of strain measures is developed in continuum mechanics courses. In contrast to stress, strain, or at least displacement, can be directly measured in experiments.

2.1 DEFINITION OF STRAIN

The analysis of stress in the previous chapter produced three of the basic equations describing deformed bodies, the stress equilibrium equations. The second set of basic equations relates the displacements and strain in a body. The definition of strain, given by Cauchy, is motivated here by comparison to the engineering strain used in strength of materials.

The strength of materials definition of normal engineering strain is based on the change in length of a line segment, Δx, on the body. The average engineering strain, ϵ_x, over Δx is the change in length of Δx divided by the original length, Δx. The generalization of this idea is to define the normal strain at a point, A, to be the limit as Δx goes to zero for all line segments Δx in the x-direction starting at the point A.

Consider uniaxial deformation of a bar. In Figure 2.1, points A and B are endpoints of a segment with length Δx on the undeformed body. The coordinate of A is x_A and of B is x_B measured from the left end of the bar. Under a deformation of the body, A moves to A' and B moves to B' (A' and B' are drawn above A and B to clarify the figure, but they probably lie on the line connecting A and B). The displacement in the x-direction of a point having coordinate x is denoted by $u(x)$.

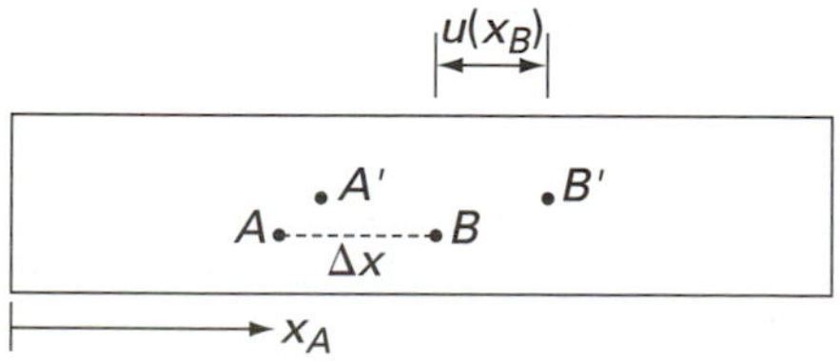

FIGURE 2.1   Displacements of the points A and B.

31

Therefore

$$x_{A'} = x_A + u(x_A);$$
$$x_{B'} = x_B + u(x_B).$$

Then the distance from A' to B' is

$$x_{B'} - x_{A'} = [x_B + u(x_B)] - [x_A + u(x_A)] = \Delta x + u(x_B) - u(x_A).$$

The change in displacement over the interval A to B is denoted by $\Delta u = u(x_B) - u(x_A)$. From strength of materials, the average strain over the length Δx is the final length minus the original length divided by the original length,

$$\frac{(x_{B'} - x_{A'}) - (x_B - x_A)}{x_B - x_A} = \frac{\Delta x + \Delta u - \Delta x}{\Delta x} = \frac{\Delta u}{\Delta x}.$$

Then the normal strain in the x-direction at the point x_A is defined to be the limit as Δx tends to zero,

$$\epsilon_x = \lim_{\Delta x \to 0} \frac{\Delta u}{\Delta x} = \frac{du}{dx}. \tag{2.1}$$

This suggests that the proper definition of the normal strain in the x-direction at a point should be the derivative of the displacement in the x-direction with respect to x evaluated at the point in question.

If $\epsilon_x > 0$, the body elongates at A in the x-direction; if $\epsilon_x < 0$, the body contracts in the x-direction at A. Notice that since the normal strain is the limit of $\Delta u / \Delta x$, which has units length divided by length, the normal strain is unitless.

A typical elastic strain in a metal is less than 0.00125. Engineers 50 years ago often wrote the strain 0.00125 as 0.00125 in./in., indicating that their measurements were made in inches. Today, engineers usually use the symbol $\mu = 10^{-6}$ and write 0.00125 as 1250 μ, which is read 1250 microstrains.

EXAMPLE 2.1   A member is mounted in a rigid wall and is axially loaded by a force, P (Fig. 2.2). The origin of the axis is at the wall.

FIGURE 2.2   Axially loaded member mounted in a wall.

Assume that for all points in the x-face, the displacement in the x-direction is given by $u(x) = kx$, where k is a constant depending on the body material and the force P. Then the strain in the x-direction is $\epsilon_x = du/dx = k$. This guess for the displacement implies that the strain in the x-direction is the same at all points of the body. A point at the wall does not move; the points at the right end of the member displace the most. While the deformation at the wall is zero, the strain is $k \neq 0$.   ∎

EXAMPLE 2.2   A member mounted to a wall is bent by a force, P, at its free end (Fig. 2.3). The length is L and the rectangular cross section has height, h, and width, w. The origin of the coordinate system is at the wall.

Assume that the displacement in the x-direction of the point (x, y, z) is given by $u(x, y, z) = kxy$, where k is a constant depending on the beam material, beam geometry, and

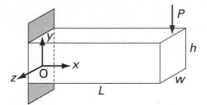

FIGURE 2.3   Cantilever beam under a point load.

the load, P. Then the strain in the x-direction is $\epsilon_x = \partial u/\partial x = ky$. Notice that the displacement at the wall is zero, but the strain is not. The strain anywhere on the x-z plane, where $y = 0$, is also zero. The strain on the top of the beam is tensile and on the bottom is compressive. The displacement is largest at the right end of the beam, but the strain at a fixed height y is the same at the free end as at the wall. Since the displacement field is a guess about the real behavior, these results must be checked by experiment. ■

This definition of strain can be generalized to two dimensions. Suppose a point $A(x, y)$ displaces to a point A'. Denote by $u(x, y)$ and $v(x, y)$ the displacements of point $A(x, y)$ in the x and y directions, respectively (Fig. 2.4). Both u and v are functions of position in the body.

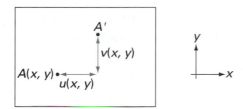

FIGURE 2.4   Cartesian displacements associated with point A.

Define the normal strains by analogy to the axial case as

$$\epsilon_x = \frac{\partial u}{\partial x} \quad \text{and} \quad \epsilon_y = \frac{\partial v}{\partial y}. \tag{2.2}$$

An examination of the definition of shear strain, γ_{xy}, given in strength of materials suggests how to define the shear strain at a point in terms of the displacements.

Definition   The average shear strain, γ_{xy}, is the change in angle between the x- and y-axes on the deformed body compared to the undeformed body. The angles are measured in radians so that the shear strain is unitless.

$$\gamma_{xy} = \frac{\pi}{2} - \beta. \tag{2.3}$$

The angle before deformation is $\pi/2$ and after deformation is β (Fig. 2.5). The subscript xy indicates the axes between which the change in angle is measured. The shear strain on a body is frequently quite difficult to measure experimentally, especially compared to normal strains which only require the measurement of lengths.

For small displacements, the definition of shear strain at a point is derived from the definition of average shear strain in terms of the change in angle between the two axes. In Figure 2.5, $\gamma_{xy} = \pi/2 - \beta = \theta_1 + \theta_2 \sim \tan(\theta_1) + \tan(\theta_2) = \Delta u/\Delta y + \Delta v/\Delta x$. To obtain the shear strain at the point O, let Δx and Δy tend to zero, so that

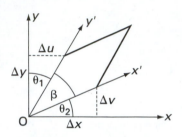

FIGURE 2.5 Shear strain in a planar deformation.

$\gamma_{xy} = \partial u/\partial y + \partial v/\partial x$. Therefore, the shear strain at a point relative to the x and y directions is defined to be

$$\gamma_{xy} = \frac{\partial u}{\partial y} + \frac{\partial v}{\partial x}. \tag{2.4}$$

To extend these ideas to the full three-dimensional case, let $u(x, y, z)$, $v(x, y, z)$, and $w(x, y, z)$ be the displacements of point $A(x, y, z)$ in the x, y, and z directions, respectively. A point displaces from position (x, y, z) to $(x+u(x, y, z), y+v(x, y, z), z+w(x, y, z))$. The strains are defined in terms of the displacements.

Definition The strains at a point, $P(x, y, z)$, in a body are obtained by evaluating the derivatives in the following expressions at point P.

$$\epsilon_x = \frac{\partial u}{\partial x};$$
$$\epsilon_y = \frac{\partial v}{\partial y};$$
$$\epsilon_z = \frac{\partial w}{\partial z};$$
$$\gamma_{xy} = \frac{\partial u}{\partial y} + \frac{\partial v}{\partial x};$$
$$\gamma_{xz} = \frac{\partial u}{\partial z} + \frac{\partial w}{\partial x};$$
$$\gamma_{yz} = \frac{\partial w}{\partial y} + \frac{\partial v}{\partial z}. \tag{2.5}$$

Notice that

$$\gamma_{xy} = \frac{\partial u}{\partial y} + \frac{\partial v}{\partial x} = \frac{\partial v}{\partial x} + \frac{\partial u}{\partial y} = \gamma_{yx}.$$

Likewise $\gamma_{xz} = \gamma_{zx}$ and $\gamma_{yz} = \gamma_{zy}$. The equations (2.5) are called the strain-displacement relations or kinematic relations. They say nothing about the cause of the displacements. The relations are valid for small displacements $u(x, y, z)$, $v(x, y, z)$, and $w(x, y, z)$.

EXAMPLE 2.3 A material in *simple extension* in the x-direction has displacements at all (x, y, z) of $u = ax$, $v = 0$, $w = 0$ for a constant a. The strains are all zero except $\epsilon_x = \partial u/\partial x = a$.

A material is in *simple shear* in the x-y plane if the displacements at all (x, y, z) are $u = by$, $v = 0$, $w = 0$ for a constant b. The strains are all zero except $\gamma_{xy} = \partial u/\partial y + \partial v/\partial x = b$. Simple shear can be thought of as a combination of the shear of Figure 2.5 and a clockwise rotation through the angle θ_2 of Figure 2.5. ∎

EXAMPLE 2.4   To simplify the notation, the coordinates are sometimes written as (x_1, x_2, x_3) rather than (x, y, z). A material is said to be in a homogeneous state of strain if all displacements are linear combinations of the positions (x_1, x_2, x_3) so that

$$u_i = \sum_{j=1}^{3} a_{ij} x_j,$$

where the nine values a_{ij} are constants. The resulting state of strain is called homogeneous because it is the same at each point of the body.   ∎

Exercise 2.1   Verify that linear strain displacement relations in the previous example induce a state of strain that is constant over all points in the body.

EXAMPLE 2.5   A prismatic rod is one with constant cross section and longitudinal faces perpendicular to the ends faces. Such a rod in torsion about its longitudinal z-axis can be described by the Cartesian displacement equations, $u(x, y, z) = ayz$, $v(x, y, z) = axz$, $w = 0$. The associated normal strains are $\epsilon_x = \epsilon_y = \epsilon_z = 0$. The shear strains are $\gamma_{xy} = \partial u/\partial y + \partial v/\partial x = 2az$. Also $\gamma_{xz} = \partial u/\partial z + \partial w/\partial x = ay$, $\gamma_{yz} = \partial v/\partial z + \partial w/\partial y = ax$.   ∎

In more complex deformations, it is difficult to find expressions for the displacements. Often simplifying assumptions are made about the strain-displacement relationship. For example, in the strength of materials analysis of bending and of torsion, the planar sections perpendicular to the longitudinal axis of the member are assumed to remain planar under the displacement.

EXAMPLE 2.6   Strains are induced when a straight beam is bent. Several different assumptions about the strain-displacement relation have been made to make the analysis simpler. Choose Cartesian coordinates with the x-axis along the longitudinal axis of the beam; assume that the cross section is constant along the beam. The y-coordinate is parallel to the bending axis in a face, and z is the coordinate perpendicular to the bending axis and the longitudinal coordinate.

If the deformed shape is the arc of a circle so that the neutral surface has constant radius of curvature ρ, the longitudinal strain at position (x, y, z) is obtained by computing the arclength cut off by an angle θ.

$$\epsilon_x = \frac{(\rho - z)\theta - \rho\theta}{\rho\theta} = -\frac{z}{\rho},$$

where z is measured from the neutral surface toward the center of curvature. This definition ensures that plane sections remain plane and perpendicular to the deformed longitudinal axis.

For a more general bending, approximate a small segment of the deformed beam as the arc of a circle and assume that there is a neutral surface which does not deform. The radius of curvature of this section, which may change from point to point, is denoted by $\rho(x)$. Then the longitudinal strain is

$$\epsilon_x(x, y, z) = -\frac{z}{\rho(x)} \sim -z\frac{d^2w}{dx^2},$$

for small displacements, where $w(x)$ is the displacement of the neutral surface transverse to the longitudinal direction of the column. Again, plane sections remain plane and transverse to the deformed longitudinal axis. The displacements are then $u(x, y, z) = -z(dw/dx)$, $v(x, y, z) = 0$, and $w(x, y, z) = w(x)$, so that u is written in terms of w. The displacement $w(x)$ must be determined from the loads and nature of the beam material. The in-plane shear is

$$\gamma_{xz} = \frac{du}{dz} + \frac{dw}{dx} = -\frac{dw}{dx} + \frac{dw}{dx} = 0.$$

Other, more complex, bending models include an approximation to a fixed strain, ϵ_c, of the centroidal surface, by superposing the strain, ϵ_c, and the bending strain to obtain the axial strain, $\epsilon_x = \epsilon_c - zw''(1 - w'^2)^{-1/2}$. In this notation, w' and w'' are derivatives with respect to the longitudinal coordinate. The bending axial strain is assumed to be proportional to the curvature of the bending neutral surface, written as $d(\arcsin(w'))/dx = w''(1 - w'^2)^{-1/2}$. This is a nonlinear expression for the curvature as opposed to the linearized version, $1/\rho = w''$, used in elementary strength of materials.

The Timoshenko beam theory drops the assumption that plane cross sections remain perpendicular to the neutral surface after bending. The plane sections remain plane but are allowed to rotate with respect to the neutral surface. The strain is a superposition of axial and pure bending. Again, the neutral surface axial strain is a constant, ϵ_c, after bending. Then at a point x, $\epsilon_x(x, y, z) = \epsilon_c - (d\phi/dx)z$ where ϕ is the angle of rotation of the column cross section at x so that $\phi = \arcsin(w/(1 + \epsilon_c))$. With this assumption, internal shear exists in the Timoshenko beam. ∎

The displacements in cylindrical coordinates (r, θ, z) are denoted by $u_r, u_\theta\ u_z$. Here the subscripts denote the coordinate direction, not partial differentiation. The strain-displacement relations are

$$\epsilon_r = \frac{\partial u_r}{\partial r};$$

$$\epsilon_\theta = \frac{1}{r}\frac{\partial u_\theta}{\partial \theta} + \frac{u_r}{r};$$

$$\epsilon_z = \frac{\partial u_z}{\partial z};$$

$$\epsilon_{r\theta} = \frac{\partial u_\theta}{\partial r} - \frac{u_\theta}{r} + \frac{1}{r}\frac{\partial u_r}{\partial \theta};$$

$$\epsilon_{z\theta} = \frac{1}{r}\frac{\partial u_z}{\partial \theta} + \frac{\partial u_\theta}{\partial z};$$

$$\epsilon_{rz} = \frac{\partial u_r}{\partial z} + \frac{\partial u_z}{\partial r}. \tag{2.6}$$

Exercise 2.2 Derive these relations from the strain-displacement relations in Cartesian coordinates. Use a small element bounded by lines in the r-direction and arcs in the θ-direction. Consider the change in dimensions of its edges, as in the case of Cartesian coordinates.

The strain-displacement equations given in equations (2.5) are linearized versions of the expressions for large displacements which involve squares of the various partial derivatives.

Definition The large displacement strains at a point, $P(x, y, z)$, are obtained by evaluating the derivatives in the following expressions at point P.

$$\epsilon_x = u_x + \frac{1}{2}(u_x^2 + v_x^2 + w_x^2);$$

$$\epsilon_y = v_y + \frac{1}{2}(u_y^2 + v_y^2 + w_y^2);$$

$$\epsilon_z = w_z + \frac{1}{2}(u_z^2 + v_z^2 + w_z^2);$$

$$\gamma_{xy} = u_y + v_x + u_x u_y + v_x v_y + w_x w_y;$$

$$\gamma_{xz} = u_z + w_x + u_x u_z + v_x v_z + w_x w_z;$$

$$\gamma_{yz} = v_z + w_y + u_y u_z + v_y v_z + w_y w_z, \tag{2.7}$$

where the subscripts on u, v, and w indicate partial differentiation.

EXAMPLE 2.7   Compare the small and large displacement strains, ϵ_x at point $(1, 0.5, 0.8)$ in the body when the displacements are given by $u(x, y, z) = 0.01x$, $v(x, y, z) = 0.2x + 0.3z$, and $w(x, y, z) = 0.5xz$.

Solution   By equation (2.5), the small strain definition produces, $\epsilon_x = 0.01$. However, by equation (2.7), the strain is defined to be $\epsilon_x(x, y, z) = 0.01 + [0.01^2 + 0.2^2 + (0.5z)^2]/2$. So at the point $(1, 0.5, 0.8)$, $\epsilon_x = 0.11005$. The small and large displacement strains at each point are distinguished by the magnitude of the displacement gradients at the point. ∎

EXAMPLE 2.8   Models of solid body displacement can be determined by choosing a large strain-displacement relation. For example, in a straight beam, choose coordinates as in the previous beam example, that is, with the x-axis along the longitudinal axis of the beam. In this case, the bending neutral surface will be allowed to be in either tension or compression. The strains are written in terms of the deflections, u and w, at the neutral surface, which are therefore functions of just x, and the radius of curvature, ρ. Again, the nonlinear relation for the curvature is used: $1/\rho = w''(1 - w'^2)^{-1/2}$. It is assumed that, in bending, the bending neutral axis remains at the centroid of the cross section. Since the curvature of the beam can contribute significantly to the strain just after bending, the quadratic displacement term in w is included.

$$\epsilon_x(x, y, z) = u' + \frac{1}{2}w'^2 - \frac{z}{\rho}$$

and $u' = \partial u/\partial x$ and $w' = \partial w/\partial x$ at the neutral surface. The axial normal strain depends only on x and z. ∎

A special case occurs when the system of strains can be viewed as two-dimensional.

Definition   A point in a body is said to be in plane strain if $\epsilon_z = 0$, $\gamma_{xz} = 0$, and $\gamma_{yz} = 0$, for some orientation of a Cartesian coordinate system at the point.

A state of plane strain can be created by constraining the deformation in one direction, perhaps by mounting a member between two parallel rigid walls. This state must not be confused with plane stress, $\sigma_z = 0$, $\tau_{xz} = 0$, and $\tau_{yz} = 0$, which occurs most typically on an unloaded surface of a body.

2.2 THE STRAIN TENSOR

The strain varies from point to point in the body, just as the stress does. The description of the deformation of a solid under stresses will be completed in Chapter 3 by writing a stress-strain relation. Since in three dimensions, the stress is most easily expressed as a tensor, we want to express the strain state at a point as a tensor as well. To write the strain tensor, it is again easier to use coordinates written as (x_1, x_2, x_3) rather than (x, y, z). The displacements in the coordinate directions are then written $u_1(x_1, x_2, x_3) = u(x, y, z)$, $u_2(x_1, x_2, x_3) = v(x, y, z)$ and $u_3(x_1, x_2, x_3) = w(x, y, z)$. So that the strain tensor components will correspond to the stress tensor components, the strain tensor, ϵ, written as a matrix has entries, ϵ_{ij}, which are defined as

$$\epsilon_{11} = \frac{\partial u_1}{\partial x_1};$$

$$\epsilon_{22} = \frac{\partial u_2}{\partial x_2};$$

$$\epsilon_{33} = \frac{\partial u_3}{\partial x_3};$$

$$\epsilon_{12} = \frac{1}{2}\gamma_{xy} = \frac{1}{2}\left(\frac{\partial u_1}{\partial x_2} + \frac{\partial u_2}{\partial x_1}\right) = \epsilon_{21};$$

$$\epsilon_{13} = \frac{1}{2}\gamma_{xz} = \frac{1}{2}\left(\frac{\partial u_1}{\partial x_3} + \frac{\partial u_3}{\partial x_1}\right) = \epsilon_{31};$$

$$\epsilon_{23} = \frac{1}{2}\gamma_{yz} = \frac{1}{2}\left(\frac{\partial u_3}{\partial x_2} + \frac{\partial u_2}{\partial x_3}\right) = \epsilon_{32}. \tag{2.8}$$

These equations can be written compactly as

$$\epsilon_{ij} = \frac{1}{2}\left(\frac{\partial u_i}{\partial x_j} + \frac{\partial u_j}{\partial x_i}\right). \tag{2.9}$$

The factor of 1/2 is included in the shear components so that the strain tensor will obey the same change of coordinate relations as the stress tensor, T. Note that the strain tensor is symmetric, $\epsilon = \epsilon^t$.

Analogously to the stress tensor, principal normal strains are defined as the eigenvalues of the strain tensor matrix. They are defined in the directions of normals to planes for which the shear strain is zero. Since the strain tensor is a real valued, symmetric matrix, there are three real eigenvalues, and the three distinct associated eigenvectors are mutually orthogonal.

Recall that α, the matrix of direction cosines, defines a change of coordinates from unprimed to primed coordinates. The stress tensor, T', written with respect to primed coordinates is related to the stress tensor written with respect to the unprimed coordinates by $T' = \alpha T \alpha^t$. Likewise, for the same change of coordinates,

$$\epsilon' = \alpha \epsilon \alpha^t, \tag{2.10}$$

where ϵ' is the strain tensor in primed coordinates.

EXAMPLE 2.9   A state of shear strain is represented in x_1-x_2-x_3 coordinates by the matrix

$$\epsilon = \begin{pmatrix} 0 & \epsilon_{12} & 0 \\ \epsilon_{12} & 0 & 0 \\ 0 & 0 & 0 \end{pmatrix}.$$

Determine the state of strain in the primed coordinate system rotated 45° about the x_3-axis counterclockwise from the given x_1-x_2-x_3 coordinates.

Solution   The rotation matrix is

$$\alpha = \begin{pmatrix} \cos 45 & \sin 45 & 0 \\ -\sin 45 & \cos 45 & 0 \\ 0 & 0 & 1 \end{pmatrix}.$$

The transformation $\epsilon' = \alpha\epsilon\alpha^t$ produces

$$\epsilon' = \begin{pmatrix} \epsilon_{12} & 0 & 0 \\ 0 & -\epsilon_{12} & 0 \\ 0 & 0 & 0 \end{pmatrix},$$

the state of principal strains.   ∎

Exercise 2.3   Repeat the previous example, but rotate in the clockwise direction about the x_3-axis.

2.3   MOHR'S CIRCLE FOR STRAIN

As in the case of stress, the strain tensor is most useful when computing three-dimensional states of strains. In plane strain problems, Mohr's circle is usually quicker to use. Mohr's circle for strain is a two-dimensional equivalent to equation (2.10). Since there is a strain Mohr's circle which is analogous to the stress Mohr's circle, a strain element can also be drawn to represent the strain components at a point. The strains on the strain element are drawn in the same sign convention as the stress vector components. However, the strains are not vector components. The normal strains are with respect to the directions indicated by the arrow, but the direction for the shear strain has no physical interpretation similar to tension and compression for normal strains. Given a strain element, it was shown in strength of materials that

$$\epsilon_{x'} = \frac{1}{2}(\epsilon_x + \epsilon_y) + \frac{1}{2}(\epsilon_x - \epsilon_y)\cos 2\theta + \frac{1}{2}\gamma_{xy}\sin 2\theta;$$

$$\gamma_{x'y'} = \gamma_{xy}\cos 2\theta + (\epsilon_y - \epsilon_x)\sin 2\theta.$$

These equations follow the same pattern as the stress equations,

$$\sigma_{x'} = \frac{1}{2}(\sigma_x + \sigma_y) + \frac{1}{2}(\sigma_x - \sigma_y)\cos 2\theta + \tau_{xy}\sin 2\theta;$$

$$\tau_{x'y'} = \tau_{xy}\cos 2\theta + \frac{1}{2}(\sigma_y - \sigma_x)\sin 2\theta,$$

letting ϵ_x correspond to σ_x, ϵ_y to σ_y, and $0.5\gamma_{xy}$ to τ_{xy}. Therefore the Mohr's circle for strain (Fig. 2.6) is plotted with $\gamma/2$ on the vertical axis and ϵ on the horizontal axis. This also helps explain the factor of one half in front of the shear terms, $\epsilon_{12} = 0.5\gamma_{xy}$, and so on in the strain tensor.

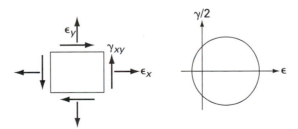

FIGURE 2.6   Mohr's circle for strain.

The principal normal strains appear on the Mohr's circle at the points where the shear strain is zero. Plane strain occurs when one of the three principal strains is zero.

EXAMPLE 2.10

A bar has several forces acting on it. At a point P on the surface, measurements have determined that $\epsilon_x = 600\mu$, $\epsilon_y = -100\mu$, and $\gamma_{xy} = -800\mu$. An x'-y' coordinate system is defined by rotating the x' axis 50° clockwise from the x-axis.

(1) Determine the strain components in x'-y' coordinates using both Mohr's circle and the strain tensor.

(2) Estimate the change in length of a line through P which is originally 2 mm long and lying in the direction of the x' axis.

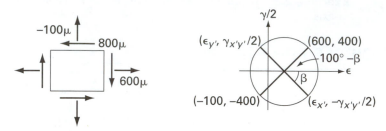

FIGURE 2.7   Strain element and associated strain Mohr's circle.

Solution

(1) First represent the strains on a strain element and draw the strain Mohr's circle (Fig. 2.7). The center is (250, 0) and the radius is $R = [(600 - 250)^2 + (400 - 0)^2]^{1/2} = 531.5\mu$. The angle $\beta = 100 - \arcsin(400/531.5) = 100 - 48.815 = 51.185°$. From right triangles inside the Mohr's circle, $\epsilon_{x'} = 250 + R\cos\beta = 250 + 333.15 = 583.15\mu$, $\epsilon_{y'} = 250 - R\cos\beta = 250 - 333.15 = -83.15\mu$, and $\gamma_{x'y'}/2 = R\sin\beta = 414.13\mu$.

While they are not needed to answer the questions posed in this example, the principal normal strains are $\epsilon_1 = 250 + R = 781.5\mu$ and $\epsilon_2 = 250 - R = -281.5\mu$.

These strains can also be computed from the strain tensor transformation. The strain tensor and the matrix of direction cosines are

$$\epsilon = \begin{pmatrix} 600 & -400 \\ -400 & -100 \end{pmatrix}; \qquad \alpha = \begin{pmatrix} \cos 50 & -\sin 50 \\ \sin 50 & \cos 50 \end{pmatrix}.$$

In constructing α, the angle in the definition of α in Chapter 1, Section 1.5 is taken to be $-50°$ since it is clockwise. Then $\epsilon' = \alpha\epsilon\alpha^t$ produces

$$\epsilon' = \begin{pmatrix} \cos 50 & -\sin 50 \\ \sin 50 & \cos 50 \end{pmatrix}\begin{pmatrix} 600 & -400 \\ -400 & -100 \end{pmatrix}\begin{pmatrix} \cos 50 & \sin 50 \\ -\sin 50 & \cos 50 \end{pmatrix} = \begin{pmatrix} 583.14 & 414.14 \\ 414.14 & -83.14 \end{pmatrix},$$

which agrees with the results obtained from Mohr's circle.

$\epsilon_{x'}$ can also be computed in a manner similar to that used to compute the normal stress component. In terms of the unprimed coordinates, a unit vector in the x'-direction is $\mathbf{n} = \cos 50\mathbf{i} - \sin 50\mathbf{j}$. The magnitude of the normal strain component, $\epsilon_{x'}$, is the magnitude of the projection of the vector $\epsilon(\mathbf{n})$ onto $\mathbf{n}$. This is obtained from the dot product of $\epsilon(\mathbf{n}) = (600\cos 50 + 400\sin 50)\mathbf{i} + (-400\cos 50 + 100\sin 50)\mathbf{j}$ and $\mathbf{n}$.

(2) The approximate change in length of a 2 mm line in the x' direction is determined by assuming that the normal strain everywhere along the line is $\epsilon_{x'} = 583.14\mu$, not just at the point P. The change in length is $\Delta L = \epsilon_{x'}L = (583.15)10^{-6}(2) = 1.16610^{-3}$ mm.   ∎

2.4 LAGRANGIAN AND EULERIAN COORDINATE SYSTEMS

In defining the stress and strain, a subtle distinction has been overlooked. The Cauchy stress tensor, T, gives the force per unit deformed area. In other words, the coordinates were chosen after the body deformed. This distinction between undeformed and deformed coordinates was also made in strength of materials. Engineering stress is the force per unit original undeformed area; true stress is the force per unit deformed area. When defining strain as it has been in this chapter, however, the strain is deformation per unit original length. The chosen coordinate system is on the undeformed body. This distinction is codified in the following definition.

Definition   A coordinate system defined on a deformed body is called Eulerian, or the current coordinate system. A coordinate system defined on an undeformed body is called Lagrangian, or a reference coordinate system.

When the displacements are small, as assumed in linear elasticity, we ignore this distinction and take the reference and spatial coordinates to be identical. However, this identification cannot be made in large displacement theories.

2.5   EQUATIONS OF COMPATIBILITY

The stress and strain vary throughout the body. Therefore the stress and strain components are all functions of position (x, y, z) in the body. Recall that for the stress component functions to be a possible valid stress field for a body, the components have to satisfy the three equations of equilibrium. Because the equations of elasticity are quite difficult to solve, one would also like criteria to check candidate functions that might describe how the strain varies in a body. If the body is deformed elastically, without fracture, the strain field must be continuous. Plastic deformations are required to create a shape with sharp corners. The derivatives of a function describing a surface are not defined at a corner. Therefore, the displacement field must be differentiable. Any set of equations that purports to describe elastic strain must be both continuous and differentiable.

To apply these conditions to the two-dimensional case of plane strain, recall that if a function $z = f(x, y)$ is continuous and has continuous first and second derivatives,

$$\frac{\partial^2 f}{\partial x \partial y} = \frac{\partial^2 f}{\partial y \partial x}.$$

Applying this to the three distinct two-dimensional strain components, $\epsilon_x = \partial u/\partial x$, $\epsilon_y = \partial v/\partial y$, $\gamma_{xy} = \partial u/\partial y + \partial v/\partial x$, one obtains

$$\frac{\partial^2 \epsilon_x}{\partial y^2} = \frac{\partial^2 (\partial u/\partial x)}{\partial y^2} = \frac{\partial^3 u}{\partial x \partial y^2};$$

$$\frac{\partial^2 \epsilon_y}{\partial x^2} = \frac{\partial^2 (\partial v/\partial y)}{\partial x^2} = \frac{\partial^3 v}{\partial x^2 \partial y};$$

$$\frac{\partial^2 \gamma_{xy}}{\partial x \partial y} = \frac{\partial^2 (\partial u/\partial y + \partial v/\partial x)}{\partial x \partial y} = \frac{\partial^3 u}{\partial x \partial y^2} + \frac{\partial^3 v}{\partial x^2 \partial y}. \qquad (2.11)$$

Combining these equations produces the two-dimensional strain compatibility equation

$$\frac{\partial^2 \epsilon_x}{\partial y^2} + \frac{\partial^2 \epsilon_y}{\partial x^2} = \frac{\partial^2 \gamma_{xy}}{\partial x \partial y}. \qquad (2.12)$$

These relations are only valid for the linearized strain-displacement relations. Much more complicated relations exist for the nonlinear strain-displacement relations.

EXAMPLE 2.11   Is the following matrix function of (x, y) a possible strain tensor for plane strain?

$$\begin{pmatrix} xy^3 & x^2 + 2xy \\ x^2 + 2xy & x^3 y \end{pmatrix} = \begin{pmatrix} \epsilon_x & \gamma_{xy}/2 \\ \gamma_{xy}/2 & \epsilon_y \end{pmatrix}.$$

Solution   Recall that $\epsilon_{12} = \epsilon_{21} = \gamma_{xy}/2$ before applying the strain compatibility equation test. The strain compatibility equation requires that $6xy + 6xy = 4$; but this relation must be true for all points (x, y) in the body. Therefore the given matrix is not a possible strain tensor. Because of the lack of strain compatibility, there are no displacements which can produce these strains. This can be checked by trying to integrate the strains to determine the associated displacements. ∎

A similar calculation produces six strain compatibility equations which are required to be satisfied by any valid three-dimensional strain field. The six components of any valid strain tensor which represents a continuous deformation satisfy

$$\frac{\partial^2 \gamma_{xy}}{\partial x \partial y} = \frac{\partial^2 \epsilon_x}{\partial y^2} + \frac{\partial^2 \epsilon_y}{\partial x^2};$$

$$\frac{\partial^2 \gamma_{xz}}{\partial x \partial z} = \frac{\partial^2 \epsilon_x}{\partial z^2} + \frac{\partial^2 \epsilon_z}{\partial x^2};$$

$$\frac{\partial^2 \gamma_{yz}}{\partial y \partial z} = \frac{\partial^2 \epsilon_y}{\partial z^2} + \frac{\partial^2 \epsilon_z}{\partial y^2};$$

$$2\frac{\partial^2 \epsilon_x}{\partial y \partial z} = \frac{\partial}{\partial x}\left(-\frac{\partial \gamma_{yz}}{\partial x} + \frac{\partial \gamma_{xz}}{\partial y} + \frac{\partial \gamma_{xy}}{\partial z} \right);$$

$$2\frac{\partial^2 \epsilon_y}{\partial x \partial z} = \frac{\partial}{\partial y}\left(\frac{\partial \gamma_{yz}}{\partial x} - \frac{\partial \gamma_{xz}}{\partial y} + \frac{\partial \gamma_{xy}}{\partial z} \right);$$

$$2\frac{\partial^2 \epsilon_z}{\partial x \partial y} = \frac{\partial}{\partial z}\left(\frac{\partial \gamma_{yz}}{\partial x} + \frac{\partial \gamma_{xz}}{\partial y} - \frac{\partial \gamma_{xy}}{\partial z} \right). \qquad (2.13)$$

Exercise 2.4   Verify equations (2.13) using a proof similar to that for equation (2.12).

These equations give information on how the strain must vary from point to point in an elastically loaded body which undergoes small displacements. Since they are a consequence of the strain-displacement equations, they cannot be used in conjunction with the strain-displacement relations to solve for additional unknowns.

2.6 ELECTRONIC RESISTANCE STRAIN GAGES

The bonded wire strain gage was invented independently by two researchers in 1938. Both had the idea of fastening a wire to the surface of a body and reading the strain by the change of resistance in the wire. One was Prof. Arthur Ruge of the Civil Engineering Department at Massachusetts Institute of Technology. Shortly later, he and colleagues presented a demonstration of a paper-backed wire gage at Baldwin-Southwark. Another man, Edward E. Simmons, Jr., while a graduate student at Caltech, had built a similar device at Caltech and had built one earlier for the Hughes Tool Co. The Simmons gage was invented slightly prior to that of Ruge. The gage was named the SR-4 after Simmons and Ruge. The patent problem was resolved by giving Simmons a royalty per gage and Ruge a percentage of the improvement patents. Simmons had to win a suit to win full royalties against the claims of Caltech for ownership of the patent early in the 1940s. He won since, as a graduate student, he was not on the Caltech payroll. His first royalty check was for $125,000. Simmons used his royalty money to enter into other business deals and never worked further on strain gages. That development was left to a company formed by Ruge and his colleague at MIT, A. V. de Forest. Simmons, when asked about his inspiration for the invention, replied that he did not know it could not be done, so he did it. He was viewed as an "attic inventor" by the Philadelphia Medal Committee of the Franklin Institute when they awarded him their 1944 Longstreth Medal. See Tatnall [5] for the full details of this fascinating story.

To experimentally determine an average normal strain, an original length must be chosen in the direction desired. This length is called the gage length. To compute the average strain over the gage length after the body deforms, the elongation of the gage length must be measured. Only average normal strain can be measured since it is physically impossible to let the gage length go to zero to obtain the strain at a point.

Strain can be measured accurately and easily using electronic resistance strain gages. It was discovered that as a length of wire is stretched or compressed, its electrical resistance changes. The resistance of a wire is $R = \rho L / A$, where L is the length, A is the cross section area, and ρ is the specific resistivity. For small A, R is large and small changes in A produce large changes in R that are easily measured. This fact can be used to measure strain by gluing a piece of wire to the surface of the body being tested so that the body and wire change length together. An electric current is passed through the wire, and the resulting resistance change, as the wire deforms, is measured. This change is calibrated so that the resistance change corresponds to the strain in the body, and the strain is read on a meter.

The strain obtained will be deformation divided by original length, where the original length is the length covered by the gage. A longer wire gives a more accurate reading. To keep the original length short, either so that the gage will fit on the body or to measure the strain at a particular location on the body, the wire is folded into many rows so that it has a long length for accuracy but still covers a short original length on the body. A typical strain gage is shown schematically in Figure 2.8. These gages can have sides as small as 1/64 of an inch or as large as desired. The small ones are made by etching the wire pattern on a metal foil.

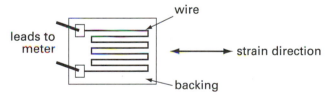

FIGURE 2.8   Strain gage.

The resistance change is measured using a Wheatstone bridge circuit. Electronically the Wheatstone bridge is composed of four resistors, R_1, R_2, R_3, R_4, with two series pairs connected in parallel and with a constant voltage applied at points A and C in Figure 2.9. In strain gage work, this voltage is typically 3 V.

In measuring strain with this circuit, strain gages replace one to four of the resistors in this circuit and act like resistors themselves. The resistance of two resistors in series is found to be the sum of their resistances. The series pair from A to B to C has, for example, resistance $R_3 + R_4$. The change in voltage (voltage drop) across a resistor is given by Ohm's law: $E = IR$, the voltage equals the current times the resistance. George S. Ohm was a German physicist who lived from 1787 to 1854.

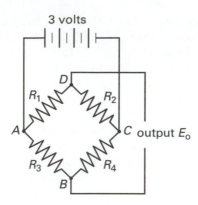

FIGURE 2.9   Wheatstone bridge circuit.

The current through ABC is $I_{ABC} = E/R_{ABC} = E/(R_3 + R_4)$. Likewise, the current through ADC is $E/(R_1 + R_2)$. The voltage drop though R_1 is $E_1 = R_1 I_{ADC} = R_1 E/(R_1 + R_2)$ and through R_3 is $E_3 = R_3 I_{ABC} = R_3 E/(R_3 + R_4)$. Therefore, taking into account the direction of current flow, $E_o = E_1 - E_3$, or

$$E_o = \frac{R_1 R_4 - R_2 R_3}{(R_1 + R_2)(R_3 + R_4)} E. \tag{2.14}$$

There is zero output voltage when the numerator is zero, $R_1 R_4 = R_2 R_3$; the system is then said to be in balance. If the system is balanced when the test begins, zero output will represent zero strain. As the test proceeds, the changing output voltage indicates the change in resistance of the gages. The manufacturer of the gage provides a number, called the gage factor, with which the measured change in resistance is converted to a strain reading. The Wheatstone bridge is used with a combination of one to four strain gages in place of the resistors in the circuit depending on the type of load, the shape of the body, and the desire to compensate for temperature effects on the gage readings. More details on strain gage technology are found in Dally and Riley [2].

EXAMPLE 2.12   The shear strain can be determined from several normal strain readings using strain gages. Strain gages can only determine strains in the plane on which they are mounted. To obtain a Mohr's circle requires three facts, for example, the coordinates of the center and radius. Therefore three strain readings are required to determine the full state of strain in the plane. Since electronic strain gages can only read normal strains, a common strategy is to measure the normal strain in three directions, say the x-direction, the y-direction, and the direction $45°$ between the x- and y-directions. These readings give information about three points on the Mohr's circle (Fig. 2.10). Since the shear strain γ_{xy} is unknown, the exact location of the three points is not known. However, the point for ϵ_x must be opposite that for ϵ_y and the point for ϵ_{45} must be $90°$ between the other two.

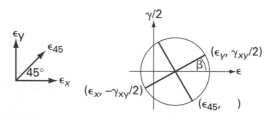

FIGURE 2.10   Mohr's circle for the strain rosette.

A relation is desired between $\epsilon_x, \epsilon_y, \epsilon_{45}$ and γ_{xy}. The center of the Mohr's circle is the point $((\epsilon_x + \epsilon_y)/2, 0)$. From the Mohr's circle geometry, if R is the radius,

$$\epsilon_{45} = \frac{1}{2}(\epsilon_x + \epsilon_y) + R\cos(90 - \beta).$$

But $\cos(90 - \beta) = \sin(\beta) = 0.5\gamma_{xy}/R$ from the Mohr's circle. Therefore, by substitution

$$\epsilon_{45} = \frac{1}{2}(\epsilon_x + \epsilon_y) + \frac{1}{2}\gamma_{xy}. \tag{2.15}$$

If the three normal strains have been measured, then γ_{xy} can quickly be calculated.   ■

2.7   OTHER STRAIN MEASUREMENT TECHNIQUES

Electronic strain gages can be used when it is permissible to attach the gage directly to the body because the presence of the gage is not expected to significantly affect the strain measurement. On delicate materials like skin, paper, and so on, a gage attached to the surface will drastically change the measurement. Noncontact methods of strain measurement are needed.

Another electronic method to measure deformation is by a linear voltage differential transformer (LVDT). This device works by correlating the change in the inductance of a coil with a deformation of the body. A cylinder containing the coil is rigidly mounted at a reference position, perhaps on a grip. An iron core that can move, parallel to the deformation direction to be measured, in and out of the center of the coil is connected to another point of the specimen, or on the grip on the other end of the specimen. As the specimen deforms, the core moves inside along the longitudinal axis of the coil so that the inductance of the coil varies.

Linear encoder devices based on moiré interference principles are also available for a similar application. The pattern is imprinted on a portion of the test stand that moves exactly as much as the specimen deforms. The scanner determines the length of the motion from variations in the resulting interference pattern.

Experimenters in several fields studying easily damaged materials, such as paper or biological tissues, have used optical noncontact strain measurement techniques. The inhomogeneity of a paper or biological specimen may make it desirable to measure local strains in addition to an average strain over a large region of the specimen. The goal of these optical techniques is to calculate local strains in a small region of the specimen rather than a single average strain for the region between the grips. Video dimension analyzers (VDA) are line scanning devices that produce a voltage proportional to the distance between lines in the video image. In this way, they measure the average normal strain in two perpendicular directions over a central region on the specimen, but not the shear strain.

Alternatively, periodically taken camera images can track markers on the specimen as the specimen deforms [3]. These techniques depend on the ability to locate the new positions of the markers after deformation. The markers are lightweight to avoid influencing the strain response and are kept small for accuracy. In previous applications, the markers have been spaced in a range of 1 to 6 mm apart and within a test area from 2.5 to 3.5 cm on a side. Others, rather than placing individual markers, spray the specimen through a sheet with holes to create closely spaced individual ink dots on the specimen. Some experimenters have tracked the position of the markers with photographs taken at discrete time intervals during the test. This method to measure the displacement is slow and risks strain rate and relaxation effects affecting the results. A less labor intensive technique automates the tracking of the markers by image processing software. If this can be done rapidly, feedback to obtain real-time experimental control, such as strain control over the loading, is possible. Software selects each video frame and locates the centroid of each of the markers by intensity measurements. Once the new marker position is determined, the instantaneous strain is approximated from geometric computations based on the strain-displacement relations. The continuous displacement fields, u and v, are obtained by fitting interpolation functions to the measured discrete marker displacements. The accuracy of the strains obtained from the partial derivatives of these displacement fields depends on the order of the interpolation functions. The fundamental difficulty is to track each marker. Two problems are that a false marker may be identified and that the specimen deformation may carry the markers out of the image processing search area.

Electronic speckle photography (ESP) is a variation on tracking markers separately. The software is designed to identify changes in the random pattern of speckles on the specimen after deformation [1]. The strains are calculated from an analysis of the two-dimensional light intensity field reflected from the deformed speckled specimen; the displacement field is deduced from the light intensity.

PROBLEMS

2.1 Compute the strains and describe the deformation of a body with strain-displacement relations $u = ax + by$, $v = 0$, $w = 0$, where a and b are constants. Draw a sketch of the deformation of a cubical body element with edges parallel to the coordinate axes.

2.2 Each of the following axially loaded members has a constant axial normal strain, ϵ, at each point. Both have the same length, L.

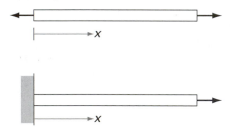

(a) For the left member, write the displacement as a function of axial position, x, from the left end with the displacement equal to zero at the center. Compute the total elongation.

(b) For the right member whose left end is mounted in a rigid wall, write the displacement as a function of axial position, x, from the left end. Compute the total elongation.

2.3 A 100 m long cable is hung vertically with a weight on its end and is fixed at the top. At each point of the cable, the weight causes a normal strain of 200 μ. In addition, at each point, the weight of the cable below each point causes a strain which is proportional to the length of the cable below the point. The total normal strain at the top of the cable is measured by a strain gage to be 450 μ. Determine the total normal strain at each point as a function of the distance of

the point from the bottom of the cable. Determine the total elongation of the cable. Ans. 0.0325 m.

2.4 The edges form a right angle at C in the triangular plate shown. Under a vertical load applied at point C, the average normal strains over CB at C are $\epsilon_x = 1000\mu$ and $\epsilon_y = 1000\mu$. Determine the final position of point C and the shear strain, γ_{xy}.

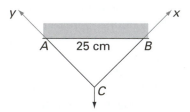

2.5 Determine the change in length of a 5 mm line through a point P making a counterclockwise angle of 35° with the x-axis. The strain at P on a body is given by

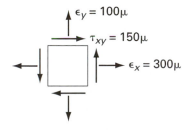

2.6 The rectangular plate shown deforms into a parallelogram. The original lengths of the sides are $PQ = 15$ cm and $PR = 30$ cm. The point Q displaces to Q', a distance of $QQ' = 0.01$ cm, and the point R displaces to R', a distance of $RR' = 0.015$ cm.

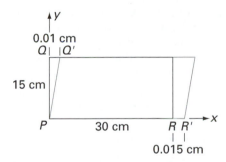

(a) Determine ϵ_x and ϵ_y at point P.

(b) Determine γ_{xy} at point P.

(c) Determine the principal strains at point P.

2.7 The edges form a right angle at C in the triangular plate shown in problem 2.4. Under a vertical load applied at point C, point C deforms vertically downward 0.05 cm.

(a) Determine the average ϵ_x and ϵ_y at point C.

(b) Determine average γ_{xy} at point C.

(c) Determine the principal strains at point C.

2.8 Determine the maximum engineering shear strain at a point with the state of strain

$$\epsilon = \begin{pmatrix} 0.0010 & 0.0012 & 0 \\ 0.0012 & 0 & 0 \\ 0 & 0 & 0 \end{pmatrix}.$$

Ans. 0.0026.

2.9 A point of a body is in plane strain with strain tensor,

$$\epsilon = \begin{pmatrix} 600 & 150 \\ 150 & -200 \end{pmatrix} \mu.$$

(a) Use Mohr's circles to determine the principal normal strains, ϵ_1 and ϵ_2.

(b) Use Mohr's circles to determine the maximum shear strain at the point.

(c) Write the rotation matrix α, needed to transform the given Cartesian coordinate system to principal strain coordinates.

(d) Verify that the trace is an invariant by showing that $\epsilon_x + \epsilon_y = \epsilon_1 + \epsilon_2$.

(e) Verify that the determinant is an invariant by showing that $\det(\epsilon) = \epsilon_1 \epsilon_2$.

2.10 A point of a body is in plane strain with strain tensor,

$$\epsilon = \begin{pmatrix} 600 & -150 \\ -150 & -200 \end{pmatrix} \mu.$$

(a) Use Mohr's circles to determine the principal normal strains.

(b) Use Mohr's circles to determine the maximum shear strain at the point.

(c) Write the rotation matrix, α, needed to transform the given Cartesian coordinate system to principal strain coordinates.

2.11 The state of plane strain at a point is $\epsilon_x = 2,500\mu$, $\epsilon_y = 4,000\mu$, and $\gamma_{xy} = -2,000\mu$. Use Mohr's circles to determine the principal strains and the maximum shear strain.

2.12 The state of plane strain at a point is $\epsilon_x = 2,500\mu$, $\epsilon_y = -4,000\mu$, and $\gamma_{xy} = 2,000\mu$. Use Mohr's circles to determine the principal strains and the maximum shear strain.

2.13 The state of plane strain at a point is $\epsilon_x = -2,500\mu$, $\epsilon_y = -4,000\mu$, and $\gamma_{xy} = -2,000\mu$. Use Mohr's circles to determine the principal strains and the maximum shear strain.

2.14 The state of strain at a point in a body is given by the strain tensor in x-y-z coordinates

$$\epsilon = \begin{pmatrix} -2,000 & 1,000 & 3,000 \\ 1,000 & 2,500 & -1,500 \\ 3,000 & -1,500 & 1,000 \end{pmatrix} \mu.$$

A set of x'-y'-z' coordinates is defined by letting the x'-axis point in the $3\mathbf{i}+2\mathbf{j}+6\mathbf{k}$ direction and the y'-axis point in the $4\mathbf{i}-2\mathbf{k}$ direction.

(a) Determine the rotation matrix, α, from the unprimed to primed coordinates.

(b) Determine the strain tensor in x'-y'-z' coordinates using the rotation matrix, α.

2.15 The displacement in a body is described in x-y-z coordinates by

$$u = (x^2 + y^2)z$$
$$v = 2xyz + x^2 z$$
$$w = 0.$$

(a) Determine the strain tensor in x-y-z coordinates at each point.

(b) Determine by a calculation whether or not compatibility holds for this strain field.

(c) What assumption about the displacements did you make in determining your answer to (a)?

2.16 Determine γ_{xy} if a strain rosette on the surface of a body measures $\epsilon_x = -100\mu$, $\epsilon_{45} = 300\mu$, and $\epsilon_y = 400\mu$.

2.17 A two-dimensional strain field has normal strains

$$\epsilon_x(x, y) = xy$$
$$\epsilon_y(x, y) = y^2.$$

What form must the shear strain have in order that the strain field satisfies the compatibility conditions? Ans. $\gamma_{xy}(x, y) = 0.5x^2 + f(x) + g(y)$.

2.18 The strain tensor at a point is

$$\epsilon = \begin{pmatrix} 100 & -200 & 0 \\ -200 & 0 & 300 \\ 0 & 300 & 0 \end{pmatrix} \mu.$$

Use a computer package such as MATLAB, Mathematica, Maple, or MACSYMA to carry out the following steps.

(a) Compute the principal strains as the eigenvalues of ϵ.

(b) Compute the eigenvectors of ϵ. Write them as unit vectors.

(c) Recall that ϵ and ϵ' are similar if $\epsilon' = P\epsilon P^{-1}$, where P is the matrix formed with rows which are the unit eigenvectors of ϵ. Construct the rotation matrix, α, in this manner. Verify that $\alpha\alpha^t = \alpha^t\alpha = I$.

(d) Verify that $\alpha\epsilon\alpha^t$ is the diagonal strain tensor whose entries are the principal strains.

(e) Determine the angle between the original x-direction and the first principal direction.

2.19 Use a symbolic computer package such as Mathematica, Maple, MATLAB, or MACSYMA to determine the angles of rotation between the original Cartesian coordinates and the principal coordinates for an arbitrary plane strain tensor. Use the method suggested in the previous problem.

2.20 The displacements, in inches, at an arbitrary point (x, y, z) in a given body are $u(x, y, z) = (x^4y + xy^3) \times 10^{-3}$, $v(x, y, z) = x^3y^3 \times 10^{-3}$, $w(x, y, z) = 0$.

(a) Determine the strains at each point, (x, y, z).

(b) Is the body in plane strain? Explain why or why not.

(c) Determine whether or not the strains are compatible.

(d) Use Mohr's circle to compute the principal normal strains at the point $(1, 2, 1)$.

(e) Use the strain tensor to compute the principal normal strains at the point $(1, 2, 1)$.

REFERENCES

[1] B. BOSTROM, New Perspectives on Test Methods for Dimensional Stability, in *Proceedings of the 3rd International Symposium: Moisture and Creep Effects on Paper, Board and Containers*, I. R. Chalmers, Ed., PAPRO, Rotorua, New Zealand 191–200, 1990.

[2] J. W. DALLY AND W. F. RILEY. *Experimental Stress Analysis*, 3rd ed., McGraw-Hill, New York, 1991.

[3] J. D. HUMPHREY, D. L. VAWTER, AND R. P. VITO, Quantification of strains in biaxially tested soft tissues, *Journal of Biomechanics*, 20(1), 59–65 (1987).

[4] L. E. MALVERN, *Introduction to the Mechanics of a Continuous Media*, Prentice Hall, Englewood Cliffs, NJ, 1969.

[5] FRANK G. TATNALL, *Tatnall on Testing*, American Society for Metals, Metals Park, OH, 1966.

CHAPTER 3

STRESS-STRAIN RELATIONS

Engineering analysis attempts to predict when a body will fail by examining the stress and strain distribution in the body and to predict the relation between the applied external loads and the resulting deformation of the body. At each point in a three-dimensional body, there are six stress components, six strain components, and three displacement components to determine, for a total of 15 unknowns. Therefore, 15 equations are required. Thus far three equilibrium equations and six strain-displacement equations have been defined. The remaining six equations are those relating the six stress components to the six strain components.

Definition   A set of equations relating the stress and strain in a material is called the constitutive equations for the material.

Once these remaining six equations have been defined, the 15 equations in addition to boundary conditions determine the stress, strain, and displacement distribution in the body. The boundary conditions depend on the shape of the body, the applied loads, and any displacement constraints. Usually, this is not a simple problem to solve.

The stress-strain, or constitutive, equations introduce the properties of the body material into the analysis. No one has yet succeeded in deriving the constitutive relations directly from knowledge of the molecular structure of the material. The relations must therefore be determined by performing experiments and fitting equations to the resulting data. Occasionally, several different types of functions can fit the same data. In this case, additional tests are required to make a choice of the form of the constitutive equations.

3.1 UNIAXIAL STRESS-STRAIN TESTS OF METALS

The simplest test to perform is an axial stress-strain test to determine how the stress in the axial direction varies with strain. This is thought of as a one-dimensional test since only one principal stress is nonzero. The test operator can either control the force on the specimen or the deformation of the specimen. For tensile tests, one of these is usually varied linearly with time. In load control, the applied force, P, is $P = rt$, where t is time and the constant r which equals the increase in force per unit time is called the load rate. The stress-strain relations obtained under different controls are not identical and the differences provide insight into the material behavior.

A uniaxial tension stress-strain curve, obtained under strain control, for mild steel (1020 steel) with a carbon content of one-fifth of 1%, 0.20%, is shown in Figure 3.1. The curve has two distinct regions, a steep initial linear part and a flatter but curved portion.

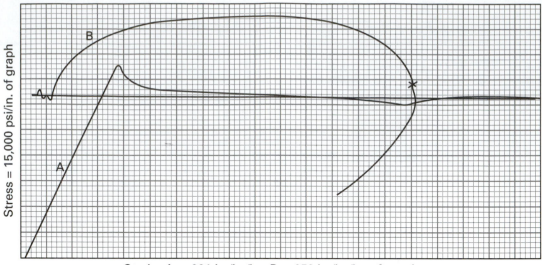

FIGURE 3.1   Stress-strain curve for 1020 steel. Portion A is the initial portion. The original pen-plotter graph was 10 inches wide and 5 inches high.

When the strain reached 0.01, the x-y plotter pen was suddenly returned to the left side of the graph and the strain scale adjusted by a factor of 50 to create portion B of the curve. The load rate was not adjusted during this continuous test.

The initial linear portion of the stress-strain curve is typical for metals, but is rarely found on the stress-strain curves for polymers. Suppose that the specimen is loaded with a slowly increasing force which is stopped at a load for which the stress is on the initial linear portion. Experiment shows that if the force is then removed so that the stress returns to zero, the strain also returns to zero. The body has recovered its original shape since the strain is zero. In this case, the specimen is said to have been elastically loaded. If the stress lies anywhere on the initial linear portion of the stress-strain curve, the metal is elastically loaded. The bonds between the molecules of the material have been stretched by the force, but once the force is removed, the bonds pull the material back to its original unloaded shape.

The equation of the initial linear portion of the uniaxial stress-strain curve is $\sigma = E\epsilon$, where E, the elastic modulus, is the slope. The stress-strain curve is linear for small strains. This result is not true for other materials such as polymers, wood, or other natural materials. The assumption of small displacements made in the linear strain-displacement relations is valid for elastically loaded metals.

If there is no initial straight line portion for a given stress-strain curve, E is often defined to be the slope of the tangent line to the stress-strain curve at the origin. These two definitions agree when the curve is straight since the tangent to a straight line is the line itself.

From $\sigma = E\epsilon$, the elastic modulus, E, must have the same units as stress. For example, E for low-carbon steel is about 30×10^6 psi or 210 GPa (giga pascals) and for aluminum is about one-third of that for steel, 10×10^6 psi or 70 GPa. In English units, E is usually written as a multiple of 10^6 to make it easy to compare the moduli for different materials.

The relative magnitudes of E for two materials indicate which material is stiffer. Since $\sigma = E\epsilon$, $\epsilon = \sigma/E$. For a fixed force, σ is constant and therefore, the larger E is, the smaller the strain is for that fixed stress. Materials with larger elastic modulus,

E, are stiffer since it is harder to strain them. Steel is stiffer than aluminum when both are elastically loaded since its elastic modulus is larger.

The idea of an elastic modulus was first introduced by Thomas Young while discussing tension and compression in his courses taught at the end of the eighteenth century in England and published in 1807. Young viewed the modulus as a physical object, as an equivalent column of material, and spoke of the weight of the modulus and the height of the modulus. His definition amounted to the current definition times the cross-sectional area of the specimen. Even though his definition is not used today, the elastic modulus is often called "Young's modulus."

The current definition of elastic modulus as the ratio of the load per unit area (stress) divided by the unit deformation (strain) produced by the load was first given by Navier, a French scientist, in his 1826 book on the strength of materials. Navier also pointed out that, to design structures, it was not sufficient to know the ultimate load of a material, which had been the primary focus of previous engineering research. He said that an engineer also needs the elastic modulus and the elastic limit. Navier's work incorrectly assumed that the relationship $\sigma = E\epsilon$ holds for all engineering materials.

The elastic region is important since, in most cases, an engineer does not want a structure to permanently change shape under forces; the structure must be elastically loaded. The structure is designed so that the maximum stress is kept in the elastic range. The maximum stress at which a material remains elastically loaded can be identified in several ways.

Definition The proportional limit, σ_{PL}, is the maximum stress on the linear portion of the stress-strain curve.

This value is the largest value for which stress is proportional to strain on the linear curve. It is possible that the material may remain elastic even after the proportional limit is exceeded, assuming such a limit exists.

Definition The elastic limit, σ_{EL}, is the largest stress at which the material remains elastic. A material is said to yield if the strain undergoes a large increase for a small increase in the stress. The yield point, σ_Y, is the stress at which the material begins to yield.

These values are sometimes also called the elastic strength or yield strength. The word "strength" used in this way is a synonym for "stress." The strengths σ_{EL}, σ_Y, are quite difficult to determine. For metals, however, they are nearly equal to σ_{PL}. Since the proportional limit is the easiest to estimate visually from the stress-strain curve for metals, it is often used as the value below which the stress must be kept in order for a material to be elastically loaded. The rate of loading does not change the elastic modulus much in a metal but can affect the proportional limit, which tends to increase with an increase in load rate.

The term "yield point" was coined in 1870 by a committee of the Institution of Civil Engineers in England [35]. While the yield point for steel is clearly visible on a stress-strain or load-deformation curve, that for copper is not (Fig. 3.2). Loading past the yield point followed by unloading changes the properties of the material. In Figure 3.2, the steel was loaded to point d, then unloaded to e. When reloaded, the curve took the path, edf, rather than egh as might have been expected. The theory of plasticity, more fully described in Chapter 9, attempts to explain this behavior.

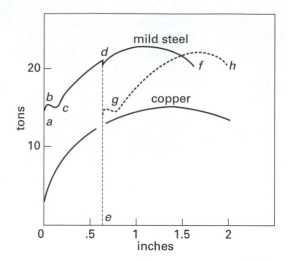

FIGURE 3.2 Tensile load versus deformation for steel and copper, adapted from [35].

3.1.1 Plastic Strain

The nonelastic stresses make up the plastic region of the stress-strain curve. On the steel curve, the plastic region is all of the curve except the initial linear portion. When stressed to these levels, the material molecules begin to flow past each other rather than just slightly separate as under elastic loading. If the stress on a material reaches the plastic region, the material is permanently strained; it changes shape and will not return to its original shape when the force on it is removed.

A long plastic region on the tensile stress-strain curve indicates that the material stretches considerably before fracturing. Such a region is typical of ductile materials like steel and aluminum. A brittle material, such as cast iron, has a very short plastic region. The brittle material fractures shortly after the elastic range of stresses is exceeded.

It is usually quite difficult to find a single formula (mathematical model) which represents the stress-strain relationship for plastic strain. Equations such as $\sigma = K\epsilon^n$, for K a constant and n a constant in the range $0 < n < 1$, are sometimes used as an approximation. The stress at yielding is σ_Y. The constants, K and n, depend on the material and so are called material constants. The constant, n, is called the strain hardening exponent.

The steel graph of Figure 3.1 was obtained under linear strain control. The dip in the graph at the end of the elastic region occurs only for steel and metals with a similar structure. This temporary decrease in stress is a consequence of the strain control; it does not occur in load control. The dip is thought to be due to slippage around carbon atoms in the steel, which is an alloy of iron and small amounts of carbon. During slip, less force is required to maintain a linear increase in the strain, and so the stress, which is the internal force divided by the original cross-sectional area, is decreased. Most materials, especially nonalloys, do not exhibit this behavior.

The most significant stress value in the plastic region is the largest stress attained.

Definition   The ultimate stress, σ_U, is the largest stress supported by the material.

The numbers E, σ_{PL}, σ_U are called material properties and can be used to assess the usefulness of various materials in the solution of a design problem. In load control, the ultimate stress is the stress at fracture since in load control the stress increases continually. However, in the strain-controlled steel tensile test curve of Figure 3.1, the ultimate stress occurs before the material fractures. Just after the material reaches its ultimate stress on a strain-controlled curve, the material begins to neck. Its cross-sectional area in one location becomes significantly smaller as the specimen is stretched. The true stress is the internal force divided by the actual cross-sectional area. But the recorded stress is the internal force divided by the larger original cross-sectional area. The tests are plotted in this manner because it is almost impossible to compute the actual cross-sectional area while the test is proceeding. The plotted stress is lower than the true stress. After necking, it takes less force to maintain a constant strain rate; the graphed stress decreases until fracture. In fact, once necking begins, the specimen can no longer be assumed to have a uniaxial stress state. In engineering design, the ultimate stress, whether obtained from a load-controlled or strain-controlled test, is often assumed to be the fracture stress.

Engineering analyses of plastic behavior are performed in terms of the true strain of the material. The true strain is defined to be the sum of all the increments in strain as the length changes, the integral from L_o to L_f,

$$\epsilon = \int_{L_o}^{L_f} \frac{dL}{L} = \ln\left(\frac{L_f}{L_o}\right). \tag{3.1}$$

The uniaxial nominal strain, $(L_f - L_o)/L_o$, sometimes called engineering strain, is defined in terms of a reference state of the body, the original length, L_o. For purposes of this discussion, the nominal strain will be denoted by e. The true strain can be written in terms of the nominal strain by

$$\epsilon = \ln\left(\frac{L_f}{L_o}\right) = \ln\left(\frac{L_f - L_o + L_o}{L_o}\right) = \ln(1 + e).$$

For example, when the nominal strain is $e = 0.02$, the true strain is $\epsilon = 0.0198$. The stress required to make the true strain 0.02 is larger than that required to make the nominal strain 0.02. The graph of the stress versus true strain lies above that of the stress versus nominal strain.

Likewise, the nominal stress, S, is defined to be the internal force divided by a reference area, usually the original area A_o, while the true stress is the internal force divided by the current area. Most data is given as a nominal stress versus nominal strain curve, but a simple relation between true and nominal stresses is obtained when the volume of the specimen remains constant. If the volume remains constant, then at each state of deformation, $AL = A_o L_o$. Therefore, $A_o/A = 1 + e$. The true stress, σ, for axial load, P, is

$$\sigma = \frac{P}{A} = \frac{P}{A_o}\frac{A_o}{A} = S(1 + e). \tag{3.2}$$

The Considère (1886) diagram (Figure 3.3) shows the relation between the nominal stress–nominal strain and the true stress–nomimal strain curves. The slope of the true stress–nomimal strain curve is the nominal stress, S, by equation (3.2). Therefore

$$\frac{d\sigma}{de} = S = \frac{\sigma}{1 + e}. \tag{3.3}$$

On the other hand, the slope of the true stress–true strain curve is σ at the ultimate tensile strength. At necking, the graph of the load P versus length L reaches a maximum;

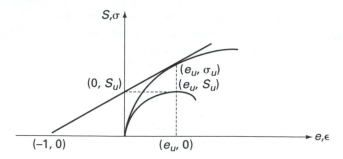

FIGURE 3.3   The Considère diagram.

so that $dP/dL = 0$. Since volume is assumed constant before necking begins and the specimen is a prism, $V = AL$, a constant, implies that $0 = dV/dL = A + L(dA/dL)$ or $dA/dL = -A/L$. By definition $P = \sigma A$. Differentiating with respect to L and evaluating at the ultimate load gives

$$0 = \frac{dP}{dL} = A\frac{d\sigma}{dL} + \sigma\frac{dA}{dL} = A\frac{d\sigma}{dL} - \sigma\frac{A}{L}.$$

Dividing out A and using the fact that $d\epsilon = dL/L$ gives $d\sigma/d\epsilon = \sigma$ at the ultimate tensile strength. The slope of the true stress-strain curve at the ultimate load is, by (3.3),

$$\frac{d\sigma}{d\epsilon} = \sigma = S_u(1 + e_u) = (1 + e_u)\left(\frac{d\sigma}{de}\right), \tag{3.4}$$

where S_u is the ultimate nominal stress on the nominal stress–nominal strain curve where necking begins and e_u is the corresponding nominal strain. The true stress–true strain curve is steeper than the nominal stress–nominal strain curve. The Considère diagram shows how to sketch the true stress–nominal strain from the nominal stress–nominal strain curve obtained from a strain-control uniaxial test. A line is drawn from the point $(-1, 0)$ to the point $(0, S_u)$. By equation (3.3), this line must be tangent to the true stress–nominal strain curve at the ultimate true stress, σ_u. This is shown by the similar triangles giving

$$\frac{\sigma_u}{1 + e_u} = \frac{S_u}{1} = S_u. \tag{3.5}$$

A plastic instability is said to occur once necking begins. At this point the material is no longer subject to a uniaxial stress state. Also, with the specimen necked, it is no longer possible to write the volume as a product of a cross-sectional area and a length. If the true stress-strain curve is given by the power law, $\sigma = K\epsilon^n$, then at the onset of the plastic instability, necking, $d\sigma/d\epsilon = \sigma$ implies that $Kn\epsilon^{n-1} = \sigma = K\epsilon^n$. The strain hardening coefficient n equals the strain at the onset of necking. This gives an experimental method of determining n without curve fitting. K and n can also be determined from a log stress versus log strain plot, $\ln(\sigma) = \ln(K) + n\ln(\epsilon)$. The log stress intercept is $\ln(K)$ and the slope is n.

3.1.2   Poisson's Ratio

During the development of the theory of deformable solid behavior in the early 1800s, experimentalists realized that their results were affected by changes in the cross-sectional area of a loaded specimen. An axially loaded member stretched in tension along its longitudinal axis is observed to become thinner. Its dimensions change in both the longitudinal (axial) and transverse directions. A transverse direction is any direction perpendicular to the axis under consideration. For example, on a circular cross-sectional member, each radius is transverse to the longitudinal axis. If a member is compressed, its transverse dimensions must increase if the volume changes only slightly. This effect is called the Poisson effect because in 1829 Siméon Denis Poisson proposed a method of monitoring it.

The Poisson effect is measured by Poisson's ratio, which is the ratio of the strain in a direction transverse to the axis under consideration, ϵ_t, to the strain in the axial direction, ϵ_a. Since the transverse strain is negative if the axial strain is positive (tension) and the transverse strain is positive if the axial strain is negative (compression), a minus sign is put in front of the ratio so that the Poisson ratio, ν, is always a positive number.

$$\nu = -\frac{\epsilon_t}{\epsilon_a}. \tag{3.6}$$

For linear elastic materials such as metals, Poisson's ratio is a constant between 0 and 0.5 . For example, for steel $\nu = 0.25$ and for aluminum $\nu = 0.33$. In other materials such as polymers, the Poisson ratio may not be a constant. Poisson had deduced from his theoretical analysis that the ratio is always 1/4. But by 1848, Wertheim had shown experimentally that the Poisson ratios for glass and for brass are not 1/4. He proposed arbitrarily that the ratio should be taken to be 1/3.

3.2 LATTICE STRUCTURE OF METALS

Metal behavior, either elastic or plastic, under loads depends on the structure of the metal as it is built up from atoms into lattices and finally into grains. These structures are held together by the bonds between atoms. External forces on a piece of material affect these atomic bonds.

3.2.1 Bonding

Atomic bonding depends on the action of the electrons in the outer shells of the bonded atoms. Each atom tries to fill its outer shell in order to reach a minimum energy state. Except for the simplest atoms, the outer shell requires eight electrons to be stable. Bonding can be classified into four general types: ionic, covalent, metallic, or van der Waals. This list is in the order of bond strength, from strongest to weakest. The melting point of solids often depends on the type of bonding, as do the heat and electrical conductivity and the behavior under forces.

Ionic bonding occurs between atoms of two different elements. If an atom with seven electrons in its outer shell nears an atom with just one electron in its outer shell, it may capture that one electron. The atom with the newly completed outer shell becomes a negatively charged ion because of the extra negative charge, while the atom giving up the electron becomes a positively charged ion. Since these opposite forces attract, the two ions become bonded. For example, this bonding forms sodium chloride (NaCl), table salt, in which the sodium gives up one electron to the chloride atom. The bonding force is nondirectional so that ions of one element surround ions of the other and vise versa. The number of atoms surrounding an ion depends on the relative size of the two ions, that is, the number of shells each ion has.

Two or more atoms can fill their outer shells by sharing some of the electrons orbiting each nucleus. This is called covalent bonding. Atoms of the same element are commonly bound in this manner. Two chlorine atoms can bond by sharing one electron to complete their outer shells to eight electrons, since a single chlorine atom has seven electrons in its outer shell. Sulfur, whose outer shell contains six electrons, can share two electrons, one each with two other sulfur atoms. This allows sulfur to form chains of covalently bonded atoms. Nonidentical atoms can also be bound covalently, for example, water (H_2O). Oxygen is short two electrons to complete its outer shell. These are provided by two hydrogen atoms, each having one electron.

Van der Waals bonding is weaker than either ionic or covalent bonding. Due to the motion of electrons around an atom, the charge of the atom can become unbalanced so that one side of the atom is temporarily negative and the other positive. Since opposite charges attract, two such unbalanced atoms can weakly bond.

Metallic bonding is described by the free-electron theory of Drude and Lorenz, proposed in 1900. They said that, in a metal crystal, the valence electrons, those in the outer shell, escape from the individual atoms and swarm in a cloud around the resulting positive ions. This structure is held together by the interaction of the positive ions and the negatively charged cloud. Because the forces involved are nondirectional, three-dimensional structures of atoms can form. The cloud structure also allows metal alloys to form. The ions of the two metals do not bond but simply mix in what is called a solid solution. The free electron theory also explains why metals are good conductors of electricity and heat. The conduction, which depends on the motion of electrons, is made easier by the free electrons in the electron cloud.

3.2.2   Formation and Structure of Crystals

As a material cools from a liquid to a solid and the energy level of the atoms is reduced, the atoms often combine in regular patterns called crystals. The crystals form most easily when the bonds between the atoms are nondirectional, either ionic or metallic. Covalent bonding, except for that involving carbon or silicon, rarely produces crystals.

When two atoms bond, the force of attraction due to the bond balances with the force of repulsion due to the intermingling of the negative electrons from each atom. The resulting energy level is at a minimum. Half the distance between the centers of the two bonded atoms in this position is taken as the radius of the atom when it is modeled as a sphere. The atoms tend to try to approach as many other atoms as possible. The atoms, thought of as spheres, try to become closely packed together. Mathematicians have abstracted this idea to study the question of the best arrangement to make spheres close-packed.

The structure of the atom, in particular its electron configuration, determines which crystal structure a solid will take. For example, the atoms, represented as spheres, could arrange themselves in layers following the pattern given in Figure 3.4.

FIGURE 3.4   Atomic organization.

If the atoms are replaced by points and lines are drawn between the points representing the bonds, one obtains a mathematical structure called a lattice, which models the crystal structure. The length of the lines represent the center to center distance between the particles represented by the points. The structure of the object is mathematically abstracted so that the structure becomes clearer and easier to analyze (Fig. 3.5).

FIGURE 3.5   Model of atomic arrangement.

A space lattice is a symmetric structure of points and connecting lines lying in three-dimensional space. A point, P, on the lattice is a center of symmetry if a line from any point on the lattice to P, when extended an equal distance through P, reaches another point on the lattice. A primitive (or basic) cell is the object whose sides are formed by three vectors A, B, and C in Figure 3.6. If these basic vectors are moved to any other point of the lattice, they will form the sides of another primitive cell with this new point as corner.

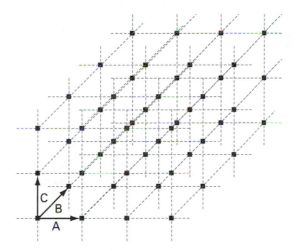

FIGURE 3.6   Space lattice.

Each point on a lattice is a center of symmetry. While a lattice extends infinitely in all directions, a crystal is modeled by just a portion of the lattice. The lattice is a geometric mathematical model for physical reality, as opposed to mathematical models composed of equations used to describe, say, the relation of stress and strain. The lattice structure of a given metal is often determined experimentally by x-ray diffraction.

Bravais proved in 1848 that there are only 14 mathematically possible lattice types. These can be classified into seven types of crystals depending on how many of the three sides of a basic lattice cell are equal and on whether or not these sides are perpendicular. Each crystal type includes from 1 to 3 of the 14 lattice types. It is important to keep in mind that the crystal is built of layers of atoms or molecules or groups of atoms, which can be viewed as lattice points, extending in all directions in space to the boundary of the crystal.

Three of these space lattices are common in metals. The stronger metals, such as chromium, alpha iron, molybdenum, and tungsten, have body-centered cubic (BCC) crystals. The cube formed by atoms in the upper and lower layer enclose a single atom in the middle layer at the center of the cube (Fig. 3.7).

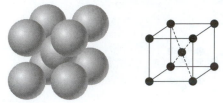

FIGURE 3.7   BCC structure.

The more ductile metals, such as aluminum, copper, gold, lead, and nickel, have a face-centered cubic (FCC) structure. Again a cube is formed by atoms in the upper and lower layer, but now four atoms in the middle layer are enclosed in the vertical faces of the cube and an additional atom is enclosed in the upper and another in the lower face. In other words, each of the six faces contains an atom at its center (Fig. 3.8).

FIGURE 3.8   FCC structure.

The metals with a hexagonal close-packed (HCP) structure, such as magnesium, titanium, and zinc, tend to be more brittle. The crystal involves three layers with the middle one skewed from the upper and lower layer so that the atoms modeled as spheres can be very closely packed. The basic cell pictured in Figure 3.9 is the repeating portion of these layers. Additional layers are built above and below in the same pattern.

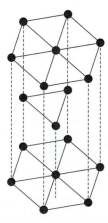

FIGURE 3.9   HCP structure.

As a material cools from its liquid to solid state, any lattice growth must be initiated by a foreign body, which forms a nucleus around which the lattice is built, even in a "pure" metal. The lattice grows in space in the directions parallel to the sides of the basic cell. Two growths approaching each other will not join unless these lattice directions happen to match in space. Since this mutual orientation rarely occurs, most of the time crystal growth is stopped when it meets another growth. The two crystals are usually separated by about the distance of two basic cells. These individual crystal growths are called grains. A material composed of many grains is said to be polycrystalline. Since it takes time for these structures to develop, slow cooling yields large grains; rapid cooling smaller grains. No structural difference is obviously apparent between the individual grain shapes of metals of the various lattice types (Figs. 3.10 and 3.11). The polycrystalline structure is an important determinant of properties such as the yield strength, fracture behavior, and creep.

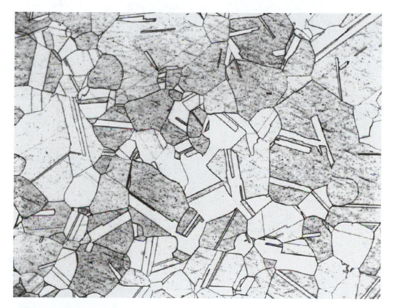

FIGURE 3.10   Conventional optical micrograph of a polycrystalline FCC copper microstructure showing in a sectioned specimen polygonal and curved grain boundaries encompassing grains that are penetrated in numerous places by planar annealing twin bands. (Courtesy of V. Ramachandran.)

Two lattice growths can meet in a small-angle grain boundary (Fig. 3.12). These growths are regarded as subgrains of a grain when the tilt angle is small, less than 15°. In this case, the boundary is stable because the atoms at the edges of the lattice are sufficiently separated. The edges of various lattice planes are defects in the perfect lattice structure.

Subgrains are visible in a reflected x-ray image because the rays are reflected differently from each of the slightly misaligned subgrain lattice structures. In Figure 3.13, the substructure is defined, first, by rather straight black or white boundaries due to displaced subgrain images because of their misorientation and, second, to finer scale boundaries within these subgrains recognizable because of the changed intensities of reflection within the finer subgrain blocks.

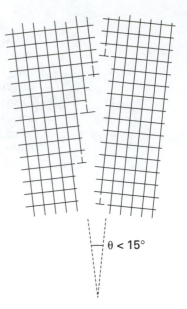

FIGURE 3.11 Optical micrograph obtained with polarized light in reflection after electropolishing and anodizing the sectioned surface of a fine-grained polycrystalline HCP alpha-titanium microstructure. (Courtesy of N. M. Madhava.)

FIGURE 3.12 The small-angle grain boundary.

FIGURE 3.13 A reflection x-ray topograph of a cross-sectional surface of a cylindrical nickel single crystal solidified from the melt and showing several levels of substructure. (Courtesy of W. J. Boettinger and M. Kuriyama.)

3.2.3 Atomic Bond Energy

By carefully examining the bond forces in the metal lattice, one can see why the elastic strain is restricted to such small values in metals. Also a physical interpretation of the elastic modulus and yield stress is possible. As is true with much of mechanics, this analysis can be formulated in terms of energy functions.

One of the first to study this problem was Charles Augustine de Coulomb (1736–1806) who determined experimentally that the attractive force in ionic bonding is inversely proportional to the square of the distance between the atoms. In this respect, it is similar to gravitational force.

In general, for a pair of atoms or molecules, the attractive force, F_a, is inversely proportional to a power of the distance, r, between either the atoms or molecules,

$$F_a = -\frac{k}{r^m}.$$

Frequently for molecules, $m = 6$. The force of attraction is inversely proportional to the square of the distance between atoms in both metallic and ionic bonds, which are both nondirectional. For example, in ionic bonding the constant $k = k_o(Z_1 q)(Z_2 q)$, where $k_o = 9 \times 10^9$ (Vm/C), $q = 1.6 \times 10^{-19}$ C and Z_1, Z_2 are the valences of the two ions.

The force of repulsion, F_r, depends on the electronic configuration of the atoms involved. If the force of repulsion is assumed to also be inversely proportional to a power of the distance between either the atoms or molecules, then it can be written as

$$F_r = \frac{B}{r^n},$$

where $n - 1$ is called the Born exponent. The exponent, n, can vary between 2 and 12 for different materials. Other researchers have proposed that

$$F_r = \lambda \exp\left(-\frac{r}{\rho}\right),$$

where λ and ρ are material constants determined from experiment.

Exercise 3.1 Sketch a graph of both models for the repulsive force to verify that they are similar in shape and are asymptotic to zero as the radius, r, tends to infinity.

The total force, F, on one of the bonded particles is the sum of the attractive and repulsive forces,

$$F = F_a + F_r = -\frac{k}{r^m} + \frac{B}{r^n}.$$

The potential energy, U, of the single bond is such that the bond force is its derivative with respect to position, $F = dU/dr$. Therefore

$$U = \frac{k}{(m-1)r^{m-1}} - \frac{B}{(n-1)r^{n-1}}.$$

The graph of a typical energy function is shown in Figure 3.14. The potential functions for metallic, ionic, covalent, and van der Waals bonding have qualitatively similar shapes.

This model of the potential energy neglects thermal vibrations of the atoms. So strictly speaking, the material is assumed to be at absolute zero (0 K). It is not yet possible to deduce the precise form of the potential energy for even two atoms from knowledge of their nuclear and electron field configuration. Therefore empirical (i.e., experimental) constants are involved.

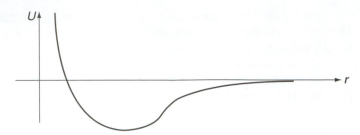

FIGURE 3.14 The potential energy for the bond force.

The atoms are in equilibrium when the attractive and repulsive forces are equal or, equivalently, when the derivative of the potential energy is zero. This equilibrium distance, a_o, is the value of r satisfying

$$\frac{dU}{dr} = -\frac{k}{r^m} + \frac{B}{r^n} = 0.$$

The equilibrium position, a_o, is the lattice parameter at the given temperature. This equilibrium must be stable; therefore the second derivative must be positive so that the equilibrium state is a minimum of the energy function.

Exercise 3.2 Verify that in order for the equilibrium to be stable, it must be true that $n > m$.

EXAMPLE 3.1 The repulsive constant for a material is found to be $B = 0.4ka_o^4$, where both B and k are measured in angstroms. Also $m = 2$ and $n = 8$. Determine the lattice parameter, a_o, for this material.

Solution At equilibrium, $r = a_o$ and the total force is zero. Therefore, the lattice parameter is the solution to

$$0 = -\frac{k}{a_o^2} + 0.4k\frac{a_o^4}{a_o^8} = \frac{k}{a_o^2}\left(-1 + \frac{0.4}{a_o^2}\right).$$

Therefore $a_o = 0.6325$ angstroms. ∎

The maximum of the binding force is the cohesive force. This force must be overcome to separate the particles to infinite distance. The force is maximum at a distance slightly larger than the equilibrium distance. The cohesive force can be computed by determining the distance r at which $dF/dr = 0$ and then finding the force at that position. F. Zwicky, in 1923, calculated the theoretical strength of a single crystal of rock salt (NaCl) at $T = 0$ to be 2000 MPa under the assumption of a perfect lattice structure, but the room temperature experimental value is only 5.31 MPa. He used $m = 1$ and $n = 9$. The discrepancy seems to be due to the assumption of a perfect crystal structure rather than to the neglect of thermal effects.

To define and compute the elastic modulus, proceed as follows. If A is the atomic cross-sectional area that is perpendicular to the bond, the stress is the force divided by the area,

$$\sigma = \frac{1}{A}\frac{dU}{dr}.$$

The stretch is defined to be $\epsilon = r/a_o$. The strain energy density, U_o, is the potential energy per unit volume of a cell, $U_o \equiv U/Aa_o$. Note that by the chain rule,

$$\frac{dU_o}{d\epsilon} = \frac{dU_o}{dr}\frac{dr}{d\epsilon} = \frac{a_o}{Aa_o}\frac{dU}{dr} = \sigma.$$

By definition, the elastic modulus is the second derivative of the strain energy density with respect to ϵ. Therefore by the previous calculation of the stress, $E = d\sigma/d\epsilon$ and by the chain rule, $E = (d\sigma/d\epsilon)|_{equil} = (a_o/A)(dF/dr)|_{equil}$. If the Born number model for the repulsive force is used with the Coulomb model for the attractive force, and taking $m = 2$,

$$E = \frac{a_o}{A}\frac{d^2U}{dr^2} = \frac{a_o}{A}\frac{dF}{dr}\bigg|_{r=equil} = \frac{a_o}{A}\left[\frac{2k}{a_o^3} - \frac{nB}{a_o^{n+1}}\right],$$

because at equilibrium, $r = a_o$. If the exponential model for the repulsive force is used,

$$E = \frac{a_o}{A}\left[\frac{2k}{a_o^3} - \frac{\lambda}{\rho}\exp\left(-\frac{a_o}{\rho}\right)\right].$$

Since equilibrium occurs at the minimum of the potential, the potential must be concave up at equilibrium; that is, E is positive.

A force function for an array of atoms or molecules in a lattice can be constructed from the analysis of a single bond. Pick one atom and place the origin of the r-axis there. The r-axis goes through a line of regularly spaced atoms in the lattice. Frequently the force between atoms in a lattice is approximated by a periodic function which is zero at the center of each atom. The atoms in an unloaded metal lattice lie at an equilibrium distance apart (Fig. 3.15).

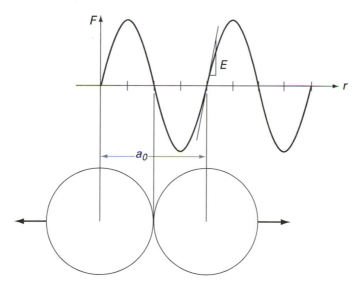

FIGURE 3.15   Periodic approximation to the force function for a one-dimensional array of atoms.

The application of external forces on the metal can push the atoms closer together, stretch the atoms slightly further apart, or can break their bonds altogether. To build a model describing the consequences of applying such force, approximate the graph of the force function in a neighborhood of equilibrium distance. This is often done with a sine function or a parabola. When the sine function is used (Fig. 3.15), the center of each atom occurs at multiples of $r = 2\pi$. The stress as r varies slightly away from zero is

$$\sigma = \sigma_o \sin\left(\frac{2\pi r}{a_o}\right).$$

But $E = (a_o/A)(dF/dr)|_{r=0} = (a_o/A)A\sigma_o(2\pi/a_o)$. Therefore, $\sigma_o = E/2\pi$.

The yield stress is taken to be the amplitude of the sine curve which fits the force function most closely at the equilibrium point. Therefore the maximum possible yield stress predicted by this model is the amplitude of the sine function, σ_o. Often, the theoretical strength is estimated as $E/10$. Experiment shows plastic flow is initiated at about $E/1,000$ in most metal or ceramic materials, a much lower stress than $\sigma_o = E/2\pi$. The failure of the experimental elastic behavior to agree with the theoretical calculation must be due to another type of bond breaking, lattice deformation, such as shear failure, or to disruptions and imperfections in the ideal lattice structure.

The deeper the potential well, the higher the predicted melting point and the elastic modulus. So materials with high melting points should be expected to have a large elastic modulus. Elasticity in an elastomer such as rubber is a very different phenomenon. There the long-chain molecules are uncurled from their minimum energy equilibrium positions as a load is applied.

3.2.4 Dislocation Theory for Metals

Crystal dislocation theory, first proposed in 1934 in terms of strength of materials by Cambridge University Professor G. I. Taylor, provides an explanation for both the ductility of metals and their ability to be cold worked. Cold working or work hardening are expressions used to describe the increasing brittleness and strength of ductile metals as they are deformed. The effects of cold working can be undone by annealing the material, that is by heating it and then allowing it to cool, to recrystallize the metal.

A crystal dislocation is a defect in the lattice structure of the metal. There are two types of line defects, a screw dislocation and an edge dislocation. An edge dislocation occurs where one level of the lattice is not a complete sheet of atoms. The line at which the sheet breaks off is an edge dislocation (Fig. 3.16). The motion of an edge dislocation under stresses is perpendicular to the edge.

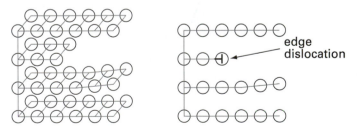

FIGURE 3.16   Edge dislocation.

A screw dislocation can be thought of as the result of shearing in the lattice parallel to a plane of lattice points. The shearing causes some of the lattice points to move through at least one lattice distance. The line indicated in Figure 3.17 is called the screw dislocation. It moves parallel to itself under stresses. Screw dislocations were postulated by F. C. Frank in 1948 to explain crystal growth. The ledge of the screw provides a place for new atoms to join the lattice as it solidifies from a solution. Most dislocations are a combination of edge and screw dislocations. Whereas the screw dislocation can move in any plane containing its line direction, the edge dislocation is

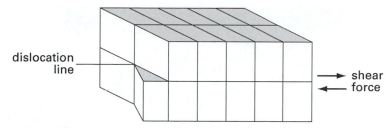

FIGURE 3.17   The screw dislocation.

restricted to move by slip in a plane containing both its line direction and its Burgers vector.

A dislocation is a line defect threading its way through the material (Fig. 3.18). The dislocation lines cannot end within the crystal lattice. They must either form a closed loop or end at grain boundaries.

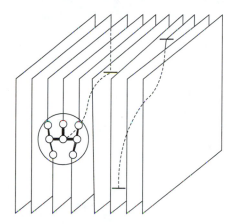

FIGURE 3.18   Schematic of crystal dislocations, including the atom positions at an edge dislocation and a curved dislocation line.

Subgrain boundaries are defined by dislocation systems (Fig. 3.19). These dislocation lines within the boundary tend to be narrower than the dislocation lines within the subgrain because of complementary overlapping strain fields.

Stresses on the lattice cause dislocation motion. When two dislocations meet, they "give birth" to hundreds more dislocations. The resulting dislocations both become entangled and repel each other causing the metal to harden as it deforms in cold working (Fig. 3.20). This process, also called strain hardening, helps explain fatigue in metal. As a metal paper clip is bent back and forth, it hardens and breaks. This effect is avoided in metals that are bent into a desired shape by then annealing to reduce brittleness.

Miller indices are a tool to describe directions and planes in a crystal. Miller indices represent only direction, while, for example, a vector has direction and magnitude. The Miller indices of a direction can be obtained from any vector in that direction. For example, the direction parallel to the vector $\mathbf{i} + \mathbf{j} + \mathbf{k}$, is written using the coefficients of the unit vectors as [1 1 1]. Any other representative of the same unoriented direction, say $\pm 5(\mathbf{i} + \mathbf{j} + \mathbf{k})$, is given the same Miller indices [1 1 1]. The set of indices representing a particular direction is always written as a set of integers by multiplying fractions by their least common denominator. A family of directions equivalent under lattice symmetry preserving rotations of the crystal is denoted by the special brackets

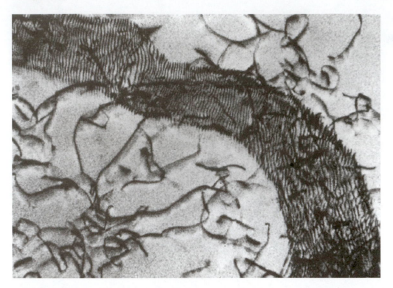

FIGURE 3.19 An x-ray projection topograph of a curved dislocation subgrain boundary within a lithium fluoride (LiF) crystal. (Courtesy of A. R. Lang.)

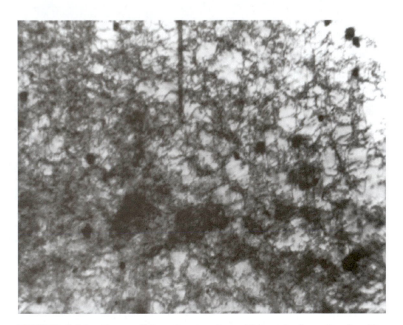

FIGURE 3.20 Dense dislocation tangles within a single grain of a silicon casting revealed by transmission x-ray topography. Also shown are carbon or carbide particles incorporated within the silicon from the graphite mold. (Courtesy of R. G. Rosemeier, K. C. Yoo, and S. M. Johnson.)

$< \cdot >$. For example, the family of directions equivalent to $\mathbf{i} - \mathbf{j}$ is written $< 1\bar{1}0 >$, where the bar is used to indicate a negative.

Miller indices are also written for planes using the fact that the direction of a normal to the plane is sufficient to describe the plane or any one parallel to it. If $\mathbf{N} = h\mathbf{i} + j\mathbf{j} + k\mathbf{k}$ is a normal to the plane, the plane is given indices, $(h\, j\, l)$. In a cube with edges having lengths a, b, c, the normal to the plane passing through the corners intersecting the coordinate axes is $\mathbf{N} = bc\mathbf{i} + ac\mathbf{j} + ab\mathbf{k}$. Therefore the Miller indices of the plane are ($bc\,ac\,ab$) or any scalar multiple thereof. For example, a possible set of indices is computed by dividing by abc to obtain ($1/a\,1/b\,1/c$). The class of all planes equivalent to ($1/a\,1/b\,1/c$) under any lattice symmetry preserving rotation of the crystal is denoted by $\{1/a\,1/b\,1/c\}$.

Definition   The plane on which a dislocation moves is called the slip plane. The dislocation Burgers vector defines the potential slip direction. The slip system is the combination of the slip plane and slip direction.

3.2.5 Ductile and Brittle Metals

A material is called ductile if it fails due to shearing. Once the stress is increased beyond the elastic range, the material begins to flow almost like a fluid and is said to yield. A ductile material has a long plastic region on a uniaxial test curve. Ductility can be measured by percent elongation or percent area reduction. Metals such as aluminum, iron, and copper, and the alloy steel are ductile. A material is brittle if it fails due to normal stresses. Such materials, for example cast iron, have very short plastic regions on their stress-strain curves.

The yield stress in a BCC metal depends strongly on strain rate and temperature as well as its atomic structure, while the yield stress in FCC metals is affected more by the strain hardening behavior. Ductility is primarily governed by the slip plane structure within the crystal. The directions of easiest slip, that is, on which the shear stress required to initiate slip is lowest, lie in planes called the slip planes. In general, the directions are those in which the atoms are closely packed. The family of slip planes for an FCC crystal is $\{1\,1\,1\}$. This group includes copper, nickel, aluminum, gold, and silver. The critical shear stress in a typical FCC metal is generally smaller than that in a BCC metal. The BCC crystal has more slip planes and systems than the FCC, 24 slip systems in BCC to 12 in FCC. Because these slip planes intersect, dislocations can transfer between planes and can easily interfere with each other's motion. The critical shear stress to cause slip is larger in the BCC materials generally because of the lattice resistance to dislocation movement. Pure iron at room temperature is BCC, as are chromium, molybdenum, and tungsten. On the other hand, there are fewer slip planes in HCP crystals. The few slip systems help explain that HCP materials tend to be brittle. HCP metals used in engineering include titanium and magnesium, with slip planes $\{1\,0\,\bar{1}\,0\}$ and $\{0\,0\,0\,1\}$, respectively.

The behavior of the uniaxial test for a ductile material can be followed on a sequence of uniaxial Mohr's circles. Shearing dominates the plastic behavior of a ductile material so that there is a shear stress, τ_m, at which the material will fail. Since the radius of the uniaxial Mohr's circle is equal to the maximum shear stress at a point, failure in a uniaxial test occurs when the uniaxial Mohr's circle hits the line $\tau = \tau_m$ on Figure 3.21. The uniaxial normal stress–normal strain curve is given next to the Mohr's circle for comparison.

3.3 GENERALIZED HOOKE'S LAW

The generalized Hooke's law is an attempt to describe the material behavior when the state of stress is three dimensional and linear elastic. Based on the uniaxial test linear results, it might be assumed that the three-dimensional constitutive equations for metals behave linearly as well. This is a leap of faith that can only be justified by showing that predictions made on the basis of this assumption agree with experimental results. Each

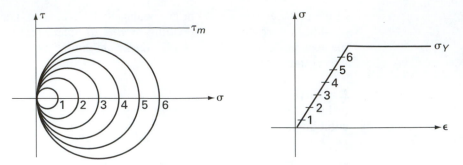

FIGURE 3.21 The growth of Mohr's circle in elastic-plastic behavior.

strain component is assumed to be a linear function of all the stress components. For example,

$$\epsilon_x = C_{11}\sigma_x + C_{12}\sigma_y + C_{13}\sigma_z + C_{14}\tau_{xy} + C_{15}\tau_{xz} + C_{16}\tau_{yz}. \tag{3.7}$$

A similar equation holds for each of the other five strain components. Each of the six equations involves six material constants, C_{ij}, for a total of 36 constants that have to be determined by experiment. This formulation holds for a linear anisotropic material, one whose material constants, such as the elastic modulus, differ when the material is tested uniaxially in different directions. Single metal crystals and any object cut from a single metal crystal are usually anisotropic. The elastic models for these crystals may be divided into nine groups, see Love [22, Chap. VI, Art. 109].

A material is called orthotropic if its material properties differ in mutually perpendicular directions and if there are three mutually orthogonal planes of material symmetry. The linear orthotropic constitutive equations are derived most easily by superposing the normal strains due to an uniaxial load in each of the three orthogonal directions. E_x is the elastic modulus in the x-direction. The various Poisson ratios are defined by $v_{yx} = -\epsilon_x/\epsilon_y$, for example. The shear modulus with respect to angle changes of the x-y axes is G_{xy}. The other constants are defined similarly. The shear strains are assumed to depend only on the corresponding shear stress.

$$
\begin{aligned}
\epsilon_x &= \frac{1}{E_x}\sigma_x - \frac{v_{yx}}{E_y}\sigma_y - \frac{v_{zx}}{E_z}\sigma_z; \\[2mm]
\epsilon_y &= -\frac{v_{xy}}{E_x}\sigma_x + \frac{1}{E_y}\sigma_y - \frac{v_{zy}}{E_z}\sigma_z; \\[2mm]
\epsilon_z &= -\frac{v_{xz}}{E_x}\sigma_x - \frac{v_{yz}}{E_y}\sigma_y + \frac{1}{E_z}\sigma_z; \\[2mm]
\gamma_{xy} &= \frac{1}{G_{xy}}\tau_{xy}; \\[2mm]
\gamma_{xz} &= \frac{1}{G_{xz}}\tau_{xz}; \\[2mm]
\gamma_{yz} &= \frac{1}{G_{yz}}\tau_{yz}.
\end{aligned}
\tag{3.8}
$$

There are apparently 12 material constants in the model, but these are reduced to nine by the symmetry requirements:

$$E_x \nu_{yx} = E_y \nu_{xy};$$
$$E_y \nu_{zy} = E_z \nu_{yz};$$
$$E_z \nu_{xz} = E_x \nu_{zx}.$$

The Poisson ratio, ν_{yx}, is due to a uniaxial stress in the y-direction so that ϵ_y is the axial strain and ϵ_x is the transverse strain. On the other hand, $\nu_{xy} = -\epsilon_y/\epsilon_x$ is due to an axial stress in the x-direction. Therefore, ν_{yx} and ν_{xy} are not reciprocals.

Wood can be approximately modeled as orthotropic, at least for small strains. The orthotropic directions are taken along the grain (parallel to the tree), in the radial direction of the tree trunk, and in the circumferential direction of the tree trunk. The properties of a wood board depend on the orientation in which it was cut from the trunk.

A material is called transversely isotropic if the material constants are the same in two of the orthotropic directions but different than those in the third direction. A drawn metal rod is transversely isotropic in the direction of the draw because the metal grain structure is deformed in that single direction. Composites made of fiber reinforcements in a matrix and also many biological tissues with internal fibers can be modeled as transversely isotropic. The distinct direction is that along the fibers; generally this is the stiffer direction. The constitutive equation is obtained from that for the linear orthotropic model by letting $E_y = E_z = E$, $G_{xz} = G_{xy} = G$, and $\nu_{xz} = \nu_{xy}$ if x is the distinct direction, that is, the y-z plane is the plane of isotropy. As shown in equation (3.12), a fourth relation is $G_{yz} = \frac{E}{2(1+\nu_{yz})}$. Transverse isotropy requires five material constants.

EXAMPLE 3.2   If the simply supported shaft of the example in the Introduction is manufactured by drawing in the longitudinal or x-direction, the shaft is transversely isotropic. The strains under the transverse load, P, are $\epsilon_x = \sigma_x/E_x$, $\epsilon_y = -\nu_{xy}\sigma_x/E$, $\epsilon_z = -\nu_{xy}\sigma_x/E$, and $\gamma = \tau/G$.

$$\epsilon_x(x,y,z) = \frac{4(L-a)P}{E_x \pi R^4 L}xy \qquad \text{if} \qquad 0 < x < a$$

$$\frac{4(L-x)aPy}{E_x \pi R^4 L} \qquad \text{if} \qquad a < x < L.$$

$$\epsilon_y(x,y,z) = \frac{-\nu_{xy}4(L-a)P}{E\pi R^4 L}xy \qquad \text{if} \qquad 0 < x < a$$

$$\frac{-\nu_{xy}4(L-x)aPy}{E\pi R^4 L} \qquad \text{if} \qquad a < x < L.$$

$$\epsilon_z(x,y,z) = \frac{-\nu_{xy}4(L-a)P}{E\pi R^4 L}xy \qquad \text{if} \qquad 0 < x < a$$

$$\frac{-\nu_{xy}4(L-x)aPy}{E\pi R^4 L} \qquad \text{if} \qquad a < x < L.$$

$$\gamma_{xy}(x,y,z) = \frac{4}{3}\frac{P(L-a)(R^2-y^2)}{G\pi R^4 L} + \frac{2Tz}{G\pi R^4} \qquad \text{if} \qquad 0 < x < a$$

$$\frac{-4}{3}\frac{aP(R^2-y^2)}{G\pi R^4 L} + \frac{2Tz}{G\pi R^4} \qquad \text{if} \qquad a < x < L.$$

$$\gamma_{xz}(x,y,z) = \frac{2Ty}{G\pi R^4}.$$

A material is isotropic if its material properties are the same in all directions on the body. All elastic moduli are the same; all shear moduli are the same; all Poisson ratios are the same. Any plane in the body is a plane of material symmetry. When the material is isotropic, it can be shown by examining rotations of the coordinates on the body that only two constants need to be determined experimentally. The following constitutive equations for linear isotropic materials can also be determined directly by the principle of superposition. So

$$\epsilon_x = \frac{1}{E}\sigma_x - \frac{\nu}{E}\sigma_y - \frac{\nu}{E}\sigma_z = \frac{1}{E}[\sigma_x - \nu(\sigma_y + \sigma_z)];$$

$$\epsilon_y = -\frac{\nu}{E}\sigma_x + \frac{1}{E}\sigma_y - \frac{\nu}{E}\sigma_z = \frac{1}{E}[\sigma_y - \nu(\sigma_x + \sigma_z)];$$

$$\epsilon_z = -\frac{\nu}{E}\sigma_x - \frac{\nu}{E}\sigma_y + \frac{1}{E}\sigma_z = \frac{1}{E}[\sigma_z - \nu(\sigma_x + \sigma_y)];$$

$$\gamma_{xy} = \frac{1}{G}\tau_{xy};$$

$$\gamma_{xz} = \frac{1}{G}\tau_{xz};$$

$$\gamma_{yz} = \frac{1}{G}\tau_{yz}.$$

(3.9)

This assumes that the shear and normal stresses and strains are completely independent and that the shear stress-strain curve is also linear. The set of six equations (3.9) taken together are called the generalized isotropic Hooke's law. They are the simplest set of constitutive equations and satisfactorily describe the material behavior of elastic, isotropic metals. They require the experimental determination of three elastic constants, E, G, and ν.

Hooke's law guarantees that for any choice of coordinates in which the shear stresses are zero, the shear strain will also be zero. This means that, in the body, the directions of the principal stresses coincide with the directions of the principal strains.

EXAMPLE 3.3   A thin plate of width, w, of length, L, and of specific weight, ρ, is hung vertically from a ceiling. Determine the displacements $u(x, y)$ and $v(x, y)$ at each point (x, y) of the plate due to the weight of the plate. Assume that $\tau_{xy}(x, y) = 0$ at each point. Also assume Hooke's law.

Solution   The coordinates are chosen so that the origin is at the center of the top edge which is w long, the x-axis is positive to the right along the top edge, and the y-axis is positive down. Assume plane stress. By the similar example (Fig. 1.11) in Chapter 1, Section 1.4, $\sigma_x(x, y) = \tau_{xy}(x, y) = 0$ and $\sigma_y(x, y) = \rho(L - y)$. Hooke's law implies that $\epsilon_x = (\sigma_x - \nu\sigma_y)/E = -\nu\rho(L - y)/E$ and $\epsilon_y = (\sigma_y - \nu\sigma_x)/E = \rho(L - y)/E$. Also $\gamma_{xy} = \tau_{xy}/G = 0$. The strain displacement relations (Chapter 2, Section 2.1) yield $\partial u/\partial x = \epsilon_x = -\nu\rho(L - y)/E$ and $\partial v/\partial y = \epsilon_y = \rho(L - y)/E$. Integration produces $u(x, y) = -\nu\rho x(L-y)/E + c_1(y)$ and $v(x, y) = \rho(2Ly - y^2)/2E + c_2(x)$.

If the plate is fixed to the ceiling along the full width, it must be true that $u(x, 0) = v(x, 0) = 0$ for all x. The conditions imply that $c_1(y) = \nu\rho xL/E$, which is impossible since $c_1(y)$ cannot be a function of x. Most likely in reality, the assumption $\tau_{xy}(x, y) = 0$ and the conclusion that $\sigma_x = 0$ is not valid because the ceiling exerts a nonuniform horizontal reaction force in response to the Poisson effect from the weight.

On the other hand, if the plate is hung from the ceiling by a pin at $(0, 0)$, then it must be true that only $u(0, 0) = v(0, 0) = 0$. In this case, $c_1(y) = 0$, which is acceptable. The condition on $v(0, 0)$ only requires that $c_2(0) = 0$ but does not determine the function $c_2(x)$. However, the shear strain must be zero. The strain displacement relation $\gamma_{xy} = \partial u/\partial y + \partial v/\partial x$

produces the relationship $0 = v\rho x/E + dc_2/dx$. Integration and using $c_2(0) = 0$ implies that $c_2(x) = -v\rho x^2/2E$. The predicted displacements are $u(x, y) = -v\rho x(L - y)/E$ and $v(x, y) = \rho(2Ly - y^2 - vx^2)/2E$. ∎

To reduce the isotropic Hooke's law from three to two experimental constants, a relation is required between E, G, and v. This is obtained by considering an elastic, isotropic material in two-dimensional pure shear and comparing the resulting stress Mohr's circle and the strain Mohr's circle (Fig. 3.22).

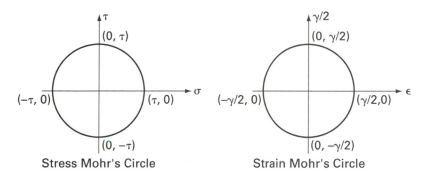

FIGURE 3.22   Mohr's circles for an isotropic material in pure shear.

The strain Mohr's circle implies that the principal strain $\epsilon_1 = \gamma/2$. Then by Hooke's law,

$$\epsilon_1 = \frac{\tau}{2G}. \tag{3.10}$$

The stress Mohr's circle shows that the principal stresses are $\sigma_1 = \tau, \sigma_2 = -\tau, \sigma_3 = 0$. Then Hooke's law says that

$$\epsilon_1 = \frac{1}{E}[\sigma_1 - v(\sigma_2 + \sigma_3)] = \frac{1}{E}[\tau - v(-\tau)] = \left[\frac{1 + v}{E}\right]\tau. \tag{3.11}$$

Comparing equations (3.10) and (3.11) yields

$$G = \frac{E}{2(1 + v)}. \tag{3.12}$$

This proof assumes that the directions of the principal stress and of the principal strain are the same. This condition is true for isotropic linearly elastic materials.

EXAMPLE 3.4   Equation (3.12) gives a good approximation of the relationship between E, G, and v as shown by those for steel: $E = 30 \times 10^6$ psi = 210 GPa; $v = 0.25$, and $G = 12 \times 10^6$ psi = 84 GPa so that $G = E/[2(1 + v)] = 12 \times 10^6$ psi. For aluminum: $E = 10 \times 10^6$ psi = 70 GPa; $v = 0.33$, and $G = 4 \times 10^6$ psi = GPa so that equation (3.12) predicts $G = E/[2(1 + v)] = 3.75 \times 10^6$ psi, which is not exactly the experimental value. ∎

To completely describe under any loading an isotropic material which behaves according to the generalized Hooke's law, it is sufficient to simply perform a single uniaxial test to obtain the material constants E and v.

The equations for the generalized isotropic Hooke's law are solved simultaneously to obtain the stresses in terms of the strains. The shear stress is easily found in terms of the shear strain, so that it is only necessary to solve the three normal strain equations for the normal stresses. This produces

$$\sigma_x = \frac{E}{1+v}\left[\epsilon_x + \frac{v}{1-2v}(\epsilon_x + \epsilon_y + \epsilon_z)\right];$$

$$\sigma_y = \frac{E}{1+v}\left[\epsilon_y + \frac{v}{1-2v}(\epsilon_x + \epsilon_y + \epsilon_z)\right]; \tag{3.13}$$

$$\sigma_z = \frac{E}{1+v}\left[\epsilon_z + \frac{v}{1-2v}(\epsilon_x + \epsilon_y + \epsilon_z)\right].$$

An examination of these three equations shows that they all follow the same pattern. The first strain component appearing corresponds to the stress component sought. The leading coefficient of each equation is $2G = E/(1+v)$. The sum of the strains, $\epsilon_x + \epsilon_y + \epsilon_z$, is the trace of the strain tensor. If one puts

$$e = \epsilon_x + \epsilon_y + \epsilon_z,$$

$$\lambda = \left(\frac{E}{1+v}\right)\left(\frac{v}{1-2v}\right) = \frac{vE}{(1+v)(1-2v)},$$

then the stress equations can be written

$$\sigma_x = 2G\epsilon_x + \lambda e;$$

$$\sigma_y = 2G\epsilon_y + \lambda e;$$

$$\sigma_z = 2G\epsilon_z + \lambda e. \tag{3.14}$$

These are called the Lamé equations and $\mu = 2G$ and λ are called the Lamé constants. Hooke's law is easiest to use when the stresses are known; the Lamé equations when the strains are known.

The number, $e = \epsilon_x + \epsilon_y + \epsilon_z$, is called the dilatation and can be given a physical interpretation. First, since $e = Tr([\epsilon])$, the trace of the strain tensor, it is independent of the choice of coordinates, as any physical quantity must be. The dilatation, e, measures the change in volume per unit volume of the material at the point (x, y, z) of the body where the strain tensor is computed.

Consider a rectangular solid around a point (x, y, z) with sides of length, Δx, Δy, Δz, so that the original volume is $V_o = \Delta x \Delta y \Delta z$. Assume that this volume deforms to a rectangular solid. Since the final length of a strained line segment is $L + \epsilon L = L(1 + \epsilon)$, the new lengths of the edges are $\Delta x(1 + \epsilon_x)$, $\Delta y(1 + \epsilon_y)$, and $\Delta z(1 + \epsilon_z)$. The final volume is

$$V_f = \Delta x(1 + \epsilon_x)\Delta y(1 + \epsilon_y)\Delta z(1 + \epsilon_z)$$

$$= \Delta x \Delta y \Delta z(1 + \epsilon_x + \epsilon_y + \epsilon_z + \epsilon_x\epsilon_y + \epsilon_x\epsilon_z + \epsilon_y\epsilon_z + \epsilon_x\epsilon_y\epsilon_z).$$

If the strains are small, all the products of strain can be neglected, so that

$$V_f \sim \Delta x \Delta y \Delta z(1 + \epsilon_x + \epsilon_y + \epsilon_z) = V_o(1 + e).$$

Therefore the change in volume is

$$\Delta V = V_f - V_o = V_o e,$$

and

$$e = \frac{\Delta V}{V_o},$$

the change in volume per unit original volume. The dilatation can be written in terms of stress when the material obeys Hooke's law.

$$e = \epsilon_x + \epsilon_y + \epsilon_z = \frac{1-2v}{E}(\sigma_x + \sigma_y + \sigma_z) = \frac{1-2v}{E}Tr(T). \qquad (3.15)$$

Exercise 3.3   Verify Equation (3.15).

EXAMPLE 3.5   If $e = 0$, there is no volume change due to the applied stress. When the sum of the normal stresses is not zero, this can only happen, by equation (3.15), if $v = 1/2$. A material that undergoes no volume change under any arbitrary stress field is called incompressible. Therefore a Poisson ratio of 1/2 is often said to be equivalent to incompressibility of the material. This analysis is based on Hooke's law. Rubber is frequently assumed to be incompressible since its Poisson's ratio is between 0.48 and 0.5; however its constitutive relation is not linear and cannot be described by Hooke's law.   ∎

EXAMPLE 3.6   Let a body be subjected to hydrostatic pressure. The only nonzero stresses are the normal stresses and they are equal; $\sigma_x = \sigma_y = \sigma_z = -p$ and $Tr(T) = -3p$. Therefore

$$e = \frac{-3(1-2v)}{E}p. \qquad (3.16)$$
∎

The change in volume per unit volume, the dilatation, is linearly proportional to the applied hydrostatic pressure, p, if the material is isotropic linearly elastic. The negative of the reciprocal of the constant of proportionality is given a special name.

Definition   The bulk modulus of elasticity, K, is the negative of the ratio of the hydrostatic pressure, p, to the dilatation, e, that the hydrostatic pressure causes. By equation (3.16),

$$K = -\frac{p}{e} = \frac{E}{3(1-2v)}. \qquad (3.17)$$

The bulk modulus for steel is approximately 24×10^6 psi (165 GPa) and for aluminum is about 10×10^6 psi (70 GPa). When $v = 0.33$, the bulk and elastic moduli are equal.

3.4   CERAMICS

Ceramics are of two major types, ionic or covalent. The ionic ceramics have a more complex lattice structure than metals. These are generally brittle because ionic bonding is much stronger than the metallic bonding of metals and because fewer slip planes exist in the crystal structure. Common ionic ceramics are sodium chloride, china, and brick. The silicate ceramics such as glass and clays are covalent ceramics. Because the ionic ceramics are brittle and the strain at failure is small, they are usually modeled as linear elastic materials.

Ceramics have quite high melting temperatures, above those of many metals, because of their crystal structure and so are used in high temperature applications. Tungsten carbide (WC) with a melting temperature of 2775°C is used as an abrasive in cutting tools. Because WC is brittle as well as very hard, it is most often used in a composite with other materials. Its density of 15.7 g/cm^3 is much greater than other ceramics. The linear thermal coefficient of expansion of most ceramics is smaller than that of most metals, but is not constant as temperature varies. For example, the linear coefficient of thermal expansion of MgO changes as the temperature is increased from room temperature to about 1300°C; its average value is about 9×10^{-6}/C. The coefficient for glass is about 5×10^{-7}/C. In nonisotropic ceramics, the linear coefficient of thermal expansion is different in different directions in the material. Graphite is made of layers of covalently bonded carbon, but the bonds between the layers are weaker van der Waals bonds. The thermal coefficient in a direction parallel to a layer is 1×10^{-6}/C but 27×10^{-6}/C perpendicular to the layers.

3.4.1   Typical Ceramic Lattice Structures

An *ion* is an atom that has gained or lost electrons so that its charge is not balanced. The ionic radius therefore differs from the atomic radius of an element. A *cation* is a positively charged metallic ion which has given up valence electrons to the negatively charged *anions*, non-metallic ions. For example, in sodium chloride, the metallic sodium cation has a positive charge of one and the chloride anion has a negative charge of one. Charge balance requires that the chemical formula for sodium chloride be NaCl.

Ionic Ceramic Crystal Structures   The simplest structures are those in which the number of cations and anions are equal. Other ceramics composed of one metal and one nonmetal may have unequal numbers of each. Even more complex ceramic structures occur for those composed of more than two elements. Anions are generally larger than the ionically bound cations so that the cations tend to take interstitial sites between the anions. Because of this, the most likely slip planes are close-packed planes of anions.

The sodium chloride structure (Fig. 3.23) is common to ceramics such as NaCl, MnS, and LiF as well as the metal oxides FeO, MgO, CoO, and CaO. The metal oxides have high melting temperatures; for MgO, $T_m = 2800$°C and for CaO, $T_m = 2570$°C. Magnesia and CaO are called refractory ceramics; since they can withstand such high temperatures. They are also quite inert in the sense that they do not react with other materials at high temperatures. Refractory ceramics are used to line furnace walls.

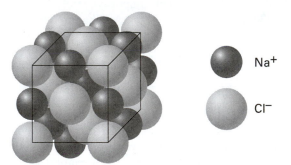

FIGURE 3.23   The sodium chloride structure.

The structure can be viewed as interlaced FCC arrangements, one of sodium and one of chloride ions. The close packed anion planes are {110} and the close packed directions are $< 1\bar{1}0 >$. These form the slip systems.

The cesium chloride (CsCl) unit cell (Fig. 3.24) has chloride anions at each of the eight corners and a cesium cation at the body center. The anions and cations are of equal number because the anions at the corners are shared by the eight adjoining unit cells. While it is superficially similar to a BCC metallic structure, it is not BCC

FIGURE 3.24   The cesium chloride structure.

because the ions are not all the same element. The cell can be thought of as a pair of interlaced simple cubic structures. Other ceramics with this structure are CsBr and CsI. There are no close-packed planes in this structure.

The zinc sulfide (ZnS), or zinc blende, structure (Fig. 3.25) has unit cell with sulfur anions in the FCC positions and zinc cations in the four interior tetrahedral positions. Ceramics with this structure also include SiC, BeO, and ZnTi.

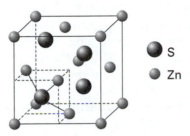

FIGURE 3.25   The zinc blende structure.

Wurtzite is formed from the HCP unit cell by adding anions at the interstitial sites. Anions are at one half the interstitial tetrahedral sites in ZnS.

Lattice Defects   Defects, or dislocations, in ionic crystals differ from those in metal structures because electrical balance is maintained. The ionic defects are paired defects. A vacancy and interstitial pair defect, or Frenkel defect, is formed when an ion leaves a lattice site open by moving to an interstitial site. Electrical balance is also maintained if a anion and cation pair is missing from the structure; this pair of vacancies is called a Schottky defect. As with metals, such disruption of the lattice structure makes slip more difficult and results in a stronger material.

Ionic ceramics are brittle because slip is difficult along the slip planes in the unit cell due to the ionic charges. The shear forces required for slip are increased because ions of like charge are brought into proximity, but then repel each other.

Covalent Ceramic Structures   A common covalent lattice structure is the diamond cubic (Fig. 3.26). It is similar to the zinc blende except that all lattice points are filled by atoms of one material. Carbon (C), germanium (Ge), and silicon (Si) can each solidify into a diamond cubic structure. If carbon does, each carbon atom is bonded to four others. This structure is relatively open, not closed packed, so that atoms have room to vibrate as temperature increases. Materials with the diamond cubic structure have low coefficients of thermal expansion, less than those of metals. Diamond, on the other hand, has twice the thermal conductivity of copper.

The silicate ceramics are built from SiO_4^{4-} tetrahedra in which the silicon and oxygen are covalently bonded (Fig. 3.27). When the tetrahedra form into large structures, they almost always bond with other tetrahedra at the corners. If two corners are shared, a linear chain structure forms as in asbestos. In this case, the oxygen to silicon ratio, O/Si = 3. The chain can form into a ring having structure $Si_3O_9^{6-}$ in

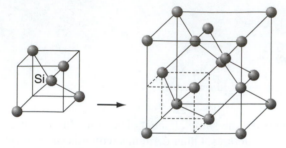

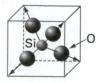

FIGURE 3.26   The diamond cubic lattice.

FIGURE 3.27   The tetrahedral structure of SiO_4^{4-}.

materials like wollastonite ($CaSiO_3$) in which three Ca are bound with each ring. A larger ring with structure, $Si_6O_{18}^{12-}$, underlies beryl ($Be_3Al_2\ Si_6O_{18}$). A layered sheet structure forms if three corners are shared, forming the unit, $Si_2O_5^{2-}$, so that $O/Si = 2.5$. An example of such a ceramic is the clay Laolinite $Al_2.Si_2O_5.OH_4$ made of two repeating layers, one of $Si_2O_5^{2-}$ and the other of $Al_2OH_4^{2+}$. When all four corners are shared, so that $O/Si = 2$, a three-dimensional structure results. Each Si bonds with four O, and each O bonds with two Si.

Most metals contract when solidified (frozen); silicon is unusual. The density of crystalline silicon is less than that of the liquid because of the covalent electron bonding condition whereby each silicon atom shares a pair of electrons with four surrounding atoms thus giving a low coordination number of four in the diamond cubic lattice structure. In the casting, shown in Figure 3.28, of relatively pure material, the individual grain orientations were found to be mutually related by a multiple twinning mechanism.

FIGURE 3.28   Longitudinal section through a large-grain-size polycrystalline silicon casting 10 cm in (horizontal) diameter and showing at the top center a reverse solidification dome that forms because of the solid expansion occurring on freezing. (Courtesy of G. M. Storti.)

3.4.2 Bending Tests for Strength

Ceramics are brittle materials so that they usually fracture at small strains. The uniaxial stress-strain curves are nearly linear. The elastic modulus ranges from the 70 GPa (10×10^6 psi) of glass to 476 GPa (68×10^6 psi) of silicon carbide. The stiffnesses of ceramics are therefore comparable to those of metals.

A tensile test of a brittle ceramic material is quite hard to perform because of the difficulty of gripping the specimen without creating cracks near the grip. A three-point or four-point bending test is used to estimate the fracture stress. A specimen is formed into a beam and simply supported at either end (two of the point loads). The third point in the the three-point load test is an increasing point load, F, applied vertically at the center of the beam with length, L, between the supports, until the specimen fractures. The internal shear stress is constant and the bending moment is linear between the supports, with a maximum of $FL/4$ under the applied load. The fracture stress, called the modulus of rupture, σ_{mr}, is usually estimated using the strength of materials bending relation, $\sigma = My/I$, because it is assumed that the beam deflection is quite small. The modulus of rupture is computed in terms of the fracture force, F_f, the length, L, and the dimensions of the specimen. When the specimen has a rectangular cross section of height h and width w, the maximum bending stress occurs under the load at the top or bottom of the beam ($y = h/2$),

$$\sigma_{mr} = \frac{3F_f L}{2wh^2}.$$

If the specimen is a rod having circular cross section with radius, R, then

$$\sigma_{mr} = \frac{F_f L}{\pi R^3}.$$

Because of the presence of the shear stress, the bending normal stress is not principal and therefore not the maximum stress in the beam in a three-point test.

The elastic modulus can also be estimated from the three-point bending test. If the beam is loaded to a small mid-beam displacement y by the vertical force F, then the elastic modulus, E, is determined from the strength of materials relation,

$$y = \frac{FL^3}{48EI}, \tag{3.18}$$

where L is the beam length and I is the beam area moment of inertia.

EXAMPLE 3.7

A 6 cm long specimen of silicon carbide is loaded in a three-point bending test to 100 N. The rectangular cross-sectional area is 0.6 cm wide and 0.1 cm thick. The measured deflection at mid-beam is 1.875 mm. Determine E.

Solution

$I = (0.006)(0.001)^3/12 = 5 \times 10^{-13} \text{m}^4$. Then from $y = FL^3/48EI$, $E = 480$ GPa. ∎

A four-point bending test is used to ensure that the bending stresses are principal over a region of the beam, because the shear stress on the vertical cross-sectional face is zero. A simply supported beam of length L is loaded with two equal and equally spaced point loads, $F/2$. A free-body diagram shows that the shear force between the inner supports is zero. This loading creates a constant, maximum, pure bending moment between the inner two applied loads, in contrast with the three-point test in which the bending moment, and therefore the normal stress, reaches a maximum at the single location under the point load. In a four-point test, then, material flaws are more likely to lie in the region of maximum stress. In a rectangular cross-sectional four-point bending specimen, strength of materials predicts that the relation between the measured fracture bending load and the stress is, from $\sigma = My/I$ and $M = Fa/2$,

$$\sigma_{mr} = \frac{3F_f a}{wh^2},$$

where a is the distance from a simply supported end to the point of application of the nearest inner load. The four-point bending test often predicts a lower modulus of rupture than the three-point test. But both are greater than that predicted by an axial test. Clearly the means of loading affects the fracture strength. The modulus of rupture can be as small as 70 MPa (10 ksi), in a glass, and as large as 1100 MPa (160 ksi) in materials such as titanium carbide.

Porosity has a strong influence on the strength and mechanical response of ceramics. Pores are created during the manufacture of a ceramic from a powder. While many of the gaps between the powder grains are filled due to diffusion during heat or pressure treatments, some voids remain. Some of these voids, or pores, are open and some are closed. The open pores are often interconnected and a liquid or gas may permeate the material through them. The closed pores are isolated from the outside of the material.

The porosity is the ratio of the volume contained in pores to the volume of the specimen. The apparent porosity, P, is the ratio of the volume of the open pores to the volume of the specimen. The apparent porosity is obtained experimentally by working in terms of equivalent volumes of water. Denote the weight per unit volume of water by ρ. Three measurements are required. D is the weight of the dry specimen in air. S is the weight of the specimen when suspended in water after time has been allowed for the open pores to fill with water. W is the specimen weight just after it is removed from the water and water has been removed from the outer surface. Then $W - D$ is the weight of the water in the open pores, and $W - S$ is the weight of water occupying the volume of the specimen. The volume fraction of pores, P, is determined by

$$P = \frac{(W - D)/\rho}{(W - S)/\rho} = \frac{W - D}{W - S}.$$

The true porosity is the ratio of the volume of all pores to the volume of the specimen. The specimen is ground to a fine powder, and the volume of the powder is measured by the volume the powder displaces in a liquid. The true porosity is $(s_g - b)/s_g$, where s_g is the ratio of D to the volume of powder and the bulk density b is the ratio of D to the volume of the specimen.

A calculation from elasticity that estimates the effect of porosity on the shear modulus, G, at porosity P in terms of the shear modulus, G_o, and the bulk modulus, K_o at zero porosity is given in [23].

$$\frac{G}{G_o} = 1 - \frac{5(3K_o + 4G_o)}{(9K_o + 8G_o)}P - AP^2, \tag{3.19}$$

where A must be determined. Because $G = E/2(1 + v)$, the bulk modulus is $K = 2G(1 + v)/3(1 - 2v)$. For example, at $v = 0.3$, $K = 2.167G$. The constant A is determined for a given v by assuming that $G = 0$ if $P = 1$. Using the fact that $G/G_o = E/E_o$ when $G = E/2(1 + v)$ yields

$$E = E_o(1 - 1.909P + 0.909P^2), \tag{3.20}$$

when the porosity is less than 50% and the Poisson ratio is 0.3. For example, for a porosity of 5%, $E/E_o = 0.907$, while for 25% porosity, $E/E_o = 0.580$. The elastic modulus decreases as the porosity increases. The relationship (3.20) was tested on alumina at room temperature [9], which has average Poisson ratio 0.3 over a range of porosities, and found to slightly overestimate E but to agree closely for G. This occurs because the relationship $G = E/2(1 + v)$ is not exactly valid experimentally. The Poisson ratio for alumina was found to increase with temperature and to vary with porosity. The elastic and shear moduli decrease with temperature. The temperature dependence for alumina was fit in [9] with the following equations.

$$\frac{E}{E_o} = 1 - \frac{1}{130}\exp\left(\frac{T}{410}\right);$$

$$\frac{G}{G_o} = 1 - \frac{1}{106}\exp\left(\frac{T}{408}\right),$$

where the temperature T is measured in Kelvins.

The modulus of rupture also decreases with porosity. This is expected because the pores are sites at which cracks leading to fracture may more easily be formed. An empirical relation is

$$\sigma_{mr} = \sigma_o \exp(-nP), \tag{3.21}$$

where σ_o and n are material constants depending on the particular ceramic. Typically $4 \le n \le 7$.

Caution is required in applying the theory of elasticity developed in subsequent chapters to ceramics because the elasticity calculations are based on the assumption that the material is continuous, that is, that there are no pores. In Table 3.1, the elastic modulus given is for 5% porosity.

TABLE 3.1   Ceramic Material Constants

Material	Symbol	E (GPa)	v	σ_{mr} (MPa)	ρ (g/cm^3)	T_m (°C)
Magnesia	MgO	207	0.36	105	3.55	2,800
Silicon carbide	SiC	470	0.19	400	3.17	2,540
Titanium carbide	TiC	462	0.19	275–430	–	3,070
Alumina	Al$_2$O$_3$	380	0.26	200–350	~ 3.90	2,050
Rock salt	NaCl	44.2	–	–	2.23	800
Diamond	C	1035	–	–	3.52	3,826
Quartz (fused silica)	SiO$_2$	69	0.25	110	2.65	1,650

3.4.3   Glass

A metal solidifies into a crystal structure as the nuclei, clusters of atoms, grow. Some materials can solidify into a noncrystalline form, called a glass. Creation of glass requires more rapid cooling, a low nucleation rate, and a slow growth rate of the nuclei due to high viscosity at temperatures near melting. For example, silica (SiO$_2$) when melted and cooled forms a glass with a neutral local structure of SiO$_4^{4-}$ tetrahedra in which each corner oxygen is shared by an adjacent tetrahedron. The glass has only a short-range order consisting of a regular arrangement of several such tetrahedra. The melting temperature of silica is 1710°C.

Some oxides such as silica can easily form into a glass. Intermediate oxides such as lead oxide or aluminum oxide which have a lower bond energy than formers may be included into the glass to adjust the mechanical properties but cannot form a glass by themselves. Modifier oxides, which have the weakest bond energy, such as Na_2O, may prevent the formation of a glass or help the glass to crystallize. If the ratio of oxygen to silicon is greater than 2.5, it is very difficult to form glass.

Materials other than silica can form a glass. For example, the metal titanium, quickly cooled into a glass, is used in the shafts of sports equipment. The ceramics boron oxide (B_2O_3) and germania (GeO_2) are good glass formers.

Viscosity of Glass The mechanical behavior of glass is temperature dependent. Shear can occur in these noncrystalline (amorphous) materials by viscous flow. The material almost acts like a fluid. The viscosity, η, of glass depends inversely on temperature and is often described by an Arrhenius-type relation,

$$\eta(T) = \eta_o \exp\left(\frac{Q}{RT}\right),$$

where η_o is a material constant, Q is the activation energy for slip, and R is the universal gas constant. If Q is in cal/mol, then $R = 1.987$; the temperature T is in Kelvins. Viscosity is measured in either poise or Pa-s. One poise is 0.1 Pa-s. Even at room temperature, glass can be considered to flow under forces, but with such a large viscosity that the flow is not visible. Glass is commonly assumed to be linearly elastic (i.e., obey Hooke's law), at room temperature.

Experimental measurement of the viscosity at two different temperatures determines the activation energy, Q, and the constant η_o if the Arrhenius relation is assumed. In a plot of the logarithm of the viscosity versus the inverse of the temperature, T, because

$$\ln\left(\frac{\eta}{\eta_o}\right) = \frac{Q}{R}\left(\frac{1}{T}\right),$$

the activation energy is obtained from the slope of the linear curve of $\ln(\eta/\eta_o)$ versus $1/T$. Notice that the calculation of the activation energy in this way from the experiments is determined by the choice of the Arrhenius-type mathematical model as well as by the material. In this sense, the value of the activation energy is not independently determined by this method.

EXAMPLE 3.8

The viscosity of a glass is 10^8 poise at 1400°C and 10^{13} poise at 900°C.

(a) Compute the activation energy for viscous flow. Assume the Arrhenius relation.

(b) Compute η_0.

(c) Compute the temperature required before the glass reaches a viscosity of 10^7 poise, at which the glass is formable.

Solution

(a) Divide the two Arrhenius equations for the temperatures given.

$$\frac{10^{13}}{10^8} = \frac{\exp\left(\frac{Q}{(1.987)(1173)}\right)}{\exp\left(\frac{Q}{(1.987)(1673)}\right)}.$$

Therefore $Q = 89.786$ kcal/mol. This compares to an activation energy for silica of 565 kJ/mol (135 kcal/mol). Another example is that the viscosity of fused silica at 1940° C is measured as 15×10^4 Pa-s.

(b) $\eta_o = 10^{13} \exp\left(\frac{-89786}{(1.987)(1173)}\right) = 1.862 \times 10^{-4}$.

(c) Repeat the process with the calculated value of Q to obtain the unknown temperature.

$$\frac{10^{13}}{10^7} = \frac{\exp\left(\frac{89786}{(1.987)(1173)}\right)}{\exp\left(\frac{89786}{(1.987)(T)}\right)}.$$

Therefore the temperature is $T = 1556°C$.   ∎

The Arrhenius equation is a two-parameter model with parameters η_o and Q. More recent experimental work shows that it does not fit the viscosity-temperature relation well. The Vogle-Fulcher-Tamman three-parameter model gives more accurate results;

$$\ln \eta = A + \frac{B}{T - T_0},$$

where A, B, and T_0 are temperature independent parameters [3].

Glass rather drastically changes behavior with temperature because of the change in viscosity. Glass blowers take advantage of this behavior by heating the glass enough that it can be easily deformed but maintains a shape and then allowing the glass to cool once the desired shape is created (10^3–$10^{6.6}$ Pa-s). When the temperature is high, but below melting, glass behaves like a viscous fluid. At lower temperature, glass acts like a solid. The transition temperature between these two responses can be found experimentally by measuring the volume of a given mass of glass as a function of temperature. As the glass cools, the volume decreases as expected. But at a certain temperature, called the glass transition temperature, T_g, there is a change in the slope of the volume-temperature curve. The change of volume is slower with respect to temperature below than it is above the glass transition. Unfortunately, the exact transition point is hard to pick out experimentally, because the transition is more gradual than sharp in most glasses. The transition point is therefore quantified by picking a viscosity that would be said to be the point at which the material changes from a viscous fluid to a glass as the temperature is decreased. Below the glass transition temperature, the atoms are frozen into relatively stable positions. Some think of deformation below T_g as elastic and that above as viscous deformation.

Definition   The glass transition, T_g, is the temperature at which the viscosity of the glass is equal to 10^{12} Pa-s.

3.5 FIBERED COMPOSITES

A fiber-reinforced material is one in which many thin, strong reinforcing fibers are embedded parallel to each other in a deformable isotropic matrix. The idea is to gain strength without creating a brittle, and therefore likely to fracture, material. The fibers give the strength, and the matrix is intended to be ductile enough to prevent crack propagation. Fibered composite material stress-strain mechanical behavior is frequently modeled as linear elastic because the internal fibers restrict the body to very small strains. Linear elasticity is applied by producing an equivalent elastic modulus, E_a, parallel to the fibers and Poisson ratios, ν_{at} and ν_{ta}, so that computations can be made as though the material were transversely isotropic. Here a refers to the axial direction parallel to the fibers and t refers to a transverse direction. The modulus parallel to the fibers is taken as an average of the modulus, E_f, for the fibers and that of the matrix, E_m, by the rule of mixtures.

The rule of mixtures for the elastic modulus is a consequence of the fact that the total tensile force, F, on the axial loaded member is the sum of that supported by the matrix, F_m, and that by the fiber, F_f. Let A be the total, A_f be the fiber, and A_m be the matrix cross-sectional areas. Because $F = F_m + F_f$ can be written in terms of the stresses as

$$\sigma_a A = \sigma_m A_m + \sigma_f A_f,$$

$$\sigma_a = \sigma_m \frac{A_m}{A} + \sigma_f \frac{A_f}{A}.$$

But since the fibers are assumed to have uniform cross sections and length, $f_m = A_m/A$ and $f_f = A_f/A$, where f_m is the volume fraction of the matrix and f_f is the volume fraction of the fibers. Notice that $f_m + f_f = 1$. Also, since each material, as well as the composite, is assumed to be linear elastic, from the stress relationship above,

$$E_a \epsilon_a = E_m \epsilon_m f_m + E_f \epsilon_f f_f.$$

But because the fibers and matrix are assumed to be rigidly attached, $\epsilon_c = \epsilon_m = \epsilon_f$ (the isostrain condition), and the rule of mixtures for the elastic modulus follows:

$$E_a = f_m E_m + f_f E_f. \tag{3.22}$$

The elastic modulus, E_t, perpendicular to the fibers is derived by assuming that the stress on the fibers and matrix is the same, the isostress condition; the strains are not equal as in the case parallel to the fibers. The total strain is assumed to be $\epsilon_t = \epsilon_m f_m + \epsilon_f f_f$. Then by substituting the linear elastic relation $\epsilon = \sigma/E$ for each summand and dividing out the equal stresses,

$$\frac{1}{E_t} = \frac{f_f}{E_f} + \frac{f_m}{E_m}; \tag{3.23}$$

or

$$E_t = \frac{E_m E_f}{f_m E_f + f_f E_m}.$$

The axial modulus, E_a, is an upper bound, and the transverse modulus, E_t, is a lower bound for the modulus in an arbitrary direction in the a-t plane.

Exercise 3.4   A composite is made of two materials having elastic moduli $E_m = 5$ GPa and $E_f = 80$ GPa. To compare the composite elastic modulus in the fiber and cross-fiber directions, sketch two graphs on the same coordinate axes: the first graph is E_a in the fiber direction versus the fiber volume fraction, f_f; the second is E_t in the transverse fiber direction versus the fiber volume fraction, f_f.

For any planar orthotropic material,

$$\nu_{at} E_t = \nu_{ta} E_a.$$

One Poisson ratio is defined by the rule of mixtures, where the matrix and fiber are each assumed isotropic,

$$\nu_{at} = \nu_f f_f + \nu_m f_m,$$

and the other Poisson ratio is

$$\nu_{ta} = \nu_{at} E_t / E_a.$$

The rule of mixtures is also used to estimate the density of a composite in terms of its constituents.

$$\rho = \rho_m f_m + \rho_f f_f. \tag{3.24}$$

EXAMPLE 3.9 Epoxy, a thermosetting polymer, is used as the matrix in high-strength fibered composites. The elastic modulus of epoxy is 6.9 GPa (1×10^6 psi). There are 65 vol% E-glass fibers, having an elastic modulus of 72.5 GPa.

(a) Compute the modulus of the composite in the fiber direction.

$$E_a = 6.9(.35) + 72.5(.65) = 49.54 \quad \text{GPa}.$$

(b) Compute the portion of an axial load, F, parallel to the fibers which is carried by the fibers.

$$\frac{F_f}{F} = \frac{\sigma_f A_f}{\sigma A} = \frac{E_f \epsilon_f A_f}{E_a \epsilon A} = \frac{E_f}{E_a} f_f = \frac{72.5}{49.54}(.65) = 0.9512.$$

The fibers carry about 95% of the load. ∎

Frequently carbon fibers with $E = 40 \times 10^6$ psi are embedded in metals such as aluminum again to gain strength while preserving ductility.

Fiber-reinforced materials seem to be favored by nature. A wood pulp fiber is made of several layers of tiny cellulosic fibrils helically wound about the open center water and food passage. Many animal and human tissues are built of layers of collagen fibers in a matrix. An important study for biomechanics is to learn in what, if any, sense these structures are optimal for carrying loads with minimal density and energy cost. Most of these biological structures cannot be modeled by linear elasticity; their response is nonlinear.

3.6 STRAIN ENERGY

Intuitively, an elastic material is one which when unloaded returns to its original stress and strain state; it recovers its original shape. George Green, an English mathematician, applied the principle of conservation of energy to elasticity through his new idea of a strain energy function. Around 1840 with the aid of the strain energy, he rederived the equations of elasticity from the principle that the equilibria are the minima of a potential energy of the system.

Elastic behavior is reversible and path independent in the sense that the current stress is assumed to depend only on the current strain, not on the deformation history of the body. In an elastic material, when the stress cycles from a particular state σ'_{ij} and then back to σ'_{ij}, the strain also cycles from ϵ'_{ij} and then back to ϵ'_{ij}. Here σ_{ij} and ϵ_{ij} denote the ijth entries in the stress and strain tensors respectively.

Thermodynamics requires that a body elastically deformed over a cycle must have zero net work performed on it. The assumption that the current stress is a function of the current strain is not sufficient to guarantee that zero work will be performed in a cycle. For example, the hypothetical elastic stress-strain curve shown in Figure 3.29 is not one to one, that is, stress is not a unique function of strain.

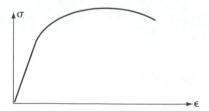

FIGURE 3.29 A stress-strain curve that is not one to one.

An additional assumption is required. Drucker's postulate of stability of a material over a cycle, which guarantees a unique strain state for each stress state, implies that over the cycle

$$\int \sum_i \sum_j \sigma_{ij} d\epsilon_{ij} = 0.$$

This is true for any path of strains chosen so that the integrand is exact. Exact means that there is some function $U_o(\epsilon_{ij})$ such that $dU_o = \sum_i \sum_j \sigma_{ij} d\epsilon_{ij}$. U_o is called the strain energy density. Integration yields

$$U_o(\epsilon_{ij}) = \int_0^{\epsilon_{ij}} \sum_i \sum_j \sigma_{ij} d\epsilon_{ij},$$

and $\partial U_o / \partial \epsilon_{ij} = \sigma_{ij}$.

Definition   A material for which a strain energy density function exists and obeys the relations, $\partial U_o / \partial \epsilon_{ij} = \sigma_{ij}$ is said to be hyperelastic.

This idea was proposed by G. Green around 1840. The name hyperelastic was not suggested until about 1960 by C. Truesdell and coworkers to distinguish such materials from possible elastic materials for which such a strain energy function does not exist.

Likewise over the cycle, it must also be true that $\int \sum_i \sum_j \epsilon_{ij} d\sigma_{ij} = 0$ so that the integrand is an exact differential $d\Psi_o = \int \sum_i \sum_j \epsilon_{ij} d\sigma_{ij}$. The function, $\Psi_o(\sigma_{ij})$, is called the complementary energy density and $\partial \Psi_o / \partial \sigma_{ij} = \epsilon_{ij}$. Both U_o and Ψ_o have units of energy per unit volume.

The total strain energy is the integral of U_o over the volume, V, of the body,

$$U = \int_V U_o dV.$$

A physical interpretation is that the work done by the internal forces in the body during an elastic deformation is stored in the body as strain energy, U. In an elastically loaded member, the strain energy might be thought of as the potential energy available to put the body back into its original shape when the load is removed.

The energy applied to a body during a deformation can be used to predict when the member will fail (Chapter 9). The behavior of the energy under the application of forces can be used to relate the deflection of the point of application of the force to the force (Chapter 7). An integral representing the energy in the body will also be the foundation of the numerical approximation techniques to be developed later (Chapter 8) to estimate solutions to the elasticity equations.

In the uniaxial case, the strain energy density integral may be interpreted as the area under the elastic portion of the stress-strain curve between zero strain and the current strain, ϵ (Fig. 3.30). The linear uniaxial stress-strain relation is one to one, $\sigma_x = E\epsilon_x$, so that U_o exists and

$$U_o = \int_0^\epsilon \sigma_x d\epsilon_x = \int_0^\epsilon E\epsilon_x d\epsilon_x = \frac{1}{2}E\epsilon^2 = \frac{1}{2}\sigma\epsilon. \qquad (3.25)$$

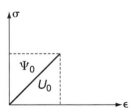

FIGURE 3.30   The strain and complementary energies as areas under a linear elastic stress-strain curve.

The uniaxial complementary strain energy density, $\Psi_o = \int_0^\sigma \epsilon_x d\sigma_x$, can be viewed as the area between the stress-strain curve and the stress axis (Fig. 3.30). If the material is linear elastic, then

$$\Psi_o = \int_0^\sigma \epsilon_x d\sigma_x = \int_0^\sigma \frac{\sigma_x}{E}d\sigma_x = \frac{1}{2E}\sigma^2 = \frac{1}{2}\sigma\epsilon. \qquad (3.26)$$

Therefore in the linear elastic case, $\Psi_o = U_o$. This is not a surprise since a straight diagonal line divides a rectangle into two equal areas. This result will be used in Castigliano's theorems in Chapter 7. If the stress-strain relation is not linear, but is monotonic, the complementary energy is still defined but is unequal to the strain energy.

The form of the uniaxial strain energy density for a linear elastic material suggests the following definition.

Definition   A Hookean material is one whose strain energy density, in terms of the strain tensor, is given by the quadratic

$$U_o = \frac{1}{2}\sum_{i=1}^6 \sum_{j=1}^6 b_{ij}\epsilon_i\epsilon_j, \qquad (3.27)$$

where ϵ_i is one of the six strain components and the b_{ij} are 36 material constants. (Here, to avoid additional subscripts on the b_{ij}, the three normal and three shear strains have been numbered from 1 to 6.)

The requirement that $\partial U_o/\partial\epsilon_x = \sigma_x$, etc. is met by choosing the coefficients, b_{ij}, so that the strain energy density can be rewritten as

$$U_o = \frac{1}{2}[\sigma_x\epsilon_x + \sigma_y\epsilon_y + \sigma_z\epsilon_z + \tau_{xy}\gamma_{xy} + \tau_{xz}\gamma_{xz} + \tau_{yz}\gamma_{yz}]. \qquad (3.28)$$

An expression for the isotropic three-dimensional complementary energy is obtained by using Hooke's law to write each strain in terms of the stresses and the fact that $U_o = \Psi_o$:

$$\Psi_o(T) = \frac{1}{2E}(\sigma_x^2 + \sigma_y^2 + \sigma_z^2) - \frac{\nu}{E}(\sigma_x\sigma_y + \sigma_x\sigma_z + \sigma_y\sigma_z) + \frac{1}{2G}(\tau_{xy}^2 + \tau_{xz}^2 + \tau_{yz}^2). \qquad (3.29)$$

Exercise 3.5 Verify equation (3.29).

Now using the Lamé equations for the linear elastic constitutive relation, equations (3.14), the isotropic strain energy density is written in terms of the strain tensor components, the dilatation, and the Lamé constants:

$$U_o(\epsilon) = \frac{1}{2}[\lambda e^2 + 2G(\epsilon_x^2 + \epsilon_y^2 + \epsilon_z^2) + G(\gamma_{xy}^2 + \gamma_{xz}^2 + \gamma_{yz}^2)]. \qquad (3.30)$$

Exercise 3.6 Verify equation (3.30).

The last two equations (3.29) and (3.30) can be used to show that in the linear elastic case, each stress component can be found as derivative of the strain energy density. For example, from equation (3.30) and the Lamé equations,

$$\frac{\partial U_o(\epsilon)}{\partial \epsilon_x} = 2G\epsilon_x + \lambda e = \sigma_x.$$

Exercise 3.7 Show that a similar procedure produces all six stress components.

Likewise from equation (3.29) and the generalized Hooke's law, each strain can be found as the derivative of the complementary strain energy by the stress. For example,

$$\frac{\partial \Psi_o(T)}{\partial \sigma_x} = \frac{1}{E}[\sigma_x - \nu(\sigma_y + \sigma_z)] = \epsilon_x.$$

Exercise 3.8 Show that a similar procedure produces all six strain components.

These results say that stress and strain are conjugate variables in the sense used in thermodynamics. Here U_o or Ψ_o play the role of thermodynamic energy functions. Recall from thermodynamics, for example, that the internal energy, $U(V, S)$, is a function of volume, V, and entropy, S. Then volume and pressure, P, are conjugate, while entropy and temperature, θ, are conjugate in the sense that

$$\frac{\partial U}{\partial V} = -P \qquad \text{and} \qquad \frac{\partial U}{\partial S} = \theta.$$

These observations indicate one direction in which the thermodynamics of fluids may be generalized to include the thermodynamics of solids.

EXAMPLE 3.10

The screw dislocation is an example of a nonuniform state of stress. The deformation (u_1, u_2, u_3) in a region around the screw dislocation is described at the point (x_1, x_2, x_3) by

$$u_1 = u_2 = 0;$$

$$u_3 = \frac{b}{2\pi} \arctan\left(\frac{x_2}{x_1}\right),$$

where b is the length of the Burgers vector. The origin is taken at the intersection of the axis of rotation (2-axis) and the slip plane (Fig. 3.31).

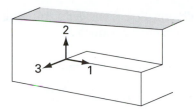

FIGURE 3.31   Coordinate system on the screw dislocation.

From the strain-displacement relations, the only nonzero strains are the elements of the strain tensor $\epsilon = (\epsilon_{ij})$,

$$\epsilon_{13} = \epsilon_{31} = \frac{1}{2}\left(\frac{\partial u_1}{\partial x_3} + \frac{\partial u_3}{\partial x_1}\right) = -\frac{b}{4\pi}\left(\frac{x_2}{x_1^2 + x_2^2}\right);$$

$$\epsilon_{23} = \epsilon_{32} = \frac{1}{2}\left(\frac{\partial u_2}{\partial x_3} + \frac{\partial u_3}{\partial x_2}\right) = \frac{b}{4\pi}\left(\frac{x_1}{x_1^2 + x_2^2}\right).$$

By Hooke's law, the corresponding stresses in the stress tensor $T = (\sigma_{ij})$, are

$$\sigma_{13} = -\frac{bG}{2\pi}\left(\frac{x_2}{x_1^2 + x_2^2}\right);$$

$$\sigma_{23} = \frac{bG}{2\pi}\left(\frac{x_1}{x_1^2 + x_2^2}\right).$$

Exercise 3.9   Draw the stress cube showing the stresses at a point where both x_1 and x_2 are positive. Use this cube and its surrounding cubes to visualize how the deformation must occur to create these shear stresses. See that Figure 3.31 is somewhat misleading because the portions both above and below the ledge are in fact distorted.

The shear stresses, σ_{13}, σ_{23}, and the shear strains, ϵ_{13}, ϵ_{23}, are both proportional to the vertical distance from the ledge and inversely proportional to the square of the straight line distance from the origin. Further, the strain displacement relations are not valid at $x_1 = 0$ because the displacement is undefined there. This analysis predicts an infinite shear strain and stress at the origin which is therefore called a singularity.

The strain energy in a region about the dislocation is

$$U = \frac{1}{2G}\int(\sigma_{13}^2 + \sigma_{23}^2)dV = \frac{1}{2}\int\frac{Gb^2}{4\pi^2}\left(\frac{1}{x_1^2 + x_2^2}\right)dV.$$

Transform this equation into cylindrical coordinates so that $dV = 2\pi r dr dx_3$, the area of a ring times the length. Then in a cylinder of radius, R, and height, L, the dislocation length in the x_3 direction, the strain energy is

$$U = \frac{Gb^2}{4\pi}\int_{r_0}^{R}\int_0^L\frac{dr}{r}dx_3 = \frac{Gb^2 L}{4\pi}\ln\left(\frac{R}{r_0}\right).$$

The lower limit, r_0, in the radial direction is taken slightly larger than 0 to avoid the singularity at the origin. This inner cylindrical region of radius, r_0, is thought of as the dislocation core. The important result of this calculation for plasticity in Chapter 9, Section 9.3 is that the strain energy of a screw dislocation is proportional to the square of the magnitude of the Burgers vector.   ■

The strain energy density can be viewed as the sum of two components, one which accounts for volume change and the other which accounts for the distortion of a volume element under a deformation. Consider a full state of stress. Let the mean normal stress and mean normal strain be defined as

$$\sigma_m = \frac{1}{3}(\sigma_x + \sigma_y + \sigma_z);$$

$$\epsilon_m = \frac{1}{3}(\epsilon_x + \epsilon_y + \epsilon_z). \tag{3.31}$$

Note that $\epsilon_m = e/3$, where e is the dilatation so that this state of strain involves only a volume change. Now consider an element on which only the mean stresses and strain act in all three coordinate directions. The stress tensor,

$$\begin{pmatrix} \sigma_m & 0 & 0 \\ 0 & \sigma_m & 0 \\ 0 & 0 & \sigma_m \end{pmatrix}$$

is called the dilatational stress tensor, since it involves only volume change. It describes a hydrostatic state of stress. The portion of strain energy density corresponding to this state is called the volumetric strain energy density, U_o^v. Then

$$U_o^v = \frac{3}{2}\sigma_m \epsilon_m.$$

In terms of stress, from the generalized Hooke's law,

$$\epsilon_m = \frac{1}{E}[\sigma_m - \nu(\sigma_m + \sigma_m)] = \frac{1 - 2\nu}{E}\sigma_m.$$

Therefore

$$U_o^v = \frac{3(1 - 2\nu)}{2E}\sigma_m^2 = \frac{\sigma_m^2}{2K} = \frac{1}{18K}(\sigma_x + \sigma_y + \sigma_z)^2, \tag{3.32}$$

since the bulk modulus $K = E/[3(1 - 2\nu)]$.

The remainder of the stress, defined by subtracting the dilatational stress from the original stress state, is called the deviatoric stress tensor, $S = T - \sigma_m I$.

$$S = \begin{pmatrix} \sigma_x - \sigma_m & \tau_{yx} & \tau_{zx} \\ \tau_{xy} & \sigma_y - \sigma_m & \tau_{zy} \\ \tau_{xz} & \tau_{yz} & \sigma_z - \sigma_m \end{pmatrix}. \tag{3.33}$$

Its trace is zero since $Tr(S) = (\sigma_x - \sigma_m) + (\sigma_y - \sigma_m) + (\sigma_z - \sigma_m) = 0$. The corresponding strain tensor

$$\begin{pmatrix} \epsilon_x - \epsilon_m & \frac{1}{2}\gamma_{yx} & \frac{1}{2}\gamma_{zx} \\ \frac{1}{2}\gamma_{xy} & \epsilon_y - \epsilon_m & \frac{1}{2}\gamma_{zy} \\ \frac{1}{2}\gamma_{xz} & \frac{1}{2}\gamma_{yz} & \epsilon_z - \epsilon_m \end{pmatrix} \tag{3.34}$$

has dilatation, its trace, $e = (\epsilon_x - \epsilon_m) + (\epsilon_y - \epsilon_m) + (\epsilon_z - \epsilon_m) = 0$. Therefore this strain tensor causes no volume change. It distorts the body without changing the volume independent of any constitutive assumption; in other words this fact is true for nonlinear materials as well. In particular, $e = 0$ for the strain tensor corresponding to the deviatoric stress in the linear elastic case. Writing the stress tensor as a sum of its dilatational and deviatoric tensors breaks the stress into a portion that causes only volume change and a portion that distorts the volume but does not change the volume. The corresponding strain energy densities are called the volumetric, U_o^v, and distortional, U_o^d, strain energy densities. Then

$$U_o = U_o^v + U_o^d. \tag{3.35}$$

For a linearly elastic isotropic material, $U_o^d = (3/4G)\tau_{oct}^2$ where the octahedral shear stress is

$$\tau_{oct} = \frac{1}{3}[(\sigma_x - \sigma_y)^2 + (\sigma_y - \sigma_z)^2 + (\sigma_z - \sigma_x)^2 + 6(\tau_{xy}^2 + \tau_{xz}^2 + \tau_{yz}^2)]^{1/2}. \tag{3.36}$$

Exercise 3.10 Verify this expression for U_o^d by using equations (3.29) and (3.32) in terms of E.

EXAMPLE 3.11 The stress tensor, T, has deviatoric stress, S:

$$T = \begin{pmatrix} -1 & 2 & -3 \\ 2 & 7 & 4 \\ -3 & 4 & 5 \end{pmatrix}. \qquad S = \begin{pmatrix} -14/3 & 2 & -3 \\ 2 & 10/3 & 4 \\ -3 & 4 & 4/3 \end{pmatrix}.$$

The trace of S is zero. ∎

EXAMPLE 3.12 For uniaxial loading in the x-direction by a stress, σ, the stress tensor at each point is

$$\begin{pmatrix} \sigma & 0 & 0 \\ 0 & 0 & 0 \\ 0 & 0 & 0 \end{pmatrix}.$$

The dilatational tensor is

$$\begin{pmatrix} \sigma/3 & 0 & 0 \\ 0 & \sigma/3 & 0 \\ 0 & 0 & \sigma/3 \end{pmatrix}.$$

The deviatoric tensor is

$$\begin{pmatrix} 2\sigma/3 & 0 & 0 \\ 0 & -\sigma/3 & 0 \\ 0 & 0 & -\sigma/3 \end{pmatrix}.$$

Therefore an axial load causes both a volume change and a distortion. ∎

Exercise 3.11 Assume that the uniaxially loaded body of the previous example is linear elastic. Show that, when $\nu = 0.25$, $U_o^v = \sigma^2/12E$ and $U_o^d = 5\sigma^2/12E$. Verify that 83% of the strain energy goes into distortion in uniaxial loading. Show that $U_o = 0.5\sigma^2/E$, independent of ν.

The deviatoric idea has at least two important applications. The deviatoric stress tensor can be used instead of the Cauchy stress tensor in the constitutive model for materials which are assumed to be incompressible. In failure theories for which it is assumed that the material fails due to volume distortion rather than change in volume, such as ductile metals, the distortional strain energy density is used as a failure criterion. Constitutive models for plasticity of metals are defined in terms of the deviatoric stress and strain when plastic behavior is assumed to be independent of the hydrostatic stress, σ_m.

3.7 CONCRETE

Concrete is a composite of cement, water, gravel, and sand, but unlike other composites it cannot be described well by a linear constitutive model. The cement, after it interacts with water by the hydration process, bonds the concrete together. The sand and gravel are called the aggregate. The amount of aggregate in the mixture ranges from about 60 to 80%. The cement is the most expensive component; the space-taking aggregate is added both to reduce the cost and to make the concrete easier to form before it sets. As the slow process of hydration continues, the concrete sets and hardens. Setting refers to the concrete holding its desired shape; hardening refers to the increased strength of concrete under forces. The proportions of the four constituent materials in the concrete mix determine its strength. The water to cement ratio is the major determinant of concrete strength. The lower the ratio, the stronger the concrete, as long as there is enough water for hydration to proceed. But extra water is often added to make the concrete easier to pour into molds; in other words, to make the concrete more workable.

Portland cement is manufactured from clay and limestone by heating the raw materials in a kiln at temperatures which range from 2300 to 3450°F. The primary constituents are lime (CaO) and silica (SiO_2) with smaller amounts of alumina (Al_2O_3) and iron oxide (Fe_2O_3). These form into the main compounds found in cement, tricalcium silicate ($3CaO.SiO_2$), dicalcium silicate ($2CaO.SiO_2$), tricalcium aluminate ($3CaO.Al_2O_3$), and tetracalcium aluminoferrite ($4CaO.Al_2O_3.Fe_2O_3$).

Hydration, the chemical combination of water and cement, is the key process in making concrete. Unfortunately for precise control, the hydration reaction is very complicated and is still not clearly understood. Each of the four major compounds in cement hydrates at different rates and the products of their hydration also interact with each other. Hydration rates increase with increasing temperature. Therefore, hydration is influenced by environmental effects such as temperature and relative humidity. Hydration releases heat which must be controlled, or the concrete structure will form cracks due to nonuniform expansion. To observe the heat release of hydration, put a small amount of quicklime in a beaker of water. The water will rapidly heat up and may even boil. Freshly poured concrete pavement is sometimes covered with hay to prevent moisture evaporation so that hydration will proceed.

The hydration reaction can continue for years. The hydrated paste, or gel, of cement and water grows outward from existing particles to fill the intervening spaces. This process forms pores of the same size as the cement particles as well as larger capillary pores in which loosely bound excess moisture can be stored. The pore structure strongly influences strength because cracks can initiate at the pores and also affects thermal conductivity. As hydration proceeds, the material properties of the concrete change. For example, concrete is usually tested 28 days after mixing to estimate its properties because it has been found that about 80% of the strength of the concrete has been developed by that time. A material whose properties change with time is said to age.

Concrete is much weaker in tension than in compression. If a concrete beam or column must carry tension, steel rods are often inserted in the concrete during pouring to carry the tensile loads. The steel rods are usually ribbed so that the concrete can grip them better. In a concrete beam, steel rods are placed longitudinally near the bottom in the longitudinal tensile stress region to carry the tension. Vertical rectangular columns can be reinforced with four steel rods, one near each corner so that a pair will always resist tensile bending stresses, irrespective of the direction in which bending occurs. A circular column can be reinforced with a helical coil of steel wire.

Prestressed concrete is a modified reinforced concrete since steel rods are also placed in the concrete. But while the concrete is poured and while it is setting, the rods are held in tension. After the concrete hardens, the forces on the ends of the rods are released; the rods relax and compress the concrete. Any tensile load must first overcome this compression before it can force the concrete into tension. Therefore, prestressed concrete can carry a larger tensile load without failure than simple reinforced concrete. Reinforced concrete can be modeled as a fibered composite.

EXAMPLE 3.13

A steel-reinforced rectangular concrete beam ($E_c = 3 \times 10^6$ psi in compression) has width $w = 10$ inches and depth 12 inches. Six steel ($E_s = 30 \times 10^6$ psi) reinforcing rods, each with cross-sectional area of $A_s = 0.2$ in^2, lie longitudinally in the beam and spaced parallel to the width two inches from the bottom of the beam (Fig. 3.32). Determine the maximum compressive stress, σ_c, in the concrete and the tensile stress in the steel rods, σ_s, in a beam face subjected to a bending moment of 20,000 ft-lb.

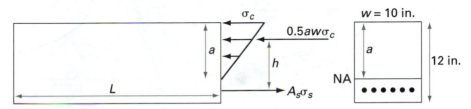

FIGURE 3.32   Beam cross section showing the stress distribution and equivalent load.

Solution

Assume that plane sections remain plane and that linear elasticity applies. Because of the presence of the rods, the location of the neutral axis is no longer through the beam face centroid. Denote the distance of the neutral axis from the top of the beam, where the concrete is in compression, by a and the radius of curvature of the neutral axis by ρ. The strain at a distance y from the neutral axis is then y/ρ. Assume that the concrete carries no load in tension. The compressive force and the tensile force on the face must be

$$\frac{1}{2}aw\sigma_c = 6A_s\sigma_s$$

so that

$$\frac{1}{2}awE_c\frac{a}{\rho} = 6A_sE_s\frac{10-a}{\rho}.$$

Therefore $a = 3.844$ inches from the top. The stresses are obtained from the fact that the moment equals either the equivalent compressive force or the equivalent tensile force times the distance, h, between them. The equivalent compressive point load is placed at $a/3$ from the top since the compressive stress distribution in the concrete is assumed linear. Therefore $h = 10 - a/3 = 8.7187$ inches.

$$\sigma_c = \frac{M}{0.5\,awh} = 1432 \text{ psi} \quad \text{and} \quad \sigma_s = \frac{M}{6A_sh} = 22{,}939 \text{ psi.} \qquad \blacksquare$$

Concrete is not a linear elastic material, although up to about 30% of the compressive strength, the uniaxial stress-strain curve is nearly linear. The typical compressive uniaxial stress-strain curve (Fig. 3.33a) increases up to a critical value σ_c, after which the stress decreases with increasing strain, possibly because of the formation of microcracks. The volume of concrete under compression contracts until the compressive strength is reached and then expands, again probably due to microcracking. Typical behavior under tension is shown in Figure 3.33b.

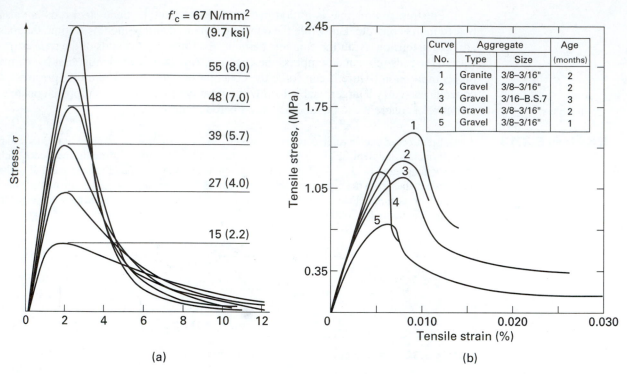

FIGURE 3.33   (a) Typical compressive and (b) tensile stress-strain curve for concrete [7, p. 256, 259]. Reprinted by permission from W.-F. Chen.

The Saenz [30] model for the uniaxial compressive stress-strain relation is

$$\sigma(\epsilon) = \frac{E_0 \epsilon}{1 + \left(\frac{E_0}{E_C} - 2 \right) \frac{\epsilon}{\epsilon_c} + \left(\frac{\epsilon}{\epsilon_c} \right)^2}, \tag{3.37}$$

where E_0 is the initial elastic modulus, ϵ_c is the strain corresponding to σ_c, and $E_C = \sigma_c / \epsilon_c$. A common estimate of E_0 in terms of the compressive strength is

$$E_0 = 33 w^{1.5} \sqrt{\sigma_c} \text{ psi}, \tag{3.38}$$

where w is the weight per cubic foot.

The Gerstle (1981) [14] empirical model for biaxial compression tests is

$$\tau_{oct} = \tau^* \left[1 - \exp \left(\frac{-G_0}{\tau^*} \gamma_{oct} \right) \right], \tag{3.39}$$

where G_0 is the initial shear modulus and τ^* is the octahedral shear strength. An estimate of G_0 can be obtained from the linear elasticity relation, $G_0 = E_0 / (2(1 + \nu_0))$, where ν_0 is the initial value of the Poisson ratio. If ν_0 is unknown, it is estimated to be $\nu_0 = 0.2$.

The Poisson ratio varies between $0.15 \leq \nu \leq 0.22$, with $\nu = 0.20$ a typical value. The Poisson ratio in concrete is not a constant; it varies with strain. One proposed relation for the change in Poisson ratio with compressive strain is that given by Elwi and Murray in 1979 [11],

$$\nu = \nu_0 \left[1.0 + 1.3763 \frac{\epsilon}{\epsilon_c} - 5.36 \left(\frac{\epsilon}{\epsilon_c} \right)^2 + 8.586 \left(\frac{\epsilon}{\epsilon_c} \right)^3 \right]. \qquad (3.40)$$

The apparent similarities between concrete response and metal plasticity have led to the development of incremental constitutive models in a manner analogous to those for metal plasticity [7].

3.8 RUBBER

Rubber is a nonlinear elastomer used extensively as an engineering material in tires, vibration dampers, rubber bands, inner tubes, children's toys, waterproofed raincoats, boots, etc. Rubber can be obtained as either a natural material from a latex from the sap of rubber trees growing in the tropics or manufactured as a synthetic polymer. Raw natural rubber is formed by treating the latex of the Havea tree with acetic acid. These trees are grown in Central America, in Brazil and in Asian countries such as India or Malaysia.

Western Europeans first discovered rubber when Columbus arrived in the Caribbean and saw the indigenous people playing a game with a rubber ball. Rubber soon became an economic motivation for the colonization of the "new world," and rubber plantations were developed in the American tropics to produce rubber for Europe. By 1823, Charles Macintosh had patented the coating of fabric with natural rubber to create rainproof coats; this is why a raincoat is sometimes referred to as a "macintosh." By the 1850s solid rubber tires were made in England. John Dunlop, an Irish veterinarian, patented the pneumatic tire, a tire containing an inner tube filled with air, in 1888, which he invented for his son's bicycle to absorb the vibrations caused by riding on rough surfaces. Automobile tires, the direct descendants of bicycle tires, were not developed specifically for cars but were merely modified bicycle tires.

3.8.1 Properties of Natural Rubber

The material properties of natural rubber are quite different from those of metals. The stress-strain curve for rubber not only is not linear (Fig. 3.34a), but also exhibits

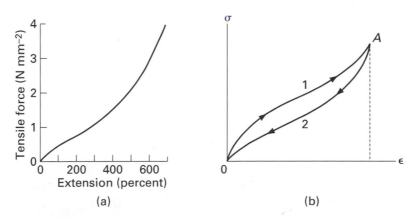

FIGURE 3.34   (a) Load-extension tensile curve for rubber [34, p. 2]. Reprinted by permission of Oxford University Press. (b) Typical loading and unloading curve for rubber (1 is the increasing load and 2 is the decreasing load).

hysteresis in the elastic range. A typical stress-strain curve showing both loading and unloading is shown in Figure 3.34b. The material is initially stiff so that a larger force is required to create stretch. Then the stiffness decreases. Finally, the stiffness increases again; this latter stage is believed due to crystallization of the rubber as the large force tends to make the rubber chain molecules line up. This increases the attractive forces between chains and stiffens the rubber. The synthetic rubbers do not crystallize as easily as natural rubber.

Notice that there is little or no initial straight line portion on the curve. The elastic modulus, E, must be defined as the slope of the tangent to the stress-strain curve at the origin. Even so, the material is still elastic on the portion of the load-extension curve shown in Figure 3.34b. The unloading portion of the stress-strain curve shown, which is not parallel to an initial straight line portion as for metals, indicates that the unloading does not follow the loading path back to the origin, as it would in an elastically loaded metal. This phenomenon, called hysteresis, occurs because time is required to relax the stretched rubber molecules back to their original unloaded shape. Therefore, the unloading curve is to the right of the loading curve in Figure 3.34b. The area inside the loading-unloading loop is a measure of the net energy lost and gained as heat.

Metals expand when they are heated and loaded. However, as discovered by J. Gough in 1805, if a strip of rubber is first loaded with a weight so that it stretches and is then heated, the rubber will contract from its stretched length. Conversely, the stretched rubber expands on being cooled. The coefficient of thermal expansion for unstretched rubber is initially larger than most metals ($\alpha = 100 \times 10^{-6}/°F$), but becomes negative once stretching begins. Gough also found that rubber heats up when it is stretched quickly. These phenomena were verified by Joule in 1859 [17] and are now often called the Gough-Joule effects.

Exercise 3.12   Hold a piece of natural rubber (containing no sulfur or carbon black) against your lips and rapidly stretch the rubber. Do you feel the heat released?

Exercise 3.13   Hang a thin strip of rubber from a support and apply a weight at the end. Perhaps using a hair dryer, carefully heat the rubber to a fixed temperature and measure the deformation. The stretched rubber should contract when heated and expand when cooled. Repeat the experiment for several different temperatures. Prepare a graph showing the temperature-strain relationship. Repeat for different loads.

The elastic modulus of natural rubber, the slope at the origin, is about 200 psi, about one ten-thousandth of the elastic modulus for most metals. Its maximum elastic strain of about 1000% is 10–20 times larger than the percent elongation of metals. This large elongation is possible since the long rubber molecules, containing about 6,000 isoprene units, straighten and line up under large loads. Natural rubber is a polymer built from chains of the isoprene monomer:

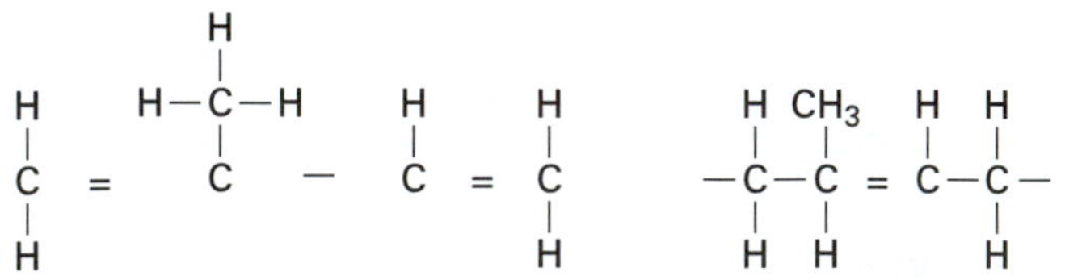

(a)   (b)

FIGURE 3.35   (a) Isoprene. (b) The isoprene monomer.

Each carbon atom is joined by four bonds, some of which are double bonds, as indicated by the double lines. To form the polymer from isoprene, the double bonds in Figure 3.35a are broken allowing a chain to form, but a double bond is created as in Figure 3.35b.

For metals, Poisson's ratio is a constant between 0.0 and 0.5 but the Poisson's ratio for rubber is a function of axial strain (Fig. 3.36). It is only for small strains and a constant volume that rubber has a Poisson's ratio of 0.5. Recall that incompressible materials which maintain a constant volume under loads are often assumed to have a Poisson's ratio of 0.5, but the calculation supporting this assumption is only valid for linearly elastic materials. Also, the relation $G = 0.5E/(1 + v)$ for linearly elastic materials does not hold unless the Poisson's ratio, v, is a constant.

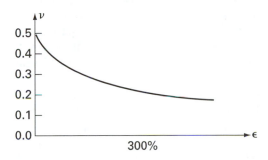

FIGURE 3.36   Poisson ratio versus stretch for rubber.

3.8.2 Kinetic Theory

The kinetic theory of rubber elasticity assumes that the mechanical behavior depends on the thermal excitation of the long-chain molecules. The chains tend to coil as heat is absorbed in the excitation of the molecules. Applying a load causes uncoiling and heat release as work is done to overcome the thermal motion. The bond lengths change under elastic deformation as do the valence angles in the chains. In the original kinetic theory, the polymer is idealized as composed of isolated and unconnected chains. The elastic properties were assumed due to the thermal motion of atoms in each chain. As a consequence, in small strains, the elastic modulus is proportional to the absolute temperature, a close approximation to the experimental behavior. Rubber was said to contract under increased temperature due to the increased thermal (Brownian) motion of the atoms in the chains. The network properties of the chains were ignored in that no motion of chains with respect to each other was assumed. This model was improved by the additional assumption that the chains are connected at entanglement junctions, so that the chains form a network structure in three-dimensional space. These junctions could not move in this model, but the atoms in the middle of the chain can move so that under thermal excitation, the chain acts almost like a jump rope held at its ends. While all chains were assumed of equal length, the distance between junctions was held to follow a Gaussian distribution. In a later improvement of the kinetic model, the entanglement junctions were allowed to move under deformation of a polymer by a load. The junctions also had a Brownian motion and their position also obeyed a Gaussian distribution. Recently, other statistical models than the Gaussian have been used along with assumptions on the number of chains meeting at a junction. No intermolecular interaction by, say van der Waals forces, is accounted for, nor is chain slippage within a junction allowed. These models provide reasonable predictions only for small strains; they fail at large deformations.

3.8.3 Vulcanization

Natural rubber is too soft for applications such as tires and is sticky at room temperature. Furthermore it is very temperature sensitive; it becomes more sticky in heat and brittle in the cold. In the first half of the nineteenth century, angry customers returned rubber-coated products because of these annoying properties; and therefore, many researchers tried to modify natural rubber to give it more useful properties.

Charles Goodyear had been trying to eliminate rubber stickiness for several years when, in 1838, Nathaniel Haywood suggested to him that adding sulfur to rubber in the presence of sunlight would make rubber less sticky; but Goodyear found that this process only affected the surface of the rubber. By accident a piece of this rubber was left on a hot stove, where it charred rather than melted. This led him to more carefully examine the process and in 1844 he received a U.S. patent for the process of vulcanization in which he heated a mixture of 20 parts sulfur, 28 parts white lead, and 100 parts rubber to 270°F, creating a thermosetting polymer with the desired hardness. Unfortunately, Thomas Hancock in England, who had heard of Goodyear's process, applied for an English patent a few weeks earlier than Goodyear, costing Goodyear his patent rights in England, the most important industrial country of that era. The first to apply rather than the first inventor was awarded the patent in the English system of that day. Hancock also made important contributions to the commercial use of the process by learning how to vulcanize rubber in steel molds in order to create in one step hardened rubber in the desired shape. The white lead is an accelerator for the vulcanization process; other materials such as litharge and magnesium oxide were also later used as accelerators.

Sulfur added to rubber increases its strength by causing cross-linking of the polymer molecules. Some of the carbon double bonds (Fig. 3.35b) are broken on the spines of pairs of molecules and reform with a sulfur atom as intermediary to link two long-chain molecules. Today rubber bands contain about 4% sulfur while very hard rubber such as used in bowling balls contain up to 40% sulfur.

Rubber is not naturally black in color. It was discovered in 1904 that adding carbon black to the vulcanized rubber increases its strength (to about 4500 psi). However, strain concentrations arise around the carbon particles. It is this carbon black which gives commercial rubber, such as that in tires, its black color. Interestingly, consumers initially rejected rubber products with carbon black filler because of what they saw as its unnatural black color.

3.8.4 Synthetic Rubber

The large effort first put into the study of polymers in the nineteenth century was partly motivated by the attempt to polymerize isoprene into a synthetic rubber. Natural rubber had been degraded to isoprene and reconstructed to rubber in the 1870s. The attempt to synthesize rubber led to the industrial use of polymers.

War and economics had a great influence on the development of synthetic rubber. During World War I, Germany was cut off from the sources of natural rubber in Central America by a shipping blockade and produced 2500 tons of synthetic rubber in a two-year period of the war. The price of natural rubber decreased drastically between 1910 and after World War I, since rubber plantations in Southeast Asia had greatly expanded, and so between the two world wars, there was little incentive for the industrial production of synthetic rubber. However, with the growing fear of war in the 1930s, rededicated efforts successfully produced synthetic rubber, particularly in Germany.

Today about half the synthetic rubber used in the United States is styrenebutadiene rubber (SBR). It is made in the proportion of 78% butadiene and 22% styrene. When reinforced with carbon black, the Young's modulus is 2000 psi, its ultimate strength is 3500 psi, and the percent elongation is 600%. It is a good insulator and is used in hose, belts, shoe soles, etc.

Butyl rubber, developed in the late 1930s, is used in inner tubes and tubeless tires, because gases do not pass through it easily. It is formed as a copolymer of isobutylene and either butadiene or isoprene. Its strength is about that of natural rubber. When reinforced by carbon black, the Young's modulus is about 900 psi, and the percent elongation is 300%.

Many other synthetic rubbers can be polymerized with differing properties, such as resistance to oil, ability to dampen vibration, tensile strength, resistance to abrasion, and reduced buildup of heat during deformation.

3.8.5   Stretch and Models for Rubber

The stretch is an alternative measure of strain, usually applied to a material allowing large deformations such as an elastomer like rubber or biological tissues.

Definition   The stretch, λ, is the ratio of a deformed length to the undeformed length.

Stretch is always positive. An undeformed body has stretch, $\lambda = 1$. The deformation is compressive if $0 < \lambda < 1$ and tensile if $\lambda > 1$. Uniaxial stretch is related to the uniaxial engineering normal strain, ϵ, by

$$\lambda = 1 + \epsilon.$$

Incompressible materials must satisfy the relation, $\lambda_1 \lambda_2 \lambda_3 = 1$, in 1-2-3 principal coordinates, as carefully proved in continuum mechanics. However, this relation can be justified by recalling the calculation made in Section 3.3 on the change of volume, which is for large deformations in 1-2-3 coordinates, $V = (1 + \epsilon_1)(1 + \epsilon_2)(1 + \epsilon_3)V_0$. Incompressible means that $V = V_0$. Here, the higher-order terms in the strains are not neglected as they were for the small strain case.

No existing mathematical stress-strain relationship proposed for rubber completely describes the behavior of rubber. The search for such a model is currently an active area of engineering and materials research. Rubber and many biological tissues are nonlinear elastic so that the linear constitutive model used for metals and ceramics is not valid.

Most of the constitutive models proposed for rubber are isothermal. The effect of temperature variation is ignored, even though rubber and some biological materials when stretched respond to temperature change in the opposite manner to metals, which expand when heated. Rubber emits heat when stretched and recovers its original temperature when unstretched. A rubber specimen held at a constant stretch contracts if heated, the opposite behavior of most other materials. This behavior is not eliminated by sulfur crosslinking of the rubber molecules when rubber is vulcanized.

Perhaps the most famous isothermal models for rubber are the neo-Hookean and the Mooney-Rivlin models. The neo-Hookean was developed from a Gaussian statistical model for the chains, suggested by the kinetic theory of rubber elasticity. The strain energy density, Ψ, is assumed to be

$$\Psi(\lambda_1, \lambda_2, \lambda_3) = G\left(\lambda_1^2 + \lambda_2^2 + \lambda_3^2 - 3\right), \tag{3.41}$$

where the λ_i are the stretches in the principal directions. A typical value for G is 0.39 N/mm^2. For a rubber sheet, assume that λ_3 is the stretch in the thickness direction. The incompressibility condition implies that $\lambda_1\lambda_2 = 1/\lambda_3$, so that the energy density function for a sheet becomes

$$\Psi(\lambda_1, \lambda_2) = G\left(\lambda_1^2 + \lambda_2^2 + \lambda_1^{-2}\lambda_2^{-2} - 3\right). \tag{3.42}$$

The isothermal Mooney-Rivlin model for an incompressible material is defined by the energy function

$$\Psi(\lambda_1, \lambda_2, \lambda_3) = C_1\left(\lambda_1^2 + \lambda_2^2 + \lambda_3^2 - 3\right) + C_2\left(\lambda_1^{-2} + \lambda_2^{-2} + \lambda_3^{-2} - 3\right). \tag{3.43}$$

For a sheet,

$$\Psi(\lambda_1, \lambda_2) = C_1\left(\lambda_1^2 + \lambda_2^2 + \lambda_1^{-2}\lambda_2^{-2} - 3\right) + C_2\left(\lambda_1^2\lambda_2^2 + \lambda_1^{-2} + \lambda_2^{-2} - 3\right), \tag{3.44}$$

where C_1 and C_2 are material constants with units of force/area. When $C_2 = 0$, the Mooney-Rivlin model reduces to the neo-Hookean.

For a uniaxial loading, $\lambda_2 = \lambda_3 = \lambda_1^{-1/2}$ to preserve the incompressibility condition. Then the strain energy is

$$\Psi(\lambda_1) = C_1\left(\lambda_1^2 + 2\lambda_1^{-1} - 3\right) + C_2\left(\lambda_1^{-2} + 2\lambda_1 - 3\right) \tag{3.45}$$

if the loading is in the 1-direction. The uniaxial stress-strain relation, for the stress per original area t versus stretch, is obtained by differentiating the strain energy.

$$t(\lambda_1) = \frac{d\Psi}{d\lambda_1} = 2C_1(\lambda_1 - \lambda_1^{-2}) + 2C_2(-\lambda_1^{-3} + 1). \tag{3.46}$$

The material parameters, C_1 and C_2, can be approximated using an experimental uniaxial stress-strain curve and equation (3.46). Because the material is incompressible, the volume is constant and $AL = A_0L_0$. Therefore the current area is $A = A_0(L/L_0) = A_0\lambda$.

$$t(\lambda_1) = 2C_1\left(1 + \frac{C_2}{C_1}\lambda^{-1}\right)(\lambda - \lambda^{-2}).$$

Because $(\lambda - \lambda^{-2}) = (\lambda-1)(\lambda^2+\lambda+1)/\lambda^2$, near $\lambda = 1$, $t(\lambda_1) = 6C_1(1+C_2/C_1)(\lambda - 1)$. As λ gets very large, $t(\lambda_1)$ is approximated by $2C_1\lambda$. The slope m at $\lambda = 1$ of the experimental curve for $t(\lambda_1)$ can be estimated from the experimental curve, although precision is rather difficult. The two facts then give

$$m = 6C_1\left(1 + \frac{C_2}{C_1}\right)$$

$$\frac{F}{A_0} = 2C_1\lambda,$$

where in the second equation λ is large, say about 2.5, and F/A_0 is the experimental stress at that value. These two equations are then solved for C_1 and C_2. Unfortunately, the values obtained will not reproduce the stresses in a test which is not axial, say a biaxial test. The Mooney-Rivlin model is better than the neo-Hookean, but still not satisfactory. These models are defined to ensure that the isothermal shear modulus, μ, is twice the sum of the material constants, $\mu = 2(C_1 + C_2)$.

Exercise 3.14   On the same set of axes, graph the uniaxial response for the Mooney-Rivlin and neo-Hookean models, when $0.8 \leq \lambda \leq 1.5$, in the case that $G = 0.39$ N/mm^2, $C_1 = 0.16$ N/mm^2, and $C_2 = 0.05$ N/mm^2.

Alexander [1] in an experiment on the inflation of an initially spherical neoprene (polychloroprene) balloon showed that the balloon can jump from its symmetrical shape to an asymmetrical shape as the pressure is increased past a critical value. The shape returns to spherical as the pressure passes a higher critical value. Alexander proposed a constitutive model to reproduce the large deformations of his experiments. In the uniaxial version of this model, the stress is given by

$$
\begin{aligned}
t(\lambda_1) &= \frac{d\Psi}{d\lambda_1} \\
&= 2c_1 \exp[k(\lambda_1^2 + 2\lambda_1^{-1} - 3)^2](\lambda_1 - \lambda_1^{-2}) \\
&\quad + 2\left(\frac{c_2}{\lambda_1^{-2} + 2\lambda_1 - 3 + \gamma} + c_3\right)(-\lambda_1^{-3} + 2).
\end{aligned} \tag{3.47}
$$

Here λ_1 is the ratio of the current radius to the original radius of the balloon. For the neoprene used by Alexander, $c_1 = 17.0, c_2 = 19.85, c_3 = 1.0, k = 0.00015$, and $\gamma = 0.735$. The neo-Hookean, Mooney-Rivlin, and Alexander models are compared to the experiment in Figure 3.37.

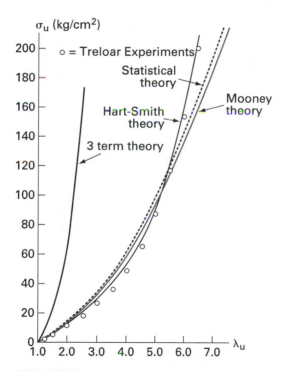

FIGURE 3.37   Experimental uniaxial curve for rubber compared to a Mooney-Rivlin, neo-Hookean (statistical), and Alexander (Hart-Smith) fit [1]. Reprinted by permission from Elsevier.

Ogden [28] built constitutive models for rubber from linear combinations of the $\varphi(\alpha) = \lambda_1^\alpha + \lambda_2^\alpha + \lambda_3^\alpha - 3$, where α is any fixed real number, so that the form of the strain energy density is

$$\Psi(\lambda_1, \lambda_2, \lambda_3) = \sum_{p=1}^{N} \frac{\mu_p}{\alpha_p} \varphi(\alpha_p), \tag{3.48}$$

for empirical constants μ_p and α_p. Twice the shear modulus equals the derivative of the biaxial $(\lambda_3 = \lambda_1^{-1}\lambda_2^{-1})$ $\partial^2 \varphi(\alpha)/\partial\lambda_1\partial\lambda_2 = \alpha^2$ when evaluated at $\lambda_1 = \lambda_2 = 1$. Therefore, the coefficients are constrained by $\sum_{p=1}^{N} \mu_p\alpha_p = 2\mu$, where μ is the shear modulus. These models with three terms, and therefore six empirical constants to work with, are capable of representing most types of rubber behavior. One such model used to fit experimental data is a three-term model ($N = 3$) with $\alpha_1 = 1.3, \alpha_2 = 4.0$, and $\alpha_3 = -2.0$, while $\mu_1 = 0.69, \mu_2 = 0.01$, and $\mu_3 = -0.0122$ (N/m^2). The α_1 term fits the data for large compressive stretches and the α_3 term fits the data for the large tensile stretches.

Exercise 3.15   Plot the uniaxial stress-strain curve for the Ogden model with the constants given in the previous paragraph over the range $0.5 \leq \lambda \leq 2.5$.

Other rubberlike materials are nitrol rubber, chloroprene rubber, chlorosulfonated polyethylene, silicone, and polyurethane.

3.9   INCOMPRESSIBLE BIOLOGICAL TISSUES

The Fung model is often used to represent incompressible biological tissue such as skin or the walls of arteries. Fung [13] proposed an energy function for the behavior of skin based on stress-strain tests and other data. The Fung strain energy per unit mass, φ, is

$$\rho_0\varphi = f(\alpha, \epsilon) + c\exp(F(a, \epsilon)), \tag{3.49}$$

where α_i, a_i, and c are constants and ρ_0 is the density.

$$f(\alpha, \epsilon) = \alpha_1\epsilon_{11}^2 + \alpha_2\epsilon_{22}^2 + \alpha_3\epsilon_{12}^2 + \alpha_3\epsilon_{21}^2 + 2\alpha_4\epsilon_{11}\epsilon_{22}; \tag{3.50}$$

$$F(a, \epsilon) = a_1\epsilon_{11}^2 + a_2\epsilon_{22}^2 + a_3\epsilon_{12}^2 + a_3\epsilon_{21}^2 + 2a_4\epsilon_{11}\epsilon_{22}. \tag{3.51}$$

Fung says that if accuracy at small strains is not imperative, then the function $f(\alpha, \epsilon)$ may be neglected.

In the uniaxial case for larger strains, the strain energy function is

$$\rho_0\varphi = c\exp(a_1\epsilon_{11}^2). \tag{3.52}$$

Biaxial stress-strain data for rabbit skin, reproduced by Fung, was modeled in terms of tensile force in grams versus the stretch ratio, λ. The constants are $\alpha_1 = \alpha_2 = 10.4, \alpha_4 = 2.59, a_1 = 3.79, a_2 = 12.7, a_4 = 0.587, c = 0.00794$. One set of tests was run with $\lambda_1 = 1$ and the other with $\lambda_2 = 1$.

Fung suggests that the equation,

$$\rho_0\varphi = c\exp(F(a, \epsilon)), \tag{3.53}$$

is also valid for the deformation of arterial walls as well as lung tissue.

The variable λ is the ratio of the current radius to the original radius of the artery viewed as a hollow cylinder. In this uniaxial case, the strain energy for the arterial wall has been assumed to be of the form

$$w(\lambda) = C\left\{\exp\left[\frac{1}{2}\Gamma(\lambda^2 - 1)^2\right] - 1\right\},\qquad (3.54)$$

where $C > 0$ has units of force per unit length, and Γ is nondimensional. The Fung model implies that the stress becomes unrealistically large for stretches larger than 1.4, a strain of 40%. A similar model can be used to represent the strain energy of a saccular aneurysm, viewed as a sphere (see Chapter 7, Section 7.5). Typical values of the constants for aneurysm walls, which are degenerate artery tissue, are $C = 0.88$ N/m, $\Gamma = 13$.

Exercise 3.16 Plot the uniaxial Fung stress strain curve, $\sigma = dw/d\lambda$ over the range $1 \leq \lambda \leq 1.4$.

Contemporary models of artery tissue view it as a layered composite in which collagen fibers are embedded [16] rather than as an isotropic material, as the Fung model assumes.

3.10 POLYMERS AND VISCOELASTIC BEHAVIOR

Polymers exist as either natural materials, such as cellulose in wood or paper, rubber, and starch, or synthetic materials, such as plastics or lubricants. Considerable effort was devoted to the study of polymers in the late nineteenth century. However, it was not until the first quarter of the twentieth century that enough was learned about their structure to permit the beginnings of commercial industrial production of useful synthetic polymers. Major breakthroughs in understanding natural polymers such as exist in the human body did not occur until even later. Today, aspects of polymer science are still not completely understood.

Polymers are built of long molecules, in contrast to the lattice crystalline structure of metals and ceramics. The polymer molecules are chains of smaller repeating structures called monomers (single *mers*). Since "poly" means "many," the name "polymer" was adopted. Polymer chains may be linear, branched, or cross-linked. Monomers are linked together to form a polymer chain molecule either by providing enough energy to break a double bond or to free an atom from the monomer, in either case to create an open bond which can be closed by linking with other monomers. This occurs under the action of a catalyst over a fixed period of time and under a certain pressure and temperature. In living animals or plants, the catalyst is called an enzyme; for the creation of synthetic polymers, the catalyst can be another molecule, light or heat.

Polymers built of covalently bonded carbon atoms are called organic. The spine of a polymer is the string of atoms or molecules holding the chain together. For example, the spine of the organic polymer, polyethylene, is a string of carbon atoms. The spine in the inorganic polymer, Silly Putty, is an alternating chain of silicon and oxygen atoms. The shape of the spine and attached atoms is determined by the balance of attractive and repulsive forces between the atoms. The atoms attached to the spine are arranged to minimize the force of attraction between atoms and the force of repulsion caused by the cloud of electrons surrounding each of these atoms.

Helical configurations, in order to reach an energy equilibrium, are taken by the spines of a molecule with larger structures attached to one side of the spine than the other. The DNA molecule, which determines the nature of a living organism, is shaped as a double helix. Two of these helical molecules, which are mirror images of each other connected by hydrogen bonds, exist in a unit.

The linear polymers have molecules that are considered one dimensional because the atoms are strung together to form a single chain. Linear polymers include polyethylene and polytetrafluoroethylene. One of the simplest mers is produced from the ethylene monomer, C_2H_4, by breaking the double carbon bond (Fig. 3.38). A double bond consists of two pairs of electrons shared by the two atoms. The resulting two open single bonds are free to bond with other mers of the same or different types. Polyethylene is formed when a chain of ethylene mers is made by bonding carbon atoms and has one of the simplest polymer structures. Polyethylene at room temperature is soft and flexible and is used to make containers. It has glass transition, $T_g = -80°C$.

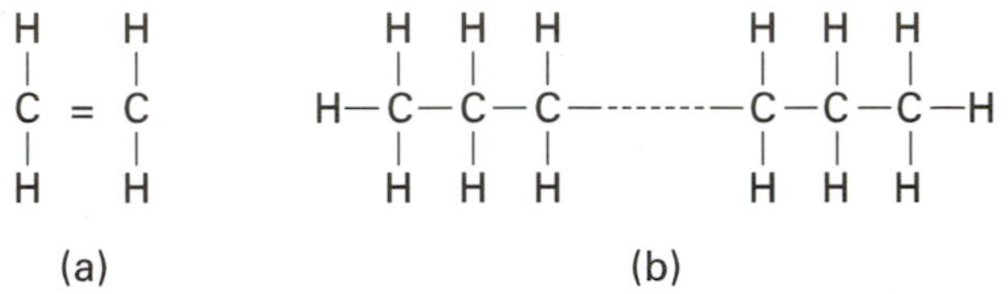

```
H   H              H  H  H           H  H  H
|   |              |  |  |           |  |  |
C = C       H — C— C— C ——————— C— C— C—H
|   |              |  |  |           |  |  |
H   H              H  H  H           H  H  H

  (a)                            (b)
```

FIGURE 3.38   (a) The ethylene monomer and (b) polyethylene.

Polymers similar in structure to polyethylene are obtained by substituting other valence one molecules or groups for the hydrogen atoms attached along the spine of the molecule. Polyvinyl chloride (PVC) is obtained by substituting a chlorine atom for one hydrogen atom on each polyethylene mer. This increases strength and raises the glass transition to $T_g = 80°C$. A plasticizer such as the toxic dioctyl phthalate is often added to PVC to make the polymer more moldable. Commonly, the plasticizer is from 20 to 50% by weight of the polymer. Aging on the order of 10 to 25 years allows the plasticizer to separate, which leads to embrittlement of the PVC. Also PVC can emit a hydrogen chloride gas which reacts with metals. Any polymer with chlorine atoms in its mer may degenerate into highly reactive components and in general should be avoided.

Polytetrafluoroethylene is formed from C_2F_4 so that the structure is analogous to that of polyethylene, but with a fluorine atom in place of every hydrogen atom, a fluoroplastic (Fig. 3.39). Polytetrafluoroethylene is better known as Teflon and has glass transition $T_g = -100°C$, melting temperature $T_m = 327°C$, and density 2.3 g/cm^3. Teflon is famous for its low coefficient of friction, about 0.04 between Telfon and most metals.

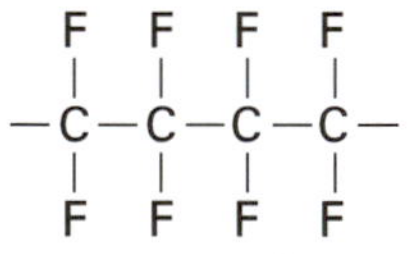

```
  F   F   F   F
  |   |   |   |
—C— C— C— C—
  |   |   |   |
  F   F   F   F
```

FIGURE 3.39   The Teflon polymer structure.

Vinyl alcohol is similar to ethylene except that one hydrogen atom is replaced by OH, the hydroxyl group. This monomer is formed into polyvinyl alcohol.

The branched polymers have a chain molecule structure that is two dimensional because linear chains are connected to a main chain at various positions along the main chain, forming a structure that resembles limbs branching off of a tree trunk. These include polymethylmethacrylate (PMMA), which is usually known as Plexiglas. This material is strong and has an amorphous chain structure. It is very brittle in tension, but a round axially loaded specimen takes a barrel shape under compression as though the material were ductile. Plexiglas, therefore, behaves differently than metals in uniaxial tests. Most metals have very similar stress-strain curves in both tension and compression, at least in the linear region. Other polymers have qualitatively distinct stress-strain curves. Figure 3.40 compares the stress-strain curves for three common polymers.

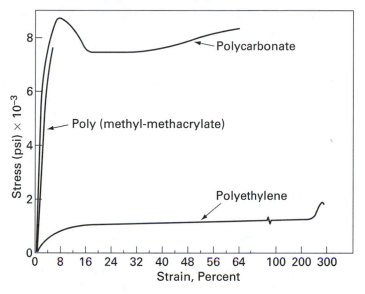

FIGURE 3.40   Stress-strain curves for three polymers loaded in tension at a strain rate of 0.3 per minute [26]. Reprinted by permission.

Vulcanized rubber is a common cross-linked polymer constructed by connecting linear chains by a bond or small chain involving sulfur along the mid section of the linear chains. Often the cross-linked molecules form a three-dimensional structure.

However, not all polymer molecules are highly stable. For example, ultraviolet radiation causes the breakdown of many polymers. If O_2 is broken into oxygen atoms by ultraviolet light, the oxygen atoms can react with nearby materials.

3.10.1 Mechanics of Polymers

Polymer deformations, which can be both large and recoverable, are a consequence of the polymer internal structure. The structure of polymer molecules was still a matter of controversy into the 1920s. H. Staudinger, a professor in Freiburg, Germany, argued correctly that polymers were large molecules, but many others believed they were aggregates of small molecules exhibiting "colloidal" behavior. A *colloid* is a gelatinous substance whose molecules, when suspended in a fluid, will not settle out. Not only are polymers not colloidal, but their molecules are not rigid. W. Kuhn suggested in 1934 that polymer molecules could be represented as random coils or helixes. Extensions would be due to the stretching of the coil. But the coil model seems too simple to explain polymer behavior. Currently, it is believed that the mechanism of polymer deformation differs from that in metals and ceramics in that it is due both to bond lengthening and to molecule rotation. Portions of a long-chain molecule may rotate in relation to other segments of the same molecule under the influence of external forces and temperature changes. Such a rotation must preserve the valance angle of covalently bonded atoms. For example, a carbon atom which is to be bonded to a given carbon must lie on the surface of a cone with an angle of 109.5° with vertex at the given carbon atom. In the aggregate, this rotational effect leads to a dimensional change in the specimen. A free volume is generated within the polymer because the molecules in their excited state cannot fit closely together. This hypothesis for the explanation for polymer behavior is very similar to the kinetic theory for rubber elasticity.

The energy of the molecule changes under rotation because of the interaction with nearby atoms that are not chemically bound with the molecules. The rotation is not free. Sections of the molecule seek configurations in which the potential energy of the molecule is minimized. The change in configuration takes time to occur so that the response of a polymer to loads is time dependent. A chain molecule with atoms attached symmetrically around the spine can take either a *trans* or *cis* configuration, and the spine is straight. For example, the polyethylene molecule can jump between the *cis* and *trans* positions. Figure 3.41 shows a top view of this arrangement, with the spine pointing out perpendicular to the page.

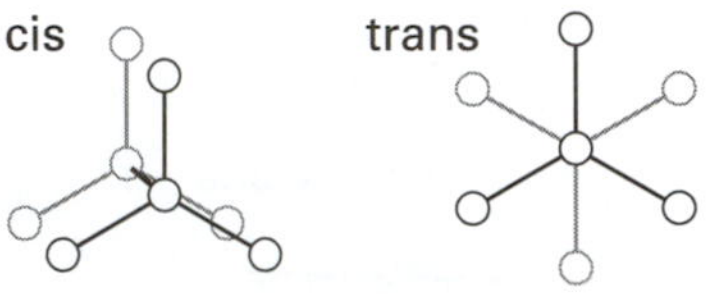

FIGURE 3.41   The trans and cis configurations.

Unloaded polyethylene has a trans configuration, with every other pair of hydrogen atoms in the same position with respect to the spine of carbon atoms. This configuration is taken because it requires the least amount of energy to hold the molecule together. If the cis and trans positions require the same amount of energy to hold a polymer molecule together, neither is favored and the molecule spine takes a randomly curved configuration. Natural and synthetic rubber, as well as other elastomers, typically have this random molecular configuration, which partially accounts for their ability to stretch.

The potential energy of a simple rotation 60° about the axis formed by the spine of the molecule has been approximated as

$$U = \frac{1}{2}U_o(1 - \cos 3\theta),$$

where θ is the angle of rotation and U_o is a measure of the molecular interactions restricting rotation. In the case of polyethylene, $U_o = 12.1$ kJ/mol.

The manner in which the individual long-chain molecules are linked in the polymeric material has great influence on the properties of the polymer, for example, strength, elasticity, translucence, and ductility. When polymer molecules are linear, they can slide easily on each other and the polymer tends to be ductile. In branched polymers, the branches can be intertwined and resist the relative motion of molecules, thereby stiffening the polymer. Molecular cross-links prevent sliding and the material is much stronger. For example, raw rubber latex contains straight chains of molecules without branches or links and is therefore soft and ductile. To harden and stiffen rubber, cross-links are created between the molecular chains by vulcanization.

An *amorphous* polymer is one whose molecules are not arranged in a definite pattern. The molecules are attracted to each other by van der Waals forces or polarity forces (positive or negative). The stronger the force, the stiffer the polymer. The stiffness can sometimes be reduced by creating branched molecules so that pairs of molecules cannot approach each other too closely and the force between them is weakened. Some molecules have a permanent polarity due to the imbalance of the atoms along their spines, giving that side a larger group of electrons and therefore a negative charge compared to the opposite side. The resulting polar forces between the molecules are quite strong and therefore these materials are stronger and harder.

Polymers may form crystals in the sense that a regular long range structure is created, but of a different type than the crystals found in metals. A plausible theory of the manner in which polymers could crystallize was not developed until the 1940s. A major contributor was P. J. Flory, who began his work in 1934 at DuPont. The thickness of a crystal is smaller than the length of the polymer molecule. Therefore the polymer molecule must fold back and forth on itself many times to form the crystal. Internal mers on the chain form the lattice points of the crystal. Linear symmetric polymer molecules are the easiest to crystallize. An example is unbranched linear polyethylene. The polymer molecule can extend outside a crystal into an amorphous area and even into another crystalline area. In a metal, the grain boundaries separate crystals. But amorphous regions separate the crystals in polymers. Other crystalline structures are also possible, such as the sphere-shaped nylon spherulitic crystals in which ribbons of folded chain molecules form rays surrounded by amorphous regions in the sphere.

Plastics are grouped into two categories. Thermosetting plastics typically form cross-links between their molecules during the initial polymerization in manufacturing of the material. If later reheated, they tend to char or flake and disintegrate. They are brittle materials. Examples are rubber, Mylar, and Bakelite, which is used as the base of electrical switches since it is a good insulator and which was one of the first commercial polymers. Thermoplastics soften when heated or reheated and are ductile. The bonds are van der Waals or polarity bonds. Examples are polystyrene, polyethylene, Nylon, Plexiglas, and Teflon.

The coefficients of thermal expansion for polymers are much greater than those of metals or ceramics. The linear polymers have the highest coefficients. For example, at room temperature that for Teflon is 99×10^{-6}/C, that of polyethylene is 180×10^{-6}/C, and the coefficient polystyrene is 63×10^{-6}/C at room temperature. The cross-linked polymers have lower coefficients of thermal expansion, but still greater than those for metals or ceramics. The coefficient for melamine-formaldehyde is 27×10^{-6}/C and that for phenol-formaldehyde is 72×10^{-6}/C. The coefficients of thermal expansion for polymers usually show a large variability with temperature, while those for metals are nearly constant around room temperature.

Most polymers are poor conductors of electricity; they make good insulators. In the 1970s, it was noticed that some polymers, such as polyacetylene, can be doped to make them conductors. Doping means that small amounts of other chemicals, such as iodine, are mixed with the polymer.

3.10.2 Methods of Building Polymers

Polymers are created by one of two processes, addition or condensation. In the addition process, the complete monomers are joined, but in condensation atoms are lost from the monomer. The extra atoms "condense" out and combine themselves to form substances such as water.

The addition process proceeds by breaking double bonds to create an open bond, and therefore only those monomers having double or triple bonds can combine by addition. For example, the double bond between carbon atoms in the monomer ethylene is broken, forming a *mer* and leaving an open bond at each end of the mer. The mers join in a chain by connecting these extra open bonds. As the double bonds are eliminated, energy is released as heat and the structure becomes more stable. Since this process, in principle, can continue indefinitely to form extremely long chains, chemical impurities are added to the solution to stop the process at some point. Energy is required to break a double bond to create a mer. Reforming single bonds to construct the long chain may require more or may emit energy. In the latter case, the process is self-sustaining.

The condensation process is used by plants to form cellulose from glucose, $C_6H_{12}O_6$, a sugar. In a water solution and under the control of an enzyme, a hydrogen atom is broken from one end of the glucose monomer and a hydroxyl group, OH, from the other end. The modified mers are then able to bond. The two groups of released atoms, H and OH, combine to form water, H_2O, the substance which is condensed out of the reaction. In wood, for example, the resulting cellulose molecule contains about 1,800 of these glucose mers. However, the cellulose chains in cotton are nearly 10 times longer and therefore stronger. The study of the properties of wood and textiles, then, is the study of polymers.

Synthetic fibers such as nylon are also formed by condensation. Nylon was invented in the 1930s by P. Schlack and patented in 1941 while Schlack worked at DuPont. These new fibers were thus available as substitutes for silk, after Far Eastern sources were cut off when the United States went to war with Japan in 1941. The shape of a stress-strain curve for nylon is shown in Figure 3.42.

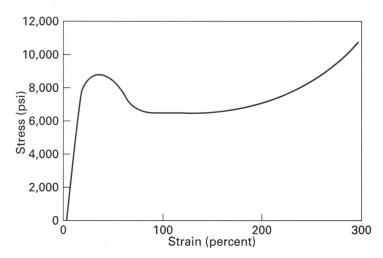

FIGURE 3.42   Typical nylon stress-strain curve.

When monomers of two or more different types are combined by addition, the resulting polymer is called a copolymer. The first successful study of copolymerization was accomplished in the late 1920s by W. O. Herrmann in Germany. For example, vinyl chloride and vinyl acetate can be copolymerized.

3.10.3   Moisture Adsorption

Several polymers, including nylon and cellulosic materials, are hydrophilic. The polymer adsorbs and bonds moisture from its environment. This process differs from taking up moisture into pores in the material, often called absorption, like a sponge. The exact mechanism by which hydrophilic polymers take moisture from the air is not fully understood. But the process depends on the relative humidity of the ambient air. The process is described by sorption isotherms. While not shown in Figure 3.43, the loss of moisture when the ambient relative humidity is reduced does not follow the same path with respect to relative humidity as the in-take of moisture does. The desorption curve typically lies above the adsorption curve.

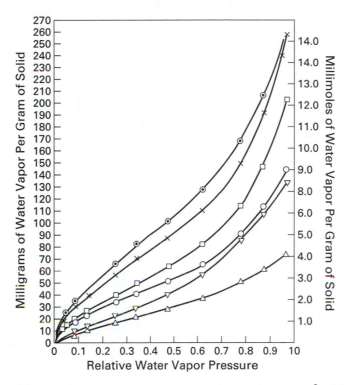

FIGURE 3.43   Adsorption isotherms of textile fibers at 25°C. (⊙) Wool. (×) Viscose. (□) Silk. (○) Cotton. (▽) Acetate. (△) Nylon. [27]. This material is used by permission of John Wiley and Sons, Inc.

The adsorption of moisture significantly changes the material properties of a hydrophilic material. Higher moisture content usually increases the magnitude of strain associated with a given stress, as shown for nylon in Figure 3.44.

3.10.4   Modeling Polymer Behavior

The time-dependent response to loads exhibited by polymers is called viscoelasticity. The behavior depends on the past history of the loading on the material, in contrast to a metal whose behavior depends primarily on the current load. The beginnings of viscoelastic modeling are described by Markovitz [25]. To capture the history dependence, Ludwig Boltzmann in 1874 used a superposition principle to produce an integral model for shear stress, which in the uniaxial case is

$$\tau(t) = \int_0^t G(t-s)\frac{d\epsilon(s)}{ds}ds = G(0)\epsilon(t) + \int_0^t \epsilon(t-s)\frac{G(s)}{ds}ds, \qquad (3.55)$$

where τ and ϵ are the shear stress and small strain, respectively. The second equality is by integration by parts with $\epsilon(0) = 0$. t is the current time, and s is past time. The function, $G(s)$, is material dependent and is to be determined by experiment. The models are called linear since they are equivalently expressed by linear differential equations.

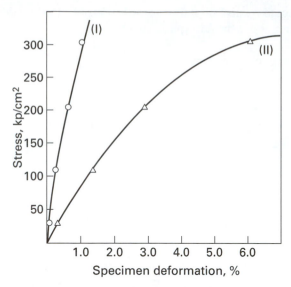

FIGURE 3.44 Stress-strain curve of a nylon 6 specimen dried at 100°C for 24 hours (I) compared to one moistened in boiling water for 5 hours (II) [6]. This material is used by permission of John Wiley and Sons, Inc.

Prior to this in about 1867, James Clark Maxwell proposed a differential equation for materials whose response depends on their past history. The equation was originally intended for a study of viscosity in gas dynamics. The spring and damper mechanical analogy was introduced by Thomson and Poynting in 1903 and led to spring and dash-pot based differential equation generalizations of the models proposed by Maxwell and by Voigt. A spring is used to account for elastic behavior. A dashpot accounts for viscosity effects. The Maxwell model has been physically interpreted as a spring and dashpot in series (Fig. 3.45).

FIGURE 3.45 Maxwell spring and dashpot in series model.

The total strain is the sum of the elastic strain, ϵ_e in the spring and the viscous strain, ϵ_v in the dashpot, while the stress, σ, is the same in both the spring and the dashpot. Since $\epsilon_e = \sigma/E$, the time derivative is $\dot{\epsilon}_e = \dot{\sigma}/E$. On the other hand, $\dot{\epsilon}_v = \sigma/c$. Therefore

$$\dot{\epsilon} = \dot{\epsilon}_e + \dot{\epsilon}_v = \frac{\dot{\sigma}}{E} + \frac{\sigma}{c}. \tag{3.56}$$

The Maxwell model is most successful for viscoelastic fluids as will be shown in Chapter 12.

Exercise 3.17 Maxwell's equation can be written in Boltzmann's integral form. Use the integrating factor $\exp(E\tau/c)$ on Maxwell's equation and assume $\sigma(0) = 0$ to show that

$$\sigma(t) = \int_0^t E \exp\left[\frac{E(\tau - t)}{c}\right]\dot{\epsilon}(\tau)d\tau, \tag{3.57}$$

where the dummy variable, τ for $0 \leq \tau \leq t$, represents past time before the current time, t.

Using integration by parts, the stress response can be written as

$$\sigma(t) = E\epsilon(t) - \frac{E^2}{c} \int_0^t \exp\left[\frac{E(\tau - t)}{c}\right] \epsilon(t)d\tau.$$

Therefore, the differential equation transforms to the Boltzmann integral forms of the constitutive models, as is always possible for linear viscoelasticity.

If the strain is held constant, the resulting process is called stress relaxation. The differential equation (3.56) becomes $\dot{\sigma}/E + \sigma/c = 0$. The solution is

$$\sigma(t) = \sigma_o \exp\left(\frac{E}{c}t\right), \tag{3.58}$$

where σ_o is the initial stress.

The Kelvin-Voigt model, proposed around 1890, represents solids much better than does the Maxwell model. The differential equation form of the Kelvin-Voigt model is

$$\dot{\epsilon} = \frac{\sigma}{c} - \frac{E\epsilon}{c}. \tag{3.59}$$

This model can be interpreted as a spring and dashpot in parallel (Fig. 3.46). The total force in the system is the sum of the forces in the spring and in the dashpot. Dividing by the area gives the Kelvin-Voigt model of equation (3.59), $\sigma = E\epsilon + c\dot{\epsilon}$.

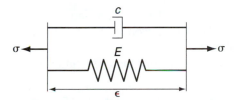

FIGURE 3.46   Kelvin-Voigt spring and dashpot in parallel model.

The stress and strain are, in general, both functions of time. In the case that the stress is held constant, creep, the strain response is determined by integrating equation (3.59) using the integrating factor, $\exp\left(\frac{E}{c}t\right)$.

$$\epsilon(t) = \epsilon(0)\exp\left(-\frac{E}{c}t\right) + \frac{\sigma}{E}\left[1 - \exp\left(-\frac{E}{c}t\right)\right].$$

Exercise 3.18   The Kelvin-Voigt equation can also be written in the integral form of Boltzmann. Use an integrating factor and assume $\epsilon(0) = 0$ to show that

$$\epsilon(t) = \int_0^t \frac{1}{c}\exp[-E(t - \tau)/c]\sigma(\tau)d\tau. \tag{3.60}$$

For many years, researchers attempted to use more and more complex combinations of spring and dashpot models to find a good history dependent material model. One of the simpler is the Zener standard linear solid. This standard linear solid viscoelastic material combines a spring with constant, E_1, in series with a Kelvin-Voigt model having constants, E_2 and c. Its differential evolution equation in terms of the stress and strain is

$$\dot{\sigma} + \frac{E_1 + E_2}{c}\sigma = E_1\dot{\epsilon} + \frac{E_1 E_2}{c}\epsilon. \tag{3.61}$$

Exercise 3.19   Verify that the standard linear solid mechanical model is represented by the differential equation (3.61). *Hint*: Let ϵ_1 be the strain in the spring, E_1, and ϵ_2 be the strain in the Kelvin-Voigt model. Then $\epsilon = \epsilon_1 + \epsilon_2$. Write the stress-strain relation for the spring E_1 and for the Kelvin-Voigt model. Eliminate ϵ_1 and ϵ_2 from these three equations.

Exercise 3.20   Write the differential equation for the standard linear solid in integral form in the case that a) the stress is constant, b) the strain is constant.

The relaxation time in the Maxwell and Kelvin-Voigt linear models is defined as $t_R = c/E$. The viscosity, c, depends on the temperature. Therefore, t_R is sometimes assumed to obey an Arrhenius relation, $t_R = t_o \exp(Q/RT)$, where Q is the activation energy for viscous flow. If Q is measured in cal/mol, then $R = 1.987$ cal/mol.K. Other researchers have defined models, especially multiple spring and dashpot combinations in series, with several relaxation times. This is done when it appears that different physical phenomena each cause relaxation. Today, the spring and dashpot model approach has been abandoned as a dead end by many workers on polymers. However, the simpler models are sometimes useful approximations to the real response of solids.

Eyring developed a nonlinear dashpot to model nonlinear polymer response. The viscosity of the dashpot is not constant but varies with the strain rate. The strain rate of this dashpot is

$$\dot{\epsilon}_v = K \sinh\left(\frac{V\sigma}{2kT}\right),$$

where K is a constant, k is the Boltzmann constant, T is the absolute temperature, and V is the Eyring volume. The Eyring volume is the volume of a polymer segment which must move to allow flow to occur [18, p. 38]. A linear spring having constant, E, in series with an Eyring dashpot satisfies the differential equation,

$$\dot{\epsilon} = \dot{\epsilon}_e + \dot{\epsilon}_v = \frac{\dot{\sigma}}{E} + K \sinh\left(\frac{V\sigma}{2kT}\right);$$

or

$$\dot{\sigma} + EK \sinh\left(\frac{V\sigma}{2kT}\right) = E\dot{\epsilon}. \tag{3.62}$$

For example, Eyring and Halsey [12] represented the tensile behavior of a cotton fiber by a three-element model.

Nonlinear viscoelasticity is a difficult subject that will be mentioned only briefly here. A class of models is a series expansion generalizing linear viscoelasticity. One form is the isothermal Green-Rivlin polynomial integral model,

$$\epsilon(t) = \int_0^t G_1(t - \tau_1)\dot{\sigma}(\tau_1)d\tau_1$$

$$+ \int_0^t \int_0^t \int_0^t G_3(t - \tau_1, t - \tau_2, t - \tau_3)\dot{\sigma}(\tau_1)\dot{\sigma}(\tau_2)\dot{\sigma}(\tau_3)d\tau_1 d\tau_2 d\tau_3 + \dots. \tag{3.63}$$

Again, the τ_i represent past time and t current time. These models require experimentally fitting the large number of time-dependent empirical functions, G_i, and so are seldom useful in practice.

Coleman [10] specialized the theory of simple materials to those with fading memory. While in principle the evolution of a particular state variable may depend on the past history of all variables, a simple material is one in which the stress depends only on the past history of the strain measure. A material is said to have fading memory if its current behavior is most strongly influenced by its behavior in the recent past.

Models including thermal effects have proved even more difficult to generate. Christensen mentions that a "coupled thermoviscoelastic theory which includes the temperature dependence of mechanical properties is necessarily nonlinear" [8, p. 99]. A possible non-isothermal model is built from the isothermal constitutive model for stress in terms of strain. The constitutive equation defining a linear thermoviscoelastic material satisfies a shift relationship in which the stress is

$$\sigma(t) = \int_{-\infty}^{t} G\big(\tau - \alpha(\theta)\big)\dot{\epsilon}(\tau)d\tau, \tag{3.64}$$

where the shift function $\alpha(\theta)$ is an increasing function of temperature θ, and τ is past time.

The difficulty in obtaining from experiment the unknown functions in these models has led to their having little influence on design to date.

3.10.5   Dynamic Loading of Polymers

Stress-strain tests for metals or ceramics are frequently performed at a small enough rate that they can be thought of as quasi-static. Polymer behavior is dependent on the rate at which the loads are applied. Viscoelastic behavior is observed when a polymer is dynamically, rather than statically, loaded. A simple periodic dynamic load illustrates the difference between elastic and viscoelastic response. The load is controlled so that the stress varies sinusoidally, a simple harmonic function. Strain, the response, lags behind the stress. This characteristic behavior of viscoelasticity is in contrast to elastic loading in which the stress and strain respond instantly to one another.

Moduli may be defined by taking the ratio of the stress and strain.

Definition   Assume that the stress and strain are out of phase by a phase shift δ. $\sigma(t) = \sigma_o \cos(\omega t)$ and $\epsilon(t) = \epsilon_o \cos(\omega t - \delta)$. In complex variables functional form, $\sigma(t) = \sigma_o \exp(i\omega t)$ and $\epsilon(t) = \epsilon_o \exp[i(\omega t - \delta)]$. The ratio is $\sigma/\epsilon = (\sigma_o/\epsilon_o)\exp(i\delta) = (\sigma_o/\epsilon_o)(\cos\delta + i\sin\delta)$. The dynamic elastic modulus, or storage modulus, is

$$J' = \frac{\sigma_o}{\epsilon_o}\cos(\delta),$$

and the dynamic loss modulus is

$$J'' = \frac{\sigma_o}{\epsilon_o}\sin(\delta),$$

and the tangent modulus is $\tan(\delta)$.

A uniaxial linear Kelvin-Voigt viscoelastic model has strain evolution in response to the sinusoidal forcing function $\sigma(t) = A\sin(\omega t)$,

$$\frac{d\epsilon}{dt} = \frac{1}{c}\left(A\sin(\omega t) - E\epsilon\right). \tag{3.65}$$

This first-order differential equation is solved by the integrating factor technique to obtain the response, where $t_r = c/E$,

$$\epsilon(t) = \frac{A}{E(1 + t_r^2 \omega^2)} \sin(\omega t) - \frac{A t_r \omega}{E(1 + t_r^2 \omega^2)} \cos(\omega t). \tag{3.66}$$

The rate dependence of the response is due to the terms involving ω.

Exercise 3.21   Verify the result of equation (3.66).

The first summand of equation (3.66) is in phase with, and proportional to, the applied stress. The constant of proportionality is the storage modulus, J'.

$$J' = \frac{1}{E(1 + t_r^2 \omega^2)}.$$

The out-of-phase term has coefficient, J'', the loss modulus.

$$J'' = \frac{t_r \omega}{E(1 + t_r^2 \omega^2)}.$$

These moduli differ from those in elasticity in that they depend on the frequency, ω, of the applied stress.

Exercise 3.22   Plot J' and J'' versus ω on the same set of axes.

The ratio $J''/J' = \tan \delta$ is called the loss tangent because δ is the phase shift in the simplification of equation (3.66) as

$$\epsilon(t) = A \sin(\omega t - \delta).$$

Here $\tan \delta = t_r \omega = c\omega/E$. This analysis shows that the strain lags the applied stress for linear viscoelastic materials. The stress and strain would be in phase for an elastic material. But it may be hard to think of this phase lag as dissipation because the amount of lag never changes after an initial transient period; this is a steady-state response.

A remarkable relationship exists between linear viscoelasticity and linear elasticity. The Laplace transform of the equations of linear viscoelasticity have the form of the equations of linear elasticity [8]. This correspondence principle means that linear elasticity techniques can be used to solve linear viscoelastic problems. First, apply the Laplace operator to transform the viscoelastic problem to an elastic problem. Solve the time-independent elastic problem and then transform the result by the inverse Laplace operator to produce a time-dependent solution for the viscoelastic problem. This relationship seems to have been first described in print by Alfrey [2] in 1944. An example calculating the time-dependent stress distribution due to a constant point load normal to the surface of a semi-infinite body is given by Lee [19] for a material described by the Voigt model.

3.11   WOOD

Wood is a bundle of nearly parallel hollow fibers, used to transport water and food. The walls of the fibers (Fig. 3.47) are composed of layers of helically wound string-like microfibrils made of cellulose. About 60% of the fiber is cellulose. Other polymers, called hemicellulose, are also present. The polymer lignin between the fibers holds the structure together. Cellulose is hydrophilic; moisture bonds to it and affects the structure of the fiber walls. Additional moisture can be trapped in the hollow fibers or in wall voids by capillary action. Therefore structures made of wood or wood products are sensitive to the ambient relative humidity and moisture content of the material. The fibers lie in the direction of tree trunk. Wood is an anisotropic material, which is frequently modeled as orthotropic.

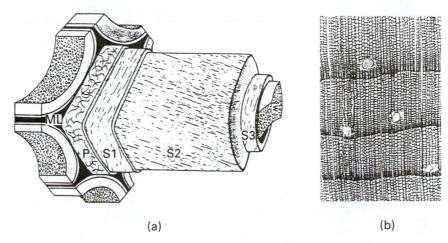

(a)	(b)

FIGURE 3.47   (a) The structure of a wood fiber and (b) a cross section of wood transverse to the fibers [32]. The dark lines in (b) are grains. Reprinted by permission from Angus Wilde Pub.

The principal directions are assumed to be the longitudinal direction parallel to the fibers, the radial direction of the trunk, and the transverse or circumferential direction of the trunk (Fig. 3.48). Wood is the most highly orthotropic material in engineering use since for small strains, the ratio of the longitudinal to the radial elastic modulus is on the order of 24.

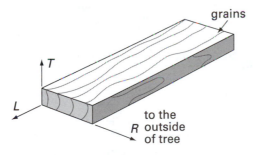

FIGURE 3.48   Orthogonal directions on a tree trunk.

The orthotropic model is not satisfactory for all types of wood. It can be tested by examining whether or not the equalities required by the orthotropic model are satisfied,

$$\frac{\nu_{LR}}{E_L} = \frac{\nu_{RL}}{E_R}; \qquad \frac{\nu_{LT}}{E_L} = \frac{\nu_{TL}}{E_T}; \qquad \frac{\nu_{RT}}{E_R} = \frac{\nu_{TR}}{E_T}.$$

Bodig and Jayne [5, p. 113] give the values in Table 3.2.

The orthotropic model [equation (3.8)] is not successful for the hardwood quaking aspen or the softwood Douglas fir. It is nearly satisfactory for oak, for beech, and for birch, all hardwoods. Wood specimens are not uniform. Some portions are clear

TABLE 3.2 Orthotropic Parameters for Selected Woods

Wood	$\dfrac{v_{LR}}{E_L}$	$\dfrac{v_{RL}}{E_R}$	$\dfrac{v_{LT}}{E_L}$	$\dfrac{v_{TL}}{E_T}$	$\dfrac{v_{RT}}{E_R}$	$\dfrac{v_{TR}}{E_T}$
Douglas fir	0.179	0.152	0.159	0.186	3.034	3.378
Oak	0.427	0.421	0.648	0.614	2.137	2.068
Quaking aspen	0.386	0.513	0.288	0.581	10.433	12.800
Beech	0.228	0.228	0.255	0.269	2.206	2.275
Birch	0.207	0.214	0.179	0.200	4.206	4.826

wood and other portions have knots, cracks, or other discontinuities. Cracks perpendicular to the grain are called checks if they occur in drying; shakes are such cracks that occur during the normal growing process. Further, latewood often has somewhat different properties than earlywood, which developed early in the growing season. This variation creates the annual rings in the trunk cross section found in trees grown in temperate regions (Fig. 3.47).

While the initial stiffness of wood, on the order of 1.2×10^6 psi with the grain, is much less than that of metals, the stiffness to specific weight ratio of wood is nearly the same as steel, aluminum, and magnesium. The stiffness to specific weight ratios are steel 0.106×10^9, aluminum 0.102×10^9, magnesium 0.098×10^9, and wood 0.100×10^9 inches. The values for copper, brass, and bronze are about half this. Wood may be the material of choice, therefore, for many types of structures.

EXAMPLE 3.14 A compressive axial load of 1,200 pounds is applied on an American beech board cut so its face lies in the radial-transverse plane. Estimate the change in length of the edges of the board assuming the stress tensor is the same at each point of the board. The dimensions are 1 inch in the radial, 2 inches in transverse, and 40 inches in the longitudinal, grain, directions. Assume a linear orthotropic model in which $E_R = 0.1901 \times 10^6$psi, $E_L = 1.895 \times 10^6$psi, $E_T = 0.0983 \times 10^6$psi, $v_{LR} = 0.37$, $v_{LT} = 0.42$, and $v_{RT} = 0.47$.

Solution $\sigma_L = -600$ psi, and $\sigma_R = \sigma_T = 0$. By equations (3.8),

$$\epsilon_L = \frac{1}{E_L}\sigma_L - \frac{v_{LR}}{E_R}\sigma_R - \frac{v_{LT}}{E_T}\sigma_T = \frac{-600}{1.895 \times 10^6} = -3.166 \times 10^{-4};$$

$$\epsilon_R = -\frac{v_{LR}}{E_L}\sigma_L + \frac{1}{E_R}\sigma_R - \frac{v_{RT}}{E_T}\sigma_T = -0.37\frac{-600}{1.895 \times 10^6} = 1.171 \times 10^{-4};$$

$$\epsilon_T = -\frac{v_{LT}}{E_L}\sigma_L - \frac{v_{RT}}{E_R}\sigma_R + \frac{1}{E_T}\sigma_T = -0.42\frac{-600}{1.895 \times 10^6} = 1.330 \times 10^{-4}.$$

If end effects are ignored, the change in length in longitudinal direction is $(-3.166 \times 10^{-4})40 = -0.0127$ inches; in the radial direction is $(1.171 \times 10^{-4})1 = 1.171 \times 10^{-4}$ inches; and in the transverse direction is $(1.330 \times 10^{-4})2 = 2.660 \times 10^{-4}$ inches. ∎

A simple nonlinear model for wood is that proposed by O'Halloran [29] in 1973,

$$\sigma = E\epsilon - A\epsilon^n. \tag{3.67}$$

The model fits the compressive uniaxial behavior of redwood in the grain direction for $n = 3$ with $E = 8.72$ GPa (1,264 ksi) and $A = 2590$ GPa (375,635 ksi) and that of Engelmann spruce for $n = 3.444495$, $E = 775, 580$ psi, and $A = 39, 304 \times 10^6$ psi.

Exercise 3.23 Plot the compressive cubic stress-strain curve for redwood. Show that the curve reaches a maximum for a strain of $\epsilon = \sqrt{E/3A} = 0.0335$ so that the model is not valid for larger strains.

3.12   PAPER

Paper is a network of polymeric composite fibers, with most of the fibers lying in the plane of the sheet (Fig. 3.49). Many different types of plant fibers have been used to make paper, but the most common today is wood. Each wood pulp fiber is itself a composite of an amorphous hemicellulose and lignin matrix surrounding largely crystalline cellulose reinforcing fibrils. The fibril crystals are of finite length when wet since they are periodically disrupted by amorphous cellulose regions along their length.

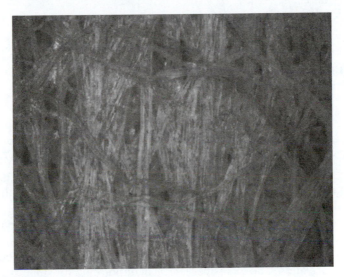

FIGURE 3.49   Wood pulp fibers at 100X magnification in corrugating medium, which is the paper used to make the flutes in corrugated boxes.

Paper is manufactured from a slurry, composed of water and 0.5–1.0% pulp fiber, distributed on a moving wire mesh. The free water drains through the mesh, and the resulting sheet is dried. A weak interfiber bond is created when two fibers approach closely enough that surface tensions can interact. As moisture is removed, the bond strength increases. Hydrogen bonding is believed to play a role in the formation of these interfiber bonds which hold the paper sheet together. Because the mesh is moving on a paper machine, the fibers tend to line up in the direction of motion, called the machine direction (MD). The direction orthogonal to the machine direction is called the cross-machine direction (CD) of the sheet. Machine-made paper therefore tends to be orthotropic, while paper made by hand, called hand sheets, is more nearly isotropic.

Paper may be the most complicated engineering material in common use. Some simple, and not entirely successful, constitutive models have been proposed for this extremely complex material. A typical stress-strain curve is shown in Figure 3.50. It is not linearly elastic, but some have suggested that, for small strains, paper can be taken to be a linear, orthotropic elastic material.

For uniaxial loading, the classical empirical hyperbolic tangent nonlinear model, first proposed in the 1950s, is

$$\sigma = c_1 \tanh\left(\frac{c_2 \epsilon}{c_1}\right) + c_3 \epsilon, \tag{3.68}$$

where c_1, c_2, and c_3 are positive constants dependent on the type of material, moisture content, and the angle of the load with respect to the machine direction. The stress for paper is given in units of N/m, ignoring the thickness, which is hard to measure.

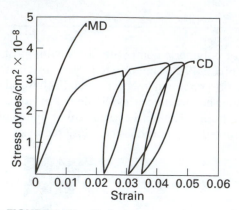

FIGURE 3.50　Typical stress-strain curves for paper [31].
Reprinted by permission from TAPPI.

Compression tests of a block of the type of paperboard used to make tubes, on which paper or film is wound, have given $c_1 = 24.96$ MPa, $c_2 = 2.93$ MPa, and $c_3 = -0.765$ MPa in the machine direction and $c_1 = 11.65$ MPa, $c_2 = 1.91$ MPa, and $c_3 = 1.648$ MPa in the cross-machine direction. The slope of the stress-strain curve at $\epsilon = 0$ may be taken as an elastic modulus, E, for this nonlinear elastic material. A calculation of $d\sigma/d\epsilon$ shows that $E = c_1c_2 + c_3$. For this paper, $E_1 = 7,240$ MPa in the machine direction and $E_2 = 2,386$ MPa in the cross-machine direction. The Poisson ratios were measured for this paper as $\nu_{12} = 0.387$ and $\nu_{21} = 0.120$. The relation $E_1\nu_{21} = E_2\nu_{12}$ which would be satisfied if this paper were linear orthotropic is within 6% of validity. So this paper is nearly orthotropic for small strains.

This model ignores the rate dependence of the stress-strain curve. An empirical modification of the hyperbolic tangent model to account for the loading rate was proposed in 1988 by Gunderson and coworkers [15]. For a constant load rate, r, they proposed that

$$\sigma = c_3 r^\beta \tanh\left(\frac{c_4\epsilon}{c_3}\right), \tag{3.69}$$

where c_3, c_4, and β all depend on the constant ambient relative humidity. This isothermal model depends only on the current rate of loading and the current relative humidity so that it is independent of the history of the loading or ambient conditions. Typical values for softwood kraft linerboard (material A) and corrugating medium (material B), which capture the fact that the strains are larger at slower rates and higher relative humidities, are given in Table 3.3 [15].

TABLE 3.3　Model Parameters for Two Types of Paper

Paper	Direction	RH %	β	c_3	c_4
A	MD	50	0.0207	7.17	2,268.
		90	0.0560	4.18	1,640.
	CD	50	0.0265	4.29	1,025.
		90	0.452	2.61	633.
B	MD	50	0.0337	4.32	1,211.
		90	0.0624	1.78	700.
	CD	50	0.0434	2.16	388.
		90	0.0475	1.18	240.

Exercise 3.24   Verify that this model predicts larger strains in the cross-machine direction of the paper than in the machine direction for a given load and load rate. Also verify that the strains in the higher relative humidity environment are predicted to be greater than in the lower relative humidity environment.

PROBLEMS

3.1   A 0.505-inch diameter tensile specimen with a 2-inch gage length begins to neck at 9,960 psi. The measured elongation at the initiation of necking is 0.22 inches. Assume that the material obeys the true stress-strain constitutive model $\sigma = K\epsilon^n$.

(a)   Determine the constants, K and n.

(b)   Determine $d\sigma/d\epsilon$ at this point.

3.2   Determine the stress and strain at which necking is initiated on the steel curve in Figure 3.1. From this information, compute the hardening coefficient, n, and the material constant K.

3.3   In the relation $\sigma = K\epsilon^n$, a copper has $n = 0.54$ and $K = 46,000$ psi. An annealed steel alloy has $n = 0.15$ and $K = 93,000$ psi. Determine the ratio of $d\sigma/d\epsilon$ for these two metals.

3.4   A tensile test of a steel specimen produces the following data. The gage length is 2 inches and the diameter of the specimen is 0.505 inches. The specimen yield load is 7,200 pounds. The necking begins at 10,400 pounds.

TABLE 3.4   Load-Elongation Data

Load (lb)	Elongation (in.)	Load (lb)	Elongation (in.)
500	0.000167	6,000	0.00199
1,000	0.000331	6,500	0.00215
1,500	0.000496	7,000	0.00232
2,000	0.000665	7,500	0.0248
2,500	0.000830	8,000	0.0412
3,000	0.000993	8,500	0.0725
3,500	0.00116	9,000	0.0950
4,000	0.00133	9,500	0.150
4,500	0.00149	10,000	0.280
5,000	0.00166	10,200	0.350
5,500	0.00183	10,400	0.600

From the data in Table 3.4, determine the ultimate true strain and true stress. Ans. $\sigma_U = 67,600$ psi.

3.5   From the data in Table 3.4 at the ultimate load, determine the value of the constant, K, in the true stress–true strain relation for the plastic region, $\sigma = K\epsilon^n$. Ans. $K = 96,079$ psi.

3.6   From the data in Table 3.4, determine $d\sigma/d\epsilon$ at the maximum load.

3.7   Based on the data in Table 3.4, what diameter specimen made of the same material would yield at 2,500 pounds?

3.8   Based on the data in Table 3.4, determine the maximum shear stress at the yield load.

3.9   Based on the data in Table 3.4, compute the volume change per unit volume at the yield load. Assume that Poisson's ratio is 0.28.

3.10   The displacement in a body is described in x-y-z coordinates by

$$u(x, y, z) = xy^2z10^{-3};$$
$$v(x, y, z) = 2xyz10^{-3};$$
$$w(x, y, z) = x^2z10^{-3}.$$

(a)   Determine the strain tensor in matrix form at each point (x, y, z).

(b)   Compute the change in volume per unit volume at point $P(1, 2, 3)$.

(c)   The material is Hookean, with Young's modulus 40 GPa and Poisson's ratio 0.25. Determine the normal stress in the z-direction at $P(1, 2, 3)$.

3.11   The stress tensor, in ksi, at a point P of a body is, in x-y-z coordinates,

$$\begin{pmatrix} 2 & -1 & 1 \\ -1 & 0 & 1 \\ 1 & 1 & 2 \end{pmatrix}.$$

(a)   Draw a cube showing the stress components on each coordinate face.

(b)   From the matrix compute the principal stresses. Do not use Mohr's circle. (*Hint*: one of the principal stresses is -1 ksi.)

(c)   Determine the stress on the face defined by $2\mathbf{i} + \mathbf{j} - 2\mathbf{k}$ in x-y-z components.

(d)   Determine the strain tensor at the point P. The material is linearly elastic, and $E = 10 \times 10^6$ psi and $\nu = 0.25$.

(e)   Compute the change in volume per unit volume at point P.

3.12   A thin plate is in plane stress. All body forces are neglected. The stress tensor at a point (x, y, z) in the plate is, in ksi,

$$\begin{pmatrix} 3y^2 & 6 & 0 \\ 6 & 3x^3 & 0 \\ 0 & 0 & 0 \end{pmatrix},$$

where $-2 \leq x \leq 2$ and $-2 \leq y \leq 2$.

(a) Determine by a computation whether or not the body is in equilibrium.

(b) The material is Hookean, with Young's modulus E and Poisson's ratio ν. Determine the normal and shear strains at point (x, y, z).

(c) Determine by a calculation whether or not the plate is incompressibly loaded at the point $(-1, 1, z)$.

3.13 A point on the surface of a body is in plane stress (MPa):

$$\begin{pmatrix} \sigma_x & \tau_{xy} \\ \tau_{xy} & \sigma_y \end{pmatrix} = \begin{pmatrix} 270 & 320 \\ 320 & -210 \end{pmatrix}.$$

(a) Determine the shear stress, $\tau_{x'y'}$, in a coordinate system rotated $30°$ counterclockwise.

(b) The material properties are $E = 21 \times 10^4$ MPa, $G = 7.6 \times 10^4$ MPa, and $\nu = 0.33$; compute ϵ_z and ϵ_{xy}.
Ans. (a) -47.84 MPa; (b) 94.3μ, 2105μ.

3.14 The principal strains at a point are $\epsilon_x = 2{,}000\mu$, $\epsilon_y = 3{,}000\mu$, and $\epsilon_z = -1{,}000\mu$. Determine the change in volume per unit volume of a small cube of material with edges parallel to the principal axes. Compare the value of the change in volume to the trace of the strain tensor.

3.15 The principal strains at a point are $\epsilon_x = 2{,}000\mu$, $\epsilon_y = -3{,}000\mu$, and $\epsilon_z = 1{,}000\mu$. Determine the change in volume per unit volume of a small cube of material with edges parallel to the principal axes. Compare the value of the change in volume to the trace of the strain tensor.

3.16 Explain why the elastic and bulk moduli for aluminum are approximately equal.

3.17 Experiments on a metal measured the bulk modulus to be 151.5 GPa and the shear modulus to be 78 GPa. Determine the elastic modulus and the Poisson ratio.

3.18 Silicon carbide (SiC) has bulk modulus $K = 191.7$ GPa and elastic modulus $E = 414$ GPa. Calculate its Poisson ratio and its Lamé constant, λ. Ans. $\nu = 0.14$, $\lambda = 70.6$ GPa.

3.19 Estimate the elastic modulus of a ceramic with 10% porosity if its elastic modulus at 0% porosity is 350 MPa.

3.20 A 5-cm-long specimen of alumina is loaded in a three-point bending test. The rectangular cross-sectional area is 0.6 cm wide and 0.2 cm high.

(a) At a load of 10N, the measured deflection is 0.01808 mm. Determine the elastic modulus.

(b) If the load is increased to 100N, the specimen fractures. Determine the modulus of rupture.

3.21 Plot the relationship between the porosity and E/E_o given by (3.20) for porosities between 0 and 50%.

3.22 Determine the relationship, analogous to (3.20), between the elastic modulus, E, and porosity for a ceramic with Poisson ratio 0.2.

3.23 The viscosity of a glass is 10^6 poise at 1000°C and 10^8 poise at 800°C.

(a) Compute the activation energy for viscous flow.

(b) Compute the temperature required so that the glass has a viscosity of 10^7 poise.

3.24 A continuous and aligned fiber-reinforced composite rod consists of 30 vol% aramid fibers ($E = 19 \times 10^6$ psi, $\sigma_U = 500{,}000$ psi) and 70 vol% polycarbonate matrix ($E = 3.5 \times 10^6$ psi, $\sigma_U = 9{,}000$ psi). The specimen has a cross-sectional area of 0.5 in^2 and is subjected to a longitudinal tensile load of 10,000 pounds parallel to the fibers.

(a) Calculate the fiber-matrix load ratio, F_f/F_m.

(b) Calculate the actual loads carried by the fiber and the matrix phases, respectively.

(c) Compute the magnitude of the stress on each of the fiber and matrix phases.

(d) Compute the strain experienced by each phase.

3.25 Determine the upper and lower possible values of the elastic modulus of a composite made of 37 vol% glass fibers ($E = 10.5 \times 10^6$ psi), 10 vol% aluminum alloy ($E = 9 \times 10^6$ psi), and the remainder nylon ($E = 0.4 \times 10^6$ psi).

3.26 The elastic modulus of S-glass is 12.6×10^6 psi while that of nylon is 0.4×10^6 psi. Fibers of S-glass are embedded longitudinally in a nylon beam. The volume fraction of fibers is 20%. Use strength of materials beam theory to estimate the beam deflection at the point just under a load of 30 pounds applied transversely at the center of the simply supported beam. The beam has a square cross section with 1-inch sides and length 6 inches. Assume the beam remains elastically loaded.

3.27 A continuous and aligned fiber-reinforced composite bar consists of 40 vol% E-glass fibers, having an elastic modulus of 72.5 GPa and 60 vol% polycarbonate matrix with elastic modulus 24 GPa. The specimen has a cross-sectional area of 4 cm^2 and is subjected to a longitudinal tensile load of 5000 N parallel to the fibers.

(a) Compute the strain experienced by each phase.

(b) Calculate the fiber-matrix load ratio, F_f/F_m.

(c) Calculate the actual loads carried by the fiber and the matrix phases, respectively.

(d) Compute the magnitude of the stress on each of the fiber and matrix phases.

3.28 A nonlinear elastic material has uniaxial constitutive equation $\epsilon = a\sigma^n$, where n is a constant. Show that the constant n is the ratio of the strain energy density, U_o, and the complementary energy density, Ψ_o, at any state of stress and strain. Notice that this model includes the linear elastic case, for which $a = 1/E$ and $n = 1$.

3.29 A machine component made of steel ($E = 30 \times 10^6$ psi, $\nu = 0.25$) has stress at a given point of

$$T = \begin{pmatrix} -1 & 2 & -3 \\ 2 & 7 & 4 \\ -3 & 4 & 5 \end{pmatrix} \text{ ksi.}$$

Compute the strain energy density, the volumetric strain energy density, and the distortion strain energy density at that point.

3.30 Compute the dilatational and deviatoric stress tensors associated with the stress, in ksi,

(a) $\begin{pmatrix} 2 & -1 & 1 \\ -1 & 0 & 1 \\ 1 & 1 & 2 \end{pmatrix}$,

(b) $\begin{pmatrix} 2 & 0 & 4 \\ 0 & 0 & 0 \\ 4 & 0 & -3 \end{pmatrix}$,

(c) $\begin{pmatrix} 3 & -10 & 0 \\ -10 & 0 & 30 \\ 0 & 30 & -27 \end{pmatrix}$.

3.31 (a) Determine the principal stresses at a point in the body where the stress state (in MPa) is

$$\begin{pmatrix} 600 & 0 & 0 \\ 0 & 160 & 320 \\ 0 & 320 & -320 \end{pmatrix}.$$

(b) Determine the corresponding principal strains for a material with $E = 30 \times 10^6$ psi and $\nu = 0.27$.

(c) Determine the corresponding deviatoric stresses in principal coordinates. Ans. (a) 600, 320, −480 MPa. (c) 1360/3, 520/3, −1880/3

3.32 The state of stress at a point on the surface of a rod is two dimensional with $\sigma_x \neq 0$, $\sigma_y = 0$, and $\tau_{xy} \neq 0$. Compute the distortion energy directly in terms of σ_x and τ_{xy}. Next use Mohr's circle to compute the principal stresses in terms of σ_x and τ_{xy}. Then recompute the distortion energy directly from the principal stresses. Verify that both computations give the same result. Ans. $(\sigma_x^2 + 3\tau_{xy}^2)/6G$.

3.33 The stress tensor, in MPa, at a point P of a body is, in x-y-z coordinates,

$$\begin{pmatrix} 1 & 3 & 0 \\ 3 & -2 & -1 \\ 0 & -1 & 1 \end{pmatrix}.$$

The material is linearly elastic. The elastic modulus is 206 GPa and the shear modulus is 70 GPa. The Poisson ratio is 0.26.

(a) Draw a cube showing the stress components on each coordinate face.

(b) From the stress tensor, compute the principal stresses. Do not use Mohr's circle. *Hint*: One of the principal stresses is 1 MPa.

(c) Determine the stress, in x-y-z components, on the face perpendicular to $-2\mathbf{i} + 5\mathbf{j} - 3\mathbf{k}$.

(d) Compute the distortion energy at this point.

(e) Compute the normal strain in the y-direction at this point.

3.34 The uniaxial compressive stress in a concrete member is assumed to obey the Saenz model. The critical stress is 8000 psi; the critical strain is 0.0033. The weight per cubic inch is 0.087 lb/in^3.

(a) Compute the stress when the strain is 0.00165.

(b) Compare this prediction to that of a uniaxial linear elastic model in which the elastic modulus is E_0, as given in the estimate for the Saenz model.

3.35 Determine the stress if $\lambda = 1.2$ in the uniaxial loading of a neo-Hookean rod with $G = 0.39$ N/mm^2. Conversely, determine the stretch when the stress is 20 kPa. (A numerical technique such as Newton's method may be necessary.)

3.36 Determine the stress if $\lambda = 2.0$ in the uniaxial loading of a Mooney-Rivlin rod with $C_1 = 0.16$ N/mm^2 and $C_2 = 0.05$ N/mm^2. Conversely, determine the stretch when the stress is 20 kPa. (A numerical technique such as Newton's method may be necessary.)

3.37 A uniaxially loaded rod is made of a Kelvin-Voigt material.

(a) Determine how the strain changes with respect to time if the axial stress is held constant.

(b) What is the strain after infinite time? Compare this result to a linear elastic material.

3.38 A 400 psi stress is required to initially hold an material obeying the Maxwell relation at a constant strain of 0.02. After 8 weeks at 20°C, the stress required to hold the 0.02 strain has been reduced to 350 psi by stress relaxation. A second test is conducted at 40°C and after 8 weeks the initial 400 psi stress is found to have dropped to 240 psi as the material is continuously held at a strain of 0.02. Assume the Arrhenius relation for the relaxation time as a function of temperature. Calculate the activation energy in cal/mol for viscous flow in the material.

3.39 Stress relaxation measurements are performed on a viscoelastic material at an elevated temperature. A specimen is suddenly pulled in tension to a strain of 0.5, while the stress necessary to maintain this constant strain is measured as a function of time, t, in seconds. Determine the viscoelastic modulus, $E_r = \sigma(10)/\epsilon$, if the initial stress is 500 psi and the stress drops to 70 psi after 30 seconds. Use the Maxwell model.

3.40 A polymer rod, which obeys the Kelvin-Voigt model, is axially loaded with a constant stress of 250 psi. The relaxation time is known to be 10 seconds. The initial strain is zero and the strain after 10 seconds is 0.158 .

(a) Determine E.

(b) Determine the predicted strain after a very long loading time.

3.41 Determine the strain response of the Kelvin-Voigt model when $\sigma(t) = at$, for a a constant. In the solution, there are three terms; two involve an exponential and one is linear. Plot the solution $\epsilon(t)$ and the linear term on the same

t-ϵ plane, when $E = 30MPa$, $c = 6$ MPa/sec and $a = 10$ MPa/sec. The initial strain is zero. What is the relationship between the two graphs?

3.42 Plot the uniaxial linear Kelvin-Voigt viscoelastic model strain evolution in response to the sinusoidal forcing function $\sigma(t) = A\sin(\omega t)$ on the same set of axes as the stress $\sigma(t)$. Assume that $E = 30MPa$, $c = 6$ MPa/sec, $\omega = 10$ rad/s, and $A = 2$ kPa. Is there a lag between the strain and stress?

3.43 Derive a differential equation describing a material whose response is modeled by a spring in parallel with a Maxwell model.

3.44 Derive the stress response under constant strain for a material whose response is modeled by a spring in parallel with a Maxwell model.

3.45 Compare the strain response when the stress is held constant for a Maxwell model in parallel with a linear spring to that of a Kelvin-Voigt model in series with a linear spring.

3.46 The Burgers model is a Maxwell model in series with a Kelvin-Voigt model.

(a) Determine the second-order differential equation which represents this model.

(b) Determine the strain response when the stress is held constant.

3.47 Use the strain transformation equations to determine the normal strains in the longitudinal-radial plane at a counterclockwise $20°$ angle to the grain in an oak board under principal stresses $\sigma_R = 8$ ksi, $\sigma_L = 15$ ksi, and $\sigma_T = 14$ ksi. Assume a linear orthotropic model in which $E_R = 0.1861 \times 10^6$psi, $E_L = 1.803 \times 10^6$psi, $E_T = 0.0956 \times 10^6$psi, $\nu_{LR} = 0.37$, $\nu_{LT} = 0.50$, and $\nu_{RT} = 0.67$.

3.48 Compute the change in volume in a rectangular block of black oak with initial edge dimensions 1 inch in the radial, 2 inches in the transverse, and 3 inches in the longitudinal (grain) directions under principal stresses $\sigma_R = 10$ ksi, $\sigma_L = 15$ ksi, and $\sigma_T = 20$ ksi. Assume a linear orthotropic model in which $E_R = 0.1861 \times 10^6$ psi, $E_L = 1.803 \times 10^6$ psi, $E_T = 0.0956 \times 10^6$ psi, $\nu_{LR} = 0.37$, $\nu_{LT} = 0.50$, and $\nu_{RT} = 0.67$.

3.49 Graph on the same set of axes the uniaxial stress-strain relation for a corrugating medium paper in the machine direction at 50 and 90% relative humidity with constants given in Table 3.3.

REFERENCES

[1] H. ALEXANDER, A constitutive relation for rubberlike materials, *International Journal of Engineering Science*, 6, 549–562 (1968).

[2] T. ALFREY, Non-homogeneous stresses in visco-elastic media, *Quarterly of Applied Mathematics*, 2, 113–119 (1944).

[3] M. BARSOUM, *Fundamentals of Ceramics*, McGraw-Hill, New York, 1997.

[4] J. M. BURGERS, Some considerations on the fields of stress connected with dislocations in a regular crystal lattice, *Proceedings Koninklijke Nederlandsche Akademie van Wetenschappen*, 42(5), 293–325 (1939).

[5] J. BODIG AND B. A. JAYNE, *Mechanics of Wood and Wood Composites*, Van Nostrand Reinhold, New York, 1982.

[6] I. BOUKAL, Effect of water on the mechanism of deformation of nylon 6, *Journal of Applied Polymer Science*, 11, 1483–1494 (1967).

[7] W.-F. CHEN AND A. F. SALEEB, *Constitutive Equations for Engineering Materials*, Wiley, New York, 1982.

[8] R. M. CHRISTENSEN, *Theory of Viscoelasticity: An Introduction*, Academic Press, New York, London (1971).

[9] R. L. COBLE AND W. D. KINGERY, Effect of porosity on the physical properties of sintered alumina, *Journal of the American Ceramics Society*, 39(11), 381, 1956.

[10] B. D. COLEMAN, Thermodynamics of materials with memory, *Archive fur Rational Mechanics and Analysis*, 17, 1–46, (1964).

[11] A. A. ELWI AND D. W. MURRAY, A 3-D hypoelastic concrete constitutive relationship, *Journal of the Engineering Mechanics Division*, ASCE, 105(EM4), 623–641 (1979).

[12] H. EYRING AND G. HALSEY, The mechanical properties of textiles, III & V, *Textile Research Journal*, 16, January and March, 13–25; 124–129; 284–285 (1946).

[13] Y. C. FUNG, *Biomechanics: Mechanical Properties of Living Tissues*, 2nd ed., Springer-Verlag, New York, 1993.

[14] K. H. GERSTLE, Simple formulation of biaxial concrete behavior, *Journal of the American Concrete Institute*, 78(1), 62–68 (1981).

[15] D. E. GUNDERSON, J. M. CONSIDINE, AND C. T. SCOTT, *Journal of Pulp and Paper Science*, 14(4), J37–J41 (1988).

[16] G. A. HOLZAPFEL, T. C. GASSER, AND R. W. OGDEN, A new constitutive framework for arterial wall mechanics and a comparative study of material models, *Journal of Elasticity*, 61, 1–48 (2000).

[17] J. P. JOULE, Some thermo-dynamic properties of solids, *Philosophical Transactions. Royal Society of London A*, 149, 91–131 (1859).

[18] A. S. KRAUZ AND H. EYRING, *Deformation Kinetics*, Wiley-Interscience, New York, 1975.

[19] E. H. LEE, Stress analysis in visco-elastic bodies, *Quarterly of Applied Mathematics*, 13, 183–190 (1955).

[20] S. G. LEKHNITSKII, *Theory of Elasticity of an Anisotropic Body*, Mir, Moscow, 1981.

[21] J. LEMAITRE AND J.-L. CHABOCHE, *Mechanics of Solid Materials*, Cambridge University Press, Cambridge, UK, 1990.

[22] A. E. H. LOVE, *A Treatise on the Mathematical Theory of Elasticity*, 4th ed., Dover, New York, 1944.

[23] J. K. MACKENZIE, The elastic constants of a solid containing spherical holes, *Proceedings, Physical Society (London)*, B63, 2–11 (1950).

[24] L. E. MALVERN, *Introduction to the Mechanics of a Continuous Media*, Prentice Hall, Englewood Cliffs, NJ, 1969.

[25] H. MARKOVITZ, Boltzmann and the beginnings of linear viscoelasticity, *Transactions of Society of Rheology* 21(3), 381–398 (1977).

[26] A. J. MCEVILY, JR., R. C. BOETTNER, AND T. L. JOHN-STON, On the formation and growth of fatigue cracks in polymers, in *Fatigue: An Interdisciplinary Approach*, J. J. Burke, N. L. Reed, and V. Weiss, Eds., Syracuse University Press, Syracuse, NY, 95–103, 1964.

[27] A. D. MCLAREN AND J. W. ROWEN, Sorption of water vapor by proteins and polymers: A review, *Journal of Polymer Science, 7*, 289–324 (1952).

[28] R. W. OGDEN, Large deformation isotropic elasticity—on the correlation of theory and experiment for incompressible rubberlike solids, *Proceedings. Royal Society of London A, 326*, 565–584 (1972).

[29] M. R. O'HALLORAN, *A Curvilinear Stress-strain Model for Wood in Compression*, Ph.D. thesis, Colorado State University, 1973.

[30] I. P. SAENZ, Discussion of "Equation for the stress-strain curve of concrete" by Desayi and Krishman, *Journal of the American Concrete Institute, 61*(9), 1229–1235 (1964).

[31] W. E. SCOTT, *Properties of Paper: An Introduction*, TAPPI Press, Atlanta, GA, 1989.

[32] G. A. SMOOK, *Handbook for Pulp and Paper Technologists*, 3rd ed., Angus Wilde Pub., Vancouver, Canada, 2003.

[33] S. P. TIMOSHENKO, *History of Strength of Materials*, Dover, New York, 1983.

[34] L. R. G. TRELOAR, *The Physics of Rubber Elasticity*, 3rd ed., Clarendon Press, Oxford, 1975.

[35] W. C. UNWIN, On the yield point of iron and steel and the effect of repeated straining and annealing, *Proceedings. Royal Society of London, 57*, 178–187 (1894).

[36] N. ZENG AND H. W. HASLACH, JR., Thermoelastic generalization of isothermal elastic constitutive models for rubberlike materials, *Rubber Chemistry and Technology 69*(2), 313–324 (1996).

[37] F. ZWICKY, Die Reissfestigkeit von Steinsalz, *Physikalische Zeitschrift, 24*(6), 131–137 (1923).

CHAPTER 4

LINEAR ELASTICITY

The 15 equations of linear elasticity have been established in the preceding chapters: the three stress equilibrium equations, the six strain-displacement equations, and the six Hookean stress-strain relations. These relations must hold for any linear elastic body. In this chapter all materials will be assumed isotropic. The behavior of different bodies is distinguished by what happens to their surfaces, the boundary conditions. The boundary conditions account for different loads and fixed displacement conditions on the surface and account for the shape of the surface. To use the 15 equations of linear elasticity to predict the behavior of a body, an engineer must be able to correctly describe the boundary conditions. The equations themselves can be simplified in some cases by using physical assumptions or mathematical tricks to reduce the number of unknowns to less than 15. Simple solutions can be combined by the principle of superposition since the theory is linear. Some cases in which a closed form solution is known will be investigated in this chapter. Later, numerical techniques will be presented to deal with the large number of other cases.

4.1 BOUNDARY CONDITIONS

The boundary of a three-dimensional body is described by equations of the form, $f(x, y, z) = 0$, and of a two-dimensional body by equations of the form $g(x, y) = 0$. The dimension of the boundary is one less than that of the body. A solid planar disk of radius R has boundary $x^2 + y^2 = R^2$, a one-dimensional curve. A solid ball of radius R has boundary $x^2 + y^2 + z^2 = R^2$, a two-dimensional surface. This boundary can also be expressed as $f(x, y, z) = x^2 + y^2 + z^2 - R^2 = 0$.

EXAMPLE 4.1 A plate with sides of length a and b is given Cartesian coordinates with origin at the center and axes parallel to the edges. The bottom boundary is $y = -b/2$, the top edge is $y = b/2$, the left side is $x = -a/2$, and the right side is $x = a/2$. ∎

Any solution to an elasticity problem, one which gives the stress $\mathbf{t}(x, y, z)$ and displacements, $u(x, y, z)$, $v(x, y, z)$, $w(x, y, z)$, must agree with the stress and displacements applied externally on the boundary. Strains are rarely given as boundary conditions since they can be obtained directly from the displacements or stresses.

The only stresses specified on the boundary are those acting on a plane tangent to the surface through a point on the surface. In a coordinate system with one axis perpendicular to the surface, one component will therefore be normal to the boundary and the other two will be shear components parallel to the plane tangent to the surface. By definition, a normal to a surface is a normal to a plane tangent to the surface at the point in question. Any or all three displacement components can be specified on the boundary. Alternatively, components might be specified in any convenient coordinate system.

If $\mathbf{n} = a\mathbf{i} + b\mathbf{j} + c\mathbf{k}$ is a unit normal to the surface at (x, y, z) and $T(x, y, z)$ is the stress tensor computed at the point (x, y, z), then $T(\mathbf{n}) = \mathbf{t}$ must be the stress on the tangent plane to the surface at (x, y, z). The unit normal can be computed from the equation $f(x, y, z) = 0$ defining the surface since the gradient vector $\nabla f = (\partial f/\partial x)\mathbf{i} + (\partial f/\partial y)\mathbf{j} + (\partial f/\partial z)\mathbf{k}$ is normal to the surface. The unit vector normal to the surface is then $\mathbf{n} = \nabla f/|\nabla f|$. For example, the unit normal at (x, y, z) on the surface of a sphere of radius R is $\mathbf{n} = (x/R)\mathbf{i} + (y/R)\mathbf{j} + (z/R)\mathbf{k}$.

EXAMPLE 4.2   Figure 4.1 shows a stress element along the boundary of a body; the stress components acting on the boundary are σ_y and τ_{yx}. The stress, σ_x, does not act on the surface of the body; it acts on a plane inside the body but near the surface.   ∎

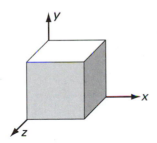

FIGURE 4.1   A boundary element for a solid body.

EXAMPLE 4.3   The stress components which must be specified on the top of a cube (Fig. 4.2) to define the stress boundary conditions are σ_y, τ_{yx}, and τ_{yz}. The components that must be specified on the right side are σ_x, τ_{xy}, and τ_{xz}.   ∎

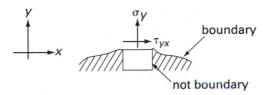

FIGURE 4.2   The boundary of a cube.

EXAMPLE 4.4   A thin disk of radius R is subjected to a uniform radial, compressive, in-plane pressure, p, applied to its outer circular surface. Specify the boundary conditions.

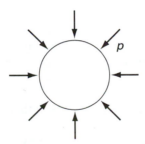

FIGURE 4.3   Disk with in-plane, constant, radial, distributed load on its boundary.

Solution   Since the external pressure, p, is represented by a vector pointing toward the center with magnitude, p, the boundary condition can be most easily expressed in polar coordinates. The boundary condition involves stresses and not displacements. The boundary condition in polar coordinates (r, θ) with origin at the disk center is that on the surface of the disk, $\sigma_r(R, \theta) = -p$ and $\tau_{r\theta}(R, \theta) = 0$. These are the two in-plane components acting on the plane tangent to the surface. No information is initially known about σ_θ, since it does not act on the plane tangent to the surface, nor about the displacements at the surface. ∎

4.2   TWO-DIMENSIONAL PROBLEMS

The system of three-dimensional elasticity equations is simplified if it can be considered to be two dimensional. For example, at a point on a free surface, one where no loads act, all stress components acting on the surface must be zero. If the coordinates are chosen so that the z-axis is perpendicular to the surface, then all stress components on the z-face are zero. The stress at the point can be thought of as two dimensional, or planar. Occasionally, the state of stress in a body can be approximated as a two-dimensional state. Recall

Definition   A state of stress at a point is said to be plane stress if a principal stress is zero.

The stress, $\sigma_z = 0$, is a principal stress. In the case of plane stress, Hooke's law becomes

$$\epsilon_x = \frac{1}{E}[\sigma_x - \nu\sigma_y];$$

$$\epsilon_y = \frac{1}{E}[\sigma_y - \nu\sigma_x];$$

$$\gamma_{xy} = \frac{1}{G}\tau_{xy}. \tag{4.1}$$

Notice that even though the stresses in the z-direction are zero, the normal strain in the z-direction is not zero, due to the Poisson effect.

$$\epsilon_z = \frac{1}{E}[\sigma_z - \nu(\sigma_x + \sigma_y)] = -\frac{\nu}{E}(\sigma_x + \sigma_y).$$

The normal strain, ϵ_z, is a principal strain since $\gamma_{zx} = \gamma_{zy} = 0$.

Thin members in which the stress on the surface perpendicular to the thickness direction remains very close to zero are often assumed to be two-dimensional bodies with a state of plane stress everywhere. This condition can only be met if the applied forces lie in the plane perpendicular to the thickness direction. A thin disk subjected to in-plane radial pressures can be assumed to be in plane stress.

In other problems, the strains in a given direction are so small that they may be neglected. Recall

Definition   A point of a body is said to be in plane strain if a principal strain is zero.

Hooke's law at a point in plane strain, $\epsilon_z = \gamma_{xz} = \gamma_{yz} = 0$, can be most simply given in terms of the strains. Since $\epsilon_z = (1/E)[\sigma_z - \nu(\sigma_x + \sigma_y)] = 0$, it follows that

$\sigma_z = v(\sigma_x + \sigma_y)$. Even though the strain in the z-direction is zero, the stress in the z-direction is not necessarily zero. The stresses in terms of the strains are, from equation (3.13),

$$\sigma_x = \frac{E}{1+v}\left[\epsilon_x + \frac{v}{1-2v}(\epsilon_x + \epsilon_y)\right];$$

$$\sigma_y = \frac{E}{1+v}\left[\epsilon_y + \frac{v}{1-2v}(\epsilon_x + \epsilon_y)\right];$$

$$\sigma_z = \frac{Ev}{(1+v)(1-2v)}(\epsilon_x + \epsilon_y)$$

$$\tau_{xy} = G\gamma_{xy}. \tag{4.2}$$

Exercise 4.1   Write these relations in terms of the Lamé constants. Then use equations (4.2) for σ_z and Hooke's law to show that $\sigma_z = v(\sigma_x + \sigma_y)$ and finally Hooke's law again to write the strains in terms of the stresses for plane strain as

$$\epsilon_x = \frac{1+v}{E}[(1-v)\sigma_x - v\sigma_y];$$

$$\epsilon_y = \frac{1+v}{E}[(1-v)\sigma_y - v\sigma_x]. \tag{4.3}$$

Plane stress and plane strain are very different states. Plane stress may have three nonzero normal strains while plane strain may have three nonzero normal stresses.

Exercise 4.2   Is it possible for a point in a body to be simultaneously in plane stress and plane strain? If so, give an example.

4.3 AIRY STRESS FUNCTION

The 15 equations of linear elasticity are very difficult to solve since they involve coupled partial differential equations. The usual solution strategy would be to simplify the system by combining some of the equations to reduce the number of unknowns. For two-dimensional problems, such a procedure can reduce the number of equations and unknowns drastically.

In two-dimensional problems, such as found on the free surface of a member, only eight equations remain in eight variables: two displacements, three strains, and three stresses. The equations are three strain-displacement, three constitutive, and two equilibrium equations. In principle, then, since there are as many equations are there are unknowns, the problem is solvable. All the variables except the stresses can be eliminated if the material is isotropic and linear elastic.

Assume plane strain. Substitute the strains from equation (4.3) into the strain compatibility equation (Chapter 2.5), $\partial^2\epsilon_x/\partial y^2 + \partial^2\epsilon_y/\partial x^2 = \partial^2\gamma_{xy}/\partial x\partial y$, which was obtained from the strain-displacement relations, to obtain

$$\frac{\partial^2}{\partial y^2}[(1-v)\sigma_x - v\sigma_y] + \frac{\partial^2}{\partial x^2}[(1-v)\sigma_y - v\sigma_x] = 2\frac{\partial^2\tau_{xy}}{\partial x\partial y}. \tag{4.4}$$

The relation $G = E/2(1+v)$ is used to simplify (4.4). This uses up the strain-displacement and constitutive equations. A second equation is obtained by differentiating the x-direction equilibrium equation by x and the y-direction equilibrium equation by y and adding

$$-\left[\frac{\partial^2 \sigma_x}{\partial x^2} + \frac{\partial^2 \sigma_y}{\partial y^2}\right] - \left(\frac{\partial F_x}{\partial x} + \frac{\partial F_y}{\partial y}\right) = 2\frac{\partial^2 \tau_{xy}}{\partial x \partial y}. \tag{4.5}$$

Equations (4.4) and (4.5) are combined by setting the right sides equal to obtain the stress compatibility equation for plane strain,

$$\left(\frac{\partial^2}{\partial x^2} + \frac{\partial^2}{\partial y^2}\right)(\sigma_x + \sigma_y) = -\frac{1}{1-\nu}\left(\frac{\partial F_x}{\partial x} + \frac{\partial F_y}{\partial y}\right). \tag{4.6}$$

Exercise 4.3 Write down the details of the calculation leading to equation (4.6).

For the case of plane stress, a similar strategy yields a slightly different stress compatibility equation,

$$\left(\frac{\partial^2}{\partial x^2} + \frac{\partial^2}{\partial y^2}\right)(\sigma_x + \sigma_y) = -(1-\nu)\left(\frac{\partial F_x}{\partial x} + \frac{\partial F_y}{\partial y}\right). \tag{4.7}$$

Exercise 4.4 Derive equation (4.7) following the pattern leading to equation (4.6).

If the body forces, $F_x\mathbf{i} + F_y\mathbf{j}$, are neglected, then these two equations are identical.

$$\nabla^2(\sigma_x + \sigma_y) = 0, \tag{4.8}$$

where $\nabla^2 = \partial^2/\partial x^2 + \partial^2/\partial y^2$ is the Laplace operator.

Exercise 4.5 Examine the stress compatibility equation in the case of plane strain when the stresses are independent of z and the body forces are zero. Show that the stress compatibility equation is satisfied if the stresses are linear functions of x and y.

Equation (4.8) and the two stress equilibrium equations form a system of three equations in three unknowns, the stress components, when the body forces are neglected. All three of these equations involve partial derivatives and so are not simple to solve. Therefore a further reduction in unknowns is sought. One equation in one unknown is possible, if a new unknown related to the stresses is introduced.

Definition The Airy stress function for either plane stress or plane strain, and zero body forces, is any function, $\phi(x, y)$, which satisfies

$$\frac{\partial^2 \phi}{\partial y^2} = \sigma_x, \qquad \frac{\partial^2 \phi}{\partial x^2} = \sigma_y, \quad \text{and} \quad \frac{\partial^2 \phi}{\partial x \partial y} = -\tau_{xy}. \tag{4.9}$$

This function for plane strain was developed by G. B. Airy in 1862.

Note that the stress in the x-direction is obtained by differentiating twice with respect to y and vice versa. The negative of the shear stress is obtained by differentiating once each with respect to x and y. The definitions are chosen so that the stress equilibrium equations are automatically satisfied.

Exercise 4.6 Substitute the relations (4.9) into the two stress equilibrium relations with zero body forces. Are these two equations satisfied?

Each problem has its own set of boundary conditions. The correct Airy stress function is the $\phi(x, y)$ which satisfies not only the relations (4.9) but also the stress boundary conditions. In terms of the Airy stress function, equation (4.8) becomes

$$\nabla^2(\sigma_x + \sigma_y) = \nabla^2\left(\frac{\partial^2\phi}{\partial y^2} + \frac{\partial^2\phi}{\partial x^2}\right)$$

$$= \frac{\partial^4\phi}{\partial x^4} + 2\frac{\partial^4\phi}{\partial x^2\partial y^2} + \frac{\partial^4\phi}{\partial y^4} = \nabla^4(\phi) = 0. \qquad (4.10)$$

Definition   The equation, $\nabla^4(\phi) = 0$, is called the biharmonic equation.

Some functions which satisfy the biharmonic equation are any polynomial in x and y of order less than 4, x^3y, xy^3, $x^4 - y^4$, $x^2y^2 - y^4/3$, $\cos x \cosh y$, $\cos y \cosh x$, $x \cos y \cosh x$, and $y \cos x \cosh y$.

The stresses obtained by equation (4.9) from any function, $\phi(x, y)$, which satisfies this equation, $\nabla^4(\phi) = 0$, automatically satisfy the three elasticity equations in stress: (4.8) and the two stress equilibrium equations. Therefore, the two-dimensional elasticity problem has been reduced to solving the biharmonic differential equation for the Airy stress function, $\phi(x, y)$. Once $\phi(x, y)$ is determined subject to the boundary conditions of the physical problem, the stresses can be obtained by taking second derivatives of $\phi(x, y)$. The strains are then obtained from the stresses using Hooke's law. Finally, the displacements are obtained from the strains by integrating the strain-displacement equations subject to displacement boundary conditions. This program sounds very straightforward to carry out. But it is seldom simple. The next section lists some techniques which might help determine $\phi(x, y)$.

EXAMPLE 4.5   Suppose that the Airy stress function for a square plate with sides a is $\phi(x, y) = cxy^3$, where the origin is at the lower left corner of the plate. Determine the stress state at each point in the plate and also the stress boundary conditions which are consistent with this Airy stress function. The stress at point (x, y) in the body is

$$\sigma_x = \frac{\partial^2\phi}{\partial y^2} = 6cxy, \qquad \sigma_y = \frac{\partial^2\phi}{\partial x^2} = 0, \quad \text{and} \quad \tau_{xy} = -\frac{\partial^2\phi}{\partial x\partial y} = -3cy^2.$$

To be consistent with this Airy function, the applied stress along the left edge where the points have coordinates $(0, y)$ must be $\sigma_x = 0$ and $\tau_{xy} = -3cy^2$. On the bottom edge where the points have coordinates $(x, 0)$, it is required that $\sigma_y = 0$ and $\tau_{xy} = 0$.   ∎

Exercise 4.7   Determine the required stress boundary conditions on the remaining edges. Do the shear stresses agree at the corners?

EXAMPLE 4.6   A thin triangular bracket is fixed in a rigid wall and supports a constant distributed load, q, across its top face (Fig. 4.4). The thickness is t and the length of the top edge is b. Let $w = q/t$. The unmounted edges form an angle β. Verify that the proposed Airy function, $\phi(x, y) = [0.5w\cot\beta/(1 - \beta\cot\beta)][xy - x^2\tan\beta + (x^2 + y^2)(\beta - \arctan(y/x))]$ produces the correct boundary conditions along the edges not attached to the wall.

Solution   Notice that the Airy function chosen depends only on the angle β not on the size of the triangular plate. The stresses at any point (x, y) are obtained from the definition of the Airy function (4.9).

$$\sigma_x = \frac{\partial^2\phi}{\partial y^2} = \frac{w\cot\beta}{(1 - \beta\cot\beta)}\left(\beta - \frac{xy}{x^2 + y^2} - \arctan\frac{y}{x}\right);$$

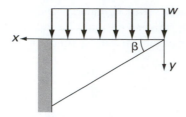

FIGURE 4.4   A thin triangular plate subject to a constant distributed load on its upper edge.

$$\sigma_y = \frac{\partial^2 \phi}{\partial x^2} = \frac{w \cot \beta}{(1 - \beta \cot \beta)} \left(\beta - \tan \beta - \frac{y}{x} + \frac{2xy}{x^2 + y^2} + \frac{y^3}{x^3 + xy^2} - \arctan \frac{y}{x} \right);$$

$$\tau_{xy} = -\frac{\partial^2 \phi}{\partial x \partial y} = -\frac{w \cot \beta}{(1 - \beta \cot \beta)} \frac{y^2}{x^2 + y^2}.$$

At the upper edge ($y = 0$), as required $\sigma_y = -w$ and $\tau_{xy} = 0$ as required. On the slanted edge $y = x \tan \beta$, the normal stress perpendicular to the edge and the shear stress parallel to the edge must be zero. In the given coordinates, along the edge,

$$\sigma_x = -\frac{w}{(1 - \beta \cot \beta)(1 + \tan^2 \beta)}$$

$$\sigma_y = -\frac{w \tan^2 \beta}{(1 - \beta \cot \beta)(1 + \tan^2 \beta)}$$

$$\tau_{xy} = -\frac{w \tan \beta}{(1 - \beta \cot \beta)(1 + \tan^2 \beta)}.$$

Putting $N = w/[(1 - \beta \cot \beta)(1 + \tan^2 \beta)]$, the stress tensor is

$$T = N \begin{pmatrix} -1 & -\tan \beta \\ -\tan \beta & -\tan^2 \beta \end{pmatrix}.$$

Rotate the coordinates through $\pi/2 - \beta$ clockwise to a primed coordinate system so that the x'-axis is perpendicular to the slanted edge. By Chapter 1.5, the rotation matrix is

$$\alpha = \begin{pmatrix} \cos(\pi/2 - \beta) & -\sin(\pi/2 - \beta) \\ \sin(\pi/2 - \beta) & \cos(\pi/2 - \beta) \end{pmatrix} = \begin{pmatrix} \sin \beta & -\cos \beta \\ \cos \beta & \sin \beta \end{pmatrix}.$$

In the primed coordinates,

$$T' = \alpha T \alpha^t = N \begin{pmatrix} 0 & 0 \\ 0 & -\sec^2 \beta \end{pmatrix}.$$

Therefore on the slanted face, the normal and shear stresses, $\sigma_{x'} = \tau_{x'y'} = 0$ as required. ∎

A similar analysis produces a stress compatibility equation for orthotropic linear elastic materials. Assume the body is in plane stress and neglect body forces. Further assume that the relation $G_{xy} = E_x E_y/[E_x + (1 + 2\nu_{xy})E_y]$ is a valid approximation. Note that this reduces to the isotropic relation $G = E/[2(1 + \nu)]$. Substitution of the orthotropic constitutive equations (3.8), when $\sigma_z = 0$, into the strain compatibility equation (2.12) gives

$$\frac{1}{E_x E_y} \left(E_y \frac{\partial^2 \sigma_x}{\partial y^2} - \nu_{yx} E_x \frac{\partial^2 \sigma_y}{\partial y^2} - \nu_{xy} E_y \frac{\partial^2 \sigma_x}{\partial x^2} + E_x \frac{\partial^2 \sigma_y}{\partial x^2} \right) = \frac{1}{G_{xy}} \frac{\partial^2 \tau_{xy}}{\partial x \partial y}.$$

Then solving this result with (4.5), and simplifying yields

$$0.5(E_x + E_y)\frac{\partial^2\sigma_x}{\partial x^2} + E_x\frac{\partial^2\sigma_y}{\partial x^2} + 0.5(E_x + E_y)\frac{\partial^2\sigma_y}{\partial y^2} + E_y\frac{\partial^2\sigma_x}{\partial y^2} = 0. \qquad (4.11)$$

Note that (4.11) reduces to the biharmonic in the isotropic case. An Airy stress function, ϕ, is defined by its relationship to the stresses as in the isotropic case.

EXAMPLE 4.7   A thin beam is loaded in bending in a manner that also induces an internal shear force, with the moment at the ends given by a distributed load, $p(L, y) = -ay$ and $p(-L, y) = ay$ for a constant $a = 1{,}000$ MPa/m (Fig. 4.5). The origin of the coordinate system is at the centroid of the beam, which has height $2b = 8$ cm and length $2L = 50$ cm. The beam is a fibered composite with fibers in the longitudinal x-direction. The fibers are graphite ($E = 280$ GPa) and the matrix is aluminum ($E = 70$ GPa). Design the volume fraction of fibers, f, so that the maximum tensile longitudinal strain, ϵ_x, in the beam is less than 0.00025.

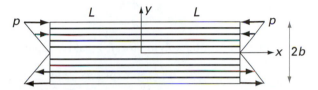

FIGURE 4.5   A fibered composite beam loaded in bending.

Solution   Because the beam is thin, assume plane stress. A suitable Airy stress function that satisfies equation (4.11) is

$$\phi(x, y) = a(3b^2xy - xy^3)/6L$$

because it is consistent with the boundary conditions if they are relaxed at the ends to require only that the shear resultant be zero while the exact pointwise boundary conditions match the Airy function in all other cases.

$$\sigma_x(x, y) = -axy/L;$$
$$\sigma_y = 0;$$
$$\tau_{xy} = -a(b^2 - y^2)/2L$$

From these equations, $\sigma_x(L, y) = -ay$, $\sigma_x(-L, y) = ay$, $\tau(x, \pm b) = 0$, $\sigma_y(x, \pm b) = 0$, and the shear resultants on the ends are zero. The maximum tensile stress, σ_x, occurs at $x = L$ and $y = -b$.

The volume fraction appears in the expressions for the orthotropic elastic moduli (Chapter 3.5). Recall that the axial modulus is $E_x = E_m(1 - f) + E_f f$ by the rule of mixtures. Also for an orthotropic material $\epsilon_x = \sigma_x/E_x - \nu_{yx}\sigma_y/E_y$. Because $\sigma_y = 0$ everywhere and the maximum axial stress is $\sigma_{x_{max}} = ab = 40$ MPa, it is necessary to solve

$$0.00025 = \epsilon_x = \sigma_x/E_x = ab/[E_m(1 - f) + E_f f],$$

which yields $f = 0.429$.   ∎

4.4   TECHNIQUES FOR SOLVING THE EQUATIONS OF ELASTICITY

The biharmonic differential equation is very difficult to solve. Particular solutions can be obtained for the biharmonic equation from solutions to the Laplace equation, $\nabla^2\phi = 0$. In Cartesian coordinates, Laplace's equation can be solved by separation of variables. Suppose that $\phi(x, y, z) = f(x)g(y)h(z)$. Then substituting into Laplace's equation

$$\frac{1}{f}\frac{d^2 f}{dx^2} + \frac{1}{g}\frac{d^2 g}{dy^2} + \frac{1}{h}\frac{d^2 h}{dz^2} = 0.$$

Each term in this sum must be a constant in order for it to add to zero for arbitrary (x, y, z). Let $a^2 = (1/f)(d^2 f/dx^2)$, $b^2 = (1/g)(d^2 g/dy^2)$, and $c^2 = (1/h)(d^2 h/dz^2)$. The three resulting differential equations are

$$\frac{d^2 f}{dx^2} = a^2 f;$$

$$\frac{d^2 g}{dy^2} = b^2 g;$$

$$\frac{d^2 h}{dz^2} = c^2 h.$$

The general solution is, after solving for f, g, and h separately and multiplying,

$$\phi(x, y, z) = \exp(\pm ax \pm by \pm cz),$$

where a, b, and c can be complex numbers.

The Laplace equation also admits particular solutions depending on the boundary conditions. For example, a particular solution could be

$$\phi_p(x, y, z) = [(x - a')^2 + (y - b')^2 + (z - c')^2]^{1/2}.$$

Particular solutions for the biharmonic include

a. Any polynomial of order three or less.

b. Any polynomial with coefficients chosen to satisfy the biharmonic.

c. If Ψ is a solution to the Laplace equation: $x\Psi$; $y\Psi$; $z\Psi$; $(x^2 + y^2 + z^2)\Psi$.

Exercise 4.8   Verify that solutions of the forms given in c) satisfy the biharmonic equation.

These results are needed only in two dimensions for the Airy stress function. The Laplace and biharmonic equations are special cases of elliptic partial differential equations studied in differential equations texts.

Often some information is available on the behavior of the stress field. For example, the stress might be known to be constant throughout the body as a good approximation. One typical strategy, using assumptions on the stress behavior, is to guess the actual Airy stress function or to guess what form it might take. The inverse method requires the engineer to guess the Airy function, ϕ, and then to verify that ϕ satisfies the equations of elasticity and that the boundary conditions are satisfied. In the semi-inverse method, the form of the Airy stress function, up to some undetermined coefficients, is guessed based on engineering insight. The coefficients are determined by requiring that the form chosen for ϕ satisfies the equations of elasticity and the boundary conditions. Any constant or linear terms in $\phi(x, y)$ may as well be neglected since they do not contribute to the stresses, which are second derivatives of ϕ.

In uniaxial loading, the stress might be assumed to be constant, σ_0 in the loading direction, zero in the other principal direction and zero shear. Since the stresses are second derivatives of the Airy stress function, the Airy function must be a second-order polynomial. The function, $\phi(x, y) = \sigma_0 y^2/2$, satisfies all equations of elasticity and the boundary conditions.

Exercise 4.9   Verify that $\phi(x, y) = \sigma_0 y^2/2$ is the Airy stress function for uniaxial loading.

Notice that the load on the boundary must be a distributed load of σ_0 force per unit area. A point load cannot satisfy the boundary stress conditions predicted by any Airy stress function. Point loads do not exist either in elasticity or in reality. Any such load would cause infinite stress at its point of application and therefore inelastic loading and material failure. Analysis by strength of materials, on the other hand, assumes that point loads can exist.

Alternatively, if one stress is known or assumed throughout the body, the Airy stress function can sometimes be recovered by integration. Suppose the loading is uniaxial and that $\sigma_x = \sigma_0$. Then by equation (4.9), $\partial^2\phi/\partial y^2 = \sigma_0$, so that integrating twice gives, $\phi(x, y) = (\sigma_0/2)y^2 + c_1(x)y + c_2(x)$. The functions $c_1(x)$ and $c_2(x)$ can be determined from the boundary conditions. Since $\sigma_y = \partial^2\phi/\partial x^2 = c_1''(x)y + c_2''(x) = 0$ on the unloaded edges $(x, \pm h)$, both $c_1''(x)$ and $c_2''(x)$ must be zero. Since $\tau_{xy} = -\partial^2\phi/\partial x\partial y = -c_1'(x) = 0$ on the unloaded edges, $c_1(x)$ is constant. So both $c_1(x)y$ and $c_2(x)$ are linear functions and do not affect the stresses deduced from the Airy function. They might as well be neglected and take $\phi(x, y) = (\sigma_0/2)y^2$.

The application of the semi-inverse method can be guided by information on the stresses gained from strength of materials.

EXAMPLE 4.8   An Airy stress function can be written for the case of pure bending of a thin beam of thickness, t. The beam has length, L, and height, $2h$ (Fig. 4.6). The origin of the coordinate system is at the centroid.

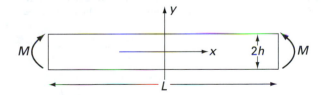

FIGURE 4.6   Bending of a thin beam.

Solution   Since the moment, M, is constant throughout the beam, the normal stress in the x-direction must be linear so that the Airy stress function must be a cubic. Assume that $\phi(x, y) = (a_3/6)x^3 + (b_3/2)x^2y + (c_3/2)xy^2 + (d_3/6)y^3$, a cubic polynomial of homogeneous order. Then $\sigma_x = \partial^2\phi/\partial y^2 = c_3x + d_3y$. The external applied moment must satisfy

$$M = \int_{-h}^{h} y\sigma_x(L/2, y)t\,dy = \int_{-h}^{h} y(c_3L/2 + d_3y)t\,dy = \frac{2}{3}d_3h^3t.$$

Since the area moment of inertia about the bending axis is $I = (2/3)h^3t$, $d_3 = M/I$.

The axial load at the end of the beam is zero, so that

$$P = \int_{-h}^{h} \sigma_x(L/2, y)t\,dy = \int_{-h}^{h} (c_3L/2 + d_3y)t\,dy = c_3htL = 0.$$

Therefore $c_3 = 0$. On the top of the beam, the shear and normal stresses are zero. The shear stress is $\tau_{xy} = -\partial^2\phi/\partial x\partial y = -b_3x - c_3y$. At any point (x, h) on the top, $\tau_{xy} = -b_3x - c_3h = -b_3x = 0$ implies that $b_3 = 0$. Also, $\sigma_y = \partial^2\phi/\partial x^2 = a_3x + b_3y$, so that at any point (x, h)

on the top, $\sigma_y = \partial^2\phi/\partial x^2 = a_3 x = 0$ and $a_3 = 0$. Therefore the Airy stress function for a two-dimensional state of pure bending is

$$\phi(x, y) = \frac{M}{6I}y^3.$$

Notice that $\sigma_x = My/I$, $\sigma_y = 0$, and $\tau_{xy} = 0$ as predicted in strength of materials.

To see if the elasticity solution is exactly equivalent to the strength of materials solution, the deflections must be computed from the equations of elasticity. From Hooke's law, $\epsilon_x = My/IE$, $\epsilon_y = -\nu My/IE$, and $\gamma_{xy} = 0$. In the coordinates chosen, M must be taken negative as drawn in Figure 4.6 so that ϵ_x is compressive on top of the beam. The strain-displacement equations give three differential equations for the displacements u and v.

$$\frac{\partial u}{\partial x} = \frac{My}{EI};$$

$$\frac{\partial v}{\partial y} = \frac{-\nu My}{EI};$$

$$\frac{\partial u}{\partial y} + \frac{\partial v}{\partial x} = 0.$$

Integrating produces

$$u(x, y) = \frac{Mxy}{EI} + u_1(y);$$

$$v(x, y) = \frac{-\nu My^2}{2EI} + v_1(x).$$

The zero shear strain condition implies that $Mx/EI + \partial u_1/\partial y = -\partial v_1/\partial x$. Because a function of x can only equal a function of y for all x and y if both functions are zero, $u_1(y) = c_1$, a constant, and $\partial v_1/\partial x = -Mx/EI$. Integration produces $v_1(x) = -Mx^2/2EI + c_2$.

$$u(x, y) = \frac{Mxy}{EI} + c_1;$$

$$v(x, y) = \frac{-\nu My^2}{2EI} - \frac{Mx^2}{2EI} + c_2.$$

The constants c_1 and c_2 are determined from the displacement boundary conditions. For example, if the origin is fixed so that $u(0, 0) = v(0, 0) = 0$, then $c_1 = c_2 = 0$. Elementary strength of materials states that at the neutral axis $(y = 0)$, $v(x, 0) = -Mx^2/2EI + c_2$, but rarely discusses $u(x, y)$. In the case of pure bending, $u(x, y)$ is linear in y for fixed x, so that plane sections remain plane. Elasticity does not imply that plane sections remain plane in arbitrary bending. ∎

Exercise 4.10   What happens to the displacement field if a beam in pure bending is cantilevered.

EXAMPLE 4.9

One cannot expect to find an Airy stress function for a cantilever beam with a point load at the top of its free end. To use elasticity, the load must be taken as a distributed shear force along the face of the free end (Fig. 4.7). If the shear force on the free end was taken as a constant, the boundary conditions would be satisfied since there was no shear on the top. The shear force therefore could not be uniformly distributed along the face.

To guess a form of the Airy stress function, recall that strength of materials predicts that stress is a quadratic function of position in the beam. The Airy stress function then must be at least a fourth order polynomial in x and y, the coordinates of points in the thin beam. It is instructive to see what happens when an invalid form is chosen for the Airy stress function. Assume that $\phi(x, y) = (a_3/6)x^3 + (b_3/2)x^2 y + (c_3/2)xy^2 + (d_3/6)y^3$, a cubic polynomial of

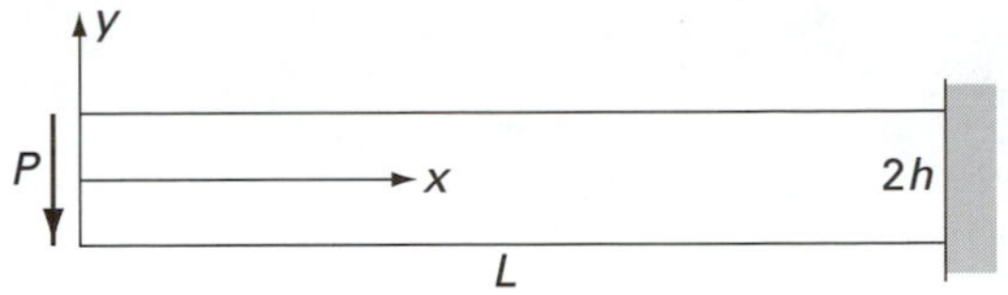

FIGURE 4.7 Cantilever beam with distributed shear on its free end.

homogeneous order, one degree lower than can be expected to satisfy the stress behavior from strength of materials. Then

$$\sigma_x = \frac{\partial^2 \phi}{\partial y^2} = c_3 x + d_3 y;$$

$$\sigma_y = \frac{\partial^2 \phi}{\partial x^2} = a_3 x + b_3 y;$$

$$\tau_{xy} = -\frac{\partial^2 \phi}{\partial x \partial y} = -b_3 x - c_3 y.$$

The equilibrium equations are satisfied, as they must be by the definition of the stresses in terms of an Airy stress function. However, the boundary conditions are not satisfied. The total shear force on the free end is for $x = 0$,

$$P = -\int_{-h}^{h} \tau_{xy} t\, dy = \int_{-h}^{h} c_3 y t\, dy = 0,$$

which is a contradiction. Therefore, the form chosen for the Airy stress function is not correct. A more successful Airy function for this situation is

$$\phi(x, y) = -(P/6I)xy^3 + (Ph^2/2I)xy,$$

(see Problem 2), where I is the area moment of inertia. ∎

These examples have all assumed that the Airy stress function is a polynomial; other functions are also possible.

If the load on a body is the superposition of loads, the Airy stress function is the sum of the Airy functions for the loads individually. Taking second derivatives to obtain the stress from the Airy function is a linear operation.

EXAMPLE 4.10 The distributed load on the free ends of a thin beam is known to be $F(y) = \sigma_0 + My/I$. From the previous examples, this is recognized as the superposition of an axial load and a pure bending moment, M. Therefore the Airy stress function must be $\phi(x, y) = (\sigma_0/2)y^2 + (M/6I)y^3$. ∎

Exercise 4.11 Compute the displacement field for this example subject to the constraint that $u(0, 0) = v(0, 0) = 0$. The origin is at the centroid of the beam.

EXAMPLE 4.11 A dam can be designed using the results of Lévy's problem (1898). Let γ be the specific weight of the fluid and p be the specific weight of the dam material. Determine the stresses under the assumption that the dam is an infinite thin wedge (Fig. 4.8).

Solution The boundary conditions are $\sigma_x(0, y) = \gamma y$, $\tau_{xy}(0, y) = 0$ and along $y = x \cot \beta$, the total stress is zero so that because the unit normal to the surface is $\mathbf{n} = \cos \beta \mathbf{i} - \sin \beta \mathbf{j}$,

$$\begin{pmatrix} \sigma_x & \tau_{xy} \\ \tau_{xy} & \sigma_y \end{pmatrix} \begin{pmatrix} \cos \beta \\ -\sin \beta \end{pmatrix} = \begin{pmatrix} 0 \\ 0 \end{pmatrix}.$$

An Airy stress function, $\phi(x, y)$, accounts for the stresses due to the external tractions. The stresses due to the body forces must be superposed on those produced by the stress function.

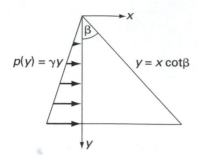

FIGURE 4.8 Thin wedge supporting a linearly varying distributed load.

Lévy assumed that the stress at (x, y) due to the weight of the dam material is a normal stress, ρy, in the y-direction. Then

$$\sigma_x = \frac{\partial^2 \phi}{\partial y^2}$$

$$\sigma_y = \frac{\partial^2 \phi}{\partial x^2} - \rho y$$

$$\tau_{xy} = -\frac{\partial^2 \phi}{\partial x \partial y}$$

Lévy assumed the stress function, $\phi(x, y) = ax^3 + bx^2y + cy^3$ so that

$$\sigma_x(x, y) = 6cy$$
$$\sigma_y(x, y) = 6ax + 2by - \rho y$$
$$\tau_{xy}(x, y) = -2bx$$

The boundary condition, $\sigma_x(0, y) = \gamma y$, implies that $c = \gamma/6$. The other two boundary conditions, $\sigma_x \cos\beta + \tau_{xy} \sin\beta = 0$, $\tau_{xy} \cos\beta - \sigma_y \sin\beta = 0$, then imply that $a = \rho \cot\beta/6$, and $b = (\gamma \cot^2\beta)/2$.

If this analysis is applied to a dam, the stress along the base is predicted by the Lévy analysis to be linear, but subsequent investigations showed that this result is not realistic for the physical dam and that numerical methods are needed to compute the stress. ∎

4.5 LINEAR THERMOELASTICITY

Changing the temperature of a body can cause it to deform. The study of the interrelation of elastic mechanical and heat effects on a body is called thermoelasticity. In a metal body, experiments show that heating the body uniformly causes any line in the body to elongate, while cooling causes contraction.

Exercise 4.12 A metal plate has a circular hole in it. If the plate is uniformly heated, does the hole get larger or smaller? Explain why. Have you done any "kitchen" experiment that might shed light on this question?

Experiment shows that the change in length, δ, of any line is proportional to the original length, L, the change in temperature, ΔT, and depends on the material. This change is given by

$$\delta = \alpha L \Delta T, \tag{4.12}$$

where α is the coefficient of linear thermal expansion. The effect of the material is contained in this term. In thermodynamics, it is shown that

$$\alpha = \frac{1}{L}\left(\frac{\partial L}{\partial T}\right)_P,$$

where the subscript indicates constant pressure. Therefore α depends on temperature. Some typical values are given in Table 4.1 for two temperatures. However, in most design work near room temperature, α for metals is assumed to be a constant since the change in the linear thermal coefficient with temperature is small in that range. But near absolute zero, the coefficient of linear expansion is proportional to the cube of temperature.

TABLE 4.1   Change in Coefficient of Linear Thermal Expansion with Temperature

Material	at $T = 0°C$ (10^{-6}/K)	at $T = 100°$ C (10^{-6}/K)	% difference
Aluminum	23	24.9	8
Copper	16.1	16.9	5
Iron	11.7	12.7	8
Nickel	12.5	14	12

The thermally induced average normal strain over a length, L, is then

$$\epsilon_t = \frac{\delta}{L} = \alpha L \Delta T \frac{1}{L} = \alpha \Delta T.$$

Is it possible to have strain without stress? The answer is yes if the body can be heated uniformly without constraints on its change in shape. If the heating is nonuniform or there are constraints, then stresses as well as strains are induced in the body.

A linearized theory of thermoelasticity is obtained by assuming that the thermal strains superpose with the mechanical normal strains. Shear strains are assumed to be unaffected by temperature. Furthermore, it is assumed that the forces on the body do not significantly affect the linear thermal coefficient, α.

EXAMPLE 4.12

A spacer of length 10 cm is to be placed between two members which are assumed rigid. This structure is subject to a temperature change of 100°C. The spacer is mounted so that thermal expansion is constrained only in the longitudinal direction. It was first suggested that the spacer could be made of an aluminum with properties $E = 70$ GPa, $\nu = 0.34$, $\alpha = 24 \times 10^{-6}/C°$, and $\sigma_Y = -95$ MPa. Then it was decided that, if possible, the spacer should be an insulator so that 99.5% dense alumina (Al_2O_3) ($E = 370$ GPa, $\nu = 0.22$, $\alpha = 5 \times 10^{-6}/C°$, $\sigma_f = -2620$ MPa) or magnesia (MgO) ($E = 280$ GPa, $\nu = 0.18$, $\alpha = 10 \times 10^{-6}/C°$, $\sigma_f = -840$ MPa) were considered. If the spacer must have cross-sectional area of 0.0001 m^2, choose the material which induces the smallest reaction force on the two separated members.

Solution

As the spacer tries to expand axially due to the increase in temperature, the walls exert a force on the bar that prevents any change in length. Assume that, since the thermal expansion is unconstrained in the y- and z-directions, $\sigma_y = \sigma_z = 0$. Then by $\epsilon_x = [\sigma_x - \nu(\sigma_y + \sigma_z)]/E + \alpha\Delta T$ and the fact that the rigid members ensure that $\epsilon_x = 0$, $\sigma_x = -\alpha E \Delta T$. The values of α given are those at room temperature (25°C). In terms of these low estimates of thermal expansion, σ_x is -168 MPa for the aluminum so that it yields and cannot be used. The value for alumina is -185 MPa and for magnesia is -280 MPa. Therefore the spacer should be made of Al_2O_3. ■

The thermoelastic modification to Hooke's law induces changes in the differential equation defining an Airy stress function for two-dimensional problems. For example, in plane stress,

$$\epsilon_x = \frac{1}{E}[\sigma_x - \nu\sigma_y] + \alpha\Delta T;$$

$$\epsilon_y = \frac{1}{E}[\sigma_y - \nu\sigma_x] + \alpha\Delta T;$$

$$\epsilon_z = -\frac{\nu}{E}[\sigma_x + \sigma_y] + \alpha\Delta T;$$

$$\gamma_{xy} = \frac{1}{G}\tau_{xy}.$$

The remaining equations of elasticity, the stress equilibrium equation and the strain-displacement equations, are not affected by the thermal behavior. Follow the same procedure used before to develop the differential equation for the Airy stress function. The result of this calculation is

$$\nabla^4\phi + \nabla^2(E\alpha\Delta T) = 0. \tag{4.13}$$

If α and E are constant throughout the body, then equation (4.13) can be written as

$$\nabla^4\phi + \alpha E\nabla^2(\Delta T) = 0.$$

Notice that if the temperature change is constant throughout the body, then equation (4.13) reduces to the biharmonic equation previously developed, $\nabla^4\phi = 0$. The Airy stress function is only affected by temperature changes when they are nonuniform since only then are there thermally induced stresses.

A body is only in thermodynamic equilibrium when the temperature is uniform throughout the body. The body apparently can be in mechanical equilibrium, in the sense that the equilibrium equation is satisfied, while the temperature is in a steady state but nonuniform throughout the body.

EXAMPLE 4.13   A thin plate is subjected to a steady state, non-equilibrium, temperature distribution. Determine the distribution of stress and strain as a function of position in the plate for mechanical equilibrium. The thin rectangular plate has small thickness, t, height, $2h$, and length $2L$. The origin is at the center of the plate and the plate lies in the x-y plane. The loading is linear elastic, with elastic modulus, E, and linear thermal coefficient of expansion, α. Assume that E and α are given. Assume plane stress. There are no external forces on the plate, but the temperature variation in the plate is given by $\Delta T(x, y) = ky^2$, with the constant k also given. This might be accomplished by putting heat sources at the upper and lower edges of the plate and a heat sink along the x-axis.

Solution   The equation, $\nabla^4\phi + \alpha E\nabla^2(\Delta T) = 0$, is too complex to solve in most situations. Since we are interested in the stress, this can be rewritten as $\nabla^2(\sigma_x + \sigma_y + \alpha E\Delta T) = 0$. The goal is to make some simplifying assumptions on the stresses that are consistent with the equations of elasticity. Since ΔT is a function of y only, the stress on horizontal cross sections of the plate should be negligible; assume that $\sigma_y = \tau_{xy} = 0$. Since lines on the plate parallel to the x-axis apparently will try to deform to different lengths, assume that $\sigma_x = \sigma_x(y)$, *i.e.* is a function of y only. Neglect body forces.

Exercise 4.13   Verify that these assumptions on the stresses are consistent with the stress equilibrium equations.

With these assumptions, only $\sigma_x(y)$ remains to be determined. The equation, $\nabla^2(\sigma_x + \sigma_y + \alpha E \Delta T) = 0$, becomes

$$\frac{d^2 \sigma_x}{dy^2} + 2\alpha k E = 0.$$

Therefore, $\sigma_x(y) = -\alpha k E y^2 + c_1 y + c_2$. The constants, c_1 and c_2, are determined from the boundary conditions on the plate. The stress along the sides of the plate is zero so that it should be true that $\sigma_x(y) = -\alpha k E y^2 + c_1 y + c_2 = 0$ for any y between $-h$ and h. This is clearly impossible. One has to use the less exact boundary condition, that the total force along the edge is zero, to obtain an approximately valid answer for $\sigma_x(y)$. The total force on the right edge in the x-direction is

$$\int_{-h}^{h} \sigma_x(y) t \, dy = t\left(-\frac{1}{3}\alpha k E 2h^3 + 2c_2 h\right) = 0,$$

so that $c_2 = \alpha k E h^2 / 3$. The total moment about the z-axis is also zero.

$$\int_{-h}^{h} y \sigma_x(y) t \, dy = \frac{t}{3}(c_1 2h^3) = 0,$$

so that $c_1 = 0$. After substituting the coefficients into the normal stress in the x-direction,

$$\sigma_x(y) = -\alpha k E y^2 + \frac{1}{3}\alpha k E h^2.$$

By Hooke's law, the strains are

$$\epsilon_x = \frac{\sigma_x}{E} + \alpha \Delta T = \frac{1}{3}\alpha k h^2,$$

$$\epsilon_y = -\frac{\nu \sigma_x}{E} + \alpha \Delta T = -\nu\left(-\alpha k y^2 + \frac{1}{3}\alpha k h^2\right) + \alpha k y^2,$$

$$\gamma_{xy} = \frac{1}{G}\tau_{xy} = 0.$$

The strain in the x-direction is constant, while the strain in the y direction varies with y. Lines parallel to the y-axis strain more towards the outer edges. ■

4.6 POLAR COORDINATES

The choice of coordinates within which to express the equations of elasticity is guided by the geometry of the body and the form of the boundary conditions. A body is geometrically axisymmetric if it is symmetric about an axis. The load is axisymmetric if it is symmetric about an axis. If the body and the load are symmetric about the same axis, the stress field is also likely to be symmetric about that axis. If the stress is axisymmetric, then polar coordinates in two dimensions or cylindrical coordinates in three dimensions might be appropriate. For example, a disk subjected to radial pressure on its edge is easier to describe in polar than in Cartesian coordinates.

The equilibrium equations, which govern the allowable variation of stress from point to point in the body, are obtained by examining a small element of the body that is in static equilibrium (Fig. 4.9). This stress diagram is converted into a free-body diagram by multiplying each stress by the area of the face on which it acts. The inner radius is r and the outer radius is $r + \Delta r$. The change in angle between the radial sides is $\Delta \theta$. The area of the face under the inner circle is the thickness, t, times the arc length, $tr\Delta\theta$. The area of the face under the outer circle is, likewise,

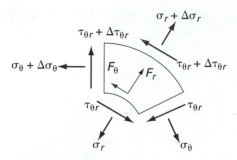

FIGURE 4.9 Stress element in polar coordinates.

$t(r + \Delta r)\Delta\theta$. The area of the sides are both $t\Delta r$. Notice that the sides of this element are not perpendicular. Therefore the forces must be broken into components in the r and θ directions. Summation of forces in the r-direction produces,

$$- \sigma_r tr\Delta\theta + (\sigma_r + \Delta\sigma_r)t(r + \Delta r)\Delta\theta - \tau_{\theta r}t\Delta r\cos(\Delta\theta/2)$$
$$+ (\tau_{\theta r} + \Delta\tau_{\theta r})t\Delta r\cos(\Delta\theta/2) - \sigma_\theta t\Delta r\sin(\Delta\theta/2)$$
$$- (\sigma_\theta + \Delta\sigma_\theta)t\Delta r\sin(\Delta\theta/2) + F_r rt\Delta\theta\Delta r = 0.$$

Summation of forces in the θ-direction produces

$$- \tau_{\theta r} tr\Delta\theta + (\tau_{\theta r} + \Delta\tau_{\theta r})t(r + \Delta r)\Delta\theta + \tau_{\theta r}t\Delta r\sin(\Delta\theta/2)$$
$$+ (\tau_{\theta r} + \Delta\tau_{\theta r})t\Delta r\sin(\Delta\theta/2) - \sigma_\theta t\Delta r\cos(\Delta\theta/2)$$
$$+ (\sigma_\theta + \Delta\sigma_\theta)t\Delta r\cos(\Delta\theta/2) + F_\theta rt\Delta\theta\Delta r = 0.$$

Divide each equation by $rt\Delta\theta\Delta r$ and let both $\Delta\theta$ and Δr tend to zero to obtain the equilibrium relations at the point (r, θ). The equilibrium equations in polar coordinates are

$$\frac{\partial\sigma_r}{\partial r} + \frac{1}{r}\frac{\partial\tau_{\theta r}}{\partial\theta} + \frac{1}{r}(\sigma_r - \sigma_\theta) + F_r = 0;$$

$$\frac{1}{r}\frac{\partial\sigma_\theta}{\partial\theta} + \frac{\partial\tau_{\theta r}}{\partial r} + \frac{2}{r}\tau_{\theta r} + F_\theta = 0. \tag{4.14}$$

The strain-displacement equations are, for u the displacement in the r direction and v the displacement in the θ direction,

$$\epsilon_r = \frac{\partial u}{\partial r};$$

$$\epsilon_\theta = \frac{1}{r}\frac{\partial v}{\partial\theta} + \frac{u}{r};$$

$$\gamma_{r\theta} = \frac{1}{r}\frac{\partial u}{\partial\theta} + \frac{\partial v}{\partial r} - \frac{v}{r}. \tag{4.15}$$

Equation (4.15) can be proved by writing the displacements, U and V, in the x and y directions (Fig. 4.10) in terms of those in the r and θ directions.
    The relation

$$U = u\cos\theta - v\sin\theta;$$
$$V = u\sin\theta + v\cos\theta.$$

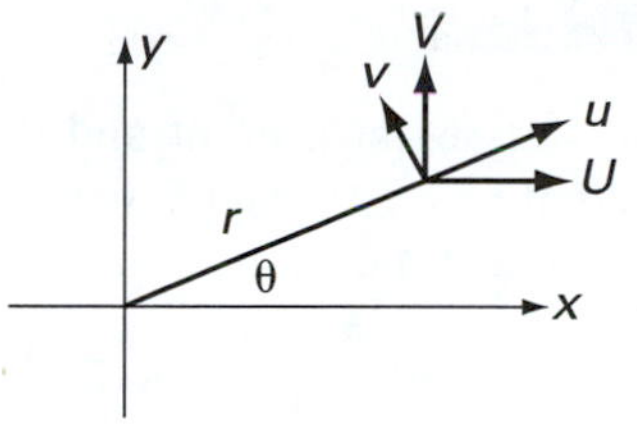

FIGURE 4.10   Relation between polar and Cartesian displacements.

is a clockwise rotation through θ from the polar to the Cartesian system. The idea is to write ϵ_x in terms of polar coordinates and then let θ vary so that ϵ_x first coincides with ϵ_r and then with ϵ_θ. By the chain rule,

$$\epsilon_x = \frac{\partial U}{\partial x} = \frac{\partial U}{\partial r}\frac{\partial r}{\partial x} + \frac{\partial U}{\partial \theta}\frac{\partial \theta}{\partial x}$$

$$= \left[\frac{\partial u}{\partial r}\cos\theta - \frac{\partial v}{\partial r}\sin\theta\right]\cos\theta$$

$$+ \left[\frac{\partial u}{\partial \theta}\cos\theta - u\sin\theta - \frac{\partial v}{\partial \theta}\sin\theta - v\cos\theta\right]\frac{(-\sin\theta)}{r},$$

using $r^2 = x^2 + y^2$ and $\theta = \arctan(y/x)$. As θ tends to 0, ϵ_x tends to ϵ_r. Therefore

$$\epsilon_r = \lim_{\theta \to 0} \epsilon_x = \frac{\partial u}{\partial r}.$$

Likewise, as θ tends to $\pi/2$, ϵ_x tends to ϵ_θ,

$$\epsilon_\theta = \lim_{\theta \to \pi/2} \epsilon_x = \frac{1}{r}\frac{\partial v}{\partial \theta} + \frac{u}{r}.$$

The shear strain is obtained from the fact that γ_{xy} tends to $\gamma_{r\theta}$ as θ tends to 0. The term $\partial U/\partial y + \partial V/\partial x$ is written in terms of polar coordinates by the chain rule and the limit is taken.

Exercise 4.14   Use the above strategy to verify the expression for the shear strain in polar coordinates.

Hooke's law, for an isotropic material in plane stress, takes the same form as the Cartesian case since the r and θ directions are perpendicular.

$$\epsilon_r = \frac{1}{E}[\sigma_r - \nu\sigma_\theta];$$

$$\epsilon_\theta = \frac{1}{E}[\sigma_\theta - \nu\sigma_r];$$

$$\gamma_{r\theta} = \frac{1}{G}\tau_{r\theta}. \tag{4.16}$$

The Poisson effect says that $\epsilon_z = (-\nu/E)(\sigma_\theta + \sigma_r)$.

4.6.1 Thick-walled Cylinders

The problem of determining the stresses at a point and the deformation of a thick-walled cylinder loaded by an internal and external pressure (Fig. 4.11) was first solved by Lamé. The internal pressure is p_i and the external pressure is p_e.

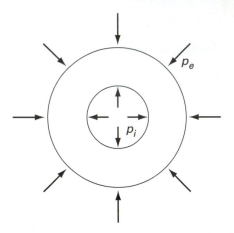

FIGURE 4.11   Thick-wall cylinder with internal and external pressure.

Lamé assumed plane stress so that the analysis applies to thin disks as well as to a cylinder with arbitrary length. The plane stress analysis applies to the cylinder, which at first glance might seem to be better viewed as in plane strain, because ϵ_z is independent of r as will be shown, so that the cylinder can be thought of as a stack of disks. Further, the solution satisfies the equations of elasticity with the given boundary conditions and is verified by experiment.

By symmetry, none of the stresses can depend on θ; further, the displacement v in the θ direction must be zero. Therefore the shear strain, $\gamma_{r\theta} = (1/r)(\partial u/\partial \theta) + \partial v/\partial r - v/r = 0$. The equilibrium equation in the r-direction is then, assuming negligible body forces,

$$\frac{\partial \sigma_r}{\partial r} + \frac{1}{r}(\sigma_r - \sigma_\theta) = 0. \tag{4.17}$$

Combining the strain-displacement relation and Hooke's law gives, since $v = 0$,

$$\epsilon_r = \frac{\partial u}{\partial r} = \frac{1}{E}(\sigma_r - v\sigma_\theta),$$

$$\epsilon_\theta = \frac{u}{r} = \frac{1}{E}(\sigma_\theta - v\sigma_r).$$

These equations are solved simultaneously for σ_r and σ_θ in terms of u and du/dr.

$$\sigma_r = \frac{E}{1 - v^2}\left(\frac{du}{dr} + v\frac{u}{r}\right)$$

$$\sigma_\theta = \frac{E}{1 - v^2}\left(\frac{u}{r} + v\frac{du}{dr}\right) \tag{4.18}$$

The results are then substituted into the equilibrium equation (4.17) to obtain a differential equation for $u(r)$,

$$\frac{d^2u}{dr^2} + \frac{1}{r}\frac{du}{dr} - \frac{u}{r^2} = 0. \tag{4.19}$$

This linear, second-order, ordinary differential equation with nonconstant coefficients is called a Euler equation. As long as $r \neq 0$, the substitution $r = e^t$ reduces the equation to a linear differential equation in t with constant coefficients. This trick works for any order Euler equation. Since $t = \ln(r)$, $dt/dr = 1/r$. By the chain rule, $du/dr = (du/dt)(dt/dr) = (1/r)(du/dt)$. By the product rule and chain rule,

$$\frac{d^2u}{dr^2} = \frac{d}{dr}\frac{du}{dr} = \frac{d}{dr}\left(\frac{1}{r}\frac{du}{dt}\right)$$

$$= -\frac{1}{r^2}\frac{du}{dt} + \frac{1}{r}\frac{d^2u}{dt^2}\frac{dt}{dr} = -\frac{1}{r^2}\frac{du}{dt} + \frac{1}{r^2}\frac{d^2u}{dt^2}.$$

Substituting these expressions for the first and second derivatives into equation (4.19) yields, after multiplying by r^2,

$$\frac{d^2u}{dt^2} - u = 0. \tag{4.20}$$

Equation (4.20) has the two independent solutions $u = e^t$ and $u = e^{-t}$ in terms of t. But since $t = \ln(r)$, the solutions in terms of r are $u = r$ and $u = 1/r$. The general solution of (4.19) is a linear combination of the two independent solutions.

$$u(r) = a_1 r + a_2 \frac{1}{r}.$$

Substitute this expression into equation (4.18) for σ_r and use the stress boundary conditions to solve for the constants, a_1 and a_2. At the inner radius, $r = a$, $\sigma_r = -p_i$ and at the outer radius $r = b$, $\sigma_r = -p_e$. The minus signs indicate that the stresses are compressive.

Exercise 4.15 Verify that the constants are $a_1 = [(1-v)/E](a^2 p_i - b^2 p_e)/(b^2 - a^2)$ and $a_2 = [(1+v)/E](p_i - p_e)a^2b^2/(b^2 - a^2)$. Then

$$u(r) = \left(\frac{1-v}{E}\right)\left(\frac{a^2 p_i - b^2 p_e}{b^2 - a^2}\right)r + \left(\frac{1+v}{E}\right)\left(\frac{(p_i - p_e)a^2b^2}{b^2 - a^2}\right)\frac{1}{r};$$

$$\sigma_r(r) = \frac{a^2 p_i - b^2 p_e}{b^2 - a^2} - \left[\frac{(p_i - p_e)a^2b^2}{b^2 - a^2}\right]\frac{1}{r^2};$$

$$\sigma_\theta(r) = \frac{a^2 p_i - b^2 p_e}{b^2 - a^2} + \left[\frac{(p_i - p_e)a^2b^2}{b^2 - a^2}\right]\frac{1}{r^2}. \tag{4.21}$$

Because of the manner in which the boundary condition was written, p_i and p_e are positive when the pressure is compressive. Observe that at any radius, the sum of σ_r and σ_θ is a constant;

$$\sigma_r(r) + \sigma_\theta(r) = 2\frac{a^2 p_i - b^2 p_e}{b^2 - a^2}. \tag{4.22}$$

Exercise 4.16 Verify using Hooke's law that, because $\sigma_z = 0$, ϵ_z is independent of r and is the constant

$$\epsilon_z = -\frac{v}{E}2\frac{a^2 p_i - b^2 p_e}{b^2 - a^2}. \tag{4.23}$$

Exercise 4.17 Choose values for the radii, a and b, and the pressures p_i and p_e. Plot the variation of σ_r and σ_θ with radius r. Do this for fixed a and b but different values of the pressures. Include the cases for which one of p_i or p_e are zero. Note that for $p_e = 0$, the magnitudes of σ_r and σ_θ are largest at $r = a$, σ_r is always compressive, and σ_θ is always tensile. But for $p_i = 0$, the magnitude of σ_r is largest at $r = b$ and that of σ_θ is largest at $r = a$, and both are always compressive.

EXAMPLE 4.14   The stress distribution in a solid rod in an external pressure, p_e, is determined from equation (4.21) by setting $a = 0$. The resulting stress state at any point is $\sigma_r = \sigma_\theta = -p_e$. Also recall that for an infinite rod, $\sigma_z = 0$ by assumption. Mohr's circle implies that the maximum shear stress in the r-θ plane is also equal to the external pressure. The radial deflection, on the other hand, depends on the distance from the center, $u(r) = -(1 - \nu)p_e r/E$. The axial strain is $\epsilon_z = 2\nu p_e/E$. ∎

EXAMPLE 4.15   An optical fiber, made of silica, can be embedded in a solid to monitor the state of strain in the body, which is often called a smart structure. The optical fiber strain sensor is designed by controlling the properties of the coatings put on the fiber. The resulting coated fiber is modeled as a thin solid cylinder encased by a thick hollow cylinder representing the coating which is made of a different material. Even though this is not yet a mature technology, such an optical fiber sensor has been designed [1] for a rod under simple loads, an axial load and an external pressure, in which the structure is viewed as in plane strain. The governing differential equation is in terms of the radial displacement in the cylinder (4.19) and is solved for the radial displacement in the optical fiber and the encasing cylinder as well as for ϵ_z in terms of the loads. This displacement is then related to the phase change of the light passing through the fiber. The phase change, after calibration, then measures the loads. One problem is that the presence of the optical fiber may affect the ambient state of strain in the body. ∎

Interference Fits   An interference fit, also called a shrink or press fit depending on how it is made, is a method to join a thick-walled cylinder around another thick-walled cylinder or a rod without using a fastener. The inner cylinder is made with an outer radius slightly larger than the inner radius of the hollow cylinder which is to go over it. The two pieces are mated by raising the temperature of the outer cylinder and/or lowering the temperature of the inner cylinder until the inner slips into the outer. Typically both cylinders are metal. As the parts come to an equilibrium temperature, the outer cylinder shrinks onto the inner cylinder and a radial pressure develops between them at their interface. The pressure not only deforms the cylinders, it also induces stresses which could cause the mated assembly to fail. The pressure also resists external loads which might tend to separate the mated pieces. To design a device using this means of fastening, the interface pressure first must be determined. The resulting internal stresses and deflections are computed from equations (4.21).

The pressure developed depends on interference between the two rods. Denote the displacement of the inner rod by $u_i(r)$ and of the outer by $u_o(r)$. The inner cylinder has inner radius, a_i, and outer radius, b_i. The outer cylinder has inner radius, a_o, and outer radius, b_o. If $b_i - a_o > 0$ the cylinders are said to interfere because they cannot fit together at room temperature. The radial interference is $\delta = b_i - a_o$.

In machine design, the types of interference fits are defined in terms of the inner radius, a_o, of the enveloping hollow cylinder. The interference for a tight fit is 0.00025 a_o, for a medium force fit is 0.0005 a_o, and for a shrink fit is 0.001 a_o.

To just allow the outer cylinder to fit over the inner one, it is necessary that $u_i(b_i) - u_o(a_o) = \delta$; the deformed radius difference must equal the interference. Denote the interface pressure by p and assume that this is the only pressure on the two cylinders. It is the external pressure for the inner rod and the internal pressure for the outer rod. For the outer rod, the radial deformation at the interface, $r = a_o$, under the interface pressure is

$$
u_o(a_o) = \left(\frac{1 - \nu_o}{E_o}\right)\left(\frac{a_o^2 p}{b_o^2 - a_o^2}\right) a_o + \left(\frac{1 + \nu_o}{E_o}\right)\left(\frac{pa_o^2 b_o^2}{b_o^2 - a_o^2}\right)\frac{1}{a_o}
$$

$$
= \frac{pa_o}{E_o(b_o^2 - a_o^2)}[(1 - \nu_o)a_o^2 + (1 + \nu_o)b_o^2].
$$

Likewise, for the inner cylinder at the interface, $r = b_i$,

$$
u_i(b_i) = \left(\frac{1 - \nu_i}{E_i}\right)\left(\frac{-b_i^2 p}{b_i^2 - a_i^2}\right) b_i + \left(\frac{1 + \nu_i}{E_i}\right)\left(\frac{-pa_i^2 b_i^2}{b_i^2 - a_i^2}\right)\frac{1}{b_i}
$$

$$
= \frac{-pb_i}{E_i(b_i^2 - a_i^2)}[(1 - \nu_i)b_i^2 + (1 + \nu_i)a_i^2].
$$

To compute the interface pressure in terms of δ, substitute these results in $u_i(b_i) - u_o(a_o) = \delta$ and solve for p. The result is negative because p is compressive.

A simpler approximation for the magnitude of p is obtained by letting $b_i \sim a_o \equiv c$,

$$
p = \frac{\delta}{c}\left[\frac{1}{E_o}\left(\frac{b_o^2 + c^2}{b_o^2 - c^2} + \nu_o\right) + \frac{1}{E_i}\left(\frac{c^2 + a_i^2}{c^2 - a_i^2} - \nu_i\right)\right]^{-1} \tag{4.24}
$$

Exercise 4.18   *Verify equation (4.24).*

This calculation of the interface pressure assumes that the two rods are both of infinite length. When a ring or disk is interference fit on a cylinder, so that the lengths are not the same, there is a stress concentration at the edge of the ring or disk. This stress concentration depends on the geometry of the ring or disk and the cylinder on which it is fit, but for design purposes a stress concentration factor of 2 applied to the results for equal length members is a conservative estimate.

A compound member made by interference fitting two hollow cylinders may be loaded by an internal and external pressure. In this case, the stress due to the interface pressure may be viewed as a residual stress. The total stress is obtained by adding to this residual pressure the stress due to the external and internal pressures. This superposition is valid because the body material is linear elastic.

The temperature difference, ΔT, between the inner and outer members which allows the outer member to slip over the inner can be computed from the radial strain expression of Hooke's law, with coefficient of thermal expansion, α,

$$
\epsilon_r(r) = \frac{1}{E}\sigma_r(r) - \frac{\nu}{E}\big(\sigma_\theta(r) + \sigma_z(r)\big) + \alpha\Delta T. \tag{4.25}
$$

However, in an unconstrained cylinder, $\sigma_z = \sigma_\theta = \sigma_r = 0$ so that $\epsilon_r(r) = \alpha\Delta T$. Therefore, the radial deformation is $u(r) = \alpha\Delta T r$.

When estimating the temperature difference required, keep in mind that the thermal coefficient of expansion given in handbooks is an average, usually over a range such as 0–100°C. In fact, the thermal expansion coefficient is a function of temperature but for a metal is nearly constant at room temperature and above. Secondly, if the temperature is raised significantly for a lengthy time, diffusion will occur in the metal. Such diffusion can both cause grain size growth and eliminate dislocations so the material properties change; for example, the material may not be as strong as it was before heating. Finally, the temperature difference should be larger than that required to just mate the members so that there is enough clearance to allow simple assembly.

EXAMPLE 4.16   A composite member is made of a hollow brass cylinder ($E = 110$ GPa, $v = 0.32$, and $\alpha = 18.7 \times 10^{-6}$ /C°) interference fit over a solid steel rod ($E = 210$ GPa, $v = 0.28$, and $\alpha = 11 \times 10^{-6}$/°C). The outer radius of the brass cylinder is 100 mm, the inner radius is 50 mm, and the outer radius of the steel rod is 50.05 mm. The length of both is the same. Compute a temperature difference required to easily slip the brass cylinder over the steel rod. Compute the interface pressure of the interference fit at room temperature.

Solution   The interference is $\delta = 0.05$ mm. Allow a clearance of, say, 0.025 mm. Therefore if the brass is heated, the temperature difference must be

$$\Delta T = \frac{u}{\alpha r} = \frac{\delta + 0.025}{\alpha r} = \frac{0.05 + 0.025}{(18.7 \times 10^{-6})50} = 80.2°\text{C}.$$

The interface pressure is, by equation (4.24) with $a_i = 0$,

$$p = \frac{\delta}{c}\left[\frac{1}{E_o}\left(\frac{b_o^2 + c^2}{b_o^2 - c^2} + v_o\right) + \frac{1}{E_i}\left(\frac{c^2 + a_i^2}{c^2 - a_i^2} - v_i\right)\right]^{-1}$$

$$= \frac{0.05}{50}\left[\frac{1}{110 \times 10^9}\left(\frac{100^2 + 50^2}{100^2 - 50^2} + 0.32\right) + \frac{1}{210 \times 10^9}(1 - 0.28)\right]^{-1} = 46.5 \text{ MPa.}\ \blacksquare$$

EXAMPLE 4.17   A 0.5-cm-thick, 12-cm-diameter disk is shrink-fit onto a 2-cm-diameter solid shaft. Both are made of an aluminum with $E = 70$ GPa and $v = 0.34$. Design the interference so that the disk can sustain a torque of $M_t = 6$ kNm without slipping when the shaft is held fixed.

Solution   A static analysis gives an approximation which can be used for the design. The interface pressure is obtained from equation (4.24), $p = 3,403 \times 10^9 \delta$ Pa. The total normal force on the inner bore of the disk is $N_p = p(2\pi c)t = p(2\pi)(0.01)(0.005) = 1.069 \times 10^9 \delta$ N. The total frictional force on the inner bore at impending slip is $f = \mu N_p = 0.428 \times 10^9 \delta$, assuming that 0.40 is the static coefficient of friction for steel on steel. The disk will not slip as long as $M_t < fc = 4.28 \times 10^6 \delta$. So the interference must be $\delta > 0.001403$ m. $\blacksquare$

Again, after assembly of a shrink fit component, the interface pressure, p, induces a "prestress" in the component. This prestress may increase the allowable service load on the assembled component. Typically the magnitude of the circumferential stress, $\sigma_\theta(r)$, jumps at the interface.

Hollow Cylinder with Closed Ends   A finite length, isotropic, hollow cylinder with closed ends has the same cross-sectional stresses, σ_r and σ_θ, as an open-ended cylinder at a point far enough from the ends. However, the pressures induce a normal stress, $\sigma_z(r)$, in the direction of the longitudinal cylindrical z- axis. Here cylindrical coordinates are used. There also may be an axial load, P, applied at the ends.

To determine $\sigma_z(r)$, observe that the summation of the forces in the longitudinal direction on the end plate implies that

$$P + p_i\pi a^2 = p_e\pi b^2 + \int_a^b \sigma_z 2\pi r\, dr. \tag{4.26}$$

The strain, $\epsilon_z(r)$, is constant over a cross-sectional face of the cylinder. The Hooke's law expression for $\epsilon_z(r)$ is

$$\epsilon_z(r) = \frac{1}{E}\sigma_z(r) - \frac{v}{E}\left(\sigma_\theta(r) + \sigma_r(r)\right) + \alpha\Delta T. \tag{4.27}$$

The integral of equation (4.27) over the cross-sectional area of the rod is

$$\epsilon_z \int_a^b 2\pi r\,dr = \frac{1}{E}\int_a^b \sigma_z 2\pi r\,dr - \frac{\nu}{E}\int_a^b (\sigma_r(r)+\sigma_\theta(r))2\pi r\,dr + \int_a^b (\alpha\Delta T)2\pi r\,dr.$$

This becomes, using the fact (4.22) that $\sigma_r(r)+\sigma_\theta(r)$ is a constant and the expression for the integral of σ_z obtained from equation (4.26),

$$\epsilon_z \pi(b^2-a^2) = \frac{1}{E}[P+p_i\pi a^2 - p_e\pi b^2] + 2\frac{\nu}{E}\pi(a^2 p_i - b^2 p_e) + \int_a^b \alpha\Delta T 2\pi r\,dr.$$

Substitution of this form of ϵ_z into (4.27) and the constant expression for $\sigma_r(r)+\sigma_\theta(r)$ yields

$$\sigma_z = \frac{a^2 p_i - b^2 p_e}{b^2 - a^2} + \frac{P}{\pi(b^2-a^2)}. \tag{4.28}$$

Notice that σ_z is independent of r so that σ_z is constant over the cross-sectional face as expected from equation (4.27). Further, the stress σ_z is independent of any temperature change.

Exercise 4.19 Convert the stress equations for a thick-walled cylinder to those for the elementary strength of materials internally pressurized thin-walled cylinder by letting $b/a \to 1$ and $p_e = 0$. Show that $\sigma_r \to 0$, $\sigma_\theta \to \sigma_{hoop} = p_i b/t$, and $\sigma_z \to \sigma_{longitudinal} = p_i b/2t$. *Hint*: Factor $b^2 - a^2$ and set $t = b - a$. Hold t fixed, while letting $a \to b$ in the remainder of the expression.

Rotating Disks and Cylinders Many machine components are designed to rotate at a constant speed. Such components include motor shafts, gears, some flywheels, etc. The rotation itself induces stresses due to the body forces irrespective of other loads on the component.

A thin constant thickness circular disk of radius b and having a center hole of radius a rotates about its center with constant angular velocity ω. The mass per unit volume of the material is ρ. Assume, as for the pressure loaded disk, that the stress at any point depends only on r, not on θ. The derivation of the governing equation follows the same pattern as that for the static disk except that the body forces are not neglected.

The body forces are determined from the accelerations written in tangential-normal components and Newton's law. The radial body force per unit volume is $F_r = \rho\omega^2 r$ and the angular body force per unit volume, F_θ, is zero because the angular velocity is constant. The equilibrium equation in the radial direction is

$$\frac{\partial\sigma_r}{\partial r} + \frac{1}{r}(\sigma_r - \sigma_\theta) = -\rho\omega^2 r. \tag{4.29}$$

As above, the combination of the strain-displacement relations and Hooke's law are solved simultaneously for σ_r and σ_θ in terms of u and du/dr to obtain equation (4.18). These results are then substituted into the equilibrium equation (4.29) to obtain a differential equation for $u(r)$,

$$\frac{d^2 u}{dr^2} + \frac{1}{r}\frac{du}{dr} - \frac{u}{r^2} = -\frac{(1-\nu^2)}{E}\rho\omega^2 r. \tag{4.30}$$

This linear, second-order, ordinary differential equation with nonconstant coefficients is a nonhomogeneous Euler equation. The solution is the sum of the solutions to the associated homogeneous equation and a particular solution, $u(r) = u_h(r) + u_p(r)$. A particular solution may be guessed of the form $u_p(r) = br^3$. Substitution shows that $b = (1 - v^2)\rho\omega^2/8E$. Therefore using the result for the homogeneous Euler equation in the static disk analysis,

$$u(r) = a_1 r + a_2 \frac{1}{r} - \frac{(1 - v^2)}{8E}\rho\omega^2 r^3. \tag{4.31}$$

Substitute this expression in the equations for the stress in terms of u (4.18) and use the boundary conditions to determine the constants a_1 and a_2.

Case 1: Solid disk. The stress at $r = 0$ cannot be infinite so that $a_2 = 0$. The condition that $\sigma_r(b) = 0$ determines a_1. The stresses are then

$$\sigma_r(r) = \frac{3 + v}{8}\rho\omega^2 \left(b^2 - r^2\right); \tag{4.32}$$

$$\sigma_\theta(r) = \frac{3 + v}{8}\rho\omega^2 \left(b^2 - \frac{1 + 3v}{3 + v}r^2\right). \tag{4.33}$$

Case 2: Disk with a hole at the center. The boundary conditions are $\sigma_r(a) = \sigma_r(b) = 0$. The stresses are then

$$\sigma_r(r) = \frac{3 + v}{8}\rho\omega^2 \left(a^2 + b^2 + \frac{a^2 b^2}{r^2} - r^2\right); \tag{4.34}$$

$$\sigma_\theta(r) = \frac{3 + v}{8}\rho\omega^2 \left(a^2 + b^2 + \frac{a^2 b^2}{r^2} - \frac{1 + 3v}{3 + v}r^2\right). \tag{4.35}$$

Note that this result reduces to that for the solid disk when a is set equal to zero. But the stress in the solid disk is not zero at $r = 0$.

Exercise 4.20   Using calculus, show that σ_r is maximum at $r = \sqrt{ab}$. Calculate the maximum σ_r.

EXAMPLE 4.18   A thin flat disk of radius 10 cm is shrink-fit onto a solid shaft of diameter 3 cm which rotates at 1,200 rpm. Both are made of a steel with $\rho = 7{,}860$ kg/m³, $E = 200$ GPa, and $v = 0.25$. Design the interference so that the disk will not separate from the shaft while the system is spinning.

Solution   Denote the interface pressure by p, by equation (4.24),

$$p = \frac{\delta}{c}\left[\frac{1}{E_o}\left(\frac{b_o^2 + c^2}{b_o^2 - c^2} + v_o\right) + \frac{1}{E_i}\left(\frac{c^2 + a_i^2}{c^2 - a_i^2} - v_i\right)\right]^{-1}$$

$$= \frac{\delta}{0.015}200 \times 10^9 \left[\left(\frac{0.1^2 + 0.015^2}{0.1^2 - 0.015^2} + 0.25\right) + (1 - 0.25)\right]^{-1} = 6{,}517 \times 10^9 \delta.$$

At separation, the radial stress must be zero at the interface. The disk and cylinder system acts approximately like a thin solid disk; but at separation the radial stress is zero at the interface $c = 0.015$ m. The stress due to the interface pressure and that due to the body forces are superposed so that

$$\sigma_r(c) = \frac{3 + v}{8}\rho\omega^2 \left(b^2 - c^2\right) - p = 4{,}928{,}983 - 6517 \times 10^9 \delta = 0.$$

Recall that 1200 rpm = 40π rad/s. Therefore, the required interference is greater than $\delta = 7.56 \times 10^{-8}$ m. The risk of the shrink fit separating is negligible in such a small radius system. However, in a large disk on a large diameter shaft system, the designer may have to compute the interference required to prevent separation. ∎

4.6.2 The Airy Stress Function in Polar Coordinates

In polar coordinates, the Airy stress function, $\phi(r, \theta)$, is one satisfying

$$\sigma_r = \frac{1}{r}\frac{\partial \phi}{\partial r} + \frac{1}{r^2}\frac{\partial^2 \phi}{\partial \theta^2};$$

$$\sigma_\theta = \frac{\partial^2 \phi}{\partial r^2};$$

$$\tau_{r\theta} = -\frac{\partial}{\partial r}\left(\frac{1}{r}\frac{\partial \phi}{\partial \theta}\right). \tag{4.36}$$

These relations can be proved using the facts that $\partial r/\partial x = \cos\theta$, $\partial r/\partial y = \sin\theta$, $\partial\theta/\partial x = -\sin(\theta)/r$, and $\partial\theta/\partial y = \cos(\theta)/r$, from $r^2 = x^2 + y^2$ and $\theta = \arctan(y/x)$, with the chain rule, $\partial\phi/\partial x = (\partial\phi/\partial r)(\partial r/\partial x) + (\partial\phi/\partial\theta)(\partial\theta/\partial x) = (\partial\phi/\partial r)\cos\theta - (\partial\phi/\partial\theta)(\sin\theta/r)$ and $\partial\phi/\partial y = (\partial\phi/\partial r)(\partial r/\partial y) + (\partial\phi/\partial\theta)(\partial\theta/\partial y) = (\partial\phi/\partial r)\sin\theta + (\partial\phi/\partial\theta)(\cos\theta/r)$. By the product rule and the chain rule,

$$\sigma_x = \frac{\partial^2 \phi}{\partial y^2} = \frac{\partial^2 \phi}{\partial r^2}(\sin\theta)^2 + \frac{2}{r}\frac{\partial^2 \phi}{\partial r \partial\theta}\sin\theta\cos\theta$$

$$- \frac{2}{r^2}\frac{\partial \phi}{\partial\theta}\sin\theta\cos\theta + \frac{1}{r}\frac{\partial \phi}{\partial r}(\cos\theta)^2 + \frac{1}{r^2}\frac{\partial^2 \phi}{\partial\theta^2}(\cos\theta)^2;$$

$$\sigma_y = \frac{\partial^2 \phi}{\partial x^2} = \frac{\partial^2 \phi}{\partial r^2}(\cos\theta)^2 - \frac{2}{r}\frac{\partial^2 \phi}{\partial r \partial\theta}\sin\theta\cos\theta$$

$$+ \frac{2}{r^2}\frac{\partial \phi}{\partial\theta}\sin\theta\cos\theta + \frac{1}{r}\frac{\partial \phi}{\partial r}(\sin\theta)^2 + \frac{1}{r^2}\frac{\partial^2 \phi}{\partial\theta^2}(\sin\theta)^2;$$

$$\tau_{xy} = -\frac{\partial^2 \phi}{\partial x \partial y} = -\frac{\partial^2 \phi}{\partial r^2}\sin\theta\cos\theta - \frac{1}{r}\frac{\partial^2 \phi}{\partial r \partial\theta}(\cos^2\theta - \sin^2\theta)$$

$$+ \frac{1}{r^2}\frac{\partial \phi}{\partial\theta}(\cos^2\theta - \sin^2\theta) + \frac{1}{r}\frac{\partial \phi}{\partial r}\sin\theta\cos\theta + \frac{1}{r^2}\frac{\partial^2 \phi}{\partial\theta^2}\sin\theta\cos\theta. \tag{4.37}$$

As θ tends to 0, σ_x tends to σ_r, σ_y tends to σ_θ, and τ_{xy} tends to $\tau_{r\theta}$. Taking the limits in the previous expressions produces equations (4.36). The expression, $\tau_{r\theta} = -(1/r)(\partial^2\phi/\partial r\partial\theta) + (1/r^2)(\partial\phi/\partial\theta)$, can be rewritten in the form $\tau_{r\theta} = -(\partial/\partial r)[(1/r)(\partial\phi/\partial\theta)]$.

Exercise 4.21   Substitute the stresses in terms of the polar form of the Airy stress function into the polar form of the equilibrium equations to verify that this definition of the Airy stress function forces the equilibrium equations to be satisfied.

EXAMPLE 4.19   Suppose the Airy stress function is $\phi(r, \theta) = Pr\theta \sin \theta$. Then the stresses are

$$\sigma_r = \frac{1}{r}\frac{\partial \phi}{\partial r} + \frac{1}{r^2}\frac{\partial^2 \phi}{\partial \theta^2}$$

$$= \frac{1}{r}P\theta \sin \theta + \frac{1}{r^2}Pr(\cos \theta + \cos \theta - \theta \sin \theta) = \frac{2}{r}P\cos \theta;$$

$$\sigma_\theta = \frac{\partial^2 \phi}{\partial r^2} = 0; \quad \text{and} \quad \tau_{r\theta} = -\frac{\partial}{\partial r}\left[\frac{1}{r}\frac{\partial \phi}{\partial \theta}\right] = 0. \qquad \blacksquare$$

The biharmonic equation in terms of polar coordinates is based on the Laplace operator, which for a function $F(r, \theta)$ is

$$\nabla^2 F(r, \theta) = \frac{\partial^2 F}{\partial r^2} + \frac{1}{r}\frac{\partial F}{\partial r} + \frac{1}{r^2}\frac{\partial^2 F}{\partial \theta^2}. \qquad (4.38)$$

This relation can be proved directly from $\nabla^2 \phi = \partial^2 \phi/\partial x^2 + \partial^2 \phi/\partial y^2$ by a change of coordinates. Alternatively, since the trace of the stress tensor is invariant under coordinate rotations, $\sigma_x + \sigma_y = \sigma_r + \sigma_\theta$ in plane stress. But by definition of the Airy stress function in Cartesian coordinates, $\nabla^2 \phi = \sigma_x + \sigma_y$. From equations (4.37), $\nabla^2 \phi = \sigma_x + \sigma_y = \partial^2 \phi/\partial r^2 + (1/r)(\partial \phi/\partial r) + (1/r^2)(\partial^2 \phi/\partial \theta^2)$, since $\cos^2 \theta + \sin^2 \theta = 1$.

Exercise 4.22   Verify that $\sigma_x + \sigma_y = \sigma_r + \sigma_\theta$ directly from equations (4.36) and (4.37).

The biharmonic equation expressed in polar coordinates is

$$\nabla^4 \phi(r, \theta) = \left[\frac{\partial^2}{\partial r^2} + \frac{1}{r}\frac{\partial}{\partial r} + \frac{1}{r^2}\frac{\partial^2}{\partial \theta^2}\right]\nabla^2 \phi(r, \theta)$$

$$= \left[\frac{\partial^2}{\partial r^2} + \frac{1}{r}\frac{\partial}{\partial r} + \frac{1}{r^2}\frac{\partial^2}{\partial \theta^2}\right]\left[\frac{\partial^2 \phi}{\partial r^2} + \frac{1}{r}\frac{\partial \phi}{\partial r} + \frac{1}{r^2}\frac{\partial^2 \phi}{\partial \theta^2}\right] = 0. \qquad (4.39)$$

EXAMPLE 4.20   The model for an edge dislocation is reported to have come to Volterra while working on a stress analysis in welded rolled sheet metal used to make piping.

The Airy stress function for an edge dislocation is chosen by analogy with the two-dimensional problem of a ring with a slit and offset d. The figure on the right (Fig. 4.12) looks like an end view of a pipe made from a rolled sheet. For this problem, if the elastic modulus of the material is E, the Airy stress function in polar coordinates is $\phi(r, \theta) = -[Ed/8\pi(1 - v^2)]r\ln(r^2)\sin \theta$. The edge dislocation moves in the x-y plane with the x-direction perpendicular to the inserted plane and the z-direction along the edge. In cylindrical coordinates (r, θ, z), the z-axis is along the dislocation line. The core of the edge dislocation is a thin cylinder surrounding the dislocation line. Inside the core, of radius r_0, the

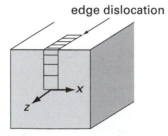

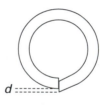

FIGURE 4.12   Edge dislocation.

distortion is so great that linear elasticity cannot be used, even as an approximation. But outside the core, the induced stresses may be described by linear elasticity. The region outside the cylinder surrounding the dislocation line is assumed to be in plane strain ($\epsilon_z = 0$). By analogy to the slit ring, the edge dislocation has Airy stress function in Cartesian coordinates, $\phi(x, y) = -[Gb/4\pi(1-\nu)]y\ln(x^2+y^2)$, where b is the length of the Burgers vector. Note that this function is not defined at the dislocation line ($x = y = 0$).

Exercise 4.23   Assume plane strain, and verify that the three normal stresses and the shear stress in the x-y plane for the edge dislocation are

$$\sigma_x = \frac{-Gb}{2\pi(1-\nu)}\frac{y(3x^2+y^2)}{(x^2+y^2)^2}; \qquad \sigma_y = \frac{Gb}{2\pi(1-\nu)}\frac{y(x^2-y^2)}{(x^2+y^2)^2};$$

$$\sigma_z = \nu(\sigma_x+\sigma_y); \qquad \tau_{xy} = \frac{Gb}{2\pi(1-\nu)}\frac{x(x^2-y^2)}{(x^2+y^2)^2}.$$

Note the singularity at the origin, as expected. These stresses may be thought of as residual stresses induced in the metal by the presence of the dislocation.

In the discussion of plasticity in Chapter 9, the question of whether or not a dislocation induces a volume change becomes relevant. Because the region is in plane strain with $\epsilon_z = 0$, $\sigma_z = \nu(\sigma_x+\sigma_y)$. By the results in Chapter 3.3, for small strains and an isotropic linearly elastic solid, the volume change per unit initial volume at each point (x, y) is, using the stresses determined in the exercise,

$$\frac{\Delta V}{V_o} = \frac{1-2\nu}{E}(\sigma_x+\sigma_y+\sigma_z) = -\frac{(1-2\nu)b}{2\pi(1-\nu)}\frac{y}{x^2+y^2}.$$

Therefore the upper half-space ($y > 0$) containing the extra half-plane of atoms decreases in volume; it is in compression as expected to force in the extra half-plane of atoms. The lower half-space where $y < 0$ increases in volume and is in tension because the atoms there are wedged apart by the half plane. However, the volume change at each point (x, y) is symmetric about the x-axis. The total volume change induced by the edge dislocation in the full volume excluding the core is therefore zero.

The strain energy induced by the edge dislocation of length L, in a large cylinder of radius, R, surrounding the core, is

$$U = \frac{Gb^2L}{4\pi(1-\nu)}\ln\left(\frac{R}{r_0}\right).$$

It is proportional to the square of the magnitude, b, of the Burgers vector, as was the case with the screw dislocation (Chapter 3, Section 3.6). This energy can be viewed as stored in the material because of the existence of the edge dislocation. Even though the calculation uses elasticity, the dislocation is not destroyed when a load is removed and so the energy is held in the material. A heat treatment, such as annealing, is one way to destroy a dislocation by diffusing atoms to complete the lattice structure. ■

4.7 PLATE WITH A HOLE

A uniaxially loaded infinite thin plate with a hole (Fig. 4.13) serves as a good example to introduce the idea of a stress concentration. Furthermore, it suggests an approach to the analysis of cracks and fracture.

A polar form for the Airy stress function for a plate without a hole is used to obtain, by analogy, the form of the Airy stress function for the plate with a hole. The boundary conditions will produce the undetermined coefficients in this form. This is the semi-inverse method of determining the Airy stress function. The plate is uniaxially

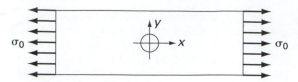

FIGURE 4.13   Axially in-plane loaded plate with a hole.

loaded with stress, σ_0, in the x-direction. The plate will be assumed to be in plane stress since it is very thin.

The behavior of the plate with the hole might be best described in polar coordinates because of the boundary conditions at the hole. The normal component of stress on the plane tangent to the hole is in the radial direction. This fact suggests using polar coordinates. The radial normal stress and the shear stress components are both zero on the surface of the hole. The plate is of infinite dimensions so that the boundary conditions at the edges can also be translated from Cartesian to polar form.

In Cartesian coordinates with x-axis parallel to the load, the stresses for the plate without a hole are $\sigma_x = \sigma_0$, $\sigma_y = \tau_{xy} = 0$. The Airy stress function for this state of stress and boundary conditions is $\phi(x, y) = \sigma_0 y^2/2$. This stress function can be transformed to polar coordinates by the substitution, $y = r \sin\theta$, where $\theta = 0$ along the x-axis, the direction of the stress, σ_0. The polar Airy stress function for uniaxial loading is

$$\phi(r, \theta) = \frac{1}{2}\sigma_0 y^2 = \frac{1}{2}\sigma_0 (r \sin\theta)^2 = \frac{1}{4}\sigma_0 r^2 (1 - \cos 2\theta). \qquad (4.40)$$

This expression is in the form of a function of r plus a function of r times $\cos 2\theta$. The stresses are

$$\sigma_r = \frac{1}{r}\frac{\partial\phi}{\partial r} + \frac{1}{r^2}\frac{\partial^2\phi}{\partial\theta^2} = \frac{1}{2}\sigma_0(1 + \cos 2\theta);$$

$$\sigma_\theta = \frac{\partial^2\phi}{\partial r^2} = \frac{1}{2}\sigma_0(1 - \cos 2\theta);$$

$$\tau_{r\theta} = -\frac{\partial}{\partial r}\left(\frac{1}{r}\frac{\partial\phi}{\partial\theta}\right) = -\frac{1}{2}\sigma_0 \sin 2\theta. \qquad (4.41)$$

Exercise 4.24   Verify these stress results from equation (4.40).

The Airy function for the plate with a hole is assumed to have the same form as that for the plate without a hole,

$$\phi(r, \theta) = f_1(r) + f_2(r)\cos(2\theta). \qquad (4.42)$$

The functions, $f_1(r)$ and $f_2(r)$, which are allowed to differ, are identical for the plate without a hole. If the functions $f_1(r)$ and $f_2(r)$ can be determined, then the stresses can be computed from $\phi(r, \theta)$. The function, $\phi(r, \theta)$, must satisfy both the biharmonic equation and the boundary conditions. Application of the biharmonic produces two ordinary differential equations, one for $f_1(r)$ and the other for

$f_2(r)$. The Laplacian is

$$\nabla^2 \phi(r, \theta) = \frac{\partial^2 \phi}{\partial r^2} + \frac{1}{r}\frac{\partial \phi}{\partial r} + \frac{1}{r^2}\frac{\partial^2 \phi}{\partial \theta^2}$$

$$= \frac{d^2 f_1}{dr^2} + \frac{d^2 f_2}{dr^2}\cos(2\theta) + \frac{1}{r}\left(\frac{df_1}{dr} + \frac{df_2}{dr}\cos(2\theta)\right) - \frac{4}{r^2}f_2\cos(2\theta).$$

The biharmonic equation can be broken into two summands, one involving $f_1(r)$ and the other $f_2(r)$.

$$\nabla^4 \phi(r, \theta) = \left[\frac{\partial^2}{\partial r^2} + \frac{1}{r}\frac{\partial}{\partial r} + \frac{1}{r^2}\frac{\partial^2}{\partial \theta^2}\right]\nabla^2 \phi(r, \theta)$$

$$= \left[\frac{\partial^2}{\partial r^2} + \frac{1}{r}\frac{\partial}{\partial r} + \frac{1}{r^2}\frac{\partial^2}{\partial \theta^2}\right]\left[\frac{d^2 f_1}{dr^2} + \frac{1}{r}\frac{df_1}{dr} + \left(\frac{d^2 f_2}{dr^2} + \frac{1}{r}\frac{df_2}{dr} - \frac{4}{r^2}f_2\right)\cos(2\theta)\right]$$

$$= 0.$$

Then, since $f_1(r)$ and $f_2(r)$ are independent, the summands must be separately zero.

$$\left[\frac{\partial^2}{\partial r^2} + \frac{1}{r}\frac{\partial}{\partial r} + \frac{1}{r^2}\frac{\partial^2}{\partial \theta^2}\right]\left[\frac{d^2 f_1}{dr^2} + \frac{1}{r}\frac{df_1}{dr}\right] = 0;$$

$$\left[\frac{\partial^2}{\partial r^2} + \frac{1}{r}\frac{\partial}{\partial r} + \frac{1}{r^2}\frac{\partial^2}{\partial \theta^2}\right]\left(\frac{d^2 f_2}{dr^2} + \frac{1}{r}\frac{df_2}{dr} - \frac{4}{r^2}f_2\right)\cos(2\theta) = 0.$$

The second equation must hold for all θ; therefore the coefficient of $\cos(2\theta)$ must also be zero. The resulting two differential equations in $f_1(r)$ and $f_2(r)$ involve only functions of r; θ is not present.

$$\left[\frac{\partial^2}{\partial r^2} + \frac{1}{r}\frac{\partial}{\partial r}\right]\left[\frac{d^2 f_1}{dr^2} + \frac{1}{r}\frac{df_1}{dr}\right] = 0;$$

$$\left[\frac{\partial^2}{\partial r^2} + \frac{1}{r}\frac{\partial}{\partial r} - \frac{4}{r^2}\right]\left(\frac{d^2 f_2}{dr^2} + \frac{1}{r}\frac{df_2}{dr} - \frac{4}{r^2}f_2\right) = 0.$$

If these equations are expanded, they can be seen to be fourth-order Euler equations. The substitution $t = \ln(r)$ converts them into fourth-order equations with constant coefficients. They have solutions

$$f_1(r) = c_1 r^2 \ln r + c_2 r^2 + c_3 \ln r + c_4;$$

$$f_2(r) = c_5 r^2 + c_6 r^4 + c_7 \frac{1}{r^2} + c_8.$$

Recall that $f_1(r)$ and $f_2(r)$ are multiples of only r^2 for the plate without a hole. The boundary conditions are in terms of stress; therefore they must be applied to the stress equations. Eight facts are required to obtain the eight unknown coefficients of $f_1(r)$

and $f_2(r)$.

$$
\begin{aligned}
\sigma_r &= \frac{1}{r}\frac{\partial\phi}{\partial r} + \frac{1}{r^2}\frac{\partial^2\phi}{\partial\theta^2} \\
&= \frac{1}{r}\left[\frac{df_1}{dr} + \frac{df_2}{dr}\cos(2\theta)\right] + \frac{1}{r^2}[-4f_2\cos(2\theta)] \\
&= c_1(1+2\ln r) + 2c_2 + \frac{1}{r^2}c_3 - \left(2c_5 + 6c_7\frac{1}{r^4} + 4c_8\frac{1}{r^2}\right)\cos(2\theta);
\end{aligned}
$$

$$
\sigma_\theta = \frac{\partial^2\phi}{\partial r^2} = c_1(3+2\ln r) + 2c_2 - \frac{1}{r^2}c_3 + \left(2c_5 + 12c_6 r^2 + 6c_7\frac{1}{r^4}\right)\cos(2\theta);
$$

$$
\tau_{r\theta} = -\frac{\partial}{\partial r}\left(\frac{1}{r}\frac{\partial\phi}{\partial\theta}\right) = \left(2c_5 + 6c_6 r^2 - 6c_7\frac{1}{r^4} - 2c_8\frac{1}{r^2}\right)\sin(2\theta). \tag{4.43}
$$

The constant c_4 does not appear so that it may be taken to be zero. All the stresses are assumed to be finite at each point (r, θ). The terms $\ln(r)$ and r^2 cannot appear in the stress expressions since they get very large as r increases. To remove them, the constants $c_1 = c_6 = 0$. Since the plate is assumed to be infinite, as r tends to infinity, σ_r tends to $(1/2)\sigma_0(1 + \cos 2\theta)$, the uniaxial stress. For large r, the only terms remaining are $\sigma_r \sim 2c_2 - 2c_5\cos(2\theta)$ so that $c_2 = \sigma_0/4$ and $c_5 = -\sigma_0/4$. This accounts for five of the eight constants. The remaining three are obtained from the boundary conditions at the hole that $\sigma_r = \tau_{r\theta} = 0$ if $r = a$ but θ is arbitrary. This gives the relations

$$
2c_2 + c_3\frac{1}{a^2} = 0, \quad \text{and} \quad 2c_5 + 6c_7\frac{1}{a^4} + 4c_8\frac{1}{a^2} = 0;
$$

$$
2c_5 - 6c_7\frac{1}{a^4} - 2c_8\frac{1}{a^2} = 0.
$$

The first pair is from $\sigma_r = 0$ and the second line from $\tau_{r\theta} = 0$. Solving these equations yields

$$
c_3 = -\frac{1}{2}a^2\sigma_0, \quad c_7 = -\frac{1}{4}a^4\sigma_0, \quad \text{and} \quad c_8 = \frac{1}{2}a^2\sigma_0.
$$

In summary then,

$$
\sigma_r = \frac{1}{2}\sigma_0\left[\left(1 - \frac{a^2}{r^2}\right) + \left(1 + 3\frac{a^4}{r^4} - 4\frac{a^2}{r^2}\right)\cos(2\theta)\right];
$$

$$
\sigma_\theta = \frac{1}{2}\sigma_0\left[\left(1 + \frac{a^2}{r^2}\right) - \left(1 + 3\frac{a^4}{r^4}\right)\cos(2\theta)\right];
$$

$$
\tau_{r\theta} = -\frac{1}{2}\sigma_0\left(1 - 3\frac{a^4}{r^4} + 2\frac{a^2}{r^2}\right)\sin(2\theta). \tag{4.44}
$$

Exercise 4.25 Graph σ_r versus r when the value of θ is fixed at $\theta = 0$ or at $\theta = \pi/2$. Do the same for σ_θ and $\tau_{r\theta}$. Also plot the stresses σ_r, σ_θ and $\tau_{r\theta}$ versus θ when $r = a$.

At the edge of the hole, $r = a$, the only stress, because of the boundary conditions, is $\sigma_\theta(a, \theta) = 0.5\sigma_0[2 - 4\cos(2\theta)]$. This stress is maximum at $\theta = \pm\pi/2$, the sides of the hole on a diameter perpendicular to the load, where it equals $3\sigma_0$. The stress concentration at the hole is obtained by comparing this stress to the corresponding stress, $\sigma_x = \sigma_0$, in the plate without a hole. So, the stress concentration induced by the presence of a hole, of any size, in an infinite plate is 3, according to this analysis.

EXAMPLE 4.21 A 20 inch by 30 inch thin plate is to be subjected to an in-plane tensile load of $F = 100{,}000$ pounds distributed perpendicularly to each of the opposite pairs of edges of length $w = 20$. It is decided to drill a circular hole in the center to allow a 0.5-inch-diameter shaft to pass through. There are two choices for the plate material, an A514 steel ($E = 29 \times 10^6$ psi, $\sigma_Y = 100$ ksi) of specific weight $\rho = 0.284$ lb/in.^{3} and 6061-T6 aluminum ($E = 10 \times 10^6$ psi and $\sigma_Y = 37$ ksi) of specific weight $\rho = 0.098$ lb/in.^{3}.

(a) Design the thickness, t, of the plate to prevent yield for each of the two choices of material if the hole is not present.

(b) Design the thickness of the plate for each of the two choices of material if the hole is present. Use the criteria that the Mises stress must be less than $\sigma_Y/2$. Assume the infinite thin-plate analysis.

(c) Would the plate with a hole at the minimum thickness be lighter if it was made of the aluminum or the steel?

Solution (a) The thickness must be $t \geq F/\sigma_Y(20)$. So for the steel, $t \geq 0.05$ inches, and for the aluminum $t \geq 0.135$ inches.

(b) The stresses for fixed θ decrease as r increases, that is, as the point is further from the hole. The maximum Mises stress [equation (1.14)] occurs at the hole, $r = a$, and is equal to the maximum σ_θ at $r = a$. Therefore the maximum Mises stress is $3\sigma_0 = 3F/20t$. Therefore, the thickness must be $t \geq 3(2)F/\sigma_Y(20)$. So for the steel, $t \geq 0.3$ inch, and for the aluminum $t \geq 0.811$ inch.

(c) At the minimum thickness, the weight of the steel plate is 51.1 pounds and the aluminum plate weighs 47.7 pounds. ∎

4.8   STRESS CONCENTRATIONS

In an axially loaded member with constant cross-sectional area, the maximum average normal stress equals the stress in the axial direction at each point in the body and occurs on the face perpendicular to the longitudinal axis of the body. The maximum stress is equal to the total internal force acting on the face divided by the area of the face. But if the axially loaded member has holes, notches, or other sharp changes in cross-sectional area, this analysis of the maximum normal stress is no longer valid.

For example, consider an axially loaded bar with two symmetric semicircular notches (or grooves) (Fig. 4.14).

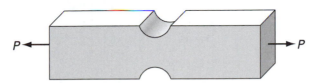

FIGURE 4.14   Axially loaded bar with symmetric notches.

Where would the bar be expected to break? Most people would guess, and experience shows, that it would first crack in the deepest point of the notch, the root of the notch. As the crack lengthens, the cross-sectional area becomes smaller, raising the stress. The bar eventually completely breaks between the two notches. The crack initially occurs, not only because the cross-sectional area is smaller here so that the average normal stress is higher, but also because stress is concentrated around the notch. The study of the mechanism of the formation of cracks, both theoretical and experimental, is called fracture mechanics and is a very active area of research today (see Chapter 10).

Stress concentrations occur in screws, around holes drilled for bolts, in sharp concave corners, and at any other abrupt change in cross section of the member. Also point loads, such as those caused by a ball bearing on a surface, create a stress concentration at the point of contact.

4.8.1   Stress Flow Model

A rough model of the stress pattern within an axially loaded bar with nonconstant cross section gives an indication of how the stress is concentrated by the shape of the bar. In axial tension, the bar will likely break first at the point where the normal stress is largest since at that point the shear stress will also be largest. Recall that a ductile material fails due to the maximum shear stress and a brittle material fails due to the maximum normal stress. Stress can be represented as a vector since it is force, which is a vector, divided by area, which is a scalar. A scalar times a vector is a vector in the same direction but with different magnitude.

In an axially loaded member with constant cross section, the axial stress vectors are all of the same magnitude and in the same direction along the axis. As a rough model, each particle of material could be viewed as a fluid particle flowing in a tube with the shape of the bar and the stress vector viewed as the velocity vector for the fluid particle. Lines, called stress trajectories, can be drawn (Fig. 4.15), showing the paths each particle would take. The velocity vectors are tangent to the lines. This model is called a stress flow model.

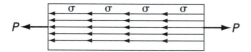

FIGURE 4.15   Stress flow model for an axial member of constant cross section.

Figure 4.16 shows a stress flow model for various cross sections of an axially loaded member: with a groove, hole, and sharp cross-sectional change.

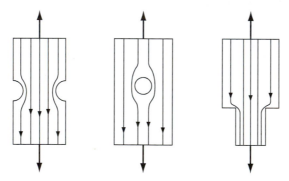

FIGURE 4.16   Stress flow models for axial members with nonconstant cross sections.

Notice where the flow lines tend to bunch together. Just as on a topographic map, these are the points where the stress increases quickly and is the largest. The points are at the root of the groove, at the sides of the hole, and at the change in cross section, respectively.

Graphs of the stress distribution along the face connecting the points of stress concentration can be drawn by placing a w-axis along the cross-sectional area, on which the w value represents the position on the width of the bar, and a stress (σ)

axis along the bar (Fig. 4.17). It might be better to draw the sigma axis perpendicular to the surface of the bar, but this three-dimensional picture is hard to draw on a two-dimensional paper. The stress at each position along the width, w, of the bar is represented by a vector whose magnitude corresponds to the magnitude of the stress. The stress vector acts at the point of the body under the tail of the vector.

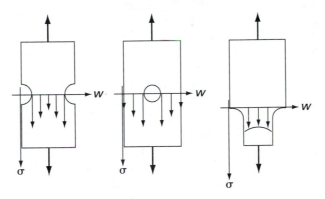

FIGURE 4.17   Stress flow near the change in cross section.

The sharp corner on the right-hand figure showing a bar with a sharp change in cross-sectional area has been rounded off to make a fillet. This rounding off lowers the stress concentration, and therefore, if possible, fillets are used in most engineering structures having concave corners created by a sharp change in cross section.

This model is not sufficient to make numerical predictions of the value of the highest stress and is therefore not completely satisfactory. The increase in stress magnitude from the average normal stress of a constant cross-sectional bar must be found experimentally.

4.8.2   Stress and Strain Concentration Factors

The increase in maximum stress is typically measured by a number called the stress concentration factor. The cross-sectional area of the constant cross section away from the drastic change in area caused by a hole, notch, and so on is called the gross area, A_g. The cross-sectional area at the point where the stress concentration is found is called the net area, A_n (Fig. 4.18).

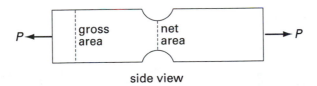

side view

FIGURE 4.18   The net and gross cross-sectional areas.

The net area is used to define the stress concentration factor. If the average normal stress is $\sigma_{ave} = P/A_n$, and σ_{max} is the maximum stress, then the stress concentration factor K_σ is defined to be

$$K_\sigma = \frac{\sigma_{max}}{\sigma_{ave}}.$$

The strain concentration factor is defined to be

$$K_\epsilon = \frac{\epsilon_{max}}{\epsilon_{ave}}.$$

If the body is elastic and axially loaded, then $\sigma = E\epsilon$ and the two concentration factors are equal, $K_\sigma = K_\epsilon$. These factors, obtained from experiment, can be found in engineering handbooks. Some handbooks define the stress concentration factor by computing the average stress in terms of the gross area. When using any handbook, be certain to check which definition of average stress is assumed. In some handbooks, the stress concentration factor K_σ is called K_t, for tensile stress concentration factor, since these results are only valid for tensile axial loads. Also, since most structures are designed to act in the elastic range, concentration factors are usually only given for the elastic case. If the stress exceeds the elastic limit (proportional limit) in a ductile material, the material tends to flow locally to smooth out the stress so that it becomes constant over the net area. A brittle material has a very short plastic stress range before fracture, and the elastic stress concentration factor can be used to estimate the maximum stress in the plastic range as well.

4.8.3 Experimental Determination of Stress Concentration Factors

It is seldom possible to measure stress in a body directly. However, techniques to measure strain, such as strain gages, allow the computation of the stress. Since the body is assumed elastic, the strain quickly gives the stress using the relation for uniaxial elastic loading, $\sigma = E\epsilon$, assuming axial loads.

EXAMPLE 4.22

Determine by experiment the stress concentration factor, K_σ, around a hole of radius 0.25 inch in a 2-inch-wide and 0.25-inch-thick aluminum bar, using nine strain gages mounted on the bar (Fig. 4.19).

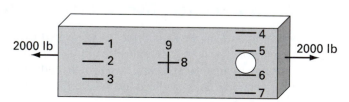

FIGURE 4.19   Placement of the strain gages for the experimental determination of the stress concentration factor.

Solution

The bar is arbitrarily loaded to 2000 pounds, a load low enough that the bar will be elastically loaded. The readings in Table 4.2 are obtained from strain gages, indicated by the short lines and numbers on the Figure 4.19.

Since gages 1, 2, and 3 are far from the hole, they can be assumed unaffected by the hole. These gages can be used to determine E, the elastic modulus. The strains in gages 1, 2, and 3 theoretically should be equal. Since they are not, the bar must not be perfectly axially loaded. These can be averaged to help eliminate this experimental error. The average normal strain is thus taken to be 405 microstrain. Since for elastic loads, $\sigma = E\epsilon$, and since $\sigma = 2000/(0.25)(2) = 4,000$ psi,

$$E = \frac{\sigma}{\epsilon} = \frac{4,000}{405(10^{-6})} = 9.88 \times 10^6 \quad \text{psi.}$$

TABLE 4.2   Strain Gage Readings

Gage #	Strain (μ)	Gage #	Strain (μ)
1	405	6	1290
2	410	7	370
3	400	8	405
4	350	9	−130
5	1,310		

Gages 8 and 9 are perpendicular to each other in the axial, a, and a transverse direction, t, and can be used to determine Poisson's ratio, v.

$$v = -\frac{\epsilon_t}{\epsilon_a} = \frac{130}{405} = 0.321.$$

Now the stress concentration factor at the hole can be determined using either gage 5 or 6. These should be the same, but again a slight experimental error has left them close but unequal.

At gage 5, $\sigma_{max} = E\epsilon_{max} = (9.88)(10^6)(1,310)(10^{-6}) = 12942.8$ psi.
At gage 6, $\sigma_{max} = E\epsilon_{max} = (9.88)(10^6)(1,290)(10^{-6}) = 12745.2$ psi.
The net area at the hole is $(1.5)(0.25) = 0.375$ and so $\sigma_{ave} = 2000/.375 = 5333$ psi.
Therefore

$$K_\sigma = \frac{\sigma_{ave}}{\sigma_{max}} = 2.43 \quad \text{at gage 5,}$$

$$K_\sigma = 2.39 \quad \text{at gage 6.}$$

To get the best experimental answer, average these to obtain $K_\sigma = 2.41$. ∎

EXAMPLE 4.23

The axially loaded bar shown in Figure 4.20 has a stress concentration factor of 2.9 at the fillet. The proportional limit of the material is 45,000 psi.

(a) Compute the average normal stress in sections AB and BC. Are these stresses in the elastic range?

(b) Will the material in the region of the fillet become plastic?

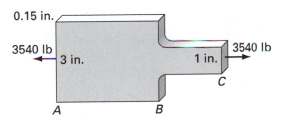

FIGURE 4.20   Stress concentration in a symmetric contraction of cross section.

Solution

(a) $\sigma_{AB} = 3540/(3)(0.15) = 7867$ psi (elastic).
$\sigma_{BC} = 3540/(1)(0.15) = 23600$ psi (elastic).

(b) $\sigma_{ave} = \sigma_{BC} = 23,600$ psi. By the definition of K_σ, $\sigma_{max} = K_\sigma\sigma_{ave} = (2.9)(23600) = 68,440$ psi (plastic). ∎

EXAMPLE 4.24

A 0.31-inch-thick flat bar, made of a material with proportional limit 57,000 psi, having a 0.125-inch-diameter hole is to be axially loaded with a force of 5,400 pounds (Fig. 4.21). The stress concentration factor at the hole is assumed to be 2.6. How wide must the bar be made to ensure it is elastically loaded throughout the bar?

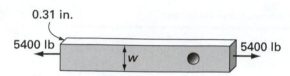

FIGURE 4.21 Bar with a hole.

Solution $\sigma_{max} \le 57{,}000$ psi for the bar to be elastic.
$57{,}000 \ge \sigma_{max} = K_\sigma \sigma_{ave} = (2.6)(5400)/[(0.31)(w - 0.125)]$. Therefore $w \ge 0.920$ inches. ∎

Experimental graphs showing the relationship between the stress concentration factor $K_t(= K_\sigma)$ and the ratios of certain dimensions for three types of flat bars are shown in Figure 4.22. Because the stress concentration factors can be given in terms of these geometric ratios, the factors for many different sized bars can be represented on a single graph. The graphs in Figure 4.22 are from the Mechanical Engineer's Reference Book [4].

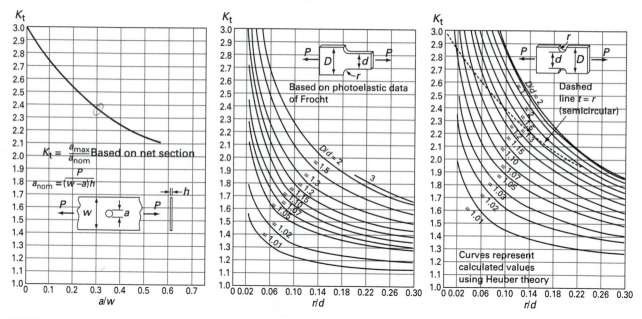

FIGURE 4.22 Reprinted by permission of Elsevier Ltd.

EXAMPLE 4.25 Determine the stress concentration factor in a 0.2 inch thick flat bar with two symmetric grooves of radius 0.3 inches and width 2.6 inches.

Solution To use the graphs given in Figure 4.22, the two ratios D/d and r/d must first be computed. D is the width and r is the groove radius. The net width, d, equals the width minus twice the groove radius. So $d = 2.6 - 2(0.3) = 2$. Therefore

$$\frac{D}{d} = \frac{2.6}{2} = 1.3 \quad \text{and} \quad \frac{r}{d} = \frac{0.3}{2} = 0.15.$$

To determine K_σ from the graphs find the curve corresponding to $D/d = 1.3$ and find $r/d = 0.15$ on the horizontal axis. Next find the intersection, C, of the curve $D/d = 1.3$ and the vertical line through $r/d = 0.15$. Read the vertical coordinate of C by following a horizontal line from C to the left-hand scale. The stress concentration factor is approximately $K_\sigma = 2.18$. ∎

4.9   CONTACT STRESSES

Two surfaces come into contact in many situations, especially within machines. The teeth on two meshed gears touch. Bearings roll under a normal load against a rotating, and perhaps vibrating, shaft. Calculation of the stresses and the corresponding displacements that are induced by contact is a difficult dynamic problem. A simpler problem is a railroad wheel rolling on a stationary track. Since there is no acceleration in the normal (vertical) direction between the wheel and the track, a static solution plus some simplifying assumptions lead to a description of the stress induced in the contact region, the shape of the contact region, and the normal displacements measuring the extent of contact deformation between the two bodies. The contact stresses between a bearing and its fixed bearing race can be similarly computed if friction is neglected. Contact problems differ from those of a single body under a fixed point load because the contact stress distribution is spread over a contact region whose size varies with the deformation of both bodies. Heinrich R. Hertz (1857–1894) first developed the static stress and deformation analysis for two elastic solids in contact [2]. He did this work in 1881, the year following the completion of his Ph.D. on electrodynamics under Helmholtz. Hertz obtained other results in elasticity, but is probably most famous for his experimental work that verified Maxwell's theoretic model for the propagation of electromagnetic waves in space. To honor these experiments, the unit of a cycle per second is now called a hertz.

Hertz proposed an elastic potential function, by analogy with electric potential theory, to compute the induced displacements and, with Hooke's law, the consequent stresses. The bodies were assumed to have infinite extent, so that the analysis is only approximately valid for real bodies of finite size. The coincident normal vectors at the point of contact on the surfaces of the bodies are assumed parallel to the direction of application of the load, P.

4.9.1   Geometric Considerations

Hertz's fundamental strategy was to write the displacements in a coordinate-free way using the curvatures of the two surfaces. He then solved the displacement equilibrium equations and finally applied Hooke's law to obtain the stresses.

A coordinate system is defined in terms of the initial point of contact and the common tangent plane to the two body surfaces at that point. The tangent plane common to both surfaces at the initial point of contact is taken as the x-y plane, with the origin at the initial point of contact, and the load and normals to the surface are in the z-direction (Fig. 4.23). The two bodies are forced together in a motion parallel to the z-axis.

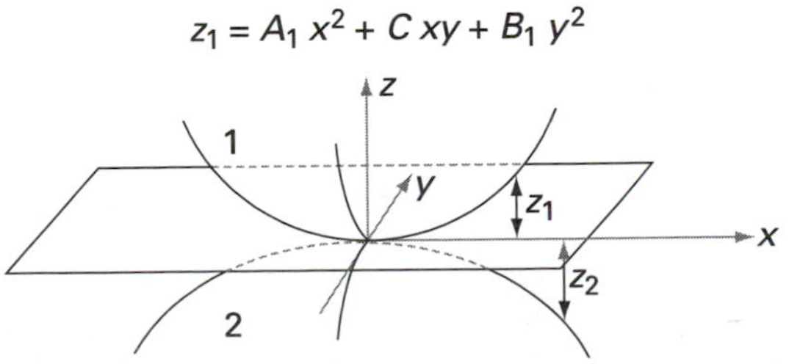

$$z_1 = A_1 x^2 + C xy + B_1 y^2$$

FIGURE 4.23   The tangent plane between two surfaces in contact.

The perpendicular distance from the common tangent plane, at $z = 0$, to one of the body surfaces is given by the z-coordinate on the body. Hertz approximated the shape of the body surfaces near the initial point of contact by quadratic functions, $z = Mx^2 + Rxy + Ny^2$. This means that the perpendicular distance, z, between a point on one surface and the common tangent plane is a homogeneous quadratic function of x and y. An orientation of the x-y axes can be chosen in the common tangent plane so that the xy terms for both surfaces have the same magnitude coefficient. In these quadratic equations for the surfaces, the subscripts refer to body 1 and body 2.

$$z_i = A_i x^2 + Cxy + B_i y^2 \qquad \text{for} \qquad i = 1, 2. \tag{4.45}$$

Then, the distance between corresponding points on the two surfaces is

$$z_1 + z_2 = Ax^2 + By^2, \tag{4.46}$$

where $A = A_1 + A_2$ and $B = B_1 + B_2$. If the positive direction of the z- axis is taken toward body 1, then both A and B are positive.

Equation (4.46) is an ellipse if the left side, $z_1 + z_2$, is a constant. This implies that all points a constant distance $z_1 + z_2$ apart on the two body surfaces lie above an ellipse in the projected contact area in the common tangent plane. The boundary of the contact area in the deformed bodies must, therefore, be an ellipse because all points on the boundary are zero distance apart. That the contact area is approximately elliptical is a consequence of the approximation of the equations representing the two surfaces near the initial contact point as quadratics. Most curved surfaces of machine parts are cylindrical or spherical, or sometimes ellipsoidal or hyperboloidal. In these cases, the Hertz model applies without problem. But if the surface is more complicated and the contact region is large, the contact region may not have an elliptic boundary.

The values of the coefficients A and B in equation (4.46) depend on the choice of coordinates, x-y. This computational difficulty can be avoided by picking a coordinate system in which A and B can be expressed in terms of the surface curvatures (see Appendix A), which are independent of the coordinates. The radius of curvature, ρ_{ij}, is the reciprocal of the curvature, κ_{ij}.

Definition   The principal plane of curvature associated to the principal curvature κ in a body is that containing the corresponding circle of curvature of radius $1/\kappa$ with center at the center of curvature. If the principal radii are not unique, then the principal planes of curvature are not unique.

The coefficients A and B can be expressed in terms of the two principal curvatures, κ_{i1} and κ_{i2}, for the two bodies, $i = 1, 2$. Let u_1-v_1 be principal Cartesian coordinates for body 1 and u_2-v_2 be principal coordinates for body 2. The distance from the tangent plane to the surface is assumed to be expressed in terms of the principal curvatures as

$$z_1 = \frac{1}{2}\kappa_{11} u_1^2 + \frac{1}{2}\kappa_{12} v_1^2; \tag{4.47}$$

$$z_2 = \frac{1}{2}\kappa_{21} u_2^2 + \frac{1}{2}\kappa_{22} v_2^2. \tag{4.48}$$

Therefore,

$$z_1 + z_2 = \frac{1}{2}\kappa_{11} u_1^2 + \frac{1}{2}\kappa_{12} v_1^2 + \frac{1}{2}\kappa_{21} u_2^2 + \frac{1}{2}\kappa_{22} v_2^2.$$

Now u_2 and v_2 are eliminated from this equation. The principal planes of curvature for the first curvature, κ_{11} and κ_{21}, in the two bodies make an angle, ω, as shown in Figure 4.24.

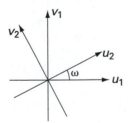

FIGURE 4.24   The relationship between the two principal coordinate systems.

Therefore the two principal coordinate systems in the two bodies at contact are related by a rotation (see Chapter 1),

$$u_2 = u_1 \cos \omega + v_1 \sin \omega;$$
$$v_2 = -u_1 \sin \omega + v_1 \cos \omega.$$

After substituting for u_2 and v_2, the sum of the vertical positions is written in the form

$$z_1 + z_2 = A'u_1^2 + C'u_1v_1 + B'v_1^2. \tag{4.49}$$

Exercise 4.26   Compute the coefficients A', B', and C' in terms of the curvatures.

For constant distance $z_1 + z_2$, equation (4.49) can be viewed as a conic. In most first-year calculus books, a technique is given to rotate the coordinates in which the equation of a conic is given to eliminate the cross term, C', in the new coordinates. The angle through which the coordinates must be rotated is β. Making the substitution to perform the rotation to x-y coordinates,

$$u_1 = x \cos \beta + y \sin \beta;$$
$$v_1 = -x \sin \beta + y \cos \beta.$$

The equation (4.49) is now in the desired form,

$$z_1 + z_2 = Ax^2 + B^2y^2,$$

where

$$A = \frac{1}{4}(\kappa_{11} + \kappa_{12} + \kappa_{21} + \kappa_{22})$$
$$- \frac{1}{4}\sqrt{[(\kappa_{11} - \kappa_{12}) + (\kappa_{21} - \kappa_{22})]^2 - 4(\kappa_{11} - \kappa_{12})(\kappa_{21} - \kappa_{22})\sin^2(\omega)};$$

$$B = \frac{1}{4}(\kappa_{11} + \kappa_{12} + \kappa_{21} + \kappa_{22})$$
$$+ \frac{1}{4}\sqrt{[(\kappa_{11} - \kappa_{12}) + (\kappa_{21} - \kappa_{22})]^2 - 4(\kappa_{11} - \kappa_{12})(\kappa_{21} - \kappa_{22})\sin^2(\omega)}.$$

Exercise 4.27   Verify these expressions for A and B in terms of the curvatures.

Then

$$2(A + B) = \kappa_{11} + \kappa_{12} + \kappa_{21} + \kappa_{22}; \tag{4.50}$$

$$2(B - A) = [(\kappa_{11} - \kappa_{12})^2 + 2(\kappa_{11} - \kappa_{12})(\kappa_{21} - \kappa_{22})\cos(2\omega) + (\kappa_{21} - \kappa_{22})^2]^{1/2}. \tag{4.51}$$

Define an angle, τ, by $\cos \tau = (B - A)/(A + B)$ so that

$$A = (1/2)[\kappa_{11} + \kappa_{12} + \kappa_{21} + \kappa_{22}] \cos^2\left(\frac{\tau}{2}\right); \tag{4.52}$$

$$B = (1/2)[\kappa_{11} + \kappa_{12} + \kappa_{21} + \kappa_{22}] \sin^2\left(\frac{\tau}{2}\right). \tag{4.53}$$

Most handbook tables for contact stresses are in terms of this angle τ introduced by Hertz.

EXAMPLE 4.26   A wheel of radius 16 inches rests on a rail whose top surface has a radius of curvature of 10 inches. Determine the angle τ.

Solution   The wheel and rail are both modeled as cylinders. The curvatures of the wheel are 0 and 1/16. The curvatures of the rail are 0 and 1/10. The planes of principal curvature are those in the wheel containing the 16-inch radius of curvature and in the rail containing the 10-inch radius of curvature since these are the largest in the two bodies. These planes are perpendicular so that $\omega = \pi/2$. By equations (4.50) and (4.51),

$$A + B = 0.5\left(\frac{1}{16} + 0 + \frac{1}{10} + 0\right) = 0.08125;$$

$$B - A = 0.5\left[\left(\frac{1}{16} - 0\right)^2 + \left(\frac{1}{10} - 0\right)^2 + 2\left(\frac{1}{16} - 0\right)\left(\frac{1}{10} - 0\right)\cos(\pi)\right]^{1/2} = 0.01875.$$

$$\cos \tau = (B - A)/(A + B) = 0.2308 \quad \text{so that } \tau = 76.66°. \quad \blacksquare$$

A special case occurs if the principal planes in the two bodies coincide ($\omega = 0$), then $2B = \kappa_{11} + \kappa_{21}$ and $2A = \kappa_{12} + \kappa_{22}$ so that

$$\tan^2\left(\frac{\tau}{2}\right) = \frac{B}{A} = \frac{\kappa_{11} + \kappa_{21}}{\kappa_{12} + \kappa_{22}}.$$

EXAMPLE 4.27   A plane of principal curvature in a sphere is any plane passing through the center. When two spheres are in contact, the two planes of principal curvature can be taken as coincident. Then since the angle ω is zero, $B/A = 1$ and the angle $\tau = \pi/2$. $\quad \blacksquare$

The sign of the curvature must be carefully chosen. The curvature is taken as positive if the center of curvature is inside the body, and negative otherwise. The curvature is positive if when one moves along the radius of curvature towards the center of curvature from the point of contact, then one immediately enters the body. For example, the principal radii of curvature for a sphere are the radii of the sphere. A sphere has positive curvature. Those of the cylinder are the radius of a circular cross section and infinity, the radius of curvature of the straight longitudinal line. Those of the torus are the radii of a vertical slice and a horizontal slice as shown in Figure 4.25.

EXAMPLE 4.28   Determine the magnitudes and the signs of the radii of curvature at the point of contact between a sphere of radius R inside a torus as shown in Figure 4.26. This configuration is a good approximation to a ball bearing inside its race.

Solution   The sphere is body 1. Then $\kappa_{11} = \kappa_{12} = 1/R$. In the race, $\kappa_{21} = -1/R_1$ and $\kappa_{22} = -1/R_2$. $\quad \blacksquare$

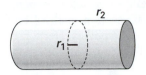

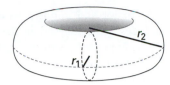

FIGURE 4.25   Principal radii of curvature of a cylinder and of a torus.

FIGURE 4.26   Sphere in contact with the interior of a torus.

4.9.2   Displacement Analysis

The displacements at a point in the body i are denoted by u_i, v_i, and w_i. Again, the z-direction is normal to the surfaces at contact. The x-y plane is tangent to the surfaces at the point of initial contact. Hertz wrote the equilibrium equations for the ith body in terms of the displacements as

$$0 = \nabla^2 u + \frac{1}{1 - 2v_i} \frac{\partial e}{\partial x};$$

$$0 = \nabla^2 v + \frac{1}{1 - 2v_i} \frac{\partial e}{\partial y};$$

$$0 = \nabla^2 w + \frac{1}{1 - 2v_i} \frac{\partial e}{\partial z}, \tag{4.54}$$

where

$$e = \frac{\partial u}{\partial x} + \frac{\partial v}{\partial y} + \frac{\partial w}{\partial z} = \epsilon_x + \epsilon_y + \epsilon_z.$$

The equilibrium equations were taken to be subject to the following boundary conditions.

1. The displacements are zero at infinity. This requires that the bodies be of infinite extent.
2. All tangential stresses are zero, even in the contact region.
3. The normal stresses outside the contact region are zero.
4. Inside the contact region, the vertical displacement difference is $(w_1 - w_2) = \alpha - z_1 + z_2$, where α is the distance the initial point of contact moves.
5. Inside the contact region, the z-component of stress is positive. Outside the contact region, the displacement difference is $(w_1 - w_2) > \alpha - z_1 + z_2$, to keep the surfaces from flowing past each other.
6. The sum of the z-direction normal stresses distributed over the projected contact area, A_p, equals the applied load, P, that is,

$$\int_{A_p} \sigma_z(x, y) dA_p = P.$$

Hertz assumed that the shape of the contact region is an ellipse centered at the initial point of contact,

$$\frac{x^2}{a^2} + \frac{y^2}{b^2} = 1. \tag{4.55}$$

The magnitudes of the major and minor axes of the ellipse, a and b, depend on the curvatures of the surfaces, as discussed in the previous section, and, of course, on the load, P, and the elastic properties of the materials of the bodies. Hertz proposed a potential function, Π, that was shown to satisfy the displacement equilibrium equations and boundary conditions,

$$\Pi = \frac{Qz}{G} + \frac{2+\lambda}{2G}\left[\int_z^\infty Q\,dz - J\right], \tag{4.56}$$

where J is a constant to make Π finite, and λ and G are the Lamé constants for the material.

Hertz chose the potential $Q(x, y, z)$, based on an analogy with an electric potential, so that the density is uniform in the ellipse.

$$Q(x, y, z) = \frac{3P}{16\pi}\int_\mu^\infty \left(1 - \frac{x^2}{a^2 + \xi} - \frac{y^2}{b^2 + \xi} - \frac{z^2}{\xi}\right)\frac{d\xi}{\sqrt{(a^2 + \xi)(b^2 + \xi)\xi}}, \tag{4.57}$$

where μ is the positive root of $\frac{x^2}{a^2+\mu} + \frac{y^2}{b^2+\mu} + \frac{z^2}{\mu} = 1$.

The potential function satisfies the equation $\nabla^2\Pi = 0$. The displacements derived from the potential satisfy both the differential equations for the displacements at equilibrium and also are zero at infinity. The displacements are obtained from

$$u = \frac{\partial\Pi}{\partial x};$$

$$v = \frac{\partial\Pi}{\partial y};$$

$$w = \frac{\partial\Pi}{\partial z} + \frac{\lambda + 2G}{2G(\lambda + G)}Q. \tag{4.58}$$

The normal stress at each point (x, y) inside the elliptical contact area is

$$\sigma_z(x, y) = \frac{3P}{2\pi ab}\left[1 - \frac{x^2}{a^2} - \frac{y^2}{b^2}\right]^{1/2}. \tag{4.59}$$

The normal stress is not a linear function of the applied force since the half lengths of the axes of the ellipse, a and b, are also functions of the load, P. Recall that the area of an ellipse is πab. At the center of the ellipse, the maximum stress is one and half times the applied load divided by the area of the elliptical contact region.

$$\sigma_z = \frac{3P}{2\pi ab}. \tag{4.60}$$

The normal stress in the contact region is often called the Hertzian contact stress.

To finish evaluating the Hertzian stresses, it remains to determine the size of the elliptical projected contact area. E_i is the elastic modulus and ν_i is the Poisson ratio for the ith body. The major and minor axes of the ellipse are calculated from

$$a = c_a\left[3P\frac{(1-\nu_1^2)/E_1 + (1-\nu_2^2)/E_2}{2(\kappa_{11} + \kappa_{12} + \kappa_{21} + \kappa_{22})}\right]^{1/3},$$

$$b = c_b\left[3P\frac{(1-\nu_1^2)/E_1 + (1-\nu_2^2)/E_2}{2(\kappa_{11} + \kappa_{12} + \kappa_{21} + \kappa_{22})}\right]^{1/3}, \tag{4.61}$$

where the constants c_a and c_b are each from an integral depending on the angle, τ [5, p. 379]. Some values are given in Table 4.3.

TABLE 4.3   Relation between τ and the Constants

Angle (τ)	c_a	c_b
0	∞	0.0
10	6.612	0.3186
20	3.7779	0.4070
30	2.7307	0.4930
40	2.1357	0.5673
50	1.7542	0.6407
60	1.458	0.7171
70	1.2835	0.8017
80	1.1278	0.8927
90	1.0000	1.0000

Notice that if the principal planes coincide so that $\tau = \pi/2$, the constants, c_a and c_b, are both 1.

EXAMPLE 4.29 Determine the maximum stress in the contact region between two spheres, each of radius 10 cm, when the load is $P = 300$ N. One sphere is 4142 steel with $E = 210$ GPa, $\sigma_Y = 1600$ MPa, and $\nu = 0.25$; the other is 7075-T6 aluminum with $E = 70$ GPa, $\sigma_Y = 470$ MPa, and $\nu = 0.3$.

Solution The principal curvatures are the inverse radii of the spheres, $\kappa_{ij} = 1/R = 1/0.10$. Also, $\tau = \pi/2$ so that $c_a = c_b = 1$. The contact region has a circular boundary of radius

$$a = \left[3PR\frac{(1 - \nu_1^2)/E_1 + (1 - \nu_2^2)/E_2}{8} \right]^{1/3} = 5.813 \times 10^{-4} \text{ m}$$

by (4.61). The maximum stress is then

$$\sigma_z = \frac{3P}{2\pi a^2} = 424 \text{ MPa}.$$

The stress at this point is below the elastic limits for the steel and the aluminum.

It also can be shown that the distance that the initial contact point moves is

$$\alpha = \left[9P^2\frac{[(1 - \nu_1^2)/E_1 + (1 - \nu_2^2)/E_2]^2}{16(1/R_1 + 1/R_2)} \right]^{1/3}. \tag{4.62}$$

(See Timoshenko and Goodier [5, p. 375].) Therefore,

$$\alpha = 9.17 \times 10^{-7} \text{ m}. \qquad \blacksquare$$

The Hertz theory assumes that the surfaces are smooth in the sense that no tangential (friction) stresses exist. Thus the analysis is a better approximation to the situation for a well-made bearing than for a tire driving on a road. Tire friction cannot be neglected since it is the force which propels the vehicle.

Hertz attempted to experimentally verify his conclusion that the contact region is an ellipse by covering the bodies with soot. After they were loaded and then unloaded, the elliptical outline of the disturbed region was clearly visible.

EXAMPLE 4.30 The hardness of a material is measured by indenting a plate of the material with a sphere. Determine the elastic contact stresses during the initial stage of the indentation test.

Solution   In the case of a sphere on a flat plate, $\tan^2(\tau/2) = 1$ and so $\tau = \pi/2$. The principal radii of curvature of a sphere are each equal to the radius, R, of the sphere; while each of the principal curvatures of the plate are zero. The Hertz calculation predicts that, for a plate and sphere even of different materials, the projected contact area is a circle with radius,

$$a = \left[3PR\frac{(1-v_1^2)/E_1 + (1-v_2^2)/E_2}{4}\right]^{1/3} \tag{4.63}$$

and the distance that the initial contact point moves is, by (4.62),

$$\alpha = \left[9RP^2\frac{(1-v_1^2)/E_1 + (1-v_2^2)/E_2}{16}\right]^{1/3}. \tag{4.64}$$

At the center of the contact region,

$$\sigma_z = \frac{3P}{2\pi a^2}. \qquad \blacksquare$$

Exercise 4.28   The bending moment in the contact region of a thin plate in contact with a sphere of radius, R, can be estimated from strength of materials by assuming the the deformed plate in this region forms the arc of a circle. The plate has width, w, and thickness, $2t$. Verify that the internal bending moment is

$$M = \frac{2Ewt^3}{3R(1-v^2)}.$$

In the case that the plate is thin compared to the contact region length, the bending stresses dominate. When the plate is thick, the stresses are the superposition of the contact stresses and the bending stress.

In handbooks, the terms representing the material properties are often expressed as

$$k_i = (1-v_i^2)/\pi E_i. \tag{4.65}$$

The case of initial contact along a line requires a slightly different analysis. Line contact is created, for example, by two cylinders in contact under a load but oriented so that their longitudinal axes are parallel. The contact region under an elastic load is a rectangle. This is in contrast to the case of crossed cylinders which have an elliptical contact region under an elastic load.

The two cylinders have radii, R_1 and R_2, and are in contact only a line of length L. The half width of the contact region under the load, P, can be shown to be

$$a = 2\sqrt{\frac{P\pi(k_1+k_2)R_1R_2}{L(R_1+R_2)}},$$

by a calculation involving the displacement of the contact line in the direction of the load.

Let σ_z be the normal stress parallel to the load at the center of the contact region. The stress distribution in the direction perpendicular to the longitudinal axis and tangent to the cylinder contact line is assumed to be a parabola so that $\sigma_z(x) = (\sigma_z/a)\sqrt{a^2-x^2}$. Then σ_z is related to the load P by equilibrium in the z-direction,

$$P = \int_{-a}^{a}(\sigma_z/a)\sqrt{a^2-x^2}dx = \frac{\pi}{2}\sigma_z aL.$$

Therefore, the maximum normal stress is

$$\sigma_z = \frac{1}{\pi}\left[\frac{P(R_1+R_2)}{R_1 R_2 L(k_1+k_2)}\right]^{1/2}. \tag{4.66}$$

The maximum normal stress under a cylinder with its longitudinal axis parallel to the plate surface is also obtained from this relation. The plate is viewed as a cylinder with an infinite radius so that $1/R_2 = 0$. Because $(R_1+R_2)/R_1 R_2 = 1/R_1 + 1/R_2$, the maximum stress is, by (4.66),

$$\sigma_z = \frac{1}{\pi}\left[\frac{P}{RL(k_1+k_2)}\right]^{1/2}, \tag{4.67}$$

where R is radius of the cylinder.

The principal stresses at the center of an elliptical contact area in terms of the principal stress, $-\sigma_z$, in the direction normal to the tangent plane to the two surfaces of contact at the the first contact point are

$$\sigma_x = -\left[2\nu + (1-2\nu)\frac{b}{a+b}\right]\sigma_z$$

$$\sigma_y = -\left[2\nu + (1-2\nu)\frac{a}{a+b}\right]\sigma_z, \tag{4.68}$$

where a is the length of the elliptical semiaxis in the x-direction and b is the length of the semi-axis in the y-direction [5, p. 380]. The largest shear stress under the contact area occurs below the surface just under the initial point of contact.

Plastic behavior under contact forces will be discussed in Chapter 9 on plasticity when hardness tests are considered.

PROBLEMS

4.1 Sketch and compare the stress Mohr's circles and the strain Mohr's circles for a state of plane stress with principal stresses $\sigma_1 = 30$ MPa, $\sigma_2 = 55$ MPa, and $\sigma_3 = 0$. The elastic modulus is $E = 210$ GPa and Poisson's ratio is 0.25.

4.2 Determine an Airy stress function for a narrow cantilever beam of thickness t and height $2h$ with a distributed shear load P on the free end by assuming the function is of the form $\phi(x,y) = c_1 x^4 + c_2 x^3 y + c_3 x^2 y^2 + c_4 xy^3 + c_5 y^4 + c_6 xy$ and determining the coefficients, c_i, from the boundary conditions. The coordinate x is measured from the free end and y from the centroidal axis. Determine the stresses, strains, and displacements as a function of (x,y). Ans. $\phi(x,y) = -(P/6I)xy^3 + (Ph^2/2I)xy$.

4.3 Determine the Airy stress function for a narrow cantilever beam of thickness t and height $2h$ with a distributed shear load P on the free end superposed with an axial load, σ_0. Compute the displacement field. Determine the wall reactions by integrating the stresses.

4.4 Determine the Airy stress function for a narrow cantilever beam of thickness t and height $2h$ with a distributed shear load P on the free end superposed with a pure bending

moment, M. Compute the displacement field. Determine the wall reactions by integrating the stresses.

4.5 For the member of problem 3, compute the principal stresses as functions of (x,y). Do the principal axes vary as x changes along the line a) $y = h$, b) $y = 0$?

4.6 Suppose that the Airy stress function for a thin square plate with sides a is $\phi(x,y) = cx^2 y^3$, where the origin is at the lower left corner of the plate.

(a) Determine the stress state at each point in the plate.

(b) Determine the stress boundary conditions which are consistent with this Airy stress function.

4.7 The thin plate shown with thickness, t, and sides a and b is linear elastic, with elastic modulus, E, and Poisson ratio, ν. An Airy stress function is proposed to be $\phi(x,y) = 4[x^2 y - xy^3/3] + 3xy^2$. Neglect body forces.

(a) Is ϕ a possible valid Airy stress function? Prove your answer by a calculation.

(b) If ϕ is valid, determine the tractions (distributed forces) on the top face of the plate that are consistent with ϕ.

(c) If ϕ is valid, draw a stress element showing the state of stress at point $(a/4, b/4)$ in the plate, in terms of the given constants.

(d) If ϕ is valid, determine the strain ϵ_x at $(a/4, b/4)$, in terms of the given constants.

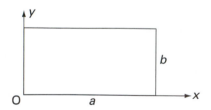

4.8 A plate of constant thickness, t, length, a, and width, b, in plane stress is assigned a stress function, $\phi(x, y) = 2x^3 y$, where the origin of the coordinates is at the plate centroid.

(a) Compute the state of stress at any point (x, y) of the plate.

(b) Sketch a graph of the shear stress distribution (τ_{xy} versus x) along the edge $y = b/2$.

(c) Determine the total resultant force along the edge $y = b/2$.

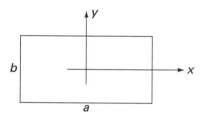

4.9 A triangular bracket (Fig. 4.4) supports a distributed load of 0.5 lb/in. along its upper edge, which is 5 inches long and 0.25 inches thick. The angle of the bracket is 45°. Determine the shear and normal stresses along the joint with the rigid wall, as a function of y.

4.10 A thin plate of constant thickness, t, length, a, and width, b, in plane stress is assigned a stress function, $\phi(x, y) = 2x^3 + 3xy - 4xy^2$, where the origin of the coordinates is at the centroid of the plate (see the figure in problem 8). It is made of an isotropic material with properties E and v.

(a) Compute the state of stress at any point (x, y) of the plate.

(b) Sketch a graph of the shear stress distribution (τ_{xy} vs. y) along the edge $x = a/2$.

(c) Determine the total resultant force along the edge $y = b/2$.

(d) Determine the state of strain at any point (x, y) of the plate.

(e) Assume the edge, $x = -a/2$, is fixed. Determine the deformation $u(x, y)$ and $v(x, y)$ of an arbitrary point in the plate.

4.11 A cantilever beam of length, L, and stiffness, EI, supports a distributed load, $w(x)$, that varies as $w(x) = px/L$ where x is measured from the free end and y from the centroidal axis. Assume that the beam is in plane stress and assume an Airy function of the form $\phi(x, y) = c_1 cy + c_2 x^3 + c_3 x^2 y + c_4 x y^3 + c_5 x^3 y^3 + c_6 x y^5$, where the coefficients, c_i, are unknown constants to be determined from the stress relations to the Airy function and the loads on the beam.

(a) Determine expressions for the stresses, σ_x, σ_y, and τ_{xy} in terms of the coefficients, c_i.

(b) Assume boundary conditions on the free end that the resultant forces are zero. Assume the exact pointwise conditions on the remaining two boundaries. Determine the coefficients c_i.

(c) Determine expressions for the displacements from Hooke's law and the strain-displacement relations.

(d) Compute σ_x and the vertical displacement of this beam from elementary strength of materials and compare to the elasticity results.

4.12 A thin beam is loaded in bending in a manner that also induces an internal shear force, with the moment at the ends given by a distributed load, $p(L, y) = -ay$ and $p(-L, y) = ay$ for a constant $a = 1,000$ MPa/m. The origin of the coordinate system is at the centroid of the beam, which has height $2b = 8$ cm and length $2L = 100$ cm. The beam is a fibered composite with fibers in the longitudinal x-direction. The fibers are E-glass ($E = 70$ GPa, $v = 0.25$) and the matrix is an epoxy ($E = 7$ GPa, $v = 0.4$). Design the volume fraction of fibers, f, so that the maximum magnitude of the compressive strain, ϵ_y, in the beam is less than 0.001.

4.13 A triangular cross-sectional retaining wall 3 meters high is to be built to hold back a pond of water of depth 3 meters. Use the Lévy analysis to make a first approximation for the design of the slope of the wall on the side away from the water. Determine the minimum angle β so that the principal compressive stress at $x = 0$ in the base has magnitude less than 10 MPa. The wall is to be made of a high-strength concrete with specific weight $\rho = 2320g$ N/m³ and $E = 30$ GPa. The specific weight of the water is $\gamma = 1000g$ N/m³. Note that this analysis assumes that concrete is linearly elastic; this is not a valid assumption but does allow the approximation of the design to be made.

4.14 A flat square aluminum plate ($E = 70$ GPa, $v = 0.34$, $\alpha = 24 \times 10^{-6}/°C$) of sides 20 cm and thickness, t, fits snugly into a grooved frame as shown. The frame is held together by four bolts. The plate is then subjected to a temperature increase of 20°C. Assume that the frame remains rigid and is negligibly affected by the temperature change. Assume that the plate does not buckle. Design the thickness, t, of the plate so that the shear force on each bolt is less than 50 kN.

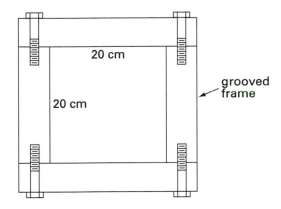

20 cm

20 cm

grooved frame

4.15 A thin rectangular plate has small thickness, t, height, $2h$, and length $2L$. Find the distribution of stress in the plate as a function of x and y. There are no external forces on the plate, but there is a temperature variation in the plate given by $T(x, y) = ky^4$. The loading is linear elastic, with elastic modulus, E, and linear thermal coefficient of expansion, α. Assume that E, α, and k are given. Assume that $\sigma_y = \tau_{xy} = 0$. Also assume that $\sigma_x = \sigma_x(y)$, that is, is a function of y only. Neglect body forces. Evaluate any other constants that appear in your answer in terms of the given constants, E, α, and k, and the dimensions of the plate.

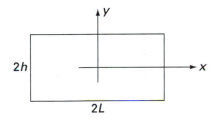

4.16 If a thin plate is unconstrained show that a temperature distribution $T(x, y) = ky$ produces zero stresses.

4.17 A steel ($E = 210$ GPa, $\nu = 0.28$, $\sigma_Y = 120$ MPa) hollow cylinder with inner radius 3 cm and outer radius 6 cm is loaded with an external pressure of 70 MPa. Would the cylinder be expected to yield at the inner surface? The material has the same magnitude yield stress in both tension and compression.

4.18 A hollow thick-walled cylinder is loaded by an internal pressure only.

(a) Compute the stresses at the outer surface $r = b$ in terms of b/a and p_i.

(b) Determine the maximum shear stress at the surface, in terms of b/a and p_i, from Mohr's circle.

(c) Graph the effect of changing the ratio b/a by plotting the maximum shear stress at the surface versus b/a.

(d) Compute the maximum allowable b/a in terms of the yield stress, τ_Y, of the material and p_i.

(e) Repeat these calculations at the inner surface $r = a$.

4.19 A composite member is made of a hollow brass cylinder ($E = 110$ GPa, $\nu = 0.32$, and $\alpha = 18.7 \times 10^{-6}/°C$) interference fit over a solid aluminum rod ($E = 70$ GPa, $\nu = 0.33$, and $\alpha = 23.6 \times 10^{-6}/°C$). The outer radius of the brass cylinder is 50 mm, the inner radius is 25 mm, and the outer radius of the aluminum rod is 25.01 mm. The length of both is the same.

(a) Compute the temperature difference required to easily slip the brass cylinder over the aluminum rod.

(b) Compute the interface pressure of the interference fit at room temperature.

(c) Use Mohr's circle to compute the maximum normal and shear stresses in the cross-sectional plane at the outer surface.

(d) Use Mohr's circle to compute the maximum normal and shear stresses in the cross-sectional plane in both the steel and aluminum at the interface of the two mated cylinders.

4.20 A composite member is made of a hollow copper cylinder ($E = 110$ GPa, $\nu = 0.34$, and $\alpha = 17 \times 10^{-6}/°C$) interference fit over a hollow steel rod ($E = 210$ GPa, $\nu = 0.28$, and $\alpha = 11 \times 10^{-6}/°C$). The outer radius of the copper cylinder is 80 mm, the inner radius is 60 mm. The outer radius of the steel rod is 60.05 mm and the inner radius is 30 mm. The lengths of both are the same.

(a) Compute the temperature difference required to easily slip the brass cylinder over the steel rod.

(b) Compute the interface pressure, p, of the interference fit at room temperature.

(c) Determine the stresses, σ_r and σ_θ, at each radial distance in the composite cylinder. Plot the results as σ_r/p versus r and σ_θ/p versus r.

4.21 A hollow steel cylinder ($E = 210$ GPa and $\nu = 0.25$) which has closed ends contains an internal pressure, $p_i = 5$ MPa. The outer radius is 100 mm and the inner radius is 90 mm.

(a) Plot the stress σ_θ at 2-mm increments of the radius.

(b) Calculate the stress σ_z.

(c) Compare these results to those obtained by using the elementary strength of materials, and thin-walled cylindrical pressure vessel equations for the hoop and longitudinal stresses.

(d) Write the stress tensor at $r = 95$ mm for the elasticity solution.

(e) Write the deviatoric stress tensor at $r = 95$ mm for the elasticity solution.

4.22 A composite member is made of a hollow steel cylinder shrink fit over a hollow steel rod ($E = 210$ GPa, $\nu = 0.28$, and $\alpha = 11 \times 10^{-6}/°C$). The outer radius of the cylinder is 6 cm, the inner radius is 2.5 cm. The outer radius of the steel rod is 2.505 cm and the inner radius is 7.5 mm. The lengths of both are the same.

(a) Compute the interfacial pressure.

(b) Compute $\sigma_r(r)$ and $\sigma_\theta(r)$ in both cylinders.

(c) Plot the results of part b) for $\sigma_r(r)$ in both cylinders on a single set of axes and the results for $\sigma_\theta(r)$ in both cylinders on a separate set of axes.

(d) Find the total stress at each point when the assembled component is subject to a torque of 1000 Nm. Use strength of materials to determine the contribution from the torque. Does the shrink fit enhance the load carrying ability of the shaft in the sense that the principal stresses for the torque alone are reduced by the presence of the shrink fit?

4.23 A hollow cylinder has end caps and contains a pressure of 8 MPa. Its inner radius is 3 cm, its outer radius is 5 cm, and its length is 50 cm. An axial tensile force of 5000 N is applied at the ends.

(a) Determine the stress state at a point that is at a radial distance of 4 cm and at 25 cm from the end.

(b) Determine the maximum shear stress at the point described in a). Describe the plane on which the maximum shear stress acts.

(c) Write the stress tensor for the point described in a).

(d) Compute the axial strain at the point described in a). The material is linear elastic with $E = 200$ MPa and Poisson ratio 0.28.

4.24 In a thin-walled cylindrical pressure vessel, scaling the inner and outer radii by the same factor leaves the hoop and longitudinal stresses unchanged.

(a) Verify this statement.

(b) Does such geometric scaling change the stresses developed in a hollow thick-walled cylinder with closed ends and containing an internal pressure?

(c) If so, determine the change in terms of the scaling factor.

4.25 Does scaling the inner and outer radii by the same factor in a press fit change the resulting interface pressure? If so, determine the change in terms of the scaling factor.

4.26 Would a thick-walled hollow cylinder loaded by both internal and external pressures be safer if it were made of steel or of aluminum?

4.27 A thin constant thickness circular disk is made of a steel with $\rho = 7,860$ kg/m^3, $E = 210$ GPa, and $\nu = 0.25$. The outer radius is 10 cm, the inner radius is 1 cm, and the disk rotates at a constant 1,200 rpm.

(a) Determine the stress state at the outer surface.

(b) Graph σ_θ versus r in the cases 1) $a = 0.2$ cm, $b = 4$ cm, and 2) $a = 0$, $b = 4$ cm. Compare the maximum values of σ_θ in the two cases and state at what value of r this maximum occurs. How does the existence of even a small hole at the center influence the stress σ_θ?

4.28 A thin flat disk of radius 12 cm is shrink-fit onto a solid shaft of diameter 4 cm that rotates at a constant speed, ω. Both are made of a steel with $\rho = 7,860$ kg/m^3, $E = 210$ GPa, $\sigma_Y = 700$ MPa, and $\nu = 0.25$. The interference is

0.0001 m. Determine the constant rotational speed at which the disk separates from the shaft. Is this mode of system failure likely to be a problem?

4.29 A thin flat disk of radius 12 cm is shrink-fit onto a hollow shaft of outer diameter 4 cm and inner diameter 2 cm that rotates at a constant speed, ω. Both are made of a steel with $\rho = 7860$ kg/m^3, $E = 210$ GPa, $\sigma_Y = 700$ MPa, and $\nu = 0.25$. The interference is 0.001 m. Determine the constant rotational speed at which the disk separates from the shaft. Is this mode of system failure likely to be a problem?

4.30 A 2-cm-thick flat disk of radius 80 cm is shrink-fit onto a solid shaft of diameter 20 cm which rotates at 1500 rpm. Both are made of a steel with $\rho = 7860$ kg/m^3, $E = 210$ GPa, $\sigma_Y = 700$ MPa, and $\nu = 0.25$. Design the interference so that the disk will not separate from the shaft while the system is spinning. Is this mode of system failure likely to be a problem?

4.31 In an arbitrary axisymmetric problem assume that no stresses depend on θ. Show that, for zero body forces, the biharmonic equation $\nabla^2[\sigma_r(r) + \sigma_\theta(r)] = 0$ reduces to

$$\left[\frac{d^2}{dr^2} + \frac{1}{r}\frac{d}{dr}\right]\left[\frac{d^2\phi}{dr^2} + \frac{1}{r}\frac{d\phi}{dr}\right] = 0,$$

which can be rewritten as

$$\frac{1}{r}\frac{d}{dr}\left\{r\frac{d}{dr}\left[\frac{1}{r}\frac{d}{dr}\left(r\frac{d\phi}{dr}\right)\right]\right\} = 0,$$

and then integrate to obtain the general form of the Airy stress function for any axisymmetric problem independent of θ:

$$\phi(r) = a\ln(r) + br^2\ln(r) + cr^2 + d.$$

The constants a, b, and c are determined by the boundary conditions.

4.32 Verify that the strain energy due to an edge dislocation at a point close to the edge dislocation is proportional to the square of the magnitude of the Burgers vector.

4.33 An infinite plate with a 0.5 inch radius hole is axially loaded with a stress of 1,500 psi.

(a) Determine the state of stress at the edge of the hole on a diameter parallel to the load.

(b) Compare the result in (a) to the state of stress at the edge of the hole on a diameter perpendicular to the load.

4.34 A 20 inch by 30 inch thin plate is to be subjected to an in-plane tensile load of $F = 100,000$ pounds distributed perpendicularly to each of the opposite pairs of edges of length $w = 20$. It is decided to drill a circular hole in the center to allow a 0.625-inch-diameter shaft to pass through. There are two choices for the plate material, a magnesium alloy containing 8.5% Al ($E = 6.5 \times 10^6$ psi, $\sigma_Y = 40$ ksi) of specific weight $\rho = 0.065$ lb/in^3 and 2014-T6 aluminum ($E = 10.6 \times 10^6$ psi and $\sigma_Y = 60$ ksi) of specific weight $\rho = 0.101$ lb/in.^{3}.

(a) Design the thickness, t, of the plate to prevent yield for each of the two choices of material if the hole is not present.

(b) Design the thickness of the plate for each of the two choices of material if the hole is present. Use the criteria that the Mises stress must be less than $\sigma_Y/2$. Assume the infinite thin-plate analysis.

(c) Would the plate with a hole at the minimum thickness be lighter if it was made of the aluminum or the magnesium alloy?

4.35 Determine the maximum tensile normal stress in an axially loaded flat bar with a hole in its center. The bar has width 2 inches and thickness 1/8 inch. The diameter of the hole is 0.5 inches. The load on the bar is 200 pounds.

4.36 An axially loaded flat bar is made of steel ($E = 210$ GPa and $\sigma_Y = 350$ MPa) and has thickness 5 mm. The large hole has diameter 3 cm and the small hole has diameter 1 cm. The width of the bar is 6 cm. Determine the maximum tensile normal stress in the bar. State clearly where it occurs. Ans. 86.4 MPa.

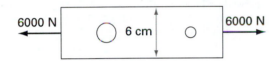

4.37 Determine the maximum normal stress in an axially loaded flat bar with 0.5 inch radius fillets and a tensile load of 2,370 pounds. The bar is 3 inches wide at one end and 2 inches wide at the other end. The thickness is 0.25 inches. The proportional limit of the material is 36 ksi. Ans. 15,550 psi.

4.38 Determine the maximum normal strain in an axially loaded 0.5 inch thick flat bar with 0.25 inch symmetric circular grooves. The width of the bar is 3 inches and the load is 4,000 pounds. The proportional limit of the material is 45 ksi.

4.39 Determine the maximum normal stress in the following 1.5-cm-thick axially loaded flat bar with two holes and fillets. The diameter of the large hole is 1.5 cm and of the small hole is 2/3 cm. The fillet radius is 1/3 cm. The width of the large end is 4 cm. The proportional limit of the material is 175 MPa. State clearly where the maximum tensile normal stress occurs. Ans. 32.8 MPa.

4.40 Determine the maximum axial tensile load, P, the following flat bar can carry with a factor of safety of 2.5. Its thickness is 0.4 inches, its width is 3 inches, the groove radius is 0.5 inches, and the hole diameter is 0.3 inches. The proportional limit of the material is 57 ksi.

4.41 A flat bar 4 inches wide with a 0.6-inch-diameter bolt hole in its center is to be axially loaded with a tensile load of 12,050 pounds. Determine the minimum thickness for the bar so that it remains elastically loaded with a factor of safety of 4. The proportional limit of the material is 60 ksi.

4.42 An axially loaded flat bar of width 3 inches and thickness 0.4 inches has a hole of diameter 0.24 inches centered on its axis. The proportional limit of the material is 65 ksi. The bar must remain elastically loaded.

(a) Compute the maximum possible load the bar could carry if the hole were not in the bar.

(b) Compute the maximum possible load the bar can carry with the hole cut in it.

(c) What percent load can the bar with hole carry compared to the same bar without the hole?

4.43 A wheel of radius 16 inches rests on a rail whose top surface has a radius of curvature of 14 inches. Determine the angle τ.

4.44 A cylinder of radius 16 inches rests on a flat plate. Determine the angle τ.

4.45 A sphere of diameter 8 inches rests on a flat plate. Determine the angle τ.

4.46 A sphere of diameter 8 cm rests on a cylinder of radius 16 cm. Determine the angle τ.

4.47 Prove that for the contact of two spheres, the normal stress at the initial point of contact is

$$\sigma_z = 0.26965 \left[\frac{P(R_1 + R_2)^2}{R_1^2 R_2^2 (k_1 + k_2)^2} \right]^{1/3}. \qquad (4.69)$$

4.48 Two spheres made of A514T1 steel with $E = 210$ GPa, $\sigma_Y = 720$ MPa, and $v = 0.25$, each of radius 10 cm, are in contact when the load is $P = 100$ N.

(a) Determine the maximum normal stress in the contact region between two spheres.

(b) Repeat the calculation, changing only the value of v to 0.3. Compare the stresses calculated in a) and b).

4.49 Two spheres made of a composite material with $E = 105$ GPa and $v = 0.25$, each of radius 10 cm, are in contact when the load is $P = 500$ N.

(a) Determine the maximum stress in the contact region between two spheres.

(b) Compare this result to the answer for problem 48 (a).

4.50 Verify that the maximum contact stress between a sphere of radius, R_1, in a spherical socket of radius, R_2, is

$$\sigma_{max} = 0.26965 \left[\frac{P(R_2 - R_1)^2}{R_1^2 R_2^2 (k_1 + k_2)^2} \right]^{1/3}.$$

4.51 A wheel of radius 16 inches rests on a rail whose top surface has a radius of curvature of 10 inches. Both are made of steel with $E = 30 \times 10^6$ psi, $\sigma_Y = 800$ ksi, and $\nu = 0.25$. The load is $P = 600$ lb.

(a) Determine the maximum normal stress in the contact region.

(b) Would it be safer to design the rail with a radius of curvature closer to that of the wheel? Compute the case for which the radius of curvature of the rail is 16 inches.

4.52 Determine the radius of the contact area, the maximum normal stress, and the deflection for a 10 mm radius sphere made of steel ($E = 210$ GPa, $\sigma_Y = 500$ MPa, $\nu = 0.25$) forced into a flat aluminum plate ($E = 70$ GPa, $\sigma_Y = 550$ MPa, $\nu = 0.3$) by a 20 N force.

4.53 Determine the contact area and maximum normal stress along the line of contact for a cylinder of radius 25 mm and length 25 cm made of steel ($E = 210$ GPa, $\sigma_Y = 250$ MPa, $\nu = 0.25$) forced into a flat 2024-T4 aluminum plate ($E = 70$ GPa, $\sigma_Y = 300$ MPa, $\nu = 0.3$) by a 2 kN force.

4.54 Verify that the maximum contact stress between a cylinder of radius, R_1, and length, L, in a cylindrical groove of radius, R_2, is

$$\sigma_{\text{max}} = 0.318 \left[\frac{P(R_2 - R_1)}{R_1 R_2 L (k_1 + k_2)} \right]^{1/2}.$$

4.55 Write the deviatoric stress tensor at the point of contact for problem 48a.

REFERENCES

[1] H. W. HASLACH, JR., AND K. WHIPPLE, Mechanical design of embedded optical fiber interferometric sensors for monitoring simple combined loads, *Optical Engineering*, *32*, 494–503 (1993).

[2] H. HERTZ, On the contact of elastic solids, *Journal für die reine und angewandte Mathematik*, *92*, 156–171 (1881).

[3] L. E. MALVERN, *Introduction to the Mechanics of a Continuous Media*, Prentice Hall, Englewood Cliffs, NJ, 1969.

[4] *Mechanical Engineer's Reference Book*, 11th ed., A. Parrish, Ed., Butterworth's, London, 1973.

[5] S. P. TIMOSHENKO AND J. N. GOODIER, *Theory of Elasticity*, McGraw-Hill, New York, 1987.

CHAPTER 5

APPLICATIONS OF LINEAR ELASTICITY AND ITS APPROXIMATIONS

The 15 equations of elasticity can rarely be solved in closed form for the structures used in engineering. Approximations of the strains or displacements are frequently required to obtain simpler solutions. Strength of materials as it is taught today usually makes geometric assumptions about the behavior of the strain that lead to linear equations to predict stresses. In this chapter, the stresses in some important engineering structures are computed in closed form, by elasticity if possible or by strength of materials. Three major load-geometry cases are considered: torsion of a solid or a thin-walled prismatic rod, bending of a beam, and a beam on an elastic foundation.

5.1 TORSION OF PRISMATIC RODS

The drive shaft on a motor is a prismatic rod under torsion. The wing of an airplane can be very roughly modeled as a thin-walled prismatic beam under a torque due to forces induced by airflow, along with other loads. A prism is a body with an arbitrary base and sides perpendicular to the base. It therefore has constant cross section and is usually described by the shape of the base. A coordinate system is chosen with origin at the centroid of the base and with one axis perpendicular to the base so that the axis is also parallel to the sides. The cross section of the prism is the area perpendicular to this axis. The applied torque, T, is a moment about the axis perpendicular to the base.

Only circular rods, those prisms with circular cross-sectional area, were considered in elementary strength of materials. The material was assumed to be isotropic, homogeneous, and Hookean. Such rods were first considered by Coulomb around the time of the U.S. war for independence in 1776. Coulomb was also the first to consider the idea of shear stress. His most important assumption, the one which ensured that all the stress and deformations would be linear functions of the torque, was that plane sections remain planar during the twist. Consequently, the shear stress at a point r units from the center on a cross-sectional face is

$$\tau = \frac{Tr}{J},$$

where $J = \pi R^4/2$ is the polar area moment of inertia about the axis perpendicular to the base for the circular cross section of outer radius, R. The angle of twist per unit length, θ, is constant if T, the cross section, and the shear modulus, G, are constant,

$$\theta = \frac{T}{JG}.$$

The angle of twist over the length, L, is

$$\theta_L = \theta L = \frac{TL}{JG}$$

if T, the cross section, and the shear modulus, G, are constant over the length, L.

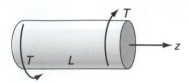

FIGURE 5.1   Torsion of a circular rod.

The assumptions are relaxed somewhat for the analysis of a torque applied to a prism of noncircular cross section. The problem of the torsion of prisms was first solved by B. de Saint-Venant in 1855 as an application of his semi-inverse method. The material is still assumed to be isotropic, homogeneous, and Hookean. The longitudinal axis of symmetry is taken to be the z-axis, and the x-y axes lie in the cross section (Fig. 5.1). The angle of twist, θ, per unit length is assumed constant throughout the prism. Saint-Venant observed that the cross sections need not remain planar under torsion; he allowed the cross section to warp out of plane. But rather than allow each cross section to deform differently, each is assumed to take the same shape so that the deflection in the z-direction in each face is independent of the location along the z-axis. In other words,

$$w(x, y, z) = f(x, y).$$

This assumption cannot be correct near the ends of the rod. For example, if the rod is mounted in the wall, the cross section at the wall has $w(x, y) = 0$, but the deflection of no other cross section satisfies this plane section condition. However, it will be assumed to be true in the following analysis. Saint-Venant in his famous principle postulated that, away from the regions at which the loads are applied, his assumptions are valid.

Differential equations governing a torsional stress function, ϕ, called the Prandtl function, will be obtained by applying some simplifying geometric assumptions to write the strain-displacement relation in terms of the angle of twist, θ, and the warping function, $f(x, y)$. The stress function permits the determination of the stresses in terms of the angle of twist. Another technique must also be developed to relate the angle of twist to the applied torque.

Assume that the base of the rod is mounted in a wall and the torque is applied at the free end. Put the origin in the face at the wall so that z is the distance from the fixed wall. The angle of twist in the face at z is $z\theta$, since the angle of twist per unit length, θ, is constant. Consider a point P at (x, y, z) in this face (Fig. 5.2) with displacement under the twist given by (u, v, w). Further assume that the cylindrical coordinates of P are (r, α, z). Under the twist, P rotates to a point P' through the angle $z\theta$. The arc cut off has length $rz\theta$. Figure 5.2 represents a typical cross section with arbitrarily oriented x-y coordinates at the center of rotation.

The arclength that the point P traces as it moves to P' is $rz\theta$. Assume that this displacement is small so that the arclength can be approximated as the line PP' which is the hypotenuse of the triangle with u and v as legs. In the right triangle, since the line PP' is approximately perpendicular to the radial line OP, the angle between v and PP' is α, the polar coordinate of P. Then u is negative since it is a displacement

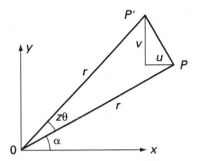

FIGURE 5.2   The rotation of point P to point P' under a torque.

in the negative x-direction, while v is positive since it is in the positive y-direction. From the right triangle,

$$u = -rz\theta \sin\alpha = -yz\theta,$$
$$v = rz\theta \cos\alpha = xz\theta,$$

by the relation between polar and Cartesian coordinates. The in-plane displacement is now written in terms of the coordinates and the angle of twist. The out-of-plane displacement is assumed to be $w = f(x, y)$. From these displacements, the strain and, then by Hooke's law, the stress can be written. At least this can be done in terms of the unknown angle of twist.

All normal strains are zero since $\partial u/\partial x = 0$, $\partial v/\partial y = 0$, and $\partial w/\partial z = 0$. The shear strains are

$$\gamma_{xy} = \frac{\partial u}{\partial y} + \frac{\partial v}{\partial x} = -z\theta + z\theta = 0;$$

$$\gamma_{xz} = \frac{\partial u}{\partial z} + \frac{\partial w}{\partial x} = -y\theta + \frac{\partial f}{\partial x};$$

$$\gamma_{yz} = \frac{\partial v}{\partial z} + \frac{\partial w}{\partial y} = x\theta + \frac{\partial f}{\partial y}.$$

In Hooke's law for the torsional prism, the normal stresses are all zero. The shear stresses are

$$\tau_{xy} = G\gamma_{xy} = 0;$$

$$\tau_{xz} = G\gamma_{xz} = G\left(-y\theta + \frac{\partial f}{\partial x}\right); \tag{5.1}$$

$$\tau_{yz} = G\gamma_{yz} = G\left(x\theta + \frac{\partial f}{\partial y}\right). \tag{5.2}$$

The stresses must satisfy the equilibrium equations; recall that the body forces are neglected. The sum of the forces in the x-, y-, and z-directions, respectively, give the following requirements for equilibrium, since the normal stresses and τ_{xy} are zero,

$$\frac{\partial \tau_{zx}}{\partial z} = 0;$$

$$\frac{\partial \tau_{zy}}{\partial z} = 0;$$

$$\frac{\partial \tau_{xz}}{\partial x} + \frac{\partial \tau_{yz}}{\partial y} = 0. \tag{5.3}$$

For these assumptions about torsion, the first two equations are automatically satisfied since τ_{xz} and τ_{yz} are not functions of z.

The term involving the displacement $w = f(x, y)$ in the stress-displacement relations can be eliminated by combining the derivatives of equations (5.1) and (5.2).

$$\frac{\partial \tau_{xz}}{\partial y} - \frac{\partial \tau_{yz}}{\partial x} = G\left(-\theta + \frac{\partial^2 f}{\partial x \partial y}\right) - G\left(\theta + \frac{\partial^2 f}{\partial x \partial y}\right) = -2G\theta. \tag{5.4}$$

Define $H = -2G\theta$.

Much as in development of the Airy stress function, a new stress function, the Prandtl stress function, $\phi(x, y)$, is defined by the following differential equations.

$$\frac{\partial \phi}{\partial x} = -\tau_{yz}; \tag{5.5}$$

$$\frac{\partial \phi}{\partial y} = \tau_{xz}. \tag{5.6}$$

Note that, in contrast to the Airy stress function, the shear stresses are first derivatives. Substitution of these relations into equation (5.4) produces Poisson's equation

$$\frac{\partial^2 \phi}{\partial x^2} + \frac{\partial^2 \phi}{\partial y^2} = \nabla^2 \phi = H. \tag{5.7}$$

In applications, Poisson's equation might be used to determine $\phi(x, y)$, which in turn determines the stresses by equations (5.5) and (5.6). Other strategies are also possible.

The Poisson equation (5.7) is a partial differential equation. Boundary conditions are required in order to solve it for a specific structure. First, since the normal stress perpendicular to a cross section is $T_z = 0$ in pure torsion, the stress function must be constant on the boundary of the cross-section. To see this, let the unit vector normal to the boundary at a point P be $\boldsymbol{n} = a\boldsymbol{i} + b\boldsymbol{j}$. The stress, $\boldsymbol{t} = T_x\boldsymbol{i} + T_y\boldsymbol{j} + T_z\boldsymbol{k}$ is given by $\boldsymbol{t} = T(\boldsymbol{n})$ so that

$$\begin{pmatrix} T_x \\ T_y \\ T_z \end{pmatrix} = \begin{pmatrix} 0 & 0 & \tau_{zx} \\ 0 & 0 & \tau_{zy} \\ \tau_{xz} & \tau_{yz} & 0 \end{pmatrix} \begin{pmatrix} a \\ b \\ 0 \end{pmatrix}. \tag{5.8}$$

Then the normal stress component is

$$T_z = a\tau_{xz} + b\tau_{yz} = 0. \tag{5.9}$$

Approximate a piece of the boundary having normal $\boldsymbol{n}$ by a straight line (Fig. 5.3).

Since $\boldsymbol{n} = \cos\alpha\,\boldsymbol{i} + \sin\alpha\,\boldsymbol{j}$, and since $\cos\alpha = \Delta y/\Delta s$, $\sin\alpha = -\Delta x/\Delta s$, by letting Δs tend to zero, the unit normal vector can be written as

$$\boldsymbol{n} = \frac{dy}{ds}\boldsymbol{i} - \frac{dx}{ds}\boldsymbol{j}.$$

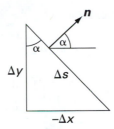

FIGURE 5.3   The normal, **n**, to the boundary of a prismatic rod.

The equation (5.9) implies, using equations (5.5) and (5.6), that

$$\frac{\partial \phi}{\partial x}\frac{dx}{ds} + \frac{\partial \phi}{\partial y}\frac{dy}{ds} = 0.$$

This expression is just the chain rule for the derivative of the two-variable function $\phi(x, y)$ with respect to s. Therefore

$$\frac{d\phi}{ds} = 0$$

along the boundary of the cross section; that is, $\phi(x, y)$ is a constant on the boundary of the cross section. If the function $\phi(x, y)$ satisfies the Poisson differential equation so does the function $\phi(x, y) + c$, where c is any constant. Therefore, the Prandtl stress function $\phi(x, y)$ can be assumed to be zero on the boundary of any cross section.

This fact can frequently be used to guide a guess for the form of a Prandtl stress function. This will be shown below in examples and the problems. This fact is also the primary tool used to relate the constant angle of twist per unit length, θ, to the applied torque.

At the ends of the rod, the normal to the surface is $\mathbf{n} = \mathbf{k}$. The stress components in the plane are

$$T_x = \tau_{xz}; \tag{5.10}$$
$$T_y = \tau_{yz}. \tag{5.11}$$

The force on the ends are zero as expected since by equations (5.10) and (5.6), the integral over the cross-sectional area, A,

$$\int\int_A T_x dx dy = \int\int_A \tau_{xz} dx dy$$

$$= \int \left(\int \frac{\partial \phi}{\partial y} dy \right) dx$$

$$= \int [\phi(x, y_2) - \phi(x, y_1)] dx = 0,$$

since (x, y_1) and (x, y_2) lie on the boundary of the cross section implies $\phi(x, y_2) = \phi(x, y_1) = 0$. Likewise, $\int\int_A T_y dx dy = 0$. The moment equation relates the applied torque, M_t, and the angle of twist. The moment about the z-axis equals the sum over the cross-sectional area, A, of the cross products of the moment arm from the z-axis and the internal forces, given as stress times area.

$$M_t \mathbf{k} = \int \int_A [(x\mathbf{i} + y\mathbf{j}) \times (\tau_{xz}\mathbf{i} + \tau_{yz}\mathbf{j})]dxdy$$

$$= \int \int_A (x\tau_{yz}\mathbf{k} - y\tau_{xz}\mathbf{k})dxdy = \int \int_A \left[-x\frac{\partial\phi}{\partial x} - y\frac{\partial\phi}{\partial y} \right]dxdy\mathbf{k}.$$

Now compute the integral

$$\int \int_A \left[-x\frac{\partial\phi}{\partial x} - y\frac{\partial\phi}{\partial y} \right]dxdy = \int dy \int -x\frac{\partial\phi}{\partial x}dx - \int dx \int y\frac{\partial\phi}{\partial y}dy$$

by parts to obtain

$$\int \int_A \left[-x\frac{\partial\phi}{\partial x} - y\frac{\partial\phi}{\partial y} \right]dxdy = -\int dy \left[x\phi|_{ends} - \int \phi dx \right]$$

$$- \int dx \left[y\phi|_{ends} - \int \phi dy \right].$$

Use the fact that $\phi(x, y)$ is zero on the boundary to obtain

$$M_t = 2 \int \int_A \phi(x, y)dxdy. \tag{5.12}$$

Since the expression for the function $\phi(x, y)$ involves the angle of twist in the term H of the Poisson equation, equation (5.12) relates the applied torque and the angle of twist.

The torque on the rod is that induced by tractions equal to the shear stresses acting on the end faces of the rod. In fact, in many numerical methods, like finite elements, torques are created by such a distributed force on the ends.

Equations (5.7) and (5.12) completely determine the torsional behavior of a prismatic rod. Even though the cross section is allowed to warp out of plane, the only stresses that this model predicts are the shear stresses on the cross-sectional face. Further, it predicts that the stress distribution in each face is the same as that in other cross-sectional faces. Both results are a consequence of the assumption that the out-of-plane warping $w = f(x, y)$ is not a function of z; that is, it is the same in all faces. This indicates that if a little more realistic model were chosen, the predicted stress fields would be much more complicated.

Saint-Venant's solution was accomplished in terms of the function, $f(x, y)$. In 1903, L. Prandtl pointed out that the mathematics of the torsion problem was analogous to that of a membrane stretched uniformly and introduced the stress function. The mathematics are also identical to the hydrodynamics problem in which $f(x, y)$ represents the velocity potential for the irrotational motion of an incompressible, inviscid fluid in a container of the same shape as the prismatic rod in which the container rotates about its axis with an angular velocity of -1 (see Love [1, p. 314]). Many elasticity problems are mathematically similar to other engineering problems. This fact motivated mathematicians to study the underlying mathematical ideas and greatly encouraged the development of several fields of mathematics.

The function $f(x, y)$, describing the warp of a face, satisfies a differential equation of the form of the Prandtl stress function. Saint-Venant noted that the equilibrium equation (5.3), when the expressions for the shear stress in terms of the function

$f(x, y)$ are substituted, becomes

$$\frac{\partial^2 f}{\partial x^2} + \frac{\partial^2 f}{\partial y^2} = \nabla^2 f = 0. \tag{5.13}$$

A function $f(x, y)$ satisfying the Laplace differential equation, $\nabla^2 f = 0$, is called a harmonic function. These functions have been extensively studied in the classical field of potential theory. An example of a harmonic function is $f(x, y) = (A \cos(nx) + B \sin(nx)) \exp(ny)$, where n, A, and B are constants.

The Prandtl stress function has a very useful geometric interpretation that gives a designer excellent insight into the shear stress distribution on a face of a prismatic rod under torsion. Because of the definition, $\frac{\partial \phi}{\partial x} = -\tau_{yz}$, $\frac{\partial \phi}{\partial y} = \tau_{xz}$, given in equations (5.5) and (5.6), the stress vector $t = \tau_{xz} i + \tau_{yz} j$ is tangent to the contour curve for ϕ. So a plot of the contour curves indicates the direction of the shear stress vector on the cross-section face. This is a consequence of the fact that the normal to the contour $\phi(x, y) = $ constant is the gradient of $\phi(x, y)$, $n = \nabla\phi = \frac{\partial \phi}{\partial x} i + \frac{\partial \phi}{\partial y} j = -\tau_{yz} i + \tau_{xz} j$. The dot product of the vectors n and t is zero and so they are perpendicular in the $x - y$ plane. Recall that the contour plot at level a for a function $\phi(x, y)$ is all points (x, y) such that $\phi(x, y) = a$. The magnitude of the total stress $|t| = (\tau_{xz}^2 + \tau_{yz}^2)^{1/2}$ increases with the magnitude of $\phi(x, y)$.

This elasticity analysis applied to a circular cross-section rod under torsion produces the same expression for the shear stress as found in elementary strength of materials. The circular and elliptical cross-sectional calculations are left to the problems at the end of the chapter.

EXAMPLE 5.1   Make a contour plot of the Prandtl stress function for a prismatic rod with a circular cross section under torsion. Here $f(x, y) = 0$ since plane sections remain plane. Therefore $\phi(x, y) = -(x^2 + y^2)/2$ and the contour lines are all circles. Therefore the stress resultants are all tangent to circles centered at the centroid. This is exactly as assumed in the strength of materials derivation of the equations for torsion of a circular rod. ∎

EXAMPLE 5.2   Determine the stress distribution in the prism with equilateral triangular cross section under torsion shown in Figure 5.4. The moment, M, acts about the z-axis.

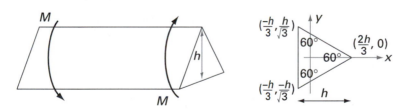

FIGURE 5.4   Equilateral triangular cross-sectional prism under torsion.

Solution   This is a good example in which the fact that the Prandtl stress function must be zero on the boundary guides a guess of the stress function. When the boundary is made up of several curves, as in this case, the Prandtl function can sometimes be taken to be a constant k times the product of the equations of the curves.

Place the origin at the centroid of the triangular cross section with the x-axis along an axis of symmetry of the triangle, as shown in Figure 5.4. The altitude of the triangle is h. The sides of the triangles then have equations

$$0 = x - \sqrt{3}y - \frac{2}{3}h;$$

$$0 = x + \sqrt{3}y - \frac{2}{3}h;$$

$$0 = x + \frac{1}{3}h.$$

A function that satisfies the requirement that it be zero on the boundary is then

$$\phi(x, y) = k\left(x - \sqrt{3}y - \frac{2}{3}h\right)\left(x + \sqrt{3}y - \frac{2}{3}h\right)\left(x + \frac{1}{3}h\right). \tag{5.14}$$

The constant k is determined from the Poisson equation. If there is no k satisfying the Poisson equation, then the guess of $\phi(x, y)$ is not the Prandtl function.

$$\frac{\partial^2 \phi}{\partial x^2} = k\left[\left(x + \frac{h}{3}\right) + \left(x + \sqrt{3}y - \frac{2}{3}h\right) + \left(x + \frac{h}{3}\right) + \left(x - \sqrt{3}y - \frac{2}{3}h\right) + 2\left(x - \frac{2}{3}h\right)\right]$$

$$= (-2h + 6x)k;$$

$$\frac{\partial^2 \phi}{\partial y^2} = k\left[-3\left(x + \frac{h}{3}\right) - 3\left(x + \frac{h}{3}\right)\right]$$

$$= -6k\left(\frac{h}{3} + x\right);$$

so that

$$\frac{\partial^2 \phi}{\partial x^2} + \frac{\partial^2 \phi}{\partial y^2} = -4kh.$$

But the Poisson equation requires that $\nabla^2 \phi = -2G\theta$. Therefore,

$$k = \frac{G\theta}{2h}.$$

The Prandtl stress function is then

$$\phi(x, y) = \frac{G\theta}{2h}\left[\frac{4}{27}h^3 - h(x^2 + y^2) + x^3 - 3xy^2\right].$$

The shear stresses at any point in the face can now be easily computed in terms of the angle of twist.

Usually one wishes to write the stress in terms of the applied torque, M, rather than the angle of twist. Note that the lower vertex is $(-h/3, -h/\sqrt{3})$. Because of the symmetry of the cross-sectional area and because the function ϕ is symmetric about $y = 0$,

$$M = 2\int\int_A \phi(x, y)dxdy = 4\int_{-h/\sqrt{3}}^{0}\int_{-h/3}^{\sqrt{3}y+\frac{2}{3}h} \phi(x, y)dxdy = \frac{G\theta h^4}{15\sqrt{3}}.$$

Therefore, $\theta = M15\sqrt{3}/Gh^4$. The shear stresses are

$$\tau_{xz} = \frac{\partial \phi}{\partial y} = k(-2hy - 6xy) \quad \text{and} \quad \tau_{yz} = -\frac{\partial \phi}{\partial x} = k(2hx - 3x^2 + 3y^2).$$

For example, consider the behavior of the shear stress along the line of symmetry, $y = 0$. The stress $\tau_{xz} = 0$. On the other hand, for $-h/3 \le x \le 2h/3$, τ_{yz} has a maximum at $x = h/3$ where $\tau_{yz} = G\theta h/6$ and a minimum at the endpoint of the interval of definition, $x = -h/3$, where $\tau_{yz} = -G\theta h/2$. Since the sign of the shear stress has no physical meaning, the maximum shear stress on the x-axis is at $x = -h/3$ and $\tau_{yz} = -G\theta h/2 = -15\sqrt{3}M/2h^3$. ∎

Exercise 5.1   Sketch the graph of the shear stress τ_{yz} when $y = 0$.

Exercise 5.2   Verify that the shear stress is zero at each vertex.

Exercise 5.3   Use a mathematical program, such as Mathematica, to make a contour plot of the Prandtl stress function for a prismatic rod under torsion with a triangular cross section. Use this graph to see which regions of the cross section support the most stress.

EXAMPLE 5.3   A prismatic rod that is under torsion has a cross section whose boundary is partially a parabola and partially a straight line defined respectively by $y = x^2$ and $y = 1$. Decide, using a computation, whether or not $\phi(x, y) = k(y - x^2)(y - 1)$, where k is a constant, is a valid Prandtl function for the prismatic rod with this cross section.

Solution $\nabla^2(\phi) = k(-2y + 4)$ so that if $\phi(x, y)$ were a Prandtl function, $k = -2G\theta/(-2y + 4)$. But k must be a constant, not a function of (x, y). Therefore, $\phi(x, y)$ is not a valid Prandtl function. ∎

The analysis of a rectangular cross-sectional prism under torsion requires the inclusion of an infinite series in the Prandtl function. The cross section has long side of length a and short side of length b. Then a Prandtl function that is zero on all four edges is

$$\phi(x, y) = \theta G \left[\frac{b^2}{4} - y^2 - \frac{8b^2}{\pi^3} \sum_{k=0}^{\infty} \frac{(-1)^k \cosh \frac{(2k+1)\pi x}{b} \cos \frac{(2k+1)\pi y}{b}}{(2k + 1)^3 \cosh \frac{(2k+1)\pi a}{2b}} \right]. \quad (5.15)$$

The infinite series is chosen so that it is zero on the edges, $y = \pm b/2$ and equal to $-(b^2/4 - y^2)$ on the edges $x = \pm a/2$. The series is obtained from a Fourier series analysis. The moment-angle of twist relation is

$$M_z = \theta G a b^3 \left[\frac{1}{3} - \frac{64b}{\pi^5 a} \sum_{k=0}^{\infty} \frac{\tanh \frac{(2k+1)\pi a}{2b}}{(2k + 1)^5} \right].$$

The maximum shear stress occurs at the middle of the longer side $(x = 0, y = \pm b/2)$ and is equal to

$$\tau_{zx}(0, \pm b/2) = \theta G b \left[1 - \frac{8}{\pi^2} \sum_{k=0}^{\infty} \frac{1}{(2k + 1)^2 \cosh \frac{(2k+1)\pi a}{2b}} \right].$$

For design purposes, the series has been evaluated, and these expressions have been written in the form

$$\tau_{zx}(0, \pm b/2) = \frac{M_z}{c_1 a b^2};$$

$$M_z = \theta c_2 G a b^3. \quad (5.16)$$

Some values for the ratio a/b are: (1) $a/b = 1$, $c_1 = 0.208$, $c_2 = 0.140$; (2) $a/b = 2$, $c_1 = 0.246$, $c_2 = 0.229$; (3) $a/b = 4$, $c_1 = 0.282$, $c_2 = 0.281$, and (4) $a/b = \infty$, $c_1 = 0.333$, $c_2 = 0.333$ [2, p. 201].

EXAMPLE 5.4   A prismatic bar has a square cross section with sides of length, s. It is made of 2014-T6 aluminum with a shear strength of 42 ksi (290 MPa) and $G = 27$ GPa. The bar is subjected to a torque $M_z = 3250$ Nm.

(a) Design the cross section so that the maximum Mises stress is less than one-fourth the shear strength.

(b) Determine the angle of twist per unit length for the designed bar.

Solution **(a)** The shear stress is maximum at the middle of any one of the edges of any cross section. At such a point, $\tau_{zx} = M_z/c_1 s^3$ and $\tau_{zy} = 0$. Therefore, at this point, the principal stresses are $0, \pm M_z/c_1 s^3$. The maximum Mises stress (1.14) is therefore $M_z/c_1 s^3$. The edge length is found from $s^3 \geq 4M_z/(\tau_Y c_1) = 4(3250)/(0.208 \times 290 \times 10^6) = 0.0002155$. The square cross section must have edges of length greater than 5.996 cm.

(b) The angle of twist $\theta = M_z/c_2 G s^4$ is therefore less than 0.06654 rad/m. ∎

EXAMPLE 5.5 A prismatic bar, which is $L_b = 0.5$ m long, has a square cross section with sides of length, s. It is made of 2014-T6 aluminum with a shear strength of 42 ksi (290 MPa) and $G = 27$ GPa. The bar is mounted in a rigid wall at one end (Fig. 5.5). At the other end, a rigid rod of length $L = 0.2$ m is attached perpendicular to the bar. A load of $P = 1000$ N is applied normal to the rod.

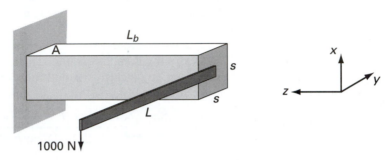

FIGURE 5.5 A cantilevered square cross-sectional bar with a rigid rod attached perpendicular to the bar.

(a) Determine the moment and load acting on the ends of the bar using statics.

(b) Determine the state of stress at point $A\left(\frac{s}{2}, 0, L_b\right)$ on the top of the bar at the wall. Write the stress tensor in terms of s. The coordinate origin is at the free end of the bar.

(c) Compute the principal stresses at point A.

(d) Design the dimension s so that the Mises stress is less than one-fourth of the shear strength.

Solution **(a)** From statics, using a free-body diagram of the rigid rod, there is a downward vertical force P and a torque of $M_t = LP$ on the free end. The bar is in combined bending and torsion.

(b) The state of stress is obtained by superposition. The torque induces a shear $\tau_{zy} = M_z/c_1 s^3$ at A. The bending stresses are obtained from the Airy stress function $\phi(x, z) = Pzx^3/6I - Ps^2xz/8I$ where z is measured from the end at which the rod is attached (problem 2, Chapter 4). The bending axial stress is $\sigma_z = Pxz/I$. But the internal shear force P induces a shear stress, $\tau_{zx} = P[x^2 - (s/2)^2]/2I$. The point A lies at $x = s/2$ and $z = L_b$. Therefore the shear stress due to bending is zero. The stress tensor at point A in terms of s and the loads is

$$T = \begin{pmatrix} 0 & 0 & 0 \\ 0 & 0 & \tau_{zy} \\ 0 & \tau_{zy} & \sigma_z \end{pmatrix} = \begin{pmatrix} 0 & 0 & 0 \\ 0 & 0 & LP/(c_1 s^3) \\ 0 & LP/(c_1 s^3) & 6PL_b/s^3 \end{pmatrix}.$$

because $I = s^4/12$.

(c) The eigenvalues of the stress tensor, which are the principal stresses, are 0, and

$$0.5\left(\sigma_z \pm \sqrt{\sigma_z^2 + 4\tau_{zy}^2}\right).$$

(d) The Mises stress is $\sqrt{\sigma_1^2 - \sigma_1 \sigma_2 + \sigma_2^2} = \sqrt{\sigma_z^2 + 3\tau_{zy}^2} = P\left[\sqrt{36L_b^2 + 3L^2/c_1^2}\right]/s^3$. To have the Mises stress less than one-fourth the shear yield,

$$s^3 \geq 4P\left[\sqrt{36L_b^2 + 3L^2/c_1^2}\right]/\tau_Y = 4.73 \times 10^{-5}.$$

So the edge length s must be greater than 3.62 cm.   ∎

5.2 BENDING OF CURVED BEAMS

Curved members loaded in bending appear as arches, as well as in chains, hooks, and other connectors. A beam is called straight if the line of centroids of all the cross sections is straight and is called curved if the curve formed by the centroids of all cross sections is not straight. The normal bending stresses in a curved beam with circular line of centroids which is in pure bending will be computed from elasticity. Then, for comparison, a strength of materials solution, the Winkler-Bach analysis, will be obtained for the longitudinal bending stress in an arbitrarily curved beam.

5.2.1 Circular Beam in Pure Bending

The beam is assumed to have constant cross section. Assume that the curve of centroids of the cross sections is the arc of a circle. The cross section is assumed to be a narrow rectangle in the sense that the thickness in the direction perpendicular to the plane containing the arc of centroids is small compared to the height, h. The beam is in pure bending with moments about the direction of the small thickness, t. The beam is also isotropic, homogeneous, and Hookean. The coordinates are cylindrical with origin at the center of curvature of the lateral boundaries of the beam and z-direction parallel to the thickness direction. In this system, the inner boundary is $r = a$ and the outer boundary is $r = b$ so that $h = b - a$ (Fig. 5.6).

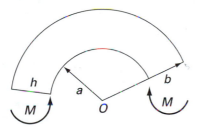

FIGURE 5.6   A curved beam with constant radius of curvature under bending.

The sign convention is that the moment, M, is positive if it decreases the radius of curvature of the beam; a positive moment tends to bend the beam further. A straight beam has infinite radius of curvature. The circumferential stresses near the top of a beam under a positive moment are therefore tensile and those near the bottom are compressive.

In other situations, assumptions were placed on the displacements to simplify the elasticity solution. Here, assumptions are made on the stress. The beam will be assumed to be in plane stress with the stress $\sigma_z = 0$ in the direction of the thickness, t. Furthermore the shear stress on each face of the beam is assumed to be zero. Therefore the only nonzero stresses on each face of the beam are σ_r and σ_θ. Neither of these stresses depend on z or θ. The problem has been reduced by the assumptions to a two-dimensional one in polar coordinates.

The polar equations of equilibrium, if the body forces are neglected, are

$$\frac{\partial \sigma_r}{\partial r} + \frac{1}{r}\frac{\partial \tau_{r\theta}}{\partial \theta} + \frac{\sigma_r - \sigma_\theta}{r} = 0;$$

$$\frac{1}{r}\frac{\partial \sigma_\theta}{\partial \theta} + \frac{\partial \tau_{r\theta}}{\partial r} + 2\frac{\tau_{r\theta}}{r} = 0.$$

Since the shear stresses are assumed zero, the second equation becomes

$$\frac{\partial \sigma_\theta}{\partial \theta} = 0,$$

so that σ_θ is only a function of r as expected. Because the stresses are functions of just r, the first equation is

$$\frac{d\sigma_r}{dr} + \frac{\sigma_r - \sigma_\theta}{r} = 0. \tag{5.17}$$

The polar equation of stress compatibility is

$$\nabla^2(\sigma_r + \sigma_\theta) = \frac{d^2(\sigma_r + \sigma_\theta)}{dr^2} + \frac{1}{r}\frac{d(\sigma_r + \sigma_\theta)}{dr} = 0. \tag{5.18}$$

These two equations can be solved simultaneously for the two unknowns, σ_r and σ_θ. If one writes $f(r) = \sigma_r(r) + \sigma_\theta(r)$, equation (5.18) becomes a version of Euler's differential equation,

$$\frac{d^2 f}{dr^2} + \frac{1}{r}\frac{df}{dr} = 0, \tag{5.19}$$

one which is linear with special nonconstant coefficients. It is solved by the transformation $t = \ln(r)$. By repeated applications of the chain rule, $df/dr = (1/r)(df/dt)$ and $d^2 f/dr^2 = (1/r^2)(d^2 f/dt^2) - (1/r^2)(df/dt)$. Substituting these into equation (5.19) gives

$$\frac{d^2 f}{dt^2} = 0.$$

The solution is $f(t) = k'' + k't$, so that $f(r) = k'' + k'\ln(r) = k'\ln(r/a) + k'''$, where the last equality follows by putting $k'' = k''' - k'\ln(a)$. Therefore equation (5.18) implies

$$\sigma_r + \sigma_\theta = k'\ln\left(\frac{r}{a}\right) + k'''. \tag{5.20}$$

Now equations (5.17) and (5.20) are solved simultaneously for σ_r and σ_θ by eliminating σ_θ to obtain

$$\frac{d\sigma_r}{dr} + \frac{1}{r}\left[2\sigma_r - k'\ln\left(\frac{r}{a}\right) - k'''\right] = 0.$$

This first-order linear differential equation in σ_r is solved using the integrating factor r^2 and gives a solution of the form

$$\sigma_r(r) = c_1 + c_2\ln\left(\frac{r}{a}\right) + \frac{c_3}{r^2}. \tag{5.21}$$

Substitution of $\sigma_r(r)$ in equation (5.17) gives

$$\sigma_\theta(r) = c_1 + c_2\left[1 + \ln\left(\frac{r}{a}\right)\right] - \frac{c_3}{r^2}. \tag{5.22}$$

The three constants, c_1, c_2, and c_3, are determined from the boundary conditions.

Exercise 5.4 An Airy function for the curved beam under pure bending is $\phi(r,\theta) = Ar^2 + B\ln r + Cr^2\ln r$, where A, B, and C are constants. Verify that this function satisfies the polar biharmonic equation (Chapter 4, Section 4.6.2). Show, using the definition of the Airy stress function in polar coordinates, that the induced stress functions $\sigma_r(r)$ and $\sigma_\theta(r)$ have the same form as equations (5.21) and (5.22) and that $\tau_{r\theta}(r) = 0$.

The boundary conditions are $\sigma_r(a) = \sigma_r(b) = 0$. Note that σ_θ is not zero at the ends. In fact the moment, M, is due to the distributed normal stresses σ_θ on the ends, but the total force on the ends is zero. So the remaining boundary conditions are

$$\int_a^b t\sigma_\theta dr = 0;$$

$$\int_a^b rt\sigma_\theta dr = M.$$

Since M is a couple, it causes the same moment about any point. Therefore in the last equation, the moment can be calculated about the origin with moment arm r to $\sigma_\theta(r)$.

The stresses are therefore

$$\sigma_r(r) = \left(\frac{4M}{tb^2N}\right)\left[\left(1-\frac{a^2}{b^2}\right)\ln\left(\frac{r}{a}\right) - \left(1-\frac{a^2}{r^2}\right)\ln\left(\frac{b}{a}\right)\right]; \tag{5.23}$$

$$\sigma_\theta(r) = \left(\frac{4M}{tb^2N}\right)\left[\left(1-\frac{a^2}{b^2}\right)\left(1+\ln\left(\frac{r}{a}\right)\right) - \left(1+\frac{a^2}{r^2}\right)\ln\left(\frac{b}{a}\right)\right], \tag{5.24}$$

where

$$N = \left(1-\frac{a^2}{b^2}\right)^2 - 4\left(\frac{a^2}{b^2}\right)\ln^2\left(\frac{b}{a}\right).$$

Notice that the stresses are linear functions of the couple, M.

Exercise 5.5 Verify that the boundary conditions produce these constants.

Exercise 5.6 Plot the stresses $\sigma_r(r)/M$ and $\sigma_\theta(r)/M$ as a function of r for a beam with $t = 0.1$, $a = 1$, and $b = 3$.

Exercise 5.7 Is there a neutral surface for the curved beam as there is for a straight beam? Is the radial or is the circumferential stress zero at the centroid of the cross-sectional area? Are these stresses tensile or compressive as r varies through the cross section?

EXAMPLE 5.6 Compute the stresses at $r = 11.5$ inches in a curved beam of outer radii, $a = 10$ inches and $b = 12$ inches subject to a pure bending moment of $M = 300$ in/lb. The cross section is rectangular with a beam thickness of 0.25 inches. Assume that the beam is elastically loaded.

Solution First, compute

$$N = \left(1-\frac{10^2}{12^2}\right)^2 - 4\left(\frac{10^2}{12^2}\right)\ln^2\left(\frac{12}{10}\right) = 1.0277 \times 10^{-3}.$$

The coefficient $4M/tb^2N = 32{,}436$. By substitution into the expressions derived for the circumferential and radial stresses,

$$\sigma_r(11.5) = 32436\left[\left(1 - \frac{10^2}{12^2}\right)\ln\left(\frac{11.5}{10}\right) - \left(1 - \frac{10^2}{11.5^2}\right)\ln\left(\frac{12}{10}\right)\right]$$

$$= (32{,}436)(-1.755 \times 10^{-3}) = -56.9 \text{ psi},$$

$$\sigma_\theta(11.5) = 32436\left[\left(1 - \frac{10^2}{12^2}\right)\left(1 + \ln\left(\frac{11.5}{10}\right)\right) - \left(1 + \frac{10^2}{11.5^2}\right)\ln\left(\frac{12}{10}\right)\right]$$

$$= (32{,}436)(2.81 \times 10^{-2}) = 911 \text{ psi}.$$

Ignore the beam curvature and use straight beam theory to compute $\sigma_x = \sigma_\theta$.

$$\sigma_x = \frac{My}{I} = \frac{(300)(0.5)}{\frac{1}{12}(0.25)(2)^3} = 900 \text{ psi}.$$

At this radius of curvature, the straight beam theory does make a satisfactory prediction of the circumferential stress. ∎

A rule of thumb is that the straight beam theory closely approximates the stress, $\sigma_\theta(r)$, when the curvature is large compared to the depth $h = b - a$. Usually the ratio of the radius of curvature at the centroid to the distance between the centroid and the outer edge is used. Such a criterion for a rectangular cross section is

$$\frac{a + \frac{h}{2}}{\frac{h}{2}} = \frac{b + a}{b - a} > 4.$$

EXAMPLE 5.7 A U-shaped frame is loaded by a force, $P = 4{,}000$ N, perpendicular to each of its free ends (Fig. 5.7). The inner radius of curvature is $a = 4$ cm. The two straight sections each have length $L = 10$ cm. The cross section is rectangular of thickness t and height $h = 8$ cm. The maximum shear stress in the curved section must be less than 40 MPa. Neglect the stresses induced by any internal shear forces. Design the thickness, t, of the member.

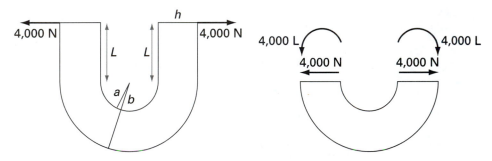

FIGURE 5.7 U-shaped frame is loaded by a force perpendicular to each of its free ends.

Solution The stresses σ_r and σ_θ are principal. Therefore the maximum shear stress acting on the various planes through any point is $0.5(\sigma_r + \sigma_\theta)$. This value is obtained from equations (5.23) and (5.24). The derivative of $0.5(\sigma_r + \sigma_\theta)$ with respect to r is positive for $a \leq r \leq b$ so that the maximum shear stress at a point increases with r. Therefore, the maximum shear stress occurs on the 45° plane at $r = b$ and is

$$0.5(\sigma_r + \sigma_\theta) = 0.5\sigma_\theta = \left(\frac{4M}{2tb^2N}\right)\left[1 - \left(\frac{a^2}{b^2}\right)\left(1 + 2\ln\frac{b}{a}\right)\right].$$

A free-body diagram of one straight section shows that at the face where the shape changes from straight to curved, there is an internal shear force equal to P and an internal moment $M = PL = 400$ Nm. The moment is positive because it decreases the radius of curvature. If the effects of the shear force are neglected, the curved section is in pure bending. From the data, $b = a + h = 12$ cm, so that $N = 0.2537$. The maximum shear stress at $r = b$ is then

$$\tau_{max} = \left(\frac{4(400)}{2t(.12)^2(0.2537)} \right) \left[1 - \left(\frac{0.04^2}{0.12^2} \right) \left(1 + 2\ln\frac{0.12}{0.04} \right) \right] = 141188/t \leq 40 \times 10^6.$$

Therefore the thickness must be $t \geq 3.53$ mm. ∎

5.2.2 Winkler-Bach Theory for Curved Beams

The analysis of the previous section was explicitly based on the equations of elasticity with an additional assumption that the stress fields are independent of θ, that is, they are the same in all faces of the beam. A strength of materials solution was developed by Winkler and Bach. In their approach, which parallels that of the elementary strength of materials solution for a straight beam, the strain is described in terms of the beam geometry and in terms of the position, y, in the beam cross section. Again, the fundamental assumption is that the cross sections remain planar under the displacement. Furthermore, the elastic modulus is assumed to be the same in tension and compression, a good assumption for metals. This solution produces the longitudinal normal bending stress, which corresponds to σ_θ in the elasticity solution. The radial stress is presumed to be zero.

This calculation of the stresses in a curved beam by strength of materials methods was devised by E. Winkler, a German engineering professor, and first presented in his strength of materials book published in 1867 while he was chairman of bridge and railroad engineering at the polytechnic institute in Prague. Timoshenko [3], in his history of strength of materials, says that, after the publication of Winkler's book, the fields of strength of materials and mathematical elasticity split apart so that subsequently the more elementary approach of strength of materials was commonly used. Winkler went on to perform many experiments to validate his analysis of curved beams. He was one of those primarily responsible for its adoption in daily engineering practice, particularly in the design of arches.

As in the elasticity analysis, a moment, M, is given a positive sign if its application decreases the radius of curvature. In a given cross-sectional face, the y-coordinate is taken from the centroid along the radius of curvature and is positive away from the center of curvature, C. The x-direction is perpendicular to the face; it corresponds to the circumferential direction, θ, in the elasticity analysis. The strength of materials method requires using geometric relations to determine the strain at each position y in an arbitrary cross-sectional face. A small section of the beam, of length Δx, is shown in Figure 5.8. As the moment is applied, the outer faces remain plane but rotate with respect to each other. Using the face BE as a reference, the face AD rotates to position $A'D'$. $O_1 O_2$ is the centroid line, which may not be the neutral axis. During the bending it is deformed by $O_1 O'$. The line $P_1 P_2$ is the line at position, y, whose strain is sought; it deforms through $P_1 P'$. Drop a line from point O' parallel to face AD until it hits line $P_1 P'$ at point H. Drop a vertical line from point O' until it hits line $P_1 P'$ at point Z.

The strain at position y is just the length $P'P_1$ divided by the length $P_1 P_2$. It remains to determine these lengths. The distance $P_1 P_2 = (R + y)\Delta\theta$ since it approximates the arc of a circle. The angle between lines $O'Z$ and $O'H$ is the original angle, $\Delta\theta$, cut off by the arclength Δx of the piece of beam being considered. The change in

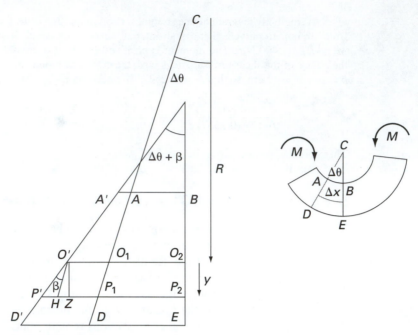

FIGURE 5.8 Section of curved beam showing approximations used to derive the Winkler-Bach equation.

this angle, β, due to the application of the moment is the angle between the lines $O'H$ and $O'P'$.

The length of the original centroidal arc, which is approximated as a line in Figure 5.8, is $O_1 O_2 = R\Delta\theta$. The strain of the centroidal arc is $\epsilon_0 = O_1 O' / O_1 O_2$ so that

$$O_1 O' = \epsilon_0 R \Delta\theta.$$

Since lines $O'H$ and AD are parallel, the distances $O_1 O'$ and HP_1 are equal. Also, using the right triangles of height y,

$$HP' = ZP' - HZ = y\tan(\Delta\theta + \beta) - y\tan(\Delta\theta) \sim y\beta.$$

Therefore, the strain at position y is

$$\epsilon_x(y) = \frac{P_1 P'}{P_1 P_2} = \frac{P_1 H + HP'}{P_1 P_2} = \frac{O_1 O' + HP'}{P_1 P_2}$$

$$= \frac{\epsilon_0 R \Delta\theta + y\beta}{(R + y)\Delta\theta}.$$

Divide top and bottom of the fraction on the right side by $\Delta\theta$ and let $\lambda = \beta/\Delta\theta$ be the angular strain, the change of angle per original angle. Then, the longitudinal normal strain, as a function of the position in the cross section is

$$\epsilon_x(y) = \frac{\epsilon_0 R + y\lambda}{R + y}. \tag{5.25}$$

This expression involves two unknowns, ϵ_0 and λ, which are determined from the equilibrium conditions. Rewrite the strain as

$$\epsilon_x(y) = \epsilon_0 + \frac{(\lambda - \epsilon_0)y}{R + y}. \tag{5.26}$$

Then by Hooke's law in the circumferential direction, the stress is

$$\sigma_x(y) = E\left[\epsilon_0 + \frac{(\lambda - \epsilon_0)y}{R + y}\right]. \tag{5.27}$$

Since the beam is assumed in pure bending, the sum of the normal forces on each cross-sectional face, of area A, must be zero.

$$\int_A \sigma_x(y)dA = \int_A E\left[\epsilon_0 + \frac{(\lambda - \epsilon_0)y}{R + y}\right]dA = 0.$$

Therefore since ϵ_0 and λ are constant over the area A,

$$\epsilon_0 A = -(\lambda - \epsilon_0)\int_A \frac{y}{R + y}dA. \tag{5.28}$$

Also the sum of the moments of the internal forces on each cross-sectional face must be the applied moment, M.

$$M = \int_A y\sigma_x(y)dA = \int_A yE\left[\epsilon_0 + \frac{(\lambda - \epsilon_0)y}{R + y}\right]dA$$
$$= E\epsilon_0\int_A ydA + E(\lambda - \epsilon_0)\int_A \frac{y^2}{R + y}dA. \tag{5.29}$$

Since y is measured from the centroid of the cross section, $\int_A ydA = 0$. A special integral appears in the equation (5.28) derived from equilibrium of forces. So define

$$Z \equiv -\frac{1}{A}\int_A \frac{y}{R + y}dA = -\frac{1}{A}\int_A \left(1 - \frac{R}{R + y}\right)dA. \tag{5.30}$$

This integral must be computed for each cross section. Its value for common cross sections is given in Table 5.1.

EXAMPLE 5.8   The value of Z for a circular cross section of radius, $r = 2$, in a curved beam whose centroidal line has constant radius of curvature, $R = 12$, is

$$Z = -1 + 2\left(\frac{R}{r}\right)^2 - 2\left(\frac{R}{r}\right)\sqrt{\left(\frac{R}{r}\right)^2 - 1} = 0.0070426.$$

Most values of Z are significantly less than one and are positive.   ■

Exercise 5.8   The value of Z for a rectangular cross section is $Z = -1 + (R/h)\ln[(R + h/2)/(R - h/2)]$. Write $Z = -1 + (R/h)\ln[(R/h + 0.5)/(R/h - 0.5)]$. Show that Z decreases as the parameter R/h increases. Plot this relation for $1 \le R/h \le 20$.

The other integral in the right-hand term of equation (5.29) can be expressed in terms of Z as follows:

$$\int_A \frac{y^2}{R + y}dA = \int_A \left[y - \frac{Ry}{R + y}\right]dA = \int_A ydA - R\int_A \frac{y}{R + y}dA = RZA, \tag{5.31}$$

TABLE 5.1 Values of Z for selected cross sections

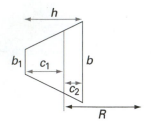

$$Z = -1 + 2\left(\frac{R}{r}\right)^2 - 2\left(\frac{R}{r}\right)\sqrt{\left(\frac{R}{r}\right)^2 - 1}$$

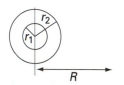

$$Z = -1 + \frac{2R}{(b+b_1)h}\left\{\left[b_1 + \frac{b-b_1}{h}(R+c_1)\right]\ln\left(\frac{R+c_1}{R-c_2}\right) - (b-b_1)\right\}$$

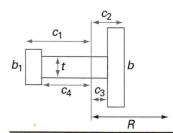

$$Z = -1 + \frac{2R}{r_2^2 - r_1^2}\left[\sqrt{R^2 - r_1^2} - \sqrt{R^2 - r_2^2}\right]$$

$$Z = -1 + \frac{R}{A}\left[b_1\ln(R+c_1) + (t-b_1)\ln(R+c_4)\right.$$
$$\left. + (b-t)\ln(R-c_3) - b\ln(R-c_2)\right]$$

since $\int_A y\,dA = 0$. Therefore, the equilibrium equations produce two equations in ϵ_0 and λ, in terms of Z,

$$\epsilon_0 = (\lambda - \epsilon_0)Z;$$
$$M = E(\lambda - \epsilon_0)RZA,$$

which are solved simultaneously to obtain

$$\epsilon_0 = \frac{M}{AER};$$
$$\lambda = \frac{1}{AE}\left(\frac{M}{R} + \frac{M}{ZR}\right).$$

Substituting these into the expression for the stress [equation (5.27)] produces the Winkler-Bach equation for the stress in a curved beam under a pure bending moment,

$$\sigma_x(y) = \frac{M}{AR}\left[1 + \frac{y}{Z(R+y)}\right]. \tag{5.32}$$

Notice that this result depends only on the geometry of the beam and the applied moment, but not on the material. It is a linear function of the moment as expected from linear strength of materials. The assumption that plane sections remain plane and the linear Hooke's constitutive model produce the linearized relation.

Exercise 5.9 Plot the normalized Winkler-Bach stress curve, $\sigma_x(y)/M$, as a function of y for a rectangular beam with height $h = 1$ and thickness 0.25 and radius of curvature, $R = 2$. Compare this curve to the one obtained from the elasticity solution.

The location, y_0 of the neutral axis where $\sigma_x(y) = 0$ can be obtained in closed form by setting equation (5.32) equal to zero.

$$y_0 = -\frac{ZR}{Z+1}. \tag{5.33}$$

The Winkler-Bach stress equation should give the elementary strength of materials beam stress when applied to a straight beam, a beam in which the radius of curvature is infinite. The Winkler-Bach equation should have My/I as its limit when $R \to \infty$. Using equation (5.31), $ARZ = \int_A y^2/(R+y)dA$, to go from the second line to the third line,

$$\sigma_x(y) = \frac{M}{AR}\left[1 + \frac{y}{Z(R+y)}\right]$$

$$= \frac{M}{AR} + \frac{M}{ARZ}\frac{y}{(R+y)}$$

$$= \frac{M}{AR} + \frac{My}{\left(\int_A \frac{y^2}{R+y}dA\right)(R+y)}$$

$$= \frac{M}{AR} + \frac{My}{\left(1+\frac{y}{R}\right)\left(\int_A \frac{y^2}{1+\frac{y}{R}}dA\right)}.$$

Then because the area moment of inertia is $I = \int_A y^2 dA$, this expression tends to $\sigma_x(y) = My/I$ as $R \to \infty$, the bending stress for a straight beam.

EXAMPLE 5.9

The tray BEFC carrying a component of weight, $W = 400$ pounds, is hung from pins B and C (Fig. 5.9). To provide clearance for the component, it is decided to bend BC into the arc of the circle shown. The inner radius of curvature must be $a = 5$ inches to clear the component. Let b be the outer radius of curvature of section BC and denote $h = b - a$. The bar ABCD has rectangular cross section h by $t = 0.375$ and is supported by pins at A and D. The members BE and CF are vertical. The straight sections AB and CD have length, $L = 10$ inches. Design the dimension, h, so that the maximum circumferential tensile stress in section BC is less than 2,500 psi. Use the Winkler-Bach analysis.

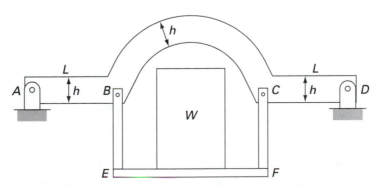

FIGURE 5.9 The tray BEFC carrying a component and hung from pins B and C.

Solution   A free-body diagram shows that the section is in pure bending with internal moment $M = -WL/2$ on each face in the section so that the maximum tensile stress occurs in each face at the inner radius, a. The coordinate of a point on the inner radius in the Winkler-Bach analysis is $y = -h/2$. Further, the centroidal radius of curvature is $R = a + h/2$. The cross-sectional area is $A = ht$. Therefore the stress is

$$\sigma_x(-h/2) = \frac{M}{ht(a+h/2)}\left(1 + \frac{-h/2}{Z(a+h/2-h/2)}\right),$$

where $Z = -1 + (R/h)\ln[(R+h/2)/(R-h/2)]$. Substituting the given values and using a numerical solution technique, such as Newton's method, the value of h is 3.98 inches.   ■

If a particular face of the curved beam is subjected to an internal normal axial load, P, as well as a bending moment, then the normal stress on that one face is obtained by superposition.

$$\sigma_x(y) = \frac{P}{A} + \frac{M}{AR}\left[1 + \frac{y}{Z(R+y)}\right]. \tag{5.34}$$

Since the beam is curved, this situation can only occur on the faces perpendicular to an external load. Other faces will additionally have shear stress on them.

EXAMPLE 5.10   The classical application of this work was to large hooks used on loading cranes in shipyards. Compute the circumferential stresses at the points Q and Q_1 on the face perpendicular to the applied load, $P = 10,000$ pounds (Fig. 5.10). The cross section is the symmetrical trapezoid shown, with the longer baseline toward the center of curvature of the hook. The bases of the trapezoid are $b = 3.5$ inches and $b_1 = 1.5$ inches; the height is $h = 5$ inches. The radius of curvature of the inner surface at the cross section, $Q_1 Q$ is 2.25 inches. Apparently, the designer expected the largest circumferential stress on the inside. The hook is shaped so that when hung from a cable, the line of action of the forces go through the center of curvature of the perpendicular section.

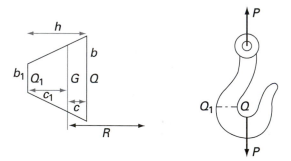

FIGURE 5.10   A hook with a trapezoidal cross section.

Solution   The internal forces on the cross section are a tensile normal load, P, at the centroid of the face and a moment $M = -RP$ about the centroidal axis of the face. The minus sign in the moment is taken since the moment tries to increase the radius of curvature by straightening out the hook. First locate the centroid, G, on the line of symmetry of the cross section from the relation giving its distance from the longer baseline,

$$c = \frac{(b+2b_1)h}{3(b+b_1)} = 2.17.$$

The distance from the center of curvature to the centroid is $R = 2.25 + c = 4.42$. The area of the trapezoid is $A = (b + b_1)h/2 = 12.5$. The distance $Q_1 G$ is $c_1 = h - c = 2.83$. From

Table 5.1,

$$Z = -1 + \frac{R}{Ah}\left\{[b_1 h + (R + c_1)(b - b_1)]\ln\left(\frac{R + c_1}{R - c}\right) - (b - b_1)h\right\} = 0.11324.$$

The position, y, on the face is positive pointing from the centroid away from the center of curvature. The position of Q is therefore, $y_Q = -2.17$.

The stress on the face at position, y, is the superposition of the axial stress due to P and the bending stress due to M.

$$\sigma_x(y) = \frac{P}{A} + \frac{M}{AR}\left[1 + \frac{y}{Z(R + y)}\right].$$

The moment happens to be equal to $-PR$; this is a special case and does not always occur. But when it does, the stress formula can be simplified to

$$\sigma_x(y) = -\frac{P}{AZ}\frac{y}{(R + y)}.$$

The stress at Q is

$$\sigma_x(-c) = -\frac{P}{AZ}\frac{-c}{(R - c)} = 6813 \text{ psi.}$$

The stress at Q_1 is

$$\sigma_x(c_1) = -\frac{P}{AZ}\frac{c_1}{(R + c_1)} = -2758 \text{ psi.}$$

As expected the stress at Q is tensile and that at Q_1 is compressive. If the hook is made of a metal-like steel, these stresses are well under the elastic limit, and so the analysis is valid.

The neutral axis for bending only is at the point, y_0, where $\sigma_x = 0$;

$$y_0 = -\frac{ZR}{Z + 1} = -0.450 \text{ inches.} \qquad \blacksquare$$

5.3 BEAMS ON AN ELASTIC FOUNDATION

Most beams studied in elementary strength of materials are either simply supported or cantilevered. These supports are rigid in both cases. In many situations, the supports are not rigid but are rather also elastic themselves. The classical situation, which seems to have motivated much of the work on this subject, is railroad rails resting on the cross ties, which are elastic. In fact, the ties rest on the ground which is also elastic. As a train passes a particular site, multiple point loads are applied to the rail by the wheels. The rail is then viewed as a long beam resting on an elastic foundation. It seems that beams on an elastic foundation were first discussed in Winkler's 1867 strength of materials book in which they were applied to predict the stresses in railroad tracks.

Such beams have been classified by whether they rest on a continuous elastic foundation, such as the ground, or on a discrete elastic foundation, such as the rails on the ties. The beams have also been distinguished by whether they are long or short because in some of the cases simplifying approximations permit an easier calculation of the displacement of the beam.

Since strength of materials is assumed, the displacement is computed first. The stresses and strains are then computed from the displacements. The analysis provides a good example of the manner in which engineers can use approximations to simplify the analysis. This permits using the resulting closed form solutions to predict the effect on a design of changing parameters. The alternative is numerical methods, but then many cases need to be computed to see the effect of changing parameters.

5.3.1   A Beam Differential Equation

In principle, the problem of a beam on an elastic foundation can be analyzed by elasticity to obtain the displacement, strains, and stresses from the 15 equations of elasticity subject to the foundation boundary conditions. Unfortunately, the magnitudes of the reaction forces on the beam from the foundation are not initially known. If the foundation is thought of as a spring, then the force exerted on the beam depends on the deformation of the spring. But this deformation is directly related to the displacement of the beam, which we are trying to determine. It seems as though the problem is circular. Of all possible deflections of the beam, the required one is that which places the beam in equilibrium with both its external loads and the elastic support.

The strength of materials assumptions are frequently made for beams on an elastic foundation. As proved in elementary strength of materials, when $v(x)$ is the vertical deflection of the neutral axis of the beam and $M(x)$ is the internal bending moment,

$$EI\frac{\partial^2 v}{\partial x^2} = M(x). \tag{5.35}$$

Differentiating twice and recalling that the second derivative of the internal moment equals the distributed load, $q(x)$, on the beam produces the relation

$$EI\frac{\partial^4 v}{\partial x^4} = q(x). \tag{5.36}$$

The elastic foundation has spring constant, $k(x)$, under each point in the beam. The constant $k(x)$ has units of force per unit area because it is the force per unit length of the beam per unit deflection of the beam. The foundation exerts a distributed load on the beam of magnitude,

$$q(x) = -k(x)v(x).$$

Therefore, the equilibrium displacement must satisfy the ordinary differential equation,

$$EI\frac{\partial^4 v}{\partial x^4} + k(x)v(x) = 0. \tag{5.37}$$

If the elastic coefficient, $k(x)$, is constant at all points along the beam, this differential equation

$$EI\frac{\partial^4 v}{\partial x^4} + kv(x) = 0 \tag{5.38}$$

is linear with constant coefficients and is relatively easy to solve.

A fourth-order linear differential equation with constant coefficients has four independent solutions of the form,

$$v(x) = e^{rx},$$

determined by four different values of r.

Substituting this into the differential equation and dividing out $\exp(rx)$, which is always positive, produces the characteristic equation,

$$EIr^4 + k = 0. \tag{5.39}$$

Therefore, the four roots are the fourth roots of $-k/EI$. The fourth roots of -1 are from DeMoivre's theorem,

$$1/\sqrt{2} + i/\sqrt{2}, \quad -1/\sqrt{2} + i/\sqrt{2}, \quad 1/\sqrt{2} - i/\sqrt{2}, \quad -1/\sqrt{2} - i/\sqrt{2},$$

where $i = \sqrt{-1}$. This can be checked by taking each of these complex numbers to the fourth power. The four roots of $-k/EI$ are then $(k/EI)^{1/4}$ times each of the complex fourth roots of -1. Put $\beta = (k/4EI)^{1/4}$. β has units of length inverse. The roots are

$$r = \beta(1 + i), \quad r = \beta(-1 + i), \quad r = \beta(1 - i), \quad r = \beta(-1 - i).$$

The general solution of the differential equation is a linear combination, a sum, of multiples of four linearly independent solutions.

$$v(x) = c_1 \exp(\beta(1 + i)) + c_2 \exp(\beta(-1 + i))$$
$$+ c_3 \exp(\beta(1 - i)) + c_4 \exp(\beta(-1 - i));$$

or

$$v(x) = e^{\beta x} [c_1 \cos(\beta x) + c_2 \sin(\beta x)] + e^{-\beta x} [c_3 \cos(\beta x) + c_4 \sin(\beta x)]. \quad (5.40)$$

Notice that the number β depends on the ratio of the stiffness of the support to the stiffness, EI, of the beam. This ratio controls the frequency of the oscillations of the deflection curve about the x-axis.

A fourth-order equation requires four boundary conditions. These determine the constants c_1, c_2, c_3, and c_4. In the following sections, several cases are calculated for different types of boundary conditions.

The boundary conditions require computation of the derivatives of the general form of the solution.

$$v'(x) = \beta e^{\beta x} [(c_1 + c_2) \cos(\beta x) + (-c_1 + c_2) \sin(\beta x)]$$
$$+ \beta e^{-\beta x} [(-c_3 + c_4) \cos(\beta x) + (c_3 + c_4) \sin(\beta x)];$$
$$v''(x) = \beta^2 e^{\beta x} [2c_2 \cos(\beta x) - 2c_1 \sin(\beta x)] + \beta^2 e^{-\beta x} [2c_3 \cos(\beta x) - 2c_4 \sin(\beta x)];$$
$$v'''(x) = \beta^3 e^{-\beta x} [(-2c_1 + 2c_2) \cos(\beta x) + (-2c_1 - 2c_2) \sin(\beta x)]$$
$$+ \beta^3 e^{\beta x} [(-2c_3 - 2c_4) \cos(\beta x) + (-2c_3 + 2c_4) \sin(\beta x)];$$
$$v''''(x) = \beta^4 e^{\beta x} [-4c_1 \cos(\beta x) - 4c_2 \sin(\beta x)] + \beta^4 e^{-\beta x} [4c_4 \cos(\beta x) + 4c_3 \sin(\beta x)].$$

Infinite Beams on Elastic Supports   A railroad rail resting on the ties can be modeled as an infinitely long beam on elastic supports. The rail is much longer than the region of application of the force from the train wheels.

The case of a single point load, P, can be calculated using the symmetry of the deflection. Choose a coordinate system with $x = 0$ at the point of application of the load. By symmetry $v(x) = v(-x)$ so that only the portion of the beam with $x > 0$ need be considered. The solution obtained will *only* be valid for positive x, not for all x. But the symmetry allows one to determine the displacements for negative x.

The displacement is assumed negligible far away from the point of application of the load, $v(x) \to 0$ as $x \to \infty$. This condition requires that the positive exponential term, $e^{\beta x}$, drop out of the expression (5.40) for $v(x)$. This is satisfied by $c_1 = c_2 = 0$, so that the form of the solution is

$$v(x) = e^{-\beta x} [c_3 \cos(\beta x) + c_4 \sin(\beta x)]. \quad (5.41)$$

The beam is concave up and has its maximum deflection under the load. The third condition is $dv/dx = \beta e^{-\beta x}[(-c_3 + c_4)\cos(\beta x) + (c_3 + c_4)\sin(\beta x)] = 0$ at $x = 0$. This implies that $c_3 = c_4$.

The load must be involved in at least one boundary condition. The last boundary condition is that the internal shear force in the beam at $x = 0$ is $-P/2$. From strength of materials, the third derivative of the deflection equals the shear force divided by EI.

$$\beta^3 e^0[(-2c_3 - 2c_4)\cos(0) + (-2c_3 + 2c_4)\sin(0)] = \frac{P}{2EI}.$$

Since $c_3 = c_4$, this implies that

$$c_3 = c_4 = \frac{P}{8EI\beta^3} = \frac{P}{2\sqrt{2}(EI)^{1/4}k^{3/4}}$$

after the expression $\beta = (k/4EI)^{1/4}$ has been substituted. Alternatively, by multiplying numerator and denominator by β,

$$c_3 = c_4 = \frac{P}{8EI\beta^3} = \frac{P\beta}{2k}.$$

The solution for positive x is

$$v(x) = \frac{P}{8EI\beta^3}e^{-\beta x}[\cos(\beta x) + \sin(\beta x)]. \tag{5.42}$$

Notice again that this solution is only valid for positive x. Substitution of negative values of x gives a meaningless answer. The calculation of the deflections on the negative x side of the beam is carried out by substituting the corresponding positive value of x in equation (5.42) and using symmetry, $v(-x) = v(x)$.

Exercise 5.10 Plot the deflection, slope, bending moment, and shear force curves as a function of x for $x \geq 0$.

(a) Verify that the deflection curve is first equal to zero at $x = 3\pi/4\beta$ and then oscillates around the x-axis.

(b) Verify that the slope is equal to zero at both the origin and at $x = \pi/\beta$.

EXAMPLE 5.11 An infinite beam on an elastic foundation ($k = 20$ MPa, $E = 210$ GPa) with square cross section is to be used to support a 100 kN point load. The maximum allowable deflection is 1 cm. Determine the cross-sectional dimensions with a safety factor of 2.

Solution The maximum deflection occurs under the point of application of the load, $x = 0$. Therefore by equation (5.42), including the factor of safety,

$$\frac{0.01}{2} = \frac{P}{8EI\beta^3}.$$

But $\beta = (k/4EI)^{1/4}$ so that $I^{1/4} = 0.034927$ and $I = 1.488 \times 10^{-6}$ m^4. If s is the length of the side of the cross section, $I = s^4/12$. The required dimension is $s = 0.0650$ m. ∎

The bending stress at the point (x, y, z), where x is again the longitudinal coordinate, and y is positive up and measured from the neutral axis, is estimated by the strength of materials relation,

$$\sigma = \frac{M(x)y}{I} = EIv''(x)\frac{y}{I} = -y\frac{P}{4\beta I}e^{-\beta x}[\cos(\beta x) - \sin(\beta x)]. \tag{5.43}$$

EXAMPLE 5.12   Verify that the beam of the previous example when $s = 0.0650$ m is elastically loaded if the yield stress is 300 MPa.

Solution   The maximum stress in an elastically supported beam on which a transverse point load is applied occurs under the point of application of the load, where $x = 0$ and at the bottom of the beam where $y = s/2$. The term, $\beta = (k/4EI)^{1/4} = 2.00$ and

$$\sigma = -\frac{-0.065}{2}\frac{P}{4\beta I} = 273 \quad \text{MPa.}$$

Because the maximum bending stress is less than the yield stress, in this uniaxial state of stress, the beam is elastically loaded.   ∎

The results for a single transverse point load can be superposed to account for beams loaded by several point loads.

EXAMPLE 5.13   Three transverse loads are applied to an infinitely long beam (Fig. 5.11). The point loads, $P_1 = 10$ kN, $P_2 = 10$ kN, and $P_3 = 10$ kN, are separated by the distance $b = 2$ m between P_1 and P_2 and $a = 3$ m between P_2 and P_3. Determine the deflection of the beam under P_3. For this beam $E = 200$ GPa, $I = 0.1 m^4$, and $k = 6$ kPa.

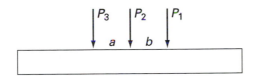

FIGURE 5.11   An infinite beam with three-point loads.

Solution   The deflection is found by superposition. The deflection expression for a point load is obtained in a coordinate system with the origin under that point load. By equation (5.42), the superposed deflection with the coordinates adjusted and using symmetry to get the deflections due to P_2 and P_1 is

$$v = \frac{P_3}{8EI\beta^3} + \frac{P_2}{8EI\beta^3}e^{-\beta a}\left[\cos(\beta a) + \sin(\beta a)\right]$$

$$+ \frac{P_1}{8EI\beta^3}e^{-\beta(a+b)}\left[\cos\left(\beta(a+b)\right) + \sin\left(\beta(a+b)\right)\right]. \tag{5.44}$$

Substituting values gives $\beta = 0.01655$ and $v = 0.0396$ m.   ∎

The analysis can be generalized using superposition to determine the deflection due to a distributed load, $w(x)$, applied to a portion of an infinitely long beam. Choose the point, Q, at which the deflection is desired. The distributed load extends a units to the left and b units to the right of Q. The idea is to superpose the deflections caused by each increment of the distributed load. To determine an expression for this deflection, first place the origin of the coordinates at point Q. All other contributions to the deflections must be written in this coordinate system. A difficulty is that the expression (5.42) previously derived for the deflection due to a point load requires the origin of the coordinate system to be under the point load. These expressions must each be translated to the coordinate system at point, Q. The approximate point load $w(x)\Delta x$ at point Q' is at position x in the coordinate system centered at Q. The contribution of this portion of the load to the deflection at Q is that which occurs at point x in the system centered at Q'. By the calculation preceding equation (5.42),

$$\Delta v = \frac{w(x)\Delta x}{2k} \beta e^{-\beta x} [\cos(\beta x) + \sin(\beta x)].$$

These increments of v can all be summed up and when $\Delta x \to 0$, the sum becomes the integral, which is the deflection at Q,

$$
\begin{aligned}
v(Q) = &\int_0^a \frac{w(x)}{2k} \beta e^{-\beta x} [\cos(\beta x) + \sin(\beta x)] \, dx \\
&+ \int_0^b \frac{w(x)}{2k} \beta e^{-\beta x} [\cos(\beta x) + \sin(\beta x)] \, dx.
\end{aligned}
\tag{5.45}
$$

The symmetry of the displacement fields was also used to obtain this expression. In general, these integrals would need to be solved numerically. If the distributed load is constant, so that $w(x) = w$, then the integrals (5.45) can be solved in closed form.

$$v(Q) = \frac{w}{2k} \left[2 - e^{-\beta a} \cos(\beta a) - e^{-\beta b} \cos(\beta b) \right]. \tag{5.46}$$

For example, the deflection at the center of a constant distributed load of length $2a$ is

$$v(0) = \frac{w}{2k} \left[2 - e^{-\beta a} \cos(\beta a) - e^{-\beta a} \cos(\beta a) \right]. \tag{5.47}$$

Notice that βa is in radians.

EXAMPLE 5.14

A constant transverse load of magnitude $w = 7.5$ kN/m is distributed over the length, $L = 4$ meters, of an infinite beam on an elastic support with constant $k = 10$ MPa. The beam is aluminum with $E = 70$ GPa and has cross-sectional dimensions, height $h = 5$ cm and width $t = 10$ cm. Determine the maximum deflection.

Solution

Place the origin of the coordinate system under the center of the distributed load. The maximum deflection, which by symmetry is at the center of the distributed load, then occurs at the origin. For this beam, $\beta = (k/4EI)^{1/4} = 0.6985$ /m. Also $\beta L = 2.794$. In the distributed load, $a = b = 2$, so that $\beta a = \beta b = 1.397$ radians or 80.05°. Applying equation (5.47), $v = 0.000717$ m. ∎

In most cases, the spring constant, k, for an elastic support must be determined experimentally. A long beam is placed on the elastic support and a point load, P, of known magnitude is applied to the center of the beam, perpendicular to the support. The deflection of the beam at a point just under the applied load is measured. Then using the relationship (5.42) for the deflection at $x = 0$, which is just under the load,

$$v_{\text{max}} = \frac{P}{8EI\beta^3} = \frac{P}{2\sqrt{2}(EI)^{1/4}k^{3/4}}, \tag{5.48}$$

the value of k is solved for. Notice the the experimentally measured values are v_{max} and the load. The spring constant k is deduced from these experimental facts under the assumptions that strength of materials is valid and that the beam is infinite. It is not really an experimental value, but rather is derived from experimental data and theoretical hypotheses. This is a common procedure in science, but one must remember that if the theoretical hypotheses are not correct, then the "experimental" values deduced from experiment may also not be correct.

In the early part of the twentieth century, a great deal of effort was expended to measure the elastic constant for a railroad bed including the ties, ballast, and underlying ground. The values were found to range between 1,400 and 2,000 psi (14 MPa). The elastic constant depends on the soil type, of course. But it also depends on the contact area between the foundation and the beam. If this analysis were applied to the slab of a roadway or airport runway, the effect of the area might become significant.

Occasionally, the spring constant of the earth is given as force per unit volume, k_e. In this case, the spring constant, k, for the elastic foundation is k_e times the width of the beam, $k = k_e w$.

5.3.2 Semi-infinite Beams

A very long beam with a downward force, P, and a counterclockwise moment, M, at one end can be approximated as a semi-infinite beam. The coordinate system is placed so that the origin is at the end where the load and moment are applied. The boundary conditions are that the deflection approaches zero as $x \to \infty$. The other two conditions are the strength of materials conditions that $EIv''(0) = M$ and $EIv'''(0) = -P$ at the end $x = 0$. Applying these conditions to

$$v(x) = e^{-\beta x} [c_3 \cos(\beta x) + c_4 \sin(\beta x)] \tag{5.49}$$

yields

$$c_3 = \frac{P + \beta M}{2EI\beta^3} \quad \text{and} \quad c_4 = \frac{-M}{2EI\beta^3}.$$

The deflection curve is therefore

$$v(x) = \frac{e^{-\beta x}}{2EI\beta^3} [P \cos(\beta x) + \beta M (\cos(\beta x) - \sin(\beta x))]. \tag{5.50}$$

This v is positive down.

The elastic supports are called discrete if they are not continuous. When the distance between the supports is small, the support can be replaced by an equivalent continuous support. Both the distance between the supports and the number β affect the error of this approximation. The rule of thumb is that the distance between any two supports must be less than $\pi/4\beta$. If each support has the same spring constant, k', and the distance l between each support is the same, then the equivalent continuous spring constant is $k = k'/l$.

PROBLEMS

5.1 Determine a general form, in terms of the Prandtl function, of the potential energy per unit length of a prism under a torque, T, according to the analysis due to Saint-Venant and Prandtl developed in this chapter. The potential energy is the sum of the strain energy and the negative of the work done by the torque.

5.2 Verify that the relationship between the Prandtl stress function, $\phi(x, y)$, and the axial displacement field-warping function, $f(x, y)$, is given by $\phi(x, y) = 0.5G[g(x, y) - \theta(x^2 + y^2)/2]$, where $g(x, y)$ is the conjugate function to $f(x, y)$. The conjugate is defined by the relations $\partial f/\partial x = $

$\partial g/\partial y$ and $\partial f/\partial y = -\partial g/\partial x$. Such a function $g(x, y)$ also satisfies the harmonic differential equation within the boundary of the cross section.

5.3 The Prandtl stress function, $\phi(x, y)$, for the torsion of a prismatic bar is zero on the outer surface. Let the cross section of a bar be circular of radius R, given by $x^2 + y^2 = R^2$, so that the z-axis is the longitudinal axis of the bar. Assume that the stress function has the form, $\phi(x, y) = A(x^2 + y^2 - R^2)$. The bar is twisted by a torque M_t. Neglect body forces. Use elasticity to answer the following.

(a)   Determine the constant A.

(b)   Determine the stresses at point (x, y) in a cross-sectional face of the bar.

(c)   Determine the angle of twist, θ, in terms of M_t.

(d)   Verify that the expression for the resultant shear stress is the same as that given in elementary strength of materials.

5.4   An elliptical prismatic rod is one whose cross-sectional area has boundary, $(x/a)^2 + (y/b)^2 = 1$. A torque, M_t, is applied at the ends.

(a)   Determine the Prandtl stress function in terms of θ, the constant angle of twist.

(b)   Relate the torque and the angle of twist.

(c)   Compute the shear stresses at each point (x, y).

(d)   Compute the ratio of the stress components at each point (x, y).

5.5   An elastic prismatic bar is subjected to pure torsion. A proposed Prandtl stress function is $\phi(x, y) = k[xy + bx^3 - ax^2 + 2by^3]$.

(a)   Perform a calculation to decide if this is a valid Prandtl stress function.

(b)   If it is not, determine conditions on the constants, a and b, which will make $\phi(x, y)$ into a valid Prandtl stress function.

(c)   In those cases that $\phi(x, y)$ is a valid stress function, determine the constant k in terms of the constant angle of twist.

(d)   In those cases that $\phi(x, y)$ is a valid stress function, determine the state of stress at an arbitrary point (x, y).

(e)   Draw a stress element showing the state of stress at the point $(1, 0)$.

5.6   A bar with the rectangular cross section shown has a torque, M_t, applied at its free end. The bar is linear elastic with shear modulus, G. Assume a constant angle of twist, θ, per unit length. Is the function $\phi(x, y) = k(a^2 - x^2)(b^2 - y^2)$, where k is a constant, a valid Prandtl stress function? Prove your answer by a calculation. Neglect body forces.

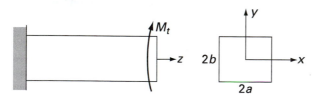

5.7   A prismatic rod that is under torsion has a cross section whose boundary is partially a semi-circle and partially a straight line defined respectively by $y^2 + x^2 = R^2$, where R is a constant, and $y = 0$. Decide, using a computation, whether or not $\phi(x, y) = k(R^2 - x^2 - y^2)y$, where k is a constant, is a valid Prandtl function for the prismatic rod with this cross section.

5.8   A 0.5-m-long prismatic bar has a square cross section with sides of length, s. It is made of 2014-T6 aluminum with

a shear strength of 42 ksi (290 MPa) and $G = 27$ GPa. The bar is mounted in a rigid wall at one end. At the other end, a rigid rod of length $L = 0.1$ m is attached perpendicular to the bar. A load of $P = 600$ N is applied normal to the rod. (See Fig. 5.5).

(a)   Determine the moment and load acting on the ends of the bar using statics.

(b)   Determine the state of stress at point A on the top of the bar at the wall (*Hint*: the bar is loaded by a combined torque and bending load). Write the stress tensor in terms of s.

(c)   Determine the state of stress at point B on the side of the bar at the wall. Write the stress tensor in terms of s.

(d)   Compute the principal stresses at point A.

(e)   Design the dimension s so that the Mises stress is less than one-fourth the shear strength. *Hint*: use point A.

5.9   A curved elastic beam has shape $a = 20$ cm, $b = 30$ cm, thickness 1 cm, and is loaded with torque $M_t = -50$ Nm.

(a)   Write the stress and deviatoric stress tensors at the point $r = a + h/4$. Compute the principal stresses and their directions.

(b)   Compute the distortion energy for the beam at that point $(G = 60$ GPa).

5.10   A flat circular beam of thickness t must have outer radius $b = 7$ inches and inner radius $a = 3$ inches. It must carry a pure bending moment of $M = 10{,}000$ in/lb. Use elasticity to design the thickness so that the maximum shear stress at $r = b$ is less than 10,000 psi.

5.11   By examining the displacement equation, $v(r)$, obtained from the strain-displacement relation in the elasticity solution for the bending of a curved beam, show that plane sections remain plane.

5.12   Verify that the centroid of a symmetrical trapezoid with longer base, b, shorter base, b_1, and height, h, is located on the axis of symmetry and a distance

$$\bar{y} = \frac{(b + 2b_1)h}{3(b + b_1)}$$

from the longer base.

5.13   Compute the expression for Z directly from the integral in the case that the cross section is rectangular with radius of curvature to the centroid, R, side b perpendicular to the radius of curvature, and height, h.

$$Z = -1 + \frac{R}{h} \ln\left(\frac{R + h/2}{R - h/2}\right).$$

Obtain the same result from the expression for Z of the trapezoidal cross section in Table 5.1.

5.14   How is the circumferential stress predicted by the elasticity and Winkler-Bach theories for a curved beam with rectangular cross-sectional area affected if the beam thickness, t, is doubled? Does the straight beam bending stress relation respond similarly?

5.15 A steady rest has three set screws 120 degrees apart to hold a workpiece. The screws each exert a 500-pound force on the workpiece. The upper section of the rest from the free end to the screw below the horizontal section A-A is circular with centroidal radius of curvature $R = 3$ inches and has a rectangular cross section with height $h = 1$ inch. The rest is made of gray cast iron with yield stress of 25 ksi in both tension and compression. Use the Winkler-Bach theory to design the width, b, of the cross section A-A so that the maximum compressive circumferential stress has magnitude less than one-third of the yield stress. Neglect the stresses induced by any internal shear force on section A-A.

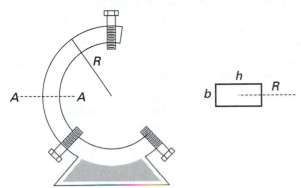

5.16 The tray BEFC (Fig. 5.9) carrying a component of weight, $W = 400$ pounds, is hung from pins B and C. To provide clearance for the component, it is decided to bend BC into the arc of the circle shown. The inner radius of curvature must be $a = 5$ inches to clear the component. Let b be the outer radius of curvature of section BC and denote $h = b - a$. The bar ABCD has rectangular cross section $h = 4$ in by the thickness, t, and is supported by pins at A and D. The members BE and CF are vertical. The straight sections AB and CD have length, $L = 15$ inches. Design the dimension, t, so that the maximum circumferential tensile stress in section BC is less than 3500 psi. Use the Winkler-Bach analysis.

5.17 A curved beam, whose centroidal line has constant radius of curvature 6 cm, has a rectangular cross section of height 4 cm and thickness 2 cm. A couple of magnitude $M = 68$ Nm is applied at the ends and in the plane of curvature of the beam.

(a) Determine the stress σ_θ at $r = 4$ and $r = 8$ using elasticity.

(b) Repeat (a) using the Winkler-Bach relation.

(c) Compare the results of (a) and (b) to the stresses computed by the straight beam relation, ignoring the beam curvature.

5.18 A curved beam, whose centroidal line has constant radius of curvature 25 cm, has a rectangular cross section of height 4 cm and thickness 2 cm. A couple of magnitude $M = 68$ Nm is applied at the beam ends and in the plane of curvature of the beam.

(a) Determine the maximum tensile and compressive σ_θ stresses in a face using elasticity.

(b) Repeat (a) using the Winkler-Bach relation.

(c) Compare the results of (a) and (b) to the stresses computed by the straight beam relation, ignoring the beam curvature.

5.19 The circular curved beam shown, which is mounted in a rigid wall at the left end, is linear elastic ($E = 210$ GPa, $v = 0.25$, and $\sigma_{PL} = 50$ MPa in tension and compression). The inner and outer radii are $a = 0.2$ m and $b = 0.3$ m. The cross section of the curved beam is circular, with radius $r = 5$ cm. Make the strength of materials assumptions that plane sections remain plane and that the radial stress and shear stresses are zero. Neglect body forces. The horizontal force $P = 2,000$ N.

(a) Using the Winkler-Bach theory, determine whether or not the beam is elastically loaded on the vertical face AB. Prove your answer by computations of the stress at A and at B.

(b) If the beam is elastically loaded on face AB, determine the location of the neutral axis in face AB. Prove your answer by a computation.

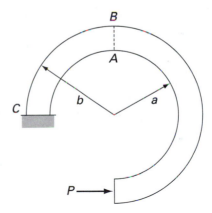

5.20 A crane hook similar to that in Figure 5.10 carries a load of 10,000 pounds and is elastically loaded. The point C is the intersection of Q_1Q and the line of action of the force P and the distance $QC = 3$ inches; the radius of the inner circular edge at Q is 2.4 inches. The cross section Q_1Q is rectangular with edges $h = 3$ inches and $b = 2$ inches.

(a) Use the Winkler-Bach analysis to compute the stress at point Q.

(b) Compute the location of the neutral axis in cross section Q_1Q.

5.21 A steel rod having square cross section, 2 inches on a side, is in the form of a semicircle having a mean radius of 12 inches. One end of the rod is fixed and the other is free, as shown. Neglect the weight of the rod. The free end is subjected to a horizontal load, P, of 2000 pounds. The proportional limit of the material is 40,000 psi. Compute the maximum and minimum stresses at the wall using the Winkler theory.

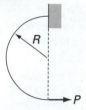

5.22 An infinite beam on an elastic foundation has a square cross section of sides 10 cm, $E = 210$ GPa, and the support is elastic with $k = 100$ MPa. Plot the displacement as a function of x when a point load of 35 kN is applied.

5.23 Determine the maximum value of the load P for which the beam of problem 23 will be elastically loaded. The proportional limit is $\sigma_{PL} = 45$ MPa.

5.24 An infinite beam on an elastic support is loaded with a transverse point load, P. Write the expression for the strain energy of the section from the point load to a face L units to the right $(x = L)$, in terms of β, P, and L.

5.25 Prove that if an infinite beam on an elastic foundation with a constant distributed load, w, of length L is such that $\beta L > 2\pi$, then the maximum deflection may be approximated as $v_{max} = w/k$.

5.26 Three transverse loads are applied to an infinitely long beam. The point loads, $P_1 = 10$ kN, $P_2 = 20$ kN, and $P_3 = 15$ kN, are separated by the distance $b = 2$ m between P_1 and P_2 and $a = 3$ m between P_2 and P_3. Determine the deflection of the beam under P_2. For this beam, $EI = 5$ MN.m^2, and $k = 12$ MPa.

5.27 An infinite beam on an elastic support with $k = 15$ MPa carries a transverse point load $P = 120$ kN. For this beam, $EI = 7.5$ MN.m^2, and $E = 200$ GPa. Assume the beam is elastically loaded. Compute the maximum tensile bending stress in the beam. The neutral bending axis is 4 cm above the elastic support. Compute the deflection of the beam at the position of application of the load.

5.28 Railroad ties are 30.5 cm apart and are 20 cm wide. The rails have $EI = 8$ MN.m^2 and the spring constant of the ties and ground is $k = 16$ MPa. Can a rail be approximated as an infinite beam on an elastic foundation?

REFERENCES

[1] A. E. H. LOVE, *A Treatise on the Mathematical Theory of Elasticity*, 4th Edition, Dover, New York, 1944.

[2] V. G. REKACH, *Manual of the Theory of Elasticity*, Mir, Moscow, 1979.

[3] S. P. TIMOSHENKO, *History of strength of materials*, Dover, New York, 1983.

CHAPTER 6

THIN-WALLED MEMBERS

Thin walled members such as I-beams and tubes are designed to increase the resistance per unit weight to a particular kind of load. They pose special problems because, while designed to resist one type of load, they often cannot resist other types of loads well. Such members are often used in place of solid cross-sectional members in order to minimize the weight of a structure. For example, the operation of a lower-weight machine generally requires less energy. A careful stress analysis is needed to ensure that such a structure will not fail. The equations of elasticity can often be considerably simplified by making approximations about the stress behavior in the thin wall. The stresses in the direction normal to the thin wall are usually assumed to be zero. Thin-walled beams in pure bending and thin-walled prismatic torsional members will be discussed after defining the principal axes and shear center in a beam with an unsymmetric load.

6.1 PRINCIPAL AXES AND PRODUCT AREA MOMENTS OF INERTIA IN A BEAM CROSS SECTION

The cross sections of a beam in pure bending are frequently assumed to remain planar during the bending. This assumption implies that the bending normal strain at a point is a linear function of the distance from the point to the neutral axis. The computation of the corresponding linear elastic normal stress involves special integrals, which are given the name area moments of inertia. While these have no intrinsic physical meaning other than geometric, they are sometimes thought of as a measure of bending stiffness since they arise in this context.

The area moment of inertia about a line $a - a$ is the integral over the area, $I_a = \int_A s^2 dA$, where s is the perpendicular distance from a point in the area to the axis $a - a$. Recall that the parallel axis theorem says that $I_x = \bar{I}_x + d^2 A$, where d is the perpendicular distance to the parallel centroidal axis. The bar over the symbol for the area moment of inertia indicates that the axis passes through the centroid. The word "area" in the area moment of inertia distinguishes it from the mass moment of inertia. Products of inertia are defined with respect to an orthogonal pair of axes. A similar parallel axis theorem holds for the products of inertia.

Definition   The product of inertia, I_{xy}, with respect to the x-y coordinate axes is

$$I_{xy} = \int_A xy\, dA.$$

The units of both the product of inertia and the area moments of inertia are length to the fourth power. If either of the axes is an axis of symmetry for the area, then the product of inertia is zero.

203

Exercise 6.1   Verify the parallel axis theorem for products of inertia,

$$I_{xy} = \bar{I}_{xy} + \bar{x}\bar{y}A,$$

where $(\bar{x}, \bar{y})$ are the coordinates of the centroid in the x-y coordinates.

EXAMPLE 6.1   The product of inertia of a rectangle with respect to axes along its edges can be computed using the parallel axis theorem.

Solution   The rectangle has base b and height h. Since it has two axes of symmetry, the centroid is at their intersection, and the product of inertia with respect to these axes is zero. Put the x'-y' coordinates along the edges with the origin at the lower left corner. The lower left corner is located at $(-b/2, -h/2)$ in the centroidal coordinates. By the parallel axis theorem,

$$I_{x'y'} = \bar{I}_{xy} + \bar{x}\bar{y}A = 0 + \frac{1}{4}b^2h^2. \qquad\blacksquare$$

The product of inertia in one x-y coordinate system is related to the product of inertia in a system, x'-y' coordinates, which is rotated through a counterclockwise angle, θ, about the origin. A point $P(x, y)$ has coordinates in the two systems that are related, using the rotation matrix (see Chapter 1, Section 1.5), by

$$\begin{pmatrix} x' \\ y' \end{pmatrix} = \begin{pmatrix} \cos\theta & \sin\theta \\ -\sin\theta & \cos\theta \end{pmatrix} \begin{pmatrix} x \\ y \end{pmatrix}.$$

Exercise 6.2   Draw the x-y coordinates and the rotated x'-y' coordinates on the same graph. Use triangles to verify the matrix relation above,

$$\begin{aligned} x' &= x\cos\theta + y\sin\theta; \\ y' &= y\cos\theta - x\sin\theta. \end{aligned}$$

With this relation between the coordinates, a calculation of $\int_A x'y'dA$ shows that

$$I_{x'y'} = I_{xy}\cos(2\theta) + \frac{1}{2}(I_x - I_y)\sin(2\theta), \qquad (6.1)$$

where $I_x = \int_A y^2 dA$ and $I_y = \int_A x^2 dA$.

The area moments of inertia can be collected in the tensor

$$(IA) = \begin{pmatrix} I_x & -I_{xy} \\ -I_{xy} & I_y \end{pmatrix}.$$

The area moments of inertia in an x'-y' coordinate system rotated $\theta°$ counterclockwise from the x-y coordinates are given by a relation similar to that obtained in Chapter 1, Section 1.5 for the stresses.

$$(IA)' = \alpha(IA)\alpha^t,$$

where α is the rotation matrix. The minus signs in front of I_{xy} in the matrix are required to make this relation valid and therefore make (IA), by definition, a tensor.

Exercise 6.3   Multiply the transformation matrices for $(IA)'$ and compare the result to equation (6.1).

Definition   A system of centroidal coordinates in which $I_{xy} = 0$ is called principal.

An axis of symmetry is always one axis of a principal system. For example, two perpendicular axes of symmetry for a cross section are principal coordinate axes. In the principal coordinates, I_x is greater than any other $I_{x'}$ and I_y is less than any other $I_{y'}$ or vice versa.

Exercise 6.4 Verify this statement about the magnitudes of the principal area moments of inertia.

Given any x-y coordinates, Equation (6.1) can be used to find the counterclockwise angle, θ, through which these coordinates must be rotated to obtain the principal coordinates in which $I_{x'y'} = 0$.

$$\tan(2\theta) = \frac{-2I_{xy}}{I_x - I_y}. \tag{6.2}$$

EXAMPLE 6.2 Determine the direction of the principal coordinates for a right-angle beam with wall thickness, t, and inner length, w, along each leg (Fig. 6.1).

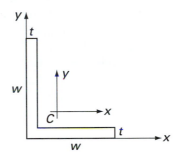

FIGURE 6.1 Coordinate systems for a right-angle beam.

Solution First determine the location of the centroid C. In coordinates along the outer edges of the beam,

$$\bar{x} = \bar{y} = \frac{\frac{1}{2}w^2 + \frac{3}{2}wt + \frac{1}{2}t^2}{2w + t}.$$

The moments of inertia are easiest to calculate in centroidal axes parallel to the edges of the rectangular sections using the parallel axis theorems. By symmetry, $\bar{I}_x = \bar{I}_y$. Therefore, equation (6.1) implies that the principal axes are at an angle $\theta = \pi/4$ from the centroidal axes parallel to the edges of the rectangular sections. Notice that one of these is an axis of symmetry. The origin is the centroid. ∎

6.2 UNSYMMETRICAL BENDING OF BEAMS

Some beam members are designed to be especially resistant to bending, but only if the load is applied in certain directions. For example, an I-beam is more resistant to bending loads applied perpendicular to the flanges than to loads applied in other directions.

To restrict attention to the effect of the unsymmetric load on bending only, the line of action of the load is assumed to pass through a special point, called the shear center, about which it induces no twisting moment. The technique to determine the shear center will be discussed in the next section. Also the beam is assumed to have a constant cross section along its length.

The strain is, as in elementary strength of materials, assumed to vary linearly with the perpendicular distance from the neutral axis. Likewise, so does the stress since the beam is made of a linearly elastic material. Consider an arbitrary point, $B(x, y)$, in the face. The stress at this point, B, because of the linear assumption is $\sigma(x, y) = ks$, where k is a constant and s is the perpendicular distance from the neutral axis. The total normal force on the cross-sectional face of area, A, is zero since the beam is subject only to a bending load.

$$0 = \int_A \sigma \, dA = k \int_A s \, dA = k\bar{s}A,$$

where $\bar{s}$ is the distance to a parallel centroidal axis. Therefore $\bar{s} = 0$. The neutral axis passes through the centroid. In general, it will not pass through the shear center, except when the shear center and centroid coincide.

In this situation, the direction of the neutral axis must be determined in order to compute the stresses. Let X-Y be a coordinate system with origin at the shear center, O, in the face (Fig. 6.2). At this stage, its orientation is arbitrary. The centroid has coordinates $(\bar{X}, \bar{Y})$ in this system. The load, **P**, passes through the shear center, O, at an angle θ with the chosen Y-coordinate. The load-induced stresses cause an internal bending moment, M, which has component $M_X = M \cos\theta$ about the X-axis and $M_Y = M \sin\theta$ about the Y-axis. Put another coordinate system x-y parallel to the system X-Y but with origin at the centroid C of the face. The neutral axis passes through the centroid at an unknown angle ϕ measured clockwise from the x-axis.

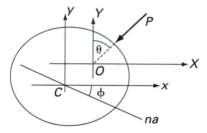

FIGURE 6.2 Coordinate systems through the centroid, C, and the shear center, O, for the cross section of a body.

The distance s in terms of the (x, y) coordinates measured from the centroidal coordinate system is

$$s = y \cos\phi + x \sin\phi.$$

The relationship between the coordinates of the point Q in the coordinate system with origin at the centroid and that with origin at the shear center is

$$x = X - \bar{X} \quad \text{and} \quad y = Y - \bar{Y}.$$

Therefore the internal bending moments due to the induced normal stresses about the x- or the X-axis are equal, $M_x = M_X$, because the axes are parallel and the moment is a couple.

$$M_X = \int_A Y\sigma \, dA = k \int_A (y + \bar{Y})[y\cos\phi + x\sin\phi]dA$$

$$= k\cos\phi \left[\int_A y^2 dA + \bar{Y} \int_A y \, dA \right] + k\sin\phi \left[\bar{Y} \int_A x \, dA + \int_A xy \, dA \right]$$

$$= k\cos\phi I_x + k\sin\phi I_{xy}. \tag{6.3}$$

The integrals $\int_A y \, dA = \int_A x \, dA = 0$ because x and y are measured from the centroid. Likewise, $M_y = M_Y$ and

$$M_Y = k\cos\phi I_{xy} + k\sin\phi I_y. \tag{6.4}$$

Equations (6.3) and (6.4) are solved simultaneously to obtain

$$\cos\phi = \frac{1}{k}\frac{M_x I_y - M_y I_{xy}}{I_x I_y - I_{xy}^2}; \tag{6.5}$$

$$\sin\phi = \frac{1}{k}\frac{M_y I_x - M_x I_{xy}}{I_x I_y - I_{xy}^2}. \tag{6.6}$$

These expressions permit the calculation of the orientation of the neutral axis and of the stresses with respect to the centroidal coordinate system. The angle ϕ is related to the known angle at which the load is applied, θ, by

$$\tan\theta = \frac{M_y}{M_x} = \frac{\cos\phi I_{xy} + \sin\phi I_y}{\cos\phi I_x + \sin\phi I_{xy}} = \frac{I_{xy} + \tan\phi I_y}{I_x + \tan\phi I_{xy}}; \tag{6.7}$$

or

$$\tan\phi = \frac{\tan\theta I_x - I_{xy}}{I_y - \tan\theta I_{xy}}. \tag{6.8}$$

If the coordinates are also principal, then, because $I_{xy} = 0$,

$$\tan\theta = \frac{I_y}{I_x}\tan\phi,$$

where the angle θ is between the applied load and the principal y-coordinate.

To complete the calculation of the stresses, recall that the magnitude of the stress is just the constant k times the perpendicular distance from the neutral axis:

$$\sigma(x, y) = k(y\cos\phi + x\sin\phi).$$

Therefore

$$\sigma(x, y) = \frac{(M_y I_x - M_x I_{xy})x + (M_x I_y - M_y I_{xy})y}{I_x I_y - I_{xy}^2}. \tag{6.9}$$

In principal coordinates, since $M_x = M\cos\theta$ and $M_y = M\sin\theta$

$$\sigma(x, y) = \frac{(M\sin\theta)x}{I_y} + \frac{(M\cos\theta)y}{I_x}. \tag{6.10}$$

The graph of the stress in (x, y, σ)-space is that of a plane passing through the origin. Therefore the stress is maximal at the largest perpendicular distance from the neutral axis.

EXAMPLE 6.3   Determine the direction of the neutral axis in a beam with rectangular cross section having a base $b = 2$ inches and height $h = 4$ inches when the load is applied at an angle of $22°$ from the vertical, and its line of action passes through the centroid.

Solution   The first step is to determine the principal coordinates. Since the area has two axes of symmetry, these are the principal coordinates. Then since the angle $\theta = 22°$ when the principal y-coordinate is known, the direction, ϕ, of the neutral axis with respect to the principal x-axis can be computed. $I_x = bh^3/12$ and $I_y = hb^3/12$.

$$\tan 22° = \frac{I_y}{I_x} \tan \phi = \frac{b^2}{h^2} \tan \phi.$$

Therefore $\tan \phi = 4 \tan(22°)$ or $\phi = 58.25°$. Notice that the neutral axis is not perpendicular to the load in this example.   ■

The longitudinal stress due to bending, σ_z, is only equal to My/I, as learned in elementary strength of materials, under certain restrictive conditions. The coordinates must be principal and the load must be along an axis of symmetry and perpendicular to the other principal axis.

The maximum stress occurs at the point in the cross section whose perpendicular distance from the neutral axis is the greatest.

EXAMPLE 6.4   The orientation of the neutral axis in a very thin beam can be determined in a similar manner. Determine the direction of the neutral axis in a beam with rectangular cross section having a base b and height h when the load is applied at an angle of β from the vertical with its line of action passing through the centroid, which is the shear center (Fig. 6.3). Examine how the direction of the neutral axis with respect to the horizontal axis changes as either $h \to \infty$ or $b \to 0$.

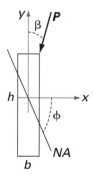

FIGURE 6.3   The neutral axis position in a thin beam determined by the angle ϕ between the principal x-axis and the neutral axis.

Solution   A rectangular area has two axes of symmetry; these are the principal coordinates. Let x be the horizontal axis. As in the previous example,

$$\tan \phi = \tan \beta \frac{I_x}{I_y} = \tan \beta \frac{h^2}{b^2}.$$

As either $h \to \infty$ or $b \to 0$, ϕ tends to $\pi/2$, no matter what β is. In other words, in a thin beam the neutral axis is nearly perpendicular to the smallest dimension edge.   ■

EXAMPLE 6.5   Determine the direction of the neutral axis in a channel beam having a horizontal base with outer length $b = 6$ inches, height $h = 12$ inches, and flange thickness of 1 inch. The load is applied at an angle of 22° from the vertical, and its line of action passes through the shear center. The beam is oriented so that the two legs of the U-shape are horizontal.

Solution   The centroid lies on the horizontal axis of symmetry. Take this axis to be the x-axis and the left side of the channel to be the y-axis. The distance from the left edge to the centroid is $41/22 = 1.864$ inches.

Next compute the area moments of inertia about the centroidal axes using the parallel axis theorem.

$$I_{xy} = (5)(1)(5.5)(3.5 - 1.864) - (5)(1)(5.5)(3.5 - 1.864) + 0 = 0$$
$$I_x = 2(1/12)(5)(1)^3 + 2(5.5)^2(5)(1) + (1/12)(1)(12)^3 = 447.3$$
$$I_y = 2[(1/12)(1)5^3 + 5(3.5 - 1.864)^2] + (1/12)(12)(1)^3 + 12(1.864 - 0.5)^2 = 70.92$$

The parallel axis theorem was not needed to show that $I_{xy} = 0$ since the x-axis is an axis of symmetry for the total area.

$$\tan \phi = \frac{\tan 22° I_x - I_{xy}}{I_y - \tan 22° I_{xy}} = 2.548,$$

so that $\phi = 68.6°$. ∎

6.3 BENDING OF THIN-WALLED BEAMS

Beams are designed with thin walls in order to increase their bending stiffness for a fixed volume of material. Much of the material in an I-beam, for example, is placed away from the neutral axis in order to increase the area moment of inertia and therefore the bending stiffness, EI, with respect to that axis. The longitudinal stress is also reduced. Such beams are frequently used to support the floors of buildings, the deck in bridges, and other structures.

EXAMPLE 6.6   The advantage of moving material away from the neutral axis can be seen by comparing the area moment of inertia of a rectangle to that of an I cross-sectional beam of the same cross-sectional area. For bending moment, M, the maximum bending stress in a rectangular cross section beam of width, w, and height h is $\sigma = M(h/2)/(wh^3/12) = 6M/wh^2$. On the other hand, a symmetrical I-beam with upper and lower flange of width w and height $h/4$ and web of width $w/4$ and height $2h$ (Fig. 6.4) has the same cross-sectional area and thus the same weight if both are made of the same material. However, as can be computed by the parallel axis theorem, its area moment of inertia is $(308/364)wh^3$, which is 9.6 times that of the rectangular beam. The maximum stress in the I-beam is $\sigma = M(5h/4)/(308/364)wh^3 = 1.558M/wh^2$. The maximum bending stress in the rectangular beam is 3.85 times that of the I-beam for the same amount of material. ∎

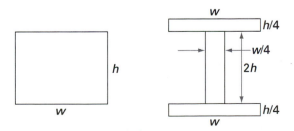

FIGURE 6.4   Dimensions of the rectangular and I cross-sectional areas.

Exercise 6.5 Verify the stress and area moment of inertia calculations in the previous example.

6.3.1 Shear Center

A thin-walled beam, such as an I-beam or a channel beam, is much less resistant to torsion than bending. It is important that the designer ensures that the line of action of the applied load passes through the shear center of the cross section if the beam is to avoid twisting while it is bending. In this way, the additional bending stiffness per unit weight is achieved without causing a disaster due to unexpected twisting. The total twisting moment on a face is the sum of that due to the external loads and that due to the internal shear stresses. The total twisting moment about the shear center must be zero when the line of action of the resultant external forces passes through it. Therefore, the twisting moment about the shear center due to the shear stresses must also be zero.

Definition The shear center is the point in the plane of the cross section about which the moment of the resultant internal shear force is zero when the line of action of the resultant external forces passes through it.

The shear center depends only on the geometry of the cross section, not on the applied loads. The center need not lie in the face itself. The shear center can be located by assuming that the line of action of the resultant external forces passes through it. The position of the shear center is that which forces the moment of the external forces to balance the moment of the resultant internal shear force.

Take the z-axis along the longitudinal axis of the beam. The shear center coordinates, (e_x, e_y), for any beam, thin-walled or not (Fig. 6.5), are defined by

$$e_y V_x + e_x V_y = \int_A (\tau_{zx} y - \tau_{zy} x) dx dy, \tag{6.11}$$

the balance of moments about the origin due to the external shear force, $\mathbf{V} = V_x \mathbf{i} + V_y \mathbf{j}$, and the shear stresses on the cross-sectional face. The two unknowns can be determined by considering two different loadings, $\mathbf{V}$, each of which produces an equation. Calculation of the shear center coordinates requires knowledge of the shear stress distribution in the cross section. Because the wall is thin, all stresses perpendicular to the beam surface are assumed to be zero. The only nonzero shear stress acts parallel to the edge of the wall.

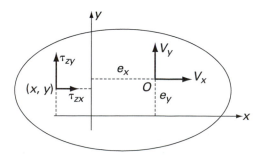

FIGURE 6.5 The external shear force $\mathbf{V} = V_x \mathbf{i} + V_y \mathbf{j}$ at the shear center and the internal shear stresses at an arbitrary point (x, y).

Definition Denote the shear force tangent to the wall cross-sectional boundary by V_s. The magnitude of V_s per unit arc length is called the shear flow, q.

An integral expression for the shear flow, q, can be obtained from the equations of equilibrium (Chapter 1, Section 1.4). The coordinate system varies from point to point. At a fixed point, the z-axis is along the longitudinal axis of the beam, and the s is tangent to the wall, while the n-axis is normal to the wall. Since the wall is thin, the stresses, $\tau_{ns} = \tau_{sn}$ and σ_s are assumed to be approximately zero. The equilibrium equation from the sum of forces in the s-direction implies that $\partial \tau_{zs}/\partial z = 0$. In other words, τ_{zs} is not a function of z. The sum of forces in the z-direction becomes

$$\frac{\partial \sigma_z}{\partial z} + \frac{\partial \tau_{zs}}{\partial s} = 0,$$

since body forces are neglected and since $\tau_{nz} = 0$. This expression can be integrated with respect to s since τ_{zs} is not a function of z. After multiplying by the thickness, $t(s)$, which is also not a function of z, the shear flow at position s is

$$q(s) = t(s)\tau_{zs} = -\int_0^l t(s)\frac{\partial \sigma_z}{\partial z}ds, \qquad (6.12)$$

where l is the length along the thin wall from $s = 0$. This result is valid if the section is open. In that case, the shear flow must be zero at the ends.

A strength of materials approximation for the shear flow is obtained from equation (6.12). The strength of materials version of beam theory implies that $\sigma_z(z, y) = M(z)y/I$, where y is the distance from the centroidal bending axis in the face of the beam and I is the area moment of inertia about the bending axis. Recall that, in linear elastic beam theory, the neutral and centroidal axes coincide. Further, from strength of materials, $dM(z)/dz = V(z)$, the shear force acting on the beam face so that $\partial \sigma_z(z)/\partial z = V(z)y/I$. Therefore, the strength of materials expression for the shear flow at the position $s = l$ is

$$q = -\int_0^l \frac{V(z)y}{I}t(s)ds = -\frac{V(z)}{I}\int_0^l y(s)t(s)ds = -\frac{V(z)}{I}\bar{y}A, \qquad (6.13)$$

where A is the cross-sectional area of the thin wall from the face where $q = 0$ ($s = 0$) to the face at arclength $s = l$ and $\bar{y}$ is the coordinate of the centroid of area A. The minus sign indicates that the direction of the shear flow is opposite that of V. In the following examples, the integral is denoted by $Q = \int_0^l y(s)t(s)ds = \bar{y}A$.

The shear center can be located by taking the moment of the shear flow about any point, O. The coordinates, (e_x, e_y) for the shear center are those in a system centered at point O. Let the wall be parametrized by the arclength, s. The perpendicular distance from O to the shear flow $q(s)$ at point s on the wall is $r(s)$. The shear center coordinates are obtained as in equation (6.11).

$$e_x V_y + e_y V_x = \int_0^l q(s)r(s)ds, \qquad (6.14)$$

where the shear flow is induced by both V_x and V_y. This equation has two unknowns, e_x and e_y. They can be determined from the two equations obtained by in turn letting V_x and V_y be zero while setting the shear flow under the integral sign to be that induced by the nonzero shear force.

The integral $\int_0^l q(s)r(s)ds$ can be written as $\int_0^l \tau(s)r(s)dA$, the moment of the shear force in the leg, but the thickness divides out of the area and shear stress so that it has become common to work in terms of the shear flow.

Alternatively, equations (6.12) and (6.9) and the fact that $dM/dz = V$ imply that

$$
\begin{aligned}
e_x V_y + e_y V_x &= \int_0^l q(s)r(s)ds \\
&= -\int_0^l \frac{\partial \sigma(x,y)}{\partial z} r(s)ds \\
&= -\int_0^l \frac{(V_y I_x - V_x I_{xy})x + (V_x I_y - V_y I_{xy})y}{I_x I_y - I_{xy}^2} r(s)ds.
\end{aligned}
\tag{6.15}
$$

The coefficients of V_x must be equal, as must those of V_y. Therefore,

$$
e_y = \int_0^l \frac{I_{xy}x - I_y y}{I_x I_y - I_{xy}^2} r(s)ds;
\tag{6.16}
$$

$$
e_x = \int_0^l \frac{-I_x x + I_{xy}y}{I_x I_y - I_{xy}^2} r(s)ds.
\tag{6.17}
$$

The shear center depends only on the shape of the cross section, and not on the applied load. The centroid of the cross section only coincides with the shear center in special cases. When the cross section has an axis of symmetry, the shear flow must also be symmetrical since the integrals, $\int_0^l xr(s)ds$ and $\int_0^l yr(s)ds$ are. The shear center therefore lies on every axis of symmetry. The centroid also lies on every axis of symmetry. Therefore, for those areas with two axes of symmetry, the shear center and the centroid coincide.

EXAMPLE 6.7 Determine the shear center of an angle beam with equal legs at an acute angle θ (Fig. 6.6).

FIGURE 6.6 Shear center, O, in an angle beam.

Solution The shear flows resulting from the stresses are directed along the legs. Irrespective of their magnitudes, the directions of these shear flows intersect at the corner. If the shear force is applied at this point, a zero twisting moment results so that the corner is the shear center. Notice however, that this point is not the centroid. The centroid lies on the horizontal axis of symmetry, as does the shear center, but in between the two legs. ∎

EXAMPLE 6.8 Determine the shear center of a thin-walled beam that has a semicircular cross section of radius R (Fig. 6.7).

Solution The term Q for the semicircle is

$$
Q(\theta) = \int x(\theta)dA_\theta = \int_0^\theta R\cos\theta\, Rt\,d\theta = R^2 t \sin\theta.
$$

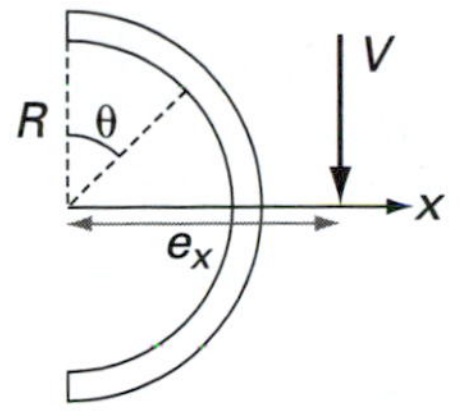

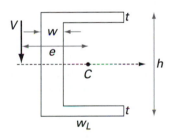

FIGURE 6.7   The shear center coordinate, e_x, in a semicircular cross section.

Notice that as expected, Q is zero at the outer ends, $\theta = 0, \pi$ and is maximal at $\theta = \pi/2$. The sum of the moments about the center of the arc is zero. The moment arm to the shear flow is R and an increment of arclength is $Rd\theta$ so that by equations (6.13) and (6.14)

$$V e_x = \int_0^\pi q R R d\theta = \int_0^\pi \frac{V Q(\theta)}{I} R^2 d\theta$$
$$= \frac{V}{I} \int_0^\pi R^2 t \sin\theta R^2 d\theta = 2\frac{V}{I} R^4 t.$$

Also the area moment of inertia for a thin-walled semicircle about its axis of symmetry is $I = \int_0^\pi (R\cos\theta)^2 t R d\theta = \pi R^3 t/2$. Therefore $e_x = 4R/\pi$, measured from the center of the arc.   ∎

EXAMPLE 6.9   Determine the shear center of a channel beam (Fig. 6.8) with equal-length legs of length w_L and midsection of height, h. The thickness of the legs is t and of the midsection is w.

FIGURE 6.8   The shear center coordinate, e, in a channel beam cross section.

Solution   The shear center must lie on the axis of symmetry, which is the horizontal centroidal axis. Assume that the resultant shear force is perpendicular to the axis of symmetry and acts at the shear center. Then the location of the shear center can be computed from the fact that the sum of the moments about any point must be zero. If this point is taken at a convenient location, the calculation can be simplified. Take the moment about one corner to reduce the computation to that of the shear force in just one leg. The area moment of inertia about the axis of symmetry, using the parallel axis theorem, is $I = 2w_L t (h/2)^2 + wh^3/12$. Here, the area moment inertia about the centroidal axis of the thin section itself has been approximated as zero since it is very thin.

Now Q is under the integral sign so that it must be taken as Q for the area from the end of the leg to the position ξ. $Q(\xi) = t\xi(h/2)$ and $dA = t d\xi$.

$$V_L = \frac{V}{It} \int_A Q dA = \frac{V}{It} \int_0^{w_L} t\xi \left(\frac{h}{2}\right) t d\xi = \frac{V t h w_L^2}{4I}.$$

The sum of the moments about the corner, keeping in mind that V_L is in the same direction as the leg, is

$$\sum M_C = Ve - V_L h = 0$$

so that

$$e = \frac{th^2 w_L^2}{4[2w_L t (h/2)^2 + wh^3/12]}$$

which is independent of the shear force. ∎

Exercise 6.6   Reduce the expression for e to

$$e = \frac{w_L^2}{2w_L + \frac{wh}{3t}}. \tag{6.18}$$

EXAMPLE 6.10   A computation of the shear flow in the I-shaped cross section of Figure 6.9 shows how the shear flow behaves at a point at which three wall sections join.

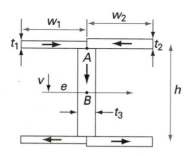

FIGURE 6.9   I-beam with one axis of symmetry.

Solution   Let point A be at the joint of vertical section and top flange. Consider a small region about A. The sum of the internal forces in the z-direction must be zero. $\tau_1 t_1 d + \tau_2 t_2 d - \tau_3 t_3 d = 0$ where d is the depth of the element in the z-direction. Therefore $q_3 = q_1 + q_2$. So, in general, at any joint of a thin-walled cross section, the incoming shear flow adds up to the outgoing shear flow.

The shear flow in the horizontal flange section 1 is

$$q_1(s) = \frac{Vh}{2I} t_1 s,$$

where s is measured from the free end. The shear flow in the horizontal flange section 2 is

$$q_2(s) = \frac{Vh}{2I} t_2 s,$$

where s is measured from the free end. At point A, $q_3(0) = q_1(w_1) + q_2(w_2)$, for s measured from point A in the vertical flange section 3. The shear flow is

$$q_3(s) = \frac{V}{I} t_3 s \left(\frac{h}{2} - \frac{s}{2} \right) + q_1(w_1) + q_2(w_2).$$

Taking moments about point B on the axis of symmetry yields, by symmetry,

$$eV = 2 \int_0^{w_1} q_1(s) \frac{h}{2} ds - 2 \int_0^{w_2} q_2(s) \frac{h}{2} ds,$$

or

$$e = \frac{h^2}{4I} t_1 w_1^2 - \frac{h^2}{4I} t_2 w_2^2 = \frac{h^2}{4I} (t_1 w_1^2 - t_2 w_2^2)$$

$$= \frac{t_1 w_1^2 - t_2 w_2^2}{2 \left(t_1 w_1 + t_2 w_2 + t_3 \frac{h}{6} \right)},$$

because $I = h^2 (t_1 w_1 + t_2 w_2 + t_3 h/6)/2$ after neglecting terms containing a thickness cubed. ∎

Exercise 6.7   Locate the shear center if the thickness is constant throughout the cross section and if $w_1 = w_2$.

Closed Sections   The shear flow in a closed section is determined by reducing the problem to that of an open section. Assume that the line of action of the applied force passes through the shear center. The technique for finding the shear center of an open section requires a point at which the shear flow is zero. In a closed section no such point is available, but one can imagine the section as separated at some point. Pick a point, A, where the arclength parameter, s, is taken to be zero. Such a point is called a cut point. The cut is imaginary, but the idea is that the cut point acts as the free end of an open section. Let q_A be the shear flow at this point. The modified shear flow, $q'(s) = q(s) - q_A$ is zero at point A. The prime does not indicate differentiation. The flow $q'(s)$ can be viewed as that in a section which is open at A since it is zero there. This method, however, requires an additional equation to solve for q_A. The shear strain $\gamma_{sz} = \partial w/\partial s + \partial u/\partial z$, where u, v, and w are the displacements in the direction of the arclength, of the normal in the face and of the longitudinal axis respectively. But $\partial u/\partial z = 0$ because there is no twist when the force is applied through the shear center. Therefore

$$\frac{q}{t} = \frac{q_A + q'}{t} = G\gamma_{sz} = G\frac{\partial w}{\partial s}.$$

Integrating the left hand equality with respect to s from point A all the way around the closed wall to A again gives

$$\oint \frac{\partial w}{\partial s} ds = 0,$$

so that

$$\oint \frac{q}{t} ds = 0. \tag{6.19}$$

Therefore in the case that the force is applied through the shear center,

$$q_A = -\frac{\oint \frac{q'}{t} ds}{\oint \frac{1}{t} ds}. \tag{6.20}$$

The analysis now proceeds as for an open section to determine q'.

EXAMPLE 6.11   Determine the shear center of the box beam with the cross section shown in Figure 6.10. Assume that the load is applied vertically along a line through the shear center so that the beam suffers no twist. The thickness $t_1 = t_3$, but $t_1 \neq t_2 \neq t_4$.

Solution   The beam cross section has a horizontal axis of symmetry so that only the horizontal position of the shear center need be computed. The shear flow can be approximated by using strength of materials. Let I_x be the area moment of inertia about the horizontal axis of symmetry, the x-axis. Neglecting terms containing the cube of a wall thickness,

$$I_x = \frac{1}{12}t_2 h^3 + \frac{1}{12}t_4 h^3 + t_1 w\frac{h^2}{4} + t_3 w\frac{h^2}{4} = \frac{1}{12}(t_2 + t_4)h^3 + t_1 w\frac{h^2}{2}.$$

because $t_1 = t_3$.

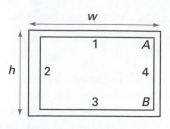
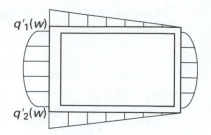

FIGURE 6.10   Box beam.

Make an imaginary cut at point A near the upper right corner so that $q = q' + q_A$. Then q' is the shear flow which is zero at point A. On the top horizontal wall, the shear flow varies linearly in s because

$$q'_1(s) = -\frac{V\,Q(s)}{I_x} = -\frac{V}{I_x}t_1 s\frac{h}{2},$$

where the arclength, s, is measured from point A. Note that

$$q'_1(w) = -\frac{V}{I_x}t_1 w\frac{h}{2}.$$

On the left vertical wall, the shear flow is parabolic in s because

$$q'_2(s) = -\frac{V}{I_x}t_2 s\left(\frac{h}{2} - \frac{s}{2}\right) + q'_1(w),$$

where the arclength, s, is measured from the top of the section. On the bottom horizontal wall, the shear flow varies linearly in s because

$$q'_3(s) = q'_2(h) + \frac{V}{I_x}t_3 s\frac{h}{2} = -\frac{V}{I_x}t_1 w\frac{h}{2} + \frac{V}{I_x}t_3 s\frac{h}{2},$$

where the arclength, s, is measured from the left of the section. On the right vertical wall, the shear flow is parabolic because

$$q'_4(s) = -\frac{V}{I_x}t_4 s\left(\frac{s}{2} - \frac{h}{2}\right) + q'_3(w) = -\frac{V}{I_x}t_4 s\left(\frac{s}{2} - \frac{h}{2}\right),$$

where the arclength, s, is measured from the bottom of the section and $q'_3(w) = 0$. Note that the moment arm involved in the Qs is taken in the coordinate system given. The shear flow is sketched in the right hand diagram in Figure 6.10. Using the computed shear flows,

$$\oint \frac{q'(s)}{t}ds = -\frac{V}{I_x}\frac{h}{2}\frac{w^2}{2} - \frac{V}{I_x}\frac{h^3}{12} - \frac{V}{I_x}\frac{t_1}{t_2}\frac{h^2}{2}w$$
$$-\frac{V}{I_x}w^2\frac{h}{2} + \frac{V}{I_x}\frac{h}{2}\frac{w^2}{2} + \frac{V}{I_x}\frac{h^3}{12}$$
$$= -\frac{Vhw}{2I_x}\left(w + h\frac{t_1}{t_2}\right).$$

Also

$$\oint \frac{1}{t}ds = 2\frac{w}{t_1} + \frac{h}{t_2} + \frac{h}{t_4}.$$

By equation (6.20),

$$q_A = \frac{Vhw}{2I_x}\frac{\left(w + h\frac{t_1}{t_2}\right)}{\left(2\frac{w}{t_1} + \frac{h}{t_2} + \frac{h}{t_4}\right)}.$$

Now take moments about the point B at the lower right corner to determine the location of the shear center, using the fact that the applied load and the shear flow oppose each other.

$$eV = \int_0^w hq_1(s)ds + \int_0^h wq_2(s)ds$$

$$= -\frac{V}{I_x}\frac{h^2}{2}t_1\frac{w^2}{2} + q_A wh - \frac{V}{I_x}wt_2\frac{h^3}{12} - \frac{V}{I_x}t_1w^2\frac{h^2}{2} + q_A wh.$$

The shear center coordinate measured from the right-hand side is

$$e = \frac{h^2}{I_x}\left[-\frac{3w^2}{4}t_1 + \frac{w^2\left(w + h\frac{t_1}{t_2}\right)}{2\frac{w}{t_1} + \frac{h}{t_2} + \frac{h}{t_4}} - \frac{wh}{12}t_2\right]$$

$$= \frac{1}{\left[\frac{1}{12}(t_2 + t_4)h + \frac{1}{2}t_1 w\right]}\left[-\frac{3w^2}{4}t_1 + \frac{w^2\left(w + h\frac{t_1}{t_2}\right)}{2\frac{w}{t_1} + \frac{h}{t_2} + \frac{h}{t_4}} - \frac{wh}{12}t_2\right]. \qquad (6.21)$$

∎

Exercise 6.8   The shear center lies on every axis of symmetry. Therefore, for a box beam with all wall thicknesses the same, the shear center lies at the centroid. Verify this fact using the expression for e of the previous example.

Note that, in Example 6.11, in the case that $t_4 = 0$ so that $q_A = 0$, the resulting expression for the C-shaped cross section, equation (6.21), agrees with that of equation (6.18) if the notation is made consistent and if w_1 is added to the result above because the distance e was measured from the left side in equation (6.18).

Exercise 6.9   Verify the agreement of equation (6.18) with equation (6.21).

Shear Center for a Multicelled Cross Section   Multicelled cross sections occur in some complex thin-walled structures, such as airplane wings. It is often necessary for designers to calculate the location of the shear center in such a structure.

The problem, again, is to determine the shear flow in each wall so that a moment can be taken. The technique applied to the box beam in the previous example can be applied to each cell, or compartment, one at a time in turn. Again, if the load is applied through the shear center, the shear flow about each cell, or compartment, must satisfy the relation

$$\oint \frac{q}{t}ds = 0.$$

An imaginary cut point is chosen in the first cell and equation (6.20) is used to calculate the shear flow at that point. The shear flow in the walls of this cell is then computed. The process is repeated for the next cell and so on.

The center is found from a moment balance, so the number of cells does not matter. It is only necessary to calculate the shear flow in the walls of each one.

EXAMPLE 6.12   The two-celled cross section of a box beam is shown in Figure 6.11. Determine the location of the shear center. The thickness of section i is t_i. But to ensure symmetry, $t_1 = t_6$ and $t_2 = t_4$.

Solution   The cross section has a horizontal axis of symmetry. Therefore, it is only necessary to compute the horizontal position of the shear center, e. Make vertical cuts at points A and C. In the

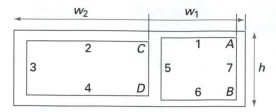

FIGURE 6.11 Two-celled box beam.

right-hand cell, section 1, the shear flow at A is q_A. In the left-hand cell, section 2, the shear flow at C is q_C. Further,

$$q_1(C) = q_5(C) + q_2(C);$$
$$q_6(D) = q_4(D) + q_5(D).$$

By symmetry, these two equations are identical.

In the right-hand cell, let $q' = q - q_A$. On the top horizontal wall, the shear flow varies as

$$q_1'(s) = -\frac{V}{I_x} t_1 s \frac{h}{2},$$

where the arclength, s, is measured from point A. On the left vertical wall, the shear flow is

$$q_5'(s) = -\frac{V}{I_x} t_5 s \left(\frac{h}{2} - \frac{s}{2}\right) + q_1'(w_1) - q_C = -\frac{V}{I_x} t_5 s \left(\frac{h}{2} - \frac{s}{2}\right) - \frac{V}{I_x} t_1 w_1 \frac{h}{2} - q_C,$$

where the arclength, s, is measured from the top of the section. On the bottom horizontal wall, the shear flow is

$$q_6'(s) = q_5'(h) + q_D + \frac{V}{I_x} t_6 s \frac{h}{2} = -\frac{V}{I_x} t_1 w_1 \frac{h}{2} - q_C + q_D + \frac{V}{I_x} t_6 s \frac{h}{2} = -\frac{V}{I_x} t_1 w_1 \frac{h}{2} + \frac{V}{I_x} t_6 s \frac{h}{2},$$

where the arclength, s, is measured from the left of the section. On the right vertical wall, the shear flow is

$$q_7'(s) = -\frac{V}{I_x} t_7 s \left(\frac{s}{2} - \frac{h}{2}\right) + q_6'(w_1) = -\frac{V}{I_x} t_7 s \left(\frac{s}{2} - \frac{h}{2}\right),$$

where the arclength, s, is measured from the bottom of the section and $q_6'(w_1) = 0$.

The left cell is treated similarly. Let $q'' = q - q_C$. On the top horizontal wall, the shear flow is

$$q_2''(s) = -\frac{V}{I_x} t_2 s \frac{h}{2},$$

where the arclength, s, is measured from point A. On the left vertical wall, the shear flow is

$$q_3''(s) = -\frac{V}{I_x} t_3 s \left(\frac{h}{2} - \frac{s}{2}\right) + q_2''(w_2) = -\frac{V}{I_x} t_3 s \left(\frac{h}{2} - \frac{s}{2}\right) - \frac{V}{I_x} t_2 w_2 \frac{h}{2},$$

where the arclength, s, is measured from the top of the section. On the bottom horizontal wall, the shear flow is

$$q_4''(s) = q_3''(h) + \frac{V}{I_x} t_4 s \frac{h}{2} = -\frac{V}{I_x} t_2 w_2 \frac{h}{2} + \frac{V}{I_x} t_4 s \frac{h}{2},$$

where the arclength, s, is measured from the left of the section. On the right vertical wall, the shear flow is

$$q_5''(s) = -\frac{V}{I_x} t_5 s \left(\frac{s}{2} - \frac{h}{2}\right) + q_4''(w_2) - q_6(0) = -\frac{V}{I_x} t_5 s \left(\frac{s}{2} - \frac{h}{2}\right) + \frac{V}{I_x} t_1 w_1 \frac{h}{2} - q_A,$$

where the arclength, s, is measured from the bottom of the section.

Application in the right-hand cell of equation (6.20), $q_A \oint \frac{1}{t}ds = -\oint \frac{q'}{t}ds$, and in the left-hand cell of $q_C \oint \frac{1}{t}ds = -\oint \frac{q''}{t}ds$ produces two linear equations in q_A and q_C.

$$q_A\left(2\frac{w_1}{t_1}+\frac{h}{t_5}+\frac{h}{t_7}\right)=\frac{Vhw_1}{2I_x}\left(w_1+h\frac{t_1}{t_5}\right)+q_C\frac{h}{t_5};$$

$$q_C\left(2\frac{w_2}{t_2}+\frac{h}{t_3}+\frac{h}{t_5}\right)=\frac{Vh}{2I_x}\left(w_2^2+w_2h\frac{t_2}{t_3}-w_1h\frac{t_1}{t_5}\right)+q_A\frac{h}{t_5}.$$

Once these are solved, one proceeds in each cell to obtain q. A check is that both cell calculations must produce the same q in the midsection 5. The moments are taken over the full cross section to obtain the shear center, e. A convenient point to take moments about is D so that if e is measured from D,

$$Ve=\int_0^h q_3(s)w_2ds+\int_0^{w_2} q_2(s)hds+\int_0^{w_1} q_1(s)hds+\int_0^h q_7(s)w_1ds. \qquad ■$$

Exercise 6.10 Verify that in the two-cell box beam of Example 6.12,

$$q_A=\frac{Vh}{2I_x}\frac{\left(w_1^2+w_1h\frac{t_1}{t_5}\right)\left(2\frac{w_2}{t_2}+\frac{h}{t_3}+\frac{h}{t_5}\right)-\left(w_2^2+w_2h\frac{t_2}{t_3}+w_1h\frac{t_1}{t_5}\right)\frac{h}{t_5}}{\left(2\frac{w_1}{t_1}+\frac{h}{t_5}+\frac{h}{t_7}\right)\left(2\frac{w_2}{t_2}+\frac{h}{t_3}+\frac{h}{t_5}\right)-\left(\frac{h}{t_5}\right)^2};$$

$$q_C=\frac{Vh}{2I_x}\frac{\left(2\frac{w_1}{t_1}+\frac{h}{t_5}+\frac{h}{t_7}\right)\left(w_2^2+w_2h\frac{t_2}{t_3}+w_1h\frac{t_1}{t_5}\right)-\left(w_1^2+w_1h\frac{t_1}{t_5}\right)\frac{h}{t_5}}{\left(2\frac{w_1}{t_1}+\frac{h}{t_5}+\frac{h}{t_7}\right)\left(2\frac{w_2}{t_2}+\frac{h}{t_3}+\frac{h}{t_5}\right)-\left(\frac{h}{t_5}\right)^2};$$

$$e=\frac{h^2}{I_x}\left[-t_3\frac{h}{6}w_2-\frac{3}{4}t_2w_2^2-\frac{1}{4}t_1w_1^2+t_7\frac{h}{6}w_1\right]+2hw_1\frac{q_A}{V}+2hw_2\frac{q_C}{V}. \qquad (6.22)$$

6.4 TORSION OF THIN-WALLED PRISMATIC MEMBERS

The cases of open and closed cross sections must be considered separately. Again, a closed thin-walled cross section is one which is a closed curve when the wall thickness is neglected. Consider a prismatic rod with an arbitrary closed thin-walled cross section. Tangential-normal coordinates are chosen in the plane with direction s tangent to and n normal to the thin cross section. The longitudinal direction is the z-axis.

The shear stress, τ_{zs}, is in the direction tangent to the wall (Fig. 6.12). Assume that this shear stress is constant across the thickness, t, at any point. This thickness need not be constant. To analyze the influence of the wall thickness on the shear stress, consider a stress element in the wall (Fig. 6.12). The normal stress in the z-direction is zero.

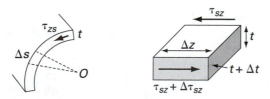

FIGURE 6.12 The shear stresses in a thin-walled cross section.

From the stress element, the sum of the forces in the z-direction must be zero at equilibrium, so that

$$(\tau_{sz}+\Delta\tau_{sz})(t+\Delta t)\Delta z-\tau_{sz}t\Delta z=0.$$

Add out the terms $\tau_{sz} t \Delta z$ and divide by $\Delta s \Delta z$. Let Δs tend to zero; as it does, $\Delta \tau_{sz}$ also tends to zero so that

$$\tau_{sz} \frac{dt}{ds} + t \frac{d\tau_{sz}}{ds} = 0,$$

or

$$\frac{d(\tau_{sz} t)}{ds} = 0.$$

Therefore the shear flow $q = \tau_{sz} t$ is a constant. If the section gets thinner, then the shear stress increases, as would be expected. The shear flow q, at each point around the circumference, can be thought of as a force per unit length vector which is tangent to the wall boundary. The shear flow multiplied by an increment of arclength of the boundary is the torsional force at that point. The sum of the moments of all these forces distributed around the circumference must equal the applied torque, T, by equilibrium.

$$\oint rq\,ds = T, \tag{6.23}$$

where the integral is taken about the closed boundary and $r(s)$ is the perpendicular distance from a fixed point in the interior, O, on the axis of rotation to the vector q. But by Figure 6.12, the wedge-shaped area increment is $dA = (1/2)r\,ds$, where A is the area of the cross section inside the wall. Since q is a constant, the torque is

$$T = \oint qr\,ds = q \oint r\,ds = q \int_A 2dA = 2qA.$$

The shear stress, as a function of s, is

$$\tau_{sz}(s) = \frac{T}{2At(s)}. \tag{6.24}$$

It remains to determine the angle of twist per unit length, θ. The displacement, w, is not a function of z. The assumption that the displacement depends only on the position in the face was also made in the study of the torsion of solid sections, discussed in Chapter 5, Section 5.1. Each face of the prismatic rod is subjected to the same out-of-plane displacement.

The shear strain is by the small strain assumption,

$$\frac{\tau_{sz}}{G} = \gamma_{sz} = \frac{\partial w}{\partial s} + \frac{\partial u}{\partial z},$$

where (u, v, w) are the displacements in the (s, n, z) directions respectively. But $\partial u / \partial z = r(s)\theta$. Substitution into the expression for the strain and integration about the cross section implies that

$$0 = \oint dw = \frac{1}{G} \oint \tau_{sz}(s)\,ds - \theta \oint r(s)\,ds.$$

where the left integral is zero since w is continuous. Recall that $\oint r(s)\,ds = 2A$. Therefore,

$$\theta = \frac{TC}{4GA^2} \int \frac{ds}{t} = \frac{q}{2GA} \int \frac{ds}{t}. \tag{6.25}$$

If the wall thickness, t, is constant, these relations are simply

$$\theta = \frac{TC}{4GA^2t},\qquad(6.26)$$

where C is the circumference.

EXAMPLE 6.13   In the case that the boundary is circular, $A = \pi R^2$ and $C = 2\pi R$. Application of equation (6.26) recovers the result from elementary strength of materials,

$$\theta = \frac{T}{2G\pi R^3 t} = \frac{T}{GJ},$$

for $J = 2\pi R^3 t$, and also

$$\tau_{sz} = \frac{TR}{J}.\qquad\blacksquare$$

EXAMPLE 6.14   A prismatic rod has a box-shaped cross section of height, h, and base, b. The wall thickness is a constant t. Determine the angle of twist when the torque, T, is applied.

Solution   The area is bh and the circumference is $2h + 2b$. By equation (6.26),

$$\theta = \frac{TC}{4GA^2t} = \frac{T(2h + 2b)}{4Gb^2h^2t}.\qquad\blacksquare$$

EXAMPLE 6.15   A square cross-sectional bar must carry a torque of $T = 800$ Nm. It is decided to make the bar thin-walled to reduce the weight of the bar. The bar is to be made of a 4130 steel with $\tau_Y = 600$ MPa and shear modulus $G = 80$ GPa. The bar must be elastically loaded with a safety factor of 2 and the angle of twist per unit length must not exceed 0.1 radians. Design the minimum edge dimension, s, of the square cross section and the minimum wall thickness, t.

Solution   Assume dimensions so that the bar is just at yield and just reaches the maximum allowable twist. Because of the safety factor, the maximum allowable shear stress is $\tau_Y/2$. Then by equation (6.24), $\tau_Y/2 = T/(2s^2t)$ and by equation (6.26), $\theta = Ts/(s^4Gt)$. Solve these equations simultaneously by eliminating t to obtain $s = \tau_Y/(\theta G) = 0.075$ m, and $t = T/(\tau_Y s^2) = 0.000237$ m. A larger thickness may need to be specified to simplify manufacturing of the bar. $\blacksquare$

If thin walls are also placed inside the prismatic cross section, they divide the cross-sectional area into regions, often called cells. Figure 6.13 shows such a rod with two cells, the left one of area, A_1, and the right cell of area, A_2. There are three independent wall sections. The left outer wall has shear flow, q_1, the right outer wall has shear flow, q_2, and the common center wall has shear flow, q_{12}. Three equations are required to determine these shear flows. A small section of length, Δz, is cut around

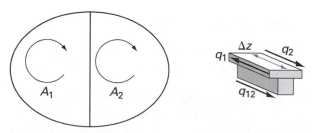

FIGURE 6.13   Two-celled thin-walled cross section with the shears at a joint of the wall sections.

the upper joint of the three wall sections. The sum of forces in the z-direction shows that

$$q_{12} = q_2 - q_1. \qquad (6.27)$$

The torque-moment of shear flow equation (6.23) combined with the relation $q_{12} = q_2 - q_1$ implies that

$$\frac{1}{2}T = q_1 A_1 + q_2 A_2. \qquad (6.28)$$

Finally, the relation between the torque and the angle of twist equation (6.25) and the requirement that the angle of twist, θ, for the full cross-sectional face must equal both the angles of twist for each cell imply that

$$\theta = \frac{1}{2GA_1}\left(q_1 \int_{C_1} \frac{ds}{t} + (q_1 - q_2)\int_{C_{12}} \frac{ds}{t}\right);$$
$$\theta = \frac{1}{2GA_2}\left(q_2 \int_{C_2} \frac{ds}{t} + (q_2 - q_1)\int_{C_{12}} \frac{ds}{t}\right), \qquad (6.29)$$

where C_1 is the left wall, C_2 is the right wall, and C_{12} is the common wall.

EXAMPLE 6.16

A vertical web of thickness 3 mm is built into a box beam with base length 1 meter and height 0.6 meters and wall thickness 2 mm, dividing it into two rectangular cells, one square with sides 0.6 meters and the other with base 0.4 meters and height 0.6 meters. The shear modulus of the material is $G = 70$ GPa. Determine the angle of twist if a torque $T = 4,000$ Nm is applied about the longitudinal axis of the beam.

Solution

First write the shear flows in terms of the angle of twist. The square box is labeled 1.

$$2G\theta = \frac{1}{(0.6)(0.6)}\left(q_1\frac{3(0.6)}{0.002} + (q_1 - q_2)\frac{0.6}{0.003}\right)$$
$$2G\theta = \frac{1}{(0.4)(0.6)}\left(q_2\frac{2(0.4)+0.6}{0.002} + (q_2 - q_1)\frac{0.6}{0.003}\right)$$

Solving simultaneously yields,

$$q_1 = 3.9158 \times 10^{-4} 2G\theta \quad \text{and} \quad q_2 = 3.5368 \times 10^{-4} 2G\theta.$$

The relation $T = 2A_1 q_1 + 2A_2 q_2$ is solved for the angle of twist to obtain $\theta = 6.0325 \times 10^{-5}$ rad/m. ∎

6.4.1 Open Section Rods

Sections that are not closed are more difficult to analyze. A common trick is to assume that the Prandtl stress function has a special form that satisfies the boundary condition, $\phi = 0$, and reduces the Poisson partial differential equation to a tractable form. A partial differential equation in two variables, such as the Poisson equation, can sometimes be reduced to a solvable ordinary differential equation by assuming that the solution separates into a product of functions, one in each variable. It is often necessary to assume that the solution, the Prandtl function, is an infinite sum of such separated functions.

To analyze the torsion of a prismatic rod with rectangular cross section of base, $2w$, and height, $2b$, work in Cartesian coordinates, and assume that the Prandtl function is

$$\phi(x, y) = \sum_{n=1,3,5,\ldots}^{\infty} a_n \cos\left(\frac{n\pi x}{2w}\right) f_n(y), \tag{6.30}$$

where the constants, a_n, and and the functions, $f_n(y)$, are to be determined from the differential equation and the boundary conditions. The Prandtl function, $\phi(x, y)$, must be zero on the boundaries of the cross section. Only odd values of n are taken to ensure that ϕ is zero on the vertical boundaries, $\phi(\pm w, y) = 0$, but not on the vertical center line.

Substituting the assumed form for ϕ into the Poisson partial differential equation (Chapter 5, Section 5.1),

$$\frac{\partial^2 \phi}{\partial x^2} + \frac{\partial^2 \phi}{\partial y^2} = -2G\theta$$

yields the ordinary differential equation for $f_n(y)$,

$$\frac{d^2 f_n}{dy^2} - \frac{n^2 \pi^2}{4w^2} f_n = -2G\theta \frac{4}{n\pi a_n} (-1)^{(n-1)/2}.$$

The solution of this second-order ordinary differential equation with constant coefficients involves hyperbolic functions since the coefficient of f_n is negative. After applying the requirement that the function be zero on the upper and lower boundaries, $f_n(\pm b) = 0$, the solution is

$$f_n(y) = \frac{32G\theta w^2}{n^3 \pi^3 a_n} (-1)^{(n-1)/2} \left[1 - \frac{\cosh(n\pi y/2w)}{\cosh(n\pi b/2w)} \right].$$

Exercise 6.11   Verify that $f_n(y)$ is zero on the upper and lower boundaries.

The stress function is therefore

$$\phi(x, y) = \frac{32G\theta w^2}{\pi^3} \sum_{n=1,3,5,\ldots}^{\infty} \frac{1}{n^3} (-1)^{(n-1)/2} \left[1 - \frac{\cosh(n\pi y/2w)}{\cosh(n\pi b/2w)} \right] \cos\left(\frac{n\pi x}{2w}\right). \tag{6.31}$$

Notice that the unknown constants, a_n, divide out so that it is unnecessary to determine them.

The vertical shear component is

$$\tau_{yz} = -\frac{\partial \phi}{\partial x} = \frac{16G\theta w}{\pi^2} \sum_{n=1,3,5,\ldots}^{\infty} \frac{1}{n^2} (-1)^{(n-1)/2} \left[1 - \frac{\cosh(n\pi y/2w)}{\cosh(n\pi b/2w)} \right] \sin\left(\frac{n\pi x}{2w}\right). \tag{6.32}$$

This shear stress is zero at the top and bottom edges, where $y = \pm b$, and along the vertical centerline $(0, y)$. The maximum vertical shear stress component is at the center of the vertical edges, $(\pm w, 0)$.

$$\tau_{\max} = \frac{16G\theta w}{\pi^2} \sum_{n=1,3,5,\ldots}^{\infty} \frac{1}{n^2} \left[1 - \frac{1}{\cosh(n\pi b/2w)} \right]. \tag{6.33}$$

Since $\sum_{n=1,3,5,\dots}^{\infty} \frac{1}{n^2} = \pi^2/8$,

$$\tau_{\text{max}} = 2G\theta w - \frac{16G\theta w}{\pi^2} \sum_{n=1,3,5,\dots}^{\infty} \frac{1}{n^2} \frac{1}{\cosh(n\pi b/2w)}.$$

These calculations can be applied to a thin rectangular section in which the horizontal dimension is small, $t = 2w$ and $h = 2b$, so that h is much greater than t. In such a case, $\cosh(n\pi h/2t)$ is large. The shear stress in a thin rectangular section is therefore

$$\tau_{\text{max}} = G\theta t. \tag{6.34}$$

The relation between the torque, T, and the angle of twist, θ, for a thin rectangular cross section is determined as for a solid cross section, using equation (6.31),

$$T = 2\int_{-t/2}^{t/2}\int_{-h/2}^{h/2} \phi \, dy \, dx = \frac{1}{3}G\theta t^3 h \left[1 - \frac{192w}{\pi^5 h} \sum_{n=1,3,5,\dots}^{\infty} \frac{1}{n^5} \tanh\left(\frac{n\pi h}{2w}\right) \right], \tag{6.35}$$

since $\sum_{n=1,3,5,\dots}^{\infty} \frac{1}{n^4} = \frac{\pi^4}{96}$. In the case of a thin rectangle, $\tanh(n\pi h/2t) \sim 1$ and

$$T = \frac{1}{3}G\theta t^3 h \left(1 - 0.630\frac{t}{h}\right). \tag{6.36}$$

For very thin members, the ratio t/h is approximately zero, and the relation becomes

$$T = \frac{1}{3}G\theta t^3 h. \tag{6.37}$$

So that the expression is of the form of that obtained in elementary strength of materials, $\theta = T/JG$, the torsional constant, J, for a thin rectangle is defined to be

$$J = \frac{t^3 h}{3}.$$

The torsional constant depends only on the area. It can be obtained for an L-cross section by adding the results for the two thin rectangles making up the L,

$$J = \frac{t_1^3 h_1 + t_2^3 h_2}{3}.$$

EXAMPLE 6.17

Determine the maximum shear stress in a constant rectangular cross-sectional beam of dimensions 25 cm by 2 cm subject to a 200 Nm torque about the longitudinal beam axis. The shear modulus is 28 GPa.

Solution

Solve equations (6.36) and (6.34) simultaneously to obtain

$$\tau_{\text{max}} = \frac{T}{\frac{1}{3}t^2 h \left(1 - 0.630\frac{t}{h}\right)}. \tag{6.38}$$

This result is independent of the shear modulus, but not of the geometry or torque. In this example, $\tau_{\text{max}} = 789.8$ kPa. ∎

EXAMPLE 6.18   Determine the angle of twist due to a 3,650 Nm torque acting on a prismatic beam with an L cross section. The wall thickness is 1.5 cm and the flanges have respective lengths of 15 cm and 22 cm. The shear modulus is 70 GPa.

Solution   The torsional constant is

$$J = \frac{1}{3}[(1.5)^3(15) + (1.5)^3(22)] = 41.625 \quad \text{cm}^4.$$

The angle of twist is

$$\theta = \frac{3,650}{(70 \times 10^9)(41.625 \times 10^{-8})} = 0.1253 \quad \text{rad.} \qquad \blacksquare$$

PROBLEMS

6.1   Verify that if either axis is an axis of symmetry for the area, then the product of inertia is zero.

6.2   Determine the angle that the principal axes make with respect to the cross-bar edge of a symmetrical T-cross section of constant thickness, t, cross-bar length, w, and leg length, h.

6.3   Determine the direction of the principal coordinates for a right-angle beam with wall thickness, 20 mm, one inner leg length, 180 mm, and the other inner leg length of 220 mm.

6.4   Determine the ratio of the sides in the rectangular cross section of a beam that would ensure that the neutral axis is perpendicular to the load for each acute angle, β, with respect to the vertical for an unsymmetric loading whose line of action passes through the shear center. Ans. $b/h = 1$

6.5   Determine the direction of the neutral axis in a beam with rectangular cross section having a base $b = 0.2$ inches and height $h = 4$ inches when the load is applied at an angle of $5°$ from the vertical and with its line of action through the centroid.

6.6   Determine the direction of the neutral axis in a channel beam having a base with outer length $b = 8$ inches, height $h = 12$ inches, and flange thickness of 1 inch. The load is applied at an angle of $22°$ from the vertical. Compare the result to that of the similar example in the text.

6.7   A cantilever beam with rectangular cross section having a base $b = 2$ inches and height $h = 4$ inches has a load of $P = 300$ pounds applied at the free end and at an angle of $25°$ from the vertical. In a face 5 inches from the free end, determine the expression for the normal stress as a function of x and y, the principal coordinates. Then use a computer program to draw a contour plot of the function $\sigma(x, y)$ over the full cross-sectional area. Estimate the location of the maximum stress from your contour plot.

6.8   Determine the shear center of a C-shaped cross section in which the upper and lower legs each have length 15 cm and thickness 0.3 cm and in which the vertical leg has length 20 cm and thickness 0.5 cm.

6.9   A constant thickness, thin-walled beam has a cross section that is the arc of a circle of radius, R, cutting off an angle of 2θ. Determine the distance from the center of the arc to the shear center. Ans. $[2R(\sin\theta - \theta\cos\theta)]/(\theta - \sin\theta\cos\theta)$.

6.10   Determine the shear center of a box beam with height 25 cm and width 30 cm. The thicknesses of the top and bottom walls are respectively 0.4 and 0.4 cm. The left side and the right side thickness are both 0.2 cm.

6.11   Determine the shear center in a square box beam with sides, w, and constant wall thickness, t, with a cut halfway up the right vertical wall. Ans. $5w/8$ from the left.

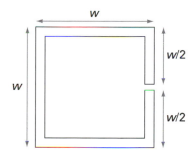

6.12   If all the wall thicknesses in the two-celled box beam example of Fig. 6.11 of Example 6.12 are equal to t and $w_1 = w_2$, determine the horizontal position of the shear center using equation (6.22). Compare the location of the shear center to that of the centroid.

6.13   A thin-walled box beam must carry a torque, $T = 1,000$ Nm. The horizontal sides are 8 cm and the vertical sides are 16 cm. The angle of twist per unit length must be less than 0.02 radians. The shear modulus is 40 GPa. Determine the minimum allowable constant wall thickness, t. Compute the shear stress for the computed minimum, t.

6.14   For functional reasons, a thin-walled bar must have a rectangular cross section of 2 cm by 3 cm and a wall thickness of 2 mm. The bar must carry a torque of 100 Nm. Prepare for material selection by determining the minimum yield stress required of the material to guarantee elastic loading. If the allowable twist per unit length is 0.1 radians, what is the minimum shear modulus required?

6.15 Determine the shear flows for an arbitrary torque, T, in a thin-walled circular cross-sectional-rod with an additional section along one diameter of the circle. The radius of the circular section is 60 mm, the thickness of the outer wall is 6 mm, and the thickness of the wall section along the diameter is 9 mm.

6.16 Determine the angle of twist for the rod in problem 15 in the case that $T = 2000$ Nm and $G = 50$ GPa.

6.17 A vertical web of thickness 5 mm is built into a box beam whose dimensions are: base length 1.2 meters, height 0.8 meters, and wall thickness 3 mm. The web divides the cross section into two rectangular cells, one with sides 0.7 meters and height 0.8 meters and the other with base 0.5 meters and height 0.8 meters. The shear modulus of the material is $G = 70$ GPa. Determine the angle of twist if a torque $T = 4,000$ Nm is applied about the longitudinal axis of the beam.

6.18 Determine the maximum shear stress in a constant rectangular cross-sectional bar with dimensions 20 cm by 1.5 cm subject to a 400 Nm torque about the longitudinal bar axis.

CHAPTER 7

ENERGY METHODS

The linear elasticity problem is to determine the equilibrium stress, strain, and displacement at each point of a body subject to given boundary conditions in terms of surface tractions or displacements. Some two-dimensional problems can be solved using a stress function, such as the Airy or Prandtl functions. Most three-dimensional problems, which require the solution of the 15 equations of linear elasticity, are so complex that approximation techniques are required.

Two traditions of equilibrium analysis exist in mechanics. The oldest and by far the most flexible is that using energy functions and variational methods. The more recent is vector analysis, as taught in Statics. Vector analysis was devised by Gibbs and Maxwell in the 1880s based on work by Hamilton on quaternions. Its primary advantages are the avoidance of complex trigonometric calculations to determine sums, projections and moments of forces and the ease in calculating the velocity and acceleration of a particle moving in three-space. Energy methods using the potential energy allow an easier calculation of equilibrium states for complex mechanical systems, such as statically indeterminate systems. Such energy method techniques can easily determine the stability of these equilibrium states, a problem nearly insolvable using statics. The Castigliano theorem, based on energy functions, permits the quick computation of the displacement at a single point in the body. More importantly, most effective approximation techniques are based on energy methods or variational principles.

Definition   A variational principle is a statement that a certain response of a mechanical system is predicted by minimizing a particular real-valued function associated with the system and the type of response.

The classical variational principle is that of virtual work, which gives a necessary condition for equilibrium of a system. Its essence was known to Leonardo da Vinci (1452–1519) and used by Galileo (1564–1642) to analyze simple machines. It was first formulated precisely in 1717 by Jean Bernoulli (1667–1748). A more complete development was published in 1760 by J. L. Lagrange (1736–1813). Euler had used it as early as 1732 to analyze the behavior of simple structures. The principle was extended to kinetics, the relation of force on a body to the motion of the body, by Hamilton in 1834. In the nineteenth and twentieth centuries, many other variational principles have been proposed and the methods have found important applications in determining the stability of the equilibrium states of a mechanical system.

Energy methods are applicable to rigid or deformable bodies; and if the bodies are deformable, energy methods are valid whether the behavior is described by strength of materials, by linear elasticity, or by any other model. In this chapter, which is concerned primarily with static, linear or nonlinear elastic, deformable bodies, attention will be restricted to the principles of virtual work, of stationary potential energy, and of complementary energy.

7.1 WORK, STRAIN ENERGY, AND COMPLEMENTARY ENERGY

Work due to a force is defined as the product of the magnitude of the force and the distance in the direction of the force that the force moves through. Work has dimensions of force times distance and units Nm, ft-lb, in-lb, etc. Suppose that the force $\mathbf{F} = F_x\mathbf{i} + F_y\mathbf{j} + F_z\mathbf{k}$ acts on a particle which moves from point P to a nearby point P'. Let the position vector from P to P' be $\mathbf{r} = r_x\mathbf{i} + r_y\mathbf{j} + r_z\mathbf{k}$ (Fig. 7.1).

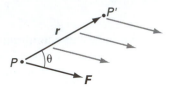

FIGURE 7.1   The force **F** on a point mass as the point mass moves from P to P'.

If θ is the angle between $\mathbf{F}$ and $\mathbf{r}$, the component of distance between P and P' in the direction of $\mathbf{F}$ is $r\cos\theta$. Therefore the work done by $\mathbf{F}$ as the particle moves from P to P' is $W = Fr\cos\theta$, where r is the magnitude of $\mathbf{r}$. But this is just the dot product of $\mathbf{F}$ and $\mathbf{r}$, $W = \mathbf{F}\cdot\mathbf{r}$. The work done by a moment, $\mathbf{M}$, as the body rotates through an angle, θ, is $W = M\theta$.

The work over a curved path, Γ, connecting points P and P' is found by breaking the path into small straight increments defined by the position vectors $d\mathbf{r} = dx\mathbf{i} + dy\mathbf{j} + dz\mathbf{k}$. The total work done by a force, which may be a function of position, $\mathbf{F}(\mathbf{r})$, is the sum of the work done on each of these small pieces of the path.

Definition   The work done by a force, $\mathbf{F}(\mathbf{r}) = F_x\mathbf{i} + F_y\mathbf{j} + F_z\mathbf{k}$, acting over a path, Γ, from point P to P' is the line integral over Γ, $W = \int_\Gamma \mathbf{F}\cdot d\mathbf{r} = \int_\Gamma (F_x dx + F_y dy + F_z dz)$.

If the value of this line integral is independent of the path between P and P', it is called exact, and the force is called conservative. The force due to gravity is conservative. In fact, any force which never changes direction as the body moves is conservative. On the other hand, a frictional force is not conservative because it always opposes the motion of the body.

Definition   The line integral, $\int_\Gamma (F_x dx + F_y dy + F_z dz)$, over a path from point P to point P' is exact if there is a real valued function $f(x, y, z)$ such that $F_x = \partial f/\partial x$, $F_y = \partial f/\partial y$, $F_z = \partial f/\partial z$.

In this case,

$$
\begin{aligned}
\int_\Gamma \mathbf{F}\cdot\mathbf{r} &= \int_\Gamma (F_x dx + F_y dy + F_z dz) \\
&= \int_\Gamma \left(\frac{\partial f}{\partial x}dx + \frac{\partial f}{\partial y}dy + \frac{\partial f}{\partial z}dz\right) \\
&= \int_P^{P'} df = f(x_2, y_2, z_2) - f(x_1, y_1, z_1),
\end{aligned}
$$

where $P = (x_1, y_1, z_1)$ and $P' = (x_2, y_2, z_2)$. The line integral representing the work due to a conservative force can be computed by evaluating $f(x, y, z)$ at the endpoints of the curve. In other words, the value of the line integral of an exact integrand only depends on the endpoints, not on the particular path between those endpoints.

EXAMPLE 7.1   Compute the work done by the force $\mathbf{F} = x\mathbf{i} + y\mathbf{j} + z\mathbf{k}$ on a point mass as the mass moves from the point (1,2,3) to the point (4,5,6).

Solution   The force is not constant, but it is conservative because the associated real valued function $f(x, y, z) = 0.5(x^2 + y^2 + z^2)$. This is checked by taking the various partial derivatives of $f(x, y, z)$. The work done is $W = f(4, 5, 6) - f(1, 2, 3) = 31.5$.   ∎

The *first law of thermodynamics*, conservation of energy, relates the work, W_e, done on the body by the external forces on a body and the heat flow into the body, Q, to the change in kinetic energy, ΔK, and the change in internal energy, U, of the body.

$$W_e + Q = \Delta K + U. \tag{7.1}$$

This conservation of energy relation is sometimes called the balance of energy assumption.

The law of kinetic energy says that if W is the work done by all forces, internal and external, then $\Delta K = W$ in a nonaccelerating coordinate system. The law of kinetic energy is equivalent to Newton's second law that the total force, $\mathbf{F}$, acting on a point mass is the product of the mass times the acceleration. To prove this, recall that the velocity is $\mathbf{v} = d\mathbf{r}/dt$ and the acceleration is $\mathbf{a} = d\mathbf{v}/dt$. Then applying Newton's second law, the increment of work over a small change in position is

$$dW = \mathbf{F} \cdot d\mathbf{r} = m(d\mathbf{v}/dt) \cdot d\mathbf{r} = m\mathbf{v} \cdot d\mathbf{v} = d\left(\frac{1}{2}mv^2\right), \tag{7.2}$$

where the square of the magnitude of $\mathbf{v}$ is $v^2 = \mathbf{v} \cdot \mathbf{v}$. If the kinetic energy is defined to be $K = mv^2/2$, then integrating equation (7.2) over the path yields $W = \Delta K$.

But the total work is the sum of the work done by the internal and external forces, $W = W_i + W_e$, so that

$$\Delta K = W_i + W_e. \tag{7.3}$$

An adiabatic system is one that is completely insulated so that there is no heat flow into or out of it; $Q = 0$. For such a system, the first law implies that since $\Delta K = W_i + W_e$,

$$W_e + 0 = W_i + W_e + U,$$

so that

$$U = -W_i. \tag{7.4}$$

This relation identifies the internal energy, in the case in which there is no heat flow in or out of the system or from point to point within it, as the potential energy of the internal forces. This energy is hard to compute since one must either identify the internal forces or find a way to circumvent such a calculation. If the body is static, not moving, the law of kinetic energy provides such a method. When $\Delta K = 0$, using equations (7.3) and (7.4),

$$0 = \Delta K = W_e + W_i = W_e - U$$

implies that

$$U = W_e. \tag{7.5}$$

Therefore in the special case that the body is both adiabatic and static, the magnitude of the internal energy can be computed from the work done by the external forces. This result (7.5) is sometimes called Clapeyron's theorem.

If the work done by the internal forces in a body is conservative, *i.e.* if the body is elastically loaded, the potential energy of the internal work is called the strain energy, U. The strain energy is often written in terms of the strain energy density, U_o, so that $U = \int_V U_o dV$, where V is the volume. In Chapter 3, Section 3.6, the strain energy density for a linear elastic material was shown to be

$$U_o = \frac{1}{2}[\sigma_x \epsilon_x + \sigma_y \epsilon_y + \sigma_z \epsilon_z + \tau_{xy} \gamma_{xy} + \tau_{xz} \gamma_{xz} + \tau_{yz} \gamma_{yz}], \tag{7.6}$$

where U_o is a function of just the strains. To compute U_o, the stresses must be written as a function of the strains. For a linear elastic material,

$$U_o(\epsilon) = \frac{1}{2}[\lambda e^2 + 2G(\epsilon_x^2 + \epsilon_y^2 + \epsilon_z^2) + G(\gamma_{xy}^2 + \gamma_{xz}^2 + \gamma_{yz}^2)], \tag{7.7}$$

where $e = \epsilon_x + \epsilon_y + \epsilon_z$, G is the shear modulus, and λ is the Lamé constant.

In the uniaxial case, all stresses are zero except $\sigma_x = E\epsilon_x$ and

$$U_o = \int_0^{\epsilon_x} \sigma_x d\epsilon_x = \int_0^{\epsilon_x} E\epsilon_x d\epsilon_x = \frac{1}{2}E\epsilon_x^2 = \frac{1}{2}\sigma_x \epsilon_x.$$

The strain energy density can be viewed as the area under the elastic portion of the stress-strain curve between $\epsilon_x = 0$ and ϵ_x (Fig. 7.2).

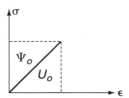

FIGURE 7.2   Strain and complementary energy for a linear elastic material.

For linear elastic materials, the area between the stress-strain curve and the σ-axis is also equal to the strain energy density since the linear stress-strain curve divides the rectangle into two equal parts. The area between the stress-strain curve and the σ-axis is called the complementary energy density, Ψ_o. So for a linear elastic material, $\Psi_o = U_o$. In uniaxial loading, $\Psi_o = \int_0^\sigma \epsilon d\sigma = \int_0^\sigma \sigma/E d\sigma = \sigma^2/2E$. The total complementary energy is $\Psi = \int_V \Psi_o dV$. For a linear elastic material, $\Psi = U$. If the stress-strain relation is not linear, but is monotonic, the complementary energy is still defined but is unequal to the strain energy.

For simple loadings such as those described in elementary strength of materials, the stresses or strains can be written in terms of the external loads or displacements. Even though, for linear elastic materials, the strain energy and the complementary energy have the same magnitude, the internal energy, U, is usually written in terms of displacements of the points at which external forces are applied, while the complementary energy, Ψ, is usually written in terms of the external forces.

EXAMPLE 7.2   Assume that the stresses are given by the strength of materials approximations, as opposed to elasticity. Write the complementary energy due to each of the types of loading considered in elementary strength of materials: axial loading, torsion, and bending.

Axial Loading: Let the body have constant cross-sectional area A, length L, and internal axial force, $N(x)$, at the face indexed by x, then $\sigma_x = N(x)/A$ and all other stress components are zero. Therefore,

$$\Psi = \int_V \Psi_o dV = \int_V \frac{\sigma^2}{2E} dV = \int_0^L \int_A \frac{N(x)^2}{A^2 2E} dA dx = \int_0^L \frac{N(x)^2}{2AE} dx. \qquad (7.8)$$

Torsion: Let the rod have constant cross-sectional area A, length L, and internal torque, $M_t(x)$, at the face indexed by x, then $\tau_{xy} = M_t(x)r/J$ at the position r from the center of the face. All other stress components are zero.

$$\Psi = \int_V \Psi_o dV = \int_V \frac{\tau_{xy}^2}{2G} dV = \int_0^L \int_A \frac{M_t(x)^2 r^2}{J^2 2G} dA dx$$

$$= \int_0^L \frac{M_t(x)^2}{J^2 2G} \left[\int_A r^2 dA \right] dx$$

$$= \int_0^L \frac{M_t(x)^2}{2JG} dx, \qquad (7.9)$$

since $J = \int_A r^2 dA$.

Bending: Let the beam have constant cross-sectional area A, length L, and shear force, $V(x)$, internal bending moment, $M(x)$, at the face indexed by x, then $\sigma_x = M(x)y/I$, $\tau_{xy} = V(x)Q/It$, and all other stress components are zero. The complementary energy density is $\Psi_o = (1/2)[\sigma_x \epsilon_x + \tau_{xy}\gamma_{xy}] = \sigma_x^2/2E + \tau_{xy}^2/2G$. Therefore, $\Psi = \int_V (\sigma_x^2/2E) dV + \int_V (\tau_{xy}^2/2G) dV$. Taking these integrals separately,

$$\int_V \frac{\sigma_x^2}{2E} dV = \int_0^L \int_A \frac{M(x)^2 y^2}{I^2 2E} dA dx = \int_0^L \frac{M(x)^2}{I^2 2E} \left[\int_A y^2 dA \right] dx = \int_0^L \frac{M(x)^2}{2IE} dx, \quad (7.10)$$

since $I = \int_A y^2 dA$.

$$\int_V \frac{\tau_{xy}^2}{2G} dV = \int_0^L \int_A \frac{V(x)^2 Q^2}{2GI^2 t^2} dA dx = \int_0^L \frac{V(x)^2}{I^2 2G} \left[\int_A \left(\frac{Q}{t} \right)^2 dA \right] dx = f_s \int_0^L \frac{V(x)^2}{2AG} dx,$$

$$(7.11)$$

where the cross-sectional factor $f_s = (A/I^2) \int_A (Q/t)^2 dA$. Note that f_s is dimensionless. ■

EXAMPLE 7.3   For a rectangular cross-sectional area, the neutral axis is parallel to the side of length t. Let the height be $2h$. Then $Q = A\bar{y} = 0.5(h^2 - y^2)t$ and $I = 2th^3/3$. The cross-sectional factor is $f_s = (A/I^2) \int_A (Q/t)^2 dA = (2th/I^2) \int_{-h}^h (Q/t)^2 t dy = 6/5$.
By an analogous calculation, $f_s = 10/9$ for a solid circular cross section. ■

7.2 CASTIGLIANO'S THEOREM

Castigliano's theorem provides a quick method of determining the unknown deflection at a point of application of a given external force. This theorem is useful when the deflection is desired at only one or a few points on the body. If the deflection must be determined at all points, other elasticity techniques are more efficient. Castigliano's theorem, which he first proved in his 1873 thesis, is also sometimes called the principle of complementary energy. Castigliano proved it only in the case of a linear elastic material, but it is also valid for a nonlinear elastic material when only small displacements are permitted. The Castigliano result was generalized by Engesser in 1889 to nonlinear elastic materials, but this result then seems to have been lost to practicing engineers until H. M. Westergaard reintroduced it in a paper published in 1942. A converse theorem gives the unknown external force causing a given displacement at its point of application. This theorem requires that the complementary energy be written in terms of the external loads on the body.

THEOREM

Castigliano's Theorem

For a linear elastic structure, the partial derivative of the complementary energy, written in terms of the external loads, with respect to the magnitude of a point load, $\mathbf{F}$, is the deflection of the body at the point of application of $\mathbf{F}$ in the direction of the force, $\mathbf{F}$.

Proof

Consider a body on which two point loads, $\mathbf{F}_1$ and $\mathbf{F}_2$, are applied at points P_1 and P_2, respectively (Fig. 7.3).

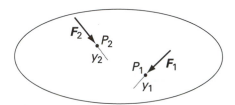

FIGURE 7.3   Forces, F_i, and the displacements, y_i, of their points of application.

When an external load, $\mathbf{F}$, whose magnitude increases from 0 to F_f, is applied to the body at point P, the point P undergoes a displacement, y, in the direction of $\mathbf{F}$. The point of application may also displace in other directions. Since the material is linear elastic, the magnitude of y and the magnitude of the force are proportional, $F(y) = cy$, for some constant, c. The work done by $\mathbf{F}$, as the magnitude increases from 0 to F_f, is then $W_e = \int_0^{y_f} F(y)dy = \int_0^{y_f} cydy = cy_f^2/2 = F_f y_f/2$, where y_f is the final displacement.

Now assume Clapeyron's theorem that $W_e = U$. Since the material is linearly elastic, $U = \Psi$, so that $W_e = \Psi$.

Assume that $\mathbf{F}_1$ increases from 0 to F_1 with a final displacement of y_1 in the direction of $\mathbf{F}_1$ and that $\mathbf{F}_2$ increases from 0 to F_2 with a final displacement of y_2 in the direction of $\mathbf{F}_2$. Then

$$\Psi = W_e = \frac{1}{2}F_1 y_1 + \frac{1}{2}F_2 y_2. \tag{7.12}$$

To finish the proof, an expression for $\Delta\Psi/\Delta F_1$ must be found in terms of y_1 and y_2. If the magnitude of $\mathbf{F}_1$ is changed by ΔF_1, then the displacements of P_1 and P_2 in the direction of $\mathbf{F}_1$ and $\mathbf{F}_2$ will change by Δy_1 and Δy_2, respectively. Changing any force on a body, in general, causes all points of the body to displace. The additional work done by increasing the force, $\mathbf{F}_1$, by an increment which changes from 0 to ΔF_1 is

$$\Delta W_e = \Delta\Psi = \frac{1}{2}\Delta F_1 \Delta y_1 + F_1 \Delta y_1 + F_2 \Delta y_2, \tag{7.13}$$

since the original force magnitudes also move though Δy_1 and Δy_2 respectively. Therefore the total work done by first applying the forces F_1 and F_2 and then increasing the magnitude of $\mathbf{F}_1$ by ΔF_1, is by adding equations (7.12) and (7.13),

$$\Psi + \Delta\Psi = \frac{1}{2}F_1 y_1 + \frac{1}{2}F_2 y_2 + \frac{1}{2}\Delta F_1 \Delta y_1 + F_1 \Delta y_1 + F_2 \Delta y_2, \qquad (7.14)$$

If the forces are applied in the opposite order, the work done must be the same as that given by equation (7.14) since the body is linear elastic. First apply the small increment ΔF_1 through Δy_1 and then apply F_1 and F_2 through the distances y_1 and y_2 as before. The work done by first applying ΔF_1 is $\Delta F_1 \Delta y_1/2$. The work done by applying F_1 and F_2 is $\Delta F_1 y_1 + F_1 y_1/2 + F_2 y_2/2$, since the already existing fixed load ΔF_1 moves through y_1. Therefore, in this order of application,

$$\Psi + \Delta\Psi = \frac{1}{2}\Delta F_1 \Delta y_1 + \Delta F_1 y_1 + \frac{1}{2}F_1 y_1 + \frac{1}{2}F_2 y_2. \qquad (7.15)$$

Setting equations (7.14) and (7.15) equal and simplifying yields

$$F_1 \Delta y_1 + F_2 \Delta y_2 = \Delta F_1 y_1. \qquad (7.16)$$

To obtain $\Delta\Psi/\Delta F_1$, divide equation (7.13) by ΔF_1;

$$\frac{\Delta\Psi}{\Delta F_1} = \frac{1}{2}\Delta y_1 + \frac{F_1 \Delta y_1 + F_2 \Delta y_2}{\Delta F_1}. \qquad (7.17)$$

Equation (7.16) is used to replace the second term on the right in equation (7.17) so that

$$\frac{\Delta\Psi}{\Delta F_1} = \frac{1}{2}\Delta y_1 + y_1. \qquad (7.18)$$

Take the limit as ΔF_1 tends to zero. Since Δy_1 also then tends to zero,

$$\frac{d\Psi}{dF_1} = y_1.$$

Therefore, if the complementary energy, written in terms of the external forces, is differentiated with respect to the magnitude of an external load, $\mathbf{F}_1$, one obtains the displacement, y_1, of the point of application of $\mathbf{F}_1$ in the direction of $\mathbf{F}_1$. ■

This theorem is also true if the load is a couple, M. In this case, the angular displacement caused by the couple, θ, is given by $\partial\Psi/\partial M = \theta$.

EXAMPLE 7.4   For a simply supported beam with a transverse point load at its center (Fig. 7.4), determine the vertical deflection under the point load by Castigliano's theorem. Compare the result to that obtained from strength of materials.

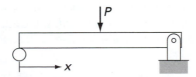

FIGURE 7.4   Simply supported beam with load, P, at center.

Solution Statics implies that

$$M(x) = \frac{Px}{2} \quad \text{if} \quad 0 \le x \le \frac{L}{2}$$

$$= \frac{Px}{2} - P(x - L) = \frac{P}{2}(L - x) \quad \text{if} \quad \frac{L}{2} \le x \le L.$$

$$\Psi = \int_0^L \frac{M^2}{2EI} dx = \frac{1}{2EI} \int_0^{L/2} (Px)^2 dx + \frac{1}{2EI} \int_{L/2}^L \frac{P^2(L - x)^2}{4} dx = \frac{P^2 L^3}{96EI}.$$

Then $\partial \Psi / \partial P = PL^3 / 48EI$ which agrees with strength of materials. ∎

EXAMPLE 7.5 Castigliano's theorem is well suited for more complex strength of materials problems, such as those involving curved beams. If the curved beam is slender enough, then as shown in Chapter 4, a strength of materials solution is obtained using the stress relations for a straight beam. Find the horizontal deflection, δ_x, of the free end of the semicircular slender rod with radius, R, shown in Figure 7.5. The load, P, applied to the free end is horizontal.

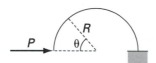

FIGURE 7.5 A curved beam with a transverse point load on the free end.

Solution Neglect any shear stress effects and write the moment at each point in the rod in terms of the angle, θ, and the load P. A free-body diagram of the curved beam section from the free end to the face defined by θ shows that the internal moment at position θ is $M = -PR\sin\theta$. The internal axial load is $N = -P\sin\theta$, and the internal shear load is $V = -P\cos\theta$. Let s be the arclength. Then $ds = Rd\theta$. Therefore, neglecting shear,

$$\Psi = \int_0^L \frac{M^2}{2EI} ds + \int_0^L \frac{N^2}{2EA} = \frac{1}{2EI} \int_0^\pi (PR\sin\theta)^2 Rd\theta + \frac{1}{2EA} \int_0^\pi (P\sin\theta)^2 Rd\theta.$$

In Example 7.4, the integral for Ψ was first solved and then the resulting Ψ was differentiated to get the deflection desired. Often it is easier to perform these operations in the opposite order, first differentiating the integral and then integrating to obtain the desired displacement. This uses the fact that $\partial \Psi / \partial P = (\partial / \partial P) \int_0^L f(P) dx = \int_0^L [\partial f(P) / \partial P] dx$. In this case,

$$\delta_x = \frac{\partial \Psi}{\partial P} = \frac{1}{EI} \int_0^L M \frac{\partial M}{\partial P} ds + \frac{1}{EA} \int_0^L N \frac{\partial N}{\partial P} ds$$

$$= \frac{1}{EI} \int_0^\pi (PR\sin\theta) R\sin\theta \, Rd\theta + \frac{1}{EA} \int_0^\pi PR\sin^2\theta \, d\theta = \frac{\pi PR^3}{2EI} + \frac{\pi PR}{2EA}. \quad ∎$$

Exercise 7.1 Redo the previous example with a rectangular cross section of height h and width w accounting for internal shear. Recall that $f_s = 6/5$ for a rectangular cross section.

Often the contributions from the axial, twist, and shear loads are smaller than that from the internal moment and may be neglected. If the curved beam is not slender, then axial loading, twist, and internal shear must also be accounted for. From equations (7.8)–(7.11),

$$\Psi = \int_0^L \frac{N^2}{2AE} dx + \int_0^L \frac{M_t^2}{2JG} dx + \int_0^L \frac{M^2}{2IE} dx + f_s \int_0^L \frac{V^2}{2AG} dx.$$

EXAMPLE 7.6   This example illustrates a common error in applying Castigliano's theorem. The problem will first be solved incorrectly, making the common error, and then correctly. Consider a simply supported beam with transverse point loads, both of magnitude, P, at $x = L/3$ and at $x = 2L/3$ (Fig. 7.6). The reactions at the supports are then both vertical and of magnitude, P. Determine the deflection at $x = L/3$, under the strength of materials assumptions and ignoring shear.

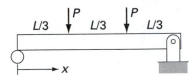

FIGURE 7.6   Simply supported beam with evenly spaced transverse equal point loads.

Solution   From free-body diagrams, the internal moment is

$$M(x) = Px \quad \text{if} \quad 0 \le x \le \frac{L}{3}$$

$$= Px - P\left(x - \frac{L}{3}\right) = \frac{PL}{3} \quad \text{if} \quad \frac{L}{3} \le x \le \frac{2L}{3}$$

$$= Px - P\left(x - \frac{L}{3}\right) - P\left(x - \frac{2L}{3}\right) = P(L - x) \quad \text{if} \quad \frac{2L}{3} \le x \le L.$$

The total complementary energy is

$$\Psi = \int_0^L \frac{M^2}{2EI}dx = -\frac{35P^2L^3}{27EI}.$$

Then apparently

$$y\left(\frac{L}{3}\right) = \frac{d\Psi}{dP} = -\frac{70PL^3}{27EI}.$$

But this derivative does not distinguish between the various loads of magnitude, P. Castigliano's theorem requires that the derivative be taken with respect to just the load P at $x = L/3$. Therefore this result is wrong.

The only way to avoid this problem is to give each of the loads with magnitude, P, a different name, say P_1, P_2, and label the reactions of magnitude P as R_1, and R_2. Then after finding Ψ, take $d\Psi/dP_1$. Finally set all loads equal to P. A shorter method is to take the derivative first as in the previous example. From free-body diagrams, the internal moment is

$$M(x) = R_1 x \quad \text{if} \quad 0 \le x \le \frac{L}{3}$$

$$= R_1 x - P_1\left(x - \frac{L}{3}\right) \quad \text{if} \quad \frac{L}{3} \le x \le \frac{2L}{3}$$

$$= R_1 x - P_1\left(x - \frac{L}{3}\right) - P_2\left(x - \frac{2L}{3}\right) \quad \text{if} \quad \frac{2L}{3} \le x \le L.$$

Then Castigliano's theorem says

$$y\left(\frac{L}{3}\right) = \frac{d\Psi}{dP_1} = \frac{1}{EI}\int_0^L M\frac{\partial M}{\partial P_1}dx$$

$$= \frac{1}{EI}\int_{L/3}^{2L/3}\left[R_1 x - P_1\left(x - \frac{L}{3}\right)\right]\left[-\left(x - \frac{L}{3}\right)\right]dx$$

$$+ \frac{1}{EI}\int_{L/3}^{L}\left[R_1 x - P_1\left(x - \frac{L}{3}\right) - P_2\left(x - \frac{2L}{3}\right)\right]\left[-\left(x - \frac{L}{3}\right)\right]dx$$

$$= \frac{1}{EI}\left(-\frac{5}{81}R_1 L^3 + \frac{1}{81}P_1 L^3\right) + \frac{1}{EI}\left(-\frac{23}{162}R_1 L^3 + \frac{7}{81}P_1 L^3 + \frac{5}{162}P_2 L^3\right).$$

Then setting $R_1 = P_1 = P_2 = P$, the correct displacement is

$$y\left(\frac{L}{3}\right) = -\frac{7PL^3}{162EI}.$$ ∎

EXAMPLE 7.7   A cantilever beam of length, L, has a couple, C, applied at its free end. Determine the slope of the rongitudinal neutral axis at the free end. Assume that the beam satisfies the assumptions of strength of materials.

Solution   Note that the couple cannot be moved without changing the internal response of the system because the beam is deformable, not rigid as was assumed in statics, when moving forces or couples. Castigliano's theorem says that if θ is the angle the couple moves through, then $\theta = \partial\Psi/\partial C$. This angle is approximately equal to the slope of the neutral axis. As the vertical face rotates, it remains perpendicular to the neutral axis since plane sections are assumed to remain planar. The neutral axis is forced to rotate through θ at the free end. An elastic deformation is assumed to be small so that $\tan\theta = \theta$. But $\tan\theta$ is just the slope of the neutral axis. For small displacements, then, the slope EIy' from strength of materials can be found as $\tan\theta = \partial\Psi/\partial C$. In this example, the internal moment is $M(x) = -C$, and so the complementary energy is $\Psi = \int_0^L (C^2/2EI)dx$. Therefore, $\tan\theta = \partial\Psi/\partial C = CL/EI$, which is unitless as expected. ∎

7.2.1   Method of Fictitious Loads

Castigliano's theorem apparently only gives the deflections at the point of application of point loads. Frequently an engineer needs to predict a deflection at a different point in the body. The trick is to temporarily apply a fictitious load, P, at the point and in the direction the deflection is needed. Once the deflection is determined in terms of P and the other loads on the body, then P is set equal to zero. The result is the desired deflection.

EXAMPLE 7.8   A cantilever beam of length, L, has a couple, C, applied at its free end (Fig. 7.7). Make the strength of materials assumptions on the stresses, neglect shear, and find the vertical deflection at the point $x = 3L/4$.

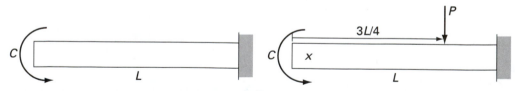

FIGURE 7.7   Cantilever beam with the fictitious load, P.

Solution   There is no load at the point $x = 3L/4$ and so put a vertical fictitious load, P, there. The internal moment is found from free-body diagrams to be

$$M(x) = -C \quad \text{if} \quad 0 \le x \le \frac{3L}{4}$$

$$= -C - P\left(x - \frac{3L}{4}\right) \quad \text{if} \quad \frac{3L}{4} \le x \le L.$$

The total complementary energy is

$$\Psi = \int_0^L \frac{M^2}{2EI} dx = \frac{1}{2EI} \int_0^{3L/4} (-C)^2 dx + \frac{1}{2EI} \int_{3L/4}^L \left[-C - P\left(x - \frac{3L}{4}\right)\right]^2 dx$$

$$= \frac{3C^2 L}{8EI} + \frac{C^2(L - 3L/4)}{2EI} + \frac{CP(L/4)^2}{2EI} + \frac{P^3(L/4)^3}{6EI}.$$

Then

$$\frac{d\Psi}{dP} = \frac{C(L/4)^2}{2EI} + \frac{P^2(L/4)^3}{2EI}.$$

Setting $P = 0$ produces $y(3L/4) = CL^2/32EI$. Alternatively,

$$y\left(\frac{3L}{3}\right) = \frac{d\Psi}{dP} = \frac{1}{EI} \int_0^L M \frac{\partial M}{\partial P} dx$$

$$= 0 + \frac{1}{EI} \int_{3L/4}^L \left[-C - P\left(x - \frac{3L}{4}\right)\right]\left[-\left(x - \frac{3L}{4}\right)\right] dx$$

$$= \frac{1}{EI} \left\{ \frac{1}{2}\left[C\left(x - \frac{3L}{4}\right)\right]^2 + \frac{P}{3}\left(x - \frac{3L}{4}\right)^3 \right\} \Big|_{3/4}^L = \frac{CL^2}{32EI} \quad \text{if} \quad P = 0. \quad ■$$

EXAMPLE 7.9   A simply supported beam of length L is loaded by a constant distributed load, p.

(a) Use Castigliano's theorem to determine the deflection at the middle of the beam.

(b) Compare this result to that obtained in elementary strength of materials.

Solution   (a) Castigliano's theorem requires differentiating the complementary energy by a point load. Differentiating with respect to a distributed load is invalid. Therefore a fictitious force, Q, must be added in the direction and at the point where the deflection is desired. To compute the complementary energy, the internal moment, $M(x)$, at each position x in the beam must be known. Here x is measured from the left end of the beam. The reactions at each support are $pL/2 + Q/2$. A free-body diagram shows that the internal moment is

$$M(x) = \left(\frac{Q}{2} + \frac{pL}{2}\right)x - \frac{px^2}{2} \quad \text{if } 0 \le x \le L/2$$

$$= \left(\frac{Q}{2} + \frac{pL}{2}\right)x - \frac{px^2}{2} - Q(x - L/2) \quad \text{if } L/2 \le x \le L.$$

Neglect shear. Then the deflection in the downward direction of the load Q is

$$y(L/2) = \int_0^L \frac{M}{EI} \frac{\partial M}{\partial Q} dx$$

after letting $Q = 0$.

$$y(L/2) = \frac{1}{EI} \int_0^{L/2} \left(\frac{pL}{2}x - \frac{px^2}{2}\right)\frac{x}{2} dx + \frac{1}{EI} \int_{L/2}^L \left(\frac{pL}{2}x - \frac{px^2}{2}\right)\left(\frac{x}{2} - x + \frac{L}{2}\right) dx$$

$$= 0.01302 \frac{pL^4}{EI}.$$

(b) By elementary strength of materials, $EIy''(x) = M(x)$, where $M(x) = pLx/2 - px^2/2$. Integrating twice and using the two boundary conditions $y(0) = y(L) = 0$ produces

$$y(x) = \frac{1}{EI} \left[\frac{pL}{12}x^3 - \frac{p}{24}x^4 - \frac{pL^3}{24}x \right].$$

Substituting the value $x = L/2$ recovers the same answer as in part (a):

$$y(L/2) = 0.01302pL^4/EI. \qquad \blacksquare$$

7.2.2 Statically Indeterminate Problems

Castigliano's theorem is frequently more efficient than traditional strength of materials procedures for determining deflections when the loads on the body are statically indeterminate. The forces on a two-dimensional statically indeterminate problem can be found by using Castigliano's theorem to determine all but three forces. Statics is then used to determine the remaining forces.

EXAMPLE 7.10

A cantilever beam of length, L, with a clockwise "point couple," C, applied at $x = L/2$ is supported by a roller at its free end (Fig. 7.8). There are four unknown forces: the three wall reactions and the vertical force, R, exerted by the roller support. The force, R, is determined by the fact that at the roller $y(L) = 0$ and at the wall $y(0) = 0$. Making the strength of materials assumptions and neglecting shear, determine R.

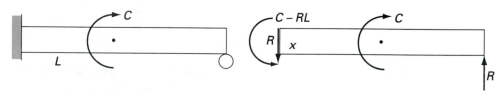

FIGURE 7.8 Free-body diagram of the cantilever beam loaded by C and supported by a roller at its free end.

Solution

The internal moment is from the free end (this R is the vertical wall reaction, which is equal to the roller support reaction),

$$M(x) = R(L - x) - C \quad \text{if} \quad 0 \le x \le \frac{L}{2}$$

$$= R(L - x) \quad \text{if} \quad \frac{L}{2} \le x \le L.$$

The total complementary energy is

$$\Psi = \int_0^L \frac{M^2}{2EI}dx = \frac{1}{2EI} \int_0^{L/2} [R(L-x) - C]^2 dx + \frac{1}{2EI} \int_{L/2}^L [R(L-x)]^2 dx$$

$$= \frac{1}{6EI} \left[\frac{7R^2L^3}{8} - \frac{9RL^2C}{4} + 3LC^3 + \frac{R^2L^3}{8} \right].$$

At the support, Castigliano's theorem implies that

$$y(L) = \frac{d\Psi}{dR} = \frac{1}{6EI} \left[\frac{7RL^3}{4} - \frac{9L^2C}{4} + \frac{RL^3}{4} \right] = 0.$$

Solving this linear equation in R produces, $R = 9C/8L$.

Alternatively, the deflection can be found by differentiating and then integrating to obtain the same expression from

$$y(L) = \frac{d\Psi}{dR} = \frac{1}{EI}\int_0^L M\frac{\partial M}{\partial R}dx = 0. \qquad \blacksquare$$

Often engineering structures, such as chains, involve a ring supporting equal and opposite loads (Fig. 7.9a). The stress and deflection analysis of such a ring is complex because the problem of finding the internal moments is statically indeterminate. The Castigliano theorem is used to determine the internal moments.

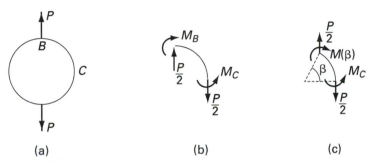

FIGURE 7.9 (a) The ring loaded by forces P. (b) The quarter section of the ring. (c) The free-body diagram of the section from face C to face β.

First neglect the distortion at the points of application of the point loads. By symmetry, only one quarter of the ring needs to be analyzed (Fig. 7.9b). The faces between B and C are labeled by the angle β. Think of face B as a rigid wall at which the moment M_B and the shear force $P/2$ act. The shear on face C must be zero because if it were not, by symmetry, the reaction shear on both sides of the face C would be in the same direction, a contradiction to the law of action and reaction. Therefore on face C there is a moment M_C and a tensile normal force $P/2$ (Fig. 7.9b). The sum of the moments on the quarter section implies that $-M_B = PR/2 - M_C$ so that the problem is statically indeterminate because both M_B and M_C are unknown.

The section C does not rotate; the Castigliano theorem then implies that the derivative of the complementary energy, $\partial\Psi/\partial M_C = 0$. At face β (Fig. 7.9c), the internal normal force is $N(\beta) = 0.5P\cos\beta$, the internal shear force is $V(\beta) = 0.5P\sin\beta$, and the internal moment is $-M(\beta) = 0.5PR(1-\cos\beta) - M_C$. Neglect the contribution of the normal and shear forces to the complementary energy so that $\Psi = \int_0^{\pi/2}(1/2EI)M(\beta)^2 Rd\beta$. Then the rotation at face C is

$$0 = \int_0^{\pi/2}\frac{M(\beta)}{EI}\frac{\partial M(\beta)}{\partial M_C}Rd\beta = -M_C + \frac{PR}{2}\left(1 - \frac{2}{\pi}\right).$$

Therefore, $-M(\beta) = 0.5PR(2/\pi - \cos\beta)$. The maximum internal moment occurs at face B where $\beta = \pi/2$ and is $-M_B = PR/\pi$. Note that, as expected, the moment at B tries to decrease the radius of curvature of the ring, while that at C tends to increase the radius of curvature of the ring. The moment is zero at $\beta = 50.46$ degrees.

EXAMPLE 7.11 Gymnasts' rings are suspended by a cable from the ceiling. The ring is required to support the dynamic load from a 150-pound gymnast with a safety factor of 4. The ring is made of a normalized 1030 steel ($E = 29 \times 10^6$ psi and $\sigma_Y = 49$ ksi). The ring is encased in a rubberlike material whose influence can be neglected. The ring has a diameter of 6 inches measured between the centroids of opposite cross sections. The cross section is to be circular of radius r.

(a) Design the dimension r so that the ring has maximum tensile stress of 4,000 psi under the force from the gymnast. Use the Winkler-Bach theory.

(b) Determine the deflection, δ_B, at the vertical cross section where the load is applied.

Solution   **(a)** Because of the safety factor, the design load is $P = 600$ pounds. In the Winkler-Bach sign convention the moment at B, in Figure 7.9a, is $M = PR/\pi$ because it decreases the radius of curvature. The largest tensile stress is at the outside of the cross section ($y = r$). By the Winkler-Bach theory after neglecting the stress due to the shear force on the face,

$$\sigma = \frac{PR}{\pi r^2 R\pi}\left[1 + \frac{r}{Z(R+r)}\right],$$

where $Z = -1 + 2(R/r)^2 - 2(R/r)\sqrt{(R/r)^2 - 1}$. Substitute the given values, set $\sigma = 4,000$, and solve numerically to obtain $r \geq 0.543$ inches.

(b) The complementary energy in the full ring is four times that in the quarter ring. By the Castigliano theorem, the deflection is

$$\delta_B = 4\int_0^{\pi/2} \frac{M(\beta)}{EI}\frac{\partial M(\beta)}{\partial P} R\,d\beta = 0.15\frac{PR^3}{EI} = 0.15\frac{PR^3}{E\pi r^4/4} = 0.001227 \text{ inches.} \quad \blacksquare$$

Castigliano's theorem works very well if the deflections at just a few points are required or some forces are unknown. The theorem has been generalized by Reissner to obtain the equilibrium stress, strain, and deflection at all points in a body from an integral involving the complementary energy. In this case, it is called the principle of complementary energy.

7.2.3   Principle of Complementary Energy for Nonlinear Elastic Materials

Castigliano proved his result only for linearly elastic materials and did so in terms of the strain energy. The generalized result, proved by Engesser, uses the complementary energy and allows for nonlinearly elastic materials. The complementary energy is not equal to the strain energy if the material is not linear elastic.

EXAMPLE 7.12   A three-member truss made of uniform straight members forms an equilateral triangular shape (Fig. 7.10). The horizontal member forming the base is supported by pins, B and D, at either end. A vertical downward point load, P, is applied to the pin, C, at the top. The truss members are made of a material with stress-strain relation $\epsilon = k\sigma^{1/3}$. Determine the vertical deflection of pin C at the top of the truss.

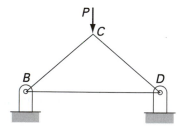

FIGURE 7.10   Three-member truss supporting a point load.

Solution   First determine the complementary energy in an axially loaded member with length L and cross-sectional area A, subject to the internal force F. The engineering strain is $\epsilon = e/L$

and the stress is F/A. The complementary energy per unit volume in terms of F is $\Psi_o = \int \epsilon d\sigma = 0.75k\sigma^{4/3} = 0.75k(F/A)^{4/3}$. So, the total complementary energy is $\Psi = AL\Psi_o = 0.75kLF^{4/3}/A^{1/3}$.

The free-body diagrams of pins C and B show that the internal force in the horizontal member BD is $P/2\sqrt{3}$ and in the other two members is $P/\sqrt{3}$. The total complementary energy of the truss is the sum of the complementary energies of the members. Therefore, for the truss, $\Psi = kL[1.5(P/\sqrt{3})^{4/3} + 0.75(P/2\sqrt{3})^{4/3}]/A^{1/3}$ and the vertical deflection at pin C is, by the generalized Castigliano theorem,

$$\frac{d\Psi}{dP} = \frac{kL}{A^{1/3}}\left(2\frac{1}{(\sqrt{3})^{4/3}} + \frac{1}{(2\sqrt{3})^{4/3}}\right)P^{1/3}.$$

Notice that the deflection is not a linear function of the load. ∎

A more subtle technique is required to apply the generalized Castigliano theorem to a straight beam made of a nonlinear elastic material. Let M be a moment applied to the beam. Then the rotation is $\theta = d\Psi/dM$. Assume that shear effects in the beam are neglected. Then $\theta = L/\rho$, where L is the length of the beam and ρ is the radius of curvature. Because θ is constant over the beam length in pure bending, the complementary energy can be written

$$\Psi = \int \theta dM = L\int \frac{dM}{\rho} = \int_0^L \int \frac{dM}{\rho} dx.$$

If a point load P is applied to the beam, the chain rule implies that the deflection of the point of application of P in the direction of P is

$$\frac{d\Psi}{dP} = \frac{d\Psi}{dM}\frac{dM}{dP} = \int_0^L \frac{1}{\rho}\frac{dM}{dP}.$$

The problem then is to write $1/\rho$ in terms of P. A similar procedure applies to find the angle of rotation under an applied point couple.

EXAMPLE 7.13

Determine the slope at the free end of a uniform, linear elastic, cantilever beam of length, L, that carries a point load, P, at the free end. The cross-sectional area is rectangular with depth $2h$ and width b. The material is such that $\sigma = K|\epsilon|^{1/2}$ if $\epsilon \geq 0$ and $\sigma = -K|\epsilon|^{1/2}$ if $\epsilon \leq 0$.

Solution

Apply a fictitious couple, C, at the free end. If x is measured from the free end, then the internal moment is $M(x) = C - Px$ and $dM/dC = 1$.

To determine $1/\rho$, recall that the internal moment is also found by integrating the internal force moments over the cross-sectional area so that $M = 2\int_0^h \sigma y b dy$. Also, $\epsilon = y/\rho$, where y is measured from the centroid of the cross-sectional area.

$$M = 2\int_0^h K\left(\frac{y}{\rho}\right)^{1/2} ybdy = \frac{4bKh^{5/2}}{5\rho^{1/2}}.$$

Therefore $1/\rho = 25M^2/16b^2K^2h^5$. Now letting $C = 0$,

$$\theta = \int_0^L \frac{1}{\rho}\frac{dM}{dC}dx = \frac{25}{(16b^2K^2h^5)}\int_0^L P^2x^2(1)dx = \frac{25P^2L^3}{48b^2K^2h^5}.$$

Again, the angle of rotation is not a linear function of P. ∎

7.3 PRINCIPLE OF VIRTUAL WORK

The principle of virtual work is a criterion for finding equilibrium states. The assumption of this principle has been tested by verifying that it makes good predictions about the behavior of mechanical systems. The idea of virtual work can be defined by thinking of the whole deformed body rather than the displacement of a single point as in Castigliano's theorem.

The configuration or shape of a body is described by indicating the position of each point in the body. The position is defined in a system of generalized coordinates. These may or may not be Cartesian coordinates. For example, a point mass moving along a line can be located by giving the Cartesian coordinate, x, of its position on the line. A point mass moving in a plane can be located by giving the Cartesian coordinates, (x, y), of its position in the plane. A point mass moving in three-space can be located by specifying the Cartesian coordinates, (x, y, z), of its position in space. These point mass systems, of dimensions one, two, and three, respectively, are described by their Cartesian coordinates. A system moving in a plane, for example, may not require two coordinates to specify its position because of constraints on the system. A pendulum is a point mass attached to a massless string that swings about a fixed pivot point. If the pendulum is constrained to move in a plane, the location of the system could be described by the Cartesian coordinates, (x, y), of the position of the point mass in the plane. However, to keep computations as simple as possible, the engineering analyst wants to use the minimum number of coordinates possible to describe the system. When the string cannot change length, the position of the point mass can be described by one coordinate, the angle, θ, between the string and a vertical reference line.

A system of generalized coordinates for a system is a set containing the minimum number of coordinates required to completely describe the configuration of the system. This need not be a Cartesian, polar, or other standard coordinate system. The pendulum on an inextensible string moving in a plane requires a single generalized coordinate, θ. It is a one-dimensional system. A pendulum moving in a fixed plane for which the string is extensible requires two generalized coordinates, θ and the length of the string. It is a two-dimensional system. The number of generalized coordinates required to describe a system is called the number of degrees of freedom.

A rigid disk contains an infinite number of points. Its configuration, when rolling on a plane, could be described by the giving the coordinates in a Cartesian coordinate system for each of the points. This strategy would require listing an infinite number of coordinates, three for each point. However, because the disk is rigid, the position can be described with a finite number of coordinates. If the disk rolls without slip and in a straight line, its position can be described by the angle, θ, between a radial line drawn on the face of the disk and a line normal to the plane on which the disk rolls. A single generalized coordinate is all that is required. It is a one degree of freedom system. Two generalized coordinates are needed to describe the position when the disk rolls with slip in a straight line, the angle θ and the horizontal distance, x, through which the center of the disk has moved. The rigid disk rolling with slip is a two degree of freedom system.

A deformable body, in general, requires an infinite number of generalized coordinates to describe its configuration, the position of each of its points. A deformable body is an infinite degree of freedom system. For example, the deformed shape of a strength of materials beam is described by the deflection of its neutral axis. The displacement of the neutral axis in the deformed beam is located by a function, $y(x)$, where x is the position along the length of the beam. An infinite number of values $y(x)$ must be given to describe the configuration of the beam.

For this beam example, the function $y(x)$ (or its graph) is thought of as the equilibrium configuration of the beam. Now fix the forces on the beam. Imagine a perturbation of the beam shape away from its equilibrium configuration; this will be called a virtual displacement, denoted δy. It does not actually occur. No forces cause it; it is merely imagined that the shape of the beam is now defined by $y(x) + \delta y(x)$ (Fig. 7.11). But as the beam goes through the imagined virtual displacement, the fixed forces on the beam do imaginary work because they would move through a distance during the virtual displacement. This will be called the virtual work, δW, done during the virtual displacement. In this case, the total work done δW is a scalar, but the virtual displacement δy is a function of x. There is one further restriction. The only virtual displacements allowed are those which preserve the constraints on the body. For a simply supported beam, the virtual displacements must be zero at the supports, so that $y(x) + \delta y(x)$ satisfies the boundary conditions. The goal is to develop a criterion that will pick out the equilibrium configuration $y(x)$ from all possible configurations of the body defined by the infinite number of possible perturbations δy.

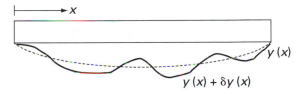

FIGURE 7.11   Schematic showing the equilibrium configuration $y(x)$ of the beam and one possible virtual displacement perturbation, $\delta y(x)$, from equilibrium.

The criterion can be motivated by recalling from Section 7.1 that the work, W, done on a system, due to both internal and external forces, equals the change in kinetic energy. This fact, again, is equivalent to Newton's law that the total force on a particle equals the mass times the acceleration of the particle and from the definition of kinetic energy as $mv^2/2$, where m is mass and v is the speed of the particle. If a body moves from equilibrium, then the work done is the change in kinetic energy, $W = mv^2/2 - 0 > 0$. Therefore, it always takes positive work to move a body from an equilibrium state. Imagine all possible motions of the system from equilibrium, the virtual displacements. If virtual work, δW, done under any and all virtual displacements is negative, then the system cannot spontaneously move from its position. Therefore the mathematician and engineer Fourier (1768–1830) proposed that a body remains in equilibrium if the virtual work done by any virtual displacement is less than or equal to zero. The statement that $\delta W \leq 0$ is often called Fourier's inequality.

For example, consider a box lying on a table. The two actual forces on it are the weight and the reaction force from the table. Of these, only the weight does work since the reaction force cannot act through a distance. Fourier's inequality is satisfied for any motion: lifting, tipping, sliding of the box, that can be imagined subject to constraint provided by the table. Therefore the box is in equilibrium on the table top.

Fourier's inequality is only a sufficient condition for equilibrium. For example, a ball balanced on the top of a hill is in equilibrium, but its weight does positive work under any virtual displacement consistent with the constraint that the ball must remain in contact with the hill. So Fourier's inequality does not predict equilibrium even though the ball is in equilibrium. An alternative condition which is both necessary and sufficient is given by the principle of virtual work. This statement is a variational principle.

Principle of Virtual Work   A body is in equilibrium if and only if the total work, δW, done on the body by all actual forces during any virtual displacement is zero.

We can easily show that this condition is equivalent to the force equilibrium condition determined in statics. Let a particle be subjected to forces $\mathbf{F}_1, \mathbf{F}_2, \ldots, \mathbf{F}_n$. For a virtual displacement $\delta \mathbf{s}$, the total virtual work is $\delta W = \sum_{i=1}^{n} (\mathbf{F}_i \cdot \delta \mathbf{s}) = (\sum_{i=1}^{n} \mathbf{F}_i) \cdot \delta \mathbf{s}$. So the virtual work is zero if the sum of the forces is zero. Conversely, $\delta W = 0$ implies that either $\delta \mathbf{s}$ is perpendicular to $\sum_{i=1}^{n} \mathbf{F}_i$ or $\sum_{i=1}^{n} \mathbf{F}_i = 0$. But the result must hold for all $\delta \mathbf{s}$, not just those perpendicular to the sum of forces, so that it must be true that $\sum_{i=1}^{n} \mathbf{F}_i = 0$. Therefore $\delta W = 0$ iff $\sum_{i=1}^{n} \mathbf{F}_i = 0$. The principle of virtual work has the same status as Newton's Laws. It is an assumption whose validity is verified by the accuracy of the predictions it makes about the behavior of mechanical systems.

The first variation of a real valued function, $F(x, y, z)$, is defined in analogy with the linear terms of a Taylor series expansion of F.

Definition   The first variation of F is

$$\delta F = \frac{\partial F}{\partial x} \delta x + \frac{\partial F}{\partial y} \delta y + \frac{\partial F}{\partial z} \delta z,$$

where δx, δy, and δz are variations of the coordinates.

A condition for the virtual work to be zero can be computed easily if the work is a function of a finite number, n, of degrees of freedom, $x_1, \ldots, x_n$. Then the variation of the work is defined to be

$$\delta W = \frac{\partial W}{\partial x_1} \delta x_1 + \ldots + \frac{\partial W}{\partial x_n} \delta x_n. \tag{7.19}$$

Since the coordinate variations, δx_i, are arbitrary and independent, the virtual work can only be zero, $\delta W = 0$, if

$$\frac{\partial W}{\partial x_i} = 0 \quad \text{for} \quad i = 1, 2, \ldots, n. \tag{7.20}$$

These n, possibly nonlinear, equations must be solved simultaneously to identify the generalized coordinates of the equilibrium state. The problem of determining the equilibrium configuration has been reduced to a simple calculus and algebra problem.

7.4   THE PRINCIPLE OF STATIONARY POTENTIAL ENERGY

A system is conservative if the work done on the system only depends on the endpoints of the displacement path. Mathematically this means that the work by a force done over the path can be represented by an exact integral. Potential energy is only defined for conservative systems, those mechanical systems on which all forces, both internal and external, are conservative. Conservative forces include spring forces and point loads which do not change direction as the system is deformed.

Definition   Suppose that the work, denoted by $W(X_1, X_2)$, done by the forces during the displacement of a system from configuration X_1 to X_2 is conservative. Then the potential energy, $\Pi(X_2)$, at configuration, X_2, with respect to configuration, X_1, is $\Pi(X_2) = -W(X_1, X_2)$.

This defines the potential energy as the negative of the work if the work is conservative.

EXAMPLE 7.14   The work done by the weight of a point mass of mass m as it is lifted through a height h is $-mgh$. Therefore the potential energy of the mass subjected only to its weight is mgh at height h.   ∎

By the principle of virtual work, the system attains an equilibrium state at those $x_1, \ldots, x_n$, where the virtual work is zero. This occurs if each $\partial W/\partial x_i = 0$ for $i = 1, 2, \ldots, n$. But since $W = -\Pi$, we can state the principle of stationary potential energy.

Principle of Stationary Potential Energy   A conservative system described by coordinates, $x_1, \ldots, x_n$, and having a potential energy $\Pi(x_1, \ldots, x_n)$ reaches an equilibrium state at those points for which $\partial \Pi/\partial x_i = 0$ for $i = 1, 2, \ldots, n$.

This principle gives n equations in n unknowns to determine the coordinates of the equilibrium states. Since the equations $\partial \Pi/\partial x_i = 0$ for $i = 1, 2, \ldots, n$ may be nonlinear, there can be more than one equilibrium state.

Castigliano's first theorem can be obtained from the principle of stationary potential energy. Suppose there is an unknown force, P, acting on a body. As it increases from zero up to P, it does the work $\int P dy$, where y is the displacement of the point of application of P in the direction of P. If the system is conservative, the potential energy is, for strain energy U,

$$\Pi = U - \int P dy.$$

Think of y as the single degree of freedom; equilibrium is obtained if $\partial \Pi/\partial y = 0$ or

$$\frac{\partial U}{\partial y} = P.$$

This equation is dual to that for Castigliano's second theorem in which $\partial \Psi/\partial P = y$. The strain energy written in terms of the displacements takes the role of the complementary energy written in terms of the loads. The roles of force and displacement are interchanged.

EXAMPLE 7.15   A rigid bar of length 6 inches is welded to the rigid wheel of radius $R = 3$ inches (Fig. 7.12). The linear spring with spring constant 100 lb/in. is initially unstretched when the bar is in the vertical position ($\theta = 0$). A 200-pound weight is hung on the free end of the rod. Use an energy method to determine the equilibrium configuration of the system.

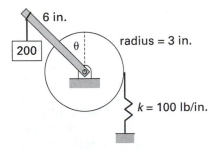

FIGURE 7.12   Wheel supported by a frictionless pin at the center and loaded by a weight and a spring.

Solution   The position (configuration) of the system is completely defined by the angle, θ. The system has a single degree of freedom. All forces on the system are conservative; the problem can be solved using the principle of stationary potential energy. Since the body is rigid, the internal energy is zero. The internal forces do no work. The potential energy of the full system is given by the sum of the potential energies of the two external forces, the 200-pound weight and the spring force.

$$\Pi(\theta) = \frac{1}{2}kR^2\theta^2 + (200)(6)\cos(\theta),$$

since the arclength, $R\theta$, cut off by the angle, θ, is the length the spring stretches. The value of θ which minimizes the potential energy is given by $d\Pi/d\theta = 0$, or

$$0 = kR^2\theta - (200)(6)\sin(\theta).$$

This equation is the same as obtained by setting the sum of the moments about the center of the wheel equal to zero. Therefore, $\sin(\theta) = 0.75\theta$. A numerical technique such as Newton's method can be used to determine that at equilibrium, $\theta = 0$ or $\theta = 1.2757 = 73.1°$. The second derivative determines the stability of these positions, $d^2\Pi/d\theta^2 = kR^2 - (200)(6)\cos(\theta)$. The equilibrium state when the bar is vertical ($\theta = 0$) is unstable because $d^2\Pi/d\theta^2 < 0$ there; the position $\theta = 73.1°$ is stable because $d^2\Pi/d\theta^2$ is positive there. The stability of an equilibrium cannot be determined from the vector methods learned in statics. The reason for this second derivative condition for stability is given in the following paragraphs.   ∎

An equilibrium configuration is stable if the system returns to that configuration after slight perturbations from the configuration. An example which uses a physical analog of the potential energy function helps make the connection between stability and the maximums and minimums of the potential energy clear. Suppose that a ball can move along the surface of a hill whose height from a reference line is given by $y(x) = a\sin(x) + b$, where x is the horizontal position. This is a conservative system since the only force which can do work as the ball moves spontaneously along the surface is the weight, mg, of the ball. It is a one-dimensional, or one degree of freedom, system since the behavior is completely described by one state variable, the horizontal position x. The potential energy is $\Pi(x) = mgy(x)$, which is the negative of the work done by the weight as the ball rises from the reference, or datum, line. The graph of $\Pi(x)$ has the same shape as the hilly surface (Fig. 7.13).

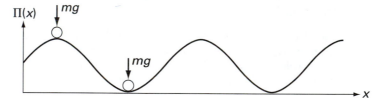

FIGURE 7.13   A ball resting on a sine-shaped hill.

The equilibria occur at those positions where the derivative of Π is zero,

$$\frac{d\Pi}{dx} = mg[a\cos(x)] = 0 \qquad \text{if} \qquad x = \frac{n\pi}{2}, \quad n = 1, 3, 5, \ldots .$$

The ball is at a hilltop when $x = \pi/2$. A slight perturbation will cause the ball to roll down the hill; it will never spontaneously return to the hilltop. This equilibrium is therefore unstable. On the other hand, if the ball is perturbed from the equilibrium at $x = 3\pi/2$ in the valley, the ball will always roll back to the equilibrium position due to the action of its weight. This is a stable equilibrium. The direction of the concavity of the potential energy function at the equilibrium point determines the stability of the equilibrium.

Definition   An equilibrium state is said to be stable if the potential energy of the equilibrium is a minimum. The equilibrium state is said to be unstable if its potential energy is a maximum.

If the potential energy depends only on a single variable and if the second derivative is nonzero, the sign of its second derivative determines stability. The case in which the second derivative is zero is discussed in Chapter 13.

EXAMPLE 7.16   A homogeneous rigid bar of length $L = 20$ inches and weight $W = 50$ pounds is attached to blocks at either end (Fig. 7.14). The blocks move without friction in guides and are connected by an inextensible cable and a spring. The spring constant is $k = 12$ lb/in., the spring is unstretched when $\theta = 0$. Determine the equilibrium positions and their stability.

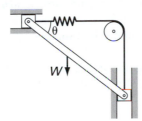

FIGURE 7.14   A rigid bar with ends constrained to move in guides.

Solution   The angle, θ, is the single degree of freedom. The stretch of the spring is $x = L \sin\theta + L \cos\theta - L$ and the weight drops a distance $0.5L \sin\theta$. Therefore the potential energy of this system is

$$\Pi(\theta) = -0.5WL \sin\theta + 0.5kL^2(\sin\theta + \cos\theta - 1)^2.$$

Then

$$\frac{d\Pi}{d\theta} = -0.5WL \cos\theta + kL^2(\sin\theta + \cos\theta - 1)(\cos\theta - \sin\theta)$$

$$= -0.5WL \cos\theta + kL^2(\cos 2\theta + \sin\theta - \cos\theta).$$

Setting the derivative of the potential energy equal to zero, substituting the values of the parameters, and solving numerically yields three equilibria when $0 \le \theta \le \pi/2$: 0.128116 rad $= 7.34°$; 0.636165 rad $= 36.45°$; and $\pi/2$ rad $= 90°$. Physically, the largest θ can be is $90°$.

The second derivative of the potential energy is needed to examine the stability of the equilibria.

$$\frac{d^2\Pi}{d\theta^2} = -0.5WL \sin\theta + kL^2(-2 \sin 2\theta + \cos\theta + \sin\theta).$$

Substitution of the values of the parameters and the equilibrium values shows that the equilibria at $7.34°$ and $90°$ are stable because the second derivative is positive, but that at $36.45°$ is unstable because the second derivative is negative.   ■

Exercise 7.2   Plot the graph of the potential energy in the previous example for $0 \le \theta \le 1.6$ radians.

7.5 INTERNALLY PRESSURIZED SPHERICAL MEMBRANES

The static equilibria of structures made of nonlinear elastic materials can often be most easily determined by energy methods.

A spherical membrane made of a nonlinear elastomer can model structures such as a balloon as well as an aneurysm in an artery of a living being. The configuration of the sphere is described by its radius at each point; it is a one degree of freedom system because the radial deformation is assumed to be the same at all points of the sphere. The relationship between the uniform internal pressure in a spherical membrane and the radius of the sphere at equilibrium is determined from the potential energy if a strain energy function for the membrane material is known.

Assume that the initial dimensions of the sphere are H and A, the undeformed thickness and radius of the membrane respectively. The state variable λ is the uniform stretch of the wall given by $\lambda = a/A$, where a is the current deformed radius, because a/A is the stretch of an equator of the sphere. The deformed thickness of the membrane is h. The elastomeric material is assumed to be incompressible. This constraint of no material volume change, $4\pi A^2 H = 4\pi a^2 h$, requires that $h = H/\lambda^2$. Only tensile stretches, $\lambda \geq 1$, are permitted because the membrane wrinkles in compression. The function $w(\lambda)$ is the strain energy per unit initial material volume of the sphere wall at the stretch λ.

The potential energy function for the spherical membrane when the internal pressure, P, is constant is the sum of the strain energy and the negative of the work done by the pressure,

$$\Pi(\lambda) = 4\pi A^2 H w(\lambda) - P\Delta V = 4\pi A^2 H w(\lambda) - P\left[\frac{4}{3}\pi a^3 - \frac{4}{3}\pi A^3\right]$$

$$= 4\pi A^2 H w(\lambda) - P\left[\frac{4}{3}\pi(A\lambda)^3 - \frac{4}{3}\pi A^3\right], \tag{7.21}$$

where ΔV is the change of volume within the sphere. This energy expression uses the fact the initial spherical surface area is $4\pi A^2$ and the volume of a sphere with radius R is $4\pi R^3/3$. By the principle of stationary potential energy, the equilibria radii occur at values of λ satisfying

$$\frac{d\Pi}{d\lambda} = 4\pi A^2 H\frac{dw}{d\lambda} - 4\pi A^3\lambda^2 P = 0.$$

Therefore, the equilibrium pressure is defined by

$$P(\lambda) = \frac{H}{A\lambda^2}\frac{dw}{d\lambda}. \tag{7.22}$$

This equilibrium relation is always a well-defined function because the potential energy depends linearly on P. The equilibria are stable if $d^2\Pi/d\lambda^2 > 0$. To relate this condition to the behavior of $P(\lambda)$, write

$$\frac{1}{4\pi A^2 H}\frac{d^2\Pi}{d\lambda^2} = \frac{d^2w}{d\lambda^2} - \frac{2A\lambda}{H}P.$$

An expression for the second derivative of w at equilibrium is obtained by differentiating $P(\lambda)$ in equation (7.22) to find that

$$H\frac{d^2w}{d\lambda^2} = A\lambda^2\frac{dP}{d\lambda} + 2A\lambda P.$$

Substitute this result in the expression for the second derivative of Π to obtain

$$\frac{1}{4\pi A^2}\frac{d^2\Pi}{d\lambda^2} = A\lambda^2\frac{dP}{d\lambda}, \tag{7.23}$$

at equilibrium. In other words, the equilibrium relation between a fixed pressure and its corresponding equilibrium stretches, $P(\lambda)$, is increasing at a stable equilibrium value. The equilibria are unstable if $P(\lambda)$ is decreasing.

This analysis can be applied to many different nonlinear elastic constitutive models for incompressible materials, such as the Mooney-Rivlin and the neo-Hookean strain energy functions (see Chapter 3, Section 3.8). The Mooney-Rivlin model is rewritten, assuming equal in-plane principal stretches $\lambda_1 = \lambda_2 = \lambda$ so that the stretch in the thickness direction is $\lambda_3 = \lambda^{-2}$, as

$$w(\lambda) = c_1\left[\left(2\lambda^2 + \frac{1}{\lambda^4} - 3\right) + \Gamma\left(\lambda^4 + \frac{2}{\lambda^2} - 3\right)\right], \tag{7.24}$$

where c_1 is a positive-valued material parameter, and $\Gamma = c_2/c_1$ is a nondimensional parameter that is zero for neo-Hookean materials and positive for Mooney-Rivlin materials.

The equilibrium pressure in the Mooney-Rivlin model is given by

$$P(\lambda) = \frac{4c_1 H}{A}(\lambda^{-1} - \lambda^{-7})(1 + \Gamma\lambda^2). \tag{7.25}$$

The stability changes at the critical points defined by $dP(\lambda)/d\lambda = 0$ because the second derivative of the potential energy changes sign or, in other terms, when

$$\Gamma = \frac{\lambda^6 - 7}{\lambda^8 + 5\lambda^2}. \tag{7.26}$$

To examine the roots of this equation, take the derivative of Γ.

$$\frac{d\Gamma}{d\lambda} = \frac{2(35 + 38\lambda^6 - \lambda^{12})}{\lambda^3(5 + \lambda^6)^2}. \tag{7.27}$$

The only positive zero of this equation is at $\lambda_c = \left(19 + 0.5\sqrt{38^2 + 4(35)}\right)^{1/6} \sim 1.84073$. The function $\Gamma(\lambda)$, equation (7.26), is monotonic increasing up to this critical value λ_c where $\Gamma = 0.214458$. Then $\Gamma(\lambda)$ decreases asymptotically to zero as λ increases. $P(\lambda)$ has no critical points for $\Gamma > 0.214458$. So, only a single stable equilibrium state exists for each value of P when $\Gamma > 0.214458$. However, $P(\lambda)$ has two critical points for $0 < \Gamma < 0.214458$. Therefore, if $0 < \Gamma < 0.214458$, there are one, two, or three equilibrium states for different values of P. Two stable equilibria lie on either side of an unstable equilibrium for those values of P having three equilibrium stretches. Figure 7.15 illustrates the equilibrium behavior in the case that $\Gamma = 0.1$. The equilibrium stretches for a particular value of p are those λ defining the intersection of the curve and a horizontal line at height p.

This case illustrates a perhaps surprising phenomena for nonlinear elastic materials. Several possible equilibria may exist for a single given load. In contrast, linear elastic materials produce a unique deformation response for a given load.

Exercise 7.3 Use a computer program to sketch the graph of $AP(\lambda)/4c_1 H$ versus λ for the case $\Gamma = 0.2$ at the equilibrium pressure $P(\lambda)$.

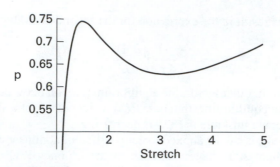

FIGURE 7.15 The equilibria load, $p = PA/(4c_1H)$, versus stretch relation for a Mooney-Rivlin material with $\Gamma = 0.1$.

In the neo-Hookean case, $\Gamma = 0$ in equation (7.25), and at equilibrium, the pressure is

$$P(\lambda) = \frac{4c_1H}{A}(\lambda^{-1} - \lambda^{-7}). \qquad (7.28)$$

Any neo-Hookean material has a change in stability behavior at $dP/d\lambda = 0$, where $d^2\Pi/d\lambda^2 = 0$ by equation (7.23), *i.e.* at $\lambda = 7^{1/6} = 1.38309$. The corresponding critical pressure is

$$P_c = 0.61973\frac{4c_1H}{A} \qquad (7.29)$$

No equilibria exist for a fixed pressure, $P > P_c$. The function $P(\lambda)$ is therefore increasing up to the limit point and is then decreasing, as λ increases. For each $0 < P < P_c$, there are two equilibrium stretches, one is stable and the other is unstable.

Exercise 7.4   Use a computer program to sketch the graph of $AP(\lambda)/4c_1H$ versus λ for a neo-Hookean sphere at the equilibrium pressure $P(\lambda)$. Compare this result to those for the Mooney-Rivlin material if $\Gamma = 0.1$ and if $\Gamma = 0.2$.

EXAMPLE 7.17   Compute the equilibrium gage pressure in an internally pressurized spherical rubber balloon made of either a Mooney-Rivlin or a neo-Hookean material at a stretch of $\lambda = 1.2$. Also determine the final thickness of the sphere. Compare the results. The Mooney-Rivlin material has $\Gamma = 0.2$. The initial radius is $A = 0.25$ meters and the initial thickness is $H = 0.01$ meters. The constant in both cases is $c_1 = 20$ MPa as measured for a natural rubber vulcanized with 8% sulfur.

Solution   The equilibrium pressure in the neo-Hookean case is $P = (4c_1H/A)(\lambda^{-1} - \lambda^{-7}) = 1.774$ MPa. The final thickness is, by the incompressibility condition, $h = H/\lambda^2 = (0.01)/(1.2)^2 = 0.00694$ meters. For the Mooney-Rivlin material, by equation (7.25), $P = 2.284$ MPa. The final thickness is also 0.00694 meters. The Mooney-Rivlin material is stiffer than the neo-Hookean because more pressure is required to produce the same deformation.   ∎

A saccular aneurysm is a biological structure that can be represented as a sphere under internal pressure, the blood pressure. A saccular aneurysm is a spherelike sac that grows from the wall of an artery due to local damage to the arterial tissue. Aneurysms frequently occur in the circle of Willis, an artery system within the lower portion of the cranium, and often at bifurcations of an artery. About 5% of the population has either unruptured or ruptured intracranial aneurysms; of these from 15 to 31% have multiple aneurysms. Intracranial aneurysms occur more frequently in women than in men.

The neck is the inlet to the saccular aneurysm at the point of attachment to the artery. The fundus is the sac itself (Fig. 7.16). An average fundus diameter is 8.4 mm. Some aneurysms can grow to a diameter of 25 mm. The cranial artery has a diameter of from 2 to 4 mm.

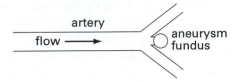

FIGURE 7.16   Schematic of a saccular aneurysm at a bifurcation in an artery.

The mechanical properties of the aneurysm walls differ from those of the artery wall to which it is attached. A healthy artery wall is a composite layered structure composed of an elastin lamella and collagen fibers, the former providing stretchability and the latter providing reinforcement. The aneurysm wall is primarily collagen; much of the original elastin from the healthy tissue has been fragmented and possibly lost.

Even though the aneurysm tissue is in reality a complex composite, it has often been approximated as an isotropic membrane. Recall (Chapter 3, Section 3.9) that a material satisfying the Fung model for isotropic biological tissue is one with strain energy per unit initial surface area,

$$w(\lambda) = C\left\{\exp\left[\frac{1}{2}\Gamma(\lambda^2 - 1)^2\right] - 1\right\}, \tag{7.30}$$

where $C > 0$ has units of force per unit length, and Γ is nondimensional. For arterial tissue, the parameters have been chosen as $C = 0.88$ and $\Gamma = 13$ in [3]. The wall thickness of an aneurysm is about $H = 0.000287$ meters. The mass per unit volume of the tissue is $\rho = 1{,}050$ kg/m^3, a little greater than that of water (1,000 kg/m^3).

The Fung potential energy for a pressurized sphere is

$$\Pi(\lambda) = C\{\exp[0.5\Gamma(\lambda^2 - 1)^2] - 1\} - \frac{1}{3}AP\lambda^3.$$

The static equilibria for the Fung model are given by setting equal to zero the derivative of this potential energy with respect to λ.

$$P(\lambda) = \frac{2C\Gamma}{A\lambda}(\lambda^2 - 1)\exp[0.5\Gamma(\lambda^2 - 1)^2]. \tag{7.31}$$

The Fung model implies that the equilibrium pressure increases monotonically for positive stretches.

This representation neglects the fact that in the body the blood pressure is not steady but rather is nearly periodic with period equal to that of the heartbeat. Medical personnel measure blood pressure in mm Hg. Blood pressure usually oscillates between about 70 (diastole) to 130 (systole) mm Hg, although this varies greatly from individual to individual. The conversion to Pascals is 133 Pa per mm Hg and to psi is 1 mm Hg = 0.01934 psi. For example, an amplitude of 30 mm Hg is about 3,999 Pa or 0.58 psi; a mean pressure of 100 mm Hg = 13,329 Pa = 1.934 psi.

EXAMPLE 7.18   A saccular aneurysm with undeformed radius $A = 0.003$ meters is subject to a blood pressure of 160 mm Hg. Assume that the Fung model with the Shah-Humphrey values of the constants describes the response.

(a) Compute the static stretched radius of the aneurysm for a blood pressure of 160 mm Hg.

(b) Compute the static stretched radius of the aneurysm for a blood pressure of 150 mm Hg.

(c) Determine the deformed radii in cases (a) and (b).

Solution

(a) The static pressure is 160 mm Hg = 21,326 Pa. The Fung model predicts a stretch of $\lambda =$ 1.2393 by numerically solving the equilibrium pressure relation [equation (7.31)].

(b) The static pressure is 150 mm Hg = 19,950 Pa. The Fung model predicts a stretch of $\lambda =$ 1.2361 again by numerically solving the equilibrium pressure relation.

(c) The deformed radius is $a = \lambda A$. In case (a) it is 3.7179 mm, and in case (b) it is 3.7083 mm.

■

7.6 INFINITE DEGREE OF FREEDOM SYSTEMS

Infinite degree of freedom systems, such as deformable bodies, require special techniques to determine their equilibrium configurations.

EXAMPLE 7.19

Determine the equilibrium position for a simply supported strength of materials beam with a constant distributed load, p (Fig. 7.17). The configuration is described by the deflected position of the neutral axis, $y(x)$, where the positive direction is up. This is an infinite degree of freedom problem since, to specify $y(x)$, values must be given at the infinite number of points between 0 and L.

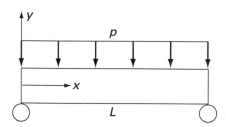

FIGURE 7.17 Simply supported beam with distributed load, p.

The beam is made of a linear elastic material. Assume the strength of materials form of the strain energy, U.

$$U = \int_0^L \frac{M(x)^2}{2EI} dx,$$

and the linearized expression for the internal moment $M(x) = y''(x)EI$, where E is the elastic modulus, I is the area moment of inertia, and $y(x)$ is the position of the neutral axis. Therefore,

$$U = \int_0^L \frac{1}{2} EI y''(x)^2 dx.$$

Because the work done by the downward external distributed force $p(x)$ (y is positive up) is

$$W_e = -\int_0^L p(x)y dx,$$

the total potential energy is

$$\Pi = \int_0^L \left[\frac{1}{2} EI y''(x)^2 + p(x)y \right] dx. \tag{7.32}$$

■

In the example, $p(x) = p$, but $p(x)$ can be any distributed load. The principle of stationary potential energy states that equilibrium occurs for the deflection $y(x)$ of the neutral axis that minimizes this integral. This means that one allows $y(x)$ to vary over all possible functions $y(x)$ satisfying the boundary conditions $y(0) = y(L) = 0$ (since the beam is simply supported) to find the one, the equilibrium deflection, which minimizes the integral.

Since this is not a finite degree of freedom problem, the equilibrium position cannot be found by solving a finite number of partial derivative equations simultaneously, but rather must be obtained by the calculus of variations, which is discussed later in Section 7.8. The calculus of variations says that the function $y(x)$ minimizing an integral of the form $\int F(x, y, y', y'')dx$ satisfies the partial differential equation

$$\frac{\partial F}{\partial y} - \frac{d}{dx}\frac{\partial F}{\partial y'} + \frac{d^2}{dx^2}\frac{\partial F}{\partial y''} = 0,$$

called the Euler-Lagrange equation. Because $F(x, y, y', y'') = 0.5EIy''(x)^2 + p(x)y$, the equation for the beam problem of Example 7.19 is the fourth-order ordinary differential equation

$$EIy'''' + p(x) = 0, \tag{7.33}$$

which is satisfied by a fourth-order polynomial. This equation is one that appeared in the strength of materials analysis of beams. The term EIy'''' is the distributed load (positive is up).

The solution can also be calculated, as in strength of materials, by integrating $EIy''(x) = M(x)$ with the boundary conditions to obtain the fourth-degree polynomial solution. Recall that the second derivative of the internal moment in a beam is the distributed load. Also recall from Chapter 1, Section 1.4, that the strength of materials solution does not satisfy the equilibrium equation for stress in the case of a simply supported beam with constant distributed load. The problem should be reformulated in terms of the elastic strain energy if an elasticity solution is desired.

7.7 RAYLEIGH-RITZ APPROXIMATION TECHNIQUE

Before discussing the calculus of variations, we will use an approximation technique to estimate the equilibrium deflection of a deformable body, such as the beam. The goal of this approximation method is to reduce an infinite degree of freedom problem, such as the beam problem, to a finite degree of freedom problem. If the finite degree of freedom version of the problem involves generalized coordinates, $x_1, \dots, x_n$, then the equilibrium state can be found by solving the equations, $\partial \Pi / \partial x_i = 0$ for $i = 1, 2, \dots, n$, simultaneously.

Rayleigh's idea was to assume a form for the approximate equilibrium displacement that satisfies the geometric boundary conditions and that involves only a finite number of coefficients, $a_1, \dots, a_n$. The coefficients, a_i, are treated as the generalized coordinates since each choice of such a set of values defines a particular displacement. The equilibrium values of the coefficients, $a_1, \dots, a_n$ are determined by first rewriting the potential energy in terms of the assumed displacement so that it is a function, $\Pi(a_1, \dots, a_n)$, of the $a_1, \dots, a_n$. The n equations, $\partial \Pi(a_1, \dots, a_n)/\partial a_i = 0$ for $i = 1, 2, \dots, n$, from the principle of stationary potential energy are solved simultaneously for the equilibrium coefficients. These determine the approximate equilibrium displacement.

EXAMPLE 7.20 Consider again the strength of materials version of the simply supported beam with constant distributed load, p (see Fig. 7.17). It has potential energy,

$$\Pi = \int_0^L \left[\frac{1}{2} EI y''(x)^2 + py \right] dx.$$

Convert this into a one degree of freedom problem by assuming that the deflection has the form, $y(x) = \alpha \sin(\pi x/L)$. This form is chosen since it satisfies the geometric boundary conditions $y(0) = y(L) = 0$ no matter what the coefficient α is. Of all the functions of the form, $y(x) = \alpha \sin(\pi x/L)$, as α is varied, the one that minimizes the approximate potential energy is the one that is the best approximation to the equilibrium deflection. The coefficient α is the single generalized coordinate so that the potential energy must be written as a function $\Pi(\alpha)$. Since $y''(x) = -\alpha(\pi/L)^2 \sin(\pi x/L)$,

$$\Pi(\alpha) = \int_0^L \left[\frac{1}{2} \alpha^2 EI \left(\frac{\pi}{L} \right)^4 \sin^2 \left(\frac{\pi x}{L} \right) + p\alpha \sin \left(\frac{\pi x}{L} \right) \right] dx$$

$$= \frac{1}{4} EI \left(\frac{\pi^4}{L^3} \right) \alpha^2 + 2 \left(\frac{L}{\pi} \right) p\alpha. \tag{7.34}$$

Differentiation yields

$$\frac{\partial \Pi}{\partial \alpha} = \frac{1}{2} EI \left(\frac{\pi^4}{L^3} \right) \alpha + 2 \left(\frac{L}{\pi} \right) p = 0,$$

so that $\alpha = -4L^4 p / EI\pi^5$. The approximate equilibrium solution is therefore

$$y(x) = - \left(\frac{4L^4 p}{EI\pi^5} \right) \sin \left(\frac{\pi x}{L} \right). \tag{7.35}$$

The approximate internal moment $M(x)$ associated with this displacement is found, in the case of strength of materials assumptions, from the relation $M(x) = EI y''(x)$. This approximation of $y(x)$ predicts that

$$M(x) = \left(\frac{4L^2 p}{\pi^3} \right) \sin \left(\frac{\pi x}{L} \right).$$

The assumed deflection approximation implies an approximation to the moment that, in this case, satisfies the simply supported boundary conditions that $M(0) = M(L) = 0$. ∎

EXAMPLE 7.21 In some cases, the Rayleigh method can produce a very good approximation to the deflection but a very poor approximation for the internal moment $M(x)$. Redo the previous example, under the strength of materials assumption, but assume that $y(x) = \alpha x(L - x)$ for some coefficient, α. Again take α to be the single degree of freedom. Notice that the form of $y(x)$ was again chosen so that the geometric boundary conditions $y(0) = y(L) = 0$ are satisfied. The beam has potential energy,

$$\Pi = \int_0^L \left[\frac{1}{2} EI y''(x)^2 + py \right] dx.$$

Substituting the assumed form of $y(x)$, which has $y''(x) = -2\alpha$,

$$\Pi(\alpha) = \int_0^L \left[\frac{1}{2} EI(-2\alpha)^2 + p\alpha x(L - x) \right] dx = 2EIL\alpha^2 + \frac{1}{6} pL^3 \alpha.$$

Then the principle of stationary potential energy, $\partial \Pi / \partial \alpha = 0$, implies that

$$\alpha = -\frac{pL^2}{24EI}.$$

The equilibrium deflection is predicted to be

$$y(x) = -\left(\frac{pL^2}{24EI}\right)x(L-x)$$

and the moment is constant,

$$M(x) = \frac{1}{12}pL^2.$$

The prediction of the moment does not satisfy the simply supported requirement that $M(0) = M(L) = 0$ and does not agree with the results from strength of materials. Since the bending stress is related to the moment, this assumed deflection will produce very wrong predictions of the stress. The order of the approximating polynomial is not large enough. Therefore this approximation of the deflection field is not as satisfactory as that involving $\sin(\pi x/L)$. ∎

The load need not be a distributed load. A point load merely changes the term accounting for the potential energy of the load in the total potential energy.

EXAMPLE 7.22   A simply supported beam of length, L, and constant bending stiffness, EI, carries a transverse downward point load, P, at its center. Assume a displacement field that satisfies the simply supported boundary conditions that the displacement is zero at both ends, $y(x) = a_1 \sin(\pi x/L)$. Again assume that the displacement is positive up. The potential energy of the load is $Py(L/2) = Pa_1$ so that the potential energy is

$$\Pi(a_1) = \int_0^L \frac{EI}{2}y''(x)^2 dx + Py(L/2) = \frac{EI}{2}a_1^2\left(\frac{\pi}{L}\right)^4\frac{L}{2} + Pa_1.$$

From the principle of stationary potential energy, $a_1 = -2PL^3/\pi^4 EI$. The maximum displacement, which is at the center, predicted by integrating the moment, $M(x)$ obtained from a free-body diagram, as in elementary strength of materials, is $y(L/2) = -PL^3/48EI$. The corresponding displacement from the sine approximation is a_1, which is approximately $-PL^3/48.7EI$, in close agreement with elementary strength of materials. The moment at the center from a free-body diagram is $M(L/2) = PL/4$, while that predicted by this approximation is $PL/4.93$, a somewhat larger error. ∎

EXAMPLE 7.23   A cantilever beam of length, L, and constant bending stiffness, EI, carries a transverse downward point load, P, at its free end. Choose a displacement function so that the displacement is zero at the wall, the slope is zero at the wall, and the moment is zero at the free end. Measure x from the wall. The function $y(x) = a_1[1 - \cos(\pi x/2L)]$ satisfies all three of these conditions. If y is positive up, the energy function is

$$\Pi(a_1) = \int_0^L \frac{EI}{2}y''(x)^2 dx + Py(L) = \frac{EI}{2}a_1^2\left(\frac{\pi}{2L}\right)^4\frac{L}{2} + Pa_1.$$

From the principle of stationary potential energy, $a_1 = -32PL^3/\pi^4 EI$. The displacement at the free end predicted by integrating the moment, $M(x)$ obtained from a free-body diagram, as in elementary strength of materials, is $y(L) = -PL^3/3EI$. The corresponding displacement from the approximation is a_1, which is approximately $-PL^3/3.044EI$, in close agreement with elementary strength of materials. The moment at the wall computed from a free-body diagram is $M(0) = PL$, while that predicted by this approximation is $8PL/\pi^2 = 0.811PL$. ∎

W. Ritz refined this idea around 1909 by taking a sequence of approximations to the deflection, $y(x)$. Let the sequence of functions, $y_n(x)$, be written in terms of another family of functions, $g_i(x)$, so that

$$y_n(x) = \sum_{i=1}^n a_i g_i(x),$$

where the a_i are scalars. The function $y_n(x)$ has one more term, $a_n g_n(x)$, than $y_{n-1}(x)$. Ritz required that these satisfy a requirement on the least squares estimation. These functions, $g_i(x)$, must be complete in the sense that the least squares error in the approximation y_n to $y(x)$ over the domain $a \leq x \leq b$

$$M_n = \int_a^b [y(x) - y_n(x)]^2 dx$$

satisfy $\lim_{n \to \infty} M_n = 0$. In other words, the least squares error M_n must go to zero as more and more terms are taken in the approximation y_n to $y(x)$.

For example, the $g_i(x)$ can be taken to be the functions $\sin(nx)$ and $\cos(nx)$ in the Fourier series approximation to $y(x)$,

$$y(x) = a_0 + \sum_{n=1}^{\infty} a_n \cos(nx) + \sum_{n=1}^{\infty} b_n \sin(nx).$$

The example above using $y(x) = \sin(\pi x / L)$ as an approximation to the equilibrium deflection of a simply supported beam with constant distributed load is the simplest example of this.

The terms of the Fourier series have the special property that

$$\int_0^{\pi} \sin(mx) \sin(nx) dx = 0 \quad \text{for} \quad n \neq m$$

$$\int_0^{\pi} \cos(mx) \cos(nx) dx = 0 \quad \text{for} \quad n \neq m$$

$$\int_0^{\pi} \sin(mx) \cos(nx) dx = 0 \quad \text{for all} \quad n, m.$$

Exercise 7.5   Verify these results using, in the first case, the trigonometric identity

$$\sin(mx) \sin(nx) = [\cos(mx - nx) - \cos(mx + nx)]/2$$

and, in the second case, a similar expression for $\cos(mx) \cos(nx)$ obtained from the sum and difference of the angle relations and likewise in the third case.

Exercise 7.6   Verify that

$$\int_0^L \sin^2\left(\frac{n\pi x}{L}\right) dx = \frac{L}{2} \quad \text{and} \quad \int_0^L \cos^2\left(\frac{n\pi x}{L}\right) dx = \frac{L}{2}$$

for all positive integers n.

This behavior of functions is codified in the following definition which is analogous to the statement that the dot product of orthogonal vectors is zero.

Definition   Two functions, $f(x)$ and $g(x)$, are said to be orthogonal over the interval $[a, b]$ if

$$\int_a^b f(x)g(x)dx = 0.$$

The approximation of the displacement, $y(x)$, by sums of orthogonal functions, as in a Fourier series, has a very important computational advantage which is shown in Example 7.24.

EXAMPLE 7.24   Repeat the beam with constant distributed load problem using the Fourier series approximation,

$$y(x) = a_1 \sin\left(\frac{\pi x}{L}\right) + a_2 \sin\left(\frac{2\pi x}{L}\right).$$

Notice that the constant term a_0 and the cosine terms had to be dropped from the Fourier series and x had to be replaced by $\pi x/L$ in order that the geometric boundary conditions $y(0) = y(L) = 0$ be satisfied by the approximation. The sine terms for higher values of n have also been dropped in this approximation.

In the strength of materials approximation, the potential energy is

$$\Pi = \int_0^L \left[\frac{1}{2}EIy''(x)^2 + py\right] dx. \tag{7.36}$$

Substitute $y(x)$ and $y''(x) = -a_1(\pi x/L)^2 \sin(\pi x/L) - a_2(2\pi x/L)^2 \sin(2\pi x/L)$ into equation (7.36) and integrate. This reduces the infinite degree of freedom problem to a two degree of freedom problem with generalized coordinates a_1 and a_2. Assume that E and I are constant over the length of the beam.

$$\Pi(a_1, a_2) = \frac{EI}{2}\int_0^L \left[a_1^2 \left(\frac{\pi x}{L}\right)^4 \sin^2\left(\frac{\pi x}{L}\right)\right.$$
$$+ 2a_1a_2 \left(\frac{\pi x}{L}\right)^2 \left(\frac{2\pi x}{L}\right)^2 \sin\left(\frac{\pi x}{L}\right)\sin\left(\frac{2\pi x}{L}\right)$$
$$\left. + a_2^2 \left(\frac{2\pi x}{L}\right)^4 \sin^2\left(\frac{2\pi x}{L}\right)\right] dx$$
$$+ p\int_0^L \left[a_1 \sin\left(\frac{\pi x}{L}\right) + a_2 \sin\left(\frac{2\pi x}{L}\right)\right] dx$$
$$= \frac{EI}{2}\left[a_1^2 \left(\frac{\pi}{L}\right)^4 \frac{L}{2} + a_2^2 \left(\frac{2\pi}{L}\right)^4 \frac{L}{2}\right] + p\left[a_1 \frac{2L}{\pi}\right].$$

Because an orthogonal sequence of approximating functions, $\sin(n\pi x/L)$, has been chosen, all the integrals involving products of the form $\sin(m\pi x/L)\sin(n\pi x/L)$ for $n \neq m$ are zero, drastically simplifying the computation.

Once the remaining integrals have been calculated, the principle of stationary potential energy produces the two equations which must be solved simultaneously for a_1 and a_2.

$$\frac{\partial \Pi}{\partial a_1} = \frac{EI}{2}a_1 \left(\frac{\pi}{L}\right)^4 L + p\frac{2L}{\pi} = 0;$$
$$\frac{\partial \Pi}{\partial a_2} = \frac{EI}{2}a_2 \left(\frac{2\pi}{L}\right)^4 L = 0.$$

The second equation implies that $a_2 = 0$ and the first that $a_1 = -4L^4 p/EI\pi^5$ so that

$$y(x) = -\frac{4L^4 p}{EI\pi^5} \sin\left(\frac{\pi x}{L}\right).$$

The identical result was obtained under the assumption that $y(x) = a_1 \sin(\pi x/L)$.

This result might lead to the guess that for the simply supported beam with a constant distributed load, sine terms with n an even number contribute nothing to the equilibrium deflection. This conjecture will be further tested in the problems for this chapter by assuming that $y(x) = a_1 \sin(\pi x/L) + a_3 \sin(3\pi x/L)$ to see if higher-order odd terms contribute to the deflection.

7.8 CALCULUS OF VARIATIONS AND THE EULER-LAGRANGE EQUATION FOR AN INTEGRAL FUNCTION

In Chapter 7, Section 7.6, it was indicated that the potential energy for an infinite degree of freedom problem is often expressed as an integral whose integrand involves the unknown displacement function. This integral is viewed as a real-valued function of functions representing the conceivable displacements. The problem of minimizing a real-valued function whose domain is a set of functions can be reduced to solving a differential equation. Conversely, the solution of a differential equation can be found numerically by finding the associated integral and then applying a numerical technique, such as the finite element method discussed in Chapter 8, to the integral.

The equilibrium displacement field for a deformable body is the displacement field which minimizes the potential energy function. In a one-dimensional case, an arbitrary displacement field might be written as $v(x)$, where x is the position along the body. The corresponding potential energy is then an integral function of $v(x)$ and its first and second derivatives. The analysis is carried out for any function of the form

$$V(x, v, v', v'') = \int_{x_1}^{x_2} F(x, v, v', v'')dx,$$

where $v'(x) = dv/dx$, and $v''(x) = d^2v/dx^2$. The problem is to view $V(x, v, v', v'')$ as a function of a single scalar variable, ϵ, so that the function may be treated by elementary calculus. This is done because it is not possible to easily indicate when a variation of the form $v(x) + \delta v(x)$ is close to $v(x)$. The trick is to think of the variation as one in the form $\epsilon \eta(x)$ for an arbitrary function $\eta(x)$; all possible variations can be expressed in such a form. Then, the most general perturbation of $v(x)$ is another function of the form $v(x) + \epsilon \eta(x)$ for $\epsilon > 0$ and $\eta(x)$ another fixed function that satisfies $\eta(x_1) = \eta(x_2) = 0$ so that the perturbation has the same boundary conditions as the equilibrium configuration. Now it is clear that if ϵ is very small, then $v(x) + \delta v(x)$ is close to $v(x)$. This is a generalization to functions of the idea that a scalar perturbation of x is $x + \Delta x$. For example, if $v(x) = \sin x$ and one chooses $\eta(x) = x^3$, the perturbation is $\sin x + \epsilon x^3$. This situation can be visualized in the case that a static beam deflection is described by the deformation of the neutral surface, $y(x)$ (Fig. 7.18).

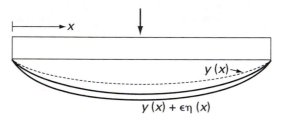

FIGURE 7.18 Two perturbations $y(x) + \epsilon \eta(x)$, for two values of ϵ and the same $\eta(x)$, of the equilibrium neutral surface, $y(x)$, of a simply supported beam.

The displacement functions are constrained by boundary conditions at x_1 and x_2 which satisfy linear relations of the form

$$a_0 + a_1 v + a_2 v' + a_3 v'' = 0,$$

for some constants a_i, some of which may be zero. The linearity of this relation guarantees that if $\eta(x)$ satisfies the end conditions, then $v(x) + \epsilon \eta(x)$ is a perturbation satisfying the end conditions, for $\epsilon > 0$.

EXAMPLE 7.25

The function $v(x) = 1 - \cos(\pi x/L)$ is an admissible displacement function for a cantilever beam because $v(0) = v'(0) = 0$. A possible perturbation function is $\eta(x) = -x^2/L^2$. ■

Exercise 7.7   On the same coordinate system, plot $v(x) = 1 - \cos(\pi x/L)$ and $v(x) + \epsilon\eta(x)$ in the cases that $\eta(x) = -x^2/L^2$, $\epsilon = \pm 0.01$, and $L = 10$.

Suppose that $y(x)$ is the function that minimizes $V(x, v, v', v'')$. Then any perturbation of $y(x)$ must produce a larger value than $V(x, y, y', y'')$. The change in value of $V(x, y, y', y'')$ when $y(x)$ is perturbed is

$$\Delta V = \int_{x_1}^{x_2} [F(x, y + \epsilon\eta, y' + \epsilon\eta', y'' + \epsilon\eta'') - F(x, y, y', y'')]dx.$$

For fixed $y(x)$ and $\eta(x)$, this expression can be viewed as a function of ϵ, a function whose domain and range are both the real numbers. A Taylor series expansion of $(\Delta V)(\epsilon)$, the change in potential energy as a function of ϵ, about $\epsilon = 0$, by expanding the function $F(x, y + \epsilon\eta, y' + \epsilon\eta', y'' + \epsilon\eta'')$ in the integrand, yields

$$V(\epsilon) = \int_{x_1}^{x_2} \left\{ F(x, y, y', y'') + \epsilon\left[\frac{dF}{dy}\eta(x) + \frac{\partial F}{\partial y'}\eta'(x) + \frac{\partial F}{\partial y''}\eta''(x) \right] + O(\epsilon^2) \right\} dx.$$

The change in V is

$$\Delta V = \int_{x_1}^{x_2} \epsilon\left(\eta\frac{\partial F}{\partial y} + \eta'\frac{\partial F}{\partial y'} + \eta''\frac{\partial F}{\partial y''} \right) dx + O(\epsilon^2),$$

where $O(\epsilon^2)$ denotes the higher-order terms in ϵ. The integral is called the first variation of V, δV. Because $\Delta V > 0$ for arbitrary ϵ, positive or negative, $\delta V = 0$ is a necessary condition at a minimum of V.

As an alternative that avoids the higher-order terms $O(\epsilon^2)$, put

$$I(\epsilon) = \int_{x_1}^{x_2} F(x, y + \epsilon\eta, y' + \epsilon\eta', y'' + \epsilon\eta'')dx.$$

By assumption, the function $y(x)$ minimizes the original integral function. Therefore, the function $I(\epsilon)$ must take a minimum at $\epsilon = 0$ or

$$\left.\frac{dI}{d\epsilon}\right|_{\epsilon=0} = \int_{x_1}^{x_2} \left(\eta\frac{\partial F}{\partial y} + \eta'\frac{\partial F}{\partial y'} + \eta''\frac{\partial F}{\partial y''} \right) dx = 0.$$

The goal is rewrite this expression in terms of the function, $\eta(x)$, but eliminating all of the derivatives of $\eta(x)$ under the integral so that the lemma below may be employed. Integration by parts gives

$$\left.\frac{dI}{d\epsilon}\right|_{\epsilon=0} = \int_{x_1}^{x_2} \left(\frac{\partial F}{\partial y} - \frac{d}{dx}\frac{\partial F}{\partial y'} + \frac{d^2}{dx^2}\frac{\partial F}{\partial y''} \right) \eta dx$$

$$+ \left[\eta\left(\frac{\partial F}{\partial y'} - \frac{d}{dx}\frac{\partial F}{\partial y''} \right)\Big|_{x_1}^{x_2} + \eta'\frac{\partial F}{\partial y''}\Big|_{x_1}^{x_2} \right] = 0. \qquad (7.37)$$

Recall that $\eta(x)$ is an arbitrary function. The key tool in arriving at the Euler-Lagrange differential equation is the fundamental lemma,

LEMMA   If $f(x)$ is continuous in $x_1 \leq x \leq x_2$ and

$$\int_{x_1}^{x_2} \eta(x)f(x)dx = 0,$$

for all $\eta(x)$ such that $\eta(x_1) = \eta(x_2) = 0$, then $f(x) = 0$ identically.

Proof

The proof is by contradiction. Assume there is a value x_3 in the interval for which $f(x_3) > 0$. Then on some smaller interval $x_4 < x_3 < x_5$, $f(x)$ is positive. Pick a continuous function $\eta(x)$ such that also $\eta(x) > 0$ on the smaller interval $x_4 < x_3 < x_5$ but zero outside it. For example, $\eta(x)$ could be the function,

$$\eta(x) = \begin{cases} 0 & \text{if } x_1 \leq x \leq x_4 \\ -(x - x_4)^3(x - x_5)^3 & \text{if } x_4 \leq x \leq x_4 \\ 0 & \text{if } x_5 \leq x \leq x_2. \end{cases}$$

Then $\int_{x_1}^{x_2} \eta(x)f(x)dx > 0$, a contradiction to the hypothesis that this integral is zero for any admissible $\eta(x)$. The assumption that $f(x)$ is not identically zero is false.   ∎

Exercise 7.8   Graph the function $\eta^*(x) = -(x - x_1)^3(x - x_2)^3$, on the interval from $x_1 = 1$ to $x_2 = 3$.

The two-dimensional case of this lemma can be proved in the same manner. If $\int\int \eta(x, y)G(x, y)dxdy = 0$ for all $\eta(x, y)$ which are zero on the boundary of the region of integration, then $G(x, y) = 0$ for all points in the region of integration.

Apply the lemma to $\frac{dI}{d\epsilon}\big|_{\epsilon=0}$ in equation (7.37). Because the functions η and η' are arbitrary, each of the three terms of equation (7.37) must be zero. In particular by the lemma, the coefficient of η under the integral must be zero. This fact produces the Euler-Lagrange differential equation which $y(x)$ must satisfy to minimize the function V,

$$\frac{\partial F}{\partial y} - \frac{d}{dx}\frac{\partial F}{\partial y'} + \frac{d^2}{dx^2}\frac{\partial F}{\partial y''} = 0. \tag{7.38}$$

Notice the alternation in signs in front of the terms. The *natural boundary conditions* are given by the other two terms of equation (7.37),

$$\eta\left(\frac{\partial F}{\partial y'} - \frac{d}{dx}\frac{\partial F}{\partial y''}\right)\Big|_{x_1}^{x_2} = \eta'\frac{\partial F}{\partial y''}\Big|_{x_1}^{x_2} = 0. \tag{7.39}$$

There also may be *forced boundary conditions* created by the supports on the body.

EXAMPLE 7.26   Determine the equilibria of the function of functions, $G(y) = \int_0^\pi [(y')^2 - y^2]dx$, where any function $y(x)$ must satisfy $y(0) = 0$ and $y'(0) = 1$. Also compute the corresponding value of the function.

Solution   The equilibrium state can be computed by solving the associated Euler differential equation. The integrand is a function of x, y, and y'. Therefore the Euler equation (7.38) is

$$0 = \frac{dF}{dy} - \frac{d}{dx}\frac{\partial F}{\partial y'} = -2y - 2y''.$$

The resulting differential equation, $y + y'' = 0$ subject to the given boundary conditions has solution $y(x) = \sin(x)$. Substituting this function into the integrand gives

$$\int_0^\pi [(y')^2 - y^2]dx = \int_0^\pi \left[\cos(x)^2 - \sin(x)^2\right]dx = 0.$$

To estimate whether the stationary state is a maximum or a minimum, calculate the value of the function $G(y)$ on the perturbation, $y_1(x) = x - x^2/\pi$, of $y = \sin(x)$. Both functions are zero at the endpoints and satisfy $y'(0) = 1$.

$$G(y_1) = \int_0^\pi \left[\left(1 - \frac{2x}{\pi} \right)^2 - \left(x - \frac{x^2}{\pi} \right)^2 \right] dx = 0.013655.$$

The function $y = \sin(x)$ probably minimizes $G(y)$ because $G(\sin(x)) = 0$.   ∎

Exercise 7.9   Graph the functions $y = \sin(x)$ and $y_1(x) = x - x^2/\pi$ for $0 \leq x \leq \pi$ on the same axes to show that one is a perturbation of the other.

EXAMPLE 7.27   A simple example which shows that the energy method produces the same result as elasticity is the uniaxial loading of a linear elastic rod, of constant cross section, A, and length, L, at either end by a force, P. The rod is fixed in a wall at its left end where $x = 0$. When the strain is written in terms of the displacement as $\epsilon_x = du/dx$, the potential energy over the full volume of the rod for a given displacement, u, is

$$\Pi(u) = \int_0^L \frac{1}{2} E \epsilon_x^2 A\, dx - P \int_0^L \frac{du}{dx} dx$$
$$= \int_0^L \frac{1}{2} E \left(\frac{du}{dx} \right)^2 A\, dx - P \int_0^L \frac{du}{dx} dx.$$

The Euler-Lagrange equation is

$$E A \frac{d^2 u}{dx^2} = 0,$$

which has solution of the form $u(x) = ax + b$ for some constants a and b to be determined by the natural boundary conditions (7.39).

$$\eta \frac{\partial F}{\partial u'} \Big|_0^L = \eta \left(E A \frac{du}{dx} - P \right) \Big|_0^L = 0.$$

The forced boundary condition that the rod is fixed in a wall at $x = 0$, requires that both $u(0) = 0$ and that $\eta(0) = 0$ since the function η must obey the forced boundary conditions of the problem. Therefore, $\eta(L)(EAa - P) = 0$, and because $\eta(L) \neq 0$, $a = P/EA$ is the constant strain expected. Also $b = 0$ because $u(0) = 0$.   ∎

EXAMPLE 7.28   A similar procedure gives the displacement field of the same rod as in the previous example, but when the material is nonlinear elastic with constitutive equation, $\sigma = E_1 \epsilon + E_2 \epsilon^3$.

$$\Pi(u) = \int_0^L \left(\frac{1}{2} E_1 \epsilon_x^2 + E_2 \frac{1}{4} \epsilon_x^4 \right) A\, dx - P \int_0^L \frac{du}{dx} dx$$
$$= \int_0^L \left[\frac{1}{2} E_1 \left(\frac{du}{dx} \right)^2 + E_2 \frac{1}{4} \left(\frac{du}{dx} \right)^4 \right] A\, dx - P \int_0^L \frac{du}{dx} dx.$$

Now the Euler-Lagrange equation is a nonlinear ordinary differential equation,

$$E_1 \frac{d^2 u}{dx^2} + 3 E_2 \left(\frac{du}{dx} \right)^2 \frac{d^2 u}{dx^2} = 0.$$

Factoring shows that either $d^2u/dx^2 = 0$ or $E_1 + 3E_2(du/dx)^2 = 0$. The latter nonlinear equation can have no solution because each term is positive. Therefore, again $u(x) = ax + b$ for some constants a and b. The natural boundary condition is, because $\eta(0) = 0$ and $\eta(L) \neq 0$,

$$\frac{\partial F}{\partial u'} \Big|_{x=L} = \left[E_1 A \frac{du}{dx} + E_2 A \left(\frac{du}{dx} \right)^3 \right] \Big|_{x=L} - P = 0.$$

In this case, the strain $\partial u/\partial x$ is a; it is constant throughout the rod. Therefore $E_1 a + E_2 a^3 = P/A = \sigma$ as expected from the constitutive model. The value of the coefficient a can be obtained by using the algebraic solution for a cubic polynomial or numerically by a root-finding algorithm like Newton's method. Again the forced boundary condition requires that $b = 0$. If it had been known that the strain is constant ahead of time, the displacement could have been obtained by integrating $du/dx = a$, but it was not known that the strain is constant until the energy analysis was performed. ∎

EXAMPLE 7.29 Determine and solve the Euler-Lagrange equation for a cantilever beam with a transverse point load at the free end, under the strength of materials assumptions. These assumptions appear in the choice of the strain energy. The potential energy is

$$\Pi(y) = \frac{EI}{2}\int_0^L y''^2 dx + Py(L).$$

The Euler equation is $y'''' = 0$, and the natural boundary conditions at the free end are $y''(L) = 0$, that is, the moment is zero, and $EIy'''(L) = 0$, that is, the shear is zero. The forced boundary conditions are that $y(0) = 0$ and $y'(0) = 0$. The solution to the fourth-order Euler differential equation is a third-order polynomial,

$$y(x) = a_0 + a_1 x + a_2 x^2 + a_3 x^3.$$

Application of the boundary conditions shows that the displacement is, as found in strength of materials,

$$y(x) = \frac{P}{6EI}x^3 - \frac{PL}{2EI}x^2.$$ ∎

EXAMPLE 7.30 An inextensible cable of length L is hung between the two points (x_1, y_1) and (x_2, y_2) in a chosen coordinate system. The specific weight of the cable is ρ and its cross-sectional area is a. Determine the equilibrium configuration, $y(x)$, of the cable.

Solution The incremental weight of a length ds of the cable is $\rho a\, ds = \rho a\sqrt{1 + y'^2}dx$, where y' is the derivative with respect to x. The potential energy of the cable with respect to the chosen datum is

$$\Pi(x) = \int_{x_1}^{x_2} y\rho a\sqrt{1 + y'^2}dx.$$

Because the cable is inextensible, the equilibrium configuration must satisfy the constraint, $L = \int_{x_1}^{x_2}\sqrt{1 + y'^2}dx$. The method of Lagrange multipliers allows inclusion of the constraint. This method states that the equilibrium subject to the constraint is obtained by optimizing the associated function, $\hat{V}(x)$, which has the same values as $\Pi(x)$, where λ is the Lagrange multiplier,

$$\hat{V}(x) = \int_{x_1}^{x_2} y\rho a\sqrt{1 + y'^2}dx + \lambda\left[\int_{x_1}^{x_2}\sqrt{1 + y'^2}dx - L\right] = \int_{x_1}^{x_2}(y\rho a + \lambda)\sqrt{1 + y'^2}dx - L.$$

Since the integrand, $F(y, y')$, depends only on y and y', the Euler equation, $\partial F/\partial y - (\partial/\partial x)(\partial F/\partial y')$ integrates to $F - y'(\partial F/\partial y') = C$, for a constant C. (Verify this statement by differentiating with respect to x and keeping in mind that the chain rule must be applied when differentiating F.) Substitution into this result of the integrand of $\hat{V}(x)$ and simplifying yields $\rho a y + \lambda = C'\sqrt{1 + y'^2}$, where $C' = C + L$. This differential equation is solved by separation of variables to obtain

$$y(x) = C''\cosh\left(\frac{x}{C''} + C_1\right) - \lambda,$$

where $C'' = C'/\rho a$, and C_1 is an integration constant. There are three unknown constants C_1, C'', and λ and so three equations are needed to determine them. The first is the constraint,

$$L = \int_{x_1}^{x_2} \sqrt{1 + y'^2} dx = C'' \sinh\left(\frac{x}{C''} + C_1\right)\Big|_{x_1}^{x_2},$$

which results from integration with the form of the equilibrium solution, $y(x)$. The other two equations are the boundary conditions, $y_1 = y(x_1)$ and $y_2 = y(x_2)$. Numerical techniques will usually be needed to simultaneously solve the three equations. ∎

EXAMPLE 7.31

An inextensible cable of length, L, and weight per length, w, is hung between two points that are at the same height (Fig. 7.19a). The horizontal distance between the two points is D. Determine the equilibrium configuration, $y(x)$, of the cable. Also express the tension in the cable at any point as a function of its height above the datum in a special coordinate system by selecting the values of C'' and λ in the solution to the previous example.

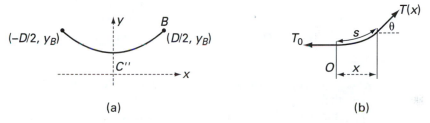

FIGURE 7.19   (a) The cable hanging from two points at the same height (shown in a special coordinate system). (b) Free-body diagram of a length s of the cable with the lowest point at the left end.

Solution

Choose a coordinate system whose origin is below the center of the cable. From the previous example, because $y(-D/2) = y(D/2)$,

$$C'' \cosh\left(\frac{-D}{2C''} + C_1\right) - \lambda = C'' \cosh\left(\frac{D}{2C''} + C_1\right) - \lambda.$$

Therefore $\cosh(-D/2C'' + C_1) = \cosh(D/2C'' + C_1)$. By symmetry of the cosh function, $C_1 = 0$.

Notice that the datum line, the location of the origin for the coordinate system, has not yet been selected. The coordinate system is chosen so that the tension in the cable at a point is related to the height of the point above the datum. Let T_0 be the tension in the cable at the lowest point, that is, at $x = 0$. In the free-body diagram (Fig. 7.19b), the sum of the horizontal forces produces $T_0 = T(x) \cos\theta$, and the sum of the vertical forces yields $ws = T \sin\theta$. The forces are related to the equilibrium condition by the fact that $dy/dx = \tan\theta$. These facts are combined to show, using $y(x)$ from Example 7.30 measured from the lowest point on the cable,

$$T(x) = T_0/\cos\theta = T_0 \sec\theta = T_0\sqrt{1 + \tan^2\theta}$$

$$= T_0\sqrt{1 + (dy/dx)^2} = T_0\sqrt{1 + \sinh^2(x/C'')} = T_0 \cosh(x/C'').$$

Therefore at any point a length s from the lowest, using the general solution for $y(x)$ from the previous example, the tension is

$$T(x) = T_0(y(x) + \lambda)/C''.$$

Now select the actual position of the coordinate system by choosing the arbitrary value C'' so that $T_0 = wC''$ and take $\lambda = 0$. Then $y(0) = C''$, and the general expression for the tension at point $(x, y(x))$ in the cable is related to its height in this coordinate system,

$$T(x) = wy(x).$$

Therefore, T_0 is the minimum tension in the cable, and the tension at the support is the maximum. The numerical value of C'' is determined from the constraint equation, $L/2 = C'' \sinh(D/2C'')$.

∎

EXAMPLE 7.32

A chain weighing $w = 0.6$ lb/ft and of length $L = 27$ inches is hung between two sprockets, which are spaced $D = 24$ inches apart. The two sprockets are mounted at the same height. Determine

(a) the maximum sag;

(b) the tension exerted on the sprockets by the chain.

Solution

Following the previous example, choose the coordinate system whose origin is C'' below the midpoint between the two sprockets. First determine C'' from the length constraint, $L/2 = C'' \sinh(D/2C'')$. A numerical equation solver applied to $13.5 = C'' \sinh(12/C'')$ yields $C'' = 14.109$. Then in the chosen coordinates the end of the chain is at $y(D/2) = C'' \cosh(D/2C'') = 14.109 \sinh(12/14.109) = 19.527$.

(a) The sag equals $y(D/2) - C'' = 5.418$ inches.

(b) The tension on the end of the chain is $wy(D/2) = (0.6/12)(19.527) = 0.977$ pounds. Also $T_0 = 0.706$ pounds.

∎

An analysis similar to that for a single integral function can be applied in the case that the potential energy is represented as a double integral, as it might for a constant thickness plate. Suppose that the function, $w(x, y)$, describing the displacement of the body is unknown. The particles making up the body are at positions (x, y) in a region, R, of the plane.

$$\Pi(w(x, y)) = \int \int_R F(x, y, w, w_x, w_y, w_{xx}, w_{xy}, w_{yy})dxdy,$$

where the subscripts on w denote partial derivatives. Perturb the solution, $w(x, y)$, which minimizes this integral to $w(x, y) + \epsilon\eta(x, y)$ for some fixed but arbitrary function, $\eta(x, y)$. Minimizing the potential energy with respect to ϵ and the application of the Taylor series around $\epsilon = 0$ yields, for $I(\epsilon)$ defined in a manner analogous to the single integral case,

$$\left.\frac{dI}{d\epsilon}\right|_{\epsilon=0} = \int \int_R \left(\eta\frac{\partial F}{\partial w} + \eta_x\frac{\partial F}{\partial w_x} + \eta_y\frac{\partial F}{\partial w_y} \right. $$
$$\left. + \eta_{xx}\frac{\partial F}{\partial w_{xx}} + \eta_{xy}\frac{\partial F}{\partial w_{xy}} + \eta_{yy}\frac{\partial F}{\partial w_{yy}} \right) dxdy.$$

The subscripts on η also indicate partial derivatives. Again the goal is to rewrite the integrand to allow use of the fundamental lemma.

Exercise 7.10   Perform the required integration by parts (see Langhaar [1, pp. 93–96]).

An application of the analogous two-dimensional lemma produces the Euler-Lagrange equation.

$$\frac{\partial F}{\partial w} - \frac{\partial}{\partial x}\frac{\partial F}{\partial w_x} - \frac{\partial}{\partial y}\frac{\partial F}{\partial w_y} + \frac{\partial^2}{\partial x^2}\frac{\partial F}{\partial w_{xx}} + \frac{\partial^2}{\partial x\partial y}\frac{\partial F}{\partial w_{xy}} + \frac{\partial^2}{\partial y^2}\frac{\partial F}{\partial w_{yy}} = 0. \quad (7.40)$$

The four natural boundary conditions are, for Cartesian coordinates and the region R a rectangle with sides parallel to the coordinate axes,

$$\eta\left(\frac{\partial F}{\partial w_x} - \frac{\partial}{\partial x}\frac{\partial F}{\partial w_{xx}} - \frac{\partial}{\partial y}\frac{\partial F}{\partial w_{xy}}\right) = 0 \text{ on edge } x = \text{constant}$$

$$\eta_x\frac{\partial F}{\partial w_{xx}} = 0 \text{ on edge } x = \text{constant}$$

$$\eta\left(\frac{\partial F}{\partial w_y} - \frac{\partial}{\partial x}\frac{\partial F}{\partial w_{xy}} - \frac{\partial}{\partial y}\frac{\partial F}{\partial w_{yy}}\right) = 0 \text{ on edge } y = \text{constant}$$

$$\eta_y\frac{\partial F}{\partial w_{yy}} = 0 \text{ on edge } y = \text{constant}.$$

These techniques for finding the equilibrium configuration can be applied to infinite degree of freedom situations for which the potential energy is defined by an integral involving the function representing the configuration. The subject of energy methods is an important part of engineering mathematics. Interested students should pursue it further in a course on energy methods in applied mechanics.

7.9 TRANSVERSELY LOADED PLATES

The study of the bending and vibration of plates goes back almost to the beginning of the development of mechanics. A prize for the solution was offered as early as 1809 by the French Institute. This prize was won in 1815 by the famous mathematician Sophie Germaine. She assumed that, as in beam theory, the analysis could be based on the curvatures of the plate. Because she was working before the full development of the theory of elasticity, her solution was not exactly correct. S. D. Poisson developed the differential equation in about 1830, but assumed that all materials have the identical Poisson ratio of 1/4. Poisson erroneously believed that three boundary conditions were needed, conditions describing the shear force, the twisting moment, and the bending moment. G. R. Kirchhoff published the first valid theory for plates in 1850, in which only two boundary conditions are required. Kirchhoff simplified the problem by writing the potential energy for a constant thickness plate in terms of the curvatures of the middle surface of the plate. To do so, he introduced the two fundamental assumptions of plate analysis. Planar cross sections normal to the flat plate surface remain planar and normal to the midplane during the deformation. Under small deflections, the midplane of the plate does not stretch even though it bends during loading transverse to the plate surface. Subsequent researchers treated the case that the middle surface can stretch in order to solve problems when the load is applied in the same plane as the plate.

The undeformed plate lies in the x-y plane with the z-coordinate perpendicular to the plate. The deformed plate is assumed to be in plane stress. The Kirchhoff assumption of plane stress implies $\tau_{xz} = \tau_{yz} = 0$, so that $\gamma_{xz} = \gamma_{yz} = 0$, however, the displacement in the z-direction is nearly independent of z because the plate is thin. Consequently, along with the Kirchhoff approximation, it is assumed that $\epsilon_z = 0$, which is not consistent with $\sigma_z = 0$ in the linear elastic case by Hooke's law. However, this problem produces only a very small error in the computations. The Kirchhoff assumption is independent of the loading and material of the plate.

When the various theories of plate deformation are developed from energy methods, they are distinguished by their assumption about the strain-displacement relations. A popular theory, which allows for large deformations and generalizes the small strain theory, is that proposed by von Kármán in the early 1900s. In the following, subscripts on the displacement components, (u, v, w), indicate partial derivatives, while subscripts on the strains or material parameters refer to the coordinate directions. The in-plane displacements at mid-thickness of the plate are $\bar{u}$ and $\bar{v}$; w is the displacement of the midplane in the z-direction.

$$\epsilon_x(x, y, z) = \bar{u}_x + \frac{1}{2}w_x^2 - zw_{xx};$$

$$\epsilon_y(x, y, z) = \bar{v}_y + \frac{1}{2}w_y^2 - zw_{yy};$$

$$\gamma_{xy}(x, y, z) = \bar{u}_y + \bar{v}_x - 2zw_{xy} + w_xw_y. \tag{7.41}$$

The strain energy function is

$$U = \frac{1}{2}\int_V \left(\sigma_x\epsilon_x + \sigma_y\epsilon_y + \tau_{xy}\gamma_{xy}\right)dxdydz. \tag{7.42}$$

After substituting the strain displacement relations (7.41), the strain energy becomes

$$U = \frac{1}{2}\int\int\left\{\frac{E}{1-v^2}\left[\left(\bar{u}_x + \frac{1}{2}w_x^2\right)^2 h + \frac{h^3}{12}w_{xx}^2 + 2vh\left(\bar{u}_x + \frac{1}{2}w_x^2\right)\left(\bar{v}_y + \frac{1}{2}w_y^2\right)\right.\right.$$

$$\left. + 2v\frac{h^3}{12}w_{xx}w_{yy} + h\left(\bar{v}_y + \frac{1}{2}w_y^2\right)^2 + \frac{h^3}{12}w_{yy}^2\right]$$

$$\left. + G\left[(\bar{u}_y + \bar{v}_x + w_xw_y)^2 h + \frac{h^3}{3}w_{xy}^2\right]\right\}dxdy. \tag{7.43}$$

7.9.1 Small Bending Deflections of Rectangular Plates

The classical small strain, small deflection theory is obtained from the following strain-displacement relations. The squared terms in the von Kármán model are neglected. The midplane itself is assumed undeformed so that $\bar{u} = \bar{v} = 0$.

$$\epsilon_x(x, y, z) = -z\frac{\partial^2 w}{\partial x^2};$$

$$\epsilon_y(x, y, z) = -z\frac{\partial^2 w}{\partial y^2};$$

$$\gamma_{xy}(x, y, z) = -2z\frac{\partial^2 w}{\partial x\partial y}. \tag{7.44}$$

The minus signs arise because z is measured so that when the second derivatives are positive the plate is in compression when z is positive and tension when z is negative. The result [equation (7.44)] is similar to beam theory in that the in-plane strains are assumed proportional to the linearized curvature of the midplane. The curvature, κ, of a curve in a plane is the reciprocal of the radius of curvature, ρ. The curvature of the deformation of a line parallel to the x-axis on the plate, described by $w(x, y)$ for a fixed value of y, is

$$\kappa = \frac{1}{\rho} = \frac{\frac{\partial^2 w}{\partial x^2}}{\left[1 + \left(\frac{\partial w}{\partial x}\right)^2\right]^{3/2}}.$$

For small deflections, the slope of the deformed line is close to zero because its tangent is nearly horizontal. The expression for small deflections in which $\partial w / \partial x$ is very close to zero is called the linearization of the curvature,

$$\frac{1}{\rho} = \frac{\partial^2 w}{\partial x^2}.$$

Likewise the linearized curvature of a line parallel to the y-axis on the deformed plate is

$$\frac{1}{\rho} = \frac{\partial^2 w}{\partial y^2}.$$

The corresponding stresses are computed (under the approximation that $\epsilon_z = 0$) in terms of the displacement in the z-direction from the assumptions on the strains [equation (7.44)] and Hooke's law for an isotropic linear elastic material.

$$\sigma_x(x, y, z) = \frac{E}{1 - v^2}(\epsilon_x + v\epsilon_y) = \frac{-zE}{1 - v^2}\left(\frac{\partial^2 w}{\partial x^2} + v\frac{\partial^2 w}{\partial y^2}\right);$$

$$\sigma_y(x, y, z) = \frac{E}{1 - v^2}(\epsilon_y + v\epsilon_x) = \frac{-zE}{1 - v^2}\left(\frac{\partial^2 w}{\partial y^2} + v\frac{\partial^2 w}{\partial x^2}\right);$$

$$\tau_{xy}(x, y, z) = G\gamma_{xy} = -2zG\frac{\partial^2 w}{\partial x \partial y}. \tag{7.45}$$

Stresses produce moments about the neutral plane and shear forces on the vertical faces of the plate of thickness h. At each point (x, y),

$$M_x = \int_{-h/2}^{h/2} \sigma_x z\, dz, \quad M_y = \int_{-h/2}^{h/2} \sigma_y z\, dz, \quad M_{xy} = \int_{-h/2}^{h/2} \tau_{xy} z\, dz. \tag{7.46}$$

$$Q_x = \int_{-h/2}^{h/2} \tau_{zx} z\, dz, \quad Q_y = \int_{-h/2}^{h/2} \tau_{zy} z\, dz. \tag{7.47}$$

Consider a rectangular plate with edges of length a and b. The plate defines a region R of the x-y plane, R: $0 \le x \le a$ and $0 \le y \le b$ (Fig. 7.20). The strain energy of the rectangular plate is

$$U = \int\int_R \left\{ \frac{D}{2}\left(\frac{\partial^2 w}{\partial x^2} + \frac{\partial^2 w}{\partial y^2}\right)^2 - D(1 - v)\left[\left(\frac{\partial^2 w}{\partial x^2}\frac{\partial^2 w}{\partial y^2}\right) - \left(\frac{\partial^2 w}{\partial x \partial y}\right)^2\right]\right\} dx\, dy, \tag{7.48}$$

where $D = Eh^3/[12(1 - v^2)]$, sometimes called the flexural rigidity of the isotropic plate. The work done by a transverse load, $p(x, y)$, is $W = \int\int_R w(x, y)p(x, y)dx\, dy$.

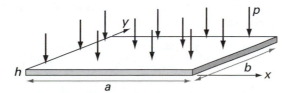

FIGURE 7.20   Flat rectangular plate with transverse constant distributed load.

The potential energy is $\Pi = U - W$. Then by equation (7.40), an isotropic rectangular plate with transverse distributed load $p(x, y)$ has Euler equation

$$\frac{\partial^4 w}{\partial x^4} + 2\frac{\partial^4 w}{\partial x^2 \partial y^2} + \frac{\partial^4 w}{\partial y^4} = \frac{p(x, y)}{D}, \tag{7.49}$$

because the second term in equation (7.48) satisfies the Euler equation for any displacement and so does not contribute. Equation (7.49) can be written as

$$\nabla^4 w = \frac{p}{D}.$$

Unfortunately, this fourth-order partial differential equation has no closed-form solution.

Exercise 7.11   Verify that equation (7.49) can be expressed as the system of two differential equations:

$$\nabla^2 w = \frac{M}{D}, \quad \text{and } \nabla^2 M = p,$$

which may be easier to solve. In applications, M is first obtained from the second equation and then w from the first equation. M is proportional to the sum of the magnitudes of the moments about the x- and y-axes.

$$M = -\frac{M_x + M_y}{1 + \nu}.$$

The natural boundary conditions along the edges are, for $x = 0$ or $x = a$,

$$\eta_x \left[\frac{\partial^2 w}{\partial x^2} + \nu \frac{\partial^2 w}{\partial y^2} \right] = 0; \tag{7.50}$$

$$\eta \left[\frac{\partial^3 w}{\partial x^3} + (2 - \nu)\frac{\partial^3 w}{\partial x \partial y^2} \right] = 0. \tag{7.51}$$

For $y = 0$ or b,

$$\eta_y \left[\frac{\partial^2 w}{\partial x^2} + \frac{\partial^2 w}{\partial y^2} \right] = 0; \tag{7.52}$$

$$\eta \left[\frac{\partial^3 w}{\partial x^3} + (2 - \nu)\frac{\partial^3 w}{\partial x^2 \partial y} \right] = 0. \tag{7.53}$$

The forced boundary conditions describe the displacement w and its slope along each edge.

Exercise 7.12   Verify the Euler equation (7.49) and natural boundary conditions.

Navier, in 1823, proposed that the solution for a simply supported plate must be periodic to automatically satisfy the boundary conditions and therefore should be taken as a double Fourier series,

$$w(x, y) = \sum_{m=1}^{\infty} \sum_{n=1}^{\infty} a_{mn} \sin\left(\frac{m\pi x}{a}\right) \sin\left(\frac{n\pi y}{b}\right). \tag{7.54}$$

The assumption that the plate is simply supported on all edges is conservative in that the deflections are largest compared to other supports.

Exercise 7.13   Verify that such a $w(x, y)$ satisfies simply supported boundary conditions on all four edges of the plate. Use the fact that the moments about the edges $x = 0$ and $x = a$ are approximately $\partial^2 w / \partial x^2$, and the moments about the edges $y = 0$ and $y = b$ are approximately $\partial^2 w / \partial y^2$.

Two equivalent strategies are possible to compute the coefficients, a_{mn}, for a given loading. This expression for $w(x, y)$ can be substituted into the expression for the potential energy, $\Pi = U - W$. The principle of stationary potential energy produces the equations $\partial \Pi / \partial a_{mn} = 0$, which are then solved for the a_{mn}. This is similar to the Rayleigh-Ritz method, except that there are an infinite number of unknowns. Alternatively, the expression (7.54) for $w(x, y)$ is substituted into the Euler equation (7.49) to obtain an infinite sequence of equations involving the a_{mn}.

$$a_{mn} = \frac{4}{\pi^4 abD \left(\frac{m^2}{a^2} + \frac{n^2}{b^2}\right)^2} \int_0^b \int_0^a p(x, y) \sin\left(\frac{m\pi x}{a}\right) \sin\left(\frac{n\pi y}{b}\right) dxdy. \quad (7.55)$$

To obtain this expression, $p(x, y)$ is written as a double Fourier series,

$$p(x, y) = \sum_{m=1}^{\infty} \sum_{n=1}^{\infty} p_{m,n} \sin\left(\frac{m\pi x}{a}\right) \sin\left(\frac{n\pi y}{b}\right) dxdy,$$

where

$$p_{m,n} = \frac{4}{ab} \int_0^b \int_0^a p(x, y) \sin\left(\frac{m\pi x}{a}\right) \sin\left(\frac{n\pi y}{b}\right) dxdy.$$

The expression for a_{mn} is then obtained by comparing the coefficients of the

$$\sin\left(\frac{m\pi x}{a}\right) \sin\left(\frac{n\pi y}{b}\right) dxdy$$

terms in equation (7.49) after substituting the Fourier series for $p(x, y)$.

Exercise 7.14   Verify this expression for a_{mn}.

EXAMPLE 7.33   Estimate the deflection at $(a/2, b/2)$ in a simply supported plate subject to a uniform transverse load, p, by using just the first term of the Navier series [equation (7.54)].

Solution   The estimate is equal to the coefficient a_{11}.

$$a_{11} = \frac{4}{\pi^4 abD \left(\frac{1}{a^2} + \frac{1}{b^2}\right)^2} \int_0^b \int_0^a p \sin\left(\frac{\pi x}{a}\right) \sin\left(\frac{\pi y}{b}\right) dxdy$$

$$= \frac{4p}{\pi^4 abD \left(\frac{1}{a^2} + \frac{1}{b^2}\right)^2} \frac{2a}{\pi} \frac{2b}{\pi} = \frac{16pa^4 b^4}{\pi^6 D(a^2 + b^2)^2}. \quad (7.56)$$

∎

Exercise 7.15   Verify that a_{mn} for a simply supported plate subject to a uniform transverse load, p is

$$a_{mn} = \frac{4p}{\pi^4 abD \left(\frac{m^2}{a^2} + \frac{n^2}{b^2}\right)^2} \frac{ab}{\pi^2 mn} \left(\cos(m\pi) - 1\right)\left(\cos(n\pi) - 1\right)$$

$$= \frac{4p}{\pi^6 nmD \left(\frac{m^2}{a^2} + \frac{n^2}{b^2}\right)^2} [(-1)^m - 1][(-1)^n - 1].$$

Therefore $a_{mn} = 0$ whenever either of m and n is even. Furthermore, the coefficients a_{mn} decrease rapidly as m and n increase.

EXAMPLE 7.34

The rectangular wall of a pressure vessel is used as a pressure sensor. The wall is a plate supported in nearly frictionless grooves so that it can be taken as simply supported. A contact is mounted adjacent to the midpoint of the plate so that the deformed plate would just hit it and complete a sensor circuit at the required pressure. The load on the plate is a constant transverse pressure, p. How far should the sensor be from the plate?

Solution

A good approximation to the deflection is given by the first term, when $m = 1$ and $n = 1$.

$$w(x, y) = \frac{16p}{\pi^6 D \left(\frac{1}{a^2} + \frac{1}{b^2} \right)^2} \sin \left(\frac{\pi x}{a} \right) \sin \left(\frac{\pi y}{b} \right).$$

At the center of the plate,

$$w \left(\frac{a}{2}, \frac{b}{2} \right) = \frac{16p}{\pi^6 D \left(\frac{1}{a^2} + \frac{1}{b^2} \right)^2}.$$

This is the distance from the undeformed plate at which the contact should be placed to signal when the desired pressure is reached in the chamber. ∎

Exercise 7.16   For the pressure vessel in the previous example, calculate the design displacement required to detect a 100 psi pressure if the wall is a 6 inches by 6 inches by 1/16-inch steel plate ($E = 30 \times 10^6$, $\nu = 0.25$).

The stresses in a simply supported plate are obtained from the double Fourier series expression [equation (7.54)] for $w(x, y)$, the strain-displacement relation (7.44), and Hooke's law.

$$\sigma_x(x, y, z) = \frac{-zE}{1 - \nu^2} \left(\frac{\partial^2 w}{\partial x^2} + \nu \frac{\partial^2 w}{\partial y^2} \right)$$

$$= \frac{zE}{1 - \nu^2} \left[\sum_{m=1}^{\infty} \sum_{n=1}^{\infty} a_{mn} \left[\left(\frac{m\pi}{a} \right)^2 + \nu \left(\frac{n\pi}{b} \right)^2 \right] \sin \left(\frac{m\pi x}{a} \right) \sin \left(\frac{n\pi y}{b} \right) \right].$$

(7.57)

Exercise 7.17   Verify that $\sigma_y(x, y, z)$ and $\tau_{xy}(x, y, z)$ are given respectively by the expressions

$$\sigma_y(x, y, z) = \frac{zE}{1 - \nu^2} \left[\sum_{m=1}^{\infty} \sum_{n=1}^{\infty} a_{mn} \left[\left(\frac{n\pi}{b} \right)^2 + \nu \left(\frac{m\pi}{a} \right)^2 \right] \sin \left(\frac{m\pi x}{a} \right) \sin \left(\frac{n\pi y}{b} \right) \right]; \quad (7.58)$$

$$\tau_{xy}(x, y, z) = -2zG \sum_{m=1}^{\infty} \sum_{n=1}^{\infty} a_{mn} \left(\frac{m\pi}{a} \right) \left(\frac{n\pi}{b} \right) \cos \left(\frac{m\pi x}{a} \right) \cos \left(\frac{n\pi y}{b} \right). \quad (7.59)$$

The maximum tensile stress at the center on the surface of a simply supported plate of thickness, h, is

$$\sigma_x \left(\frac{a}{2}, \frac{b}{2}, \frac{h}{2} \right) =$$

$$\frac{hE}{2(1 - \nu^2)} \left[\sum_{m=1}^{\infty} \sum_{n=1}^{\infty} a_{mn} \left[\left(\frac{m\pi}{a} \right)^2 + \nu \left(\frac{n\pi}{b} \right)^2 \right] \sin \left(\frac{m\pi}{2} \right) \sin \left(\frac{n\pi}{2} \right) \right]$$

if the length a is the short side. An approximation is obtained by taking the stress to one term, $m = n = 1$.

$$\sigma_x \left(\frac{a}{2}, \frac{b}{2}, \frac{h}{2} \right) = \frac{hE}{2(1 - v^2)} a_{11} \left[\left(\frac{\pi}{a} \right)^2 + v \left(\frac{\pi}{b} \right)^2 \right].$$

EXAMPLE 7.35   A 5-inch by 8-inch plate of thickness 0.125 inch is simply supported and made of an aluminum ($E = 10 \times 10^6$, $v = 0.33$) whose yield stress is $\sigma_Y = 80$ ksi. Using the one-term approximation, estimate the largest uniform transverse load the plate can carry without either principal normal stress at the surface at midplate exceeding yield.

Solution   Put the x-axis along the 8-inch edge and the y-axis along the 5-inch edge. Then the midplate is at $(x, y) = (4, 2.5)$. The stresses on the surface at midplate are

$$\sigma_x = \frac{hE}{2(1 - v^2)} a_{11} \left[\left(\frac{\pi}{a} \right)^2 + v \left(\frac{\pi}{b} \right)^2 \right] = \frac{96a^4b^4}{h^2\pi^6(a^2 + b^2)^2} p \left[\left(\frac{\pi}{a} \right)^2 + v \left(\frac{\pi}{b} \right)^2 \right] = 587.6p;$$

$$\sigma_y = \frac{hE}{2(1 - v^2)} a_{11} \left[\left(\frac{\pi}{b} \right)^2 + v \left(\frac{\pi}{a} \right)^2 \right] = \frac{96a^4b^4}{h^2\pi^6(a^2 + b^2)^2} p \left[\left(\frac{\pi}{b} \right)^2 + v \left(\frac{\pi}{a} \right)^2 \right] = 920.5p;$$

$$\tau_{xy} = 0,$$

where a_{11} is given by equation (7.56). Note that the shear stress is zero only at midplate. σ_y is the largest principal stress, and the maximum allowable constant transverse load is $p = 86.9$ psi. Notice that the value of the elastic modulus does not influence the calculation. Further, keeping each stress below the yield stress may not guarantee that yield does not occur (see Biaxial Yield Criteria, Chapter 9, Section 9.6). ∎

More complex edge conditions require a different assumption for the displacement, w. One that is satisfactory for plates simply supported on the two opposite edges parallel to the y-axis and free on the other two edges is

$$w(x, y) = \sum_{m=1}^{\infty} f(y) \sin \left(\frac{m\pi x}{a} \right),$$

where the function, $f(y)$, is to be determined from the differential equation. Again, this function automatically satisfies the boundary conditions. Assumption of this form for $w(x, y)$ is often called the Lévy solution.

7.9.2   Bending of Circular Plates

A circular plate can be given polar coordinates with origin at the center of the plate (Fig. 7.21) so that it defines a region R of the r-θ plane.

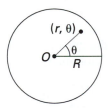

FIGURE 7.21   The polar coordinate system for a circular plate.

When the plate is given polar coordinates, the transverse deflection is $w(r, \theta)$; the displacements are $u = -z \frac{\partial w}{\partial r}$ in the r-direction and $v = -z \frac{1}{r} \frac{\partial w}{\partial \theta}$ in the θ-direction.

The strain displacement relations are, by Chapter 2, Section 2.1,

$$\epsilon_r(r, \theta, z) = -z \frac{\partial^2 w}{\partial r^2};$$

$$\epsilon_\theta(r, \theta, z) = -z \left(\frac{1}{r} \frac{\partial w}{\partial r} + \frac{1}{r^2} \frac{\partial^2 w}{\partial \theta^2} \right);$$

$$\gamma_{r\theta}(r, \theta, z) = 2z \left(\frac{1}{r^2} \frac{\partial w}{\partial \theta} - \frac{1}{r} \frac{\partial^2 w}{\partial r \partial \theta} \right). \tag{7.60}$$

The strain energy of the circular plate with transverse deflection, $w(r, \theta)$, is obtained from that of a rectangular plate by the transformation $x = r \cos \theta$ and $y = r \sin \theta$. Again, D is the flexural rigidity.

$$U = \int \int_R \frac{D}{2} \left(\frac{\partial^2 w}{\partial r^2} + \frac{1}{r} \frac{\partial w}{\partial r} + \frac{1}{r^2} \frac{\partial^2 w}{\partial \theta^2} \right)^2$$

$$- D(1 - v) \left[\frac{\partial^2 w}{\partial r^2} \left(\frac{1}{r} \frac{\partial w}{\partial r} + \frac{1}{r^2} \frac{\partial^2 w}{\partial \theta^2} \right) + D(1 - v) \left(\frac{1}{r} \frac{\partial^2 w}{\partial r \partial \theta} - \frac{1}{r} \frac{\partial w}{\partial \theta} \right)^2 \right] r \, dr \, d\theta. \tag{7.61}$$

Consider an isotropic, circular plate of radius, R, and thickness, h, which has clamped edges as the boundary conditions and is subjected to a transverse pressure, $p(r, \theta)$. The Euler equation is

$$\nabla^4 w = \left(\frac{\partial^2}{\partial r^2} + \frac{1}{r} \frac{\partial}{\partial r} + \frac{1}{r^2} \frac{\partial^2}{\partial \theta^2} \right) \left(\frac{\partial^2}{\partial r^2} + \frac{1}{r} \frac{\partial}{\partial r} + \frac{1}{r^2} \frac{\partial^2}{\partial \theta^2} \right) w = \frac{p(r, \theta)}{D}. \tag{7.62}$$

Assume that the transverse load is uniform, $p(r, \theta) = p$. The load and the boundary conditions are symmetric so that the displacement is a function of only the radius, $w = w(r)$. Therefore, the derivatives involving θ in the Euler differential equation are zero so that the governing differential equation is

$$\frac{d^4 w}{dr^4} + \frac{2}{r} \frac{d^3 w}{dr^3} - \frac{1}{r^2} \frac{d^2 w}{dr^2} + \frac{1}{r^3} \frac{dw}{dr} = \frac{p}{D}.$$

This is a Euler-type differential equation and can be solved by first multiplying through by r^4 and then making the substitution $t = \ln(r)$. The homogeneous solution must have the form $w_h(r) = A + Br^2 + C_1 \ln r + C_2 r^2 \ln r$. The particular solution is $w_p(r) = pr^4/64D$. The deflection has the form

$$w(r) = A + Br^2 + C_1 \ln r + C_2 r^2 \ln r + \frac{pr^4}{64D}.$$

The displacement must not be infinite at the center of the plate ($r = 0$). Therefore $C_1 = C_2 = 0$. The forced boundary conditions are that $w(R) = 0$ and $w'(R) = 0$ since the edges are clamped. The first condition implies that

$$A + BR^2 + \frac{pR^4}{64D} = 0,$$

and the second that

$$2BR + \frac{pR^3}{16D} = 0.$$

Solving simultaneously implies that

$$w(r) = \frac{pR^4}{64D} - \frac{pR^2}{32D}r^2 + \frac{pr^4}{64D} = \frac{p}{64D}(R^2 - r^2)^2. \tag{7.63}$$

When the plate is instead simply supported, the boundary conditions are that $w(R) = 0$ and the moment, M_r at the edge must be zero, where

$$M_r = \int_{-h/2}^{h/2} z\sigma_r dz = -D\left[\frac{d^2w}{dr^2} + v\left(\frac{1}{r}\frac{dw}{dr} + \frac{1}{r^2}\frac{d^2w}{d\theta^2}\right)\right],$$

by Hooke's law. The isotropic, circular plate of radius R and thickness h subjected to a constant transverse pressure, p, over the full surface has deflection

$$w(r) = \frac{pR^4}{64D}\left(\frac{5+v}{1+v}\right) - \frac{R^2}{32D}\left(\frac{3+v}{1+v}\right)r^2 + \frac{pr^4}{64D}. \tag{7.64}$$

Exercise 7.18 Verify that this expression satisfies the differential equation and the boundary conditions.

Next, consider an isotropic, simply supported, circular plate of radius, R, and thickness, h, with point load, F, applied transversely at its center. The Euler equation is

$$\frac{d^4w}{dr^4} + \frac{2}{r}\frac{d^3w}{dr^3} - \frac{1}{r^2}\frac{d^2w}{dr^2} + \frac{1}{r^3}\frac{dw}{dr} = 0. \tag{7.65}$$

The natural boundary conditions are

$$v\frac{dw}{dr} + r\frac{d^2w}{dr^2} = 0 \quad \text{if} \quad r = R$$

$$\lim_{r\to 0}\left(r\frac{d^3w}{dr^3} + \frac{d^2w}{dr^2} - \frac{1}{r}\frac{dw}{dr}\right) = \frac{F}{2\pi D}.$$

The forced boundary conditions are that $w(R) = 0$ and $M_r = 0$. Again, the solution to equation (7.65) is

$$w(r) = A + Br^2 + C_1 r^2 \ln r + C_2 \ln r.$$

The displacement is found from the forced and natural boundary conditions to be

$$w(r) = \frac{F}{16\pi D}\left[\frac{3+v}{1+v}(R^2 - r^2) - 2r^2 \ln\left(\frac{R}{r}\right)\right]. \tag{7.66}$$

The deflection under the concentrated point load, F, by taking the limit of $w(r)$ as r tends to zero, is

$$w(0) = \frac{FR^2}{16\pi D}\frac{3+v}{1+v}.$$

However, there is a singularity in the second derivative at $r = 0$ because of the presence of the concentrated load at that location. The second derivative of $w(r)$ is discontinuous there. In contrast, the second derivative of the deflection for beam theory is continuous at a concentrated load.

The stresses in any circular plate are obtained in terms of polar coordinates by a calculation similar to that for rectangular coordinates because Hooke's law is valid for any orthogonal coordinate system. The three-dimensional coordinate system in which the plate lies is cylindrical and again the approximation is made that $\epsilon_z = 0$.

Exercise 7.19   Verify that the expressions for σ_r, σ_θ, and $\tau_{r\theta}$ in terms of $w(r, \theta)$ are

$$\sigma_r(r, \theta, z) = \frac{-zE}{1 - \nu^2}\left[\frac{\partial^2 w}{\partial r^2} + \frac{\nu}{r}\frac{\partial w}{\partial r} + \frac{\nu}{r^2}\frac{\partial^2 w}{\partial \theta^2}\right];$$

$$\sigma_\theta(r, \theta, z) = \frac{-zE}{1 - \nu^2}\left[\nu\frac{\partial^2 w}{\partial r^2} + \frac{1}{r}\frac{\partial w}{\partial r} + \frac{1}{r^2}\frac{\partial^2 w}{\partial \theta^2}\right];$$

$$\tau_{r\theta}(r, \theta, z) = 2zG\left[\frac{1}{r^2}\frac{\partial w}{\partial \theta} - \frac{1}{r}\frac{\partial^2 w}{\partial r \partial \theta}\right]. \tag{7.67}$$

EXAMPLE 7.36   An isotropic, circular plate of radius, R, and thickness, h, which has clamped edges, is subjected to a uniform transverse pressure, p. Determine the principal stresses and the maximum shear stress on the surface at the center of the plate.

Solution   The deflection is given by equation (7.63). The derivatives of $w(r)$ with respect to r are

$$w'(r) = -(pr/16D)(R^2 - r^2) \quad \text{and} \quad w''(r) = (-p/16D)r(R^2 - 3r^2).$$

Therefore, the stresses on the surface at the center are, by equation (7.67),

$$\sigma_r(0, \theta, h/2) = \frac{3p(1 + \nu)R^2}{8h^2};$$

$$\sigma_\theta(0, \theta, h/2) = \sigma_r(0, \theta, h/2);$$

$$\tau_{r\theta}(0, \theta, h/2) = 0.$$

Again, the result is independent of the elastic modulus. Because $w(r)$ is not a function of θ, $\tau_{r\theta}$ is always zero, and σ_r and σ_θ are the principal stresses at each point. At the center, the maximum shear stress, by Mohr's circle, is

$$\tau_{max} = \frac{1}{2}(\sigma_r - \sigma_\theta) = 0. \qquad\blacksquare$$

7.9.3 Orthotropic Plates

A linear elastic orthotropic plate is one in which the material properties, such as the elastic modulus or Poisson's ratio, differ in orthogonal directions in the plane of the undeformed plate. The orthotropic plate analysis is applicable for rolled metal sheets in which the rolling process elongates the grains in a preferred direction so that the metal is no longer isotropic. Such a plate can often be modeled as an orthotropic material having a coordinate axis in the direction of rolling. Likewise the model can be applied to fibered composite plates with fibers running parallel the length of plate and embedded in a matrix or to a natural composite such as wood in which the parallel cells forming the grain are viewed a fibers. Generally, to apply an orthotropic model, the material properties are taken as averages in the respective directions.

Assume small strains, small deflections, and that the constitutive model for the material is linear orthotropic.

$$\sigma_x = \frac{E_x}{1 - \nu_x \nu_y}(\epsilon_x + \nu_y \epsilon_y);$$

$$\sigma_y = \frac{E_y}{1 - \nu_x \nu_y}(\epsilon_y + \nu_x \epsilon_x);$$

$$\tau_{xy} = G\gamma_{xy}. \tag{7.68}$$

In the notation of Chapter 3, Section 3.3, $\nu_y = \nu_{yx}$ and $\nu_x = \nu_{xy}$. Put $\mu = 1 - \nu_x \nu_y$ and $\bar{E} = E_y/E_x$. After substituting the strain displacement relations equation (7.41), the strain energy becomes

$$U = \frac{1}{2} \int \int \left\{ \frac{E_x}{\mu} \left[\left(\bar{u}_x + \frac{1}{2}w_x^2 \right)^2 h + \frac{h^3}{12} w_{xx}^2 + 2\nu_y h \left(\bar{u}_x + \frac{1}{2}w_x^2 \right) \left(\bar{v}_y + \frac{1}{2}w_y^2 \right) \right. \right.$$
$$\left. + 2\nu_y \frac{h^3}{12} w_{xx} w_{yy} + \bar{E} h \left(\bar{v}_y + \frac{1}{2}w_y^2 \right)^2 + \bar{E}\frac{h^3}{12} w_{yy}^2 \right]$$
$$\left. + G \left[(\bar{u}_y + \bar{v}_x + w_x w_y)^2 h + \frac{h^3}{3} w_{xy}^2 \right] \right\} dxdy. \tag{7.69}$$

Exercise 7.20 Determine the Euler equation for an orthotropic plate in the case of the nonlinear strain-displacement assumptions (7.41).

The strain energy for the small displacement–small strain assumption of equation (7.44) is

$$U = \frac{1}{2} \int \int \left[\frac{E_x}{\mu} \left(\frac{h^3}{12} w_{xx}^2 + 2\nu_y \frac{h^3}{12} w_{xx} w_{yy} + \bar{E}\frac{h^3}{12} w_{yy}^2 \right) + G\frac{h^3}{3} w_{xy}^2 \right] dxdy. \tag{7.70}$$

In the case of the small strain-displacement assumption, assume $E_x \nu_y = E_y \nu_x$. Put

$$D_x = \frac{h^3 E_x}{12(1 - \nu_x \nu_y)}, \qquad D_y = \frac{h^3 E_y}{12(1 - \nu_x \nu_y)},$$

$$D_{xy} = \frac{h^3 E_x \nu_y}{12(1 - \nu_x \nu_y)} = \frac{h^3 E_y \nu_x}{12(1 - \nu_x \nu_y)}, \quad \text{and} \quad G_{xy} = \frac{h^3 G}{12}.$$

Then the Euler equation for small displacements and for transverse load, $p(x, y)$, is

$$D_x \frac{\partial^4 w}{\partial x^4} + 2(D_{xy} + 2G_{xy})\frac{\partial^4 w}{\partial x^2 \partial y^2} + D_y \frac{\partial^4 w}{\partial y^4} = p(x, y). \tag{7.71}$$

Typically the partial differential equation is solved by guessing a form of the solution that automatically satisfies the boundary conditions. This form is substituted into the differential equation to determine unknown coefficients or functions.

The partial differential equation (7.71) can be solved, if the plate is simply supported, by making the Navier assumption for the form of the displacement,

$$w(x, y) = \sum_{m=1}^{\infty} \sum_{n=1}^{\infty} w_{mn} \sin\left(\frac{m\pi x}{a}\right) \sin\left(\frac{n\pi y}{b}\right).$$

Exercise 7.21 Verify that $w(x, y)$ automatically satisfies the simple support forced boundary conditions, irrespective of the choices of the w_{mn}.

Assume further that the transverse load can be expressed as a Fourier series,

$$p(x, y) = \sum_{m=1}^{\infty} \sum_{n=1}^{\infty} p_{mn} \sin\left(\frac{m\pi x}{a}\right) \sin\left(\frac{n\pi y}{b}\right).$$

Then

$$w_{mn} = \frac{p_{mn}}{D_x \left(\frac{m\pi}{a}\right)^4 + 2(D_{xy} + 2G_{xy})\left(\frac{m\pi}{a}\right)^2 \left(\frac{n\pi}{b}\right)^2 + D_y \left(\frac{n\pi}{b}\right)^4},$$

where

$$p_{mn} = \frac{4}{ab} \int_0^b \int_0^a p(x, y) \sin\left(\frac{m\pi x}{a}\right) \sin\left(\frac{n\pi y}{b}\right) dx dy.$$

If $p(x, y) = p$ is a constant, then

$$p_{mn} = \frac{4p}{mn\pi^2}[(-1)^m - 1][(-1)^n - 1].$$

Exercise 7.22 Reduce the expression for w_{mn} to the isotropic case and compare to previously obtained results.

If the plate is not simply supported on all four edges, but is simply supported only on two edges, the Lévy form of the solution may be used.

$$w(x, y) = \sum_{n=1}^{\infty} f_m(x) \sin\left(\frac{n\pi y}{b}\right).$$

The stress computation in the small strain, small deflection case depends on equation (7.44) and on Hooke's law for a linear orthotropic material [equation (7.68)].

$$\sigma_x(x, y, z) = \frac{E_x}{1 - \nu_x \nu_y}(\epsilon_x + \nu_y \epsilon_y) = \frac{-zE_x}{1 - \nu_x \nu_y}\left(\frac{\partial^2 w}{\partial x^2} + \nu_y \frac{\partial^2 w}{\partial y^2}\right); \quad (7.72)$$

$$\sigma_y(x, y, z) = \frac{E_y}{1 - \nu_x \nu_y}(\epsilon_y + \nu_x \epsilon_x) = \frac{-zE_y}{1 - \nu_x \nu_y}\left(\frac{\partial^2 w}{\partial y^2} + \nu_x \frac{\partial^2 w}{\partial x^2}\right); \quad (7.73)$$

$$\tau_{xy}(x, y, z) = G\gamma_{xy} = -2zG\frac{\partial^2 w}{\partial x \partial y}. \quad (7.74)$$

EXAMPLE 7.37 Determine the principal stresses and the maximum shear stress on the surface at the midplane of a simply supported linear orthotropic plate with sides a and b. The plate is loaded with a transverse uniform pressure, p. Use the one-term approximation.

Solution   Choose a coordinate system with the x-axis along the edge of length a and the y-axis along the edge of length b. The stresses are

$$\sigma_x(a/2, b/2, h/2) = \frac{-hE_x}{2(1 - \nu_x\nu_y)} \left(\frac{\partial^2 w}{\partial x^2} + \nu_y \frac{\partial^2 w}{\partial y^2} \right)$$

$$= \frac{hE_x}{2(1 - \nu_x\nu_y)} w_{11} \left[\left(\frac{\pi}{a}\right)^2 + \nu_y \left(\frac{\pi}{b}\right)^2 \right]; \qquad (7.75)$$

$$\sigma_y(a/2, b/2, h/2) = \frac{-hE_y}{2(1 - \nu_x\nu_y)} \left(\frac{\partial^2 w}{\partial y^2} + \nu_x \frac{\partial^2 w}{\partial x^2} \right)$$

$$= \frac{hE_y}{2(1 - \nu_x\nu_y)} w_{11} \left[\left(\frac{\pi}{b}\right)^2 + \nu_x \left(\frac{\pi}{a}\right)^2 \right]; \qquad (7.76)$$

$$\tau_{xy}(a/2, b/2, h/2) = 0, \qquad (7.77)$$

where

$$w_{11} = \frac{16p}{\pi^2 \left[D_x \left(\frac{\pi}{a}\right)^4 + 2(D_{xy} + 2G_{xy}) \left(\frac{\pi}{a}\right)^2 \left(\frac{\pi}{b}\right)^2 + D_y \left(\frac{\pi}{b}\right)^4 \right]}.$$

The principal stresses at the midplate are therefore $\sigma_x(a/2, b/2, h/2)$ and $\sigma_y(a/2, b/2, h/2)$. The maximum shear stress is $0.5[\sigma_x(a/2, b/2, h/2) + \sigma_y(a/2, b/2, h/2)]$. ∎

7.10 SOLUTION OF DIFFERENTIAL EQUATIONS USING ENERGY METHODS

The energy method produces differential equations for the displacements, the Euler equations, which may be quite difficult to solve. An alternative might be to work directly with the integral for the potential energy to find a solution numerically. This strategy is that taken by the finite element method described in Chapter 8. In fact, one can begin with a differential equation model for a system, recover the integral for which the original differential equation is the Euler equation, and then apply a numerical technique to the integral as a means of solving the original differential equation.

Suppose a differential equation is written in the form, $\mathcal{L}(y) = f(x)$, where $\mathcal{L}(y)$ is a linear differential operator involving only even-order derivatives. A solution of the differential equation is $y(x)$. An integral function that has Euler equation equal to the original differential equation is

$$I(y) = \int_a^b [\mathcal{L}(y)y - 2yf(x)]dx.$$

There may be other integral functions with the same Euler equation, but this one often has a physical interpretation when the differential equation describes the static behavior of a structure.

EXAMPLE 7.38   The differential equation for a simply supported, strength of materials beam of length L (not the operator) is $EIy'''' = P$ with boundary conditions $y(0) = y(L) = y''(0) = y''(L) = 0$. The integral function for which it is the Euler equation is

$$I(y) = \int_0^L (EIy''''y - 2Py)dx.$$

To return this integral equation to the standard form for a beam, integrate the first term twice by parts.

$$I(y) = \int_0^L (EIy''''y - 2Py)dx = EIy'''y\big|_0^L - \int_0^L (EIy'''y' - 2Py)dx$$

$$= 0 - EIy''y'\big|_0^L + \int_0^L (EIy''y'' - 2Py)dx = \int_0^L (EIy''^2 - 2Py)dx.$$

This integral is twice the integral for the potential energy of the beam, but it has the same minimizing function, $y(x)$. Verify this result by computing the Euler equation of the integral function. ∎

PROBLEMS

7.1 The cantilever beam has length, L, and is loaded by a vertical force, P, on the free end. Work in the coordinate system shown. Make the strength of materials assumptions and neglect shear.

(a) Use Castigliano's theorem to determine the deflection at the free end of the cantilever beam.

(b) Use Castigliano's theorem to determine the slope of the free end of the cantilever beam.

(c) Verify your answers using the strength of materials technique of integrating the moment.

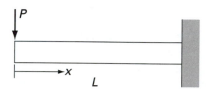

7.2 The cantilever beam has length, L, and is loaded by a couple, C, on the free end. Work in the coordinate system shown.

(a) Use Castigliano's theorem to determine the deflection at the free end of the cantilever beam. Make the strength of materials assumptions and neglect shear.

(b) Use Castigliano's theorem to determine the deflection at $x = L/2$ of the cantilever beam. Make the strength of materials assumptions and neglect shear.

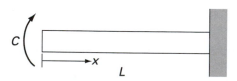

7.3 A cantilever beam has length L and is loaded by a constant distributed load, p. Make the strength of materials assumptions and neglect shear.

(a) Use Castigliano's theorem to determine the vertical deflection of the free end in terms of p, L, E, and the area moment of inertia I.

(b) Use Castigliano's theorem to determine the slope of the free end in terms of p, L, E, and the area moment of inertia I. Work in a coordinate system from the free end.

7.4 The cantilever beam shown has length $2L$ and is loaded by a constant distributed load on its right-hand half. Use Castigliano's theorem to determine the vertical deflection of the free end in terms of w, L, E, and the area moment of inertia I. Make the strength of materials assumptions and neglect shear.

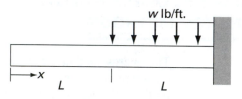

7.5 Use Castigliano's theorem to determine the deflection of the free end of the following cantilever beam. The beam has length, L, and is loaded by a constant distributed load, w, and by a couple, $wL^2/6$, on the free end. Make the strength of materials assumptions and neglect shear.

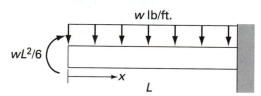

7.6 A cantilever beam of length L and constant EI has a roller placed under its free end so that the free end of the beam just touches the roller. A constant transverse distributed load, p, is applied to the beam pushing it against the roller. Determine the reaction force the roller exerts on the beam.

7.7 By Castigliano's theorem, derive a formula for the deflection, q, at the center of a simply supported linear elastic beam of rectangular cross section with a single load at the center, taking shear deformation into account.
Ans. $q = PL^3/48EI + 0.3PL/GA$.

7.8 A simply supported beam with overhang is supported by a pin at A and a roller at B. It is elastically loaded by a constant vertical distributed load p. Use Castigliano's theorem to determine the change in slope (from undeformed to deformed state) at the support B. Neglect shear.

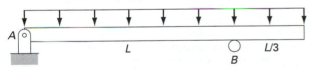

7.9 A ring of radius, $R = 3$ inches to the cross-sectional centroid, and cross-sectional radius of $r = 0.25$ inches is pulled on opposite sides by forces $P = 200$ pounds. The proportional limit for the material of the ring is $\sigma_{PL} = 56$ ksi in both tension and compression.

(a) In the face of the ring perpendicular to the applied loads, determine the maximum tensile and compressive stresses.

(b) Repeat this calculation for the case that the loads are compressive rather than tensile.

(c) Which loading, tensile or compressive, is safer?

7.10 For the ring of problem 9, locate the neutral axis in the face of the ring perpendicular to the applied loads if

(a) the loading is tensile,

(b) the loading is compressive.

7.11 A chain is made of flat circular links of outer radius 1 inch and inner radius of 0.5 inches. Design the thickness t so that the maximum tensile stress in the face perpendicular to the line of action of the load $P = 500$ pounds is less than 40,000 psi.

7.12 Gymnast's rings are suspended by a cable from the ceiling. The ring is required to support the dynamic load from a 150-pound gymnast with a safety factor of 4. The ring is made of a 2024-T4 aluminum with $E = 10.8 \times 10^6$ psi and $\sigma_Y = 46$ ksi. The ring is encased in a rubberlike material whose influence can be neglected. The ring has a diameter of 6 inches measured between the centroids of opposite cross sections. The cross section is to be circular of radius r.

(a) Design the dimension r so that the deflection, δ, is less than 0.016 inches at the vertical cross section where the load is applied.

(b) Determine the maximum tensile stress under the force from the gymnast. Use the Winkler-Bach theory and verify that the ring is elastically loaded.

7.13 A 3-inch-diameter circular hole is cut in the center of a 5-inch-diameter, 0.5-inch-thick steel plate ($G = 12 \times 10^6$, $E = 30 \times 10^6$). A cut is also made along one radius. How much force must be applied normal to the faces along the radial cut to separate the outer edges of the cut faces 0.001 inches at the centroid of the face of the cut?

7.14 For a uniform straight member loaded axially by force, P, having length L, cross-sectional area, A, and made of a

material having stress-strain relation $\sigma = E\epsilon^n$, ($n \neq -1$ and usually $0 < n < 1$), write the total strain energy in terms of the load and the axial displacement, e. (*Note:* The case $n = 1/3$ has been used as an approximation for rubber.)

7.15 For a uniform straight member loaded axially by force, P, having length L, cross-sectional area, A, and made of a material having stress-strain relation $\epsilon = (1/K) \sinh(K\sigma/E)$, write the total complementary energy in terms of the load and the displacement, e. (This relation has been used to model paper.)

7.16 For a uniform straight member loaded axially by force, P, having length L, cross-sectional area, A, and made of a material having stress-strain relation $\sigma = E_1\epsilon + E_2\epsilon^3$, write the total strain energy in terms of the load and the displacement, e. (This relation has been used to model wood.)

7.17 Determine the deflection, q, at the free end of a uniform, linear elastic, cantilever beam of length, L, that carries a point load, P, at the free end. The cross-sectional area is rectangular with depth $2h$ and width b. The material is such that $\sigma = K\epsilon^3$.

7.18 A simply supported beam with rectangular cross section of depth h and width b carries a concentrated load, P, at its center. The length is L. The stress-strain relation of the material is $\sigma = K\epsilon^{1/3}$. Neglecting shear deformation, derive a formula for the deflection, q, of the center by Castigliano's theorem. Ans. $q = 343P^3L^5/(2160b^3K^3h^7)$.

7.19 A one degree of freedom, conservative system has potential energy $V(x) = 2x^3 - 3x^2 - 36x + 2$. Determine the equilibrium states in terms of the single coordinate x and determine whether the state is stable or unstable.

7.20 A rigid bar of length $L = 10$ cm is initially in a vertical position and is supported by a pin at the bottom. A torsional spring at the pin offers resistance to rotation of the bar; its spring constant is $K = 100$ Nm. A vertical load, $P = 1,200$ N, is applied at the top of the bar.

(a) Determine the equilibrium states in terms of the angle from the vertical. Decide whether the states are stable or unstable.

(b) Determine the equilibrium states and their stability if $P = 200$ N.

7.21 A rigid bar of weight W and length L is supported on the left end by a frictionless pin and on the right end by a vertical spring. The spring constant is k. Assume that the spring remains perpendicular to the bar. The spring is unstretched when the bar is horizontal.

(a) Use the principle of stationary potential energy to determine the equilibrium position of the bar.

(b) How far is the spring stretched in the equilibrium position?

(c) Verify your answer to (a) using static equilibrium methods.

(d) Is this equilibrium position stable? Prove your answer.

7.22 A rigid bar of weight W and length L is supported on the left end by a frictionless pin. It is supported at the middle by a spring with constant, k_1, and on the right end by a vertical spring with constant, k_2. Assume that the springs remain perpendicular to the bar. The springs are unstretched when the bar is horizontal.

(a) Use the principle of stationary potential energy to determine the equilibrium position of the bar.

(b) How far are the springs stretched in the equilibrium position?

(c) Verify your answer to (a) using static equilibrium methods.

(d) Is this equilibrium position stable? Prove your answer.

7.23 A steel bar is designed to be supported by a horizontal pin at its center and a vertical spring at each end. The weight of the bar is represented as a constant distributed load of w. Since the specific weight of steel is 0.283 lb/in.³ and the rectangular cross section is 4 inches by 5 inches, $w = 4.245$ lb/in. Both linear springs have spring constant, $k = 0.2$ lb/in. The manufacturing process cannot be expected to locate the pin exactly at the center. The bar is allowed to be at most 0.05 radians off horizontal at equilibrium.

(a) Use the method of stationary potential energy to determine the maximum allowable ratio, $r = L_l/L_s$, of the length L_l of the longer side to the length L_s of the shorter side after manufacturing. Assume that the rod is rigid. (The ratio determines the design tolerance.) Since the displacement is small, the approximation $\sin\theta \sim \theta$ may be used.

(b) Determine whether or not the equilibrium position is stable.

7.24 A spherical membrane is pressurized and is made of a neo-Hookean material. Use the potential energy [equation (7.21)] to describe the equilibria. Assume that $p = PA/4c_1H = 0.5$.

(a) Plot $\Pi(\lambda)/4\pi c_1 A^2 H$ for $1 \le \lambda \le 5.5$.

(b) Determine the two equilibria stretches λ and decide which is stable and which is unstable.

7.25 A spherical membrane is pressurized and is made of a Mooney-Rivlin material. Use the potential energy (7.21) to describe the equilibria. Assume that $p = PA/4c_1H = 0.7$ and that $\Gamma = 0.1$.

(a) Plot $\Pi(\lambda)/4\pi c_1 A^2 H$ for $1 \le \lambda \le 5.5$.

(b) Determine the three equilibria stretches λ and decide whether they are stable or unstable.

7.26 A cantilever beam of length, L, and constant bending stiffness, EI, supports a constant distributed load of w N/m. A Rayleigh-Ritz approximation is made using the vertical displacement $y(x) = a[1 - \cos(\pi x/2L)]$, where x is measured from the wall.

(a) Determine the coefficient, a.

(b) Determine the internal moment function, $M(x)$, predicted by this analysis.

(c) Compare the moment at the wall predicted by (b) to that obtained from statics.

(d) Compare the deflection of the free end predicted by this Rayleigh-Ritz approximation to that obtained by elementary strength of materials.

7.27 A simply supported beam of length $L = 10$ with constant E and I is loaded by a constant distributed load, p. On a computer, plot the curves $EIy(x)/p$ on the same graph and plot $M(x)/p$ on the same graph for the following four cases. Discuss the differences and similarities between the four cases.

(a) Use statics to find $M(x)$ and integrate as in strength of materials to determine $y(x)$.

(b) Use the Rayleigh-Ritz method assuming that $y(x) = a_1 \sin(\pi x/L)$.

(c) Use the Rayleigh-Ritz method assuming that $y(x) = a_1 x(L - x)$.

(d) Use the Rayleigh-Ritz method assuming that $y(x) = a_1 \sin(\pi x/L) + a_3 \sin(3\pi x/L)$.

Notice that case (d) is the same as (b) except for an additional term. This is a consequence of the fact that $\sin(\pi x/L)$ and $\sin(3\pi x/L)$ are orthogonal functions. In both cases (b) and (d) both the deflection and the moment boundary conditions are satisfied. However, in case (c), while the deflection boundary conditions are satisfied, the moment boundary conditions are not satisfied and the moment is a very bad approximation.

7.28 A simply supported beam of length, L, and constant bending stiffness, EI carries a transverse downward distributed load defined by $p(x) = \alpha x$, where x is measured from the left end and α is a constant. Use a one-term sine function to approximate the deflection, $y(x)$.

7.29 A simply supported beam of length, L, and constant bending stiffness, EI carries a transverse downward load, P, at $x = L/4$ where x is measured from the left end. Use a one-term sine function to approximate the deflection, $y(x)$.

7.30 A cantilever beam of length, L, and constant bending stiffness, EI carries a transverse downward load, P, at $x = L/2$ where x is measured from the fixed support. Use a one-term trigonometric function to approximate the deflection, $y(x)$.

7.31 Estimate the equilibrium deflection of each point on the beam of problem 4 (looked at from the other side to make the computations easier). Make the strength of materials assumption that the internal moment is given by $M(x) = EIy''(x)$. Assume that the deflection is of the form $y(x) = a_0 + a_1 x + a_2 x^2 + a_3 x^3$. Use the coordinate system shown on the figure.

(a) Use the boundary conditions to determine some of the coefficients a_i.

(b) Use the Rayleigh-Ritz method to determine the remaining coefficients, a_i.

(c) Compute the predicted deflection at the free end and compare it to your answer in problem 4. Discuss the result.

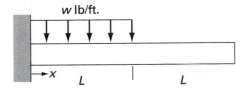

w lb/ft.

L L

7.32 Use the Rayleigh-Ritz method to approximate the deflection of a cantilever beam of length, L, loaded by a constant distributed force, p.

(a) Assume a deflection of the form, $v(x) = a_0 + a_1x + a_2x^2 + a_3x^3$ and calculate the approximate deflection at each point x. Make the strength of materials assumptions.

(b) Determine the percent error between the actual moment at the wall and the moment at the wall derived from the assumed displacement.

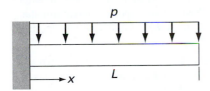

p

L

x

7.33 A uniform linearly elastic simply supported beam carries a uniformly distributed load, p (lb/in.). The beam has length L and constant EI. Use the assumptions of elementary strength of materials. Derive the natural boundary conditions. Obtain the equation of the equilibrium deflection curve by the Euler equation. Eliminate arbitrary constants of integration. Ans. If y is positive down, $y(x) = px(L - x)(L^2 + Lx - x^2)/24EI$.

7.34 A uniform linearly elastic beam that is clamped at the ends carries a linearly distributed load that varies from zero at one end to p_o at the other end. The beam has length L and constant EI. Use the assumptions of elementary strength of materials. Obtain the equation of the equilibrium deflection curve by the Euler equation. Eliminate arbitrary constants of integration. Ans. If y is positive down, $y(x) = p_o x^2(L - x)(2L^2 - Lx - x^2)/120EIL$.

7.35 The strain energy of a special type of beam is

$$U = k \int_0^L \left(\frac{y}{a^2} + \frac{y'}{a} + y'' \right)^2 dx,$$

where k and a are constants. The end $x = 0$ is clamped, and the end $x = L$ is free. The beam carries a distributed lateral load, $p(x)$. State the forced boundary conditions. Derive the natural boundary conditions and the Euler equation which governs the deflection $y(x)$. Ans. At $x = L$, $y''' - y/a^3 = 0$ and $y'' + y'/a + y/a^2 = 0$; $y'''' + y''/a^2 + y/a^4 = p/2k$, assuming y is positive down.

7.36 Determine the stationary configuration, $y(x)$, for the function $G(x, y) = \int_0^2 [(y')^2 + yy' + 4y]dx$ subject to the initial conditions, $y(0) = 3$ and $y'(0) = 0$. Ans. $y(x) = x^2 + 3$.

7.37 Determine the stationary configuration, $y(x)$, for the function $G(x, y) = \int_0^\pi yy''dx$ subject to the initial conditions, $y(0) = 0$ and $y'(0) = 2$.

7.38 Determine the curve whose configuration gives the shortest distance between two points (a, A) and (b, B) by finding the stationary state of the distance function, $G(x, y) = \int_a^b \sqrt{1 + (y')^2}dx$. Ans. $y(x) = [(B - A)/(b - a)]x + (Ab - Ba)/(b - a)$, a straight line.

7.39 As a model for a static bicycle chain, consider a chain weighing $w = 0.6$ lb/ft and of length $L = 19$ inches hung between two sprockets, which are spaced $D = 18$ inches apart. The two sprockets are mounted at the same height. Determine

(a) the maximum sag;

(b) the tension exerted on the sprockets by the chain.

7.40 The strain energy of a two-dimensional body is assumed to be is given by

$$U = K \int \int (w_{xx}^2 + w_{xx}w_{yy} + w_{yy}^2)dxdy,$$

where K is a constant and $w(x, y)$ is the deflection perpendicular to the body. The body supports an arbitrary transverse distributed load, $p(x, y)$. Determine the differential equation governing $w(x, y)$.

7.41 Show that if the load on an isotropic, linear elastic, simply supported, rectangular plate with edges of length a and b is

$$p(x, y) = p_o \sin\left(\frac{m\pi x}{a}\right) \sin\left(\frac{n\pi y}{b}\right),$$

then

$$w(x, y) = \frac{p_o a^4 b^4}{\pi^4 D(a^2 + b^2)^2} \sin\left(\frac{m\pi x}{a}\right) \sin\left(\frac{n\pi y}{b}\right).$$

7.42 Estimate the deflection at $(a/2, b/2, h/2)$ in a simply supported isotropic plate subject to a uniform transverse load, p, by using just the first three nonzero terms of the Navier series [equation (7.54)].

7.43 A 5-inch by 8-inch plate of thickness 0.125 inches is simply supported and made of an aluminum ($E = 10 \times 10^6$, $v = 0.33$). It is given a coordinate system with the x-axis along the 8-inch edge and the y-axis along the 5-inch edge. A uniform pressure of 50 psi is applied transversely to the plate. Determine the principal stresses and the maximum shear stress at the point $(2, 2.5)$ on the surface. Use the one-term approximation.

7.44 A 5-inch by 8-inch plate of thickness 0.125 inches is simply supported and made of a steel ($E = 30 \times 10^6$, $\nu = 0.26$). It is given a coordinate system with the x-axis along the 8-inch edge and the y-axis along the 5-inch edge. A uniform pressure of 100 psi is applied transversely to the plate. Determine the principal stresses and the maximum shear stress at the point $(2, 1.25)$ on the surface. Use the one-term approximation.

7.45 Determine the deflections of an isotropic, linear elastic, rectangular plate that is simply supported on the two opposite edges parallel to the x-axis, is free on the other two edges, and that supports a constant distributed load, p. Make the Lévy approximation with $f(y) = 1$.

7.46 Verify that the transverse deflection, $w(r, \theta)$, of a circular plate of radius, R, that supports a constant transverse pressure, p, and that has simply supported edges is independent of θ and is given by equation (7.64):

$$w(r) = \frac{pR^4}{64D}\left(\frac{5+\nu}{1+\nu}\right) - \frac{R^2}{32D}\left(\frac{3+\nu}{1+\nu}\right)r^2 + \frac{pr^4}{64D}.$$

7.47 An isotropic, circular plate of radius, R, and thickness, h, that has clamped edges, is subjected to a uniform transverse pressure, p. Determine the principal stresses and the maximum shear stress on the surface at $r = R/2$.

7.48 An isotropic, circular plate of radius, R, and thickness, h, that is simply supported, is subjected to a point load, F, at its center.

(a) What value for the stress σ_r at the center is predicted by the displacement equation (7.66)?

(b) Plot $Dw(r)/F$ over the range $0 \le r \le R$ when $R = 2$ and $\nu = 0.3$.

7.49 An isotropic, circular plate of radius, R, and thickness, h, that has clamped edges is subjected to a point load, F, at its center. Verify that the displacement is

$$w(r) = \frac{F}{16\pi D}\left[2r^2 \ln\left(\frac{r}{R}\right) + R^2 - r^2\right],$$

and

$$w(0) = \frac{FR^2}{16\pi D}.$$

Hint: The natural boundary condition is

$$\lim_{r \to 0}\left(r\frac{d^3w}{dr^3} + \frac{d^2w}{dr^2} - \frac{1}{r}\frac{dw}{dr}\right) = \frac{F}{2\pi D}.$$

7.50 A simply supported linear orthotropic plate with sides a and b is loaded with a transverse uniform pressure, p. The coordinate system is chosen with the x-axis along the edge of length a and the y-axis along the edge of length b. Determine the principal stresses and the maximum shear stress on the surface at $(a/4, b/2, h/2)$. Use the one-term approximation.

7.51 Write the differential equation $y'' = x$ with $y(0) = 0$ and $y(2) = 8$ as an integral function for which it is the Euler equation. Compute the Euler equation of the result to verify it.

7.52 Verify that the integral functions $I(x) = \int_0^T (0.5m\dot{x}^2 - 0.5kx^2)dt$ and $J(x) = \int_0^T (m\ddot{x}x + kx^2)dt$ have the same Euler equation.

REFERENCES

[1] H. L. LANGHAAR, *Energy Methods in Applied Mechanics*, Wiley, New York, 1962.
[2] S. G. LEKHNITSKII, *Anisotropic Plates*, 2nd ed., Gordon and Breach, New York, 1968.
[3] A. D. SHAH AND J. D. HUMPHREY, Finite strain elastodynamics of intracranial saccular aneurysms. *Journal of Biomechanics 32*, 593–599 (1999).

CHAPTER *8*

INTRODUCTION TO THE FINITE ELEMENT METHOD FOR NUMERICAL APPROXIMATIONS

The finite element method (FEM), for solids, has the same goal as elasticity, to calculate the stress, strain, and displacement fields in a body. In elasticity, these are determined from the three equilibrium equations, the stress-strain relations, and the strain-displacement relations. The finite element method is an approximation technique often derived from a variational principle such as virtual work or stationary potential energy in place of the equilibrium equations of elasticity. It estimates the stress, strain, and displacement at certain points in the body, called nodes. From these results, the behavior throughout the body can be numerically approximated.

The goal of this description is to show a connection between the finite method and energy methods for solids, as developed in Chapter 7, to give insight into some of the approximations made, and to show how the method relates to elasticity and stress analysis. Such insight is needed to judge the validity of the results so that the method is not a magical black box to the user. The method can produce incorrect answers. Anyone who suggests that the method can be used to check experimental results has it exactly backward. The finite element method is an approximation to a given mathematical model, and its results should be verified experimentally. Like any numerical method, the internal algorithms can introduce unexpected errors. In this chapter, only displacement models will be considered. Many other types of elements are possible so that this short introduction to the FEM is by no means the full story. Some of these other elements cannot be rationalized as clearly as those based on a variational principle. A full course on this technique is recommended for those who intend to use the finite element method professionally.

The FEM was first developed in the 1940s and 1950s. However, some of the ideas had previously appeared in work like that of Leibnitz in 1696 on the brachistochrone problem or of Schelbach in 1851 on the plateau problem. The foundations of the FEM were laid in Hrenikoff's [3] study in 1941 of structures made of bars and beams, in Courant's work in 1943 on torsion, and in the work of Argyris (1957) often called a primitive FEM. The method took its current character in work done in 1953, but published in 1956, by Turner, et al. [6] on linear plates. It appears that the term "finite element" was first used by Clough in 1960.

The FEM involves three components. Its foundation is a variational principle such as the virtual work or stationary potential energy. Once this principle is chosen, an approximation is made. This approximation is the source of the term "finite element." Finally, the equations developed in the first two steps must be solved on a computer.

The finite element method, applied to displacement models for solids, can be viewed as a refinement of the Rayleigh-Ritz approximation. In the Rayleigh-Ritz method, an assumed form of an approximate displacement function for the complete body is inserted into the potential energy function. The unknown coefficients are determined by minimizing the resulting potential energy function. These coefficients usually have no physical interpretation. In the finite element method, the approximations are made to small pieces of the body. These approximations are then assembled to give an approximation for the full body. The assumed approximations are chosen so that the coefficients in the approximation have a physical interpretation.

The building blocks of the finite element method are the elements. The set of elements is called the mesh. In a solids problem, each element is used to model the behavior of a small piece of the body. But elements are not actually pieces of the body. An element can deform in ways the body cannot. For example, a gap could occur between two adjacent elements that does not exist on the real deformed body. Each element is defined by its shape and the choice of the function describing the displacement of points in the element. The displacement function is typically taken as a polynomial. A key idea of FEM is to write the unknown coefficients in the assumed displacement function in terms of the displacements of the points in the element taken as nodes. This means that there is a relation between the number of nodes and the order of the displacement function. There must be as many nodal displacement components as there are undetermined coefficients in the assumed displacement field. The nodes can be chosen anywhere inside the element, but, for computational efficiency, are often chosen at the corners of the element. In the FEM, the nodal displacements are the initial set of unknowns to be solved for.

Once the elements are defined, the system of elements must be assembled to represent the full body. The external boundary conditions are then applied and the resulting system of equations is solved for the nodal displacements. Other quantities needed are found from the nodal displacements.

Current thought divides finite element methods into two types and a combination of the two. In the "h-version," the approximating displacement function is fixed. Accuracy is obtained by making the diameter, h, of each element smaller. The "p-version" fixes the mesh and obtains increased accuracy by changing the approximating polynomial to one with higher degree. For example, the commerical FEM package, ANSYS, is an h-version. In each element type, the approximating function is fixed. Accuracy is improved by modifying the mesh. Most published analysis has concentrated on the h-version. Perhaps only 10% of current research is devoted to the more recent p-version.

In order to ensure a reasonable approximation by the FEM, the elements are usually required to satisfy certain conditions. The following list is given in Cook [2].

1. The displacement field within an element must be continuous.

2. When nodal displacements are given values corresponding to constant strain, the displacement field must produce the constant strain state throughout the element.

   This is so that, as elements are taken smaller (in the h-version) and the actual strain becomes nearly constant within them, the finite element method will reproduce this state of constant strain. In this manner, the finite element solution converges to the exact solution as the finite element mesh is made finer.

3. The element must accurately represent rigid body motion. When the nodal displacements correspond to rigid body motion, the element must produce zero strain and zero nodal forces.

4. Compatibility must exist between elements. The assumed displacement field must not make the elements separate or overlap.

   This condition is frequently violated; sometimes the values on an overlap cancel out in a useful manner. But in all cases, this condition should be satisfied for nodal displacements representing constant strain states. Compatibility may not be satisfied between elements but is enforced at the nodes by "joining" the elements at nodal locations (by requiring that two nodes that are in different elements but which represent the same point in the body have the same displacement).

5. Invariance: The element should have no preferred directions. Under a set of loads having a fixed orientation with the element, the response should be independent of how the element is oriented in the global coordinate system.

For example, an assumed displacement field in the x-y plane of the form

$$a_1 + a_2 x + a_3 y + a_4 xy,$$

for constants a_i, would be considered to be balanced, but a field of the form,

$$a_1 + a_2 x + a_3 y + a_4 x^2,$$

would not satisfy this requirement.

Like many rules, these are not always followed. For example, efficient plate elements have been devised which are either incompatible in some nodal displacements or overlap after loading.

In the FEM based on the principle of stationary potential energy, the equilibrium equations of elasticity are usually not satisfied within elements. However, they are satisfied within the element called the constant strain triangle. Equilibrium is usually not satisfied between elements. This is illustrated by the constant strain triangle in which the stress is also constant, but the stress usually differs between elements. Equilibrium of nodal forces and moments is satisfied from the principle of stationary potential energy. Finite element methods based on other variational principles may have other properties.

8.1   THE FUNDAMENTAL EQUATIONS FOR THE FEM

The fundamental equations of elasticity are the strain-displacement relations, the constitutive law, and the equilibrium equations. In the finite element method, the equations of equilibrium are replaced by equations arising from a variational principle. Furthermore, in order to easily implement the finite element method on a computer, the equations are usually desired in a linear form. Computers are very efficient at solving systems of linear equations, since all a computer can do, in the final analysis, is add numbers. In this sense, the development of the FEM was driven by, and is a consequence of, the invention of the high-speed digital computer.

Since the unknowns to be solved for are the nodal displacements, the assumed displacement field within each element must be written, in the element coordinate system, as a linear function of the nodal displacements. For each element, if the nodal displacements are u_i, v_i, w_i, for $i = 1, \dots, q$ (q nodes), the equations take the form

$$\{f\}_e = \{u(x, y, z), v(x, y, z), w(x, y, x)\}^t = N\{u_i, v_i, w_i\}^t, \tag{8.1}$$

where N is an $3 \times m$ matrix if m is the number of nodal coordinates. If each node has r coordinates and there are q nodes, then $m = rq$ and $\{u_i, v_i, w_i\}^t$ is a column vector with m entries. The transpose of a row vector is a column vector. The matrix N is a function of (x, y, z), the position in the element.

Next the element strains, $\{\epsilon\}_e$, can be written in terms of the nodal displacements by using the strain-displacement relations. In order to implement this on a computer, a linear relation of the form

$$\{\epsilon\}_e = B\{d\}_e \tag{8.2}$$

is desired, where B is a function of position in the element and $\{d\}_e$ is the column vector of nodal displacements. The matrix, B, is obtained by differentiating the matrix, N, in a manner corresponding to the strain-displacement equations to obtain the strains from the displacements. In two dimensions, for example, N is $2 \times m$, and B is $3 \times m$,

$$B = \begin{pmatrix} \frac{\partial}{\partial x} & 0 \\ 0 & \frac{\partial}{\partial y} \\ \frac{\partial}{\partial y} & \frac{\partial}{\partial x} \end{pmatrix} N.$$

EXAMPLE 8.1   Consider a one-dimensional bar element of length, L, with two nodes, one at either end (Fig. 8.1). If all displacements are in the direction of the axis of the element, then there are two nodal displacements, $u_1 = u(0)$ and $u_2 = u(L)$, in the element coordinate system, x, with origin at the left node.

FIGURE 8.1   One-dimensional bar element.

Assume that the displacement of each point in the element is given by a linear function of the form $u(x) = a_1 + a_2 x$. Now write the nodal displacements: $u_1 = u(0) = a_1$ and $u_2 = u(L) = a_1 + a_2 L$. Solving these two equations simultaneously yields $a_2 = (u_2 - u_1)/L$. Therefore the assumed displacement field can be written in terms of the nodal displacements, in matrix form, as

$$u(x) = \begin{pmatrix} 1 - \frac{x}{L} & \frac{x}{L} \end{pmatrix} \begin{pmatrix} u_1 \\ u_2 \end{pmatrix}.$$

In this case, $N = \begin{pmatrix} 1 - x/L & x/L \end{pmatrix}$ is a function of x. To obtain the strain in the x-direction, the small strain displacement equation says that $\epsilon_x = \partial u/\partial x = (u_2 - u_1)/L$. In matrix form,

$$\epsilon_x(x) = \begin{pmatrix} \frac{-1}{L} & \frac{1}{L} \end{pmatrix} \begin{pmatrix} u_1 \\ u_2 \end{pmatrix}.$$

In this case, the 1×2 matrix $B = \begin{pmatrix} -1/L & 1/L \end{pmatrix}$ is the termwise derivative of N with respect to x, since the strain-displacement relation is $\epsilon_x = \partial u/\partial x$.   ∎

The one-dimensional example above is a constant strain element, since $\epsilon_x = \partial u/\partial x = a_2 = (u_2 - u_1)/L$, a constant. If the assumed displacement field were of any order higher than linear, then the element would not be constant strain.

EXAMPLE 8.2   Other displacement functions can be assumed for the one-dimensional element. In global one-dimensional coordinates, let x_i be the coordinate and let u_i be the displacement of the ith node. A linear displacement field can be defined in terms of the u_i rather than the x_i using the "chapeau functions,"

$$f(x) = u_i \frac{1 - (x - x_i)}{(x_{i+1} - x_i)} + u_{i+1} \frac{x - x_i}{(x_{i+1} - x_i)}, \quad x_i \le x \le x_{i+1}. \tag{8.3}$$

Then $f(x_i) = u_i$ and $f(x_{i+1}) = u_{i+1}$. They are called chapeau (the French word for hat) since the graphs look like hats. The graph of $f(x)$ between x_i and x_{i+1} is the sum of those shown in Figure 8.2 when they are multiplied by the proper u_i, the nodal displacement of the node at x_i.

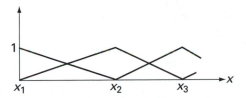

FIGURE 8.2   The chapeau functions.

These equations give a simple method of writing an assumed linear displacement field in global rather than element coordinates. ∎

Exercise 8.1   Plot the chapeau functions for $x_1 = 0$, $x_2 = 2$, $x_3 = 3$, and $u_i = 1$.

Constitutive relations are required between the stress and the strain, in each element. Put

$$\{\epsilon\}_e = \begin{pmatrix} \epsilon_x \\ \epsilon_y \\ \gamma_{xy} \end{pmatrix} \quad \text{and} \quad \{\sigma\}_e = \begin{pmatrix} \sigma_x \\ \sigma_y \\ \tau_{xy} \end{pmatrix}.$$

If the relation is two-dimensional linear elasticity, the equation is linear of the form

$$\{\sigma\}_e = D\{\epsilon\}_e, \tag{8.4}$$

where, from the equations representing plane stress in terms of the strains,

$$D = \frac{E}{1-v^2} \begin{pmatrix} 1 & v & 0 \\ v & 1 & 0 \\ 0 & 0 & \frac{1-v}{2} \end{pmatrix} \tag{8.5}$$

for plane stress. This expression uses the fact that the shear modulus is

$$G = \frac{E}{2(1+v)} = \frac{E}{1-v^2} \frac{1-v}{2}.$$

Exercise 8.2   Determine the matrix D that appears in equation (8.5) in the case of plane strain.

8.2   COMBINED EQUATIONS FOR A SINGLE ELEMENT

Once the body is represented by finite elements in which the displacements, and so the stress and strain, depend on the nodal displacements through an assumed displacement field, the nodal displacements must be related to the external forces. The nodes and elements must be chosen so that each external force is applied at a node. The variational principle chosen makes the connection between the external loads and the nodal displacements.

The process will be carried out for an elastic body loaded by conservative forces so that the principle of stationary potential energy is valid. Recall that the principle of stationary potential energy is the variational principle of virtual work in the case of a conservative system.

Let $\mathbf{F}(x, y, z) = F_x(x, y, z)\mathbf{i} + F_y(x, y, z)\mathbf{j} + F_z(x, y, z)\mathbf{k}$ be the body force at the point (x, y, z) of the body in a global Cartesian coordinate system for the body. Also, let $\mathbf{t}(x, y, z) = T_x(x, y, z)\mathbf{i} + T_y(x, y, z)\mathbf{j} + T_z(x, y, z)\mathbf{k}$ be the surface traction at the point (x, y, z) on the surface, S, of the body.

By definition, the variation of a function $f(x)$ is $\delta f = (\partial f/\partial x)\delta x$. The variation of the uniaxial strain energy density, $U_o = \sigma\epsilon/2 = E\epsilon^2/2$, is $\delta U_o = (\partial U_o/\partial\epsilon)\delta\epsilon = E\epsilon\delta\epsilon = \sigma\delta\epsilon$. In the same way, the variation in total strain energy in the three-dimensional case is $\delta U = \int_V \delta U_o dV = \int_V (\sigma_x\delta\epsilon_x + \sigma_y\delta\epsilon_y + \sigma_z\delta\epsilon_z + \tau_{xy}\delta\gamma_{xy} + \tau_{xz}\delta\gamma_{xz} + \tau_{yz}\delta\gamma_{yz})dV$. The virtual work of a force $\mathbf{F}$ through an arbitrary virtual displacement $\delta\mathbf{r} = \delta u\mathbf{i} + \delta v\mathbf{j} + \delta w\mathbf{k}$ is $F_x\delta u + F_y\delta v + F_z\delta w$. Thus, the variation in work due to the body force is $F_x\delta u + F_y\delta v + F_z\delta w$ and the variation of the surface traction is $T_x\delta u + T_y\delta v + T_z\delta w$.

Suppose that the body is divided into n elements. The variation of total potential energy, $\delta\Pi$, for the whole body, due to a virtual displacement $(\delta u(x, y, z), \delta v(x, y, z), \delta w(x, y, z))$ is the sum of that on each element

$$\delta\Pi = \sum_{i=1}^{n}\int_V (\sigma_x\delta\epsilon_x + \sigma_y\delta\epsilon_y + \sigma_z\delta\epsilon_z + \tau_{xy}\delta\gamma_{xy} + \tau_{xz}\delta\gamma_{xz} + \tau_{yz}\delta\gamma_{yz})dV$$

$$- \sum_{i=1}^{n}\int_V (F_x\delta u + F_y\delta v + F_z\delta w)dV - \sum_{i=1}^{n}\int_S (T_x\delta u + T_y\delta v + T_z\delta w)dS,$$

$$(8.6)$$

where the integrals are taken over the volume, V, or surface, S, of each element.

Recall that the principle of stationary potential energy states that the body is in equilibrium if $\delta\Pi = 0$. Next make an approximation by applying the Rayleigh-Ritz method. Assume a displacement field, $\{f(x, y, z)\}$, in each element, written in terms of the nodal displacements, $\{d\}$. The displacement can then be given in the matrix form,

$$\{f\} = N\{d\},$$

where the entries in the matrix, N, are all functions of (x, y, z). This produces an approximation to the variation in potential energy, $\delta\Pi$, which will be denoted by $\delta\Pi_a$. To complete the development of the equations relating the external loads to the nodal displacements, $\delta\Pi_a$ is written in a more convenient form and set equal to zero. The principle of virtual work is applied to the approximate potential energy to obtain the stationary state. It is assumed that the stationary state for the approximate potential energy is a good approximation for the stationary state of the exact potential energy and therefore is a good approximation to the equilibrium state.

This equation allows the determination of approximate nodal displacements. The displacements throughout each element are then computed from the assumed displacement field. The approximate strains are obtained from equation (8.2). Finally, the stress-strain relation, for example Hooke's law, is used to compute the approximate stresses.

This program can be simply carried out if the approximate potential energy variation is written as

$$\delta\Pi_a = \sum_{i=1}^{n}\int_V [\{\delta\epsilon\}^t\{\sigma\} - \{\delta f\}^t\{F\}]dV - \sum_{i=1}^{n}\int_S \{\delta f\}^t\{T\}dS = 0. \qquad (8.7)$$

This uses the fact that column multiplication on the right becomes row multiplication on the left if one changes the matrix to its transpose. For example,

$$\begin{pmatrix} a & b \\ c & d \end{pmatrix}\begin{pmatrix} u \\ v \end{pmatrix} = (u \quad v)\begin{pmatrix} a & c \\ b & d \end{pmatrix} = \begin{pmatrix} u \\ v \end{pmatrix}^t \begin{pmatrix} a & b \\ c & d \end{pmatrix}^t.$$

To get equation (8.7) in terms of the nodal displacements, substitute the expressions (8.2) and (8.1) for the strain and the displacement. In the case that the constitutive (stress-strain) relation is linear, $\{\sigma\} = D\{\epsilon - \epsilon_o\}$, where the term, $\{\epsilon_o\}$, contains the initial, or residual, strain and the thermal strain, if any. The fact that $\{\epsilon\} = B\{d\}$ implies that $\{\sigma\} = DB\{d\}$. Also $\{\epsilon\}^t = \{d\}^t B^t$ implies that

$$\{\delta\epsilon\}^t\{\sigma\} = \{\delta d\}^t B^t DB\{d\} - \{\delta d\}^t B^t D\{\epsilon_o\}.$$

Now define the stiffness matrix for each element to be

$$k_e = \int_V B^t DB dV. \tag{8.8}$$

Notice that this depends on the assumed displacement field through D and the stress-strain relation through B. The advantage of this formulation is that once an element is defined, the stiffness matrix, k_e, can be computed just once and stored in the finite element computer program. The user can then call this matrix whenever choosing the element to model a problem, without being required to reprogram it.

Also since $\{f\} = N\{d\}$, $\{\delta f\}^t\{F\} = \{\delta d\}^t N^t\{F\}$, and $\{\delta f\}^t\{T\} = \{\delta d\}^t N^t\{T\}$. Set the nodal force matrix, $\{Q\}_e$, to be

$$\{Q\}_e = \int_V [N^t\{F\} + B^t D\{\epsilon_o\}]dV + \int_S N^t\{T\}dS. \tag{8.9}$$

Then the approximate potential energy becomes

$$\delta\Pi_a = \sum_{i=1}^{n}\{\delta d\}^t[k_e\{d\}_e - \{Q\}_e] = 0. \tag{8.10}$$

This equation must be true for any possible displacement variation, $\{\delta d\}^t$. Therefore, it must be true that $[k_e\{d\}_e - \{Q\}_e] = 0$, or

$$k_e\{d\}_e = \{Q\}_e. \tag{8.11}$$

This linear relation defines the nodal displacements in each element in terms of the nodal forces, the thermal strains, the body forces, and tractions.

The element stiffness matrix, k_e, is a symmetric matrix since it is its own transpose. $(B^t DB)^t = B^t D^t (B^t)^t = B^t DB$ since D is symmetric ($D = D^t$). This agrees with the Maxwell reciprocity theorem which states that the deflection at node i due to a unit force at node j is equal to the deflection at node j caused by a unit force at node i.

EXAMPLE 8.3   A one-dimensional axial element has two nodes, one at each end, with an assumed displacement field which is linear. The coordinate along the element is x. Assume that the body force per unit volume is $\mathbf{F} = -\gamma\mathbf{i}$, and that there is a traction, $\mathbf{T} = p\mathbf{i}$, applied at $x = 2L/3$. Compute the element stiffness matrix, k_e, and the nodal force matrix, $\{Q\}_e$.

Solution   The displacement field is $u(x) = a_1 + a_2 x$. From the calculations above, the displacement matrix is

$$\{f\} = \left(1 - \tfrac{x}{L} \quad \tfrac{x}{L} \right) \left(\begin{matrix} u_1 \\ u_2 \end{matrix} \right) = N\{d\}.$$

The strain in terms of the displacements $\{d\}$ is

$$\{\epsilon\} = \left(-\tfrac{1}{L} \quad \tfrac{1}{L} \right) \left(\begin{matrix} u_1 \\ u_2 \end{matrix} \right) = B\{d\}.$$

The stress-strain relation, since the matrix D in one dimension is just the elastic modulus E, is

$$\{\sigma\} = E\{\epsilon - \epsilon_o\} = \{\sigma\} = D\{\epsilon - \epsilon_o\},$$

where the initial strain plus the strain due to a change in temperature is $\epsilon_o = \Delta L/L + \alpha\theta$ if α is the coefficient of thermal expansion and θ is the change in temperature. The element stiffness matrix is

$$k = k_e = \int_0^L B^t D B dV = \int_0^L \left(\begin{matrix} -\tfrac{1}{L} \\ \tfrac{1}{L} \end{matrix} \right) E \left(-\tfrac{1}{L} \quad \tfrac{1}{L} \right) A dx = \frac{EA}{L} \left(\begin{matrix} 1 & -1 \\ -1 & 1 \end{matrix} \right). \tag{8.12}$$

Notice that k_e is a symmetric matrix. The nodal force matrix, Q, is found from

$$\{Q\}_e = \int_V N^t\{F\}dV + \int_V B^t D\{\epsilon_o\}dV + \int_S N^t\{T\}dS.$$

For a one-dimensional element, the surface is composed of the endpoints. Therefore $\int_S N^t\{T\}dS$ is a sum over the two endpoints, $\sum N^t\{T\}$. If the force, P, is applied at a node, this sum is easy to compute. However, if P is not applied at a node, equivalent weighted components of P, the statically equivalent forces, are applied at each node.

$$\{Q\}_e = \int_0^L \left(\begin{matrix} 1 - \tfrac{x}{L} \\ \tfrac{x}{L} \end{matrix} \right) (-\gamma) A dx + \int_0^L \left(\begin{matrix} -\tfrac{1}{L} \\ \tfrac{1}{L} \end{matrix} \right) E \left(\frac{\Delta L}{L} + \alpha\theta \right) A dx + P \left(\begin{matrix} 1 - \tfrac{2}{3} \\ \tfrac{2}{3} \end{matrix} \right)$$

$$= -\gamma A \left(\begin{matrix} \tfrac{1}{2} \\ \tfrac{1}{2} \end{matrix} \right) + AE \left(\frac{\Delta L}{L} + \alpha\theta \right) \left(\begin{matrix} -1 \\ 1 \end{matrix} \right) + P \left(\begin{matrix} \tfrac{1}{3} \\ \tfrac{2}{3} \end{matrix} \right). \tag{8.13}$$

Therefore $k_e\{d\}_e = \{Q\}_e$ becomes

$$\frac{EA}{L} \left(\begin{matrix} 1 & -1 \\ -1 & 1 \end{matrix} \right) \left(\begin{matrix} u_1 \\ u_2 \end{matrix} \right) = \{Q\}_e. \tag{8.14}$$

∎

Exercise 8.3 Write FORTRAN code to represent the element stiffness matrix k_e found in the above example.

EXAMPLE 8.4   A one-dimensional axial element has two nodes, one at each end, and an assumed displacement field $f(x) = u(x) = a_1 + a_2 x^3$ and represents a rod mounted in a rigid wall at the left end. The rod is loaded by an axial force P at the free end, node 2 (Fig. 8.3).

Solving the displacement field in terms of the nodal displacements, u_1 and u_2, produces

$$u(x) = \left(1 - \tfrac{x^3}{L^3} \quad \tfrac{x^3}{L^3} \right) \left(\begin{matrix} u_1 \\ u_2 \end{matrix} \right);$$

$$\epsilon(x) = \left(-\tfrac{3x^2}{L^3} \quad \tfrac{3x^2}{L^3} \right) \left(\begin{matrix} u_1 \\ u_2 \end{matrix} \right);$$

and

$$k = \frac{9EA}{5L} \left(\begin{matrix} 1 & -1 \\ -1 & 1 \end{matrix} \right).$$

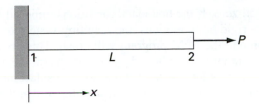

FIGURE 8.3   A rod represented as a single one-dimensional bar (axial) element.

Since the rod is mounted in a rigid wall at the left end, $u_1 = 0$. But there is an unknown wall reaction R at the wall. The goal is then to determine R and u_2 assuming that P is known. The element stiffness equation, $k\{d\} = \{Q\}$, is

$$\frac{9EA}{5L}\begin{pmatrix} 1 & -1 \\ -1 & 1 \end{pmatrix}\begin{pmatrix} u_1 \\ u_2 \end{pmatrix} = \begin{pmatrix} R \\ P \end{pmatrix}.$$

This produces the equations

$$\frac{9EA}{5L}(u_1 - u_2) = R;$$

$$\frac{9EA}{5L}(-u_1 + u_2) = P.$$

But $u_1 = 0$, therefore the second equation implies that

$$u_2 = \frac{5PL}{9AE}.$$

This is different from the deformation of the rod obtained in strength of materials because the assumed displacement field produces non-constant strain in the rod. The wall reaction $R = -P$ is obtained from the first equation; this result is as expected from static equilibrium. Show that, for example, the strain at $x = L/2$ is $5P/12AE$, not P/AE as in the constant strain rod. ∎

8.3   ASSEMBLY OF THE FINITE ELEMENT EQUATIONS

The behavior of each element in terms of its nodal displacements, $\{d\}$, is described by the matrix equation $k_e\{d\}_e = \{Q\}_e$. The equations for all elements must now be combined into one very large system of equations which give the behavior of the body in terms of the displacements, $\{d\}$, of all nodes, $K\{d\} = Q$, where K is called the global stiffness matrix. K must be symmetric since each of the element stiffness matrices is symmetric. By determining the inverse of K, the displacements of all nodes are computed. Frequently, because K is hard to invert, to solve this system of equations, the program must apply numerical techniques to approximate even further the solutions of the equation $K\{d\} = Q$, that is, the nodal displacements, $\{d\}$. At this point, two sets of approximations have been made: one to solve $K\{d\} = Q$ and the other when the approximate displacement field was chosen. It is extremely difficult to estimate the error introduced by these approximations. A good way to understand the procedure for assembling the large number of element equations is to examine an example.

The element stiffness matrix, k_e, and the assembled body stiffness matrix, K, are in general singular matrices. Physically, this is a consequence of the fact that a rigid motion of the element is not prohibited by the general formulation. A rigid translation occurs if the corresponding nodal displacement components at each node are identical. A rigid rotation occurs, for example, if a single node is fixed so that its displacement components are all zero. If the boundary conditions imposed to restrict the nodal displacements allow rigid body motions, the resulting system of equations will not have a unique solution and the FEM program will crash. This problem can be avoided by imposing a sufficient number of boundary conditions on the nodal displacements. For example, in a plane problem, at least three nodal displacement components must be specified. Two can be given at a single node and the third at another, or one component at each of three nodes can be given.

At each node either the nodal displacement or the nodal force in each coordinate direction is known. Both the global displacements and external forces can be grouped into the known set and the unknown set. Let d_u and Q_u be the unknown nodal displacements and forces respectively. Q_u will include the unknown reactions at the supports. Let d_k and Q_k be the known nodal displacements and forces respectively. The resulting global stiffness matrix K can be broken into corresponding subblocks. Then the stiffness equation for the body, $K\{d\} = R$, can be partitioned into

$$\begin{pmatrix} K_{11} & K_{12} \\ K_{21} & K_{22} \end{pmatrix} \begin{pmatrix} d_u \\ d_k \end{pmatrix} = \begin{pmatrix} Q_k \\ Q_u \end{pmatrix}, \tag{8.15}$$

where K_{ij} is a block submatrix of K. Then

$$K_{11}\{d_u\} + K_{12}\{d_k\} = \{Q_k\}; \tag{8.16}$$
$$K_{21}\{d_u\} + K_{22}\{d_k\} = \{Q_u\}. \tag{8.17}$$

In this situation, K_{11} must be a nonsingular matrix so that its inverse K_{11}^{-1} exists. Then equation (8.16) can be solved first for the unknown nodal displacements, $\{d_u\}$.

$$\{d_u\} = K_{11}^{-1}(K_{12}\{d_k\} - \{Q_k\}). \tag{8.18}$$

This may have to be done numerically to obtain K_{11}^{-1} on the computer, by techniques such as Gauss elimination, etc. Once $\{d_u\}$ is computed, then the unknown forces, $\{Q_u\}$, can easily be computed, with no further numerical approximation, from equation (8.17). The reaction forces are required to keep the displacements at the values imposed by the boundary conditions.

EXAMPLE 8.5   A straight member is mounted in a wall at its left end and has an axial load, P, applied at its right end. The cross-sectional area, A, is constant, and the material has elastic modulus, E.

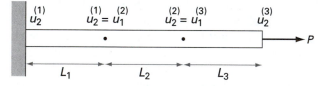

FIGURE 8.4   A rod represented by three one-dimensional bar elements.

Create a finite element model by dividing the member into three one-dimensional elements of lengths L_1, L_2, and L_3, respectively (Fig. 8.4). Choose the nodes at the ends of the elements. Make these constant strain elements by choosing the approximate displacement function to be of the form $f(x) = u(x) = a_1 + a_2 x$ in each element. It is extremely important to realize that each element has its own coordinate system in which x varies from zero at the left end to L_i at the right end. Number the displacement of the nodes by placing the node number as a subscript and the element number as a superscript. Reading from left to right, the first element has nodes u_1^1 and u_2^1; the second has nodes u_1^2 and u_2^2; while the third has nodes u_1^3 and u_2^3. In order to connect the elements, a compatibility, or interelement continuity, condition is applied. Let u_1,

u_2, u_3, and u_4 be the displacements, respectively, of the four points corresponding to the nodes in the global (as opposed to the element) coordinate system. The compatibility condition for nodal displacements requires that

$$u_1^1 = u_1, u_2^1 = u_1^2 = u_2, u_2^2 = u_1^3 = u_3, \quad \text{and} \quad u_2^3 = u_4. \tag{8.19}$$

Now the element governing equations can be rewritten in terms of the nodal displacements, u_1, u_2, u_3, and u_4, in the global coordinates. Notice that the dimension of K_e for this formulation of the system is four by four, while the dimension of k_e for each element is two by two.

Element 1: There is an unknown external reaction force Q_u acting on element 1 at node 1 and no other external forces.

$$\frac{EA}{L_1} \begin{pmatrix} 1 & -1 & 0 & 0 \\ -1 & 1 & 0 & 0 \\ 0 & 0 & 0 & 0 \\ 0 & 0 & 0 & 0 \end{pmatrix} \begin{pmatrix} u_1 \\ u_2 \\ u_3 \\ u_4 \end{pmatrix} = \begin{pmatrix} Q_u \\ 0 \\ 0 \\ 0 \end{pmatrix}.$$

Element 2: $\{Q\} = \{0\}$ on element 2.

$$\frac{EA}{L_2} \begin{pmatrix} 0 & 0 & 0 & 0 \\ 0 & 1 & -1 & 0 \\ 0 & -1 & 1 & 0 \\ 0 & 0 & 0 & 0 \end{pmatrix} \begin{pmatrix} u_1 \\ u_2 \\ u_3 \\ u_4 \end{pmatrix} = \begin{pmatrix} 0 \\ 0 \\ 0 \\ 0 \end{pmatrix}.$$

Element 3: Now Q is no longer zero since P acts on the right end at node 4.

$$\frac{EA}{L_3} \begin{pmatrix} 0 & 0 & 0 & 0 \\ 0 & 0 & 0 & 0 \\ 0 & 0 & 1 & -1 \\ 0 & 0 & -1 & 1 \end{pmatrix} \begin{pmatrix} u_1 \\ u_2 \\ u_3 \\ u_4 \end{pmatrix} = \begin{pmatrix} 0 \\ 0 \\ 0 \\ P \end{pmatrix}.$$

These systems of equations are now assembled to give a system for the full body by adding all the equations.

$$AE \begin{pmatrix} \frac{1}{L_1} & -\frac{1}{L_1} & 0 & 0 \\ -\frac{1}{L_1} & \frac{1}{L_1} + \frac{1}{L_2} & -\frac{1}{L_2} & 0 \\ 0 & -\frac{1}{L_2} & \frac{1}{L_2} + \frac{1}{L_3} & -\frac{1}{L_3} \\ 0 & 0 & -\frac{1}{L_3} & \frac{1}{L_3} \end{pmatrix} \begin{pmatrix} u_1 \\ u_2 \\ u_3 \\ u_4 \end{pmatrix} = \begin{pmatrix} Q_u \\ 0 \\ 0 \\ P \end{pmatrix}.$$

The matrix, K, on the left is called a banded matrix since all nonzero terms form a band that is symmetrical around the main diagonal. The computer memory required to solve these equations can be reduced by storing only those values along the band.

Finally, the boundary conditions restricting the displacements, u_1, u_2, u_3, and u_4, must be imposed before solving for the displacements. In this case, the left end is fixed so that $u_1 = 0$. The stiffness matrix is partitioned into blocks corresponding to the known and unknown forces and displacements, where K_{11} is 1×1, K_{12} is 1×3, K_{21} is 3×1, and K_{22} is 3×3. The unknown displacements are obtained from equation (8.16), which produces the three equations in the three unknowns, u_1, u_2, and u_3, given by the matrix equation,

$$AE \begin{pmatrix} \frac{1}{L_1} + \frac{1}{L_2} & -\frac{1}{L_2} & 0 \\ -\frac{1}{L_2} & \frac{1}{L_2} + \frac{1}{L_3} & -\frac{1}{L_3} \\ 0 & -\frac{1}{L_3} & \frac{1}{L_3} \end{pmatrix} \begin{pmatrix} u_2 \\ u_3 \\ u_4 \end{pmatrix} = \begin{pmatrix} 0 \\ 0 \\ P \end{pmatrix}.$$

Now solve this matrix equation, either by hand or numerically on a computer. Each row of the matrix equation represents a separate algebraic equation.

$$\left(\frac{1}{L_1} + \frac{1}{L_2}\right) u_2 - \frac{1}{L_2} u_3 = 0;$$

$$-\frac{1}{L_2} u_2 - \left(\frac{1}{L_2} + \frac{1}{L_3}\right) u_3 - \frac{1}{L_3} u_4 = 0;$$

$$-\frac{1}{L_3} u_3 + \frac{1}{L_3} u_4 = \frac{P}{AE}.$$

These are solved simultaneously to obtain

$$u_2 = \frac{PL_1}{AE}; u_3 = \frac{P(L_1 + L_2)}{AE}; \quad \text{and} \quad u_4 = \frac{P(L_1 + L_2 + L_3)}{AE}.$$

These displacements agree with those of strength of materials for an axial load since the one-dimensional elements were chosen to be constant strain elements, as underlies the strength of materials equation, $\delta = PL/AE$. This can be verified by showing that the strain in each element is $\epsilon = P/AE$.

Application of equation (8.17) gives the reaction force at the wall as

$$-\frac{AE}{L_1} u_2 = -\frac{AE}{L_1} \frac{PL_1}{AE} = -P,$$

as expected from a free body-diagram.   ∎

Exercise 8.4   Write FORTRAN code to assemble the element stiffness matrices into the global stiffness matrix found in the above example. Do this by calling a subroutine containing the element stiffness matrix.

8.4   A BEAM ELEMENT

The deflection of a beam, in strength of materials, is assumed to be represented by the transverse displacement of the neutral axis of the beam. An element whose generalized displacements are the transverse displacement and the rotation of the neutral axis can capture the response of such a beam. The uniform one-dimensional beam element of length, L, has flexural stiffness, EI. There are two nodes, one at each end, with generalized displacements a transverse displacement, v_i, and rotation, θ_i, at the ith node (Fig. 8.5). The cross section of the beam is accounted for only in the area moment of inertia portion, I, of the flexural stiffness. In this formulation, the rotation is the derivative of the transverse displacement, $\theta = dv/dx$, where x is the element coordinate.

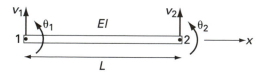

FIGURE 8.5   A strength of materials beam element.

To have a complete element, the displacement must depend on as many undetermined coefficients as there are nodal displacements in the element, four. The simplest such displacement assumption is the cubic,

$$v(x) = a_1 + a_2 x + a_3 x^2 + a_4 x^3,$$

with the as-yet undetermined coefficients, a_i. In this case,

$$\theta(x) = a_2 + 2a_3 x + 3a_4 x^2.$$

The nodal displacements must satisfy

$$\{d\} = \begin{pmatrix} v_1 \\ \theta_1 \\ v_2 \\ \theta_2 \end{pmatrix} = \begin{pmatrix} 1 & 0 & 0 & 0 \\ 0 & 1 & 0 & 0 \\ 1 & L & L^2 & L^3 \\ 0 & 1 & 2L & 3L^2 \end{pmatrix} \begin{pmatrix} a_1 \\ a_2 \\ a_3 \\ a_4 \end{pmatrix} = (DA)\{a\}.$$

By multiplying on the left by the inverse matrix $(DA)^{-1}$, $v(x)$ is written in terms of the nodal displacements, $\{d\}$,

$$v(x) = \begin{pmatrix} 1 & x & x^2 & x^3 \end{pmatrix} \begin{pmatrix} 1 & 0 & 0 & 0 \\ 0 & 1 & 0 & 0 \\ -\frac{3}{L^2} & -\frac{2}{L} & \frac{3}{L^2} & -\frac{1}{L} \\ \frac{2}{L^3} & \frac{1}{L^2} & -\frac{2}{L^3} & \frac{1}{L^2} \end{pmatrix} \begin{pmatrix} v_1 \\ \theta_1 \\ v_2 \\ \theta_2 \end{pmatrix},$$

and

$$\frac{d^2\theta}{dx^2} = B\{d\} = \begin{pmatrix} 0 & 0 & 2 & 6 \end{pmatrix} (DA)^{-1}\{d\}.$$

The element stiffness is

$$k_e = \int_0^L B^t EI B dx = \frac{EI}{L^3} \begin{pmatrix} 12 & 6L & -12 & 6L \\ 6L & 4L^2 & -6L & 2L^2 \\ -12 & -6L & 12 & -6L \\ 6L & 2L^2 & -6L & 4L^2 \end{pmatrix}. \qquad (8.20)$$

Exercise 8.5   Verify this expression for the element stiffness.

EXAMPLE 8.6   Determine the displacement field for a strength of materials cantilever beam of length, $2L$, with transverse point loads of P_1 at the center and P_2 at the free end (Fig. 8.6). Use two of the elementary beam elements and compare the result to the strength of materials solution.

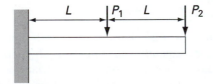

FIGURE 8.6   A cantilever beam of length, $2L$, with transverse point loads of P_1 at the center and P_2 at the free end.

Solution   The global displacements are obtained from the assembled stiffness relation, $K\{D\} = \{F\}$,

$$\frac{EI}{L^3}\begin{pmatrix} 12 & 6L & -12 & 6L & 0 & 0 \\ 6L & 4L^2 & -6L & 2L^2 & 0 & 0 \\ -12 & -6L & 12+12 & -6L+6L & -12 & 6L \\ 6L & 2L^2 & -6L+6L & 4L^2+4L^2 & -6L & 2L^2 \\ 0 & 0 & -12 & -6L & 12 & -6L \\ 0 & 0 & 6L & 2L^2 & -6L & 4L^2 \end{pmatrix}\begin{pmatrix} V_1 \\ \Theta_1 \\ V_2 \\ \Theta_2 \\ V_3 \\ \Theta_3 \end{pmatrix} = \begin{pmatrix} 0 \\ 0 \\ -P_1 \\ 0 \\ -P_2 \\ 0 \end{pmatrix}.$$

The forced boundary conditions for a cantilever beam are that $V_1 = \Theta_1 = 0$. Crossing out the first two rows and columns produces the system which must be solved. The loads are negative since they act in the negative V direction.

$$\frac{EI}{L^3}\begin{pmatrix} 24 & 0 & -12 & 6L \\ 0 & 8L^2 & -6L & 2L^2 \\ -12 & -6L & 12 & -6L \\ 6L & 2L^2 & -6L & 4L^2 \end{pmatrix}\begin{pmatrix} V_2 \\ \Theta_2 \\ V_3 \\ \Theta_3 \end{pmatrix} = \begin{pmatrix} -P_1 \\ 0 \\ -P_2 \\ 0 \end{pmatrix}.$$

Multiplying on the left by K^{-1} produces

$$\begin{pmatrix} V_2 \\ \Theta_2 \\ V_3 \\ \Theta_3 \end{pmatrix} = \frac{L^3}{EI}\begin{pmatrix} \frac{1}{3} & \frac{1}{2L} & \frac{5}{6} & \frac{1}{2L} \\ \frac{1}{2L} & \frac{1}{L^2} & \frac{3}{2L} & \frac{1}{L^2} \\ \frac{5}{6} & \frac{3}{2L} & \frac{8}{3} & \frac{2}{L} \\ \frac{1}{2L} & \frac{1}{L^2} & \frac{2}{L} & \frac{2}{L^2} \end{pmatrix}\begin{pmatrix} -P_1 \\ 0 \\ -P_2 \\ 0 \end{pmatrix} = \begin{pmatrix} -(\frac{1}{3}P_1 + \frac{5}{6}P_2)\frac{L^3}{EI} \\ -\frac{1}{2}P_1\frac{L^2}{EI} \\ -(\frac{5}{6}P_1 + \frac{8}{3}P_2)\frac{L^3}{EI} \\ -(\frac{1}{2}P_1 + 2P_2)\frac{L^2}{EI} \end{pmatrix}.$$

Therefore the displacement for $0 \le x \le L$ is

$$v(x) = \begin{pmatrix} 1 & x & x^2 & x^3 \end{pmatrix}\begin{pmatrix} 1 & 0 & 0 & 0 \\ 0 & 1 & 0 & 0 \\ -\frac{3}{L^2} & -\frac{2}{L} & \frac{3}{L^2} & -\frac{1}{L} \\ \frac{2}{L^3} & \frac{1}{L^2} & -\frac{2}{L^3} & \frac{1}{L^2} \end{pmatrix}\begin{pmatrix} 0 \\ 0 \\ -(\frac{1}{3}P_1 + \frac{5}{6}P_2)\frac{L^3}{EI} \\ -\frac{1}{2}P_1\frac{L^2}{EI} \end{pmatrix}.$$

Exercise 8.6   In a similar manner determine the displacement field for $L \le x \le 2L$.

The exact strength of materials solution is

$$v(x) = -\frac{1}{2}(P_1L + 2P_2L)x^2 + \frac{1}{6}(P_1 + P_2)x^3 \quad \text{if} \quad 0 \le x \le L$$

$$= -\frac{1}{2}(P_1L + 2P_2L)x^2 + \frac{1}{6}(P_1 + P_2)x^3 - \frac{1}{6}P_1(L - x)^3 \quad \text{if} \quad L \le x \le 2L.$$

The finite element method gives the exact solution because the element displacements chosen were cubic polynomials, just as is the exact solution.   ∎

8.5   CONSTANT STRAIN TRIANGLE

Two-dimensional structures such as plates or other thin members are often modeled by triangular or rectangular elements. The simplest such element is the constant strain triangular element with nodes at the vertices of the triangle (Fig. 8.7). The element will be taken to have a small constant thickness. This will be used to model two-dimensional problems in the sense that the element is two-dimensional and each node will be allowed to displace only in the plane (not perpendicular to the plane). The

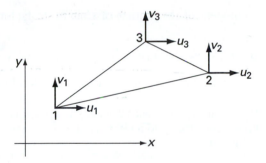

FIGURE 8.7   The constant strain triangle with nodal displacements.

displacement at the ith node can be resolved into two components, $u_i\mathbf{i} + v_i\mathbf{j}$ in a global coordinate system.

Since there are three nodes in the triangle, there will be six nodal displacement components to determine. These will be the undetermined coefficients of the assumed displacement field, $u(x, y)$ and $v(x, y)$. Therefore, these must both be linear, or they could be of higher order as long as they only involve six unknown coefficients.

To create a constant strain element, the displacement fields must be linear. Assume that

$$
\begin{aligned}
u(x, y) &= \alpha_1 + \alpha_2 x + \alpha_3 y; \\
v(x, y) &= \alpha_4 + \alpha_5 x + \alpha_6 y.
\end{aligned}
\tag{8.21}
$$

The relations (8.21) must be written in terms of the nodal displacements, u_1, u_2, u_3, v_1, v_2, v_3, by solving the six equations obtained from evaluating equations (8.21) at the nodal points, (x_1, y_1), (x_2, y_2), (x_3, y_3). In matrix form, this displacement equation is written as

$$
\begin{pmatrix} u_1 \\ u_2 \\ u_3 \\ v_1 \\ v_2 \\ v_3 \end{pmatrix} =
\begin{pmatrix}
1 & x_1 & y_1 & 0 & 0 & 0 \\
1 & x_2 & y_2 & 0 & 0 & 0 \\
1 & x_3 & y_3 & 0 & 0 & 0 \\
0 & 0 & 0 & 1 & x_1 & y_1 \\
0 & 0 & 0 & 1 & x_2 & y_2 \\
0 & 0 & 0 & 1 & x_3 & y_3
\end{pmatrix}
\begin{pmatrix} \alpha_1 \\ \alpha_2 \\ \alpha_3 \\ \alpha_4 \\ \alpha_5 \\ \alpha_6 \end{pmatrix}.
\tag{8.22}
$$

This can be viewed as a matrix equation

$$
\{d\}_e = \begin{pmatrix} C & 0 \\ 0 & C \end{pmatrix} \{\alpha\}_e,
$$

where

$$
C = \begin{pmatrix}
1 & x_1 & y_1 \\
1 & x_2 & y_2 \\
1 & x_3 & y_3
\end{pmatrix}.
\tag{8.23}
$$

To solve for $\{\alpha\}_e$ in terms of the nodal displacements, the matrix C must be inverted. Recall from linear algebra that the inverse of a matrix C is given by

$$
C^{-1} = \frac{1}{\det(C)} adj(C),
$$

where the ijth component of the adjoint of C is the cofactor of the entry C_{ji} in C. Here,

$$\det(C) = x_2y_3 - x_3y_2 + x_3y_1 - x_1y_3 + x_1y_2 - x_2y_1.$$

$\det(C)$ is positive because the nodes are numbered counterclockwise.

Exercise 8.7 Prove that $\det(C)$ is twice the area, A, of the triangle. *Hint*: First show that if $\mathbf{r}_1$ is the position vector from node 1 to node 3 and if $\mathbf{r}_2$ is the position vector from node 1 to node 2, then $A = 0.5|\mathbf{r}_1 \times \mathbf{r}_2|$.

The inverse is then expressed as

$$C^{-1} = \frac{1}{2A} \begin{pmatrix} x_2y_3 - x_3y_2 & x_3y_1 - x_1y_3 & x_1y_2 - x_2y_1 \\ y_2 - y_3 & y_3 - y_1 & y_1 - y_2 \\ x_3 - x_2 & x_1 - x_3 & x_2 - x_1 \end{pmatrix}. \tag{8.24}$$

This allows the displacement field to be written in terms of the nodal displacements, $\{f\}_e = N\{d\}_e$.

$$\begin{pmatrix} u(x,y) \\ v(x,y) \end{pmatrix} = \begin{pmatrix} 1 & x & y & 0 & 0 & 0 \\ 0 & 0 & 0 & 1 & x & y \end{pmatrix} \begin{pmatrix} C^{-1} & 0 \\ 0 & C^{-1} \end{pmatrix} \{d\}_e. \tag{8.25}$$

The strain displacement relation, $\{\epsilon\}_e = B\{d\}_e$, in terms of the nodal displacements, is

$$\begin{pmatrix} \epsilon_x(x,y) \\ \epsilon_y(x,y) \\ \gamma_{xy}(x,y) \end{pmatrix} = \begin{pmatrix} \frac{\partial}{\partial x} & 0 \\ 0 & \frac{\partial}{\partial y} \\ \frac{\partial}{\partial y} & \frac{\partial}{\partial x} \end{pmatrix} N\{\alpha\}_e$$

$$= \begin{pmatrix} \frac{\partial}{\partial x} & 0 \\ 0 & \frac{\partial}{\partial y} \\ \frac{\partial}{\partial y} & \frac{\partial}{\partial x} \end{pmatrix} \begin{pmatrix} 1 & x & y & 0 & 0 & 0 \\ 0 & 0 & 0 & 1 & x & y \end{pmatrix} \{\alpha\}_e$$

$$= \begin{pmatrix} 0 & 1 & 0 & 0 & 0 & 0 \\ 0 & 0 & 0 & 0 & 0 & 1 \\ 0 & 0 & 1 & 0 & 1 & 0 \end{pmatrix} \begin{pmatrix} C^{-1} & 0 \\ 0 & C^{-1} \end{pmatrix} \{d\}_e. \tag{8.26}$$

Therefore, B is constant since it is independent of (x,y), but it does depend on the nodal coordinates.

$$B = \frac{1}{2A} \begin{pmatrix} y_2 - y_3 & y_3 - y_1 & y_1 - y_2 & 0 & 0 & 0 \\ 0 & 0 & 0 & x_3 - x_2 & x_1 - x_3 & x_2 - x_1 \\ x_3 - x_2 & x_1 - x_3 & x_2 - x_1 & y_2 - y_3 & y_3 - y_1 & y_1 - y_2 \end{pmatrix}.$$

The sum of the entries in each row is zero. This fact can be used as a check in numerical examples.

Exercise 8.8 Show that

$$\begin{pmatrix} \frac{\partial}{\partial x} & 0 \\ 0 & \frac{\partial}{\partial y} \\ \frac{\partial}{\partial y} & \frac{\partial}{\partial x} \end{pmatrix} \begin{pmatrix} 1 & x & y & 0 & 0 & 0 \\ 0 & 0 & 0 & 1 & x & y \end{pmatrix} \{\alpha\}_e$$

gives the correct strains in the original displacement assumption, that is, that $\epsilon_x = \alpha_2$, $\epsilon_y = \alpha_6$, and $\gamma_{xy} = \alpha_3 + \alpha_5$.

Finally, the element stiffness matrix can be computed once a choice of the stress-strain displacement relation is made. Typically, the stress-strain matrix, D, represents either plane stress or plane strain.

$$k_e = \int_V B^t D B \, dV = B^t D B w A, \tag{8.27}$$

where w is the constant element thickness and A is the element area.

In the case of plane stress for a linear elastic material,

$$D = \frac{E}{1-\nu^2} \begin{pmatrix} 1 & \nu & 0 \\ \nu & 1 & 0 \\ 0 & 0 & \frac{1-\nu}{2} \end{pmatrix}.$$

Write B in the form

$$B = \frac{1}{2A} \begin{pmatrix} b_1 & b_2 & b_3 & 0 & 0 & 0 \\ 0 & 0 & 0 & a_1 & a_2 & a_3 \\ a_1 & a_2 & a_3 & b_1 & b_2 & b_3 \end{pmatrix}.$$

Performing the multiplication yields the symmetric matrix,

$$k_e = \frac{Ew}{4A(1-\nu^2)} \times \tag{8.28}$$

$$\begin{pmatrix} b_1^2 + a_1^2(1-\nu)/2 & b_1 b_2 + a_1 a_2(1-\nu)/2 & b_1 b_3 + a_1 a_3(1-\nu)/2 & a_1 b_1(1-\nu)/2 + a_1 b_1 \nu & a_1 b_2(1-\nu)/2 + a_2 b_1 \nu & a_1 b_3(1-\nu)/2 + a_3 b_1 \nu \\ & b_2^2 + a_2^2(1-\nu)/2 & b_2 b_3 + a_2 a_3(1-\nu)/2 & a_2 b_1(1-\nu)/2 + a_1 b_2 \nu & a_2 b_2(1-\nu)/2 + a_2 b_2 \nu & a_2 b_3(1-\nu)/2 + a_3 b_2 \nu \\ & & b_3^2 + a_3^2(1-\nu)/2 & a_3 b_1(1-\nu)/2 + a_1 b_3 \nu & a_3 b_2(1-\nu)/2 + a_2 b_3 \nu & a_3 b_3(1-\nu)/2 + a_3 b_3 \nu \\ & & & a_1^2 + b_1^2(1-\nu)/2 & a_1 a_2 + b_1 b_2(1-\nu)/2 & a_1 a_3 + b_1 b_3(1-\nu)/2 \\ & & & & a_2^2 + b_2^2(1-\nu)/2 & a_2 a_3 + b_2 b_3(1-\nu)/2 \\ & & & & & a_3^2 + b_3^2(1-\nu)/2 \end{pmatrix}.$$

Only the upper triangular elements are shown; the missing elements are found by symmetry.

Exercise 8.9 Compute k_e if D represents plane strain for a linear elastic material. Use a symbolic mathematics program to do the computation.

$$D = \frac{E}{(1+\nu)(1-2\nu)} \begin{pmatrix} 1-\nu & \nu & 0 \\ \nu & 1-\nu & 0 \\ 0 & 0 & \frac{1-2\nu}{2} \end{pmatrix}.$$

EXAMPLE 8.7 A thin beam of length 10 inches and height 1 inch, which is fixed at one end, is axially loaded at its free end by a distributed load of $p = 4{,}000$ lb/in^2. The beam is made of steel ($E = 30 \times 10^6$ psi and $\nu = 0.25$) and is 0.2 inches thick. Use a two-constant strain triangle finite element model to determine the deflections and stresses in the member. Assume plane stress.

Solution The choice of elements and the element node numbering is shown in Figure 8.8. The global nodes are 1:(0,1), 2:(0,0), 3:(10,0), 4:(10,1). In element 1, $(x_1, y_1) = (0, 1)$, $(x_2, y_2) = (0, 0)$,

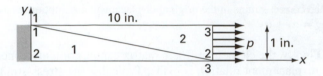

FIGURE 8.8   The two-element finite element model.

$(x_3, y_3) = (10, 0)$. Therefore $a_1 = x_3 - x_2 = 10$, $a_2 = x_1 - x_3 = -10$, $a_3 = x_2 - x_1 = 0$, $b_1 = y_2 - y_3 = 0$, $b_2 = y_3 - y_1 = -1$, $b_3 = y_1 - y_2 = 1$. From equation (8.28), the stiffness matrix is

$$k_e = 320{,}000 \begin{pmatrix} 37.5 & -37.5 & 0 & 0 & -3.75 & 3.75 \\ -37.5 & 38.5 & -1 & -2.5 & 6.25 & -3.75 \\ 0 & -1 & 1 & 2.5 & -2.5 & 0 \\ 0 & -2.5 & 2.5 & 100 & -100 & 0 \\ -3.75 & 6.25 & -2.5 & -100 & 100.375 & -0.375 \\ 3.75 & -3.75 & 0 & 0 & -0.375 & 0.375 \end{pmatrix}.$$

In element 2, $(x_1, y_1) = (0, 1)$, $(x_2, y_2) = (10, 0)$, $(x_3, y_3) = (10, 1)$. Therefore $a_1 = 0$, $a_2 = -10$, $a_3 = 10$, $b_1 = -1$, $b_2 = 0$, $b_3 = 1$.

Exercise 8.10   Compute k_e for element 2.

The elements must next be assembled to obtain the stiffness matrix for the structure. There are four nodes, each with two displacement components. Therefore, the global stiffness matrix is 8×8. Due to the node numbering chosen, it is not quite as easy to assemble the stiffness matrix in this example as in the bar examples above. At global nodes 1 and 3, there is a contribution from both elements. At global node 2, element 1 contributes and at global node 4, element 2 contributes. The global stiffness, K, is obtained by adding the rows corresponding to a given node, but placing the entries in the proper column for the global node number. For example, the first row is the sum of the u components at node 1. The superscript indicates the element.

$$(k_{11}^{(1)} + k_{11}^{(2)}, k_{12}^{(1)}, k_{13}^{(1)} + k_{12}^{(2)}, k_{13}^{(2)}, k_{14}^{(1)} + k_{14}^{(2)}, k_{15}^{(1)}, k_{16}^{(1)} + k_{15}^{(2)}, k_{16}^{(2)}).$$

Because only element 1 involves global node 2, the second row is

$$(k_{21}^{(1)}, k_{22}^{(1)}, k_{23}^{(1)}, 0, k_{24}^{(1)}, k_{25}^{(1)}, k_{26}^{(1)}, 0).$$

The third row is

$$(k_{31}^{(1)} + k_{21}^{(2)}, k_{32}^{(1)}, k_{33}^{(1)} + k_{22}^{(2)}, k_{23}^{(2)}, k_{34}^{(1)} + k_{24}^{(2)}, k_{35}^{(1)}, k_{36}^{(1)} + k_{25}^{(2)}, k_{26}^{(2)}).$$

Because only element 2 involves global node 4, the fourth row is

$$(k_{31}^{(2)}, 0, k_{32}^{(2)}, k_{33}^{(2)}, k_{34}^{(2)}, 0, k_{35}^{(2)}, k_{36}^{(2)}).$$

The remaining rows follow the same pattern. Alternatively, because global node 4 is not in element 1, its stiffness matrix is adjusted by inserting zeros in the fourth and eighth rows and columns. Likewise, because global node 2 is not in element 2, its stiffness matrix is adjusted by inserting zeros in the second and sixth rows and columns. The resulting two matrices are added to assemble the total stiffness matrix. Therefore,

$$K = 320{,}000 \begin{pmatrix} 38.5 & -37.5 & 0. & -1. & 0. & -3.75 & 6.25 & -2.5 \\ -37.5 & 38.5 & -1. & 0 & -2.5 & 6.25 & -3.75 & 0 \\ 0. & -1. & 38.5 & -37.5 & 6.25 & -2.5 & 0. & -3.75 \\ -1. & 0 & -37.5 & 38.5 & -3.75 & 0 & -2.5 & 6.25 \\ 0. & -2.5 & 6.25 & -3.75 & 100.375 & -100. & 0. & -0.375 \\ -3.75 & 6.25 & -2.5 & 0 & -100. & 100.375 & -0.375 & 0 \\ 6.25 & -3.75 & 0. & -2.5 & 0. & -0.375 & 100.375 & -100. \\ -2.5 & 0 & -3.75 & 6.25 & -0.375 & 0 & -100. & 100.375 \end{pmatrix}.$$

The force array is obtained by dividing the distributed load into two equal point loads applied at global nodes 3 and 4 in the u-direction. This approximation will introduce an error into the results.

The rows and columns, 1, 2, 5, 6 corresponding to global nodes 1 and 2 in the stiffness equation $K\{D\} = \{F\}$ can be deleted because of the boundary condition that $u_1 = v_1 = u_2 = v_2 = 0$. This equation is then solved for the displacements of global nodes 3 and 4 by multiplying on the left by K^{-1} to obtain

$$
\begin{pmatrix} U_3 \\ U_4 \\ V_3 \\ V_4 \end{pmatrix} = \begin{pmatrix} 1.647 \times 10^{-6} & 1.572 \times 10^{-6} & 3.94 \times 10^{-7} & 3.56 \times 10^{-7} \\ 1.572 \times 10^{-6} & 1.660 \times 10^{-6} & -4.19 \times 10^{-7} & -4.62 \times 10^{-7} \\ 3.94 \times 10^{-7} & -4.19 \times 10^{-7} & 8.23 \times 10^{-6} & 8.24 \times 10^{-6} \\ 3.56 \times 10^{-7} & -4.62 \times 10^{-7} & 8.24 \times 10^{-6} & 8.28 \times 10^{-6} \end{pmatrix} \begin{pmatrix} \frac{P}{2} \\ \frac{P}{2} \\ 0 \\ 0 \end{pmatrix}
$$

$$
= \begin{pmatrix} 1.6098 \times 10^{-6} P \\ 1.6164 \times 10^{-6} P \\ -1.261 \times 10^{-8} P \\ -5.307 \times 10^{-8} P \end{pmatrix}. \tag{8.29}
$$

where $P = pw = 800$ pounds is the total axial force applied. The lack of symmetry shown in the unequal U_1 and U_2 displacements and the non-zero V_1 and V_2 at the two nodes is due to the fact that the chosen mesh is not symmetric with respect to the direction of the load.

To examine the error in this example, compare the computed values of the U displacements to the expected displacement at global nodes 3 and 4, $\delta = PL/AE = 1.667 \times 10^{-6} P$. Here, A is the cross-sectional area of the beam. ∎

8.5.1 Area Coordinates

The previous formulation, once the nodal displacements are computed, allows the computation of the displacements, stresses, and strain inside the element in terms of the global coordinate system. Frequently, it is easier to compute the stiffness matrix in terms of an element coordinate system rather than the global coordinates. Number the nodes counterclockwise as shown in Figure 8.9, which represents a triangle in the x-y plane.

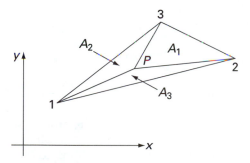

FIGURE 8.9   The triangle in area coordinates.

Each point, P, inside the triangle and on its boundary is given coordinates (L_1, L_2, L_3) by drawing lines from the nodes, the vertices of the triangle, to the point P. These lines divide the triangle into areas A_1, A_2, and A_3, where A_i is opposite node i. The coordinates, called area coordinates, are defined by

$$
L_1 = \frac{A_1}{A}, \quad L_2 = \frac{A_2}{A}, \quad \text{and} \quad L_3 = \frac{A_3}{A}. \tag{8.30}
$$

While it appears that a point in the plane is being given three coordinates, only two are independent. Since $A = A_1 + A_2 + A_3$, the three L_i are related by $L_1 + L_2 + L_3 = 1$. Observe that node 1 has coordinate $(1, 0, 0)$, node 2 is $(0, 1, 0)$, and node 3 is $(0, 0, 1)$. The line from node 1 to node 3 has coordinates of the form $(L_1, 0, L_3)$, from node 1 to node 2, $(L_1, L_2, 0)$ and from node 2 to node 3, $(0, L_2, L_3)$.

The area coordinates are related to the global Cartesian coordinates by writing the areas, A_i, in terms of the Cartesian coordinates. The position vector from node i to P is denoted $\mathbf{r}_i$ and is written in terms of the Cartesian nodal coordinates (x_i, y_i) and those of $P(x, y)$ as $\mathbf{r}_i = (x - x_i)\mathbf{i} + (y - y_i)\mathbf{j}$. Then

$$A_1 = 0.5|\mathbf{r}_2 \times \mathbf{r}_3|, \quad A_2 = 0.5|\mathbf{r}_1 \times \mathbf{r}_3|, \quad \text{and} \quad A_3 = 0.5|\mathbf{r}_1 \times \mathbf{r}_2|. \tag{8.31}$$

It can then be shown, using equation (8.24), that

$$\begin{pmatrix} L_1 \\ L_2 \\ L_3 \end{pmatrix} = \frac{1}{2A} C^{-t} \begin{pmatrix} 1 \\ x \\ y \end{pmatrix}, \tag{8.32}$$

where $\{u\} = C\{\alpha\}$.

The assumed linear element displacements can be written as

$$\begin{aligned} u(L_1, L_2, L_3) &= L_1 u_1 + L_2 u_2 + L_3 u_3, \\ v(L_1, L_2, L_3) &= L_1 v_1 + L_2 v_2 + L_3 v_3. \end{aligned} \tag{8.33}$$

Writing $\{f\}_e = N\{d\}_e$ implies that

$$N = \begin{pmatrix} L_1 & L_2 & L_3 & 0 & 0 & 0 \\ 0 & 0 & 0 & L_1 & L_2 & L_3 \end{pmatrix}. \tag{8.34}$$

Take derivatives with respect to the global coordinates (x, y) to obtain the strains, $\{\epsilon\}_e = B\{d\}_e$, as in equation (8.26). The derivatives of the L_i in N are functions only of the nodal coordinates. For example, since $L_1 = A_1/A$ and since A depends only on the nodal coordinates,

$$A_1 = 0.5|\mathbf{r}_2 \times \mathbf{r}_3| = 0.5[(x - x_2)(y - y_3) - (x - x_3)(y - y_2)],$$

$$\frac{\partial L_1}{\partial x} = \frac{1}{2A}|(y - y_3) - (y - y_2)| = \frac{1}{2A}|y_2 - y_3|.$$

Finally the element stiffness matrix is found from B and the stress-strain relation D as in equation (8.27).

Writing the assumed displacement field in terms of the area coordinates, which are element coordinates, allows one to determine the strains and element stiffness matrix directly in terms of the global coordinates of the nodes. This is true whether or not the assumed displacement field is linear.

8.6 ISOPARAMETRIC ELEMENTS

An affine element is one whose approximating polynomials are linear. In this case, the edges remain straight. Less computation is needed if the edges remain straight. To analyze the behavior of bodies with curved edges or faces, elements whose edges can curve are required. An isoparametric family of elements is one using polynomial approximation functions of order greater than the linear. The use of higher-order polynomials may allow the edges to be curved after loading. One of the earliest publications on this topic was that by Irons in 1966 [4].

A planar isoparametric quadrilateral element with four nodes requires displacement equations with eight undetermined coefficients for completeness. The element has local coordinates.

$$u(\zeta, \eta) = a_1 + a_2\zeta + a_3\eta + a_4\zeta\eta;$$
$$v(\zeta, \eta) = b_1 + b_2\zeta + b_3\eta + b_4\zeta\eta.$$

The isoparametric equations are given in local coordinates, $0 \leq \eta \leq 1$ and $0 \leq \zeta \leq 1$ (Fig. 8.10). These coordinates are analogous to the area coordinates for the constant strain triangle and can be related to the global coordinates.

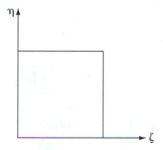

FIGURE 8.10 The rectangular isoparametric element and its local coordinates.

A straight edge is created with these elements if $\eta = 0$, $\eta = 1$, $\zeta = 0$, or $\zeta = 1$ because the displacement relation is linear along each boundary. For example, when $\eta = 0$, $u = a_1 + a_2\zeta$ and $v = b_1 + b_2\zeta$. This element generalizes the affine element in this sense. If this element is joined with other straight edge elements, like the affine, overlap is avoided. This property is particularly useful in the interior of a body.

Since the element is planar, there are two displacement components u and v at each node for a total of eight. The element is complete because eight coefficients must be determined.

Along the boundary of body, an FEM practitioner may want elements which have a boundary which can deform into a curved shape. Such an element is the 10 degree of freedom planar isoparametric quadrilateral element formed with a node at each corner and a fifth node at the center. The parametric equations for this element are

$$u(\zeta, \eta) = a_1 + a_2\zeta + a_3\eta + a_4\zeta\eta + a_5\zeta\eta(1 - \zeta);$$
$$v(\zeta, \eta) = b_1 + b_2\zeta + b_3\eta + b_4\zeta\eta + b_5\zeta\eta(1 - \zeta).$$

The boundaries for $\eta = 0$, $\zeta = 0$, or $\zeta = 1$ are straight because the displacement relation is linear. But the boundary for $\eta = 1$ is a quadratic and so is curved.

8.7 ERROR ESTIMATION

The finite element method is a numerical algorithm which approximates the solution to a chosen mathematical model for a system. The results depend on the choice of the size of the elements and on the type of element, including the number of nodes and the order of the approximating functions.

The finite element program can crash or produce unexpectedly small displacements for a given loading. The condition of producing drastically smaller displacements is called locking by FEM practitioners. Locking is a common danger when dealing with incompressible materials like rubber, called volume locking, when solving a Timoshenko beam model, called shear locking, and when solving a membrane model, called membrane locking.

A common cause of locking is that one of the matrices, often the stiffness matrix, is ill-conditioned. The condition number associated with a matrix gives an indication of the behavior of its determinant and therefore of its inverse. Recall that the inverse of a matrix is found by dividing its adjoint matrix by the determinant.

Definition   The condition number for a square $n \times n$ matrix, A, is

$$cond(A) = ||A|| \cdot ||A^{-1}||,$$

where the matrix norm for $A = (a_{ij})$ is the maximum over the rows of the sum of the absolute values of the entries in a row,

$$||A|| = \max_i \left(\sum_{j=1}^{n} |a_{ij}| \right), \quad i = 1, \dots, n.$$

For example, if A is diagonal with entries a_{ii}, then $cond(A) = a_{\max}/a_{\min}$, the ratio of largest to the smallest eigenvalue. Because any real symmetric matrix can be diagonalized, this expression of the condition number is often the most convenient in mechanics. The condition number is always greater or equal to one; the identity matrix has condition number equal to one.

EXAMPLE 8.8   The Hilbert matrix is

$$H = \begin{pmatrix} 1 & \frac{1}{2} & \frac{1}{3} & \cdots & \frac{1}{n+1} \\ \frac{1}{2} & \frac{1}{3} & \frac{1}{4} & \cdots & \frac{1}{n+2} \\ \cdots & & & & \\ \frac{1}{n+1} & \frac{1}{n+2} & \frac{1}{n+3} & \cdots & \frac{1}{2n+1} \end{pmatrix}.$$

For the simplest case of $n = 1$, $cond(H) = (3/2)(18) = 27$. As n gets larger, the condition number of H increases. ∎

A stiffness matrix K is said to be ill-conditioned if small changes in K produce large changes in the solution, $\{D\}$, to the equation, $K\{D\} = \{F\}$. If the condition number of K is very large, then K is ill-conditioned. Some insight into why this is so depends on the fact that in multiplication of a matrix A by a column b, so that $A \cdot b = c$, then $||c|| \leq ||A|| \, ||b||$.

Exercise 8.11   Verify the inequality, $||c|| \leq ||A|| \, ||b||$, for the matrix norm.

In the stiffness matrix equation, $F = K \cdot D$, perturb the force to $F + \Delta F$. The stiffness equation becomes $F + \Delta F = K \cdot (D + \Delta D)$ so that $\Delta F = K \cdot \Delta D$. By Exercise 8.11, $||F|| \leq ||K|| \, ||D||$ and $||\Delta D|| \leq ||K^{-1}|| \, ||\Delta F||$. Divide the second of these equations by $||D||$ and then use the first equation.

$$\frac{||\Delta D||}{||D||} \leq ||K^{-1}||\frac{||\Delta F||}{||D||} \leq ||K|| \, ||K^{-1}||\frac{||\Delta F||}{||F||} = cond(K)\frac{||\Delta F||}{||F||}.$$

The relative change in the displacement is bounded by the condition number of the stiffness matrix times the relative change in the force. So, if the condition number is very large, the displacement can undergo a large change for a small change in the relative force.

An ill-conditioned matrix may be interpreted by the computer as having determinant zero, due to rounding. In this case, the attempt to numerically find the inverse leads to division by zero and crashes the program. Ill-conditioning can occur when the values in two rows are close to each other.

This mathematical problem arises when a practitioner of the finite element method unintentionally produces the locking effect. For example, in plate-bending problems with the transverse shear strains included, these strains are very small. The stiffness matrix is then very stiff, or ill-conditioned, creating locking. Special efforts must be made in the integration of the terms in the stiffness matrix. Ill-conditioning can arise both from the physical form of the problem or from the manner in which the computer manipulates the numerical matrix equations. Sometimes ill-conditioned matrices in the finite element method produce accurate results, but often this is not the case.

The condition number can also be used to estimate the relationship between the number of correct decimals k in the numerical entries of the stiffness matrix and the number of correct decimals d in the displacement solution. One rule of thumb is that this truncation error is given by $\log_{10} cond(K)$ so that $k \geq d - \log_{10} cond(K)$.

Approximation error in the finite element method refers to a measure of how closely the assumed displacement field, $u(x)$, fits the exact displacement field, $u_e(x)$, obtained from the variational principle, written in global (not element) coordinates. A special effort is required to define what closeness means. A measure of the error between $u(x)$ and $u_e(x)$ could be invented by somehow combining the pointwise differences between the two functions, $u(x) - u_e(x)$. For example, the largest such difference could be used as an error estimate, error $= \max_{a \leq x \leq b} |u(x) - u_e(x)|$. The least squares error estimate, S, provides such a measure if only a finite number of points are considered, $S = \sum_{i=1}^{n}(u(x_i) - u_e(x_i))^2$, where the error is squared so that positive and negative errors do not add out and give a false reading of the total error. This measure is generalized to an infinite number of points between two limits, $a \leq x \leq b$, by taking the error estimate to be the integral

$$S = \int_a^b (u(x) - u_e(x))^2 dx.$$

This is made mathematically precise by defining the L_2-norm of a function to be $||u||_0 = \left(\int_a^b u(x)^2 dx \right)^{1/2}$. In higher dimensions, this is a generalization of the usual norm, or distance squared function. The L_2 error estimate is the L_2-norm of the difference, $S^{1/2} = ||u(x) - u_e(x)||_0$.

The fact that many finite element approximations are related to an energy function has lead to the invention of the energy norm as a measure of the difference between two functions. This norm depends on the derivatives of the displacements because the strain energy is a function of the displacement derivatives through the strain-displacement relations (Chapter 2, Section 2.1). The energy norm over the interval $a \leq x \leq b$ for an elasticity problem satisfying second-order differential equations is defined for square integrable functions, $u(x)$.

$$||u||_1 = \left\{ \int_a^b \left[u^2 + \left(\frac{du}{dx} \right)^2 \right] dx \right\}^{1/2}.$$

The error estimate is then obtained as

$$||u(x) - u_e(x)||_1 = \left\{ \int_a^b \left[(u(x) - u_e(x))^2 + \left(\frac{du}{dx} - \frac{du_e}{dx} \right)^2 \right] dx \right\}^{1/2}.$$

The energy of the approximation is greater than that of the exact solution because the exact solution minimizes the potential energy.

For an interpolation polynomial of degree p, the error in the energy norm is bounded,

$$||u(x) - u_e(x)||_1 < ch^p,$$

where c is a constant and h is a characteristic length of the elements. The exponent p is also called the rate of convergence. The rate of convergence for the energy norm is usually smaller than that of the L_2-norm.

EXAMPLE 8.9   Calculate the error in the L_2-norm for the one-dimensional bar modeled by a single element (Fig. 8.3).

Solution   The exact displacement in global coordinates is $u(x) = u_2 x/L$ for $0 \le x \le L$. The approximate displacement field is $u(x) = u_2 x^3/L^3$.

$$||u(x) - u_e(x)||_0 = \left[\int_0^L (u(x) - u_e(x))^2 dx \right]^{1/2} = \left[\int_0^L u_2^2 \left(\frac{x^3}{L^3} - \frac{x}{L} \right)^2 dx \right]^{1/2}$$

$$= \left[u_2^2 \left(\frac{L}{7} - \frac{2L}{5} + \frac{L}{3} \right) \right] \sim L^2,$$

because u_2 is proportional to L.

On the other hand, the constant strain element gives the exact displacement so there would be zero error in that case. ∎

Only displacement elements have been discussed in this chapter, in order to show how they can be derived from a variational principle like stationary potential energy. Many other types of elements have been used. Elements based on an assumed stress field, in which the unknowns are nodal forces, can be derived from the complementary energy. Any variational principle can be used to generate an associated element type. Mixed elements based on a combination of forces and displacements as the nodal unknowns are possible. Hybrid elements have been constructed with an assumed displacement on the boundaries, but an assumed stress field on the interior of the element.

8.8   FINITE ELEMENT METHOD PROJECTS

The following projects can be carried out on commercial FEM packages, such as ANSYS, to gain experience with such packages and to begin to understand the approximations made. Each is a two-dimensional problem. The mesh should be constructed directly by the operator in each case rather than making use of the automatic meshing available in the preprocessor of some packages. Each of these problems can be modeled with two-dimensional, isoparametric quadrilateral elements with four corner nodes of the type typically used for solids. Element 42 in ANSYS is of this type. An important point of each project is practice in validating FEM results by comparing them to known analytical solutions or to experimental results.

8.8.1 Project 1: Thick-Walled Cylinders

A body that is geometrically symmetric with respect to an axis and on which the loads are also symmetric to the same axis is called axisymmetric. A thick-walled cylinder is axisymmetric with respect to the longitudinal axis, the z-axis. The Lamé elasticity analysis predicts the radial displacement, u, the radial normal stress, σ_r, and the circumferential (tangential) stress, σ_θ, for an axisymmetric loading. The cylinder has inner radius $r = a$ and outer radius $r = 3a$ (Fig. 8.11). The load is an external distributed compressive pressure, p_o. The cylinder is made of aluminum so that $\nu = 0.33$ and $E = 70$ GPa.

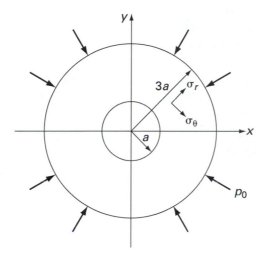

FIGURE 8.11   A disk with a hole and in-plane radial loads on the edges.

A thick-walled cylinder cannot be modeled by the thin-walled approximation of elementary strength of materials, which assumes the cylinder radius is at least 10 times the wall thickness. For the thin-walled cylinder, the longitudinal normal stress is $\sigma_a = pR/2t$ and the circumferential normal stress is $\sigma_c = pR/t$, where R is the radius, p is the internal pressure, and t is the wall thickness. These two stresses are principal. The radial stress, σ_r, is assumed to be zero since the wall is thin.

Perform the following tasks.

1. Use the elasticity solution obtained by Lamé for thick-walled cylinders to calculate expressions for u, σ_r, and σ_θ. Discuss the differences in the deformation and stress assumptions made for the thick-walled cylinder and those made for thin-walled pressure vessels.

2. Perform a finite element analysis with at least 10 elements in the radial direction for the following four cases. Take advantage of the symmetry of the structure.

   a. The number of elements is approximately 60 and the applied external pressure is modeled by forces applied to the nodal positions on the outer boundary.

   b. The number of elements is approximately 60 and the applied external pressure is modeled by distributed pressures across each element on the outer boundary.

    c. The number of elements is approximately 300 and the applied external pressure is modeled by forces applied to the nodal positions on the outer boundary.

    d. The number of elements is approximately 300 and the applied external pressure is modeled by distributed pressures across each element on the outer boundary.

Discussion

    1. Plot the elasticity and FEM solutions for σ_r/a versus r/a in each of the four cases. Compare the results on the same graph. Repeat for plots of σ_θ/a versus r/a.

    2. Plot the elasticity and FEM solutions for the radial displacement, u, in the form of $u(r)/a$ versus r/a in each of the four cases. Compare the results on the same graph.

    3. How accurate is the finite element method in stresses, in displacements? Does the number of elements affect the accuracy? What is the effect of using nodal forces rather than distributed forces?

8.8.2 Project 2: Thick Members

The object of this project is to compare the computed stresses, strains, and displacements for FEM and analytical solutions for two differently shaped bodies. FEM and analytical solutions are to be produced for a straight cantilever beam and for a curved cantilever beam, both subjected to a constant distributed stress, q, normal to the top surface. Each beam is made of aluminum ($E = 70$ GPa, $\nu = 0.33$). The cross section of each is rectangular of width $t = 0.1$.

    For each beam, perform two finite element approximations, first with 60 elements and second with 300 elements. In each of these cases, put 10 elements in the vertical or the radial direction.

Straight Beam

    1. Assume that the stress function for the straight beam (Fig. 8.12) is

$$\phi(x, y) = \frac{q}{2I}\left(-\frac{1}{6}x^2y^3 + \frac{1}{30}y^5 - \frac{1}{15}c^2y^3 + \frac{1}{2}c^2x^2y - \frac{1}{3}c^2x^2\right),$$

where I is the area moment of inertia, c is half the height, and y is measured down from the centroid of the cross-sectional area.

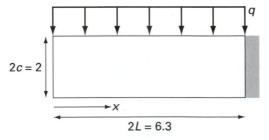

FIGURE 8.12   Straight beam with transverse distributed load.

    a. Determine the expressions for $\sigma_x(x, y)$, $\sigma_y(x, y)$, and $\tau_{xy}(x, y)$.

    b. Verify that the boundary conditions are satisfied on all edges.

2. Plot σ_x/q versus y on the same graph for the 60-element FEM, 300-element FEM, analytic elasticity, and strength of materials solutions. Do three graphs; one at $x = 0$, one at $x = L$, and one at $x = 2L$.

3. Plot σ_y/q versus y on the same graph for the 60-element FEM, 300-element FEM, analytic elasticity, and strength of materials solutions. Do three graphs; one at $x = 0$, one at $x = L$, and one at $x = 2L$.

4. Plot τ_{xy}/q versus y on the same graph for the 60-element FEM, 300-element FEM, analytic elasticity, and strength of materials solutions. Do three graphs; one at $x = 0$, one at $x = L$, and one at $x = 2L$.

5. Compare by computing percent errors and discuss the results obtained. How closely do the finite element solutions agree with the analytical? Does the number of elements affect the result?

Curved Beam   The inner radius is $a = 3$ and the outer radius is $b = 5$ (Fig. 8.13). Compute the stresses at the wall using the Winkler-Bach theory. Present your results in a table.

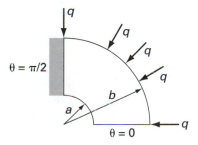

FIGURE 8.13   Curved beam with radial in-plane load on its edges.

1. Plot σ_r/q versus r on the same graph for the 60-element FEM, 300-element FEM, and the Winkler-Bach results at $\theta = \pi/2$. Repeat for just the FEM results at $\theta = \pi/4$.

2. Plot σ_θ/q versus r on the same graph for the 60-element FEM, 300-element FEM, and the Winkler-Bach results at $\theta = \pi/2$. Repeat for just the FEM results at $\theta = \pi/4$.

3. Plot $\tau_{r\theta}/q$ versus r on the same graph for the 60-element FEM, 300-element FEM, and the Winkler-Bach results at $\theta = \pi/2$. Repeat for just the FEM results at $\theta = \pi/4$.

Discussion   Compare your results for the straight and curved beams at the cross sections on the beams for which the graphs were prepared. Notice that the length of the straight beam is approximately the same as the arclength of the inner surface of the curved beam at $r = a$.

8.8.3   Project 3: Beams with and without a Notch

A notch introduces a stress concentration into a beam. This project is to perform finite element computations of the displacements, strains, and stresses in a cantilever beam, loaded with a transverse point load at its free end, with and without a notch. The results are compared to the strength of materials prediction for the beam without a notch and to the experimental stress concentration for the beam with a notch. The computations are made with two different numbers of elements to indicate the effect of changing the mesh size.

The beam is assumed to be very thin so that it is in plane stress. The beam is made of steel with $E = 210$ GPa and $v = 0.25$. The beam height is 10 cm, the thickness is $t = 5$ mm, and the length is $L = 60$ cm. The point load at the top of the free end is $P = 600$ N (Fig. 8.14). Assume that the beam is elastically loaded. Neglect the weight of the beam.

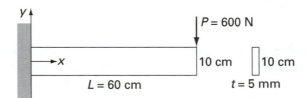

FIGURE 8.14   Cantilever beam without a notch.

Beam without a Notch

1. Determine the stress, strain, and displacements at each point in the beam using strength of materials. Find the vertical displacement using $EIv'' = M(x)$. Also try to find u and v from the strength of materials strains using the strain-displacement relations.

2. Use a 60-element mesh with 6 elements in the horizontal direction and 10 elements in the vertical direction to determine the stress, strain, and displacements at the nodes.

3. Use a 300-element mesh with 30 elements in the horizontal direction and 10 elements in the vertical direction to determine the stress, strain, and displacements at the nodes.

4. Plot graphs with each of the 10 vertical nodal positions on the horizontal axis and the computed value on the vertical axis. Do this at three cross-sectional faces for the strength of materials, the 60-element mesh, and the 300-element mesh results.

    a. σ_x at the wall, at 30 cm from the wall, and at the free end (three graphs, each containing three curves).

    b. σ_y at the wall, at 30 cm from the wall, and at the free end (three graphs, each containing three curves).

    c. The displacement, u, in the longitudinal direction, at 30 cm from the wall, and at the free end (two graphs, each containing three curves).

    d. The displacement, v, in the vertical direction, at 30 cm from the wall, and at the free end (two graphs, each containing three curves).

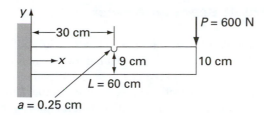

FIGURE 8.15   Cantilever beam with a notch.

Beam with a Notch   A notch is cut in the top at the midpoint of the beam
(Fig. 8.15), so that the root is 30 cm from the wall. The notch has dimensions $h = 1$ cm,
radius $a = 0.25$ cm, and the height from the root of the notch to the bottom of the beam
is $b = 9$ cm. The thickness $t = 5$ mm. The length is $L = 60$ cm.

1. Estimate the stress concentration in the root of the notch using a Neuber diagram. Use the bending version by assuming the moment applied is the same as
   the internal moment induced at the root of the notch by the transverse load, P,
   at the free end.

2. Apply the FEM to determine the stress, strain, and displacements at the nodes.
   Use a mesh with sufficient elements around the curved portion of the notch, say
   10. Try to get the smallest elements in the region of the largest change in stress.
   Try to have either about 60 elements or about 300 elements to compare with the
   beam without a notch.

3. Plot graphs with the nodal positions on the horizontal axis and the computed
   value on the vertical axis.

   a. σ_x at the wall, at 30 cm from the wall, and at the free end (three graphs).

   b. σ_y at the wall, at 30 cm from the wall, and at the free end (three graphs).

   c. The displacement, u, in the longitudinal direction, at 30 cm from the wall,
      and at the free end (two graphs).

   d. The displacement, v, in the vertical direction, at 30 cm from the wall, and at
      the free end (two graphs).

Discussion

1. For the solid beam compare the longitudinal stresses and displacements at the
   wall, midplane, and free end. Do the elasticity results agree with the strength of
   materials predictions? How does the number of elements in the mesh affect the
   results? Do the stresses at the various locations agree with your expectations?
   Does the elasticity analysis by the finite element method predict any normal
   stresses in the vertical direction in the beam? Compare the FEM prediction of
   stress near the point of application of the load with that obtained from strength
   of materials.

2. Compare the longitudinal stress in the root of the notch with the prediction of
   the Neuber diagram.

3. Compare the stresses in the beam with a notch with those in the solid beam.
   How does σ_x vary around the arc of the notch?

8.8.4   Project 4: Beam with Fillet

The response of irregularly shaped bodies to loads is difficult to analyze using strength of materials or a closed-form elasticity calculation. The member shown in Figure 8.16 is made of aluminum ($E = 70$ GPa and $v = 0.33$) and is considered to be in plane stress. The thickness in the direction perpendicular to that shown in Figure 8.16 is 0.5 cm. The member is mounted in a rigid wall at its left end. A vertical distributed force is applied along the top. The curved portion of the bottom boundary is a quarter-circle of radius 5 cm. The change in cross-sectional area should introduce a stress concentration. The length from the rigid wall to the free end is 60 cm. The length of the straight portion is 50 cm. The height at the wall is 10 cm and at the free end is 5 cm.

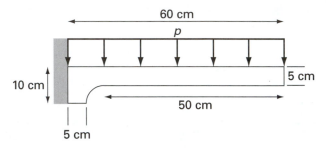

FIGURE 8.16   Cantilever beam with a fillet and distributed load.

Assume that the member is elastically loaded. Neglect its weight. Perform a FEM analysis to determine the stress and displacements at each node.

1. Use a 90-element mesh with 5 elements along the bottom of the straight portion of the member and 4 along the curved boundary. Form 10 elements in the direction transverse to the portion of the bottom boundary on which you just placed the 5 elements.

2. Use a 360-element mesh with 20 elements along the bottom of the straight portion of the member and 16 along the curved boundary. Form 10 elements in the direction transverse to the portion of the boundary on which you just placed the 36 elements.

3. Plot the graphs listed below with each of the transverse nodal positions on the horizontal axis and the computed value normalized by dividing by the distributed load, p, on the vertical axis. Do not forget to account for the thickness of the member.

   a. σ_x at four locations: halfway around the quarter-arc of the circle and at left end, middle, and the free end of the straight portion (four graphs, each containing two curves).

   b. σ_y at four locations: halfway around the quarter-arc of the circle and at left end, middle, and the free end of the straight portion (four graphs, each containing two curves).

   c. τ_{xy} at four locations: halfway around the quarter-arc of the circle and at left end, middle, and the free end of the straight portion (four graphs, each containing two curves).

   d. The displacement, u, at four locations: halfway around the quarter-arc of the circle and at left end, middle, and the free end of the straight portion (four graphs, each containing two curves).

e. The displacement, v, at four locations: halfway around the quarter-arc of the circle and at left end, middle, and the free end of the straight portion (four graphs, each containing two curves).

Discussion

1. Discuss the effect of changing the number of elements in the mesh.
2. Design and present a strength of materials calculation, using a cantilever beam or some other idea, to estimate the stresses in this member. Compare and discuss the FEM and strength of materials computations near the middle of the straight portion and near the free end.
3. Estimate the stress concentration predicted by the FEM analysis. This requires you to make and justify a calculation of the average (nominal stress) in the member at the point where the stress is concentrated. Discuss how the stresses tangent to the surface vary around the fillet.

PROBLEMS

8.1 Consider the one-dimensional element shown, with origin at the center of the element and with length $2L$. Assume that the nodes, numbered 1 and 2, are at the right and left end, respectively. Assume the element has constant cross-sectional area, A. The assumed displacement field in the element is $u(x) = \alpha_1 x + \alpha_2 x^2$.

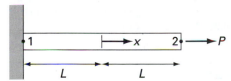

(a) Write $u(x)$ in terms of the nodal displacements, u_1 and u_2.

(b) Write strain $\epsilon(x)$ in terms of the nodal displacements, u_1 and u_2.

(c) Is this a constant strain element? Explain your answer.

(d) Write the stiffness matrix for this element using integrals in terms of x.

(e) Evaluate the stiffness matrix integral.

(f) Assume that the element is loaded at node 2 by an axial force, P, and that node 1 is fixed (*i.e.* $u_1 = 0$). Compute the displacement at node 2. Is the resulting answer reasonable? Why or why not? Ans. (f) $u_2 = 6PL/7AE$.

8.2 Compute the stiffness matrix for a one-dimensional element of length, L, having three nodes: one at $x = 0$, one at $x = L/2$, and one at $x = L$. Assume a displacement field of the form $u(x) = a_1 + a_2 x + a_3 x^2$. Is this a constant strain element?

8.3 The one-dimensional element shown has origin at the left end of the element and length L. There are nodes at each end. The element has constant cross-sectional area, A. The assumed displacement field is $u(x) = \alpha_1 + \alpha_2 x^3$.

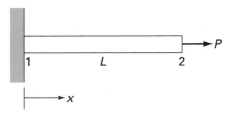

(a) Write $u(x)$ in terms of the nodal displacements, u_1 and u_2.

(b) Write the strain $\epsilon(x)$ in terms of the nodal displacements, u_1 and u_2.

(c) Is this a constant strain element? Explain your answer.

(d) Write the stiffness matrix for this element using integrals in terms of x.

(e) Evaluate the stiffness matrix.

(f) Assume that the element is loaded at node 2 by an axial force, P, and that node 1 is fixed (i.e., $u_1 = 0$). Compute the displacement at node 2.

(g) Determine the strain at $x = L/2$.

8.4 The one-dimensional element has origin at the left end of the element and length L. There are nodes at each end. The element has constant cross-sectional area, A. The assumed displacement field is $u(x) = \alpha_1 + \alpha_2 x^4$.

(a) Write $u(x)$ in terms of the nodal displacements, u_1 and u_2.

(b) Write the strain $\epsilon(x)$ in terms of the nodal displacements, u_1 and u_2.

(c) Is this a constant strain element? Explain your answer.

(d) Write the stiffness matrix for this element using integrals in terms of x.

(e) Evaluate the stiffness matrix.

(f) Assume that the element is loaded at node 2 by an axial force, P, and that node 1 is fixed (i.e., $u_1 = 0$). Compute the displacement at node 2.

(g) Determine the strain at $x = L/2$. Ans. (f) $u_2 = 7PL/16AE$; g) $\epsilon(L/2) = 7P/32AE$.

8.5 A bar of constant cross section, A, and made of a material with elastic modulus, E, is mounted on a fixed and rigid wall at its left end and is axially loaded by a force, P, at its right end.

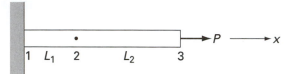

Model the bar with two one-dimensional, constant strain elements of lengths L_1 and L_2 respectively.

(a) Assemble the element stiffness matrices and write the finite element governing equations, $K\{d\} = Q$ for this model.

(b) Solve the equations for the nodal displacements and then determine the strain in each element. Are your answers reasonable? Why or why not?

(c) Compare your results to those obtained for a three-element model.

8.6 Determine the displacement field for a strength of materials problem, cantilever beam of length, L, with a transverse point load of P at the free end. Use one elementary beam element and compare the result to the strength of materials solution.

8.7 Determine the displacement field for a strength of materials problem, cantilever beam of length, $L = L_1 + L_2$, with transverse point loads of P_1 at a distance of L_1 from the fixed end and P_2 at the free end. Use two of the elementary beam elements and compare the result to the strength of materials solution.

8.8 Determine the displacement field for a strength of materials, cantilever beam of length, $L = L_1 + L_2$, with a constant transverse distributed load of magnitude w over the full length. Use two of the elementary beam elements and compare the result to the strength of materials solution. Compare the error between the cubic polynomial finite element result and the fourth-order polynomial strength of materials result.

8.9 Determine the stiffness matrix for a constant strain triangular two-dimensional element with nodes at $(0, 0)$, $(0, 1)$, and $(1, 0)$ in the global body coordinates. Assume plane stress.

8.10 Determine the stiffness matrix for a constant strain triangular two-dimensional element with nodes at $(0, 0)$, $(0, 2)$, and $(1, 1)$ in the global body coordinates. Assume plane stress.

8.11 Compute the matrix C for the constant strain triangular two-dimensional element with nodes at $(0, 0)$, $(0, 2)$, and $(1, 1)$ in the global body coordinates and verify that the $\det(C)$ is twice the area of the triangle.

8.12 A thin beam of length 10 inches and height 1 inch, which is fixed at one end, is axially loaded on its free end by a shear load of $p = 4,000$ lb/in. The beam is made of steel ($E = 30 \times 10^6$ psi and $\nu = 0.25$) and is 0.2 inches thick. Use a two-element constant strain triangle model to determine the deflections and stresses in the member. Compare the computed nodal displacements to displacements calculated by strength of materials.

8.13 A thin beam of length 10 inches and height 1 inch, which is fixed at one end, is axially loaded on its free end by a tensile load $P = 400$ pounds at the upper node and by a compressive load $P = 400$ pounds at the lower node. The beam is made of steel ($E = 30 \times 10^6$ psi and $\nu = 0.25$) and is 0.2 inches thick. Use a two-element constant strain triangle model to determine the deflections and stresses in the member. Compare the computed nodal displacements to displacements calculated by strength of materials.

8.14 Determine the area coordinates of the point of intersection of the medians in an equilateral triangular two-dimensional element. Ans. $(1/3, 1/3, 1/3)$.

8.15 Determine the area coordinates of the centroid in the triangular two-dimensional element with nodes at $(0, 0)$, $(14, 0)$, and $(9, 12)$.

8.16 Determine the area coordinates of the midpoint of the unequal side in an isosceles triangular two-dimensional element.

8.17 Determine the area coordinates of the midpoint of the height to the unequal side in an isosceles triangular two-dimensional element.

8.18 Compute the condition number of the Hilbert matrix when $n = 2$.

8.19 Compute the condition number of the global stiffness matrix, K^{-1}, in equation (8.29).

REFERENCES

[1] I. BABUŠKA AND A. K. AZIZ, Survey lectures on the mathematical foundations of the finite element method, in *The Mathematical Foundations of the Finite Element Method with Applications to Partial Differential Equations*, A. K. Aziz, Ed., Academic Press, New York, 1972, pp. 3–363.

[2] R. D. COOK, *Concepts and Applications of Finite Element Analysis*, Wiley, New York, 1974, pp. 86–91.

[3] A. HRENIKOFF, Solution of problems in elasticity by the framework method. *Journal of Applied Mechanics Transactions ASME*, *8*, 169–175 (1941).

[4] B. M. IRONS, Engineering applications of numerical integration in stiffness methods, *AIAA Journal*, *4* (11), 2035–2037.

[5] J. T. ODEN, Finite elements: An introduction, in *Handbook of Numerical Analysis II. Finite element methods (Part 1)*, P. G. Ciarlet and J.-L. Lions, Eds., North Holland, Amsterdam, 1991, pp. 3–12.

[6] M. J. TURNER, R. M. CLOUGH, H. C. MARTIN, AND L. J. TOPP, Stiffness and deflection analysis of compex structures, *Journal of Aeronautical Science 23*, 805–823 (1956).

CHAPTER 9

PLASTICITY

9.1 INTRODUCTION

Plastic behavior of metal differs from elastic behavior in that the metal is permanently deformed. Under the right combination of stresses transmitted within a body, the metal begins to flow almost like a fluid. The flow units are dislocation defects in the crystal lattice. As in a fluid, shear stresses would be expected to be very important, and this is true. The energy applied in elastic loading is recoverable since elastic loading is conservative. Energy is dissipated during plastic loading, and so some of the work done on the body by the applied loads is not recoverable. Most, but not all, of the energy of plastic deformation is transformed to heat which raises the temperature of the body to a minor extent during the plastic deformation. An elastically deformed body changes volume slightly; however, volume is assumed to be the same before and after a plastic deformation, once the load is removed. Finally, the elastic stress-strain relation for metals is linear, while the relation for metal plastic behavior is nonlinear and depends on the more complicated underlying material behavior at the atomic or molecular level.

9.1.1 Product versus Processing

In most product design applications, the goal is to ensure that the body remains elastically loaded. This requires criteria to predict under what combination of stresses plastic deformation is initiated so that the structure can be designed to keep the stresses safely away from this yield limit. For example, permanent changes in shape cannot be tolerated in machines with close-fitting parts. In manufacturing a shaped product, an engineer wishes to permanently deform a piece of metal to create a useful part. This process must be carefully controlled so that the resulting piece has dimensions within an acceptable tolerance. The energy required to deform the body must be minimized to reduce manufacturing costs.

Several types of manufacturing processes are available to plastically shape metal bodies. A rolling mill (Fig. 9.1) is used to create thinner sheets from existing flat metal plates. The thicker plate is forced between two rollers, which squeeze the plate thinner. The loading between the work rolls pressing on the plate and the larger backup rollers is elastic. As the metal is squeezed between the working rolls, it speeds up as does a fluid passing through an orifice because the volume is constant in plastic deformation. This flow produces anisotropic material properties in the newly flattened sheet and gives the sheet a texture. The design problem is to force the sheet through the working rollers as quickly as possible to increase production. But this must be done without generating a nonuniform temperature distribution in the sheet, which would distort the sheet, and

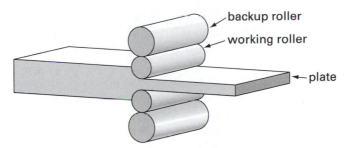

FIGURE 9.1   Schematic of a rolling mill to thin a metal sheet.

without producing roller vibrations transverse to the sheet, which would create a non-uniform sheet thickness. Such vibration is called chatter. There is a small expansion in thickness just after the material passes through the working rollers because of recovery of the elastic portion of the strain when the load is removed.

Many parts are made by cold-worked sheet forming, either by bending or by punching. The sheet is most formable when it is bent so that the bend line is perpendicular to the texture along the rolling direction. The metal stretches upon bending and because of volume preservation becomes thinner in the bend. As a rule of thumb, the bend radius must be greater than the thickness of the sheet. Again the part must be bent slightly more than required in the final result since the elastic portion of strain is recovered when the bending load is removed.

Automobile and aircraft metal panels are formed by forcing a sheet into a die. The recovery of elastic strains must be accounted for in shaping the die so that the part after elastic spring-back is the desired shape. The volume is preserved so that, in some portions of the sheet during forming, the metal is in compression and others in tension. If the compressive stresses are too high a local buckling causes wrinkles in the formed product. If the tensile stresses are too high, tearing can result. Both are serious problems in the formation of aluminum automobile panels. In either case, the part will have to be rejected.

Holes of diameter greater than the thickness of the sheet can be punched. Smaller holes are usually drilled. Unfortunately, the sides of the holes formed by the plastic shear action of the cut are not perpendicular to the sheet because of the plastic flow of the metal.

A rod of metal can be given a different cross section by forcing it through a die or hole which has the shape of the desired cross section. This process is called extrusion. A T cross-section bar could be formed from a circular rod, for example (Fig. 9.2). Because high temperatures are generated at the die due to plastic deformation energy dissipation and die friction, the surface of the extruded rod can form hot tears. These are small fractures periodically along the surface of the material and make the product useless.

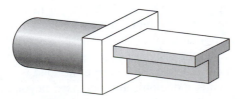

FIGURE 9.2   Extrusion of stock into a T cross section.

The dimensions of a part, say a rod, can be decreased or modified by simply stretching it until it is plastically loaded. Such operations are called drawing. Another type of drawing is the formation of a cup shape by forcing a punch into a sheet without tearing the sheet.

A distorted product results if these processes are not carefully done. Mathematical models are needed to describe and optimize the manufacturing procedures. Research is continuing to answer this need and to supplement manufacturers' qualitative understanding of these processes.

9.1.2 Engineering Fundamentals

Barré de Saint Venant is often given credit for initiating the study of plasticodynamics with his 1871 paper on bending of prismatic bars and 1872 paper on the plastic deformation of hollow circular cylinders subjected to an internal pressure. Saint Venant assumed that volume is preserved during plastic deformation and that the principal strain directions coincide with the principal stress directions. Saint Venant may have been interested in this subject by his contact with Tresca's 1868 papers on the flow of metals under high stresses.

A major impetus to the modern field of plasticity was Prandtl's 1921 paper on a two-dimensional pressure field acting on the surface of an infinite body. He did this while applying his primary efforts to aerodynamics, at the University of Gottingen. Apparently Prandtl was able to take advantage of the similarities of metal flow to fluid flow in developing his ideas.

Another impetus came from the advent of crystal dislocation theory in 1934, given in separate papers by Taylor, Orowan, and Polanyi. Taylor was concerned with explaining the low shear strength of metal crystals, while Orowan and Polanyi examined the temperature dependence of the rate of plastic flow.

Engineers, in order to control the plastic response of metals, need first to know what happens and why it happens. Then in order to predict the plastic response to loads and use it in design, mathematical models for plastic behavior must be developed. The bulk experimental response of metals to various types of loading is first presented. A materials science model explains this behavior in terms of the dislocations within the atomic lattice of the metal. As a first step in the mathematical prediction of plastic behavior, uniaxial loading is modeled. Then the idea of a yield surface is defined. Finally, mathematical models for plastic behavior under arbitrary stress states are given. The more realistic case of rate-dependent models is included under viscoplasticity.

9.2 PLASTIC BEHAVIOR

Researchers have investigated simpler plastic loads than those used in manufacturing in order to try to understand the plastic behavior of metal bodies. As is very usual, many of these tests are performed under uniaxial loading. An attempt is then made to generalize the results to multiaxial loading. These experiments describe the response of the bulk material. Later the response of the metal crystal lattice will be considered. The material is assumed to be isotropic before the load is applied. Some phenomena occur as a uniaxial load is monotonically increased beyond a threshold stress called the yield stress or elastic limit. Others are a consequence of uniaxial cyclic loading in which the maximum loads exceed the elastic limit.

9.2.1 Constant Volume

Tresca in his 1865 experiments on lead found that the volume after loading into the plastic region and then unloading was the same as the initial volume. He assumed that plastic behavior of metals is isochoric (volume preserving). Measurement of small changes in volume during plastic loading by Bauschinger in 1877 showed this assumption to be false. In 1949, P. W. Bridgman measured a decrease in volume of copper and mild steel while these metals were plastically loaded. Many have neglected this data and assumed incompressibility when constructing mathematical models of plastic behavior. On the other hand, experiments by Bridgman in the early 1950s show that the yield stress is independent of hydrostatic pressure.

During a loading causing plastic deformation, the volume of the material can temporarily change. But after the load is removed, this volume change is recovered. Bridgman said that the volume change is largely recoverable and reversible on the release of the stress [6]. The volume change effect, although very small, is magnified just before fracture. This led Bridgman to suggest that some reversible process associated with the recoverable volume change prepares the way for fracture. Taylor and Quinney pointed out in 1934 that most of the work of permanent deformation in plastic loading is converted to heat. Bridgman had to account for this by holding his plastic load constant for 15 or 20 minutes to allow the heat to dissipate. Bell [5] confirmed in the 1970s that the material is not incompressible during plastic loading.

9.2.2 Strain Hardening

Hardening occurs within a solid as the load on the solid is monotonically increased beyond the elastic limit of the material. In a fluid, once the material begins to flow, the constant stress at which flow began is sufficient to cause an increasing shear strain in the material. A solid usually responds differently. As the load passes the elastic limit, a normal stress-strain curve shows that once the material begins to flow, a constant stress does not increase the strain. Instead a larger and larger stress is needed to increase the strain. This phenomenon is called hardening, or strain hardening or work hardening.

The uniaxial load-unload experiment is to increase the load beyond the elastic limit, decrease it to zero, and then again increase the load. There is an apparent new elastic limit on the second increase of load just past the flow stress of the initial plastic loading. The curve for the second increase in load has the same elastic slope as the initial straight line elastic portion of the initial stress-strain curve (Fig. 9.3). The material shows a new elastic limit on reloading that is essentially equal to the flow stress reached before the initial unloading, and this is higher than the original one. The overall material response is attributed to strain hardening, or work hardening, produced by the prior deformation. The internal microstructure of the material has been changed.

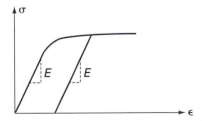

FIGURE 9.3 Unloading parallel to the initial straight line portion of a metal stress-strain curve.

Such behavior presents a problem for the design engineer because the new material yield stress has been obtained at the cost of reduced ductility of the material.

Pure aluminum, a face-centered cubic (FCC) material, exhibits little work hardening at room temperature and at a conventional deformation rate. Therefore its stress-strain curve can be approximated as elastic-perfectly plastic (Fig. 9.4). On the other hand, copper, also an FCC material, exhibits considerable work hardening under the same conditions. Such hardening influence on the apparent yield stress of deformed copper and copper alloy materials such as brass is indicated in the commercial market, where copper materials are designated as being quarter-hard or half-hard when purchased. Its stress-strain curve can be approximated by a linear elastic followed by a linear plastic response (Fig. 9.5); the linearity of the plastic portion leads this to be called linear hardening. Until the 1940s, most analysis was of perfect plasticity since the response of the primary engineering material, mild steel, could be approximated as being perfectly plastic in that the strain hardening appears small compared to the initial yield point magnitude. The increasing use, in the 1950s, of aluminum alloys, which work harden, required the study of hardening models.

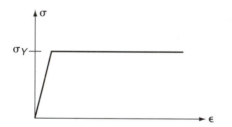

FIGURE 9.4   Perfectly plastic stress-strain curve.

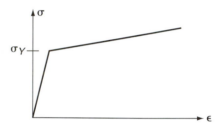

FIGURE 9.5   Linear hardening plastic behavior.

Body-centered cubic (BCC) iron, when alloyed with carbon to make steel, exhibits the yield point behavior produced by the carbon. This behavior is characteristic of BCC metals such as chromium, molybdenum, tungsten, niobium, vanadium, and tantalum, in each case because of the interaction on the atomic level between interstitial solute atoms and the dislocation "flow" units responsible for plastic deformation.

After a material has been work hardened, its stress-strain behavior is no longer isotropic. Anisotropy is observed in the direction of loading of a material which is uniaxially loaded into the plastic region.

A permanent strain is created when a metal is loaded into the plastic region. If the material is loaded to a stress σ_0 that is in the plastic region, the specimen will be permanently deformed. The strain remaining after the stress is removed is called the permanent strain, ϵ_p, caused by loading the material to the stress σ_0. No force acts externally on the material and yet there is a nonzero strain remaining.

If the material is loaded to a strain, ϵ_0, and stress, σ_0, before unloading, then the strain ϵ_0 can be divided into elastic and plastic portions. The permanent strain, ϵ_p, remains constant if the material is then unloaded. The strain recovered during the unloading is called the elastic strain, ϵ_e. The strain, ϵ_0, before unloading can be written as the sum $\epsilon_0 = \epsilon_e + \epsilon_p$.

EXAMPLE 9.1   A material has the stress-strain curve shown and is loaded to a stress of $\sigma_0 = 43,000$ psi (Fig. 9.6). Determine the magnitudes of the elastic and permanent portions of the strain at that stress.

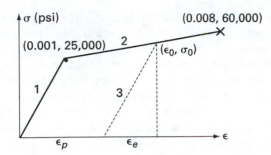

FIGURE 9.6   The relationship between the elastic and plastic portions of strain.

Solution   Lines 1 and 2 are the elastic and plastic portions of the stress-strain curve. Line 3 is the elastic curve obtained when unloading the materials from the stress, σ_0. The equation of the plastic portion under load is $\sigma - 25,000 = 5 \times 10^6(\epsilon - 0.001)$ by the point-slope formula for a line. Therefore substituting $\sigma_0 = 43,000$ yields $\epsilon_0 = 0.0046$. The elastic modulus from line 1 is $E = 25 \times 10^6$. The slope of line 3 equals E since parallel lines have the same slope. But the slope of line 3 can also be written as σ_0/ϵ_e. Therefore $\epsilon_e = \sigma_0/E = 43,000/25,000,000 = 0.00172$. The fact that $\epsilon_0 = \epsilon_e + \epsilon_p$ implies that $\epsilon_p = \epsilon_0 - \epsilon_e = 0.0046 - 0.00172 = 0.00288$. The material, after being plastically loaded to a stress of 43,000 psi and being unloaded, is then capable of elastically supporting a 43,000 psi stress.   ∎

9.2.3   Cyclic Phenomena

Cyclic loadings into the plastic region, which involve complete reversals from tension to compression or vice versa, can cause various material responses. Cyclic softening, cyclic hardening, shakedown, ratchetting, and relaxation of the mean stress can occur.

The Bauschinger effect is observed when a material is first uniaxially loaded plastically in tension and then the loading system is reversed to determine the plastic response in compression. The yield stress in compression after first plastically loading in tension is smaller in absolute value than the yield obtained by first loading in compression (Fig. 9.7). This response was noted in some materials by Bauschinger in 1886.

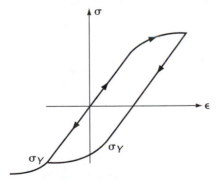

FIGURE 9.7   The Bauschinger effect.

These strain-influenced yield stresses continue to change if the load on the material is repeatedly cycled into the tensile and compressive plastic regions.

Definition   A cyclic loading is said to be in stress (strain) control if the stress (strain) is forced to vary between two fixed limits. The strain (stress) responds to this external control.

When subjected to a cyclic load into the plastic region, the material can either harden or soften on subsequent cycles. Hardening means that the yield stress magnitude increases on each subsequent cycle. Softening means the yield stress amplitude decreases. Frequently after several cycles, the response stabilizes in the sense that the yield stresses in tension and compression remain essentially constant on following cycles. This response can be observed in either stress control or strain control.

A material is cyclically loaded in strain control if on each cycle just enough stress is applied to bring it to the same strain in tension and compression. If the material is softening, the amplitude of the stress on each subsequent cycle decreases (Fig. 9.8b). The material hardens if the stress amplitude increases (Fig. 9.9b).

The material softens in a cyclic load that is stress controlled if the strain increases on each cycle as the material is brought to the same stress. In this case, the loops in stress-strain space expand (Fig. 9.8a). The opposite behavior occurs if the material hardens in either strain or stress control (Fig. 9.9a). The loops narrow. Softening behavior is atypical of age-hardened materials. Generally, a material cyclically hardens to the level of the applied strain (or stress).

Two special phenomena can occur in cyclic stress control loading, depending on the geometric relation of subsequent cycles in strain-stress space. The behavior is called shakedown if the mean stress applied is zero and if after a few cycles, the graphs of all subsequent cycles lie on top of each other. The material ceases to harden. The

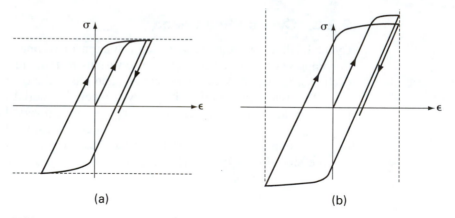

FIGURE 9.8   (a) Softening in stress control and (b) strain control.

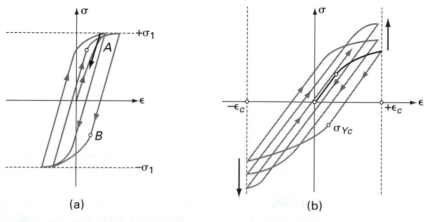

FIGURE 9.9   (a) A hardening material in stress control and (b) strain control.

material is said to elastically harden to the level of the applied stress if the cyclic loop closes to a straight line. Then, there is no energy loss per cycle.

In some stress control situations, the applied stress cycles are not symmetric about $\sigma = 0$. The plastic behavior is called ratchetting if each subsequent response cycle moves further to the right in strain-stress space than the previous cycle. The increment of strain, which can be the same between each pair of cycles, is called the ratchetting strain (Fig. 9.10).

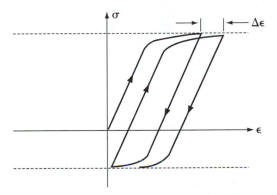

FIGURE 9.10   Ratchetting in plastic behavior.

A strain-controlled cyclic loading can produce either relaxation in which the mean stress response is zero or nonrelaxation in which the mean stress is nonzero.

The response is called kinematic hardening if in each cycle the elastic region remains the same size but its center translates in stress space in subsequent cycles. The mean elastic stress, that at the center, is called the backstress. It is the internal stress in the body when the external stress is at its neutral position.

9.3 MATERIAL MODELS FOR THE FLOW OF METALS

The bulk material behavior described in the previous section is explained in crystalline materials, such as structural metals, at the level of the crystal or grain size of the material. The grain size is generally measured in tens or hundreds of micrometers, hence, it is designated as the microstructural level. Recent advanced materials research investigates the properties of materials having nanostructural grain sizes (10^{-9} m). The plastic response of a metal is largely determined by the behavior of dislocations within these individual crystals or grains and by the interaction of dislocations with crystal boundaries (or grain boundaries). Under a plastic deformation, the crystalline structure of the metal is preserved because dislocation movement results in the translation of two crystal parts by a unit lattice period distance; see F in Figure 9.11. Only specific atoms have changed position within the crystal. The local stress is nonuniform. A much larger uniform stress would be required to move a whole plane of atoms as one unit. The number of dislocations increase with plastic straining and fragment the grains into a mosaic of smaller, more misoriented subgrains. Each grain has its own crystal orientation relative to the coordinate axes of the test specimen. The lattice directions in adjacent grains do not in general match.

The symmetry of the different lattices provides different smallest periodic separations of atoms in different lattice directions, and hence some directions are preferred for dislocation origin and movement. Slip generally occurs in the close-packed directions. The lattice can be viewed as a layered structure of atoms. The atoms in each

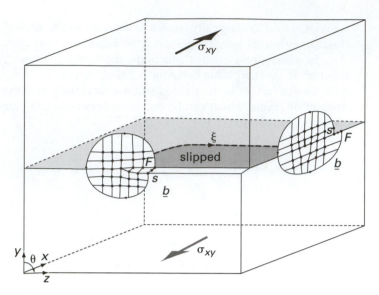

FIGURE 9.11 Plastic slip deformation of a metal lattice structure due to the shear σ_{xy}. The Burgers vector length is b. A screw dislocation geometry is at the left and an edge dislocation geometry is at the right, both connected by the dislocation line vector ξ. Their slip direction is indicated by s. The dislocation characters along the dislocation line are combinations of screw and edge components.

adjacent layer are situated in the depressions between atoms of the layer below in an attempt to be as closely packed as possible. Figure 9.12 shows the situation for the FCC case. The planes along which the atoms slide are called slip planes. The number of possible slip planes depends on the symmetry properties of the crystal lattice.

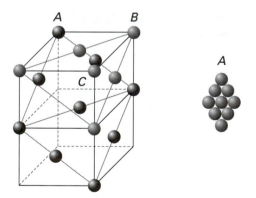

FIGURE 9.12 An FCC lattice showing the stacking sequence of close-packed (111) planes.

The slip distance in any direction must be an integer multiple of the repeat distance, b, in the metal structure, giving a quantized character to the slip step displacement at the atomic level. Several types of dislocation reactions are possible in an FCC lattice because of the large number of slip systems. Recall that there is an ABCABC... stacking sequence in an FCC lattice. The close-packed directions in FCC are the $\langle 110 \rangle$ directions. So a possible Burgers vector is $(a/2)[110]$ of smallest repeat length $\sqrt{2}a/2$, where a is the lattice parameter. However, slip does not always occur in a straightforward manner along the slip direction. Cross-slip allows the dislocation to move onto another slip plane. Atoms within the close-packed planes, the $\{111\}$ planes, of the FCC lattice can most easily slide along the atom layers in specific $\langle 112 \rangle$ directions along the lines of interstices (holes), moving from one depression between atoms to a neighboring depression.

A two-stage slip displacement in $\langle 112 \rangle$ directions produces a net displacement in the close-packed $\langle 110 \rangle$ directions to give slip in a manner that corresponds to the minimum strain energy required for the dislocation motion. By the examples in Chapter 3, Section 3.6, and Chapter 4, Section 4.6.2, the strain energy induced by either an edge or screw dislocation is seen to be proportional to the square of the magnitude of the Burgers vector, b. For example, a dislocation, instead of being forced to connect a B-level site to a B-level site in a $\langle 110 \rangle$ direction, can first be forced to connect from a B-level site to a neighboring C-level site and then from the C-level site to a B-level site. Thus, the unit slip distance in the FCC structure can be comprised of $(a/6)[\bar{1}2\bar{1}] + (a/6)[\bar{2}11] = (a/2)[\bar{1}10]$, in terms of the Burgers vectors, $\vec{b}_2 + \vec{b}_3 = \vec{b}_1$, where the vector addition is componentwise. The strain energy for the unit dislocation from the B-level site to a B-level site in a $\langle 110 \rangle$ direction is proportional to $|b_1|^2 = a^2/2$. The strain energy for the two component partial dislocation is proportional to $|b_2|^2 + |b_3|^2 = a^2/6 + a^2/6$. Because $|b_1|^2 > |b_2|^2 + |b_3|^2$, the two-stage FCC slip characterization is more favorable.

Each stage of the two-stage slip is called imperfect because it does not produce an identity translation of the lattice. A stacking fault is created when partial slip has occurred. The width of the stacking fault is the distance determined by the distance separating the partial dislocations. The further apart the partial dislocations the smaller the strain energy they induce in the lattice, however, the greater the energy present in the extent of the stacking fault. A material-dependent stacking-fault energy is required to form a stacking fault. These faults inhibit cross-slip, which is the transfer of dislocation slip to an intersecting slip plane. A greater applied shear stress is needed to recombine the partial $\langle 121 \rangle$ dislocations so that a single $\langle 110 \rangle$ type dislocation can be formed. The relatively high shear stress required to accomplish constriction of the partials to a single unit dislocation is responsible for the "stage II"–type strain hardening characteristic of FCC metals. The recombined dislocation may cross-slip if it is of screw type that is always contained in another $\{111\}$ plane. The Burgers vector and the dislocation line must lie in both planes involved in cross-slip in order for the dislocation to cross-slip. The Burgers vector and the dislocation line for a screw dislocation are parallel so that cross-slip may occur at a screw dislocation. However, because for an edge dislocation the Burgers vector and dislocation line are perpendicular, it is geometrically impossible for an edge dislocation to slip in two crystallographically distinct planes. The material dependence of cross-slipping is evidenced by the fact that it is much more common in aluminum than in copper because of the high stacking fault energy of aluminum.

Exercise 9.1   Show that there are 12 slip systems in an FCC lattice (four planes of $\{111\}$-type and three directions of $\langle 110 \rangle$ in each plane).

The $\langle 111 \rangle$ directions are close-packed in the BCC lattice. There are no close-packed planes, but the slip planes are $\{110\}$, $\{112\}$, and $\{123\}$ in which the $\langle 111 \rangle$ directions lie. The unit slip is $(a/2)[111]$ of length $\sqrt{3}a/2$. There are a total of 48 BCC slip systems, although some are activated only at high temperature. In BCC structures, there are three partial dislocations associated with a unit $\langle 111 \rangle$ type dislocation but the indication is that the stacking fault separation between the partials is very small and cross-slip relatively uninhibited. The slip systems in an HCP lattice are generally of $\{0001\}\langle 11\bar{2}0 \rangle$ type but other slip systems occur as well.

Calculations have estimated the force needed to move an edge dislocation in a cubic lattice. Peierls in 1940 [16] and Nabarro in 1947 [15] used the sine approximation to the interatomic bond energy, discussed in Chapter 3, Section 3.2.3, to compute the shear stress required to move a dislocation between adjacent equilibrium positions,

$$\tau = C_1 \exp\left(\frac{-C_2 d}{b}\right),$$

where C_1 and C_2 are material constants, d is the distance between adjacent slip planes, and b is the magnitude of the Burgers vector. This intrinsic shear stress for dislocation movement is called the lattice friction stress or the Peierls-Nabarro stress. The calculation indicates that the shear stress required to move an edge dislocation increases with the Burgers vector magnitude and decreases with slip plane separation, as expected physically. In part, because the Burgers vector for a ceramic crystal is generally larger than that for metals, a larger lattice friction stress is expected in a ceramic, and likewise for ordered intermetallic crystals. Slip should be easiest on the close-packed planes and in the close-packed directions. Experimentally, the Peierls-Nabarro stress is known to be lower than the measured yield stress in FCC metals, but is closer to the yield stress in BCC metals.

When a unidirectional stress is applied to a crystal, shear stresses are generated on all possible planes including the potential slip planes. The resolved shear stress, the shear stress component, along a slip plane in a given direction can be computed from the normal stress, σ, applied to the lattice using a rotation of the stress tensor, as in Chapter 1, Section 1.5. Put coordinates on the crystal so that the 3-direction is parallel to the direction of the applied stress. The 1- and 2-directions can be any mutually orthogonal directions that are also orthogonal to the 3-direction. In this system, the stress tensor components are $T_{33} = \sigma$ and $T_{ij} = 0$ otherwise. Orient the rotated coordinates so that the 3'-direction is orthogonal to the slip plane and so that the 1'-direction is in the slip direction. In this system, the stress tensor is $T' = \alpha T \alpha^t$, where α is the orthogonal matrix of direction cosines. The angle λ is that between the load and the slip direction so that $\alpha_{13} = \cos\lambda$. The angle ϕ is between the load and the normal to the plane (Fig. 9.13) so that $\alpha_{33} = \cos\phi$. The desired resolved shear stress is $\tau_r = T'_{31}$. Matrix multiplication shows that $T'_{31} = \alpha_{13}\alpha_{33}T_{33}$. Therefore,

$$\tau_r = \sigma \cos\phi \cos\lambda. \tag{9.1}$$

Alternatively, by geometry, the applied force, F, has shear component along the slip direction equal to $F \cos\lambda$. The area of the plane normal to the applied force is A; then the area of the slip plane is $A/\cos\phi$. The shear stress is the shear force component divided by the area: $\tau_r = \sigma \cos\phi \cos\lambda$. A vector method to make this calculation, called Schmid's law, was given in Chapter 1, Section 1.1.

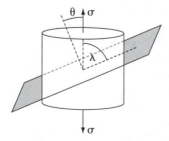

FIGURE 9.13 The angles between the load and the slip system.

EXAMPLE 9.2   The applied stress of 900 psi is parallel to the [001] direction in an FCC crystal. Determine the resolved shear stress acting on the (111) slip plane in the [$\bar{1}$10] direction and in the [01$\bar{1}$] direction.

Solution   The cosine of the angle ϕ is $1/\sqrt{3}$. The cosine of the angle λ for the [$\bar{1}$10] direction is 0 so that there is no resolved shear stress and no reason for slip to occur in this direction. The cosine of the angle λ for the [01$\bar{1}$] direction is $-1/\sqrt{2}$. Therefore the resolved shear stress in this direction is $\tau = (900)\cos(54.76°)\cos(135°) = -367$ psi.   ∎

Definition   The critical resolved shear stress, τ_{crss}, is the minimum shear stress on a slip plane in a slip direction required to initiate slip.

For example, at room temperature the experimental critical shear stress for pure copper (FCC) in the (111)[1$\bar{1}$0] slip system is about 0.65 MPa. The critical shear stress for pure iron (BCC) in the (110)[$\bar{1}$11] slip system is 27.5 MPa and that in pure zinc (hexagonal closed-packed [HCP]) in the (0001)[11$\bar{2}$0] slip system is 0.18 MPa. These values are all significantly lower than the shear yield stress values for the bulk material. So it must be possible to have local slip at stresses below commonly accepted bulk yielding stress values.

The theoretical value of the shear stress required to initiate slip in a perfect lattice can be calculated from the atomic forces and their relation to the elastic moduli of the lattice. The computed value is an appreciable fraction of the shear modulus and, thus, is much larger than observed in experiments. In fact, individual crystals are generally weaker than polycrystals; these also have lower strengths than are calculated for the perfect crystal lattice. This was one factor that led researchers to postulate that real crystal lattices are not perfect. X-ray diffraction measurements also indicated that crystals are not perfect but contain a mosaic structure. The imperfections involved in shear deformations are crystal dislocations. They or, more accurately, their strain fields are easily detected today with a variety of observational techniques.

The shear stress needed to propagate slip with a crystal dislocation is small mainly because the slip is accomplished one atomic repeat step at a time. In a perfect lattice, every atom along a given slip plane must be moved simultaneously to induce slip. Thus, the critical shear stress is much lower in a real lattice with dislocations than in the theoretical perfect lattice. Near to the dislocation center, or core, the atoms are in partly sheared positions relative to the perfect lattice and so dislocation movement adds to the required displacement. Thus crystal dislocations provide the physical mechanism by which permanent deformation occurs. The yield stress depends on the dislocation substructure in the lattices of grains that make up a conventional polycrystalline material.

The speed of dislocation motion depends on the effective shear stress, the imposed deformation rate, and the temperature. A viscoelastic description based on these factors implies that the limiting velocity is near the speed of sound in the metal. As mentioned, Polanyi and Orowan were led to postulate the existence of dislocations for explaining the strength of real materials from a viscoplasticity viewpoint.

Plastic deformation by slip produces a system of steps on the surface where the dislocations exit from within the crystal or grain volume. The slip dislocation is a line defect in the slip plane. However, the slip steps are produced by groups of dislocations having emerged from a single slip plane. Adjacent multiple steps occur on closely parallel planes as a consequence of the manner in which a dislocation line length increases, or the dislocations multiply, as embryonic slip spreads across a slip plane and spreads to an adjacent plane by a mechanism of cross-slipping.

Definition   A slip band is a closely grouped system of parallel slip planes.

These surface steps are evidence of the orientations of preferred slip planes in the crystal lattice. Such steps appearing as lines on the surface of crystals or grains within a polycrystal are easily observed at minor magnifications. The observation of slip band structures at varying macroscopic strains established that many dislocations move on closely parallel planes. The increasing size of the slip bands at larger strains is a measure of the increase in dislocation activity induced by the larger strains.

The lines of dislocations are generally curved. The number of dislocations is measured in terms of the total length of lines in a volume of the material. The units are therefore centimeters of line length per cubic centimeter, $cm\ cm^{-3}$, or number of dislocation lines per square centimeter, cm^{-2}. If a metal is in an annealed condition, a typical density of dislocations is of the order of $10^6\ cm^{-2}$. Plastically deforming the metal can increase the dislocation density by a factor of a thousand. Electron microscope photographs have verified that the dislocations multiply during plastic deformation.

F. C. Frank and W. T. Read, in 1950 [9], proposed a planar mechanism for dislocation creation that, in a modified form, was subsequently verified by experimental observation. This is now called a Frank-Read, or F-R, source. It was proposed that the ends of a dislocation line segment become pinned in the lattice so that the ends can no longer move. Further deformation of the material bends the dislocation line around the pinned ends until it meets itself in one part as a closed loop. A segment is pinched off and returns to the original bowing position of the line and then it is able to bow out again as in the original bowing operation. The other loop surrounding the segment spreads outward as if it were the section of a surface of a bubble. The same process is repeated to create more dislocation loops. In most cases, the F-R mechanism is thought to operate in modified form through the dislocation loops cross-slipping onto parallel planes, called the double cross-slip mechanism for dislocation multiplication. Figure 9.14 shows that a dislocation loop lying in a (111) plane of an FCC lattice is permitted to cross-slip in a $(1\bar{1}1)$ plane. Once cross-slipping a certain distance, the dislocation cross-slips back into the original plane and so parts of the original dislocation loop lie in two parallel (111) planes giving the result that a line segment lies in each plane. The portion lying in the $(1\bar{1}1)$ plane may become pinned. Then the original description given by Frank and Read is operative in each plane so that new dislocation loops are created in the parallel (111) planes (Fig. 9.14).

Work hardening of materials is explained by the mutual interaction of dislocations that leads to their blocking each other from moving. As the dislocation density increases during plastic deformation, it is harder for a given dislocation to find a free path on which to move. Therefore, as observed experimentally in uniaxial tests, it takes more applied stress to continue the deformation. The material is stronger. Some mathematical models, therefore, use dislocation density as a variable. An introduction to dislocation mechanics is given in Chapter 4 of [14].

A dislocation pile-up occurs when many dislocations accumulate against an obstacle, such as a grain boundary. The pile-up requires repeated activation of dislocations from a source in a slip plane. The pile-up may be blunted by dislocation sources on secondary slip planes or by cross-slip. A pileup exerts a backstress on the source to eventually limit the emitted dislocations. The local lattice structure exerts a drag stress on the dislocations in the pile-up (ref. [20], Section 9.47).

Most engineering materials are polycrystalline. At room temperature, the grain boundaries act to prevent dislocation movement outside of the grains in which they originate. As a consequence, the dislocations pile up at the grain boundary. For this reason, the grain size of a material has a very important influence on the yield stress.

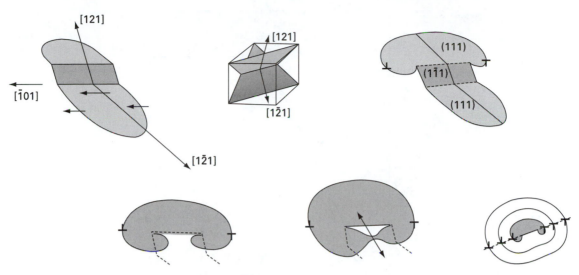

FIGURE 9.14 Frank-Read mechanism for dislocation creation.

The grain size is controlled by heat treatment of the metal. If a material is annealed, it tends to have large grains. Recall that in annealing, the metal is heated above the recrystallization temperature and allowed to cool slowly. However, if the material is cooled quickly, the grains are smaller. Of two identically shaped bodies made of the same material, the one with smaller grains should be the stronger. The smaller grain size provides a smaller distance to pile-up dislocations and thereby strengthens the material. E. O. Hall in 1951 and then N. J. Petch proposed a relationship between the yield stress, σ_Y, and the average grain diameter, l, based on dislocation pileups [2].

$$\sigma_Y = \sigma_0 + kl^{-1/2}, \tag{9.2}$$

where σ_0 and k are material constants. This is also called the Hall-Petch equation. The grain size dependence was originally established for mild steel but was extended later to all metals. The importance of the relationship is illustrated by the three stress-strain curves for brass in Figure 9.15. The equation expressed in terms of dislocation pile-ups at the grain boundaries takes the form

$$\sigma_Y = m[\tau_0 + k_s l^{-1/2}], \tag{9.3}$$

where m is an orientation factor related to the slip system geometry, k_s is the slip band stress intensity factor for overcoming the grain boundary resistance, and τ_0 is the appropriately averaged critical resolved shear stress. So, σ_0 represents the resistance to dislocation motion within the crystal (grain). The factor, k_s, plays a role similar to the stress intensity factor for fracture mechanics (Chapter 10).

The properties of dislocation pile-ups were intimately involved in the establishment of the Hall-Petch equation [3]. The effective applied stress, τ, that is, the difference between the applied shear stress and the frictional stress opposing dislocation movement in the slip band, is expressed from a pile-up model as

$$\tau = \left(Gb \frac{\tau^*}{\pi a} \right)^{1/2} x^{-1/2}, \tag{9.4}$$

where τ^* is the locking stress for the dislocation at the pile-up tip, the constant a depends on the nature of the dislocation, b is the magnitude of the Burgers vector, G is the shear modulus, and x is the pile-up diameter.

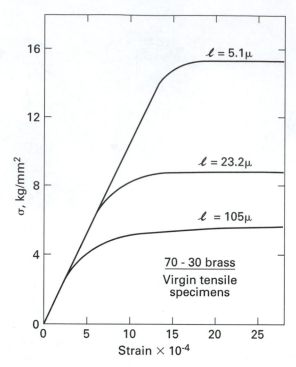

FIGURE 9.15 Tensile stress-strain curves for three polycrystalline 70-30 alpha-brass specimens with average grain diameters of 5.1, 23.2, and 105 μm. (Courtesy of W. L. Phillips.)

Experimental values of the material constant k in equation (9.2) are given in Table 9.1 [1].

TABLE 9.1 Typical Tensile Hall-Petch Stress Intensities

Material	k (MN-m$^{-3/2}$)	k (MPa-mm$^{1/2}$)	Lattice
Aluminum	0.063	2.0	FCC
Copper	0.158	5.0	FCC
Iron	0.206	6.5	BCC
Steel	0.780	23.4	BCC
Molybdenum	1.768	55.9	BCC
Zinc	0.220	7.0	HCP
Magnesium	0.279	8.8	HCP
Titanium	0.403	12.8	HCP

EXAMPLE 9.3   Estimate the yield stress in a steel with an average grain diameter of 75 μm and $\sigma_0 = 100$ MPa.

Solution   The Hall-Petch equation implies that the yield stress is

$$\sigma_Y = \sigma_0 + kl^{-1/2} = 100 + (23.4)(0.075)^{-1/2} = 185 \text{ MPa.} \qquad \blacksquare$$

High temperatures strongly affect the behavior of metals. The first result is the exciting of the atomic lattice to a more energetic equilibrium state. The thermally activated dislocations move more easily. But higher temperatures also facilitate mass transport by atomic diffusion within the lattice. The diffusion tends to remove lattice defects such as dislocations and lower the strength of the metal. Annealing is one such heat treatment that is commonly used to reverse the consequences of cold working.

The yield stress can also depend on the strain rate. It has been shown in uniaxial tests of titanium, for example, that a higher strain rate increases the yield stress. However, the elastic modulus is unaffected by the strain rate.

Twinning contributes to plastic deformation in materials with few slip systems, such as HCP materials. Engineering HCP materials include magnesium, titanium, and zinc. Twinning is relevant to engineering design because it is a deformation process. A twin is a shear displacement over multiple lattice planes so that the lattice on one side of the boundary is a mirror image of the lattice on the other side. The twin is therefore characterized by an invariant shear strain [3]. This is in contrast to slip which is a translation characterized by the unit slip distance, the magnitude of the Burgers vector. Slip is analogous to translation, while twinning is analogous to rotation. In the former, the Burgers vector is an invariant. In twinning, the shear strain, a change in angle, is invariant. This classification is similar to the idea that every motion can be split into a translation and a rotation. Twins due to annealing are more likely in FCC materials, while mechanical twins are observed in BCC and HCP materials.

Twinning by reorienting the crystal structure can create new slip systems, which may be now oriented in such a way that the applied load more easily causes slip (i. e., induces a higher shear stress on the reoriented slip plane for a given load). On the other hand, twin occurrence promotes strain hardening because it acts as an effective grain size refinement. The boundary between the twins can block dislocation motion, but it is not as effective a blocker as a grain boundary because the planar interface between the twin and matrix may have at least one common slip direction. Stacking faults in aluminum or copper have approximately one-half of the coherent surface energy of the twin. The large shear strains accompanying twinning lead to its being frequently associated with brittle fracture. Furthermore, a plastic region resulting from slip often surrounds a twin to accommodate the local twinning deformation. Twinning is easier in HCP metals, like zinc, where the lower symmetry of the HCP lattice allows smaller twinning shear strains.

9.3.1 Techniques for Strengthening a Metal

All common methods of strengthening a metal depend on blocking dislocation motion. Deformation strengthening, also called strain hardening or work hardening, is a consequence of dislocation multiplication and interactions as the metal is plastically deformed, as discussed above. Empirical graphs that relate the percent of cold working (the deformation occurs at room temperature) to the resulting tensile strength are available in metals handbooks. As the percent of cold working is increased, the yield stress and the tensile strength increase until they converge to the same value so that all ductility of the metal is lost. Further, the metal grains elongate producing anisotropic mechanical behavior of the metal. Also, a nonuniform residual stress is created by the energy stored in the dislocations. The strain energy stored in a screw dislocation was computed in Chapter 3, Section 3.6, and that stored in an edge dislocation in Chapter 4, Section 4.6.

Solidification strengthening refers to controlling the grain size in the metal produced during casting. One method controls the rate of cooling. The faster the metal cools, the greater number of crystal nuclei and the less time available for atoms to diffuse to a grain and increase its size. Consequently, the grains are smaller than they would be under slow cooling. The smaller grains increase the grain boundary present to limit dislocation movement within the grains. A metal with small grains is stronger than one with larger grains. The number of grains, and thus their size, can also be controlled by adding foreign particles to the molten metal. The particles serve as additional nucleation sites as the metal solidifies. A larger number of sites produces more grains. The existence of more grains implies a smaller grain size and thus more grain boundary to block dislocation motion.

Solid solution strengthening involves the introduction of point defects in the metal lattice to block slip by distorting the lattice. The defects can either be substitutional due to the presence of an atom of a different size in a lattice position or be interstitial due to the presence of an atom in an interstitial site in the lattice. Copper-nickel alloys are substitutional at all concentrations; in this case both pure metals have an FCC structure. A reasonable amount of zinc can be substituted into the FCC copper lattice to produce brass even though zinc is HCP. Steel is made by introducing carbon into interstitial positions in the iron lattice.

Dispersion strengthening uses the boundaries of particles added to the solvent metal to block slip. The strengthening is similar to that of particles produced by precipitation. For example, an equilibrium second phase is created in a aluminum-magnesium alloy if the solubility limit is exceeded. A second phase also strengthens 4% copper-aluminum or a solder that is an eutectic tin-lead alloy. Cooling techniques can create nonequilibrium hard phases such as martensite in steel; their boundaries also block dislocation motion. Age hardening by heating above the solvus temperature, quenching, and then aging at a temperature below the solvus can produce finer dispersions of precipitates that strengthen the alloy. Not all alloys are age-hardenable. For example, the 2000, 6000, 7000, and 8000 series of aluminum are age-hardenable. The symbol T on an aluminum specification such as 6061-T6 indicates that the aluminum has been age hardened.

9.4 MODELING UNIAXIAL PLASTIC BEHAVIOR

Most plasticity models assume that the total strain at any equilibrium load beyond the elastic region is the sum of an elastic component and a plastic component.

$$\epsilon = \epsilon_e + \epsilon_p. \tag{9.5}$$

This is not a superposition assumption based on some linear relation between elastic and plastic behavior; it follows instead from the uniaxial load-unload experiment. The elastic component, ϵ_e, is due to the stretching of atomic bonds between the atoms in the crystal lattice, and the plastic portion, ϵ_p, is a consequence of slip on crystalline planes and the subsequent reformation of bonds. The specimen is loaded beyond its elastic limit. When the material is unloaded, it is found that there is a permanent strain, ϵ_p, and that some strain has been recovered. The recovered portion is said to be the elastic component, ϵ_e (Fig. 9.6). There is no creep, for example, or any other time-dependent response. Therefore this model is not realistic at temperatures above half the melting temperature.

Since the recovery stress-strain graph is a straight line parallel to the initial straight-line portion of the uniaxial stress-strain curve, the elastic component at stress σ can be computed from the slope of this curve.

$$\epsilon_e = \frac{\sigma}{E}, \tag{9.6}$$

where E is the elastic modulus. This elastic portion, for most plastic loads, is much smaller than the plastic component.

A general model for uniaxial plasticity has the following form. Let σ_Y be the stress at which plastic flow begins as the stress is increased from a reference state.

$$\epsilon_e = \frac{\sigma}{E};$$

$$\epsilon_p = \begin{cases} g(\sigma) & \text{if } \sigma \geq \sigma_Y \\ 0 & \text{if } \sigma < \sigma_Y. \end{cases} \tag{9.7}$$

The function, $g(\sigma)$, gives the threshold stress, σ, for plastic flow when $\epsilon_p = g(\sigma)$.

Uniaxial plastic constitutive models can be obtained in this format by letting the material first be at a reference state. Then as the load is steadily increased, it reaches its first yield at the stress, σ_Y. Subsequent unloading and reloading will produce new yield stresses according to the function, $g(\sigma)$.

9.4.1   Perfect Plasticity

The plastic extrusion experiments on lead done by H. Tresca in 1865 showed that the stress remained constant if the strain was further increased after reaching an ultimate stress at about a 0.35 strain. B. Saint Venant and M. Lévy in 1871, possibly to simplify the analysis, assumed instead that the stress is constant after the end of the linear region of the stress-strain curve. Even though in 1878, Tresca objected to this use of his data, the idea has continued to be used. A material in which the stress at a point remains constant once yielding occurs at the point is called perfectly plastic. Other points in the body can remain elastic. The uniaxial stress-strain curve, which is shown in Fig. 9.4, also has an elastic portion. Therefore, such a material is sometimes called elastic-perfectly plastic. Once the yield stress is reached, the material can continue to strain with no increase in stress. No additional force is required to maintain the flow after the flow has been initiated. Such a material is idealized in the sense that no strain hardening occurs. The yield stress remains the same no matter how often the material is loaded into the plastic region and unloaded into the elastic region. The yield stress is identical to the ultimate stress. This simplest possible model of plastic behavior is also called the perfectly plastic or the elastoplastic model.

$$\epsilon = \begin{cases} \epsilon_e = \frac{\sigma}{E} & \text{if } |\sigma| < \sigma_Y \\ \epsilon_e + \epsilon_p, & \text{if } |\sigma| = \sigma_Y, \end{cases} \tag{9.8}$$

where ϵ_p is arbitrary. This behavior is often represented by a spring and constant coefficient friction block schematic diagram (Fig. 9.16). Once σ is large enough, the block on the left slips, creating plastic strain. The spring accounts for the elastic strain. The behavior of aluminum is sometimes approximated by the perfect plasticity model since aluminum exhibits slight hardening.

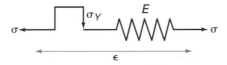

FIGURE 9.16   Spring and friction block mechanical analog for perfect plasticity.

This model represents the behavior at a point in a loaded body. Some portions of the body may be elastically loaded while other portions are plastically deformed under a given load. The perfectly plastic model can be used to estimate the load required to yield portions of a rod under torsion or a beam in bending.

Torsion   A circular rod of radius, R, is subjected to a torque on its ends. The external torque, T, is related to the stress, τ, on a face of area, A, by

$$T = \int_A \tau \rho dA = \int_0^R \tau \rho 2\pi \rho d\rho, \qquad (9.9)$$

where the area of a thin ring about the center is $dA = 2\pi \rho d\rho$. In order to evaluate this integral, the shear stress, τ, must be viewed as a function of the radius, ρ. The strength of materials assumptions imply that the strain is a linear function of radius. In the elastic region, the stress is a linear function of strain. A combination of these implies that the stress is a linear function of radius in the elastic region. The body is elastically loaded in its center and out to a radius, ρ_Y, where the stress is the shear yield stress, τ_Y (Fig. 9.17). Since the stress is zero at the center, the linear stress versus radius relation must be

$$\begin{aligned} \tau(\rho) &= \frac{\tau_Y \rho}{\rho_Y}, \quad \text{if } \rho \le \rho_Y \\ &= \tau_Y, \qquad \text{if } \rho \ge \rho_Y. \end{aligned} \qquad (9.10)$$

Substitute this expression into the integral for the torque to obtain

$$T = \int_0^{\rho_Y} \rho \left(\frac{\tau_Y \rho}{\rho_Y} \right) 2\pi \rho d\rho + \int_{\rho_Y}^R \tau_Y \rho 2\pi \rho d\rho. \qquad (9.11)$$

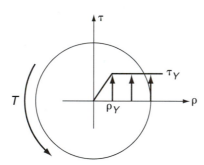

FIGURE 9.17   Torsional member in perfectly plastic loading.

EXAMPLE 9.4   Determine the minimum torque required to bring each point on the face to a plastic state.

Solution   The condition that the torque be minimum is that $\rho_Y = 0$. Therefore by equation (9.11), the fully plastic torque required is

$$T_{fp} = \int_0^R \tau_Y \rho 2\pi \rho d\rho = \frac{2}{3} \tau_Y \pi R^3. \qquad (9.12)$$

■

EXAMPLE 9.5   A rod of radius, $R = 2$ cm, and length, $L = 20$ cm, is loaded by a torque, $T = 480\pi$ Nm. The material is perfectly plastic with $\tau_Y = 100$ MPa and $G = 70$ GPa. Determine the smallest radius at which yielding occurs.

Solution   The torque relation, equation (9.11), is

$$T = \int_0^{\rho_Y} \rho \left(\frac{\tau_Y \rho}{\rho_Y} \right) 2\pi \rho d\rho + \int_{\rho_Y}^{R} \tau_Y \rho 2\pi \rho d\rho$$

$$= \frac{1}{2}\pi \tau_Y \rho_Y^3 + \frac{2}{3}\pi \tau_Y R^3 - \frac{2}{3}\pi \tau_Y \rho_Y^3.$$

Substitution of the given values implies that $\rho_Y = 0.01474$ m. ∎

When the torque is removed, residual stresses remain. Also there is elastic spring-back which leaves a permanent angle of twist. The release of the torque is equivalent to applying the original torque, but in the opposite direction, to the deformed rod. Under the release, the material remains elastically loaded until τ_Y is reached in the opposite direction. The recovered elastic stress, which varies linearly with radius but in the opposite direction, is superposed on the original shear stress. The residual stress on the outer radius is

$$\tau_Y - \frac{TR}{J}.$$

The recovered elastic stress at radius, ρ_Y, is $T\rho_Y/J$ by linearity, so that the residual stress is

$$\tau_Y - \frac{T\rho_Y}{J}.$$

The permanent angle of twist over the length, L, is the the difference between the angle of twist after the initial loading and the elastically recovered angle of twist. A relation between the angle of twist, θ, and the strain, γ, is only known in the linear elastic region. The angle of twist after the initial loading is found, using the strain at the radius ρ_Y. From $\gamma L = \rho_Y \theta$,

$$\theta = \frac{\gamma L}{\rho_Y} = \frac{\tau_Y L}{\rho_Y G}.$$

The recovered angle of twist is

$$\theta_e = \frac{TL}{JG}.$$

The permanent angle of twist is the difference.

EXAMPLE 9.6   Compute the residual stress distribution and permanent angle of twist of one end of the rod with respect to the other if the torque is removed in the previous example.

Solution   At the outer radius the residual stress is

$$\tau_Y - \frac{TR}{J} = 100 - 120 = -20 \text{ MPa}.$$

At radius ρ_Y, the residual stress is

$$\tau_Y - \frac{T\rho_Y}{J} = 100 - 88.4 = 11.6 \text{ MPa}.$$

The sign is positive if the residual shear stress is in the same direction as that due to the original loading and negative if in the opposite direction. The stress varies linearly between the center and radius ρ_Y and linearly outside the radius ρ_Y. A sketch of the residual stress distribution is given in Figure 9.18.

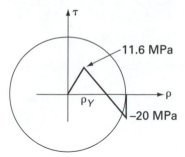

FIGURE 9.18 The residual stress in an unloaded elastic-perfectly plastic rod.

The permanent angle of twist is

$$\theta_p = \theta - \theta_e = \frac{\tau_Y L}{\rho_Y G} - \frac{TL}{JG} = 0.01939 - 0.01714 = 0.00225 \text{ rad.} \quad \blacksquare$$

Bending A similar analysis can be given for bending of a perfectly plastic beam. A portion of the material in a cross-sectional face may be plastically loaded. The internal moment, M, on a face with area A of the beam can be expressed, in terms of the longitudinal bending stress, as

$$M = \int_A \sigma y \, dA. \tag{9.13}$$

The beam is assumed to have rectangular cross section with width, t, and height, h, for ease in dealing with the area so that $dA = t\,dy$. The beam is to be elastic from the centroidal axis to a distance $\pm y_e$ away from the centroidal axis. The strength of materials assumption again implies that the stress is a linear function of y so that (Fig. 9.19)

$$
\begin{aligned}
\sigma(y) &= -\frac{\sigma_Y y}{y_e}, && \text{if } 0 \le y \le y_e \\
&= -\sigma_Y, && \text{if } y_e \le y \le h/2 \\
&= \frac{\sigma_Y y}{y_e}, && \text{if } -y_e \le y \le 0 \\
&= \sigma_Y, && \text{if } -h/2 \le y \le -y_e.
\end{aligned}
\tag{9.14}
$$

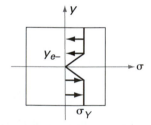

FIGURE 9.19 Stress distribution in a perfectly plastic member in bending.

Then if the cross section is symmetric about the centroidal axis, by equation (9.13),

$$M = 2\left[\int_0^{y_e}\left(\frac{\sigma_Y y}{y_e}\right)ytdy + \int_{y_e}^{h/2}\sigma_Y ytdy\right] = \frac{2}{3}\sigma_Y ty_e^2 + \sigma_Y t\left[\left(\frac{h}{2}\right)^2 - y_e^2\right].$$

(9.15)

EXAMPLE 9.7 Determine the minimum bending moment that must be created in a face so that it is fully plastic. When the face is fully plastic, it is called a plastic hinge. The stress is

$$\sigma(y) = \sigma_Y \quad \text{if } -\frac{h}{2} \le y \le 0$$
$$= -\sigma_Y \quad \text{if } 0 \le y \le \frac{h}{2}.$$

The minimum moment, M_{fp}, is then

$$M_{fp} = \int_{-h/2}^0 \sigma_Y ytdy + \int_0^{h/2}(-\sigma_Y)ytdy = \frac{1}{4}\sigma_Y th^2.$$

(9.16)

∎

Elastic spring-back occurs when the moment is removed, leaving a permanent bending deformation. This spring-back must be accounted for when manufacturing components by bending so that the final curvature of the bend meets specifications. Empirical relations exist to estimate the bending spring-back for hardening materials. From this starting point, some trial and error is often needed to ensure that the final radius of curvature is the desired one. However, the residual stress and the final radius of curvature of the beam can be computed for the idealized perfectly plastic materials.

The release of the moment is accounted for by superposing the negative of the linear elastic stress due to the moment, My/I, onto the stress in the loaded metal because the stress for the unloading is linear with respect to distance from the centroid. The resultant is the final residual stress in the cross section of the beam. Let h be the height of the beam perpendicular to the bending axis.

EXAMPLE 9.8 A steel beam ($E = 210$ GPa, $\sigma_Y = 150$ MPa) is plastically loaded by a couple $M = 6,000$ Nm. The height of the rectangular cross section of the beam is $h = 8$ cm and the thickness is $t = 3$ cm.

(a) Determine y_e after the moment is applied.

(b) Determine the radius of curvature of the midplane of the beam after the couple is released.

Solution (a) By (9.15), $M = \sigma_Y t(h/2)^2 - \sigma_Y ty_e^2/3$, the distance $y_e = 0.02828$ m.

(b) The residual stress at y_e is $\sigma_x = \sigma_Y - My_e/I = 17.417$ MPa. Because this point is elastically loaded, the corresponding strain is $\epsilon_x = \sigma_x/E = 0.00008294$. From the strain relation for an elastic beam, $\epsilon_x = y_e/\rho$, where ρ is the radius of curvature of the centroidal plane. Therefore, $\rho = y_e/\epsilon_x = 341$ m.

∎

9.4.2 Hardening Models

A uniaxial, initially isotropic, hardening, elastoplastic solid can be described by the Ramberg-Osgood model. Again, the strain is the sum of elastic and plastic components, $\epsilon = \epsilon_e + \epsilon_p$. In this model,

$$\epsilon_e = \frac{\sigma}{E}$$

(9.17)

$$\epsilon_p = \left\langle\frac{|\sigma| - \sigma_Y}{K}\right\rangle^M sgn(\sigma).$$

(9.18)

Two special functions are used in the definition of ϵ_p. The function $sgn(x)$ is defined by

$$sgn(x) = \begin{cases} 1 & \text{if } x > 0 \\ -1 & \text{if } x < 0. \end{cases} \tag{9.19}$$

The function $\langle x \rangle$ is defined by

$$\langle x \rangle = \begin{cases} x & \text{if } x > 0 \\ 0 & \text{if } x < 0. \end{cases} \tag{9.20}$$

The model can be inverted to write the monotonic curve of threshold stresses as

$$\sigma = \sigma_Y + K\epsilon_p^{1/M}. \tag{9.21}$$

A schematic graph is shown in Figure 9.20.

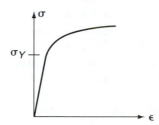

FIGURE 9.20   Stress-strain curve for the Ramberg-Osgood material.

The material constant, M, is the hardening exponent, and the material constant, K, is the coefficient of plastic resistance. In the case of zero hardening, $1/M = 0$, this model reduces to the perfectly plastic model. The hardening exponent, $n = 1/M$, mentioned in Chapter 3, Section 3.1.1, is about 0.5 for FCC metal such as copper or austenitic steel. It is about 0.15 for BCC metals such as steel, and only 0.05 for HCP metals such as titanium. The larger the hardening exponent, the easier the metal is to cold work. Experiments have shown that K is proportional to the square root of the density of dislocations in the material.

Recall that if a material which hardens is loaded to stress, σ, in the plastic region and then unloaded, the yield stress on a subsequent loading equals the stress σ to which the material was originally loaded. The Ramberg-Osgood model therefore gives the new yield stress if a hardening metal is plastically loaded.

EXAMPLE 9.9   Determine the axial stress required in a 316L steel rod to create a permanent strain of 0.015 in the axial direction. Typical values of the constants for 316L steel at room temperature are $\sigma_Y = 138$ MPa, $K = 435$ MPa, and $M = 4.55$.

Solution   The Ramberg-Osgood model implies that the required stress is

$$\sigma = 138 + 435(0.015)^{1/(4.55)} = 311 \text{ MPa.} \qquad \blacksquare$$

Manufacturing processes such as rolling or drawing are often described by the concept of the percent coldworking, which measures the change in cross-sectional area of the workpiece. The percent coldworking is defined as $\%CW = (A_o - A_f)/A_o \times 100$, where A_o is the original cross-sectional area and A_f is the cross-sectional area after coldworking. Because the volume of the material is assumed constant, $A_f L_f = A_o L_o$. Therefore in percent, the engineering strain perpendicular to the cross section is equal to the percent coldworking times the ratio of the cross sections,

$$\epsilon = \frac{(L_f - L_o)}{L_o} \times 100 = \frac{(A_o - A_f)}{A_f} \times 100 = \%CW(A_f/A_o).$$

EXAMPLE 9.10   Drawing is used to reduce the diameter of a copper wire 0.25 inches in diameter (Fig. 9.21). The copper parameters are $\sigma_Y = 10$ ksi (before loading), $1/M = 0.54$, and $K = 46,000$. Determine the largest possible reduction without causing yielding in the reduced section of the wire after it passes through the die.

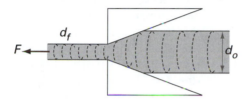

FIGURE 9.21   A schematic of wire drawing.

Solution   First, neglecting friction, the force, F, needed to pull the wire through the die is that sufficient to yield the original cross section. The force is $F = \sigma_Y A_o = 10,000\pi(0.125)^2 = 491$ pounds.

The stress in the reduced wire cross section is equal to F/A_f. To have this equal to the yield stress in the reduced wire, the Ramberg-Osgood relation requires

$$\frac{F}{.25\pi d_f^2} = \sigma_Y + K\left(\frac{(d_o^2 - d_f^2)}{100 d_f^2}\right)^{1/M}.$$

Substitution of the known values gives a nonlinear equation in d_f. This is solved numerically to obtain the minimum final diameter possible, $d_f = 0.236$ inches.

More dimensional reduction can be obtained by alternately coldworking and annealing in stages to end up with the desired yield stress.   ∎

9.5   MODELING MULTIAXIAL PLASTIC BEHAVIOR

Attempts to mathematically model the plastic behavior of metals can be traced back to Tresca, who around 1864 proposed the maximum shear stress failure criteria to predict at what three-dimensional state of stress a material is plastically loaded. Saint Venant and Lévy examined isotropic flow in 1871. Modern efforts to model plasticity got their major impetus just after World War II in Prager's examinations of anisotropic kinematic hardening.

A theory of plasticity must satisfy several conditions. The response is irreversible in a thermodynamic sense since there is dissipation and a permanent plastic strain. The response is independent of the loading rate. The volume of the body is assumed to remain constant, even though experiment indicates that the volume is the same before and after plastic loading but is not quite constant during plastic loading. A threshold, defined in terms of the stresses, is assumed at which plastic flow of the material begins. Further, the magnitude of the hydrostatic stress at a point has negligible effect on the yield condition. The response, in the case of hardening, depends on the loading history of the body.

9.5.1 The Yield Surface

Experiment apparently shows that if the stress at a point exceeds a threshold, then material plastic flow begins at that point of the body. In a uniaxial test, this simply means that the uniaxial stress exceeds the yield stress. Such a transition is especially clear in steel but not so clear in FCC metals such as aluminum. Plastic deformation is due to dislocation motion. Small plastic strains on the order of 10 microstrains may be created at stresses lower than what is commonly taken as the yield stress. Such small strains would be hard to detect in the typical uniaxial stress-strain test. Most classical descriptions of plasticity postulate a yield stress. However, a few contemporary theories of viscoplasticity, or time-dependent plasticity, do not make such an assumption.

Such a threshold is more difficult to define in a general three-dimensional state of stress. Many combinations of stress can produce plastic behavior. While the yield surface is graphed in principal stress space, the yield condition and the evolution of the plastic strain in a three-dimensional small strain plasticity model are usually described in terms of the deviatoric stress, $\mathbf{S}$. The threshold is defined in terms of a real valued function, $f(\mathbf{S})$, of the deviatoric stress tensor. Typically, the function, f, is defined so that if $f(\mathbf{S}) < 0$, then the state of stress is elastic and if $f(\mathbf{S}) = 0$, the state of stress is plastic. The graph of $f(\mathbf{S}) = 0$ is called the yield surface. During a plastic deformation, the state of stress remains on the yield surface.

EXAMPLE 9.11   A spherical yield surface of radius, R, is represented by $s_1^2 + s_2^2 + s_3^2 = R^2$ where the s_i are the principal stresses for the deviatoric stress, $\mathbf{S}$. In this case, $f(\mathbf{S}) = s_1^2 + s_2^2 + s_3^2 - R^2$.   ∎

Recall that the stress tensor, $\mathbf{T}$, and the deviatoric stress tensor are related by $\mathbf{T} = \mathbf{S} + p\mathbf{I}$, where $p = Tr(\mathbf{T})/3$ is the hydrostatic component of stress. The deviatoric accounts for volume distortion and the hydrostatic tensor $p\mathbf{I}$ accounts for volume change (Chapter 3, Section 3.6) under small strains. The volume change at a point due to the stress state represented by the tensor, $\mathbf{T}$, is proportional to the trace of $\mathbf{T}$ if the material is linearly elastic (Chapter 3, Section 3.3). The deviatoric stress, $\mathbf{S}$, state causes no volume change if the material is linearly elastic because its trace is zero, so the deviatoric state of stress causes only volume distortion at the point. Working in terms of the deviatoric stress represents the assumption that the threshold or yield is independent of the hydrostatic pressure and that the evolution of plastic deformation occurs under constant volume in a small strain theory.

If the material is isotropic, the function, f, must give the same value when the stress tensor is expressed in any other rotated coordinate system. The function, f, must be of such a form that

$$f(\alpha \mathbf{S} \alpha^T) = f(\mathbf{S}),$$

where α is an orthonormal tensor, as was defined in Chapter 1, Section 1.5. In this case, the yield condition is completely defined by f and the principal stresses, s_1, s_2, s_3, of $\mathbf{S}$.

The yield surface is a surface in the three-dimensional principal stress space whose interior is the set of elastic states of stress (Fig. 9.22). The yield surface must bound a convex set. A set is convex if any two points in the set can be connected by

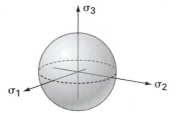

FIGURE 9.22   Schematic yield surface in principal stress space.

a straight line which also lies completely within the set. This implies that, for a convex set, the tangent plane at any point on the surface only intersects the surface at the point of tangency. A loading process is represented by a curve in principal stress space, called the process path.

If the material has the same yield threshold in tension as in compression, the yield function must then also be symmetric in the sense that

$$f(s_{11}, s_{12}, s_{13}, s_{22}, s_{23}, s_{33}) = f(\pm s_{11}, \pm s_{12}, \pm s_{13}, \pm s_{22}, \pm s_{23}, \pm s_{33}).$$

Hardening is accounted for by forcing the yield surface to change during plastic flow to ensure that the path in stress space representing the plastic process stays on the yield surface.

9.6   THRESHOLD CRITERIA FOR BIAXIAL LOADING

The classical strategy in the case of plane stress was to merely identify the yield surface obtained under various assumptions. The yield surface in the plane stress case is a closed curve in the principal stress plane. The material is said to be elastic under the particular set of yield assumptions if the point (σ_1, σ_2) in principal stress space lies inside the region enclosed by the yield surface and plastic otherwise. No attempt was made to describe the evolution of the plastic strain in the case of ductile behavior. If the material is brittle and fractures rather than yields, the fracture stress is taken as the threshold stress. In what follows, the failure stress and strain will be denoted by σ_f and ϵ_f, respectively. These values can refer to the stress and strain at which either yielding or fracture occurs.

The stress and strain at which a material either begins to yield or fractures due to a uniaxial load can be measured relatively easily. But for an arbitrarily shaped body under arbitrary loads, the prediction of either yield or fracture is very difficult. Some criterion is needed to make predictions without testing every material in every single possible shape under every possible loading. An ideal criterion would be one that predicts yielding or fracture for an arbitrary body from the data obtained from a simple uniaxial test. Usually these classical theories assume that isotropic brittle materials fail due to normal stress or strain and isotropic ductile materials fail due to shear stress or a measure of the volume distortion.

As a tool to predict failure, the classical theories use normal stress, normal strain, shear stress, the strain energy, or the distortion energy, among other possibilities. Each of these reaches its failure value at the same load in a uniaxial test, but this is no longer true if the state of stress is either two- or three-dimensional. In the axial test, the maximum shear stress is $\tau_f = \sigma_f/2$. In a pure shear test, the maximum normal stress and maximum shear stresses have the same magnitudes. The following Mohr's circles indicate this (Fig. 9.23).

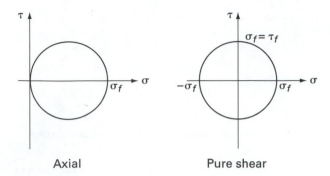

FIGURE 9.23 Mohr's circles for axial and pure shear loads.

A point on the surface of a body is in a state of plane stress if there is no stress on the surface at that point. In many types of loading, the stress is maximum at the surface. Therefore one can expect yielding to begin at the surface. This is true in both pure torsion and pure bending loading. There are two nonzero principal stresses, σ_1 and σ_2; the third principal stress is zero.

The state of stress at each point in a body in plane stress is defined by its two principal stresses. These stresses can be plotted in a plane with coordinates, σ_1 and σ_2, as the point, (σ_1, σ_2). The failure criterion determines all possible combinations of stress, σ_1 and σ_2, at which failure can occur. These stress pairs also can be plotted in the σ_1-σ_2 plane. The resulting set of curves is called the yield surface or the failure boundary. If a state of stress, represented by a point, (σ_1, σ_2), lies inside the boundary the loading is safe; if the point falls outside this region, failure is expected.

9.6.1 Maximum Normal Stress Criterion

This theory was proposed by W. J. M. Rankine (1820–1872), an engineering professor at the University of Glasgow in Scotland. Rankine suggested the condition that a body fails at a point if the maximum principal stress there exceeds the normal failure stress, σ_f, from a uniaxial test. The safe, or elastically loaded region can be determined from Mohr's circles; the case of one stress positive and the other negative is shown in Figure 9.24a. The safe region is bounded by the lines, $\sigma_1 = \pm\sigma_f$ and $\sigma_2 = \pm\sigma_f$. The failure boundary is a square (Fig. 9.24b).

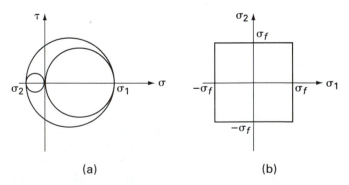

FIGURE 9.24 (a) Mohr's circle for opposite sign principal stresses. (b) Rankine failure boundary.

The criterion is not satisfactory for ductile materials, which can withstand very high hydrostatic stresses, and thus large maximum principal stress, without yielding. The criterion does work well for brittle materials however. It is one of the most accepted for design practice using brittle materials.

9.6.2 Maximum Shear Stress Criterion

This criterion has a history that goes back to Coulomb in 1773, but the primary credit is usually given to H. E. Tresca due to his experiments published in 1872. Barré de Saint Venant developed the mathematics of the criterion in 1871. J. J. Guest made some improvements in 1900. Today the criterion is often called the Tresca yield condition.

The maximum shear stress criterion states that a material fails at a point in a multiaxial state of stress if the maximum shear stress on some plane through the point exceeds the failure shear stress obtained in a uniaxial test of the material.

The Mohr's circle for a uniaxial test shows that $\tau_f = \frac{1}{2}\sigma_f$. To plot the yield boundary in the σ_1-σ_2 plane, each quadrant has to be considered separately since the sign of the stress affects the result. If both principal stresses for a certain state of stress in the body have the same sign, the largest of the three planar Mohr's circles has a diameter equal to the largest of the principal stresses (Fig. 9.25).

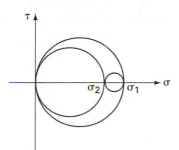

FIGURE 9.25   Mohr's circle for two positive and one zero principal stresses.

Therefore in the first and third quadrants, the yield surface for the Tresca condition is the same as that for the maximum normal stress criterion. On the other hand, if the two principal stresses existing at a point in the body have opposite signs, then the diameter of the largest Mohr's circle is $|\sigma_1 - \sigma_2|$ and its radius is $\tau_{max} = \frac{1}{2}|\sigma_1 - \sigma_2|$ (Fig. 9.24a). The boundary in the second and fourth quadrants is determined by

$$\frac{1}{2}|\sigma_1 - \sigma_2| = \tau_f = \frac{1}{2}\sigma_f. \tag{9.22}$$

In the second quadrant, the graph is

$$\sigma_2 = \sigma_1 + \sigma_f$$

and in the fourth quadrant is

$$\sigma_2 = \sigma_1 - \sigma_f.$$

Therefore the safe region has the hexagonal boundary shown in Figure 9.26. The tests to be discussed in Section 9.6.3 show that the Tresca condition is conservative for ductile materials, and so it is used in design codes.

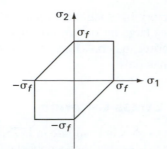

FIGURE 9.26 Boundary for the Tresca failure condition.

9.6.3 Maximum Principal Normal Strain Criterion

Saint Venant proposed that the material fails at a point if the maximum normal strain at the point is greater or equal to the normal failure strain observed in a uniaxial test, $\epsilon_{max} \geq \epsilon_f = \sigma_f/E$. This idea superseded the maximum normal stress criterion favored by Lamé and Rankine.

The calculation of the failure boundary in the σ_1-σ_2-plane requires Hooke's law to determine the stress-strain relationship. Because the elastic material is linear and isotropic, the principal strain directions are the same as the principal stress directions. Recall that $\sigma_3 = 0$. Four cases must be considered.

a. When $\sigma_1 > 0$ and ϵ_1 is the maximum principal strain so that $\epsilon_1 = \sigma/E$. Then Hooke's law

$$\epsilon_1 = \frac{1}{E}(\sigma_1 - \nu\sigma_2) \tag{9.23}$$

implies that

$$\frac{1}{E}\sigma = \frac{1}{E}(\sigma_1 - \nu\sigma_2) \tag{9.24}$$

or

$$\sigma_2 = \frac{1}{\nu}(\sigma_1 - \sigma_f). \tag{9.25}$$

b. If ϵ_2 is the maximum principal strain, and $\sigma_2 > 0$, then

$$\epsilon_2 = \frac{1}{E}(\sigma_2 - \nu\sigma_1) \tag{9.26}$$

implies that

$$\sigma_2 = \nu\sigma_1 + \sigma_f. \tag{9.27}$$

c. When $\sigma_1 < 0$ and ϵ_1 is the maximum principal strain so that $\epsilon_1 = \sigma/E$. Then Hooke's law

$$\epsilon_1 = \frac{1}{E}(\sigma_1 - \nu\sigma_2) \tag{9.28}$$

implies that

$$\sigma_2 = \frac{1}{\nu}(\sigma_1 + \sigma_f). \tag{9.29}$$

d. If ϵ_2 is the maximum principal strain, and $\sigma_2 < 0$, then

$$\epsilon_2 = \frac{1}{E}(\sigma_2 - \nu\sigma_1) \tag{9.30}$$

implies that

$$\sigma_2 = \nu\sigma_1 - \sigma_f. \tag{9.31}$$

These four lines enclose a diamond-shaped region which is symmetric about the $\sigma_1 = \pm\sigma_2$ lines and with σ_1-intercepts at $\pm\sigma_f$ and σ_2-intercepts also at $\pm\sigma_f$ (Fig. 9.27). The vertices are obtained by solving the intersecting pairs simultaneously. The vertices occur at $\left(\frac{\sigma_f}{1-\nu}, \frac{\sigma_f}{1-\nu}\right)$, $\left(\frac{-\sigma_f}{1-\nu}, \frac{-\sigma_f}{1-\nu}\right)$, $\left(\frac{\sigma_f}{1+\nu}, \frac{-\sigma_f}{1+\nu}\right)$, and $\left(\frac{-\sigma_f}{1+\nu}, \frac{\sigma_f}{1+\nu}\right)$.

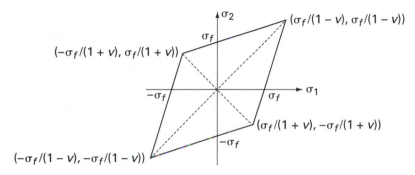

FIGURE 9.27   Boundary for the maximum principal normal strain criterion.

Exercise 9.2   Verify the coordinates of the vertices by solving the equations of the appropriate pairs of lines simultaneously.

9.6.4 Maximum Distortion Energy Criterion

Many researchers contributed to the development of this criterion, including Maxwell, M. T. Huber (1904), and von Mises (1913). Mises developed it mathematically by fitting a smooth curve around the hexagonal Tresca yield curve. The smooth surface is easier to use in calculations than the Tresca surface. It is is frequently called the von Mises yield theory today. This criterion has superseded the maximum strain energy theory because yielding is independent of the hydrostatic stress. Since the portion of strain energy producing a volume change, therefore, does not affect yielding, the important component is the distortion energy, U_d.

Recall from Chapter 3, Section 3.6, that the strain energy for uniaxial loading is

$$U = \frac{\sigma_f^2}{2E} = \frac{E\epsilon_f^2}{2}. \tag{9.32}$$

The distortion energy is written in terms of the octahedral shear stress as

$$U_d = \frac{3}{4G}\tau_{\text{oct}}^2, \tag{9.33}$$

where

$$\tau_{\text{oct}} = \frac{1}{3}\left[(\sigma_x - \sigma_y)^2 + (\sigma_x - \sigma_z)^2 + (\sigma_y - \sigma_z)^2 + 6(\tau_{xy}^2 + \tau_{xz}^2 + \tau_{yz}^2)\right]^{1/2}. \tag{9.34}$$

Then, since $G = E/2(1 + \nu)$, for axial loading,

$$U_d = \frac{1 + \nu}{3E}\sigma_f^2. \qquad (9.35)$$

Exercise 9.3 Verify that this expression for the distortion energy due to an axial load is correct.

The von Mises criterion says that yielding begins when the distortion energy at a point exceeds that value of distortion energy obtained at yield in a uniaxial test. The condition is therefore, by equations (9.33) and (9.35), that

$$\left[(\sigma_x - \sigma_y)^2 + (\sigma_x - \sigma_z)^2 + (\sigma_y - \sigma_z)^2 + 6(\tau_{xy}^2 + \tau_{xz}^2 + \tau_{yz}^2)\right] \geq 2\sigma_f^2. \qquad (9.36)$$

The distortion energy magnitude is independent of the coordinates. Therefore, it is easiest to work in principal coordinates to compute the yield boundary in the σ_1-σ_2-plane. The yield boundary in a three-dimensional state of stress is

$$U_d = \frac{1 + \nu}{6E}\left[(\sigma_1 - \sigma_2)^2 + (\sigma_1 - \sigma_3)^2 + (\sigma_2 - \sigma_3)^2\right] = \frac{1 + \nu}{3E}\sigma_f^2. \qquad (9.37)$$

The condition simplifies in the biaxial, plane stress case to

$$\sigma_1^2 + (\sigma_1 - \sigma_2)^2 + \sigma_2^2 = 2\sigma_f^2 \qquad (9.38)$$

or

$$\sigma_1^2 - \sigma_1\sigma_2 + \sigma_2^2 = \sigma_f^2. \qquad (9.39)$$

The graph of the yield boundary is an ellipse with axes on the $\sigma_1 = \sigma_2$ and the $\sigma_1 = -\sigma_2$ lines (Fig. 9.28). The intercepts on both axes occur again at $\pm\sigma_f$. The major and minor axes of the ellipse lie on the lines $\sigma_2 = \pm\sigma_1$.

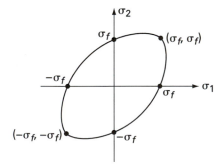

FIGURE 9.28   Boundary for the maximum distortion energy criterion.

Note that yield can occur even if both σ_1 and σ_2 have magnitude less than the uniaxial yield stress, $\sigma_f = \sigma_Y$, in the second and fourth quadrants. Notice also that in the uniaxial case, when $\sigma_2 = 0$, the criterion reduces to $\sigma_1 = \sigma_f$, as expected. Experiments show that this is an excellent criterion for the yielding of ductile materials.

9.6.5 Octahedral Shear Stress Criterion

The octahedral shear stress criterion, which is given in terms of the stress, is an alternative to the distortion energy criterion, which is defined in terms of energy.

The octahedral shear stress criterion predicts yield at a point in the material when the magnitude of the octahedral shear stress at a point exceeds that observed at yield in an axial test.

In terms of principal coordinates, this criterion predicts a yield boundary

$$\tau_{oct}^2 = \frac{1}{9}\left[(\sigma_1 - \sigma_2)^2 + (\sigma_1 - \sigma_3)^2 + (\sigma_2 - \sigma_3)^2\right] = \frac{2}{9}\sigma_f^2. \tag{9.40}$$

Then, failure occurs if

$$\tau_{oct} = \frac{\sqrt{2}}{3}\sigma_f = 0.471\sigma_f. \tag{9.41}$$

9.6.6 Comparison of the Various Failure Criteria

These criteria can be compared by superimposing all the failure boundaries on one principal stress plane. The smaller the safe region, the more conservative the theory. Figure 9.29 shows that the maximum shear stress and distortion energy criteria are both more conservative than the maximum normal stress criterion.

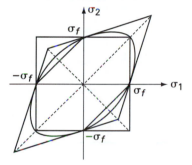

FIGURE 9.29   Overlay of the boundaries of the maximum shear stress, distortion energy, maximum normal stress, and maximum normal strain criteria.

The Tresca (maximum shear stress) and the Mises (maximum distortion energy) criteria for yielding are empirical models that extend uniaxial failure criteria to biaxial and triaxial loadings. Even though they are empirical, some justification for them is offered by dislocation theory.

The Tresca condition corresponds to the idea that plastic flow of a metal is due to dislocation motion. Dislocation glide is a result of shear stresses acting on the slip plane for the dislocation. See Chapter 1, Section 1.1, for a method of calculating the shear stress on a slip plane induced by an applied stress. While it is likely that some dislocations move under stresses less than the bulk measured yield stress, it is very plausible that the larger yield stress is necessary to move a large enough number of dislocations that the plastic flow can be observed on the bulk material level.

The Mises distortion energy criterion assumes that volume change can be neglected so that only the distortional energy is relevant. Again, since plastic behavior is due to dislocation motion, the question becomes whether or not a dislocation induces a volume change in a metal. Recall from Chapter 3, Section 3.6, that in an isotropic linearly elastic material like a metal, the volume change per unit volume is proportional to the sum of the normal stresses (the trace of the stress tensor). Since a screw dislocation only induces shear stresses (Chapter 3, Section 3.6), there is no volume change. Likewise, as discussed in Chapter 4, Section 4.6.2, while the edge dislocation does induce local volume change, the total volume change around an edge dislocation is zero. Therefore, this analysis implies that because the dislocations cause no volume change, the volumetric strain energy is zero and only the distortional strain energy can change during dislocation flow or dislocation multiplication. This suggests that plastic behavior is volume preserving, an assumption often made in analyzing plastic behavior. However, as discussed in Chapter 9, Section 9.2.1, experiment does not fully support such an assumption that plastic flow is volume preserving.

Results that can be tested by experiment are obtained by computing for each criterion the predicted relationship between the failure stress in an axial test and that in a torsion test. Draw the Mohr's circle for the uniaxial test and for pure shear. Note that in the pure shear or torsion test, $\sigma_1 = -\sigma_2 = \tau_f$. Assume that the principal normal stress, σ_1, is the same in both tests. Table 9.2 shows the predicted value of the failure magnitude first in an axial test, then in a torsion test. The right-hand column shows the relationship if the failure conditions from the axial and torsion test are equal, as they must be.

TABLE 9.2 Comparison of Failure Criteria

Criterion	Axial	Torsion	Relationship
Maximum normal stress	$\sigma_{max} = \sigma_f$	$\sigma_{max} = \tau_f$	$\tau_f = \sigma_f$
Maximum normal strain	$\epsilon_{max} = \sigma_f/E$	$\epsilon_{max} = (\sigma_1 - \nu\sigma_2)$ $= \frac{1+\nu}{E}\tau_f$	$\tau_f = \frac{1}{1+\nu}\sigma_f$
Maximum shear stress	$\tau_{max} = 0.5\sigma_f$	$\tau_{max} = \tau_f$	$\tau_f = 0.5\sigma_f$
Maximum distortion energy	$U_d = \frac{1+\nu}{3E}\sigma_f^2$	$U_d = \frac{1+\nu}{E}\tau_f^2$	$\tau_f = \frac{1}{\sqrt{3}}\sigma_f = 0.577\sigma_f$
Maximum octahedral stress	$\tau_{oct} = \frac{\sqrt{2}}{3}\sigma_f$	$\tau_{oct} = \frac{\sqrt{2}}{\sqrt{3}}\tau_f$	$\tau_f = \frac{1}{\sqrt{3}}\sigma_f = 0.577\sigma_f$

Experiments published by G. I. Taylor and H. Quinney in 1931 showed that for ductile metals, the value of τ_f from a torsion test on a thin circular cylinder (which is presumed to be in pure shear) is from 0.55 to 0.60 of the value of σ_f from a uniaxial tensile test. The average value over several ductile materials was about 0.57. Therefore their results indicate that the distortion energy or octahedral shear stress criteria, which are equivalent, are the best for ductile materials. The maximum shear stress theory prediction that $\tau_f = 0.5\sigma_f$ is about 15% less than measured. Therefore the maximum shear stress theory is conservative. It is often used in design instead of the distortion energy theory because it is easier to compute for two-dimensional states of stress. The maximum normal stress or maximum normal strain criteria are often used for brittle materials under the assumption that for them failure occurs due to normal stress rather than shear stress. Fracture occurs for a brittle material before any significant yielding happens. The discussion in Chapter 10 shows that in fact fracture is more complex than this strategy assumes.

EXAMPLE 9.12 The uniaxial yield stress for a material is $\sigma_Y = 750$ MPa. Using first the Tresca criterion and then the von Mises criterion, determine whether yielding will have occurred under combined tension and torsion with $\sigma_x = 270$ MPa and $\tau_{xy} = -360$ MPa.

Solution   **The Tresca Criterion:**   Because the radius of the largest Mohr's circle is $0.5\sqrt{\sigma_x^2 + 4\tau_{xy}^2}$ and $\tau_Y = \sigma_Y/2$,

$$(2\tau_{max})^2 = \sigma_x^2 + 4\tau_{xy}^2 = 270^2 + 4(360)^2 = 591,300 \quad \text{MPa}^2 > \sigma_Y^2 = 562,500 \quad \text{MPa}^2.$$

Note that $2\tau_{max} = 768.96$ MPa. Yielding occurs by the Tresca criterion.

The von Mises Criterion: Using the criterion comparing the distortion energies in the form of equation (9.36),

$$\sigma_x^2 + 3\tau_{xy}^2 = 270^2 + 3(360)^2 = 461,700 \quad \text{MPa}^2 < \sigma_Y^2 = 562,500 \quad \text{MPa}^2.$$

Yielding does not occur according to the von Mises criterion. The Tresca criterion is more conservative for design.

Alternatively, the criteria can be verified in principal coordinates if the principal stresses are first calculated: $\sigma_1 = 519.48$, $\sigma_2 = 0$, and $\sigma_3 = -249.48$ MPa. For the Tresca criterion, $2\tau_{max} = \sigma_1 - \sigma_3 = 768.96 > \sigma_Y = 750$ MPa. For the von Mises criterion, $\sigma_1^2 - \sigma_1\sigma_3 + \sigma_3^2 = 461700 < \sigma_Y^2$. ∎

9.6.7   Applications of the Biaxial Threshold Criteria in Design

Combined loading problems from strength of materials provide simple examples to show the application of the threshold criteria in design. A typical problem is to determine the largest load a given body can support without yielding. More often in design, the load which must be carried is known. The dimensions of the structure must then be determined to ensure that the structure will remain elastically loaded.

The point in the body at which the largest stresses occur must first be determined. In complex loadings and/or geometries, a finite element analysis might be attempted. However, this requires numerical values to be assumed for all unknowns and reduces the decision to a trial-and-error approach. If an analytic expression is available for the stress at each point, the unknowns can be solved directly. This, in a nutshell, is the reason that mathematical models, if they exist, are more useful in design than numerical methods such as the finite element method. Unfortunately, the models are not always available. Strength of materials models exist since enough approximations have been made that closed-form expressions can be written for the stress state.

On the other hand, if the loads and dimensions for a design are known, numerical methods can easily predict the state of stress at each point in the body. This procedure is sometimes called design analysis.

EXAMPLE 9.13   Determine the maximum value of P for the axial load, P, bending load, $F = 3P$, and torque, $T = P/4$, that can be applied to the rod of radius, $R = 1$ inch, and length, $L = 10$ inches (Fig. 9.30), while avoiding failure by yielding. The material is aluminum with proportional limit $\sigma_{PL} = 64$ ksi and Poisson ratio $\nu = 0.3$. Assume the elementary strength of materials models for the stresses.

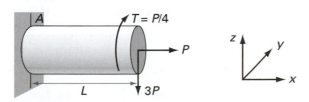

FIGURE 9.30   A rod in combined axial, torsional, and bending loading.

Solution

The first step is to locate the point in the body at which the stress is expected to be largest. This rod is subjected to axial, torsion, and bending loads. The principal stresses due to the axial load are the same throughout the body. The torsional shear stress is largest on the outer surface of the rod. The bending normal stress, which is in the axial direction in the rod, is largest at a point at the greatest distance from the neutral axis and lying in the face on which the largest bending moment acts. Since the axial load is tensile, rather than compressive, the largest normal stress occurs where it adds with the largest tensile normal bending stress. The point A on the rod is likely to be the one at which the principal and maximum shear stresses are largest. At A, the shear stress due to bending is zero, and

$$\sigma_x = \frac{P}{\pi R^2} + \frac{3PLR}{0.25\pi R^4},$$

$$\tau_{xy} = \frac{0.25PR}{0.5\pi R^4}.$$

The von Mises criterion in the form [equation (9.36)] implies that

$$\sigma_x^2 + 3\tau_{xy}^2 = \left(\frac{P}{\pi R^2} + \frac{3PLR}{0.25\pi R^4}\right)^2 + 3\left(\frac{0.25PR}{0.5\pi R^4}\right)^2 = 64{,}000^2.$$

Therefore $P = 1662$ pounds. ∎

In plane stress biaxial loading, the distortion energy yield surface is given by $\sigma_1^2 - \sigma_1\sigma_2 + \sigma_2^2 = \sigma_Y^2$, where σ_1 and σ_2 are the in-plane principal stresses (the other one is zero). This has led engineers to define a Mises stress, σ_m, (see Chapter 1, Section 1.7) by

$$\sigma_m = \sqrt{\sigma_1^2 - \sigma_1\sigma_2 + \sigma_2^2}. \tag{9.42}$$

Then yield occurs in plane stress when the Mises stress, σ_m, equals the uniaxial yield stress. Many finite element programs produce the Mises stress as output so that the user can quickly check whether the material at the point has yielded under the given loading conditions, at least if the distortion energy yield criterion is assumed.

A more complex design problem is presented by spinning disks. In this example, the Tresca yield criterion is applied.

EXAMPLE 9.14

Determine the constant rotational speed, ω rad/s, at which a thin spinning disk becomes fully plastic. The disk has a central hole of radius a and outer radius, b. The material is assumed not to harden so that it is modeled as perfectly plastic. The material has mass density, ρ. Assume the Tresca yield condition.

Solution

From Chapter 4, Section 4.6.1.2, because $\tau_{r\theta} = 0$, the principal stresses under elastic loading are known to be $\sigma_r(r)$, $\sigma_\theta(r)$, and $\sigma_z = 0$. The stress $\sigma_\theta(r)$ was shown to be greater than $\sigma_r(r)$ so that $0 \leq \sigma_r(r) \leq \sigma_\theta(r)$, and also $\sigma_\theta(r)$ decreases as r increases, where r is the radial coordinate of the point in question. The maximum shear stress on any plane through this point is $\tau_{max} = 0.5(\sigma_\theta - 0) = \sigma_\theta/2$. The Tresca criterion for yielding at this point requires that $\tau_{max} = \sigma_Y$, the uniaxial yield stress. Therefore the yield condition is $\sigma_\theta(r) = \sigma_Y$.

Because $\sigma_\theta(r)$ decreases as r increases, yield is initiated at $r = a$ the speed, ω, obtained from the expression for $\sigma_\theta(r)$ in Chapter 4, Section 4.6.1.2, by setting $\sigma_\theta(a) = \sigma_Y$ and $r = a$.

$$\omega^2 = \frac{4\sigma_Y}{\rho(3+\nu)b^2}.$$

Note that if the disk is solid, the expression for $\sigma_\theta(r)$ in Chapter 4, Section 4.6.1.2, by setting $\sigma_\theta(a) = \sigma_Y$ and $r = a$ is

$$\omega^2 = \frac{8\sigma_Y}{\rho(3+\nu)b^2}.$$

Therefore the existence of an extremely tiny central hole reduces the speed that initiates yield by a factor of $1/\sqrt{2}$, according to the Tresca condition.

The fully plastic loading speed is found by assuming that $\sigma_\theta(r) = \sigma_Y$ for all r. The stress equilibrium equation (4.14) with body force per unit volume $F_r = \rho\omega^2 r$ from Newton's law is

$$\frac{d\sigma_r}{dr} + \frac{1}{r}(\sigma_r - \sigma_\theta) + \rho\omega^2 r = 0.$$

Writing this expression with $\sigma_\theta(r) = \sigma_Y$ as

$$r\frac{d\sigma_r}{dr} + \sigma_r - \sigma_Y + \rho\omega^2 r^2 = \frac{dr\sigma_r}{dr} - \sigma_Y + \rho\omega^2 r^2 = 0$$

and integrating yields

$$\sigma_r = \sigma_Y - \frac{1}{3}\rho\omega^2 r^3 + C.$$

The boundary condition $\sigma_r(a) = 0$ implies that $C = \rho\omega^2 a^3/3 - \sigma_Y$. This expression for C substituted into the boundary condition $\sigma_r(b) = 0$ gives the fully plastic constant rotational speed,

$$\omega^2 = \frac{3\sigma_Y(a-b)}{\rho(a^3 - b^3)} = \frac{3\sigma_Y}{\rho(a^2 + ab + b^2)}.$$

Setting $a = 0$ produces the fully plastic constant rotational speed for a solid disk. ∎

9.7 THRESHOLD CRITERIA FOR TRIAXIAL LOADING

The triaxial criteria are based on those classical failure theories that are appropriate to plastic behavior, which is due to a shear flow, such as the maximum shear stress and the maximum distortion energy theories. Since, according to the Bridgman experiments, the yield stress is independent of the hydrostatic pressure, these criteria are expressed in terms of the deviatoric stress tensor, $\mathbf{S}$, rather than the Cauchy stress tensor, $\mathbf{T}$ (see Chapter 1 and Chapter 3, Section 3.6).

A general yield condition at a point of an isotropic body is defined by a function of the deviatoric stress state at that point, $f(\mathbf{S}, k)$, where k is a material constant. The constant, k, should be related to the uniaxial yield stress, σ_Y, so that the uniaxial test can be used to predict yield in a three-dimensional state of stress. If the material is anisotropic, additional material constants are required. The criterion is

$$\begin{aligned} f(\mathbf{S}, k) &< 0 \quad \text{when the material is elastically loaded} \\ f(\mathbf{S}, k) &= 0 \quad \text{when the material is plastically loaded.} \end{aligned} \tag{9.43}$$

To represent the plane stress criteria by such a function f, for each criterion, the boundaries of the safe regions pictured in Figure 9.29 must be the graph of the equation $f = 0$. Each of these boundaries is the yield surface for the corresponding theory in plane stress.

When the three-dimensional deviatoric stress is written in principal coordinates, the function, $f(\mathbf{S}, k)$, depends on four terms, s_1, s_2, s_3, and k. The dependence can be reduced to two terms so that the function $f(\mathbf{S}, k)$ is dimensionless. First, the idea is to rewrite the function in terms of the stress invariants, $I_\mathbf{S}$, $II_\mathbf{S}$, and $III_\mathbf{S}$. This notation differs from that of Chapter 1, Section 1.6, so that the stress tensor involved may be indicated. These invariants are the coordinate independent coefficients of the characteristic equation for the eigenvalues, λ, of the tensor, $\mathbf{S}$, $\det(\mathbf{S} - \lambda\mathbf{I}) = -\lambda^3 + I_\mathbf{S}\lambda^2 - II_\mathbf{S}\lambda + III_\mathbf{S}$. Since the trace of the deviatoric stress tensor is always 0 and since

$I_S = Tr(\mathbf{S}) = 0$, only two of the invariants appear in the function. Therefore, the yield criterion can be written as the function, $f(II_S, III_S, k)$. Now requiring the material constant k to have units of stress and rewriting the function as $f(II_S/k^2, III_S/k^3)$ makes the yield criterion dimensionless and dependent on only two terms.

The stress invariants, which are the same no matter in what coordinate system the tensor is written, can be most simply expressed in the principal coordinates. Let s_1, s_2, and s_3 be the principal stresses of the tensor $\mathbf{S}$. In principal coordinates, $\mathbf{S}$ has the principal stresses as the diagonal entries and zeros elsewhere. The invariants are

$$I_S = s_1 + s_2 + s_3;$$
$$II_S = s_1s_2 + s_1s_3 + s_2s_3;$$
$$III_S = s_1s_2s_3.$$

Exercise 9.4   Verify these expressions by writing out the characteristic equation.

Two of the invariants are easily identified; $I_S = Tr(\mathbf{S})$ and $III_S = \det(\mathbf{S})$. The second invariant also satisfies a useful relation. For any tensor, $\mathbf{S}$,

$$II_S = \frac{1}{2}\left(I_S^2 - Tr(\mathbf{S}^2)\right). \tag{9.44}$$

Exercise 9.5   Verify relation (9.44) by working in principal coordinates and substituting the expressions for the invariants into equation (9.44). Note that $Tr(\mathbf{S}^2) = s_1^2 + s_2^2 + s_3^2$.

9.7.1 von Mises Criterion

This criterion is the three-dimensional generalization of the maximum distortion energy failure theory for an isotropic material. It assumes that plasticity is due to slip governed by shear stress. The criterion is that, at yield, the distortional energy at a point in a three-dimensional state of stress is equal to the distortion energy at yield in the uniaxial case. This criterion cannot be expected to be exact because there is no reason to assume that the material behaves exactly the same under a three-dimensional state of stress as it does under a one-dimensional state of stress.

The von Mises yield surface for a three-dimensional state of stress with principal stresses, σ_1, σ_2, and σ_3, is defined by

$$\sigma_Y = \frac{1}{\sqrt{2}}\left[(\sigma_1-\sigma_2)^2 + (\sigma_2-\sigma_3)^2 + (\sigma_1-\sigma_3)^2\right]^{1/2}. \tag{9.45}$$

The Mises criterion can also be expressed in terms of the stress invariants of the deviatoric stress, $\mathbf{S}$. The Mises criterion for yield is

$$f\left(\frac{II_S}{k^2}, \frac{III_S}{k^3}\right) = \frac{II_S}{k^2} + 1 = 0. \tag{9.46}$$

This criterion is independent of the third invariant.

Denote the principal stresses associated with the deviatoric by s_1, s_2, and s_3. Further, as is verified by writing out the characteristic equation for $\mathbf{S}$,

$$II_S = \frac{1}{2}\left(I_S^2 - Tr(\mathbf{S}^2)\right) = -\frac{1}{2}Tr(\mathbf{S}^2) < 0, \tag{9.47}$$

since $I_{\mathbf{S}} = Tr(\mathbf{S}) = 0$. So that the von Mises criterion [equation (9.46)] says that the material yields at a point under a deviatoric stress $\mathbf{S}$ if

$$Tr(\mathbf{S}^2) = 2k^2. \tag{9.48}$$

The value of the material parameter k is determined from a uniaxial test. In uniaxial loading, the entries of the Cauchy stress tensor are $T_{11} = \sigma_0$ and $T_{ij} = 0$ otherwise. The entries of the deviatoric stress are all zero except that $s_1 = \frac{2}{3}\sigma_0$, $s_2 = -\frac{1}{3}\sigma_0$, and $s_3 = -\frac{1}{3}\sigma_0$. Then

$$Tr(\mathbf{S}^2) = s_1^2 + s_2^2 + s_3^2 = \frac{2}{3}\sigma_0^2.$$

Therefore for the von Mises criterion, where $\sigma_0 = \sigma_Y$,

$$k = \frac{\sigma_Y}{\sqrt{3}}.$$

Expanding equation (9.45) produces equation (9.48) written in terms of σ_Y.

EXAMPLE 9.15   In simple shear where $T_{12} = T_{21} \neq 0$ and $T_{ij} = 0$ otherwise, the condition is computable from the fact that the Cauchy and deviatoric stress tensors are equal. The only nonzero entries are the in-plane shears, $T_{12} = T_{21}$. Then the

$$Tr(\mathbf{S}^2) = 2T_{12}^2 = 2k^2.$$

Therefore, the condition implies that $k = T_{12}$, and for the von Mises criterion the normal stress at yield is related to the shear stress at yield by

$$\frac{1}{\sqrt{3}}\sigma_Y = T_{12}. \tag{9.49}$$

This reproduces the result obtained in the previous section where only two-dimensional states of stress were considered.   ∎

In principal stress space with coordinates, σ_1, σ_2, and σ_3, the values of stress that make the yield function $f = 0$, the yield boundary, is an inclined right circular cylinder with radius $R = \sqrt{2}\sigma_Y/\sqrt{3}$ (Fig. 9.31). The axis of the cylinder passes through the origin and is parallel to the vector, $\mathbf{i} + \mathbf{j} + \mathbf{k}$. The cross-sectional planes have equations of the form $\sigma_1 + \sigma_2 + \sigma_3 = d$, where d is a constant.

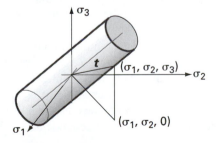

FIGURE 9.31   The von Mises yield surface in principal stress space.

To verify these statements, recall that any stress tensor, $\mathbf{T} = (T_{ij})$, is the sum of the deviatoric stress tensor, $\mathbf{S}$, and the hydrostatic pressure tensor, $p\mathbf{I}$, where $p = Tr(\mathbf{T})/3 = (\sigma_1 + \sigma_2 + \sigma_3)/3$. Assume that the tensor, $\mathbf{T}$, is given in principal coordinates so that it is diagonal. Let $(\sigma_1, \sigma_2, \sigma_3)$ be a point on the yield surface. The vector from the origin to this point is $t = (\sigma_1\mathbf{i} + \sigma_2\mathbf{j} + \sigma_3\mathbf{k})$, where $\mathbf{i}$, $\mathbf{j}$ and $\mathbf{k}$ are the unit vectors defining the principal axes. Because each $\sigma_i = s_i + p$, the stress can be written as

$$\mathbf{t} = (s_1\mathbf{i} + s_2\mathbf{j} + s_3\mathbf{k}) + p(\mathbf{i} + \mathbf{j} + \mathbf{k}).$$

In fact, these two summands are orthogonal. Their dot product is $pTr(\mathbf{S}) = 0$, since the trace of any deviatoric stress tensor is zero. The square of the magnitude of the vector, $s_1\mathbf{i} + s_2\mathbf{j} + s_3\mathbf{k}$, equals $s_1^2 + s_2^2 + s_3^2 = Tr(\mathbf{S}^2)$; recall that $\mathbf{S}$ is diagonal. Now assume that the vector $\mathbf{t}$ represents a state of plastic deformation so that its tail is at the origin and its tip is on the yield surface. The yield surface is the von Mises surface defined by $Tr(\mathbf{S}^2) = 2k^2$, a constant. The stress vector $\mathbf{t}$ has therefore been broken into a sum of a vector lying along the axis of the cylinder, $p(\mathbf{i} + \mathbf{j} + \mathbf{k})$, and a perpendicular vector that always has a fixed magnitude. The surface must therefore be an infinite cylinder of radius $\sqrt{2}k = \sqrt{2}\sigma_Y/\sqrt{3}$.

Exercise 9.6   Write $\mathbf{p} = p(\mathbf{i} + \mathbf{j} + \mathbf{k})$. Verify that if $|\mathbf{t} - \mathbf{p}|$ is the magnitude of the vector, then the equation of the cylinder $|\mathbf{t} - \mathbf{p}| = \sqrt{2}\sigma_Y/\sqrt{3}$ reduces to equation (9.45).

The intersection of the cylinder with any one of the three coordinate planes is the ellipse forming the yield surface for plane stress of Section 9.6.4 because in each case one of the principal stresses is zero.

A loading that begins at zero stress and varies can be plotted as the path that the tip of the vector $\mathbf{t} = (\sigma_1\mathbf{i} + \sigma_2\mathbf{j} + \sigma_3\mathbf{k})$ sketches in principal stress space. The path passes through the elastic region contained inside the yield surface, and the path may eventually intersect the yield surface. If it does, the material yields. Some theories to be discussed are most effective in describing behavior in the case of proportional loading.

Definition   A loading is said to be proportional (increased proportionally) if the principal stresses vary linearly in the principal stress space as the parameter α changes with time,

$$\mathbf{t} = (\sigma_1\mathbf{i} + \sigma_2\mathbf{j} + \sigma_3\mathbf{k}) = \alpha(t)(a\mathbf{i} + b\mathbf{j} + c\mathbf{k}), \tag{9.50}$$

where a, b, and c are fixed constant stresses.

The path in stress space traced by a proportional load is a straight line and, if $\alpha(t)$ increases, must eventually intersect the yield surface, say, at α_o, unless the path is parallel to $\mathbf{i} + \mathbf{j} + \mathbf{k}$, the center line of the cylinder. If, instead of increasing each stress proportionally to some level, the path is taken first in the σ_1-σ_2 plane as $\mathbf{t} = \alpha(a\mathbf{i} + b\mathbf{j})$ as α varies to α_o and then in the σ_3-direction to $\alpha_o c\mathbf{k}$ to reach the same state of stress as loading proportionally, then a different yield point will occur. The path in the σ_1-σ_2 plane intersects the yield cylinder at a lower value of α. In fact, by the plane stress Mises criterion, the yield occurs when α reaches the value such that $\sigma_1^2 - \sigma_1\sigma_2 + \sigma_2^2 = \sigma_Y^2$. In this sense, the criterion is loading path dependent.

Exercise 9.7   Sketch in principal coordinate space containing a von Mises yield surface the loading path, $P\mathbf{i} + 2P\mathbf{j} + 3P\mathbf{k}$ as the stress parameter, P, increases. Compute, in terms of σ_Y, the value P_Y at which the path intersects the yield surface. Compare this to the loading path obtained by moving first along the σ_1-axis a distance P_Y, then along the σ_2-axis a distance $2P_Y$, and finally along the σ_3-axis a distance $3P_Y$. Does this path intersect the yield surface before it is completed?

The Mises threshold criterion is most often used to predict whether a state of stress induces yielding by computing the criterion directly rather than checking geometrically whether or not the point representing the state of stress in principal stress space lies within the Mises cylindrical yield surface.

EXAMPLE 9.16   The state of stress, in ksi, at a point is

$$\mathbf{T} = \begin{pmatrix} 20 & 20 & -20 \\ 20 & 30 & 0 \\ -20 & 0 & 10 \end{pmatrix} \quad \text{ksi.} \tag{9.51}$$

The uniaxial yield stress is $\sigma_Y = 52$ ksi. Does the Mises criterion predict yield?

Solution   Several forms of the Mises criterion are available. The one which is computationally most convenient should be chosen.

Method 1:

In x-y-z coordinates, compare the distortion energy for $\mathbf{T}$ with the uniaxial distortion energy at yield, $(1 + \nu)\sigma_Y^2/3E$. By equation (9.36),

$$U_d = \frac{1+\nu}{6E}[(20-30)^2 + (20-10)^2 + (30-10)^2 + 6(20^2 + (-20)^2 + 0^2)$$

$$= \frac{1+\nu}{6E}5400 < \frac{1+\nu}{3E}\sigma_Y^2 = \frac{1+\nu}{6E}5408.$$

The criterion predicts that yield will not occur.

Method 2:

Work in principal coordinates for $\mathbf{T}$. The eigenvalues of $\mathbf{T}$ are the principal stresses, $\sigma_1 = 50$, $\sigma_2 = 20$, $\sigma_3 = -10$ ksi. By the simplified form of the criterion equation (9.37),

$$(50-20)^2 + (50+10)^2 + (20+10)^2 = 5400 < 2\sigma_Y^2 = 5408 \text{ ksi}^2.$$

The criterion predicts that yield will not occur.

Method 3:

Work with the deviatoric stress, $\mathbf{S}$, in x-y-z coordinates. Use the criterion in the form $Tr(\mathbf{S}^2) = 2k^2 = 2\sigma_Y^2/3$.

$$\mathbf{S} = \begin{pmatrix} 0 & 20 & -20 \\ 20 & 10 & 0 \\ -20 & 0 & -10 \end{pmatrix}; \quad \mathbf{S}^2 = \begin{pmatrix} 800 & 200 & 200 \\ 200 & 500 & -400 \\ 200 & -400 & 500 \end{pmatrix}$$

$Tr(\mathbf{S}^2) = 800 + 500 + 500 = 1800 < 2\sigma_Y^2/3 = 1802.7 \text{ ksi}^2$. The criterion predicts that yield will not occur.

Method 4:

Work with the deviatoric stress, $\mathbf{S}$, in principal coordinates and use $Tr(\mathbf{S}^2) = 2k^2 = 2\sigma_Y^2/3$. The principal deviatoric stresses are the eigenvalues of $\mathbf{S}$; $s_1 = 30$, $s_2 = 0$, and $s_3 = -30$ ksi. In principal coordinates, $\mathbf{S}$ is diagonal, as is $\mathbf{S}^2$, and

$$Tr(\mathbf{S}^2) = 30^2 + 0^2 + (-30)^2 = 1,800 < 2\sigma_Y^2/3 = 1,802.7 \text{ ksi}^2.$$

$Tr(\mathbf{S}^2)$ has the same value in both x-y-z and principal coordinates because it is an invariant of $\mathbf{S}$. The criterion predicts that yield will not occur.

Working with the deviatoric stress usually requires more computational effort. The deviatoric stress was introduced above primarily to help identify the shape of the Mises yield surface. ∎

Whenever the distortion energy of the applied stress exceeds the uniaxial distortion energy at yield, the material deforms plastically. Hardening must occur. However, the threshold criteria give no means to predict the evolution of the plastic strain as the material hardens. Such a prediction of the plastic strain evolution is made by a multiaxial plasticity model. Plasticity models are often written in terms of the deviatoric stress (See Chapter 9, Section 9.8).

9.7.2 Tresca Criterion

The Tresca maximum shear stress criterion can also be generalized to a three-dimensional state of stress. The yield stresses in uniaxial tension and simple shear are related as in the two-dimensional state of stress case. If k is the yield stress in pure shear,

$$\frac{1}{2}\sigma_Y = T_{12} = k.$$

The yield conditions in terms of principal stresses are then that one of the following conditions must be met.

$$0.5(\sigma_1 - \sigma_2) = \pm k; \qquad 0.5(\sigma_1 - \sigma_3) = \pm k; \qquad 0.5(\sigma_2 - \sigma_3) = \pm k.$$

Squaring and multiplying these three equations produces the general Tresca yield condition in the form of the single equation $f(\mathbf{S}, k) = 0$,

$$[(\sigma_1 - \sigma_2)^2 - 4k^2][(\sigma_1 - \sigma_3)^2 - 4k^2][(\sigma_2 - \sigma_3)^2 - 4k^2] = 0.$$

Denote $\bar{I}I_{\mathbf{S}} = Tr(\mathbf{S}^2)$ and $I\bar{I}I_{\mathbf{S}} = Tr(\mathbf{S}^3)$. The Tresca condition is then rewritten in terms of these invariants as

$$\frac{1}{2}\bar{I}I_{\mathbf{S}}^3 - 3I\bar{I}I_{\mathbf{S}}^2 - 9k^2\bar{I}I_{\mathbf{S}}^2 + 48k^4\bar{I}I_{\mathbf{S}} - 64k^6 = 0 \qquad (9.52)$$

at yield.

This produces a hexagonal cylinder (Fig. 9.32) that fits inside the right circular Mises cylinder with the vertices touching the cylinder. The two-dimensional analog is obtained by cutting the cylinder by a plane parallel to the coordinate plane.

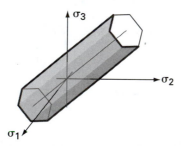

FIGURE 9.32   The Tresca yield surface in principal stress space.

EXAMPLE 9.17   Reconsider the example to which the Mises criterion was applied, but instead use the Tresca condition to predict whether or not the material yields at a point with state of stress, in ksi,

$$\mathbf{T} = \begin{pmatrix} 20 & 20 & -20 \\ 20 & 30 & 0 \\ -20 & 0 & 10 \end{pmatrix} \text{ ksi.} \qquad (9.53)$$

The uniaxial yield stress is $\sigma_Y = 52$ ksi.

Solution   The material yields at the point if the maximum shear stress is greater than the uniaxial shear yield stress. The principal stresses of **T** are $\sigma_1 = 50$, $\sigma_2 = 20$, $\sigma_3 = -10$ ksi. The uniaxial yield stress in shear is $\tau_Y = \sigma_Y/2 = 26$ ksi.

$$(\sigma_1 - \sigma_2)/2 = (50 - 20)/2 = 15 < \tau_Y;$$
$$(\sigma_2 - \sigma_3)/2 = (20 + 10)/2 = 15 < \tau_Y;$$
$$(\sigma_1 - \sigma_3)/2 = (50 + 10)/2 = 30 > \tau_Y.$$

The Tresca criterion predicts that the material yields at the point because τ_{max} obtained from the largest Mohr's circle is greater than τ_Y.   ∎

The yield surfaces for both the von Mises and the Tresca theories are convex. Drucker called such materials stable. Prager (1945) and Drucker (1949) proposed another yield criterion:

$$f\left(\frac{II_\mathbf{S}}{k^2}, \frac{III_\mathbf{S}}{k^3}\right) = II_\mathbf{S}\left(1 - c\frac{III_\mathbf{S}^2}{II_\mathbf{S}^3}\right) + k^2 = 0, \tag{9.54}$$

where c and k are material constants. This criterion generalizes the von Mises criterion [equation (9.46)] by including the effect of the hydrostatic stress, which appears in the term involving $III_\mathbf{S}$. Yielding in materials like rocks does depend on the hydrostatic stress.

9.7.3   Isotropic and Kinematic Hardening Yield Surfaces

If a yield surface is assumed, once a varying state of stress reaches the yield surface in principal stress space, it can only move on that surface or return to the inside of the convex surface. These are the only two allowable possibilities in plasticity models for the subsequent behavior of the loading path if the stress state is on the yield surface. Physically, in the case that the next states of stress are inside the yield surface, the material has been unloaded to an elastic state. Alternatively, the following states of stress can lie on the yield surface. In this case, the material remains plastically loaded. However, and this is the key idea, the yield surface is not fixed over time. The surface can move in principal stress space as the material is loaded plastically because of changes in the dislocation structure. The yield surface is defined by the criterion, $f\left(\frac{II_\mathbf{S}}{k^2}, \frac{III_\mathbf{S}}{k^3}\right) = 0$, which depends on parameters that can change as the material hardens.

In isotropic hardening the radius of the yield surface changes on subsequent loadings into the plastic region. In kinematic hardening, the center translates in principal stress space. Both phenomena can occur in the same material.

The translation of the yield surface in principal stress space because of kinematic hardening is represented by a new plasticity variable, **X**, with units of stress, which acts like a center to locate the current position of the yield surface in principal stress space. The yield criterion for kinematic hardening has the form $f(\mathbf{S} - \mathbf{X}) - k = 0$. This model can represent the ratchetting shown by some materials under cyclic loading and the Bauschinger effect. The center of each cycle is located by the variable, **X** (Fig. 9.33) which is called the backstress. The backstress tensor gives the state of stress at the center. In principal coordinates, the three diagonal elements can be viewed as the components of a vector in principal stress space. The backstress tensor represents the residual stress which builds up due to the clustering of dislocations of the same polarity (see Section 9.3 on dislocation pile-up). Each dislocation induces a stress field; if they have the same polarity, the stresses are additive. The sum is the backstress. If two dislocations are not of the same polarity, they annihilate each other when they meet.

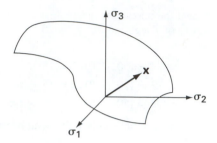

FIGURE 9.33   The backstress, **X**, locating the translation of a yield surface in principal stress space.

9.8   THREE-DIMENSIONAL PLASTICITY MODELS

Models that relate the total strain, both the elastic and plastic portions, to the stress are called *deformation theories*. Incremental models that relate a change in plastic strain, $d\epsilon^p$, to the change in stress and to the current stress are called *flow theories*. The latter are the most successful and are amenable to numerical methods. The deformation theories work best for proportional loading.

Three-dimensional generalizations of the uniaxial perfectly plastic and hardening models require the choice of a model for a yield surface in stress space to predict the stresses at which plastic deformation is initiated and the postulation of a system of evolution equations to predict how plastic deformation proceeds. Again, the evolution of the loading process can be recorded by plotting the stress components in stress space at each time to form a path as a function of time which represents the process. While the plasticity response to a loading process is path dependent, it is not rate dependent.

Most plasticity theories postulate a yield surface in stress space enclosing the elastic stress states. The yield surface may not be fixed in stress space. Hardening as the material is plastically loaded requires that the yield surface expands or translates in stress space. Different plastic flow laws are required if the hardening is isotropic or kinematic. Most laws assume that the material is linear elastic, that the hardening does not affect the elastic properties such as the elastic modulus, E, and the Poisson ratio, ν, and that the total strain can be partitioned into elastic and plastic portions, $\epsilon = \epsilon_e + \epsilon_p$.

Three-dimensional plasticity models are a current research topic. One difficult problem, for example, is to devise a model which correctly predicts the response to cyclic plastic loading of metals. The application of three-dimensional plasticity models usually requires sophisticated numerical techniques. This section is a brief introduction to this challenging problem.

9.8.1   Perfect Plasticity

A material is perfectly plastic if it does not work harden. Perfectly plasticity does not admit a one-to-one correspondence between stress and strain. Examination of a uni-axial, perfectly plastic, stress-strain curve shows that the plastic strain and the plastic strain rate are indeterminate after yielding. The relationship between the various rate of change strain components at a given plastic strain can be affected by varying the stress, as represented by the Lévy-Mises model.

The Lévy-Mises plasticity model appears to have been devised by analogy with the linear Newtonian viscous fluid equation that says that the shear stress is proportional to the rate of deformation. At any time in the plastic deformation, the deviatoric stress, **S**, is assumed to be the product of a viscosity-like term and the rate of strain tensor,

$$\mathbf{D} = \left(\frac{d\epsilon_{ij}}{dt} \right),$$

to predict the rate of change of the strain under the given state of stress, **S**,

$$s_{ij} = \frac{k}{\sqrt{-II_\mathbf{D}}} \mathbf{D}_{ij}, \tag{9.55}$$

where k is a constant determined by the yield surface, and $II_\mathbf{D}$ is the second invariant of **D**. In a further analogy with fluids, which are incompressible when the trace of the deformation tensor is zero, a material obeying this constitutive model can be said to be plastically incompressible since $Tr(\mathbf{D}) = Tr(\mathbf{S})\sqrt{-II_\mathbf{D}} = 0$.

The only yield surface consistent with this model is the von Mises. To see this, assume that the coordinates are principal so that **D** is diagonal. Since k is a constant, it is the same for any stress and rate of deformation. Because $Tr(\mathbf{D}) = 0$, the application of equation (9.44) to **D** implies that $II_\mathbf{D} = -0.5 Tr(\mathbf{D}^2)$. Squaring the Lévy-Mises equation (9.55) and taking the trace of the result implies that $-II_\mathbf{D} Tr(\mathbf{S}^2) = k^2 Tr(\mathbf{D}^2)$. Substitution of $II_\mathbf{D} = -0.5 Tr(\mathbf{D}^2)$ and simplifying shows that it must be true that $Tr(\mathbf{S}^2) = 2k^2$, which is the Mises criterion. Therefore, $k = \sigma_Y/\sqrt{3}$. Experiments have shown the Mises yield criterion to be approximately valid for aluminum, copper, nickel, iron, medium carbon steel (0.25–0.6% carbon), and strain hardened mild steel (< 0.25% carbon).

The model of Lévy and Mises is called rigid-plastic since it neglects the elastic strain and assumes that the material immediately begins to plastically deform, $\epsilon^e = 0$. This assumption is justified by the fact that the elastic portion of the strain is usually quite small compared to the plastic portion. Neglecting the elastic strain is common in applications which involve large plastic strains.

This is a rate-independent model since the multiplication of **D** by a constant, to change the deformation rates uniformly, does not affect the stress. Other changes in the stress which leave the stress on the yield surface change the ratio of the various pairs of strain rates. In other words, the evolution of the plastic strain is changed.

Exercise 9.8   Verify that $II_{a\mathbf{D}} = a^2 II_\mathbf{D}$ for any real number a. Substitute $a\mathbf{D}$ into the Lévy-Mises model, and verify that the stress is identical to that corresponding to **D**.

When the stress tensor is diagonal, that is, in principal coordinates, the Lévy-Mises model implies that the ratios of a deviatoric stress component to its corresponding strain rate component are all equal at a given time,

$$\frac{s_{11}}{\frac{d\epsilon_{11}}{dt}} = \frac{s_{22}}{\frac{d\epsilon_{22}}{dt}} = \frac{s_{33}}{\frac{d\epsilon_{33}}{dt}}.$$

But the value of the equal ratios may be different at different times. This idea was suggested first by Lévy in 1871 and later by von Mises in 1913.

EXAMPLE 9.18   In the uniaxial case, say in a rod with axial stress σ, the deviatoric tensor is diagonal with $s_{11} = \frac{2}{3}\sigma$, $s_{22} = -\frac{1}{3}\sigma$, and $s_{33} = -\frac{1}{3}\sigma$. Since the Lévy-Mises model implies that the ratio of the deviatoric stress to its corresponding strain rate is constant, the ratio of transverse to axial normal strain rates is $-1/2$. This strain rate ratio is reminiscent of the negative of the Poisson ratio. Recall that the Poisson ratio for an incompressible material is also 1/2. But the Lévy-Mises material is plastically incompressible, so that the analogy is complete. ∎

EXAMPLE 9.19   A rectangular plate is made of a rigid-plastic material which obeys the Lévy-Mises model. The uniaxial yield stress is σ_Y. The plate is biaxially loaded in the plane by σ_x and σ_y. Determine the relationship of σ_x and σ_y at yield.

Solution   Determine the deviatoric stress.

$$\mathbf{S} = \begin{pmatrix} \frac{2}{3}\sigma_x - \frac{1}{3}\sigma_y & 0 & 0 \\ 0 & -\frac{1}{3}\sigma_x + \frac{2}{3}\sigma_y & 0 \\ 0 & 0 & -\frac{1}{3}\sigma_x - \frac{1}{3}\sigma_y \end{pmatrix}.$$

The yield surface must be the von Mises surface so that $\frac{1}{2}Tr(\mathbf{S}^2) = \sigma_Y^2/3$. This implies that $\sigma_x^2 - \sigma_x\sigma_y + \sigma_y^2 = \sigma_Y^2$. This is exactly the biaxial distortion energy yield condition, discussed in Section 9.6.4, as expected. The strain rates are undetermined, but their ratio is the same as the ratio of the corresponding deviatoric components.   ∎

An incremental plasticity theory describes how small changes in variables such as the plastic strain depend on small changes in the stress and other variables. Such incremental theories are useful in stepwise numerical methods to predict plastic behavior. The strains are true strains, those measured with respect to the current configuration. From the current state, the increment of strain is computed. This model permits one to design an incremental numerical method to compute the current strains over increments of time or stress.

The rigid-plastic Lévy-Mises model was extended to elastic-plastic loadings by Prandtl and Reuss. The model was suggested for two-dimensional problems by L. Prandtl in 1924 and in 1930 by E. Reuss for general loadings. This is a flow or incremental theory that says that the change in plastic strain is governed by

$$d\epsilon_{ij}^p = s_{ij}d\lambda, \tag{9.56}$$

where s_{ij} are the entries of the deviatoric stress, $\mathbf{S}$, and $d\lambda$ is a positive function of proportionality. The multiplier, $d\lambda$, must be determined from the yield condition. Summing products of the components of the Prandtl-Reuss equation (equation 9.56) implies that the second invariant of the plastic strain increment tensor is a multiple of the second invariant of the deviatoric tensor.

$$II_{\mathbf{d\epsilon}^p} = (d\lambda)^2 II_{\mathbf{S}}.$$

In the case that the yield surface is defined by the von Mises yield condition, $II_{\mathbf{S}} = -k^2$. Therefore,

$$d\lambda = \frac{1}{k}\sqrt{-II_{\mathbf{d\epsilon}^p}},$$

so that

$$d\epsilon_{ij}^p = s_{ij}d\lambda = s_{ij}\frac{1}{k}\sqrt{-II_{\mathbf{d\epsilon}^p}}.$$

This is similar to the Lévy-Mises (equation 9.55) when the material is perfectly plastic, in which case k is a constant.

The Prandtl-Reuss model can be written as a system of coupled ordinary differential equations so that the evolution of the plastic strains can be determined numerically,

$$\frac{d\epsilon_{ij}^p}{dt} = s_{ij}\frac{d\lambda}{dt},$$

where

$$\frac{d\lambda}{dt} = \frac{1}{k}\sqrt{-II_{D^p}},$$

and $\mathbf{D}^p = (d\epsilon_{ij}^p/dt)$ is the rate of plastic strain tensor so that $II_{D^p} = -\frac{1}{2}\frac{d\epsilon_{ij}^p}{dt}\frac{d\epsilon_{ij}^p}{dt}$ where the index notation means the sum of the squares of each of the nine elements of $\mathbf{D}$. The evolution of the plastic strain for a given state of stress, $\mathbf{S}$, is

$$\frac{d\epsilon_{ij}^p}{dt} = \frac{s_{ij}}{k}\sqrt{-II_{D^p}}. \tag{9.57}$$

Because the term $d\epsilon_{ij}^p/dt$ appears on both sides of the equation, an implicit numerical method must be used. The simplest of these is the backward Euler algorithm. The Newton-Raphson method is usually applied to find the roots of the resulting equation.

9.8.2   Internal State Variables and Dissipation

Materials that exhibit hardening, and so are not perfectly plastic, require variables that describe the progress of the hardening. Hardening is due to the cumulative behavior of many dislocations in the material. It would be difficult to write equations to account for each dislocation individually, to measure the change in position of each dislocation individually, or to pinpoint dislocation multiplication at a particular site. Internal state variables are defined as those whose magnitude varies as the bulk dislocation behavior changes. These might be thought of as quantities that measure the consequences of the sum of the behavior of all the dislocations and other structural disturbances in the plastic material.

Definition   An internal state variable, used to describe the state of a system, measures the effects of a physical phenomenon which cannot itself be directly measured.

For example, the total strain at a position on a body can be directly measured by a strain gage; it is a state variable. However, the plastic strain cannot be measured directly. It can only be computed as the difference of the total strain and the elastic strain, both of which can be directly measured. The plastic strain might be thought of as an internal state variable. The components of the rate of plastic strain tensor, $\dot{\epsilon}_{ij}^p$, are often taken as a set of nine (or three if in principal coordinates) internal state variables. However, many contemporary plasticians doubt that the plastic strain should be taken to be a state variable because it is not unique for given conditions at a chosen time.

Other internal state variables are used to account for the loading history of the material. Odqvist in 1933 proposed the cumulative plastic strain, β, as a measure of the past history of plastic loading. The cumulative plastic strain is a function of time, t.

$$\beta(t) = \int_0^t [\frac{2}{3}\dot{\epsilon}_{ij}^p(\tau)\dot{\epsilon}_{ij}^p(\tau)]^{1/2} d\tau. \tag{9.58}$$

The expression under the integral is the square root of a sum of nine terms.

Each internal state variable has a corresponding generalized force, called an *affinity*, as in thermodynamics. An energy function, Ψ, is defined for plasticity which depends on both the internal and external state variables. The affinity, A_i, corresponding to the internal state variable, a_i, is by definition, $A_i = \partial\Psi/\partial a_i$. The plastic energy function is intentionally analogous to the elastic strain energy, U, for which $\partial U/\partial\epsilon_{ij} = \sigma_{ij}$. This definition makes the affinity, A_i, and the internal state variable, a_i, conjugate thermodynamic variables, in the same sense that, say, temperature and entropy are conjugate. This conjugacy relationship is not sufficient to account for the dissipation in plasticity. A set of equations must be produced that describe how the internal state variables evolve, that is, change values, over time.

In 1928, von Mises presented the idea of a plastic potential. A potential determines how the thermodynamic variables evolve in a plastically deforming material. As in the idea of conjugate variables determined by the energy function, the partial derivative of the potential function with respect to one variable gives information about its conjugate variable, but in this case determines the rate of change of the variable.

Definition A function, $\phi(A_1, \ldots, A_n)$ of the generalized forces is called a dissipation function, or a plastic potential if the rate of change of each of the internal state variables satisfies

$$\frac{da_i}{dt} = \dot{\lambda}\frac{\partial\phi}{\partial A_i},$$

where the proportionality coefficient $\dot{\lambda}$ is also a scalar function of the generalized forces. $\dot{\lambda}$ satisfies $\dot{\lambda} > 0$ if the yield function $f = 0$ and if the state of stress is elastic, $f < 0$, then $\dot{\lambda} = 0$. These evolution equations should be regarded as the plasticity constitutive model.

The rates of change of the plastic variables are given by a plasticity model, but the theory is still rate independent since the dissipation is a function of the current state variables and not of their time derivatives.

A material is plastically stable if the work done in producing an increment of plastic strain is positive, and in addition, if the work done in a loading-unloading cycle is positive. In this case, the dissipation potential, ϕ, is exactly the yield surface function, f. This was proved by D. C. Drucker in 1951 [8].

Definition A plasticity theory is called associated if the plastic potential and the yield surface function are the same. Otherwise, the plasticity theory is called nonassociated.

Associated theories have been successful in predicting the plastic behavior of metals. Other material, such as soils, require a nonassociated theory.

EXAMPLE 9.20 Determine the response of a perfectly plastic material that has the von Mises yield function, $f(\mathbf{S}) = s_{ij}s_{ij} - 2k^2$ as its plastic potential.

Solution The partial derivatives of the potential with respect to the deviatoric stress give the rate of change the plastic strain, which is assumed to be the conjugate internal state variable to the deviatoric stress in some theories.

$$\frac{d\epsilon_{ij}}{dt} = \dot{\lambda}\frac{\partial f}{\partial s_{ij}} = 2s_{ij}\dot{\lambda}.$$

In terms of increments, this result is the Lévy-Mises relation, if the factor 2 is absorbed into $d\lambda$,

$$d\epsilon_{ij} = s_{ij}d\lambda.$$

Previously, the Lévy-Mises equations were shown to imply the von Mises yield criterion. ∎

In this context of internal state variables and plastic potentials, three-dimensional models of isotropic hardening and of kinematic hardening can be written.

9.8.3 Isotropic Hardening

The Prandtl-Reuss isotropic hardening model is defined by an energy function, Ψ, which depends on both the plastic strain and the accumulated plastic strain, β. The von Mises yield criterion is $k = \sqrt{-II_S}$. Then $k = \sigma_Y/\sqrt{3}$ leads to the concept of an equivalent stress, often seen in commercial plasticity finite element programs. In the case that the yield surface is assumed to be that of the von Mises criterion, the Mises stress equation (9.42) written in terms of principal stresses can also be written in terms of the deviatoric stress.

$$\sigma_Y = \left[\frac{3}{2}s_{ij}s_{ij}\right]^{1/2} = \sqrt{\frac{3}{2}Tr(\mathbf{S}^2)}. \tag{9.59}$$

Exercise 9.9 Verify the expression (9.59) using (9.42). *Hint*: do the uniaxial case first.

Definition The equivalent von Mises stress is

$$\sigma_{eq} = \sqrt{\frac{3}{2}Tr(\mathbf{S}^2)} = \sqrt{-3II_S}, \tag{9.60}$$

where $\mathbf{S}$ is the deviatoric stress associated to the stress $\mathbf{T}$.

Therefore, the three-dimensional stress state is equivalent to the axial stress, σ_{eq}, a scalar. The yield surface for isotropic hardening is defined by

$$f(S, R, k) = \sigma_{eq} - R - \sigma_Y. \tag{9.61}$$

The scalar variable $R = \rho\frac{\partial\Psi}{\partial\beta}$ accounts for the change in yield surface radius as the plastic flow continues. The radius is a function of the internal state variable, the accumulated plastic strain β, so that $R = h(\beta)$ where $h(0) = 0$.

Isotropic hardening can be accounted for in the Prandtl-Reuss model by allowing the multiplier $d\lambda$, which depends on the flow surface, to vary with the plastic deformation. The precise value of $d\lambda$ is determined by the choice of an isotropic hardening yield surface and by the fact that a path in state space representing plastic deformation must remain on the yield surface.

The flow, or evolution, equations are obtained from the yield surface function, f. The plastic strain and the applied stress are conjugates as are the accumulated plastic strain, β, and the yield surface radius, R.

$$\frac{d\epsilon_{ij}^p}{dt} = \dot{\lambda}\frac{\partial f}{\partial s_{ij}}$$

$$\frac{d\beta}{dt} = -\dot{\lambda}\frac{\partial f}{\partial R}.$$

Therefore the incremental constitutive model corresponding to equation (9.61) is

$$d\epsilon_{ij}^p = \frac{3}{2}d\lambda\frac{s_{ij}}{\sigma_{eq}}$$

$$d\beta = d\lambda.$$

In plastic flow, the path must remain on the yield surface; therefore $\dot{f} = df/dt = 0$. This and equation (9.61) imply, since $R = h(\beta)$, that $(d\sigma_{eq}/dt) - h'(\beta)(d\beta/dt) = 0$. Then, solving for the increment $d\beta$ produces

$$d\lambda = d\beta = H(f)\frac{d\sigma_{eq}}{h'(\beta)}, \qquad (9.62)$$

where the Heaviside step function is $H(f) = 0$ if $f < 0$ and $H(f) = 1$ if $f \geq 0$. But equation (9.62) is only valid if the equivalent stress is increasing so that $d\sigma_{eq} > 0$. Therefore the Prandtl-Reuss flow equations for isotropic hardening can be expressed in terms of the applied stresses by

$$d\epsilon = d\epsilon^e + d\epsilon^p;$$

$$d\epsilon_{ij}^e = \frac{1+\nu}{E}ds_{ij} - \frac{\nu}{E}dTr(\mathbf{T})\delta_{ij}; \qquad (9.63)$$

$$d\epsilon_{ij}^p = \frac{3}{2}H(f)\frac{\langle d\sigma_{eq}\rangle}{h'(\beta)}\frac{s_{ij}}{\sigma_{eq}},$$

where δ_{ij} is the Kronecker delta and the second equation is Hooke's law.

To identify R, consider uniaxial loading with axial stress, σ, so that $\sigma = \sigma_{eq}$. Under uniaxial loading, the uniaxial strain hardening law $\epsilon^p = g(\sigma)$ implies that

$$d\epsilon^p = g'(\sigma)d\sigma,$$

where the prime indicates differentiation by σ, so that the tangent modulus at a point on the stress strain curve beyond yielding is

$$\frac{d\sigma}{d\epsilon^p} = \frac{1}{g'(\sigma)}.$$

By equation (9.62), it also follows that

$$\frac{d\sigma}{d\epsilon^p} = \frac{d\sigma}{d\beta}\frac{d\beta}{d\epsilon^p} = h'(\beta)\frac{d\beta}{d\epsilon^p} = h'(\epsilon^p),$$

so that $1/g'(\sigma) = h'(\epsilon^p)$. Therefore R can be computed as

$$R = h(\beta) = \int_0^\epsilon h'(\epsilon^p)d\epsilon^p$$
$$= \int_{\sigma_Y}^\sigma \frac{1}{g'(\sigma)}d\epsilon^p(\sigma) = \int_{\sigma_Y}^\sigma d\sigma = \sigma - \sigma_Y. \qquad (9.64)$$

The radius R is just the distance from the initial yield stress to the current yield stress (Fig. 9.34).

EXAMPLE 9.21   Let the uniaxial strain hardening be given by the power law, $g(\sigma) = \sigma_Y + K(\epsilon^p)^n$, where K and n are material constants. Then as a function of plastic strain, $R = K(\epsilon^p)^n$. ∎

Proportional Loading   In most cases, the solution for the plastic strain determined by a plasticity model cannot be obtained in closed form for a general multiaxial loading. A closed form solution is possible in one special case in which the material satisfies the Prandtl-Reuss isotropic hardening model with the von Mises yield surface.

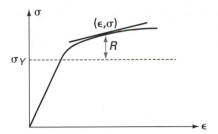

FIGURE 9.34   The terms σ_Y and R in the Prandtl-Reuss model.

The plastic strain integral produced by the model can be evaluated if the initial plastic strain is zero and if the loading is proportional and monotonically increasing. In other situations numerical methods are required.

The fact that the loading is proportional can be expressed in terms of a constant tensor, $\mathbf{T}'$. The loading is proportional if there is an increasing scalar function of time, $a(t)$, such that the stress is $\mathbf{T} = a(t)\mathbf{T}'$. The principal directions are unchanged as the load varies over time. A similar statement is true for the deviatoric stress, $\mathbf{S} = a(t)\mathbf{S}'$.

Under all the above assumptions, the differential for the plastic strain assumed in the Prandtl-Reuss model, equation (9.63), can be integrated to obtain the plastic strain as a function of the stress state. The yield surface is defined by $f = \sigma_{eq} - R - \sigma_Y$ so that in the plastic region, the Heaviside step function $H(f) = 1$. Since $\epsilon^P(0) = 0$ and σ_{eq} is positive, $h'(\epsilon^P) = 1/g'(\sigma)$, equation 9.63 implies that

$$\frac{d\epsilon^P}{dt} = \frac{3}{2}\frac{1}{h'(\beta)}\frac{s_{ij}}{\sigma_{eq}}\frac{d\sigma_{eq}}{dt};$$

(9.65)

$$\epsilon_{ij}^p = \frac{3}{2}\int_0^t g'(\sigma_{eq})\frac{d\sigma_{eq}}{\sigma_{eq}}s_{ij} = \frac{3}{2}\int_0^t g'(\sigma_{eq})\frac{d\sigma_{eq}}{\sigma_{eq}}a(t)s'_{ij}.$$

But $\sigma_{eq} = \sqrt{-3II_{\mathbf{S}}} = \sqrt{-3II_{a(t)\mathbf{S}'}} = a(t)\sqrt{II_{\mathbf{S}'}}$. Solve this expression for $a(t)$ and substitute into the previous integral to obtain

$$\epsilon_{ij}^p = \frac{3}{2}\frac{s'_{ij}}{\sqrt{-3II_{\mathbf{S}'}}}\int_0^t g'(\sigma_{eq})d\sigma_{eq},$$

(9.66)

where the terms independent of σ_{eq} have been removed from under the integral. The integral can now be simply evaluated to be $g(\sigma_{eq})$ at time t, the uniaxial hardening law. After multiplying the numerator and denominator by $a(t)$ and using $a(t)\sqrt{II_{s'}} = \sigma_{eq}$, the plastic strain is

$$\epsilon_{ij}^p = \frac{3}{2}\frac{g(\sigma_{eq})}{\sigma_{eq}}s_{ij}.$$

(9.67)

In the uniaxial case, this reduces to the uniaxial hardening law. For this reason, σ_{eq} is called the equivalent stress since it gives the same result as a uniaxial stress of the same magnitude.

EXAMPLE 9.22   A 316L steel rod of radius, r, fixed at one end is subjected to a tensile axial load, P, and a torque, M_t, which are increased from zero in a fixed ratio, $M_t = nP$. Determine the plastic strain on the surface of the rod in terms of the load, P. Assume a rigidly plastic material and the Prandtl-Reuss model with the Mises yield criterion. The hardening law is a power law with $\sigma_Y = 133$ MPa, $K_Y = 435$, and $M_Y = 4.5$.

Solution   As $P(t)$ increases with time, it plays the role of $a(t)$ in the preceding discussion. From elementary strength of materials, the uniform axial stress is $\sigma = P/\pi r^2$ and the shear stress at the surface is $\tau = M_t/.5\pi r^3 = 2n\sigma/r$. Therefore the loading is proportional and the preceding analysis applies. Compute the deviatoric, $\mathbf{S}$, for this state of stress. The equivalent stress, $\sigma_{eq} = (-3II_\mathbf{S})^{1/2} = (\sigma^2 + 3\tau^2)^{1/2}$. Therefore,

$$\sigma_{eq} = \sigma\sqrt{1 + 12\frac{n^2}{r^2}}.$$

The power law for the hardening model is

$$g(\sigma_{eq}) = \left(\frac{\sigma_{eq} - 133}{435}\right)^{4.5}.$$

The previous two values are substituted in

$$\epsilon_{ij}^p = \frac{3}{2}\frac{g(\sigma_{eq})}{\sigma_{eq}}s_{ij},$$

where

$$\mathbf{S} = \begin{pmatrix} \frac{2}{3}\sigma & \tau & 0 \\ \tau & -\frac{1}{3}\sigma & 0 \\ 0 & 0 & -\frac{1}{3}\sigma \end{pmatrix}. \qquad \blacksquare$$

9.8.4 Kinematic Hardening

Kinematic hardening is represented by a translation of the yield surface in stress space over time as the material is plastically loaded. The shape of the yield surface does not change if there is no isotropic hardening. Again, such a model can approximately predict the Bauschinger effect. The state of stress at the center in possibly non-principal coordinates is viewed as an internal state variable, the backstress tensor $\mathbf{X}$ (Fig. 9.33), which is conjugate to an unidentified internal state variable, the tensor α. The conjugacy is expressed in terms of an energy function Ψ for which $\mathbf{X} = -\partial\Psi/\partial\alpha$ and Ψ is assumed quadratic in α so that $d\mathbf{X}/d\alpha = c$, for a constant c. The general form of a kinematic hardening model for the yield surface is $f = f_y(\mathbf{S} - \mathbf{X}) - k$ for some function f_y which depends on the chosen yield condition. The deviatoric stress $\mathbf{S} = (s_{ij})$. W. Prager first proposed in 1955 a model which assumes that the center of yield surface changes linearly with the plastic strain, $\mathbf{X} = h(\epsilon_{ij}^p)$. To formulate this idea in terms of the yield surface $f = 0$, it is assumed that $f = |\mathbf{S} - \mathbf{X}| - k$. The model is associated. Then $d\epsilon_{ij}^p/dt = \dot{\lambda}(\partial f/\partial s_{ij}) = -\dot{\lambda}(\partial f/\partial X_{ij}) = d\alpha_{ij}/dt$ so that in this case the internal state variable α must be the plastic strain tensor.

The flow laws of the Prager kinematic model arise from the conjugate relations in which the change in location of the center of the yield surface is assumed to be proportional to the increment of plastic strain. The incremental form is

$$d\epsilon_{ij}^p = d\lambda\frac{\partial f}{\partial s_{ij}};$$
$$dX_{ij} = cd\epsilon_{ij}^p.$$

The term, $d\lambda$, is determined from the fact that $df = 0$ during plastic flow because the process path must remain on the yield surface. Therefore, using the index summation notation and s_{ij} for the elements of **S**,

$$df = \frac{\partial f}{\partial s_{ij}} ds_{ij} + \frac{\partial f}{\partial X_{ij}} dX_{ij}$$

$$= \frac{\partial f}{\partial s_{ij}} ds_{ij} - c \frac{\partial f}{\partial s_{ij}} d\epsilon_{ij}^p = \frac{\partial f}{\partial s_{ij}} ds_{ij} - c \frac{\partial f}{\partial s_{ij}} \frac{\partial f}{\partial s_{ij}} d\lambda = 0.$$

The last equation in this sequence is solved for $d\lambda$. The form of the flow equations depends on the particular choice of the yield surface f appropriate to the material of the body.

Exercise 9.10   From uniaxial loading, show that the constant c is the hardening modulus, the slope of the linearly increasing hardening curve, $h(\epsilon^p)$. The uniaxial curve has the form of Figure 9.5. Assume that the strain at yield is approximately zero. Then,

$$c = \frac{\sigma_u - \sigma_Y}{\epsilon_u},$$

where σ_u is the ultimate stress.

In kinematic hardening, if the translation of the yield surface obeys the Prager model, the reduction to plane stress does not do so. In fact, under this type of kinematic hardening, the Tresca yield surface in plane stress can deform as well as translate. On the other hand, a Mises yield surface in principal coordinates moves in the direction of the radial vector connecting the center of the yield surface to the current state of stress, **S**, on the yield surface. Therefore, Ziegler [22] in 1959 proposed a hardening evolution equation for the backstress, **X**, of the form

$$\dot{X}_{ij} = \dot{\lambda}(s_{ij} - X_{ij}), \tag{9.68}$$

where $\dot{\lambda} > 0$ is determined by the requirement that the current state of stress remain on the yield surface. The backstress evolution equation in the Prager model is

$$\dot{X}_{ij} = c\dot{\epsilon}_{ij}^p = c\dot{\lambda}\frac{\partial f}{\partial s_{ij}}.$$

9.9   HARDNESS TESTS

An indentation test is used to measure the hardness of a material. A ball or pointed indenter test instrument is pushed into the specimen by a specified, fixed force to produce an indentation in the specimen. The load is usually taken large enough that the stress is far into the plastic region and so the indentation is permanent. Heinrich R. Hertz developed the theory of contact stresses in elastic bodies, discussed in Chapter 4, Section 4.9, into a definition of the hardness of a material. Hertz's measure of hardness was obtained from the contact of a sphere of fixed radius onto a flat plate made of the material being tested. The use of a spherical indenter forces the contact region to be circular. He defined the hardness as the mean pressure on the elastic contact area.

The conventional hardness test measures greater deformations than those at the onset of plastic yielding and corresponds to a single point on a stress-strain curve. For example, a material with a lower yield stress will often show greater strain hardening and vice versa. Hence an early consideration in plasticity was to measure the flow stress at a relatively large value of strain, say, near the maximum load point, to characterize the tensile strength of a material. Thus early work on hardness testing was concerned, not with elastic contact, but with the mean pressure required to produce a substantial indentation within the material surface layer.

A common technique for comparing the material toughness of two materials is to compare the indentation depth a ball indenter produces in their surface under identical loads. The Meyer hardness of a material is the stress obtained by applying a transverse force to push a hard steel ball directly into a flat plate of the test material. The Meyer hardness, MH, is simply the force divided by the projected area of the indentation, that is, an analogous hardness to Hertz's elastic description;

$$MH = \frac{W}{\pi a^2} = \frac{4W}{\pi d^2}.$$

The diameter, $d = 2a$, of the plastic indentation depends on the radius, R, of the ball, the moduli of the ball and the test material, and the applied force. Meyer proposed this measure of hardness in 1908. Frequently, a plastic indentation corresponding to $a/R = 0.375$ is taken as the standard to measure the Meyer hardness.

The Brinell hardness, proposed by J. A. Brinell in 1900, attempts to account for strain hardening occurring while the material is plastically deformed during an indentation test. Rather than dividing the load by the projected area of the indentation as in the Meyer hardness, Brinell divided by the surface area of the residual indentation assumed to be a spherical cap. In a common standard test machine, the Brinell hardness is measured by obtaining the diameter of the indentation due to pressing a hardened steel sphere of diameter 10 mm into a plate of the test material at a load equivalent to 3,000 kg. This diameter D of the sphere is generally required to be in the range, $0.25 \leq d/D \leq 0.5$, to obtain the Brinell hardness. The surface area, A_s, of a spherical cap is calculated using the calculus expression for the surface area of a volume of revolution of a curve, $y(x)$, about an axis. Let x be the distance measured from the indenter sphere center along the diameter that intersects the center of the indentation. Let $y(x)$ be the perpendicular distance from that central diameter to the edge of the indentation at height, x. For the spherical indentation of depth, h, $y(x) = \sqrt{R^2 - x^2}$, the increment of arclength is $\sqrt{1 + y'^2}dx$, and so $A_s = \int_{R-h}^{R} 2\pi y \sqrt{1 + y'^2}dx = \pi Dh$. The Brinell hardness number is

$$BHN = \frac{P}{A_s} = \frac{P}{\pi Dh} = \frac{P}{\pi \frac{D^2}{2}\left[1 - \sqrt{1 - \left(\frac{d}{D}\right)^2}\right]}, \quad (9.69)$$

where $h = 0.5D\left[1 - \sqrt{1 - \left(\frac{d}{D}\right)^2}\right]$ by problem 45. The Brinell and Meyer hardness numbers are usually given in kg/mm^2. The result of an indentation test on a silicon crystal is shown in Fig. 9.35.

A Rockwell testing machine measures hardness by assigning a number to the penetration depth, rather than the diameter, of the indentation. The indenter is either a small steel sphere for relatively soft materials or a diamond cone for harder materials. The Rockwell hardness, R_B, is not related to the yield strength, but the test is commonly used, for one reason, because it involves a convenient pre-load reference penetration depth from which the greater penetration under full load is measured. An empirical relation between the Rockwell hardness employing a 1/16 inch diameter ball and the Brinell hardness is

$$BHN = \frac{7300}{130 - R_B}.$$

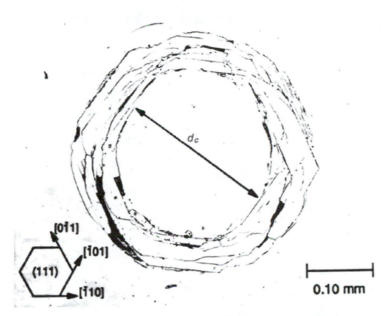

FIGURE 9.35   Crystallographically modified, supposedly Hertzian-type, multiple ring cracks resulting from a steel ball indentation on a (111) silicon crystal surface.

EXAMPLE 9.23   A steel plate is indented under a force of 3,000 kg by a 10-mm ball. The diameter of the plate indentation is 4 mm. Compute the Brinell hardness number and the Meyer hardness for this steel.

Solution $d = 4$ and $D = 10$, and $P = 3,000$ kg, so that

$$BHN = \frac{P}{\pi \frac{D^2}{2}\left[1 - \sqrt{1 - \left(\frac{d}{D}\right)^2}\right]} = \frac{3,000}{\pi \frac{10^2}{2}\left[1 - \sqrt{1 - .4^2}\right]} = 229 \text{ kg/mm}^2.$$

The Meyer hardness is

$$MH = \frac{3,000}{\pi 2^2} = 239 \text{ kg/mm}^2. \qquad \blacksquare$$

The Brinell hardness number is relatively constant at lower loads, but then decreases appreciably at higher loads. This implies that the Brinell hardness is not in fact a material constant. However, the Brinell hardness number remains in use because it seems to be proportional to the ultimate tensile strength in some metals. For example, in many steels, the ratio of the ultimate tensile stress to the Brinell hardness, in terms of kg/mm², is about 1/3. A common rule of thumb is that, for steel, the tensile strength in psi is about 500 times the BHN in kg/mm²; although some use the factor 470.

On the other hand, the Meyer hardness is relatively independent of the load used to impress the indenting sphere into the test material. Meyer suggested, in what has become known as Meyer's law, that the load, W, is related to the indentation diameter, d, by

$$W = kd^n, \qquad (9.70)$$

where n is a material constant. Meyer found that n is about 2.5 in fully annealed metals and near 2 if the metal is fully work-hardened. The coefficient, k, depends on the ball indenter diameter, D. Meyer's experiments indicated that $k = A/D^{n-2}$, for some constant A. This means that Meyer's law can be written in a form similar to the Meyer hardness,

$$\frac{W}{d^2} = A \left(\frac{d}{D} \right)^{n-2}. \tag{9.71}$$

The Meyer hardness is independent of the load, W, if the ball indenter diameter, D, is adjusted to keep the left side of equation (9.71) constant. Meyer's law, which is an empirical relation, is only valid for $d/D \geq 0.1$.

The need to investigate the behavior of the hardness numbers in terms of the applied load or ball indenter diameter leads naturally to a discussion of the curve that can be obtained if a ball of fixed diameter is used to indent a plate under different loads. The graph of the load versus the indentation diameter is called a continuous indentation test [4].

In Meyer's law, equation (9.71), the left side is $\pi/4$ times the mean stress. If the ratio d/D is thought of as proportional to the true strain, then Meyer's law is analogous to the classical plasticity hardening curve (equation 9.21),

$$\sigma = K(\epsilon_p)^{n'}. \tag{9.72}$$

Therefore $n - 2$ is the hardening exponent. In the fully annealed case, it is 0.5, as assumed in some hardening laws (Chapter 3, Section 3.1.1). In the fully work-hardened case, it is zero, that is, perfectly plastic. Tabor [19] predicted that the equivalent compressive strain $\epsilon = 0.2 d/D$ and verified that the plastic stress versus true strain curve in compression can be obtained from the Meyer test for annealed copper and mild steel, but the estimate from the Meyer test does not hold for all metals.

The mean hardness contact pressure under elastic loading, from Hertz, is

$$\sigma_H = \bar{\sigma}_3 = \frac{P}{\pi a^2} = \frac{4}{3\pi} \left[\frac{(1 - v_1^2)}{E_1} + \frac{(1 - v_2^2)}{E_2} \right]^{-1} \left(\frac{d}{D} \right), \tag{9.73}$$

where d and D are the contact area diameter and the ball diameter, respectively. The material constants E_1, v_1 and E_2, v_2 are those of the indenter and the specimen, respectively. The factor d/D corresponds to elastic strain [7].

It is possible to relate the Meyer hardness number to the grain size by an expression analogous to the Hall-Petch equation, $\sigma = \sigma_0 + k l^{-1/2}$, where l is the grain diameter and k and σ_0 are constants. The expression for the Meyer hardness in terms of grain diameter is then

$$MH = MH_0 + K_H l^{-1/2},$$

where MH_0 and K_H are positive constants. This relation demonstrates that materials with smaller grain sizes are harder, as is known for ceramics.

The Brinell and Rockwell tests, with their larger indenters, are often used to determine an average hardness of a material. Because the hardness is averaged over a larger region, it is plausible that the measurement may also be related to the overall strength of the material. The Vickers and the Knoop tests use much smaller area indenters and may therefore be used to determine the microhardness. With careful experimentation, it may be possible to measure the hardness, for example, of the precipitates in a steel.

The Vickers test and Brinell test produce similar readings for homogeneous materials. The indenter used in the Vickers test is a square-based pyramid usually made of diamond. The angle between the opposite faces is commonly $\alpha = 136°$. The length, d, of the diagonal of the square-shaped indentation is measured under a load, P, which may range from 1 to 120 kg. A microscope may have to be used to make the measurement. The Vickers number is the load divided by the surface area of the indentation,

$$HV = \frac{2P}{d^2} \sin(0.5\alpha) \quad \text{kg/mm}^2. \tag{9.74}$$

In contrast to indentations by a sphere, the indentations by the Vickers pyramid are geometrically similar irrespective of their depth.

Exercise 9.11   Verify that the choice of the angle, $\alpha = 136°$, guarantees that the depth h of the indentation is 1/7 of the diagonal, d.

The load in the Vickers test is smaller than used in the Brinell test. Therefore the surface must be prepared by polishing or other technique so that the diagonal of the small indentation can be measured. Care must be taken that the surface preparation does not affect the results. The large range of loads permitted in the Vickers test means that more different types of materials can be tested and compared by this hardness test. The Vickers test is commonly used for the brittle ceramics because its smaller load requirement makes the ceramic material less likely to shatter.

Some approximate values for the Vickers hardness of the micro-constituents of steels are pearlite 250 kg/mm², bainite 450 kg/mm², martensite 800 kg/mm², and cementite 1,100 kg/mm². Recall that different heat treatments are needed to create most of these microconstituents. These values compare to the Vickers hardness of pure iron, which is about 70 kg/mm².

TABLE 9.3   Vickers Hardness for Some Ceramics

Material	Vickers hardness (kg/mm²)
NaCl	21
Fused SiO_2	540
MgO	660
Al_2O_3	2,100–2,370
SiC	2,400–3,300
Diamond	8,000–9,000

Some researchers have conjectured that the yield stress along slip planes in ceramics is one-third of the Vickers micro-hardness. The compressive strength in ceramics without many flaws ranges from about one-half to three-fourths of this yield stress. More research is needed on this issue, especially because cracking is generally associated with micro-hardness indentations in ceramics. Such cracking has been analyzed based on indentation fracture mechanics to describe, for example, the ring cracking shown in Figure 9.35 for silicon and a consequent $c^{-3/2}$ dependence on load for the crack diameter, c [18].

A continuous indentation test gives a load-deformation relation locally. The small Vickers indenter permits testing of non-homogeneous materials without getting averaged results as when one tests a bulk specimen. The indentation strain hardens the material locally. In an x-ray topographic image such as Figure 9.36 of an MgO cleavage surface on which hardness indentations have been placed, the dark regions surrounding each indentation are due to the induced strain fields of dislocations. Separate indentations are made within the grain and on the boundaries to investigate their relative hardness. The subgrains in this crystal are outlined by light lines where the individual subgrain images are separated or dark lines where they overlap. Cleavage steps are visible as dark lines due to localized strains across the subgrains.

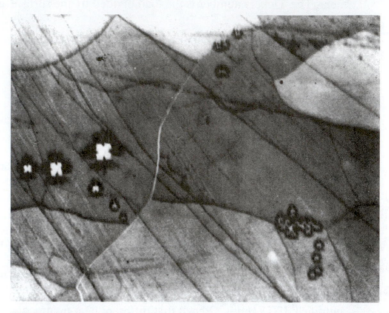

FIGURE 9.36   An x-ray reflection topograph of microindentation hardness impressions within the subgrain volumes and on the subgrain boundaries on the cleavage surface of an MgO crystal. (Courtesy of A. C. Raghuram.)

The local strain hardening near an indentation is tested by making small indentations around the initial larger indentation (Fig. 9.37).

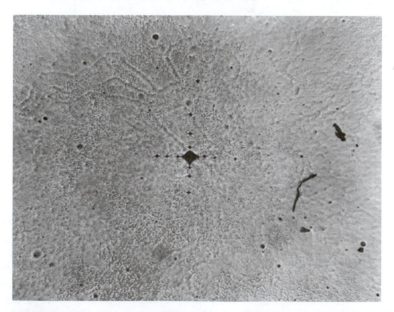

FIGURE 9.37   Tiny diamond pyramid indentations put along the vertical and horizontal diagonal axes outside on a large central indentation on a (001) copper crystal surface. (Courtesy of K. C. Yoo and B. Roessler.)

The slip and cracking behavior near an indentation aligned along the crystallographic axes is striking for MgO (Figure 9.38a). The dislocations formed by the indentation create troughs in the ⟨100⟩ directions as well as usually difficult cleavage cracking on the {110} planes. Evidence of significant elastic recovery is seen in the distorted shape of the indentation. A dislocation model description for the indentation is in Figure 9.38b.

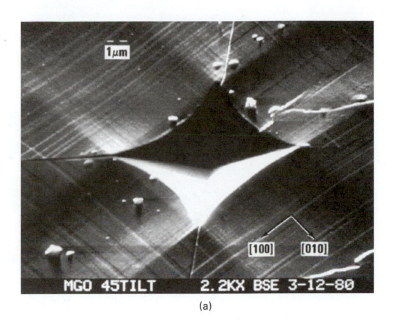

(a)

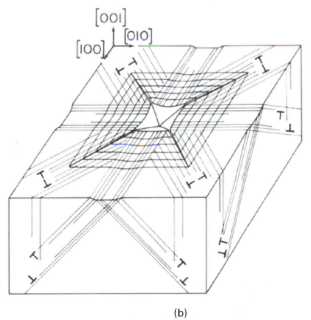

(b)

FIGURE 9.38   (a) Scanning electron micrograph of a diamond pyramid indentation aligned along crystallographic axes on the cleavage surface of an MgO crystal. (Courtesy of C. Cm. Wu.) (b) Dislocation model of the deformation field and induced cleavage cracking at an aligned diamond pyramid indentation on an MgO crystal surface, involving ⟨100⟩{110} slip systems.

Plastic anisotropic behavior is tested using a knife-like Knoop indenter, which produces an elongated diamond-shaped indentation. In Figure 9.39, relatively large Knoop indentations test for plastic anisotropy across the grains, while small diamond pyramid Vickers indentations, within individual grains, test for the local strength differences. The fact that the Vickers indentations are similar indicates that different grains have essentially the same plastic strength.

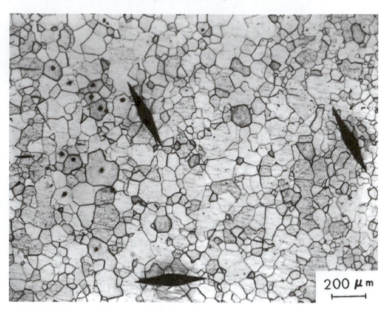

200 μm

FIGURE 9.39 Polycrystalline grain structure in BCC tantalum subjected to Vickers and Knoop indentations. (Courtesy of V. Ramachandran.)

If l is the length of the long diagonal of the elongated diamond-shaped Knoop indentation and if A the projected area of the diamond indentation, then the Knoop hardness is

$$HK = \frac{P}{A} = 14.2\frac{P}{l^2},\tag{9.75}$$

which is again usually measured in kg/mm^2. The constant 14.2 may vary with the exact design of the indenter.

Hardness tests other than those using an indenter have been devised. A scratch test uses a pointed tool to attempt to scratch the surface of the test specimen. Tabor found in 1956 that the tool scratches as long as its hardness is at least 20% greater than that of the specimen. The shape of the point of the tool does not affect the result. Mohs in 1822 had produced a scratch test for minerals. He chose 10 minerals as his scale in ascending hardness: talc, gypsum, calcite, fluorite, apatite, orthoclase, quartz, topaz, corundum, and diamond. These were assigned Mohs' numbers, M_0 equals 1 through 10. If the specimen is scratched by one mineral but not by the next one on the scale, then it is assigned a Mohs' hardness equal to that of the mineral that did not cause a scratch. Because scratching and indenting both involve plastic deformation of the specimen, there is a relation between the two types of tests. The least hard of Mohs' minerals was talc. If this is assigned the Vickers hardness, HV_1 in units of kg/mm^2, then the Vickers hardness of a specimen is given by the relationship

$$HV = HV_1(1.6)^{M_o-1}.\tag{9.76}$$

Each of the minerals used in the Mohs scale has Vickers hardness 60% greater than the one next lower on the scale, except for diamond which does not satisfy the relationship (9.76). For example, the Vickers hardness of quartz is 1,100 kg/mm^2, while that of diamond is 8,000 kg/mm^2 (refer to Table 9.3).

9.10 VISCOPLASTICITY

Plasticity, by convention, models only rate-independent behavior. If the material response depends on the rate of application of the load or some other rate-dependent parameter, its behavior is called viscoplastic. Viscoplastic behavior is a nonequilibrium response of the material. Creep of metals under a constant load, for example, is a viscoplastic behavior (see Chapter 12 for a detailed discussion of creep). In reality, all material response is rate-dependent to some extent. Plasticity theories can only provide an approximation of material response. The yield stress, for example, can depend on the rate that the load is applied. The path that a viscoelastic material follows in principal stress space need not remain on the yield surface as it must in plastic behavior. Plasticity is a special, limiting case of viscoplasticity. Since the 1980s, much emphasis has been placed on developing design models to represent viscoplastic behavior.

It is often assumed, again, that the elastic component, ϵ_e, of the strain is not affected by viscoplastic behavior. Experiment shows that if the material is unloaded at a high rate, the elastic strain is not affected by the viscoplastic strain. Therefore the strain can be decomposed into an elastic and a viscoplastic component, $\epsilon = \epsilon_e + \epsilon_{vp}$. In the 1950s, as a generalization of the separation of strain into elastic and plastic components, it was proposed that the strain separates into three components so that $\epsilon = \epsilon_e + \epsilon_p + \epsilon_{vp}$. This model is no longer thought to be valid with the exception of a few materials like stainless steel in which creep is very slow. However, some experiments indicate that, in this approach, the hypothesized plastic strain and viscoplastic strain components are coupled. So the separation of the strain into three components is probably not valid even in this case.

The viscoplastic hardening curves have the same qualitative shape as those of plasticity, but are strain rate dependent. Typically, if the strain rate is higher, more stress is required to produce the same strain. Therefore if the uniaxial hardening stress-strain curves at two different constant strain rates are plotted on the same graph, the one at the higher rate will lie above the one with the lower strain rate.

Uniaxial models with particular application to high strain rate loadings which distinguish between BCC and FCC materials have been developed by Zerilli and Armstrong [21]. The plastic stress-strain relation is for negligible stress recovery, is

$$\sigma = \sigma_Y + K\epsilon^{1/2},$$

where K is the strain hardening coefficient and σ_Y is the yield stress. The yield stress for BCC materials such as iron and steel depends on the absolute temperature, T, strain rate, and the average grain diameter, l.

$$\sigma_Y = B\exp[-\beta T] + \sigma_G + kl^{-1/2},$$

where $\beta = \beta_0 - \beta_1 \ln\dot{\epsilon}$ and B, σ_G, k, β_0, and β_1 are material constants. The athermal stress, σ_G, depends on the solutes in the metal and the initial dislocation density. In contrast, the yield stress for an FCC material such as copper varies according to a thermally dependent strain hardening. For the FCC case,

$$\sigma_Y = \sigma_G + kl^{-1/2},$$

where, in the preceding stress-strain relation, $K = B_0 \exp[-\alpha T]$ and $\alpha = \alpha_0 - \alpha_1 \ln \dot{\epsilon}$ for material constants, B_0, α_0, and α_1. At large strains, the $\epsilon^{1/2}$ term is replaced by $\epsilon_r(1 - e^{-\epsilon/\epsilon_r})$, where ϵ_r is a constant [3]. This term is approximately $\epsilon^{1/2}$ when $\epsilon/\epsilon_r < 1.0$. Because the coefficients in this model depend on strain rate, this is a viscoplasticity model. Otherwise, it has the same form as the Ramberg-Osgood model.

More recent evolution models have suggested that metal creep occurs with the application of any load (e.g., [12]). At small stresses, the stress-strain curve appears to be linear elastic only because the creep strain is very small. In this case, the idea of a distinct plastic yield surface is dropped. The apparent yield occurs when the plastic component of strain becomes much larger than the elastic component. In the classical model, a distinct yield surface exists because it is assumed that a critical stress must be attained to break atomic bonds and allow slipping of neighboring atomic planes in the lattice. In these recent models, plastic strain is assumed to occur because of changes in the dislocation structure and associated slipping. The changes are initiated with the application of any size load. Such models remain to be fully validated.

The viscoplastic response of metals depends on the behavior of the dislocations in the crystal lattice. The application of load or variations of the ambient temperature in general induce changes in the dislocation structure of metals. Such changes cause energy losses so that the system is dissipative rather than conservative as in elastic loading. A primary task in modeling viscoplastic behavior is correctly representing this dissipation.

A metal is not homogeneous. For example, the dislocations are not evenly distributed in the material. Often internal state variables are used to model the material microstructure variation during a nonequilibrium process in many contemporary thermodynamically based viscoplasticity models. Internal variables are sometimes called hidden state variables because they cannot be directly measured.

A typical set of internal variables includes the backstress tensor, B_{ij}, and the scalars the elastic yield stress, Y, and the drag stress, D. The backstress tensor, $\boldsymbol{B}$ (called $\mathbf{X}$ above), accounts for kinematic hardening. It measures stiffness change as the material plastically deforms and is caused by the non-homogeneous nature of the dislocation structures. The backstress, introduced by Nowick and Machlin in 1947, represents the inner stress field induced by stuck dislocations. The scalars drag stress, D, and the elastic yield stress, Y, account for isotropic hardening due to the accumulation of plastic deformation. The drag stress, a scalar strength parameter, is proportional to the square root of the dislocation density. In other theories, the overstress, $\boldsymbol{\sigma} - \boldsymbol{B}$, has been given the primary role of an independent internal variable. Another possibility is to represent the dislocation structure with a finite or infinite set of dislocation arrangement tensors, $\Gamma^{(k)}$.

Because viscoplastic response is a nonequilibrium behavior, it is modeled by time dependent differential equations, called evolution equations. The evolution equations for the response in the nonequilibrium thermodynamics model give the rate of change of the state variables. As an example, the state variables are the strain, ϵ, the backstress tensor, $\boldsymbol{B}$, the drag strength, D, the yield strength, Y, and the entropy, η. The temperature is θ, and the deviatoric stress is $\boldsymbol{S}$. The plastic strain tensor is denoted, $\boldsymbol{\epsilon}^p$.

A viscoplastic model requires an empirical description of the variation of the plastic strain with time. This is usually expressed in terms of a norm of the plastic strain tensor. The norm of the plastic strain tensor is a number defined by

$$||\dot{\boldsymbol{\epsilon}}^p|| = \sqrt{\frac{2}{3}\sum_{i=1}^{3}\sum_{j=1}^{3}(\dot{\epsilon}_{ij}^p \dot{\epsilon}_{ij}^p)}. \tag{9.77}$$

The description also requires a special mathematical function, $\langle x \rangle$, which is always positive,

$$\langle x \rangle = \begin{cases} x & \text{if } x \geq 0 \\ 0 & \text{if } x < 0. \end{cases} \tag{9.78}$$

The evolution of plastic strain is

$$\dot{\epsilon}_{ij}^p = ||\dot{\boldsymbol{\epsilon}}^p|| \frac{1}{2} \frac{S_{ij} - B_{ij}}{||\boldsymbol{S} - \boldsymbol{B}||}, \tag{9.79}$$

where the norm of the plastic strain rate has an empirically obtained form as a function of the external load and of the internal variables, such as

$$||\dot{\boldsymbol{\epsilon}}^p|| = f(\theta) A \left(\left\langle \frac{||\boldsymbol{S} - \boldsymbol{B}|| - Y}{D} \right\rangle \right)^n, \tag{9.80}$$

where $n > 0$, $A > 0$ are material constants and $||\boldsymbol{S} - \boldsymbol{B}|| - Y \leq D$. Typically $f(\theta) = \exp(-Q/R\theta)$, the Arrhenius form.

In the Freed-Walker model [10] proposed in 1993, the evolution of the backstress, $\boldsymbol{B}$, is given by

$$\dot{B}_{ij} = 2H_B \left(\dot{\epsilon}_{ij}^p - \frac{B_{ij}}{2L} ||\dot{\boldsymbol{\epsilon}}^p|| - f(\theta) R_b \frac{B_{ij}}{2||\boldsymbol{B}||} \right), \tag{9.81}$$

where $H_B > 0$, L is the limiting backstress, and $R_b > 0$ are moduli and $f(\theta)$ accounts for thermal recovery. The drag strength evolution is given by

$$\dot{D} = h \left(||\dot{\boldsymbol{\epsilon}}^p|| - \frac{D}{l_d} ||\dot{\boldsymbol{\epsilon}}^p|| + f(\theta) r_d \right), \tag{9.82}$$

where $h > 0$, $l_d > 0$, and $r_d \geq 0$. The evolution of the yield strength is

$$\dot{Y} = h_y \left[||\dot{\boldsymbol{\epsilon}}^p|| - \frac{(Y - Y_o)}{l_y} ||\dot{\boldsymbol{\epsilon}}^p|| - f(\theta) r_y \right], \tag{9.83}$$

where Y_o is the annealed yield strength and $h_y > 0$, $l_y > 0$, and $r_y > 0$. Here Y_0, the annealed yield, could be written in terms of the Hall-Petch empirical relation to the grain size. In these three evolution equations, the first term is strain hardening, the second is strain softening or dynamic recovery, and the third is thermal recovery, which is independent of the mechanical dissipation. The thermal softening (recovery) moduli, R_b, r_d, and r_y, are functions of state. This type of model is the subject of current research.

PROBLEMS

9.1   A material has the stress-strain curve shown. Compute

(a) E,

(b) σ_{PL},

(c) $\sigma_{0.2\%}$, and

(d) ϵ_p and ϵ_e for a strain of $0.003 = 3,000\ \mu$.

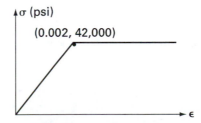

9.2 Compute, but do not estimate, the 0.2% yield stress on the following stress-strain curve. Ans. 37,500 psi.

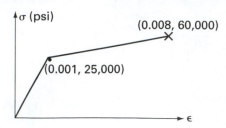

9.3 A material has the stress-strain curve shown. Compute

(a) E,

(b) σ_{PL},

(c) $\sigma_{0.2\%}$, and

(d) ϵ_p and ϵ_e for a stress of 495 MPa.

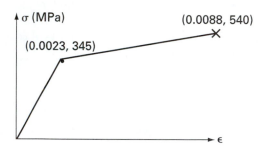

9.4 A material has the following stress-strain curve. Determine

(a) the elastic modulus,

(b) the proportional limit,

(c) the 0.2% yield strength,

(d) the elastic strain and the plastic strain if an axially loaded member made of the material is loaded to 40 ksi.

Ans. (a) 12×10^6 psi, (b) 24,000 psi, (c) 41,143 psi, (d) $\epsilon_p = 1,870\mu$, $\epsilon_e = 3,330\mu$.

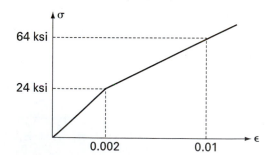

9.5 A material has the stress-strain curve shown here. Compute

(a) E,

(b) σ_{PL},

(c) $\sigma_{0.2\%}$, and

(d) ϵ_p and ϵ_e for a stress of 70,000 psi.

Ans. (a) 17.5×10^6 psi, (b) 28,000 psi, (c) 48,293 psi, (d) $\epsilon_e = 0.004$; $\epsilon_p = 0.006$.

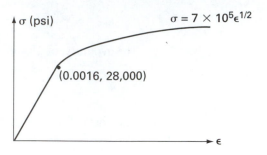

9.6 The uniaxial stress-strain curve for a metal is qualitatively the same as in the curve in Problem 5 except that the transition from linear occurs at (0.003, 240). The curve for the hardening portion is of the form, $\sigma = k\epsilon^{0.55}$. The stress is measured in MPa.

(a) Determine the material constant, k.

(b) Determine the elastic and plastic components of the total strain when the stress reaches 260 MPa.

(c) How much does a 30 cm long axially loaded rod made of this material elongate when the stress reaches 260 MPa? How much of this elongation is recovered when the load is removed?

9.7 Consider a single crystal of aluminum oriented so that a tensile stress is applied along a [001] direction. If slip occurs on a (111) plane and in a [$\bar{1}$01] direction, and is initiated at an applied tensile stress of 164 psi, compute the critical resolved shear stress.

9.8 Suppose that a single crystal of an FCC metal is oriented so that the [101] direction is parallel to an applied stress of 3,000 psi. Calculate the resolved shear stress acting on the (111) slip plane in the [0$\bar{1}$1] direction.

9.9 A dislocation begins to slip in the (101)[$\bar{1}$11] slip system when a stress of 8,000 psi is applied in the [021] direction of a BCC crystal.

(a) Calculate the critical resolved shear stress.

(b) Assuming the same critical resolved shear stress, will a dislocation in the (211)[1$\bar{1}\bar{1}$] slip system move under the same 8,000 psi stress? Support your answer with a calculation.

9.10 The critical resolved shear stress for an α-iron is approximately 10 MPa on the (1$\bar{1}$1)[110] slip system. Determine the approximate tensile stress in the [211] direction required for slip on the (1$\bar{1}$1)[110] slip system.

9.11 A titanium has Hall-Petch constants $\sigma_o = 60$ ksi and $k = 20$ psi/in.$^{1/2}$. A designer wishes to specify a heat treatment to produce a yield stress of 100 ksi. What average diameter grain sizes should be produced?

9.12 The radius of a steel rod is $R = 2$ cm. The yield stress is $\tau_Y = 110$ MPa. Determine the torque required so that all portions of the body further than $0.6R$ from the axis of symmetry are plastically loaded. Assume the material is perfectly plastic.

9.13 A hollow shaft made of aluminum has inner radius 2 cm and outer radius 3 cm. The yield stress is $\tau_Y = 40$ MPa. Assume the material is perfectly plastic. Determine the minimum torque that must be applied to a face to make it fully plastic.

9.14 Determine the maximum torque, T_y, that can be applied so that a steel rod ($\tau_Y = 90$ MPa) of radius 1.2 cm remains fully elastic.

9.15 Show that the minimal torque required to bring each point to a perfectly plastic state of stress is $T_{fp} = \frac{4}{3}T_y$ for a circular rod.

9.16 Determine the torque which must be applied and then released to produce a permanent twist of $\theta = \pi/9$ rad with respect to the ends. The rod is steel ($G = 12 \times 10^6$ psi, $\tau_Y = 30$ ksi), the radius is 0.75 inches, and the length is 12 inches. Assume the material is perfectly plastic.

9.17 A steel beam has a cross section that is a 2 cm square. The yield stress is $\sigma_Y = 200$ MPa. Determine the internal moment required so that all portions of the body a distance from the centroidal axis greater than 0.8 cm are plastically loaded. Assume the material is perfectly plastic.

9.18 A beam made of aluminum has width 2 cm and height 3 cm. The yield stress is $\sigma_Y = 180$ MPa. Determine the minimum internal moment which must be applied to a face to make it fully plastic. Assume the material is perfectly plastic.

9.19 Determine the maximum internal moment, M_Y, which can be applied so that a beam remains fully elastic. Show that $M_{fp} = 1.5M_Y$.

9.20 A steel beam ($E = 210$ GPa, $\sigma_Y = 150$ MPa) is plastically loaded by a couple $M = 6,600$ Nm. The height of the rectangular cross section of the beam is $h = 8$ cm and the thickness is $t = 3$ cm. Determine the radius of curvature of the midplane of the beam after the couple is released. Assume the material is perfectly plastic.

9.21 A steel beam ($E = 210$ GPa, $\sigma_Y = 150$ MPa) is plastically loaded by a bending couple $M = 3,000$ Nm about an axis parallel to the thickness dimension. The height of the rectangular cross section of the beam is $h = 5$ cm and the thickness is $t = 4$ cm. Assume the material is perfectly plastic.

(a) Determine the shortest distance y_e from the neutral axis at which the material is plastically loaded after the moment is applied.

(b) Determine the radius of curvature of the midplane of the beam after the couple is released.

9.22 A steel beam ($E = 210$ GPa, $\sigma_Y = 150$ MPa) is plastically loaded by a couple M. The height of the rectangular cross section of the beam is $h = 8$ cm and the thickness is $t = 3$ cm. The radius of curvature of the midplane of the beam after the couple is released is to be 40 meters. Assume the material is perfectly plastic. Determine the required bending load M. *Hint*: Solve two equations simultaneously and get y_e as well as M.

9.23 Determine the permanent axial strain in a rod of radius 2 cm under a load of $F = 500$ kN. The material obeys the Ramberg-Osgood isotropic hardening constitutive model with initial yield stress $\sigma_Y = 140$ MPa, hardening coefficient $M = 4.5$, and $K = 400$ MPa.

9.24 A rod of radius, $r = 2$ cm, is to have its radius permanently reduced 5% by stretching under a uniform axial load. Determine the required load if

(a) the material is perfectly plastic with $E = 70$ GPa, $\nu = 0.33$, $\sigma_Y = 100$ MPa;

(b) the material obeys the Ramberg-Osgood isotropic hardening constitutive model with initial yield stress $\sigma_Y = 140$ MPa, hardening coefficient $M = 4.5$, and $K = 400$ MPa.

9.25 Drawing is used to reduce the diameter of a copper-zinc wire 0.4 inches in diameter. The copper-zinc parameters are $\sigma_Y = 12$ ksi (before loading), $1/M = 0.5$, and $K = 130,000$. Determine the largest possible reduction without causing yielding in the reduced section of the wire after it passes through the die. A numerical solution will be necessary.

9.26 An austenitic steel plate 0.25 inches thick is to be manufactured by rolling a thicker plate. The austenitic steel parameters are $\sigma_Y = 30$ ksi (before loading), $1/M = 0.52$, and $K = 220,000$. The required yield stress in the manufactured plate is 100 ksi. Can the Ramberg-Osgood uniaxial relation be used to estimate what thickness plate the process should start with? Neglect changes in the width of the plate due to the rolling process.

9.27 A thick-walled cylinder with inner radius $a = 2$ inches and outer radius $b = 3$ inches is loaded by an internal pressure, p. The cylinder is made of mild steel ($E = 30 \times 10^6$ psi, $\nu = 0.25$, $\sigma_Y = 40$ ksi).

(a) Use the Tresca criterion to determine the maximum allowable pressure to avoid yielding on the inner surface.

(b) Repeat this analysis using the von Mises criterion.

9.28 A thick-walled cylinder with inner radius a inches and outer radius $b = 3$ inches is loaded by an internal pressure, $p = 2000$ psi. The cylinder is made of mild steel ($E = 30 \times 10^6$ psi, $\nu = 0.25$, $\sigma_Y = 40$ ksi).

(a) Use the Tresca criterion to determine the maximum radius a which avoids yielding on the inner surface.

(b) Repeat this analysis using the von Mises criterion.

9.29 A curved beam with inner radius $a = 6$ cm and outer radius $b = 12$ cm is in pure bending in its plane of curvature due to an applied moment, M. The thickness is $t = 1$ cm. It is made of aluminum ($E = 70$ GPa, $v = 0.33$, $\sigma_Y = 60$ MPa).

(a) Determine its state of stress at the lower surface ($r = a$) in terms of M using elasticity.

(b) Use the Tresca criterion to determine the maximum allowable moment to avoid yielding at the lower surface ($r = a$).

(c) Repeat this analysis using the von Mises criterion.

9.30 A steel rod ($E = 200$ GPa, $v = 0.29$, $\sigma_Y = 180$ MPa) of radius 2 cm and length 25 cm is loaded by a torque of $M_t = 200\pi$ Nm and and axial load of $P = 64\pi$ kN. Calculate the stresses using elementary strength of materials. *Note*: $J = \pi R^4/2$.

(a) Determine whether or not the Tresca criterion predicts yielding at a point on the surface of the rod.

(b) Determine whether or not the maximum normal stress criterion predicts yielding at a point on the surface of the rod.

9.31 A rod of length 14 inches and radius 2 inches is mounted in a rigid wall. It is loaded with a torque of $M_t = -(32,000\pi)\mathbf{i}$ in.-lb about the x-axis and a bending moment of $M = -(24,000\pi)\mathbf{k}$ in.-lb about the z-axis.

(a) Write the stress tensor at the point $(6, 1, 0)$ in the rod as a matrix. (Use strength of materials to compute stress components.)

(b) Determine the resultant stress at point $(6, 1, 0)$ acting on the plane with normal $\mathbf{N} = 2\mathbf{i} - 5\mathbf{j} + 3\mathbf{k}$.

(c) Using determinants, compute the principal stresses at point $(6, 1, 0)$.

(d) Use the maximum shear stress criterion to predict whether or not the rod will yield at point $(6, 1, 0)$. An axial test shows that the material of the rod yields at 10,600 psi.

(e) Use the distortion energy criterion to predict whether or not the rod will yield at $(6,1,0)$.

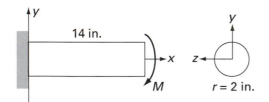

9.32 A rod of length 12 inches and radius 2 inches is mounted in a rigid wall. A rigid vertical bar of length 6 inches is attached to the centroid of the rod at the free end of the rod. The rigid bar is loaded with a force of $1,000\pi\mathbf{i}$ (lb) at its free end. A torque $M_t = -8,000\pi\mathbf{i}$ (in.-lb) is also applied at the free end. An axial test of the material of the rod shows that it yields at 6,000 psi. (The origin of the coordinate system is at the wall.)

(a) Write the stress tensor, as a matrix, at the point $(4, 2, 0)$ in the top of the rod. Use strength of materials to compute the stress components.

(b) Determine the resultant stress vector at point $(4, 2, 0)$ acting on the plane with normal $\mathbf{N} = \mathbf{i} - 2\mathbf{k}$.

(c) Using determinants, compute the principal stresses at point $(4, 2, 0)$.

(d) Use the maximum shear stress criterion to predict whether or not the rod will yield at the point $(4, 2, 0)$.

(e) Use the distortion energy criterion to predict whether or not the rod will yield at the point $(4, 2, 0)$.

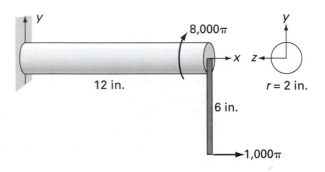

9.33 A rod of length 24 inches and unknown diameter, D, is mounted in a rigid wall. A rigid vertical bar of length 32 inches is attached to the centroid of the rod at the free end of the rod. The rigid bar is loaded with a force of $-1000\pi\mathbf{k}$ (lbs) at its free end. (The origin of the coordinate system is at the wall.) The axial yield stress is found by experiment to be 36 ksi. The designer wants a factor of safety of 2.5 so that the design axial failure stress is taken to be 14.4 ksi. The Poisson ratio is 0.30. Assume the largest stress is at point $A(0, 0, D/2)$.

(a) Use the maximum shear stress criterion to determine the minimum diameter, D, so that the body will not fail by slip (yielding).

(b) Use the distortion energy criterion to determine the minimum diameter, D, so that the body will not fail by slip (yielding).

(c) Now assume a 4-inch diameter and carry out the following calculations.

(1) Write the stress tensor, as a matrix, at the point $(4, 0, 2)$ on the front of the rod. Use strength of materials to compute the stress components.

(2) Determine the resultant stress vector at point $(4, 0, 2)$ acting on the plane with normal $\mathbf{n} = \mathbf{i} - 2\mathbf{j}$.

(3) Using determinants, compute the principal stresses at point $(4, 0, 2)$.

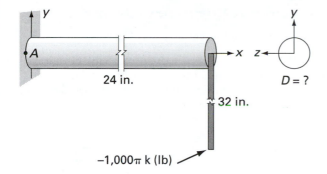

24 in.

32 in.

$D = ?$

$-1,000\pi$ k (lb)

9.34 Compare in a spinning disk without a hole the constant rotational speed for the initiation of yield and that for the fully plastic state as predicted by the Tresca yield condition in a perfectly plastic material. Do this by computing the ratios of the speeds.

9.35 A thin disk is attached to a shaft 2 cm in diameter which will rotate at 3,600 rpm. The disk is made of a 1100-H14 aluminum with mass density 2,719 kg/m², elastic modulus 70 GPa, Poisson ratio of 0.3, and uniaxial yield stress of 95 MPa. Assume perfectly plastic behavior and use the Tresca criterion to design the maximum outer radius of the disk to avoid yielding at any point.

9.36 A steel machine part ($E = 200$ GPa, $\nu = 0.29$, $\sigma_Y = 180$ MPa) is proportionally loaded so that the principle stresses at a point of interest are $a20$, $a30$, and $-a40$ MPa, where a is the proportionality constant (which changes with time as the load changes).

(a) At what value of a does the Mises criterion predict yield will occur at this point?

(b) Write the stress tensor in terms of a.

(c) Write the deviatoric stress tensor in terms of a.

(d) Compute the value of the hydrostatic stress at this point when a is the yield value determined in (a).

9.37 The uniaxial yield stress for a steel material is $\sigma_Y = 950$ MPa. Determine, using first the Tresca criterion and then the von Mises criterion, whether or not yielding occurs at a point with stress state,

$$\begin{pmatrix} 0 & 0 & 300 \\ 0 & -400 & 0 \\ 300 & 0 & -800 \end{pmatrix} \quad \text{MPa.} \quad (9.84)$$

9.38 The uniaxial yield stress for a material is $\sigma_Y = 185$ MPa. Determine using first the Tresca criterion and then the von Mises criterion whether yielding occurs at a point with stress state,

$$\begin{pmatrix} -50 & 0 & 0 \\ 0 & -80 & 80 \\ 0 & 80 & -200 \end{pmatrix} \quad \text{MPa.} \quad (9.85)$$

9.39 Prove that $II_{\mathbf{T}} = \frac{1}{2}(I_{\mathbf{T}}^2 - Tr(\mathbf{T}^2))$ for any tensor, $\mathbf{T}$.

9.40 A rectangular plate is made of a rigid-plastic material that obeys the Lévy-Mises model. The coordinate directions are along the plate edges with the origin at a corner. The uniaxial yield stress is $\sigma_Y = 80$ MPa. Determine whether the plate is elastically or plastically loaded when

(a) the plate is biaxially loaded in the plane by $\sigma_x = 80$ MPa and $\sigma_y = 40$ MPa;

(b) the plate is biaxially loaded in the plane by $\sigma_x = 80$ MPa and $\sigma_y = 80$ MPa.

9.41 A 316L steel rod of radius, R, and length, L, is fixed at one end. The rod is subjected to a tensile axial load, P, and a transverse bending load, N, at the free end which are increased from zero in a fixed ratio, $N = nP$. Determine the plastic strain at a point of maximal tensile strain in terms of P. Assume a rigidly plastic material and the Prandtl-Reuss model for isotropic hardening with the Mises yield criterion. The hardening law is a power law with $\sigma_Y = 133$ MPa, $K = 435$, and $M = 4.5$.

9.42 A steel plate is indented under a force of 3,000 kg by a 10-mm ball. The diameter of the plate indentation is 3.8 mm.

(a) Determine the indentation depth.

(b) Compute the Brinell hardness number and the Meyer hardness for this steel. Ans. 255 kg/mm²; 265 kg/mm².

9.43 A plate is indented under a force of 3,000 kg by a 10-mm ball. The diameter of the plate indentation is 2.05 mm.

(a) Determine the indentation depth.

(b) Compute the Brinell hardness number and the Meyer hardness for the material of the plate. Ans. 899 kg/mm²; 909 kg/mm².

9.44 Assume that $n = 2$ in Meyer's law and that the ball diameter is $D = 10$ mm. Plot W/A versus d for $0.1 \leq d/D \leq 1.0$.

9.45 Show that the depth, h, of an indentation of diameter, d, caused by a ball of diameter D is

$$h = \frac{D}{2} - \sqrt{\left(\frac{D}{2}\right)^2 - \left(\frac{d}{2}\right)^2} = \frac{D}{2}\left[1 - \sqrt{1 - \left(\frac{d}{D}\right)^2}\right].$$

Then show that when d/D is near zero, that

$$h \sim \frac{D}{4}\left(\frac{d}{D}\right)^2.$$

Hint: You may need to consider the Taylor series near zero of

$$\sqrt{1 - \left(\frac{d}{D}\right)^2}.$$

Of course, when d/D is near 1, then the depth approximately equals the radius of the ball, not $D/4$.

9.46 A material with BHN = 260 kg/mm² is indented by a 10-mm diameter sphere under a 3,000 kg load.

(a) What depth indentation will be made by the sphere?

(b) What is the corresponding Meyer hardness index?

9.47 In a manufacturing process, it is desired to make permanent circular indentations in the flat surface of a machine part by pressing a hard steel, 10-mm-diameter ball into the surface. The Brinell hardness of the material is 280.

(a) What force (in kg) must be applied to the ball to create the desired permanent indentation of depth 0.4 mm?

(b) Compute the Meyer hardness of the material.

9.48 Compute the Vickers hardness of talc.

9.49 If made of the same material, which is likely to be stronger, a screw that is cast or one whose threads are machined? A screw of the same dimensions is produced by both processes. Explain your choice.

REFERENCES

[1] R. W. ARMSTRONG, The influence of polycrystal grain size on mechanical properties, in *Advances in Materials Research*, Vol. 4, H. Herman, Ed., Wiley Interscience, 1970, pp. 101–146.

[2] R. W. ARMSTRONG, I. CODD, R. M. DOUTHWAITE, AND N. J. PETCH, The plastic deformation of polycrystalline aggregates, *Philosophical Magazine 7*, 45–58 (1962).

[3] R. W. ARMSTRONG, Dislocation mechanics description of polycrystal plastic flow and fracturing behavior, in *Mechanics of Materials*, M. A. Meyers, R. W. Armstrong, and H. O. K. Kirchner, Eds., Wiley, New York, 1999.

[4] R. W. ARMSTRONG AND W. H. ROBINSON, Combined elastic and plastic deformation from a continuous indentation hardness test, *New Zealand Journal of Science*, *17*, 429–434 (1974).

[5] J. F. BELL, The decrease of volume during loading to finite plastic strain, *Meccanica*, *31*, 461–472 (1996).

[6] P. W. BRIDGMAN, *Studies in Large Plastic Flow and Fracture*, McGraw-Hill, 1952, p. 213.

[7] W. J. COUSINS, R. W. ARMSTRONG, AND W. H. ROBINSON, Young's modulus of lignin from a continuous indentation test, *Journal of Materials Science*, *10*, 1655–1658 (1975).

[8] D. C. DRUCKER, A more fundamental approach to plastic stress-strain relations, *Proceedings of the 1st U.S. National Congress of Applied Mechanics*, AMSE, New York, 487–491 (1951).

[9] F. C. FRANK AND W. T. READ, Multiplication processes for slow moving dislocations, *Physical Reviews*, *79*, 772 (1950).

[10] A. D. FREED AND K. P. WALKER, Viscoplasticity with creep and plasticity bounds, *International Journal of Plasticity*, *9*, 213–242 (1993).

[11] E. O. HALL, The deformation and ageing of mild steel: III. Discussion of results, *Proceedings of the Physical Society of London*, *B64*, 747 (1951).

[12] H. W. HASLACH, JR., A non-equilibrium thermodynamic geometric structure for thermoviscoplasticity with maximum dissipation, *International Journal of Plasticity*, *18*, 127–153 (2002).

[13] L. E. MALVERN, *Introduction to the Mechanics of a Continuous Media*, Prentice Hall, Englewood Cliffs, NJ, 1966.

[14] F. A. MCCLINTOCK AND A. S. ARGON, *Mechanical Behavior of Materials*, Addison-Wesley, Reading, MA, 1966.

[15] F. R. N. NABARRO, Dislocations in a simple cube lattice. *Proceedings of the Physical Society*, *59*, 256–272 (1947).

[16] R. PEIERLS, On the size of a dislocation, *Proceedings of the Physical Society*, *52*, 34–37 (1940).

[17] N. J. PETCH, The cleavage strength of polycrystals. *Journal of Iron Steel Institue*, *174*, 25 (1953).

[18] H. SHIN, Y. L. TSAI, J. J. MECHOLSKY, AND R. W. ARMSTRONG, Elasticity, plasticity and cracking at indentations in single crystal silicon, *Journal of Materials Science Letters*, *12*, 1274–1276 (1993).

[19] D. TABOR, *The Hardness of Metals*, Clarendon Press, Oxford, UK, (1951).

[20] P. VEYSSIÉRE, Dislocations and the plasticity of crystals. In *Mechanics of Materials*, M. A. Meyers, R. W. Armstrong, and H. O. K. Kirchner, Eds., Wiley, New York, 1999.

[21] F. J. ZERILLI AND R. W. ARMSTRONG, "Constitutive Relations for the Plastic Deformation of Metals," *High Pressure Science and Technology–1993*, S. C. Schmidt, J. W. Shaner, G. A. Samara, and M. Ross, Eds., American Institue of Physics, New York Conf. Prof. *309(2)*, 989 (1994).

[22] H. ZIEGLER, A modification of Prager's hardening rule. *Quarterly Applied Mathematics*, *17*, 55–65 (1959).

CHAPTER 10

FRACTURE

10.1 INTRODUCTION

A body is said to fracture when it separates into two or more pieces. The separation creates additional surface area in the body. Atoms behave differently on a surface than they do in the interior of a body since they are no longer surrounded by other atoms, and this behavior is responsible for the surface energy. Most materials used in engineering structures contain tiny cracks or other discontinuities. Fracture is usually preceded by a rapid crack extension in the material. The study of fracture therefore focuses primarily on the lengthening of a crack, and the resultant growth in surface area, as the load on the body is increased. The load is restricted to monotonically increasing loads. Fracture under cyclic or random variable loading is called fatigue and is discussed in Chapter 11. Plane stress and plane strain problems are considered in this chapter. The behavior of cracks in a three-dimensional state of stress is the subject of current research.

The engineering study of fracture mechanics does not emphasize how a crack is initiated; the goal is to develop methods of predicting how a crack propagates, that is, how it lengthens. Crack propagation models can be developed for fracture of brittle or of ductile materials. Most classical analyses are restricted to brittle fracture in which the crack propagation is not accompanied by large plastic strains. Since the material is assumed brittle, the theory of linear elasticity can be used. Such materials include glasses, ceramics, and low-temperature metals. However, in special cases, the theory can be applied to mild steel. For example, when holes are punched in a metal plate, there is often enough plastic work hardening in the material around the hole that the material can be approximated as brittle. Brittle fracture theories can then be used to estimate the critical loads which will propagate cracks near such holes. Ductile fracture and its underlying material foundations are reviewed at the end of this chapter.

In a material without cracks, the strain and the displacement fields vary continuously over the body. However, at a crack these fields are discontinuous, and this fact can be used to mathematically describe a crack. There is no cohesion between the material on opposite faces of a crack, but there could be sliding friction between the two faces. The propagation of a crack depends on the behavior at the tip of the crack, also called the crack root. The line along the root of a three-dimensional crack is often called the crack front.

The simplest, and oldest, tool to predict fracture is the ultimate normal stress obtained from a uniaxial test. Until the early part of the twentieth century, engineers used the maximum normal stress or the maximum normal strain criterion to predict the fracture of bodies loaded in multiaxial states of stress and made of brittle materials, those which fracture after negligible plastic behavior. However, these criteria were known to be unsatisfactory. The groundbreaking and extremely important work of the British aeronautical engineer A. A. Griffith around 1920 produced a new criterion to repair this difficulty and initiated the modern era of fracture mechanics. Griffith used an energy approach to predict the fracture of linear elastic, brittle materials such as glass. Almost 30 years later, an American engineer, George Irwin, attacked the problem of fracture by analyzing the stress field near a crack tip and eventually, with others, created a method of predicting fracture for both brittle materials and ductile metals. In ductile metals, a local plastic zone is generated near the crack tip; methods for describing this zone for perfectly plastic materials were developed. These techniques examine the problem of fracture from a large-scale point of view. Contemporary materials scientists examine the plastic zone and its relation to fracture on a microstructural scale (micromechanics) in an attempt to clearly describe the mechanisms which produce fracture. Success in this effort would lead to more accurate criteria by which designers could predict fracture.

The Griffith surface energy approach and the Irwin stress intensity approach are the foundations for the prediction of brittle crack propagation. Both make essentially the same prediction for the critical stress required to initiate brittle fracture. These are equilibrium theories in the sense that they assume that the crack grows to relieve a buildup of stress in order to return the body to an equilibrium state. Neither the Griffith nor the Irwin analyses give any information about the microstructural mechanisms of fracture.

10.2 ENERGY AND THE GRIFFITH CRITERION FOR BRITTLE FRACTURE

Brittle fracture is that preceded by very little plastic deformation. Brittle fracture is initiated by flaws or cracks in the material. It can occur in ductile materials at average stresses much smaller than the yield stress. The formation of the larger crack surface resulting in fracture is a consequence of the breaking of atomic bonds, but can occur when the average stress in the body is as small as one hundredth of the bond strength. Many experimentalists had observed, prior to 1900, that a size effect exists in the strength of materials. A smaller body is stronger than a larger body made of the same material. A. A. Griffith made precise the idea that the strength of a body is related to the size of cracks in the body. Griffith's original goal was to study the effect of surface treatments, like grinding or polishing, on metal machine parts subject to fluctuating loads in service. Surface scratches were known to lead to rupture under stresses smaller than the elastic limit. This phenomenon is now considered part of the fatigue problem discussed in Chapter 11.

In his famous 1920 paper [11], Griffith proposed an equilibrium energy criterion for crack growth initiation that results in isothermal brittle fracture. Griffith considered crack growth in a homogeneous body which remains in equilibrium, so that his criterion is said to apply to equilibrium crack propagation. Griffith implicitly assumed that the system is conservative. Therefore the loading must be elastic to the point of fracture, and there is no dissipation of energy due to temperature increases or sound emission, for example. The load should be viewed as increasing quasi-statically. The criterion only applies to brittle fracture because of these assumptions.

As a crack grows, its surface area increases. Griffith assumed that the crack is formed by the annihilation of the tractions (forces per unit area) acting on the fracture plane. Work must be done to break the intermolecular bonds along the plane which is to become the new crack surface. The work or energy per unit area required to separate the surface faces further than the radius of molecular action is measured as the surface tension, γ, also called the surface energy. The total potential energy is the sum of the strain energy, the potential energy of the load, and the surface energy of the crack faces.

Recall that the potential energy of a conservative force is the negative of the work it does from a datum.

Griffith's key idea was that the sum of the potential energy of the forces and the strain energy cannot increase by the creation of a crack with traction-free surfaces. Griffith postulated that the mechanical potential energy decreases as the crack grows. By the first law of thermodynamics, conservation of energy,

$$\frac{dW}{dt} = \frac{dU}{dt} + \frac{dT}{dt} + \frac{d\Gamma}{dt},$$

where W is the work done by the conservative forces, U is the strain energy, T is the kinetic energy, and Γ is the energy used in increasing the crack surface area, A. Since the system is quasi-static, the change in kinetic energy is negligible. All terms only depend on the crack area, so that by the chain rule, dW/dt is

$$\frac{dW}{dA}\frac{dA}{dt} = \frac{dU}{dA}\frac{dA}{dt} + \frac{d\Gamma}{dA}\frac{dA}{dt}.$$

Since dA/dt is nonzero during crack propagation,

$$\frac{dW}{dA} = \frac{dU}{dA} + \frac{d\Gamma}{dA}.$$

Define the mechanical potential energy of the system to be $\Pi = U - W$. Griffith's basic relation for the initiation of crack propagation is then

$$-\frac{d\Pi}{dA} = \frac{d\Gamma}{dA}.$$

In other words, the mechanical potential energy decreases as the surface energy of the crack surface increases. The total potential energy is then $\Pi + \Gamma$ so that crack propagation is initiated if

$$\frac{d(\Pi + \Gamma)}{dA} = 0. \tag{10.1}$$

Griffith viewed this as a new stationary principle for crack propagation. Recall (see Chapter 7) that a stationary principle states that a particular phenomenon occurs if an associated function has all derivatives with respect to the independent variables equal to zero.

The strain energy is one of the components of the total potential energy of a system. Griffith assumed that if a hole is made in the body without disturbing the state of stress in the remainder of the body, the strain energy must decrease by an amount equal to the strain energy of the material formerly in the hole. This strain energy loss is computed by integrating the strain energy density over the volume of the hole. Some of this strain energy becomes surface energy on the surface of the hole. Thus as the sum of the strain energy and potential energy of the applied loads, the mechanical energy Π, decreases, the surface energy increases. These energies find their balance when the system is in equilibrium as determined by the principle of stationary total potential energy. The mechanical and surface energies are implicitly assumed to be independent of each other in this analysis. Some recent experiments hint that they can be coupled, but these will not be discussed in this text.

For a given crack length, the system is in equilibrium for a critical stress state. At this stress, the crack is at the point of extending itself. Therefore the equilibrium stress for a crack of a given length is also the critical stress required to extend it. Any higher stress would require a new equilibrium state in the body which can only be achieved by lengthening the crack.

Griffith proposed a new failure criterion. The crack continues to extend as long as the mechanical energy release rate is greater than the energy per unit surface area required to form new crack surfaces.

$$\frac{d(\Pi + \Gamma)}{dA} > 0.$$

The critical point at which crack extension begins is that at which the energy release rate equals the energy required to form new crack surfaces.

Today the first term in equation (10.1) is given a special symbol, G, in honor of Griffith. Irwin [15] described this in the 1950s by the concept of the mechanical energy released during an incremental crack extension. The rate of energy release, G, is the rate at which the total potential energy decreases. If, for a given crack size, G exceeds a critical value, the crack will grow. The critical value must be determined experimentally. The energy release rate, G, is sometimes thought of as the force driving the fracture. G depends on the loads on, and the geometry of, the body, as well as the orientation of the crack. Irwin's analysis assumes that the material is a continuum so that elasticity theory may be applied. It ignores the lattice structure of metals with all the voids between atoms.

Definition   For an arbitrary crack, the energy release rate is

$$G = -\frac{d\Pi}{dA},$$

where A is the crack interfacial area. G has units of energy per area.

If the crack is assumed to have unit thickness, it is common to define

$$G = -\frac{d\Pi}{dc},$$

where c is the crack length and where the total potential energy per unit width Π depends on the crack length and the applied load. Here G has units of energy per area per width of the crack surface. This version of G can also be viewed as a force on the crack line.

The energy for crack growth must come from the stored strain energy in *deformation control* when no external work is done on the body (that is, deformation fixed as well as controlled). On the other hand, in load control, the forces do work on the body which may be used in propagating the crack.

Deformation control can be achieved in what are called hard supports by which the rate of deformation is controlled. Another example is stress relaxation, in which the deformation of the specimen is fixed and the subsequent change in reaction force at the support with respect to time is monitored by a load cell at the support. The stress relaxation process is most easily observed in polymers because the change in stress observed may be quite large.

In load control, the load rate is controlled. This situation is sometimes called soft supports. A creep test in which an axial dead load is applied and the resulting deformation over time is measured is an example of a load control test.

EXAMPLE 10.1   If the cracked specimen is loaded between fixed grips, hard supports, the work done by the force at the grips is zero. The specimen is in deformation control because the grip separation determines the deformation. In this case, $\Pi = U$,

$$G = -\frac{dU}{dA}.$$

All the energy provided for crack propagation comes from a reduction in the strain energy of the body. This justifies calling G the energy release rate. An alternate formulation for G is

$$G = -\frac{dU}{dc}.$$ ∎

EXAMPLE 10.2   If a cracked specimen is deformed by a dead load, soft supports, the Clapyron theorem (see Chapter 7, Section 7.1) implies that the work done by the loads is twice the increase in potential energy.

$$G = -\frac{d\Pi}{dA} = \frac{dW}{dA} - \frac{dU}{dA} = 2\frac{dU}{dA} - \frac{dU}{dA} = \frac{dU}{dA}. \tag{10.2}$$

In this case, all the energy to propagate the crack is provided by the work done by the load. G is thought of as the crack driving force. The specimen in this situation is under load control. ∎

EXAMPLE 10.3   A uniaxial member is loaded by P under load control. Assume that the crack growth is quasi-static so that, during the growth, the total complementary energy, Φ, is by definition the negative of the total strain energy, U, plus the work done by the load, P, as the crack growth lets the load move through a deformation, u. The total complementary energy is $\Phi = -U + Pu$. But then Φ is the negative of the total mechanical potential energy. If P is held fixed during the crack growth, then by definition, $G = d\Phi/dA = -dU/dA + P(du/dA)$. But since $G = dU/dA$ in load control from the previous example, $G = 0.5P(du/dA)$ and so $dU/dA = 0.5P(du/dA)$. Therefore in the special case of quasi-static crack growth, which is an idealization of reality, as the crack propagates and u increases, half the work done on the body by the constant force, P, goes into the crack driving force and half into increasing the strain energy. ∎

The fact that $d\Gamma/dA = 2\gamma$, since there are two surfaces involved in a crack, implies

$$G = -\frac{d\Pi}{dA} = \frac{d\Gamma}{dA} = 2\gamma. \tag{10.3}$$

An axially loaded infinite plate with an elliptical hole was viewed by C. E. Inglis [13] as a model for the opening of a crack under a uniaxial load. Griffith [11] had available the Inglis analysis of the stress distribution in such a plate published just a few years earlier in 1913. The elliptical hole is assumed to be oriented so that the major axis is perpendicular to the applied axial load. The major axis has length $2a$ and the minor axis has length $2b$. The axial stress on the plate is σ_0. As the point at the tip of the crack is approached, the normal stress in the y-direction (Fig. 10.1) at the tip tends to

$$\sigma_{max} = \sigma_0 \left(1 + 2\frac{a}{b}\right),$$

(see Chapter 10, Section 10.2.2). For very small b, a thin elliptical crack, this value is very large, and it tends to infinity as b tends to zero. The elliptical hole becomes a crack in the limit as b tends to zero. The major axis $2a$ is then the crack length. The crack is said to propagate, or grow, if the crack length, $2a$, changes during the loading. Griffith calculated the critical stress in the case of a thin infinite plate having a crack of length $2c$ parallel to one edge of the plate and an in-plane constant stress σ

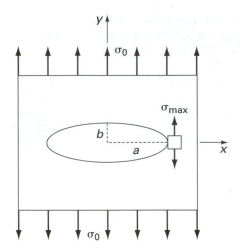

FIGURE 10.1   Plate with an elliptical hole.

applied perpendicular to the crack surface. The plate is assumed to have constant unit thickness.

Griffith assumed that the surface of a crack is stress free and that, under this assumption, the theory of elasticity gives the correct stresses in the body. If the crack is formed by the sudden annihilation of the tractions which were on its newly formed surface, the potential energy is no longer at a minimum. It must in general decrease. The available potential energy, which Griffith approximated as the strain energy of the elliptical region which models the crack, must break tractions at the crack tips until equilibrium is reached. Griffith argued that the decrease in potential energy to reach equilibrium is equal to amount of strain energy in the elliptical crack region due to loading the body less the increase in surface energy.

Since there are two crack surfaces of length $2c$, the increase in surface energy for formation of the crack in an unbroken plate is $4c\gamma$, where γ is the surface energy per unit width per unit length. The change in potential energy per unit width is then

$$\Delta\Pi = 4c\gamma - \Delta U.$$

Using Inglis's calculations for a biaxially loaded plate with an elliptical hole, Griffith assumed, on the basis of example strain energy computations for the biaxially loaded plate and one which showed that a crack parallel to a uniaxial load does not influence the stress field (this latter assumption has been more recently challenged but the argument is unresolved), that the change in strain energy from the uncracked linear elastic, thin infinite plate to one having a very narrow crack of length $2c$ is, per unit width,

$$\Delta U = \frac{\pi c^2 \sigma^2}{E'}, \tag{10.4}$$

where σ is the stress applied to the plate. $E' = E$ for plane stress and $E' = E/(1-\nu^2)$ for plane strain. The total change in potential energy per unit width is then

$$\Delta\Pi = 4c\gamma - \frac{\pi c^2 \sigma^2}{E'}. \tag{10.5}$$

The Griffith criterion is that this energy is stationary, $d\Delta\Pi/dc = 0$, at the critical stress, σ_c, for a crack of length $2c$. At this stress, σ_c, the crack may extend. The growth is determined entirely by the stress perpendicular to the crack in this analysis. The changes that Griffith made in his argument between his 1921 and 1924 papers are described by Atkins and Mai [5, pp. 194–195].

For plane stress,

$$\sigma_c = \sqrt{\frac{2E\gamma}{\pi c}};$$

(10.6)

and for plane strain,

$$\sigma_c = \sqrt{\frac{2E\gamma}{\pi(1 - \nu^2)c}}.$$

(10.7)

Substitution of the critical stress into the strain energy expression (10.4) shows that equilibrium is achieved when the strain energy change is twice the surface energy.

One important observation from the Griffith criterion is that the critical stress is proportional to the inverse of the square root of the existing crack length. The smaller the crack, the harder it is to initiate crack propagation. Griffith checked this result by experiments on a hard glass; he claimed to have verified his model within a 10% experimental error and noted that the error might be due to plastic behavior at the crack tip. Glass, which responds to cracks of sufficient size by brittle fracture, was chosen because at room temperature it behaves linear elastically until it fractures and so appears to satisfy the assumptions of the Griffith rupture criterion. Later, Marsh [21] argued that glass behaves plastically because, among other issues, hardness indentations resemble those of steel and because the fracture energy of a glass can be 50 times the surface energy predicted by the brittle fracture theory. Marsh suggested that, since glass does not work harden, it should fracture catastrophically at its yield stress (flow stress), in a manner similar to a brittle material. The higher experimental fracture energy is accounted for by plastic work done on the glass and by blunting of the crack tip.

A direct experimental measurement of the surface tension of glass at room temperature is not easy. Griffith measured the surface tension at temperatures between 745° and 1,110°C where glass is less viscous, and then assuming that the surface tension is approximately a linearly decreasing function of temperature, extrapolated a value at room temperature.

EXAMPLE 10.4   The surface tension for glass at 15°C calculated by Griffith was 0.0031 lb-in. The elastic modulus was $E = 9.01 \times 10^6$ psi, and the Poisson ratio $\nu = 0.251$. The tensile strength was 24,900 psi. Griffith tested thin wall tubes with scratched surfaces by pressurizing the inside until the tubes burst (Fig. 10.2). Compute the critical stress for a crack of half-length $c = 0.01$ inches along the longitudinal direction.

Solution   Griffith argued that the member can be assumed to be in plane stress if the radius is large compared to the wall thickness. The wall is viewed as a thin plate with a crack. The crack is opened

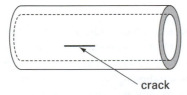

crack

FIGURE 10.2   Longitudinal crack in a thin-walled cylinder.

by the circumferential, or hoop, stress; this stress is at least an order of magnitude greater than the internal pressure. The critical stress is therefore given by

$$\sigma_c = \sqrt{\frac{2E\gamma}{\pi c}} = 1334 \text{ psi},$$

about one-twentieth of the tensile strength.   ∎

EXAMPLE 10.5   Determine the critical stress for propagation of a crack of length 10 mm in a thin rectangular aluminum plate uniaxially loaded perpendicular to the crack surface. The crack is parallel to the loaded edge of the plate. The surface energy for this aluminum is about 1.20 J/m^2. Its elastic modulus is 70 GPa.

Solution   This is a crack opening problem that may be analyzed using the Griffith criterion if it is also assumed that the plate behaves linearly elastically up to the critical stress. By equation (10.6), the critical stress is

$$\sigma_c = \sqrt{\frac{2(70 \times 10^9)(1.2)}{\pi(0.005)}} = 3.27 \text{ MPa}. \tag{10.8}$$

This stress is certainly within the elastic range for aluminum. In fact, the stress is extremely small, 500 psi in English units. This indicates that the Griffith theory by itself is not realistic for ductile metals.   ∎

The ambient environment influences the surface energy. Glass can "fatigue" until it fractures under a constant stress that is much less than the fracture stress under rapid loading. Experiment shows that the fracture stress under rapid loading is about 3.3 times larger than the constant stress, when the specimens are required to break in less than 1,000 hours. Orowan [23] explained this strange result by an adsorption-diffusion effect. A slowly growing crack allows air and moisture to diffuse to the crack root, thereby lowering the surface energy. If fracture occurs rapidly, the adsorbed film of air or moisture cannot penetrate fast enough to keep up with the surface growth at the root. For rapid growth, the surface energy in a vacuum must be used in the Griffith equation to predict the critical stress. The surface energy for mica is 4,500 ergs/cm^2 in vacuum and 375 ergs/cm^2 in air. Therefore, the Griffith equation predicts that the critical stress in a vacuum is 3.46 times that in air. Orowan assumed that a similar ratio would hold for glass.

Griffith assumed that all the energy causing fracture goes into atomic bond separation as represented by the surface energy. But for polycrystalline materials, such as most engineering metals, many other small-scale phenomena are involved in fracture. There might, for example, be dissipation by slip at the grain surfaces; energy may be used in various types of dislocation behavior near the crack tip. These are, today, all combined into a material parameter called the resistance to fracture, R. The Griffith fracture criterion is then rewritten to state that fracture occurs if $G \geq R$.

10.2.1 Stability

A crack that continues to grow under a constant load is said to be unstable. An unstable crack grows abruptly. A crack is said to be stable if it propagates a small length under a given load but no further. A crack under an applied increasing displacement is stable if the crack ceases growing when the applied displacement is held constant. The crack is stable, equilibrium is attained, and the crack propagates no further, if the stress is held constant at σ_c, when the total potential energy is a minimum with respect to the crack length, c. The energy can be a maximum at the critical value. In such cases, the crack propagation is unstable. The example computed by Griffith, equation (10.5), has negative second derivative, $d^2\Pi/dc^2$, if $c < 1$, and therefore, the crack is unstable. The crack will grow continuously as soon as the the stress is increased above σ_c.

The stability behavior of crack growth under load control differs drastically from that under deformation control. Often deformation controlled loading produces a stable crack growth, while load control often produces unstable crack propagation.

EXAMPLE 10.6   Compute G for the double-cantilever beam test specimen under a constant wedging condition of displacement, h (Fig. 10.3). The separation of the split ends is $2h$. Is the crack propagation stable?

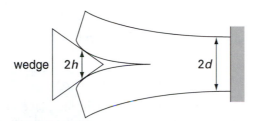

FIGURE 10.3   Wedge opening a crack in a double-cantilever test specimen.

Solution   View each piece as a cantilever beam fixed at the crack tip. The change in strain energy due to the formation of the crack equals the strain energy due to the bending of the two beams. If P is the reaction force between the wedge and the specimen, then the displacement h of the free end is $h = Pc^3/3EI$, where $I = wd^3/12$. Solve for P. The internal moment at the point x in each cantilevered section is then $M = Px = 3hEIx/c^3$, where x is measured from the left end of the beam. The elastic strain energy from beam theory is $U = 2\int_0^c (M^2/2EI)dx$ (Chapter 7, Section 7.1) so that $U = wEd^3h^2/4c^3$ where the crack length c is the length of the beam from the crack tip to the free end, w is the beam width, and $2d$ is the beam height. The work done by the reaction force from the wedge is zero. The wedge controls the deformation of the beams, and so the specimen is in deformation control. Therefore,

$$G = -\frac{dU}{dA} = -\frac{1}{w}\frac{dU}{dc} = \frac{3Ed^3h^2}{4c^4}.$$

Notice that $d^2\Pi/dc^2 = d^2U/dc^2 > 0$. Therefore the behavior is stable. The crack will not continue to propagate unless the wedge is driven in further.   ■

10.2.2   Appendix: Plate with an Elliptical Hole

C. E. Inglis, in 1913, was aware that cracks are a precursor of fracture and that plastic yield at the crack tip has an important influence on crack propagation. He viewed an elliptical hole in an infinite plate as a model for a crack in the limiting case that the minor axis tends to zero. His goal was to determine the state of stress around such a hole predicted by elasticity. His analysis of a plate having an elliptical hole was carried out in elliptical coordinates, (α, β), to more easily determine the stresses normal and parallel to the surface of the hole (Fig. 10.4). Such coordinates are an example of curvilinear coordinates, whose directions change from point to point in the body. Polar coordinates are another example of such curvilinear coordinates. The coordinate axes are given by the ellipse, for α constant,

$$\frac{x^2}{c^2\cosh^2(\alpha)} + \frac{y^2}{c^2\sinh^2(\alpha)} = 1,$$

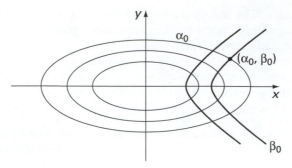

FIGURE 10.4 Elliptical coordinates.

and the hyperbola intersecting the x-axis, for β constant,

$$\frac{x^2}{c^2 \cos^2(\beta)} - \frac{y^2}{c^2 \sin^2(\beta)} = 1,$$

which is orthogonal to the ellipses. A point in the plane in elliptical coordinates is represented by a pair (α_0, β_0) corresponding to the unique ellipse, $\alpha = \alpha_0$, and the unique hyperbola, $\beta = \beta_0$, which intersect at that point. The ellipse and hyperbola have perpendicular tangents at this point of intersection so that they determine orthogonal coordinates. The normal stresses in the two directions, parallel to the curves at the point of intersection, are therefore perpendicular to each other.

Elliptical coordinates are related to Cartesian coordinates by

$$x = c \cosh(\alpha) \cos(\beta)$$
$$y = c \sinh(\alpha) \sin(\beta).$$

Exercise 10.1 Graph the ellipse corresponding to $\alpha = \pi/4$ and the hyperbola corresponding to $\beta = \pi/3$. Show that, at their point of intersection, their tangents are perpendicular to each other.

The boundary of an elliptical hole is defined by $\alpha = $ constant (Fig. 10.5). If $\alpha = 0$, then the ellipse is a straight line of length $2c$ along the x-axis. Therefore, $\alpha = 0$ is used to represent the crack in the plate.

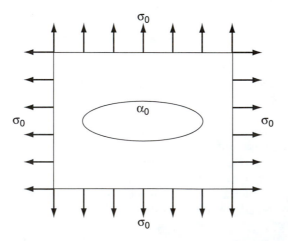

FIGURE 10.5 Biaxially loaded plate with elliptical hole.

The general solution obtained by Inglis is in the form of an infinite series. The case of a plate loaded by an equal biaxial stress, σ_0, on the edges has a relatively simple form.

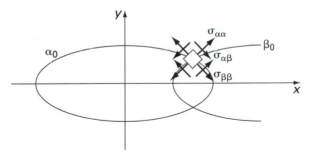

FIGURE 10.6   Stress element in elliptical coordinates.

Inglis calculated the normal stress in the α direction, perpendicular to the ellipse $\alpha = \alpha_0$ (Fig. 10.6), to be

$$\sigma_{\alpha\alpha} = \sigma_0 \frac{\sinh 2\alpha (\cosh 2\alpha - \cosh 2\alpha_0)}{(\cosh 2\alpha - \cos 2\beta)^2}.$$

The stress in the β direction perpendicular to the hyperbola, $\beta = \beta_0$,

$$\sigma_{\beta\beta} = \sigma_0 \frac{\sinh 2\alpha (\cosh 2\alpha + \cosh 2\alpha_0) - 2\cos 2\beta}{(\cosh 2\alpha - \cos 2\beta)^2}.$$

The shear stress is

$$\sigma_{\alpha\beta} = \sigma_0 \frac{\sin 2\beta (\cosh 2\alpha - \cosh 2\alpha_0)}{(\cosh 2\alpha - \cos 2\beta)^2}.$$

The normal displacements are $u_\beta = 0$, and

$$u_\alpha = \frac{hc^2\sigma_0}{8\mu}[(\kappa - 1)\cosh 2\alpha - (\kappa + 1)\cos 2\beta + 2\cosh 2\alpha_0],$$

where μ is the shear modulus, and $\kappa = 3 - 4\nu$ in plane strain and $\kappa = (3 - \nu)/(1 + \nu)$ in plane stress. The fact that the coordinates are curvilinear are accounted for in the factor,

$$h = \sqrt{\frac{2}{c^2(\cosh 2\alpha - \cos 2\beta)}}.$$

Exercise 10.2   Verify that in biaxial loading the normal stress at the edge of the elliptical hole is

$$\sigma_{\beta\beta} = 2\sigma_0 \frac{\sinh 2\alpha_0}{\cosh 2\alpha_0 - \cos 2\beta}.$$

Griffith estimated the strain energy lost by removing the material of the elliptical hole by integrating over an elliptical volume, α, containing an elliptical hole, α_0,

$$\frac{1}{2}\int_0^{2\pi}\left[\frac{u_\alpha}{h}\sigma_{\alpha\alpha} + \frac{u_\beta}{h}\sigma_{\alpha\beta}\right]d\beta,$$

where $u_\beta = 0$ on the boundary α_0. The change in strain energy due to the creation of the elliptical hole $\alpha = \alpha_0$ was computed to be

$$U_{\alpha_0} = \frac{(\kappa + 1)\pi c^2 \sigma^2}{8\mu} \cosh(2\alpha_0).$$

Let $\alpha_0 \to 0$ to obtain a narrow crack. The change in strain energy due to the presence of the narrow crack, U, for plane stress is then, using $\mu = E/[2(1 + \nu)]$,

$$U = \frac{(\kappa + 1)\pi c^2 \sigma^2}{8\mu} = \frac{\pi c^2 \sigma^2}{E}. \tag{10.9}$$

Inglis also wrote the stress in terms of the radius of curvature of the elliptical hole. At the vertex of the ellipse on the major axis, the maximum tensile stress is perpendicular to the major axis of the ellipse, in the β direction,

$$\sigma_{\max} = \sigma_0 \left(1 + 2\frac{a}{b}\right),$$

where a is the major axis half length, and b is the minor axis half length (Fig. 10.1). The radius of curvature of an ellipse at the vertex is $\rho = b^2/a$. In terms of the radius of curvature, the maximum stress is

$$\sigma_{\max} = \sigma_0 \left(1 + 2\sqrt{\frac{a}{\rho}}\right).$$

As the root of the crack gets sharper so that the radius of curvature tends to zero, the stress induced is larger. Consequently, a crack with a small radius of curvature at the root is more likely to propagate than one with a larger radius of curvature.

Inglis had therefore shown that the stress at the crack tip is proportional to the square root of the half crack length and is inversely proportional to the square root of the radius of curvature at the crack tip. Inglis first presented his results at a meeting of the Institution of Naval Architects in England. One of the questioners after Inglis's talk rejected the significance of this work because it was not clear to him that this mathematics applies to real cracks, especially in shafts made of ductile steel. But even so, the application of Inglis's analysis in the Griffith mathematical model has inspired much work on predicting fracture, even if only through the attempts to improve the Griffith model.

10.3 STRESS CRITERIA

Little additional work was done on fracture mechanics until near the end of World War II. A large number of tankers and merchant ships built for service in the war developed fractures within a few years of being commissioned. Most of these were brittle fractures, occurring rapidly and accompanied by a loud sound. The problem of such catastrophic fractures was attacked from a new point of view by George R. Irwin, an engineer at the Naval Research Laboratory near Washington, D.C. Irwin and others investigated the stress behavior in a neighborhood of the crack tip. Irwin proposed the two important ideas of the energy release rate, G, and the stress intensity factors, K. Irwin [14] defined G at the onset of fast fracturing or at crack arrest as the magnitude, per unit extension of the crack, of the energy transferred from mechanical or strain energy to other forms of energy in the vicinity of the crack, including heat. Irwin also developed a new fracture criterion in terms of the stress intensity factors.

Irwin categorized fractures into three types, called mode I, mode II, and mode III, the building blocks for a general fracture (Fig. 10.7). These correspond to the fact that on a plane, such as the crack surface, there are three orthogonal coordinate directions, one normal and two parallel to the face. Each mode describes a different way the two crack faces can move with respect to each other. Mode I fracture, called the opening mode, occurs when the two faces move transversely away from each other, parallel to the normal stress direction on the face of the crack. Mode I cracking is caused by the symmetric displacement normal to the crack surface. A crack transverse to a normal stress responds in mode I. Shearing of the crack surface occurs in the other two modes. Mode II fracture is an asymmetric displacement normal to the crack front and is called the sliding mode. In a plate subjected to shear loads at opposite edges, a crack parallel to the loads propagates in mode II. Mode III, the tearing mode, is an asymmetric displacement parallel to the crack front. Mode III crack loading occurs, for example, on a circumferential crack on the surface of a torsionally loaded circular shaft. In many real structures with more complex loads, a crack propagates by a combination of these modes.

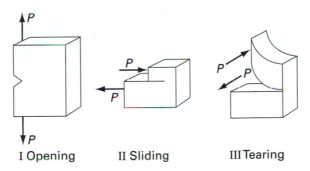

I Opening   II Sliding   III Tearing

FIGURE 10.7   The three modes of fracture.

The Griffith analysis takes a global viewpoint based on energy functions for the full body. A local examination near the crack requires an investigation of the stress field around the crack tip. Irwin [14] hoped that an understanding of this stress field would explain many of the mysteries of crack extension. As suggested by taking the limit as the minor axis tends to zero in the Inglis stress computation for an infinite plate with an elliptical hole, the stress at the crack tip becomes infinite. The stress function is said to have a singularity at the crack tip.

Definition   If a function value tends to infinity as the value in the domain approaches a point, P, then the point P is called a singularity of the function.

EXAMPLE 10.7   **(a)** The function $f(r) = r^{-1/2}$ has a singularity at $r = 0$.

(b) The function $f(r) = \ln(r)$ also has a singularity at $r = 0$. This is called a logarithmic singularity to describe the manner in which the function grows toward infinity as r tends to zero from the positive side.   ∎

Exercise 10.3   Plot the functions $f(r) = r^{-1/2}$ and $f(r) = \ln(r)$ on the same system of coordinates for $0 < r < 10$. Compare their graphs near $r = 0$.

The crack tip is a singularity of the stress function in the elliptical hole model. This fact guides the analysis of crack propagation. The existence of a singularity at the crack root is not enough to guarantee that the crack will propagate since the singularity exists for any crack length and under any load. Every elastic stress field near a crack tip has a singularity at the tip, but not all cause the crack to grow. An additional condition is needed to formulate a criterion for crack propagation. Irwin suggested that the magnitude, or intensity, of the stress field in a neighborhood of the crack tip determines when the crack will grow.

The stress field at each point is assumed to obey the equations of elasticity, but at the crack tip the stress must be infinite. This singularity is not physically reasonable as will be discussed later. M. L. Williams [32] presented, in 1957, an Airy stress function for a body with a notch, or reentrant corner (Fig. 10.8). The stress function, ϕ, must satisfy the biharmonic equation, $\nabla^4 \phi = 0$ (Chapter 4, Section 4.6.2). Polar equations are chosen with the origin at the crack tip. The angle, θ, is measured from the x-axis passing through the line of symmetry of the crack. The singularity must occur at $r = 0$. The boundary conditions are that the stresses are zero on the crack face; $\sigma_\theta = \tau_{r\theta} = 0$ at $\theta = \pm\alpha$, where α is the angle from the positive x-axis, the line of symmetry of the notch, to the face. The limiting case of a very narrow notch, when $\alpha = \pi$, is taken as a model for a crack.

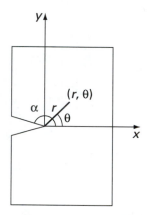

FIGURE 10.8   Notch model for a crack.

An Airy stress function is guessed which separates into a product of functions of r and of θ. The simplest such function that will produce a singularity in the stress at $r = 0$ has the form

$$\phi(r, \theta) = r^{\lambda+1} f(\theta).$$

This function is substituted into the biharmonic to determine the function $f(\theta)$. The resulting ordinary differential equation is

$$\frac{d^4 f}{d\theta^4} + 2(\lambda^2 + 1)\frac{d^2 f}{d\theta^2} + (\lambda^2 - 1)^2 f = 0, \tag{10.10}$$

and the boundary condition is $f = df/d\theta = 0$ at $\theta = \pm\alpha$.

Exercise 10.4   Verify this result by making the substitution of ϕ in the biharmonic written in polar form; see Chapter 4, Section 4.6.2.

The differential equation is a fourth-order linear equation with constant coefficients. Its solution has the form

$$\begin{aligned} f(\theta) = {} & C_1 \cos((\lambda - 1)\theta) + C_2 \sin((\lambda - 1)\theta) \\ & + C_3 \cos((\lambda + 1)\theta) + C_4 \sin((\lambda + 1)\theta). \end{aligned} \tag{10.11}$$

Exercise 10.5   Verify that this form satisfies the differential equation for f.

The possible values for λ are determined by substituting this form for $f(\theta)$ into the boundary conditions written in terms of f. The only acceptable solutions are those in which not all of the C_i are zero. The equations for the constants, C_i, can be written in matrix form.

$$\begin{pmatrix} \cos\left((\lambda-1)\alpha\right) & \cos\left((\lambda+1)\alpha\right) \\ (\lambda-1)\sin\left((\lambda-1)\alpha\right) & (\lambda+1)\sin\left((\lambda+1)\alpha\right) \end{pmatrix} \begin{pmatrix} C_1 \\ C_3 \end{pmatrix} = \begin{pmatrix} 0 \\ 0 \end{pmatrix};$$

$$\begin{pmatrix} \sin\left((\lambda-1)\alpha\right) & \sin\left((\lambda+1)\alpha\right) \\ (\lambda-1)\cos\left((\lambda-1)\alpha\right) & (\lambda+1)\cos\left((\lambda+1)\alpha\right) \end{pmatrix} \begin{pmatrix} C_2 \\ C_4 \end{pmatrix} = \begin{pmatrix} 0 \\ 0 \end{pmatrix}.$$

To avoid the unstressed solution in which all $C_i = 0$, the determinants of the two by two matrices in the previous two equations must both be zero so that at least one of the coefficients, C_i, is nonzero. This requirement produces the equations

$$\lambda \sin(2\alpha) + \sin(2\alpha\lambda) = 0;$$
$$-\lambda \sin(2\alpha) + \sin(2\alpha\lambda) = 0.$$

These equations when added require that $\sin(2\alpha\lambda) = 0$. If the crack is a narrow slit, then $\alpha = \pi$ and the equation determining acceptable values of λ becomes

$$\sin(2\pi\lambda) = 0.$$

The possible values are then

$$\lambda = \frac{1}{2}n,$$

for any positive or negative integer n. The negative values of n can be rejected since they produce infinite displacements near the crack root. The case $n = 0$ gives an unbounded strain energy near the root, which is also impossible. The integer, n, must be positive. The only value producing a stress singularity is $n = 1$. The terms for other positive values of n do not affect the singularity at, nor significantly change the stresses near, the crack tip. Therefore, the Airy stress function,

$$\phi(r, \theta) = r^{3/2} f(\theta),$$

with the correct functional form of f based on $\lambda = 1/2$ is used to compute the stresses in terms of the constants, C_i. From Chapter 4, Section 4.6.2,

$$\sigma_r = \frac{1}{r}\frac{\partial \phi}{\partial r} + \frac{1}{r^2}\frac{\partial^2 \phi}{\partial \theta^2}; \quad \sigma_\theta = \frac{\partial^2 \phi}{\partial r^2}; \quad \tau_{r\theta} = -\frac{\partial}{\partial r}\left(\frac{1}{r}\frac{\partial \phi}{\partial \theta}\right).$$

The values of the constants, C_i, are determined by the boundary conditions on the body away from the crack. The boundary conditions allow the constants, C_i, to be written in terms of constants, K_I and K_{II}, called the stress intensity factors. Again, these are determined from the far-field boundary conditions on the cracked body and its geometry.

Only mode I and mode II crack loadings are possible in plane stress or strain. In plane stress or plane strain, the only conditions where the Airy stress function is valid, the dominant terms of the stress and strain are as follows. These expressions are valid only in a neighborhood of the crack tip. For this reason, they are often called the near-field solutions of Irwin. While Irwin derived them by another technique based on the

1939 work of the elastician Westergaard using a complex variables representation of the Airy stress function, they agree with simplifications of the infinite series solutions given above.

$$\sigma_r(r, \theta) = \frac{1}{4\sqrt{2\pi r}}\left[K_I\left(5\cos\left(\frac{1}{2}\theta\right) - \cos\left(\frac{3}{2}\theta\right)\right)\right.$$
$$\left. + K_{II}\left(-5\sin\left(\frac{1}{2}\theta\right) + 3\sin\left(\frac{3}{2}\theta\right)\right)\right]; \tag{10.12}$$

$$\sigma_\theta(r, \theta) = \frac{1}{4\sqrt{2\pi r}}\left[K_I\left(3\cos\left(\frac{1}{2}\theta\right) + \cos\left(\frac{3}{2}\theta\right)\right)\right.$$
$$\left. + K_{II}\left(-3\sin\left(\frac{1}{2}\theta\right) - 3\sin\left(\frac{3}{2}\theta\right)\right)\right]; \tag{10.13}$$

$$\tau_{r\theta}(r, \theta) = \frac{1}{4\sqrt{2\pi r}}\left[K_I\left(\sin\left(\frac{1}{2}\theta\right) + \sin\left(\frac{3}{2}\theta\right)\right)\right.$$
$$\left. + K_{II}\left(\cos\left(\frac{1}{2}\theta\right) + 3\cos\left(\frac{3}{2}\theta\right)\right)\right]. \tag{10.14}$$

The displacements are

$$u_r(r, \theta) = \frac{1}{4\mu\sqrt{2\pi}}r^{1/2}\left[K_I\left((2\kappa - 1)\cos\left(\frac{1}{2}\theta\right) - \cos\left(\frac{3}{2}\theta\right)\right)\right.$$
$$\left. + K_{II}\left(-(2\kappa - 1)\sin\left(\frac{1}{2}\theta\right) + 3\sin\left(\frac{3}{2}\theta\right)\right)\right]; \tag{10.15}$$

$$u_\theta(r, \theta) = \frac{1}{4\mu\sqrt{2\pi}}r^{1/2}\left[K_I\left(-(2\kappa + 1)\sin\left(\frac{1}{2}\theta\right) + \sin\left(\frac{3}{2}\theta\right)\right)\right.$$
$$\left. + K_{II}\left(-(2\kappa + 1)\cos\left(\frac{1}{2}\theta\right) + 3\cos\left(\frac{3}{2}\theta\right)\right)\right], \tag{10.16}$$

where μ is the shear modulus and

$$\kappa = \begin{cases} \frac{3-\nu}{1+\nu} & \text{plane stress} \\ 3 - 4\nu & \text{plane strain.} \end{cases}$$

Notice that at the crack tip ($r = 0$) the stresses have a singularity in this theory. Therefore, the result cannot be accurate at the crack tip since no material can support an infinite stress. Along the line of the direction of crack propagation ($\theta = 0$), the normal stresses depend only on the mode I stress intensity factor, while the shear stress depends only on the mode II factor.

$$\sigma_r(r, 0) = \sigma_\theta(r, 0) = \frac{1}{\sqrt{2\pi r}}K_I; \tag{10.17}$$

$$\tau_{r\theta}(r, 0) = \frac{1}{\sqrt{2\pi r}}K_{II}. \tag{10.18}$$

Furthermore, the mode I terms in the normal stresses are symmetric about the line $\theta = 0$, the x-axis, while the mode II terms are asymmetric about $\theta = 0$.

Exercise 10.6   Sketch a graph of σ_r and of $\tau_{r\theta}$ with respect to r for the case that $\theta = 0$.

Because the Irwin theory is linear elastic, fracture due to combined modes may be analyzed by superposition of the separate effects of each mode.

EXAMPLE 10.8

Determine the near field stress expressions at the crack tip of length $2c$ in a thin infinite plate biaxially loaded by σ_x and σ_y (Fig. 10.9). The crack is oriented at an angle, ψ, with the y-axis.

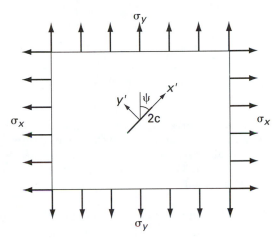

FIGURE 10.9   A crack at angle, ψ, in a biaxially loaded plate.

Solution

Write $k = \sigma_x/\sigma_y$ so that the biaxial stresses can be viewed as σ in the y-direction and $k\sigma$ in the x-direction. For an uncracked plate, σ and $k\sigma$ are principal stresses so that the Cauchy stress tensor, T, is diagonal. Place new coordinates x' and y' parallel and perpendicular to the crack, respectively. Apply the orthogonal rotation tensor, α, (see Chapter 1, Section 1.5) to obtain the stress tensor in the primed coordinates, $T' = \alpha T \alpha^t$.

$$\sigma_{x'} = \frac{k+1}{2}\sigma - \frac{k-1}{2}\sigma \cos(2\psi);$$

$$\sigma_{y'} = \frac{k+1}{2}\sigma + \frac{k-1}{2}\sigma \cos(2\psi);$$

$$\tau_{x'y'} = -\frac{k-1}{2}\sigma \sin(2\psi).$$

Since the plate is infinite, it can now be thought of as an infinite rectangular plate with edges parallel and perpendicular to the crack. The normal stresses in the prime directions can be viewed, by superposition, as the sum of a biaxial stress, $\sigma_{y'}$, and in the x'-direction, $\sigma_{x'} - \sigma_{y'}$. The biaxial stress causes a crack opening singularity and the shear stress causes a mode II singularity. These can be superposed along with the constant contribution to $\sigma_{x'}$ from the nonbiaxial stress in the x' direction. In this case,

$$K_I = \frac{1}{2}[k + 1 + (k-1)\cos(2\psi)]\sigma\sqrt{\pi c};$$

$$K_{II} = -\frac{k-1}{2}\sin(2\psi)\sigma\sqrt{\pi c};$$

and

$$\sigma_{x'} = \frac{K_I}{\sqrt{2\pi r}} \cos\left(\frac{1}{2}\theta\right)\left[1 - \sin\left(\frac{1}{2}\theta\right)\sin\left(\frac{3}{2}\theta\right)\right]$$
$$- \frac{K_{II}}{\sqrt{2\pi r}} \sin\left(\frac{1}{2}\theta\right)\left[2 + \cos\left(\frac{1}{2}\theta\right)\cos\left(\frac{3}{2}\theta\right)\right] - (k-1)\sigma\cos(2\psi); \tag{10.19}$$

$$\sigma_{y'} = \frac{K_I}{\sqrt{2\pi r}} \cos\left(\frac{1}{2}\theta\right)\left[1 + \sin\left(\frac{1}{2}\theta\right)\sin\left(\frac{3}{2}\theta\right)\right]$$
$$+ \frac{K_{II}}{\sqrt{2\pi r}} \sin\left(\frac{1}{2}\theta\right)\cos\left(\frac{1}{2}\theta\right)\cos\left(\frac{3}{2}\theta\right); \tag{10.20}$$

$$\tau_{x'y'} = \frac{K_I}{\sqrt{2\pi r}} \sin\left(\frac{1}{2}\theta\right)\cos\left(\frac{1}{2}\theta\right)\cos\left(\frac{3}{2}\theta\right)$$
$$+ \frac{K_{II}}{\sqrt{2\pi r}} \cos\left(\frac{1}{2}\theta\right)\left[1 - \sin\left(\frac{1}{2}\theta\right)\sin\left(\frac{3}{2}\theta\right)\right]. \tag{10.21}$$

∎

Exercise 10.7 Verify the above expressions for the stresses parallel and normal to the slanted crack by first transforming the given stresses in the unprimed system to the primed system. Separate the stress in the x'-direction into $\sigma_{y'} + (\sigma_{x'} - \sigma_{y'})$. Then use superposition of the stress field due to the equibiaxial loading, $\sigma_{y'}$, and that due to the shear, $\tau_{x'y'}$. Account for $(\sigma_{x'} - \sigma_{y'})$ in the normal stress in the x'-direction.

Mode III crack loadings can occur in torsionally loaded circular cross-sectional shafts with a circumferential crack (Fig. 10.10). The uncracked shaft has a standard cylindrical coordinate system with z-axis along the longitudinal axis, but to examine the crack, a cylindrical coordinate system is used instead in a radial plane containing a rod diameter and the longitudinal axis. The crack forms a notch in this plane which is surrounded by polar coordinates as in Figure 10.8, and z is perpendicular to this plane and so is not the longitudinal axis. The shear conditions are that the displacements and stresses near the crack satisfy $u_r = u_\theta = \sigma_r = \sigma_\theta = \sigma_z = \tau_{r\theta} = 0$. In pure shear with stresses τ_{rz} and $\tau_{z\theta}$, the strain-displacement equations, combined with Hooke's law, $\tau_{rz} = \mu(\partial u_z/\partial r)$ and $\tau_{\theta z} = (\mu/r)(\partial u_z/\partial \theta)$, and the equilibrium equation $\partial(r\tau_{rz})/\partial r + \partial\tau_{\theta z}/\partial\theta = 0$, combine to form the harmonic differential equation (Chapter 5, Section 5.1), $\nabla^2 u_z = 0$. The boundary conditions are $\tau_{z\theta} = 0$ if $\theta = \pm\alpha$, where α is the crack angle as in Figure 10.8. The shear stresses $\tau_{\theta z}$ create the tearing so that the crack propagates in the radial direction. A longitudinal crack that propagates in the radial direction is similarly Mode III except that the crack plane is now a cross section.

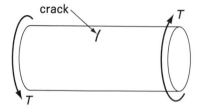

FIGURE 10.10 Circumferential crack in a shaft under a torque, T.

Assume that the displacement in the z direction is a separable function of the form

$$u_z(r, \theta) = r^\lambda f(\theta).$$

Substitution into the harmonic partial differential equation,

$$\nabla^2 u_z = \frac{\partial^2 u_z}{\partial r^2} + \frac{1}{r}\frac{\partial u_z}{\partial r} + \frac{1}{r^2}\frac{\partial^2 u_z}{\partial \theta^2} = 0,$$

and dividing out the terms involving r reduces the problem to the ordinary differential equation,

$$\frac{d^2 f}{d\theta^2} + \lambda^2 f = 0,$$

with the boundary condition, $df/d\theta = 0$ if $\theta = \pm\alpha$.

Exercise 10.8 Verify that the given differential equation for $f(\theta)$ follows from the harmonic partial differential equation.

The solution of the differential equation for f is of the form

$$f(\theta) = D\sin(\lambda\theta).$$

The boundary condition implies that $\cos(\alpha\lambda) = 0$. A crack that is a slit is obtained when $\alpha = \pi$, so that, again, $\lambda = \frac{1}{2}n$ for any integer n. Again, only the positive integers n are acceptable and the solution with $n = 1$ dominates. The coefficient, D, is rewritten in terms of the mode III stress intensity factor, K_{III}, and the Prandtl stress function analysis of Chapter 5, Section 5.1, is applied to obtain

$$\tau_{rz}(r, \theta) = \frac{1}{2}\sqrt{\frac{2}{\pi r}}K_{III}\sin\left(\frac{1}{2}\theta\right); \tag{10.22}$$

$$\tau_{\theta z}(r, \theta) = \frac{1}{2}\sqrt{\frac{2}{\pi r}}K_{III}\cos\left(\frac{1}{2}\theta\right); \tag{10.23}$$

$$u_z(r, \theta) = \frac{1}{\mu}\sqrt{\frac{2}{\pi}}K_{III}r^{1/2}\sin\left(\frac{1}{2}\theta\right), \tag{10.24}$$

where μ is the shear modulus.

Exercise 10.9 Draw a stress cube that shows that one or both of τ_{rz} and $\tau_{\theta z}$ cause tearing.

The stress intensity factors, K_I, K_{II}, and K_{III} are not the same as stress concentration factors because they depend on the magnitude of the load as well as the shape of the body. A stress concentration factor depends only on the geometry of the body. Each stress intensity factor is of the form,

$$K = Y\sigma\sqrt{\pi c},$$

where Y is a function depending on the shape of the body.

Figure 10.11 contains the Figures for Table 10.1.

Irwin postulated a different, but equivalent, fracture criterion from that of Griffith. Irwin's fracture criterion states that if the stress intensity factor, which depends on the external loads on the body, exceeds some critical value, K_c, then the crack propagates. The critical value, K_c, can be determined by experiment for each mode separately. The material property, K_{Ic}, is computed from the experimental curve representing the applied load versus the displacement across a notch at the specimen edge as described

TABLE 10.1 Stress Intensity Factors

Loading	Factor, K
1. General	$K_I = Y\sigma\sqrt{\pi c}$
2. Disk rotating at constant velocity	$K_I = 1.12\sigma\sqrt{\pi c}$
3. Plate of width $2w$, crack length $2c$	$K_I = \left(\dfrac{1-\frac{c}{2w}+0.326\frac{c^2}{w^2}}{\sqrt{1-\frac{c}{w}}}\right)\sigma\sqrt{\pi c}$
4. Symmetric edge cracks, c, width $2w$	$K_I = \left(\dfrac{1.12-0.61\frac{c}{w}+0.13\frac{c^3}{w^3}}{\sqrt{1-\frac{c}{w}}}\right)\sigma\sqrt{\pi c}$
5. Asymmetric edge crack, c, width w	$K_I = \left(\dfrac{1.12-0.23\frac{c}{w}+10.6\frac{c^2}{w^2}-27.1\frac{c^3}{w^3}+30.4\frac{c^4}{w^4}}{\sqrt{1-\frac{c}{w}}}\right)\sigma\sqrt{\pi c}$
6. Circumferential crack c, cylinder	$K_I = \sigma\sqrt{\pi c}\left(\frac{D}{d}+\frac{1}{2}-\frac{3d}{8D}-0.36\frac{d^2}{D^2}+0.73\frac{d^3}{D^3}\right)\frac{1}{2}\sqrt{\frac{D}{d}}$
7. Circumferential crack c, cylinder	$K_{III}=\frac{16T}{\pi D^3}\sqrt{\pi c}\left(\frac{D^2}{d^2}+\frac{D}{2d}+\frac{3}{8}+\frac{5d}{16D}+\frac{35}{128}\frac{d^2}{D^2}+0.21\frac{d^3}{D^3}\right)\frac{3}{8}\sqrt{\frac{D}{d}}$
8. Long beam	$K_1 = \frac{2\sqrt{3}Pc}{h^{3/2}B}$

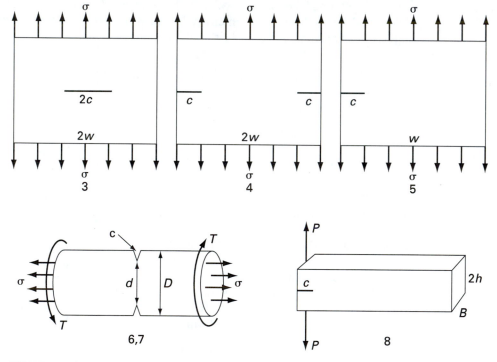

FIGURE 10.11 Dimensions used in the expressions for the stress intensity factors given in Table 10.1.

in ASTM Test Standard E399 [1]. The value computed depends on both the temperature and the load rate. In E399, a load P_Q is defined on this curve; this load generally falls just above the initial linear portion of the applied load versus the displacement across the notch curve. The property, K_{Ic}, is then computed using a relation, unique to the specimen shape, involving the load P_Q, the specimen dimensions and measurements from the crack length at fracture. This protocol is valid only for plane strain in which the plastic zone is small. The plastic zone at the crack tip that is due to the large stresses there must be small in the sense that $2.5(K_{I_c}/\sigma_{YS})^2$ is less than or equal to the specimen thickness and to the crack length, where σ_{YS} is the 0.2% yield strength. The calculation underpinning this criterion is discussed in Section 10.4.

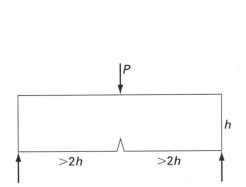

FIGURE 10.12   Crack opening in three-point bending.

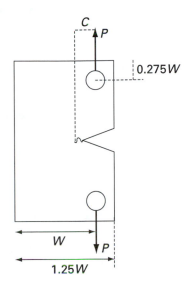

FIGURE 10.13   Crack opening in a tensile specimen.

The critical values, K_I, for crack opening in the linearly elastic fracture mechanics theory are obtained from tests such a three-point bending (Fig. 10.12) or a compact tension specimen (Fig. 10.13). The goal is to keep the plastic region at the tip very small with respect to the specimen thickness so that the measured fracture toughness, K_{I_c}, is a material property. Therefore, large specimens are often required to determine the fracture toughness. The intensity factor decreases with increasing thickness for smaller thicknesses due to the plastic zone at the crack tip. It is a much more difficult problem to determine a critical intensity factor when combined loads of several types of modes act on the body.

TABLE 10.2   Critical Stress Intensity Factors

Material	K_c (MPa-m$^{1/2}$)	Yield stress (MPa)
AISI 4340 steel	47.3	345
Carbon steel	219.8	240
2014-T4 aluminum	28.6	448
7075-T651 aluminum	29.7	545
Al_2O_3	2.0–6.0	
MgO	2.5	
TiC	3.0–5.0	
WC	6.0–20.0	
Soda lime silica glass	0.82	

EXAMPLE 10.9   Determine the critical stress for an axially loaded plate made of AISI 4340 steel and of width, $2w = 60$ mm, with a crack of length $2c = 0.145$ mm transverse to the load.

Solution   The stress intensity factor from Table 10.1 is

$$K_I = \left(\frac{1 - \frac{c}{2w} + 0.326\frac{c^2}{w^2}}{\sqrt{1 - \frac{c}{w}}} \right) \sigma \sqrt{\pi c}.$$

Substitute the critical stress intensity from Table 10.2 and the dimensions to obtain the critical stress, $\sigma = 3134$ MPa. Notice that the coefficient involving dimensions is approximately one. This stress is about 10 times the yield stress. The plate will yield before brittle crack propagation is initiated. ∎

EXAMPLE 10.10   A steel disk must rotate at $N = 2,720$ rpm. The bore has diameter $a = 2$ inches. Determine the minimum outer diameter, b, of the disk so that a crack of length $c = 0.1$ inches will not propagate. It is known that the critical mode I stress intensity is 38 ksi-in$^{1/2}$, Poisson's ratio is 0.3, the yield stress is $\sigma_Y = 80$ ksi, and the specific weight is 0.283 lb/in^3.

Solution   The stress intensity factor is

$$K_I = 1.12\sigma\sqrt{c\pi}. \tag{10.25}$$

Therefore the maximum allowable stress is $\sigma_{all} = 60.533$ ksi by equation (10.25). From Chapter 4, Section 4.6.1.3, the maximum stress is a circumferential tensile stress on the inner surface of the disk,

$$\sigma_{max} = \frac{3+\nu}{4}\rho\omega^2 c^2 \left[1 + \frac{1-\nu}{3+\nu}\left(\frac{a}{b}\right)^2 \right],$$

where $\rho = 0.0870$ is the mass density and ω is the angular velocity. Note that in this case, $\omega = 2\pi N/60 = 284.5$ radians/s. Substitute the expression for the allowable stress and solve for $b = 4.5$ inches. ∎

An important application of such mode III crack loading is in torsionally loaded shafts with a radial surface crack, such as might occur in the shaft of a motor.

EXAMPLE 10.11   Determine the critical torque, T, for a shaft of diameter, $D = 24$ mm, with a circumferential notch of depth, $c = 2$ mm (Fig. 10.14). It must rotate at $N = 5,000$ rpm. The critical stress intensity has been experimentally determined as $K_{IIIc} = 60$ MPa-m$^{1/2}$.

Solution $d = D - 2c = 20$ mm. From Table 10.1,

$$K_{III} = \frac{16T}{\pi D^3}\sqrt{\pi c}\left(\frac{D^2}{d^2} + \frac{D}{2d} + \frac{3}{8} + \frac{5d}{16D} + \frac{35}{128}\frac{d^2}{D^2} + 0.21\frac{d^3}{D^3} \right)\frac{3}{8}\sqrt{\frac{D}{d}}.$$

Substitute the critical stress intensity and other values and solve for $T = 1,452$ Nm. ∎

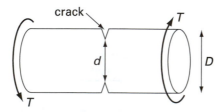

FIGURE 10.14   Torsional shaft with a circumferential notch of depth, $\frac{1}{2}(D-d)$.

There is a relationship between G and Irwin's stress intensity factors. In an infinite plate with a mode I crack, by the Griffith analysis and by equation (10.3), $\sigma_c\sqrt{\pi c} = \sqrt{E'G}$, for a mode I crack loading. Also $K_{I_c} = \sigma_c\sqrt{\pi c}$ ($Y = 1$). Therefore

$$K_{I_c} = \sqrt{E'G}, \tag{10.26}$$

where $E' = E$ in plane stress and $E' = E/(1 - v^2)$ in plane strain. In the general situation, the variation in strain energy as the crack length undergoes a variation δc is

$$\delta U = 2\int_{c+\delta c}^{c} 0.5(\sigma_y u_y + \tau_{xy} u_x + \tau_{yz} u_z)dx, \tag{10.27}$$

where the displacement field is held constant. Substituting the stress expressions for $\theta = 0$ and the strain expressions for $\theta = \pi$ from Irwin's field equations produces, as $\delta c \to 0$,

$$G = \frac{1}{E'}(K_I^2 + K_{II}^2) + \frac{1}{2\mu}K_{III}^2, \tag{10.28}$$

where μ is the shear modulus and $E' = E$ in plane stress and $E' = E/(1 - v^2)$ in plane strain.

In his original analysis for plane stress [14], Irwin interpreted his coefficient by computing the work done by the stress normal to the crack face in closing a small increment, α, of the crack extension. Irwin assumed that the shape of the opening, v, near the crack tip is parabolic so that $v(x) = (2K/\sqrt{\pi}E)\sqrt{2(\alpha - x)}$. By equation (10.13), the normal stress along the $\theta = 0$ line is $\sigma(x) = K/\sqrt{2\pi x}$. Then the work is

$$\int_0^\alpha \sigma(x)v(x)dx = \int_0^\alpha \frac{2K^2}{E\pi}\sqrt{\frac{(\alpha - x)}{x}}dx = \frac{2K^2}{E\pi}\frac{\alpha\pi}{2} = \alpha\frac{K^2}{E} = \alpha G.$$

Therefore G is a generalized force. In crack opening, the work is negative and so G is the crack opening force.

10.4 NONBRITTLE FRACTURE

The linear elastic fracture theory does not apply to many engineering metals. Ceramics, which are ionically or covalently bonded, generally undergo elastic fracture but face-centered cubic (FCC) metals undergo ductile fracture. Body-centered cubic (BCC) metals such as steel fracture by an intermediate means. In elastic fracture, it is expected that a tensile stress pulls atoms apart. However, in the plastic zone shear stresses induce slip and do plastic work. Experiments on most engineering metals show that the critical stress is larger than predicted by the linear elastic model. This additional work required is attributed to work done in the plastic region.

Orowan generalized the Griffith criterion in 1952 to include the plastic work, γ_p, in addition to the surface energy, γ, in order to make the model more realistic for the fracture of metals.

$$\sigma_c = \sqrt{\frac{2(\gamma + \gamma_p)E}{\pi c}}. \tag{10.29}$$

The plastic work is on the order of 100–1,000 J/m^2, while the surface energy is about 2 J/m^2. Therefore the term γ is often neglected.

EXAMPLE 10.12   A plain carbon steel ($E = 200$ GPa) has $G = 25$ kN/m.

(a) Compute the critical stress intensity, K_c, and the combined surface energy and plastic work term.

(b) If $\sigma_Y = 350$ MPa, what length of internal crack can be tolerated within the material to ensure that general yielding will occur before brittle fracture?

Solution   **(a)** Assuming plane stress, $K_c = \sqrt{GE} = 70.7$ MPa-m$^{1/2}$.

$$\gamma + \gamma_p = \frac{1}{2}G = 12.5 \quad \text{kJ/m}^2.$$

(b) $\sigma_Y = 350 \times 10^6 < \sqrt{GE/\pi c}$.

$$c < \frac{GE}{\pi \sigma_Y^2} = 0.013 \text{ m.} \qquad \blacksquare$$

10.4.1 Plastic Zone at the Crack Tip

The Irwin theory of stress intensity factors does not describe the actual mechanisms of crack growth in many materials because of the unrealistic stress singularity it predicts at the crack tip. This singularity results from the assumption that the crack is a slit so that the tip is extremely sharp. The stress equations are not valid at the crack tip. When the theory is applied to metals, the large stresses near the crack tip should result in plastic behavior there.

Irwin, in his 1961 paper [16], developed a technique to estimate the influence on the stress field near the crack tip by the existence of the plastic zone. The presence of a plastic zone at the crack tip changes the stress field near the crack and requires a modification of the brittle fracture analysis. To keep the form of the theory the same and yet account for this yielded region, Irwin proposed, for a perfectly plastic material, adding an adjustment, r_y, to the crack length in the calculation of the stress-intensity factor. His idea was to estimate the additional crack length required so that the stress field for the adjusted, or effective, crack length coincides with the stress field for the real crack beyond the plastic zone. The added crack length should be smaller than the plastic zone because stress relaxation in the plastic zone requires that more stress be carried beyond r_y.

From experiments on determining the stress-intensity factors, Irwin concluded that for plane stress, the correction term should be in the range $0.8r_y$–$1.3r_y$, and for plane strain in the range $0.3r_y$–$0.5r_y$, where $r_y = (K/\sigma_Y)^2/2\pi$. He suggested that in plane stress, the effective additional crack length should be r_y and in plane strain, the effective additional length should be $r_y/2\sqrt{2}$. In plane strain, the effective crack length is smaller than in plane stress because there are increased elastic constraints around the crack. Then the Irwin relation becomes $K = Y\sigma\sqrt{\pi c_{\text{eff}}}$, where $c_{\text{eff}} = c + r_y$.

Irwin said that the adjustment not only describes the effect of plastic strains on the crack edge stress field, but also on a fracture mode transition. Irwin had noted that for a crack in a plate the fracture mode changes from flat-tensile to oblique-shear as the plate thickness is reduced. In tests on 7075-T6 aluminum, this transition occurred when the plate thickness was $2r_y$.

The Irwin expression for the plastic zone correction can be written in terms of the applied stress.

$$r_y = \frac{1}{2\pi}\left(\frac{K}{\sigma_Y}\right)^2 = \frac{c}{2}\left(\frac{Y\sigma}{\sigma_Y}\right)^2,$$

under the assumption that $K = Y\sigma\sqrt{\pi c}$. The effective stress-intensity factor is then

$$K_e = Y\sigma\sqrt{\pi(c+r_y)} = Y\sigma\sqrt{\pi c}[1+(Y\sigma/\sigma_Y)^2/2].$$

The stress in the plastic zone is σ_Y because the material is assumed perfectly plastic, while the stress along the crack centerline outside the plastic zone, in the elastic region, obeys the relation $\sigma_y(x) = K_e/\sqrt{2\pi x}$. Here x is measured from the tip of the effective crack.

EXAMPLE 10.13   An aluminum alloy has a yield stress of 50 ksi. A large thin rectangular plate made of this aluminum with a center crack of full length 0.5 inches is observed to fracture under an in-plane axial 30 ksi stress normal to the crack. Determine the value of the stress intensity predicted by

(a) elastic fracture theory, and

(b) the elastic theory adjusted by Irwin's plasticity correction.

Solution   The plate is in plane stress. The half-crack length is $c = 0.25$.

(a) Assuming that $Y = 1$, $K_c = \sigma\sqrt{\pi c} = 30,000\sqrt{0.25\pi} = 26,587$ psi-in$^{1/2}$.

(b) $r_y = c(\sigma/\sigma_Y)^2/2 = 0.25(30/50)^2/2 = 0.045$ inches. So,

$$K_c = \sigma\sqrt{\pi(c+r_y)} = 30,000\sqrt{(0.25+0.045)\pi} = 28,880 \text{ psi-in}^{1/2}.\qquad\blacksquare$$

But in fact K changes with the existence of the plastic zone. The effective stress-intensity factor should be found without assuming $K = Y\sigma\sqrt{\pi c}$ in the expression for r_y. The stress intensity factor in r_y should be the effective K_e,

$$K_e = Y\sigma\sqrt{\pi\left[c + \frac{1}{2\pi}\left(\frac{K_e}{\sigma_Y}\right)^2\right]}.$$

Square both sides and solve for K_e.

$$K_e = Y\sigma\frac{\sqrt{\pi c}}{\sqrt{1-\frac{1}{2}\left(\frac{Y\sigma}{\sigma_Y}\right)^2}}.\qquad(10.30)$$

This value of the stress intensity does not necessarily satisfy the relation $K_e^2 = EG$.

EXAMPLE 10.14   Compare the latter [equation (10.30)] effective stress intensity factor, K_e, to the one calculated in the previous example. Assume the shape factor $Y = 1$.

Solution $K_e = 30,000\dfrac{\sqrt{0.25\pi}}{\sqrt{1-\frac{1}{2}\left(\frac{30}{50}\right)^2}} = 29,360$ psi-in$^{1/2}$. $\blacksquare$

The Irwin estimate can be related to the extent of the plastic zone in a perfectly plastic material for a mode I crack loading using a force equilibrium argument. The normal stress, σ_y, in the direction of the applied stress in the elasticity model has a singularity at the crack tip when examined along the line extending the crack, $\theta = 0$ or the x-axis. The stress is by equation (10.17)

$$\sigma_y = \frac{K_I}{\sqrt{2\pi x}}.\qquad(10.31)$$

This approximation uses just the first terms of the elasticity expression, those in which the singularity dominates. Therefore, this expression is only valid in a small neighborhood of the crack tip. Since the stress cannot actually be infinite as predicted by elasticity, a perfectly plastic region forms in which the stress is $k\sigma_Y$, where, σ_Y is the uniaxial yield stress and k is a constant which depends on the failure theory that applies. Recall that there is an equibiaxial state of stress near the crack tip along the crack direction.

A first estimate of the extent of this region is made by simply computing the position x at which the assumed stress distribution reaches the yield, $k\sigma_Y$.

$$k\sigma_Y = \frac{K_I}{\sqrt{2\pi x}}.$$

This implies that the plastic zone extends from the crack tip, $x = 0$, to $x = r_1$,

$$r_1 = \frac{K_I^2}{2\pi \sigma_Y^2 k^2}.$$

But this simple estimate of the extent of the plastic region cannot be correct since the resulting stress field is no longer in equilibrium for a perfectly plastic material. The shaded region in Figure 10.15 measures the amount of stress that, while balanced in the elastic model, is not balanced in the elastic–perfectly plastic model. The stress must redistribute to attain equilibrium.

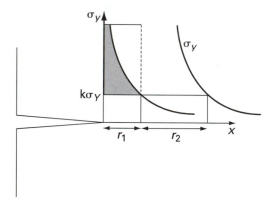

FIGURE 10.15   Plastic zone and stress state near a crack tip.

The elastic curve given by equation (10.31) merely translates to the right a distance, r_2, extending the perfectly plastic region, until stress equilibrium is achieved. The fact that force equilibrium is maintained by this translation is expressed in the relation setting the shaded area equal to the area under $k\sigma_Y$ and over r_2 in Figure 10.15,

$$\int_0^{r_1} \left(\frac{K_I}{\sqrt{2\pi x}} - k\sigma_Y \right) dx = r_2 k\sigma_Y. \tag{10.32}$$

Since the elastic curve is assumed to translate to its new position,

$$\frac{K_I}{\sqrt{2\pi r_1}} = k\sigma_Y.$$

Solving this expression for r_1 and substituting in equation (10.32) after integration shows that $r_1 = r_2$ and therefore the extent of the plastic zone is

$$r_1 + r_2 = \frac{K_I^2}{k^2 \pi \sigma_Y^2}. \tag{10.33}$$

The Tresca yield condition requires that $k = 1$ for plane stress, which is satisfactory for thin plates. In plane strain, on the other hand, it is now common to take $k = \sqrt{3}$ from the Mises criterion. In this case the plastic region extends a distance

$$\frac{K_I^2}{3\pi \sigma_Y^2}$$

from the crack tip.

EXAMPLE 10.15   Determine the Irwin type estimate of the plastic zone extent around a crack of length $2c = 2$ mm in a thin plate of width $2w = 80$ mm made of steel with a yield stress of 130 MPa. The applied stress is $\sigma = 50$ MPa.

Solution   From Table 10.1,

$$K_I = \left(\frac{1 - \frac{c}{2w} + 0.326 \frac{c^2}{w^2}}{\sqrt{1 - \frac{c}{w}}} \right) \sigma \sqrt{\pi c}.$$

$K_I = 1.000286 \sigma \sqrt{\pi c} = 2.803$ MPa-m$^{1/2}$. The plate is in plane stress so that, for $k = 1$,

$$r_1 + r_2 = \frac{K_I^2}{\pi \sigma_Y^2} = \frac{2.501}{130^2} = 0.1480 \text{ mm.} \qquad \blacksquare$$

To apply the plane strain fracture criterion, the size of the plastic zone must be small compared to the specimen dimensions, say in a ratio of about 1/50. In particular, the crack length must be much larger than the extent of the plastic zone and the uncracked dimensions from the crack to the specimen edges must be large. Further, the specimen must be thick enough that the body is mostly in plane strain. An experimentally determined condition that ensures these conditions are met is that the thickness, B (a commonly used notation in fracture analysis), crack half-length, c, the distance, h, from the crack to the edge on which the load is applied, and $w - c$, where w is the specimen width, of the specimen satisfy,

$$B, w - c, h, c \geq 2.5 \left(\frac{K_I}{\sigma_Y} \right)^2. \tag{10.34}$$

Because of Irwin's experiments, some say that if a plate thickness is less than $2r_y$, then the plane stress assumption is satisfactory.

10.4.2   The Crack Tip Opening Displacement

The relation $K_e^2 = GE$ is not proved and so cannot be used to predict a critical fracture stress in materials with a large plastic zone. An alternative material parameter is needed. The crack tip opening displacement is used for materials with significant plastic flow.

Dugdale carried out an analysis of the plastic zone that differed from that of Irwin. The analysis assumed plane stress and an applied stress, σ, such that σ/σ_Y is small. Dugdale defined an effective crack that extends to the end of the plastic zone, that is, to the point at which the internal stress near the crack along the midline of the crack equals the yield stress. In Irwin's analysis, such an effective crack half-length would be $c + r_1 + r_2$.

If one imagines a crack of this length superimposed on the original crack of half-length c, the imagined crack has an opening, at the position of the original crack tip, of length δ from the centerline (Fig. 10.16).

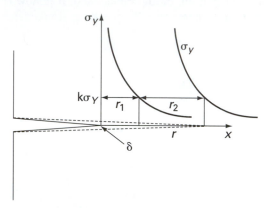

FIGURE 10.16   Effective crack length, with width δ at the actual crack tip.

Definition   The crack opening displacement (COD), δ, is the separation width of the effective crack measured at the position of the tip of the actual crack. This is also called the crack tip opening displacement (CTOD).

This imagined opening displacement might be used, for example, as a measure of how far the stress intensity attempts to separate the molecules at the tip where the crack growth occurs. In tests, the CTOD is the actual crack surface separation displacement measured at the location of the tip of the initial crack; it results from both elastic and plastic deformation.

The Irwin equation (10.16) for the displacement u_θ at $\theta = -\pi$ and r arbitrary implies that the displacement for crack opening only is

$$u_\theta = \frac{1}{4\mu\sqrt{2\pi}} r^{1/2} K_I (2\kappa + 2).$$

Apply this to the effective crack. Put the origin at the tip of the effective crack. Then the tip of actual crack is at $r = r_1 + r_2$ and $\theta = \pm\pi$. The crack opening displacement, δ, is twice the displacement, u_θ. Using Irwin's expression $r_1 + r_2 = K_I^2/(\pi \sigma_Y^2)$ produces for plane stress

$$\delta = 2u_\theta = \sqrt{2}\frac{4}{\pi E}\frac{K_I^2}{\sigma_Y},$$

for a crack open to the end of the effective tip. This value is too large compared to experiment. Irwin used Westergaard's elasticity calculation for an infinite plate to estimate

$$\delta = \frac{4}{\pi E}\frac{K_I^2}{\sigma_Y}. \tag{10.35}$$

Dugdale [10] in 1960 analyzed the extent of the plastic region without the Irwin restriction to perfectly plastic materials. But he only considered a crack opening load on a thin infinite plate. This is therefore a plane stress solution. He assumed an effective crack length that was the sum of the actual physical crack and the length of the plastic zone. He also assumed that all the plastic behavior is concentrated along the center line in the plastic zone. According to the Dugdale model, the Irwin plastic zone extent calculation was underestimated by 20%, and the distance is

$$\frac{\pi K_I^2}{8\sigma_Y^2}.$$

This estimate of the size of the plastic zone depends on the size of the crack and on the external loading through the "strength," K_I, of the singularity.

EXAMPLE 10.16

Compute the size of the plastic zone according to the Dugdale model for a crack of length $2c = 2$ mm in a thin plate of width $2w = 80$ mm made of steel with a yield stress of 130 MPa. The applied stress is $\sigma = 50$ MPa.

Solution

The Dugdale estimate is $\pi^2/8 = 1.234$ times that of Irwin. From Example 10.15, the zone extent is $\pi^2(0.1480)/8 = 0.1826$ mm. ∎

The Dugdale model, for mode I behavior of a crack in a thin plate in plane stress, predicts a crack opening displacement of

$$\delta = \frac{K_I^2}{E\sigma_Y}. \tag{10.36}$$

No analytical expression exists for the crack tip opening displacement in a case of plane strain.

EXAMPLE 10.17

For a crack of length $2c = 2$ mm in a thin plate of width $2w = 80$ mm made of steel with a yield stress of 130 MPa and $E = 200$ GPa, compute the crack opening displacement predicted by the Irwin and Dugdale models. The applied stress is $\sigma = 50$ MPa.

Solution

By Example 10.15, $K_I = 2.803$ MPa-m$^{1/2}$. The Dugdale estimate is

$$\delta = \frac{K_I^2}{E\sigma_Y} = 0.000302 \text{ mm.}$$

The Irwin estimate is $4/\pi$ times the Dugdale, $\delta = 0.000385$ mm. ∎

The crack tip opening criterion is used in cases of nonbrittle fracture. The critical CTOD, δ_{I_c}, is the value when slow stable crack extension begins. It is a material property even if there is a large plastic zone. On the other hand, δ_c, is the value when unstable crack extension begins. It may depend on the specimen size and geometry. For example, the CTOD for a single-edge bend specimen (Fig. 10.12) at a fixed temperature is calculated from experiment by

$$\delta = \frac{K^2(1 - \nu^2)}{2\sigma_Y E} + \frac{r_p(W - c_o)v_{pl}}{r_p(W - c_o) + c_o + z}, \tag{10.37}$$

where K is the stress intensity factor for the original crack length, $r_p = 0.44$, W is the specimen width, v_{pl} is the plastic component of the crack mouth opening displacement, c_o is the initial crack length, and z is the distance of the knife edge measurement point from the notched edge on the single-edge bend specimen. Here the plastic component of the crack mouth opening displacement is computed by subtracting $P \times C$ from the value of the crack mouth opening displacement, v, where P is the load and the compliance, $C = \Delta v / \Delta P$ on the load versus load-line displacement curve. The compliance can also be computed in terms of the crack length using relationships dependent on the specimen shape (see ASTM E1820, Annex 1 [1]).

The crack opening displacement predicted by the Dugdale model is less than that predicted by the Irwin model (10.35), by $\pi/4 = 0.7854$. In either case, a critical value of the crack opening displacement, δ_{I_c}, correlates with the critical value of K_I for brittle crack propagation. In the Dugdale model for small values of σ/σ_Y, $\delta_{I_c} = K_{I_c}^2 / E\sigma_Y$ and the energy release rate relation, $G_I = K_{I_c}^2/E$, imply that

$$G_{I_c} = \delta_c \sigma_Y. \tag{10.38}$$

Experiment shows that the critical crack tip opening displacement at fracture is a material property. The idea of a critical crack opening fracture criterion was suggested independently by A. H. Cottrell and A. A. Wells in 1961.

10.4.3   The Parameter, *J*, for Nonlinear Elastic Materials

Cracks in nonlinear elastic materials are studied by the parameter J, where

$$J = \frac{1}{B}\frac{dU}{dc}, \tag{10.39}$$

B is the specimen thickness, and U is the total potential energy for the cracked and nonlinear elastic system. J generalizes the term G introduced by Irwin. It has also been applied to plastic loadings by ignoring dissipative energy losses and not allowing unloading. It is commonly used in cases of large plastic regions around the crack tip.

Today, fracture analysis is often carried out in terms of a line integral form of J called the J-integral proposed by J. R. Rice in 1968 [27]. This path independent integral along a contour, Γ, around the crack tip stress singularity is roughly the integral of the difference between the nonlinear elastic strain energy (work per unit volume done by the load), W, and the work input through the contour done by the stresses. This work difference, since $dx_2 = n_1 ds$,

$$J = \int_{\Gamma} \left(W n_1 - (\mathbf{T} \cdot \mathbf{n}) \cdot \mathbf{u}_1 \right) ds, \tag{10.40}$$

where the parameter, s, is the arclength along Γ. The term $W n_1$ is the work per unit volume done by the load; n_1 is the component of $\mathbf{n}$ in the 1-direction. The vector $\mathbf{n}$ is the unit vector perpendicular to Γ at s, $\mathbf{T} \cdot \mathbf{n}$ is the stress vector on the $\mathbf{n}$-face, and $\mathbf{T}$ is the stress tensor (Chapter 1). The stress is dotted with $\mathbf{u}_1$, the derivative with respect to x_1 of the displacement vector, $\mathbf{u}$, at s along Γ to get the work per unit volume in the 1-direction passing through the contour, Γ. The integral tends to G as the contour shrinks to the singularity. It is proportional to the product, $\sigma_Y \delta$.

J is measured experimentally using the area under the load versus line-load displacement curve (see ASTM Testing Standard E1820). Experiment shows that J varies with the specimen thickness. It increases from its value for a thin member in plane stress until the thickness is large enough that the specimen is in plane strain. It remains constant for larger thicknesses once plane strain is achieved. The value in plane strain is taken as a material constant.

10.4.4 Crack Extension Resistance

The effective stress intensity describes the elastic stress field surrounding the plastic zone. It cannot be used to predict a critical stress for propagation as in a brittle fracture. A single parameter, K_c, is sufficient in brittle fracture; otherwise a function of the crack extension called the *crack extension resistance*, R, is commonly used to determine the critical stress for the propagation of stable cracks. Presumably in materials with a large plastic zone, some work is performed at the crack tip to propagate the crack while additional work is done in the plastic zone. The resistance is the rate of energy dissipation in performing these two types of work when the crack growth is stable. Work is required, for example, to extend the plastic zone as the crack propagates. Some call the work to create new surfaces the essential work of fracture. In elastic fracture, all the work done is essential work of fracture. An experimental method to determine R is defined in ASTM Test Standard E1820. R is computed in terms of the stable crack extension from a load versus load-line displacement curve. In a typical $J - R$ curve, the resistance increases with the crack extension, but does not depend on the initial crack size. The $J - R$ curve is a plot of J versus crack extension below upper limits on J and on the crack extension; the limits depend on the specimen dimensions and the crack length. The stable crack growth becomes unstable when $G = R$.

The stress intensity is often related to the resistance factor, R, by

$$K_I = \sqrt{E'R}. \tag{10.41}$$

10.5 DISLOCATION BEHAVIOR NEAR A CRACK TIP—BRITTLE AND DUCTILE MATERIALS

The physical mechanisms of crack propagation occur within the plastic zone at the crack tip. One of the mysteries of elementary strength of materials is the full meaning of the distinction between brittle and ductile materials. It is easy enough to say that brittle materials fail due to normal stress and ductile materials fail due to shear stress. But there are all shades of behavior in between.

The previous analyses all are based on a continuum model of the material. Such a model omits the details of the fracture mechanisms since the metal grains are built of an often imperfect atomic lattice with point, line, and area defects. Consequently, the continuum models may fail to make accurate fracture predictions.

Elasticity continuum models exhibit a singularity both at cracks and at dislocations. The elastic models for an edge and for a screw dislocation were discussed in Chapter 3, Section 3.6, and Chapter 4, Section 4.6.2. Not only can these singularities interact mathematically, but there is also a physical relationship between cracks and dislocations. Dislocations can be generated by the large stresses at the crack tip. These may facilitate cracking. But existing dislocations further away can also interact with the crack by presenting a "shield" for the crack tip and inhibit crack propagation. In this sense, the dislocations may stabilize the crack. The crack propagation is said to be arrested. The material toughness is the energy absorbed in propagating a crack a unit length. Existing dislocations can increase toughness by shielding, through the dislocation singularity, the crack tip from the influence of applied external stress. This type of interaction between crack tips and dislocations determines whether a material fracture is judged to be brittle or ductile.

Dislocation emission from a crack tip is related to the shape of the crack tip. The larger the radius of curvature at the tip the smaller the stress magnitude. A rounded tip may create a stress sufficient to generate a plastic zone around the tip, but a sharp tip is required for a singularity. As the crack propagates, dislocations are created at the crack tip and move away from the emission surface. The dislocations are said to be emitted. The dislocation motion tends to blunt the crack tip. Elastic fracture and blunting are sometimes viewed as competing mechanisms under which the material undergoes a sharp fracture if the local stress at the tip exceeds the fracture stress and blunting if the local stress exceeds the critical shear stress and this is less than the fracture stress. The blunted crack tip is no longer a stress singularity, and some analyses show that the maximum stress at the tip may be reduced to three times the yield stress. This idealized classification ignores the effects of mechanisms such as hardening in the plastic zone. These mechanisms compete in BCC metals. However, especially in steel, more complex microstructural behavior occurs. The carbide precipitates can be quite brittle. Nearby dislocations in the plastic zone induce stress concentrations sufficient to cause some precipitate particles ahead of the crack tip to fracture. This forms a sharp microcrack in the plastic region which itself may propagate ahead of the blunted crack tip if enough energy is supplied to separate the surrounding atoms. Some experiments show the required energy to be on the order of 9-14 J/m^2, which is greater than the surface energy of 2 J/m^2 [19]. Crack blunting does not occur in ceramics.

Brittle failure is due to unstable crack formation and growth. Brittle fracture is of two types. In cleavage fracture, two adjacent lattice planes in a crystal separate. Separation occurs more easily along certain planes in the lattice, for example, {001} in BCC metals and {0001} in HCP metals. As the crack reaches another grain, it changes direction to follow a cleavage plane in the lattice structure of the new grain. This change requires an added amount of energy due to plastic tearing of the connecting material. If the fracture surface is inspected (for example, Fig. 10.17), the local cleavage fractures can be identified by the shiny, smooth surface in each grain. The upper limiting threshold stress for such a fracture is

$$\sigma_t = \sqrt{\frac{E\gamma}{a}},$$

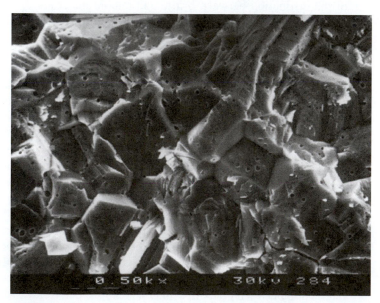

FIGURE 10.17   A scanning electron fractograph of cleavage fracturing among polycrystal grains of MgO, also showing significant porosity within the material. (Courtesy of J. D. Williams.)

where *a* is the unloaded lattice parameter. In mixed mode loading, there is a balance between cleavage planes and shear induced dislocation emission surfaces. The cleavage plane is the plane which forms the crack surface as it propagates into the material. However, this plane changes if the crack propagation changes direction in the body.

The second major type of brittle fracture is intercrystalline fracture. Brittle cracks can occur along grain boundaries. The crack follows the grain boundaries. This mechanism tends to occur most often in materials with weakly bonded grains. Cracks can also form around precipitate particles at the grain boundaries.

Ductile failure is mostly evidenced by necking in a tensile test. In fully ductile failure, the material necks until the cross section is just a few atoms wide. The local stress is then sufficiently increased to break the remaining bonds. Ductile fracture requires more energy than brittle failure because of the work associated with plastic behavior. The surface of a ductile fracture is usually dull. The dislocations are more mobile in ductile fracture, and the crystal slips along the critical shear planes. An FCC material has multiple {111} slip planes, and essentially no cleavage planes, both of which account for the observed ductility of FCC metals. In one type of ductile fracture, the surface is fibrous where local plastic deformation occurs.

A second important type of ductile fracture creates a dimpled pattern on the fracture surface. Often, the metal has crystallized around a nucleus formed by a higher temperature stabilized phase precipitate of the metal. Decohesion of the particles leads to crack initiation at these inclusions in each grain creating an indentation or dimple in each grain at the center of which is the nucleating particle. In a sense, these fractures are the result of void growth. Disjoint particles can be left within each hole. During fracture the holes link up which leads to the desription of ductile fracturing as a hole-joining process (Fig. 10.18, Fig. 10.19). Ductile crack growth is relatively stable compared to cleavage.

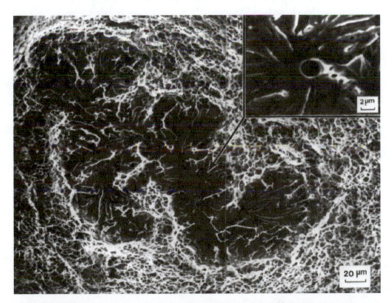

FIGURE 10.18   A scanning electron fractograph obtained from the fracture surface of a pressure vessel steel tested near the upper end of the Charpy V-notch ductile to brittle transition, showing a localized cleavage region spanning several grains, with a surrounding area of ductile fracturing by hole-joining. The inset indicates that the fracture origin was at a broken inclusion particle that triggered cleavage in one grain that then spread to neighboring grains. (Courtesy of X. J. Zhang and G. R. Irwin.)

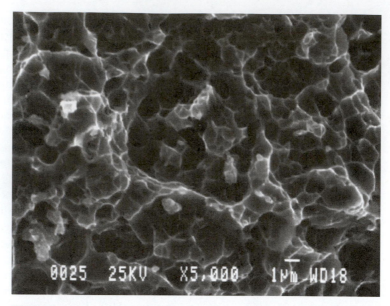

FIGURE 10.19 A high-magnification scanning electron fractograph of hole-joining linkages on the ductile fracture surface obtained in a tensile test of an AerMet 100 high-strength steel. (Courtesy of X. J. Zhang.)

Armstrong in 1966 [2] proposed the parameter $\beta = \mu b / \gamma$, where μ is the shear modulus, b is the magnitude of the Burgers vector, and γ is the surface energy, as a measure of the intrinsic ductility of a pure material (originally given as $(\gamma / \mu b)^{1/2}$). A metal was proposed to be ductile if $\beta < 10$. This parameter clearly distinguishes between materials that cleave and those that do not.

In (crack-containing) fracture mechanics tests, the stress intensity, K, for cleavage fracture in steel has a Hall-Petch dependence on the average grain diameter, l.

$$K = cs^{1/2}[\sigma_0 + kl^{-1/2}], \qquad (10.42)$$

where c is a material parameter, s is the effective length of the local plastic zone associated with unstable crack growth, σ_0 is a friction stress for dislocation motion within the polycrystalline grains, and k is the microstructural stress intensity [3, 4].

10.6 CRACK INITIATION

One of Griffith's fundamental ideas was that flaws within a material can grow into dangerous cracks. These flaws might be introduced during the processing of the materials or by later exposure to chemical, variable thermal, or other environmental effects. Many of the latter effects operate primarily on the surface of the body.

Crack nucleation can also be induced by dislocation behavior within the material. The flow processes of local plasticity can cause a pile-up of dislocations at a barrier such as a grain boundary, an inclusion of a particle of another phase of the material, or at an existing pile-up. The resulting stress concentration can initiate a crack. In this process, the shear stress component along the slip planes of the lattice is as important as the tensile stresses attempting to separate the crack planes. The dislocation pile-ups form slip bands.

Suppose that two adjacent barriers are separated by a distance d. The lattice friction stress is τ_0. As the dislocations pile up against the barriers, the stress relaxes from the applied stress, τ_{xy}, to τ_0 so that there is a recovered elastic strain of $(\tau_{xy} - \tau_0)/\mu$, where μ is the shear modulus. There is a plastic strain nb/d, where b is the magnitude of the Burgers vector. These two strains are equal in equilibrium so that the equilibrium number of dislocations is

$$n = (\tau_{xy} - \tau_0)\frac{d}{\mu b}.$$

The force $(\tau_{xy} - \tau_0)b$ is exerted by the applied load on each dislocation in the pile-up. Therefore crack nucleation occurs when

$$(\tau_{xy} - \tau_0)n = p_t, \tag{10.43}$$

where p_t is the tensile cohesive strength. The cohesive strength model implies that $p_t = \gamma/b\pi$, where γ is the surface energy.

Substitution of the expression for the tensile cohesive strength in equation (10.43) results in the Petch relation, with d equal to the grain diameter

$$\tau_{xy} = \sigma_f + \sqrt{\frac{\mu\gamma}{\pi d}}. \tag{10.44}$$

This relation implies that brittle solids may be strengthened by refining them to lower the minimum value of d by reducing the material grain size. The dependence on the inverse square root of the grain size is very similar to the Griffith dependence on the crack length. This theory does not account for the inhibition of cracking by dislocation shielding and residual stresses.

This relationship has led authors such as Bilby, Cottrell, and Swinden [7] to propose stress fields based on dislocation patterns. The Bilby, Cottrell, Swinden, Dugdale (BCSD) crack occurs in plane strain. The plastic zone near the tip is viewed as a region on the crack plane, so that it is one dimensional. The dislocation field lies in this plane.

A Griffith model mode I crack as described in this chapter closes after the stress is removed. A Zener-Stroh-Koehler (ZSK) crack does not. In the plastic region, dislocation pile-ups, especially of edge dislocations, at a grain boundary or other obstacle can initiate a microcrack. Cottrell proposed that pile-ups may also occur at the intersection of two slip planes and cause microcrack nucleation. The applied stress drives the dislocations together in such a manner as to form a crack nucleus. The crack tip in these cases is sharp at the lattice scale and does not close when the applied stress is removed. Weertman [30] has written the stress field equations for such cracks. Also see [22, Chapter 17].

10.7 DESIGN

A designer must specify the geometry, dimensions, and material of a member, as well as the loads on it, to ensure that the member will not fail in service. The dependence of the fracture strength on crack size requires that design strategies based on yielding alone be modified. Design only to prevent yielding of a metal component assumes that the material of the body is without defects. In practice, this is rarely the case. Defects can be introduced in the production of the materials, in manufacturing transformations such as rolling, cutting, milling, etc., and by environmental actions on the member in service. The designer must control the microstructure by specifying a heat treatment, cold working, etc. for the material. The finish on a machined part must also be carefully specified to control the surface flaws. A designer must size the body to prevent crack induced fracture, yielding, and geometric instabilities such as buckling. Instabilities are treated in Chapter 13.

EXAMPLE 10.18   Determine the maximum allowable load on a uniaxially loaded thin plate of width $w = 250$ mm and thickness $t = 5$ mm made of a material with yield stress $\sigma_Y = 450$ MPa and critical crack opening intensity $K_c = 30$ MPa-m$^{1/2}$. The allowable crack length is 2 mm. Assume center cracks.

Solution   From Table 10.1,

$$K_I = \left(\frac{1 - \frac{c}{2w} + 0.326 \frac{c^2}{w^2}}{\sqrt{1 - \frac{c}{w}}} \right) \sigma \sqrt{\pi c}.$$

Substitute the given values and solve for the allowable stress. Again, the coefficient, Y, is essentially 1. The allowable stress is $\sigma = 378.5$ MPa. The crack will begin to propagate at a load less than the yield stress. A design that protected against just yielding would be likely to fail. ■

The yield criteria such as the Tresca or von Mises are pointwise criteria. The fracture criteria such as the Griffith or stress intensity are global in the sense that they make a prediction that a crack will propagate from the loads on, and the shape of, the body.

A member with a geometric discontinuity causing a stress concentration, such as a plate with a hole, may develop a crack at the boundary of the discontinuity. A designer must decide whether the stress concentration or the stress intensification due to the crack dominates the behavior. For example, consider an axially loaded plate with a hole of radius, R, in which a crack of length, c, perpendicular to the load has occurred at the hole (Fig. 10.20) [8, p. 299].

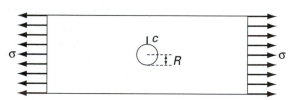

FIGURE 10.20   Axially loaded plate with a hole having a crack at the hole.

If the crack is small, the upper portion of the plate above the hole can be viewed as a plate with a crack at its lower edge. By Table 10.1, if the plate is very wide compared to the crack, the stress intensity is $K = 1.12(\text{normal stress})\sqrt{\pi c}$. But the stress at the crack is 3σ because of the stress concentration factor of 3 due to the hole. The stress intensity is therefore

$$K_I = 1.12(3\sigma)\sqrt{\pi c}.$$

If the crack is large, the hole can be considered part of the crack. The stress intensity for a plate with a center crack, assuming a shape factor $Y = 1$, is

$$K_I' = \sigma \sqrt{\pi (c + R)}.$$

The graphs of these two stress intensities can be plotted as $K_I / (\sigma \sqrt{R})$ versus c/R for fixed σ. For small cracks, $K_I < K_I'$ and for large cracks, $K_I > K_I'$. Their intersection is found by setting $K_I = K_I'$ and solving to find $c/R = 0.0972$. The curve for K_I approximates the exact stress intensity relation for a plate with a hole at small crack lengths and K_I' approximates it at large crack lengths. For cracks smaller than the critical ratio, a designer can make estimates using K_I and for larger cracks using K_I' if the exact expression for the stress intensity K is not available. For a plate with a hole,

$$K = 0.5(3 - r)\left(1 + 1.243(1 - r)^3\right)\sigma\sqrt{\pi c},$$

where $r = c/(R + c)$.

Exercise 10.10   Plot $K_I/\sqrt{c}$ and $K_I'/\sqrt{c}$ versus c/R on the same coordinate axes over the range $0 \le c/R \le 0.2$. Verify algebraically that their intersection occurs at $c/R = 0.0972$.

EXAMPLE 10.19   A large uniaxially loaded, 4340 steel plate with a hole of radius $R = 0.5$ inches has a crack at the edge of the hole of length $c = 0.01$ inches.

(a) Estimate at what stress the crack will propagate.

(b) If the plate must support an in-plane axial tensile load of $P = 8,000$ lb/in perpendicular to the crack, determine the minimum required plate thickness, t, to avoid crack propagation.

Solution   (a) The ratio $c/R = 0.02$ so that an estimate may be made from $K_I = (1.12)3\sigma\sqrt{\pi c}$. From Table 10.2, $K_{I_c} = 47.3$ MPa-m$^{1/2} \sim 42,490$ psi-in$^{1/2}$. Solving for σ yields a critical stress of $\sigma = 71,345$ psi.

(b) The stress is $\sigma = P/t$. Therefore, $t \ge 8,000/71,345 = 0.1121$ inches.   ∎

10.8   TEARING OF RUBBER

Rubber researchers have developed a rupture criterion which is similar to the Griffith criterion for linear elastic materials. The expression for rubber, which is a nonlinear elastic material, is given in terms of the strain energy, W, per unit volume at fracture. This stored energy at break is $W = W_b - W_d$, where W_b is rupture energy per unit volume and W_d is the energy dissipated prior to rupture. Roughly, if these energies are measured in erg/cm^3, $W_b = 4.1W_d^{2/3}$. The energy required to extend the fracture by a unit area is denoted by 2θ; the 2 is used because there are two surfaces to the crack. The energy $\theta \sim r\bar{W}_b$, where r is the tip radius of the tear and $\bar{W}_b$ is the breaking energy per unit volume when no flaw is present. For hydrocarbons, θ can be as small as 10 J/m^2 (10^4 erg/cm^2). However, experimental values are typically in the range of 10^2 to 10^5 J/m^2. The energy θ has been experimentally found to be dependent on temperature and on the rate of tearing. For example, in a vulcanized styrene-butadiene rubber (SBR), the log of θ varies approximately linearly with the log of the tearing rate. If c is the length of the initial crack, the rupture criterion is

$$4cW \ge 2\theta. \tag{10.45}$$

Because the measured activation energy for rubber fracture is very similar to that for rubber viscous flow, some researchers believe that the dominant failure mechanism in rubber is the breaking of intermolecular bonds. The initial flaws in manufactured rubber components typically have lengths on the order of 0.001 cm. Behavior at the crack tip raises the critical value of θ as much as to 10^5 J/m^2. For example, a natural rubber crystallizes under high strain. Such crystallization at a crack tip can greatly increase tearing resistance.

The tear resistance of natural rubber is higher than that of most synthetic rubbers. Although, crystallizing synthetics do have high tear resistance. Vulcanizates of styrene-butadiene rubber (SBR) are less tear resistant than natural rubber but their resistance is still high compared to other synthetic rubbers. SBR is the second most common industrial rubber behind natural rubber; it is often used in tires. SBR was developed by E. Tchunker and A. Bock in 1929, using a 3 to 1 ratio of butadiene to styrene, and went into large-scale production in Germany in 1937. Vulcanized chloroprene rubber as well as the polyester urethane and polyether urethane rubbers are also highly tear resistant in comparison to other elastomers. Epichlorohydrin rubbers are frequently used in the automobile industry because of their high tear resistance and good mechanical properties. Tear resistance can be increased by cross-linking or the use of fillers like carbon black. Cracks form easily in natural rubber but do not grow as quickly as those in SBR, in which crack initiation is slower.

The uniaxial failure envelope is a curve in strain-stress space that relates the uniaxial stress at fracture to the uniaxial strain at fracture, ϵ_B, for different temperatures, different strain rates, or other parameters. Harwood and Payne in 1968 [25] showed a relation between the strain energy (work done) to break, U_B and ϵ_B, to be $U_B = A\epsilon_B^2$, where the coefficient A accounts for temperature, fillers, cross-link density, and so on. The experiments were for SBR gum rubber and SBR filled with carbon black. The relation seems to be also valid for natural rubber and for NBR; the former crystallizes and the latter is amorphous. Notice that this result does not assume any initial flaw size; although the manufactured specimen is very likely to have as-received flaws.

Cracking in natural and most synthetic rubbers is strongly encouraged by oxidation from exposure to automobile exhaust. Ozone in particular accelerates crack growth when the rubber is under a stress. Fluororubber, fluorosilicone rubber, and nitrile rubber are synthetics which have high ozone resistance. Rubber with lightly colored fillers which is exposed to sunlight for long periods of time can develop a series of cracks on its surface. This effect is much reduced with dark filler such as carbon black.

10.9 POLYMER FRACTURE

Polymer fracture is initiated more easily in the amorphous than in the crystalline regions due to breaking the weaker intermolecular van der Waals bonds between the long-chain molecules rather than the covalent bonds between the mers in the long-chain polymer molecule.

The full separation of a polymer into two surfaces is the end of a somewhat complicated process of molecular scission in which the chains are broken. The thermofluctuational theory for polymer fracture says that cracks in polymers are initiated by thermal fluctuations of the atoms which increase their vibration, which causes the rupture of chemical bonds. The loads on the body reduce the activation energy required to break these bonds. The local stresses also make it less likely that bonds will reform once broken. Zhurkov, in the 1950s, showed that volatiles are given off by a polymer as it is stressed; he conducted tensile tests of polymers held in the vacuum chamber of a mass-spectrometer. The polymers tested included PMMA (polymethylmethacrylate), polyethylene, and other commonly used polymers. In many cases, the composition of the volatiles formed was the same as those formed during thermal degradation. The chains, or portions of chains, or free radicals created by scission of covalent bonds, break off the body and are detectable as a gas by various imaging techniques, such as infrared spectroscopy.

For polymers to fracture, not only is the magnitude of the stress important, but also the time during which it is applied because the mechanical behavior of a polymer is time-dependent. The Griffith model fails to account for the time-dependent behavior of polymers and, in particular, does not account for the thermal motion of atoms. The

Russian experimenter S. N. Zhurkov and coworkers measured the fracture strength instead by the durability, the time from the initial application of the load until fracture. They proposed that the relation between the durability, τ, and a fixed applied stress is temperature-dependent.

$$\tau = \tau_0 \exp\left(\frac{U_0}{kT}\right) \exp\left(-\frac{\nu\sigma}{kT}\right), \tag{10.46}$$

where τ_0 is a constant in range of 10^{-12} to 10^{-13}, U_0 is an activation energy which is nearly the energy of the chemical bonds, k is the Boltzmann constant, T is the absolute temperature, and ν is a material constant. However, the expression is valid only over a range of stresses above zero and below $\sigma = U_0/\nu$. This molecular model for fracture does not involve the idea of a critical crack length. Further, the prediction of this constant stress model that the radicals are produced at a constant rate has not been confirmed when the stress is varied [18].

G. M. Bartenov modified this theory by proposing that fracture in polymers involves first microcrack initiation and then two stages of crack growth, the first slow and the second more rapid. In the slow stage, when the stress is below a critical stress, the broken bonds in a microcrack may reform and close the crack. When the stress is larger than a critical stress, the bonds are prevented from reforming and the crack grows rapidly [26, p. 244*ff*].

The preconditions for scission depend on the type of polymer. The two primary mechanisms are shear and crazing. Excessive shear stress may cause molecules to slide relative to each other in analogy with the shear of metals along slip planes. Shearing depends on the deviatoric component of stress. The hydrostatic stress, σ_h, plays a large role by lowering the critical shear. The classical metal failure criteria are modified to reflect this. In the case of the maximum shear stress criterion, failure occurs if

$$\tau_{\max} \geq \tau_0 - k\sigma_h,$$

where τ_o is the critical value from a uniaxial test and k is a material constant. The octahedral stress theory is modified similarly. Failure occurs if

$$\tau_{\mathrm{oct}} \geq \tau_0 - k\sigma_h.$$

Crazing is observed in tensile monotonic loading of uncross-linked polymers at high enough stresses. Crazing looks like a region of thin cracks. In contrast to cracks, however, the volume between the apparent crack surfaces is filled with polymer material of lower density than that of the bulk material. Crazing is a result of the failure of intermolecular bonds, which in amorphous polymers are weaker than the covalent bonds within a molecule. In the crazed region, higher stress causes a bundle of molecules to line up in what are called fibrils, each forming a denser region. The crazed region appears shiny because it has an index of refraction lower than that of the bulk material. Crazes are induced by stress concentrations at flaws and impurities in the material and form before cracking begins. Microvoids can form between the fibrils and initiate cracks. Crazes form in thin layers oriented perpendicularly to the maximum tensile stress. Crazing can occur in the region ahead of the crack tip, allowing the tip to propagate. The craze layer ahead of the crack may separate from the bulk polymer under high crack propagation rates or in higher temperatures [29, p. 202].

The banding or striations observed in amorphous polymers but not in crystalline polymers are due to the extension of the crack by peeling of the craze wedge. The separation of the craze wedge lowers the local stress. The crack propagation pauses because a new craze region must then form before the crack can continue to grow. A lower local stress is required for craze formation than fracture. Each subsequent new craze wedge peels away as the process repeats. Therefore the fracture is not brittle.

Polymers with higher stiffness and subjected to a higher hydrostatic stress tend to craze at a lower strain. Fracture occurs when the fibrils break. In fact, cracks developing from a craze region near an existing crack may propagate in the opposite direction of the original crack propagation and join that crack, forming a region of branched cracks.

The Griffith condition is not satisfactory for polymers primarily because of the fractures in advance of the crack tip. The existence of crazing also contradicts the Griffith model. The crazes have very sharp tips so that they themselves cannot be modeled by narrow ellipses as in the Griffith theory. In fact, the initiation of a craze by breaking intermolecular bonds is similar to a brittle failure as in the Griffith theory, but the subsequent sliding of fibrils leading to craze failure is a propagation rate-dependent, viscous process to which the Griffith model does not apply. This two-step alternation causes the work of fracture to oscillate.

Berry in 1964 [6] experimentally determined the critical tensile stress for a wide range of initial crack sizes in both PMMA and polystyrene (PS). Both materials fail to follow the Griffith criterion for small crack sizes. For initial cracks of less than 0.5 mm in PS and less than 1 mm in PMMA, fracture is caused by a much smaller stress than predicted by the Griffith model. Berry suggested that perhaps flaws in the craze region near the crack were larger than the crack and so were responsible for the fracture. PMMA is more susceptible to crazing than PS. In this way, he avoided challenging the Griffith criterion. R. W. Armstrong, on the other hand, in 1972 plotted Berry's critical stress data versus $c^{-1/2}$, where c is the initial crack length; see [3]. The resulting curve was closely represented by the Bilby, Cottrell, and Swinden model for the critical fracture stress, σ_c [7], which accounts for a plastic zone at the crack tip (see Chapter 10, Section 10.5),

$$\sigma_c = \frac{2\sigma_Y}{\pi} \arccos\left(\frac{c}{s+c}\right) \sim \sigma_Y \sqrt{\frac{s}{s+c}},$$

where σ_Y is the yield stress and s is the length of the plastic zone. This fit is evidence that the failure of the Griffith criterion for small cracks in polymers, the small crack effect, is related to the existence of a relatively large plastic zone at the crack tip. The small crack effect is also observed in metals [4].

The Griffith condition, (equation 10.6), has been modified to account for the time-dependent behavior of polymers. The critical stress, σ_c, for a crack of length $2c$ in an infinite plane is

$$\sigma_c^2 = \frac{2E\gamma}{c\pi(1-\nu^2)},$$

where the elastic modulus analog, E, is a function involving the load history at the boundary of the specimen; the relaxation modulus, ν, is the limit value of the Poisson ratio in the rubbery or glassy state; and the fracture energy, γ, is time dependent rather than constant as for metals. However, this extension of the Griffith criterion to polymers does not always succeed in predicting the critical stress.

For example, in 1974 J. G. Williams and G. P. Marshall [31] proposed a rate- and temperature-dependent form of the modulus for PMMA,

$$E = E_0 t^{-n} \exp\left[\frac{nH}{R}\left(\frac{1}{T} - \frac{1}{T_0}\right)\right], \qquad (10.47)$$

where E_0 is the modulus at time $t = 1$, n is a function of time (which can be related to the tangent modulus of the material), R is the gas constant, T_0 is the original temperature, and T is the current temperature. The activation energy $H = 10^4 R^{-1}$ for PMMA. This model applied at an original temperature of 20°C predicts a temperature rise of about 20°C and a crack speed of about 50 mm/s when the crack begins to grow unstably; the latter value is in close agreement with experiment. Unfortunately, the results at extremes of temperature are not so close to experiment.

Crack growth can be encouraged by environmental influences. Chemically active gases can degrade the crack tip region, which is under a high strain, and therefore increase the rate of crack propagation by making it easier to fracture the long-chain molecules. Likewise, solvents may also increase the rate of crack propagation, but they seem to act on the electrostatic bonds between long-chain molecules. Therefore, solvents, depending on their rate of diffusion, can greatly increase crack growth in polymers which are not highly cross-linked.

10.10 PAPER FRACTURE

Wood pulp paper is made of wood fibers bonded to each other by hydrogen bonds. Paper fracture involves the competing processes of bond and fiber failure. The dominant mechanism of tensile fracture is believed to be primarily fiber-fiber debonding with perhaps some fiber fracture (Fig. 10.21). Once the interfiber bonds are broken, the crack grows by fiber pull-out. Shallhorn [28], by testing papers with different bonding strength, found that the fracture resistance increases with bonding strength. From this, he deduced that fiber pull-out is more responsible for energy dissipation during fracture than is fiber fracture. Shallhorn offered the explanation for the increase in fracture resistance that longer fibers can participate in pull-out once the bond strength is reduced.

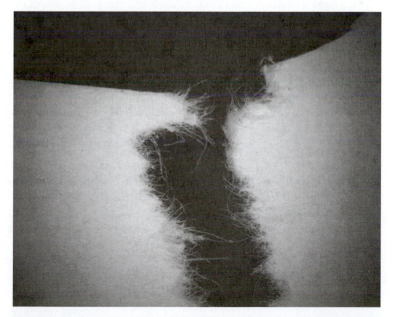

FIGURE 10.21   The fracture surface of a 20-pound bond paper tensile specimen showing fiber pull-out—8X magnification. (Courtesy of Bjorn Becker and H. Haslach.)

10.11 FLAW DETECTION

The evaluation of a part for flaws without further damaging the part is called nondestructive evaluation (NDE). Nondestructive testing initially used ultrasonics, eddy currents, x-rays, dyes, and magnetic particles. Because materials can survive with flaws if the flaws are small enough, starting around the 1970s, the problem became not one of detecting the presence of flaws, but their size. Today computerized signal analysis is a key in the success of these methods. Statistics plays a large role in assessing the results of the interaction of the interrogating fields with flaws. Many systems are now commercially available to experimentally measure the flaws in a given piece of material. These include sound, optical, thermal, and electrical nondestructive tests.

One of the simplest methods that is often part of the manufacturing process uses continuous scanning of products, such as pipes or the paper web on a paper-making machine, passing though infrared light beams to determine changes in dimensions. This system, which senses the disruptions in the infrared beams caused by flaws, allows high speed identification of defect lengths down to about 0.5 mm. Other optical techniques use lasers to detect dimensional changes as small as 3 μm. The laser beams are delivered by optical fibers, a method that permits transmitting the beams into areas that are hard to access.

Flaws in conducting materials can be detected by eddy currents. A probe containing a coil carrying an electrical current is placed near the surface. The magnetic field induced by the coil creates eddy currents in the material. This phenomenon was apparently first noticed by Hughes in 1879. These eddy currents, in a feedback process, increase the impedance of the coil. Flaws interrupt the eddy currents and thereby further increase the impedance of the coil. The resulting voltage increase across the coil indicates the presence of the flaw. Some manufacturers of this type of sensing equipment say that it can detect defects as small as 0.05 mm deep.

Thermography measures the thermal emission of a body. In this application, it is a heat transfer method that is used for quick flaw inspection of large structures. The idea is to create a thermal gradient along the material surface. A flaw impedes the conduction of thermal energy, allowing detection of the flaw by the spike it causes in the thermal image. Coatings may affect the emissivity reading; this problem may be overcome by creating a heat flow in two opposite directions. In that case, a crack has a different signature than that produced by a nonuniform coating. A technical problem with this method is to develop a scanning technique rather than a stationary one.

Ultrasonic test equipment monitors the interaction of the sound wave with a defect by measuring the change in phase or amplitude in the reflected wave. Sound waves propagate through a solid due to the vibrations of atoms within the solid so that a flaw disrupts the propagation. The ultrasonic methods can be used to test nonconducting materials like polymers. The size of the flaw that can be detected in ultrasonic testing depends on the chosen frequency of the transducer. In industrial inspections, it is assumed that the technique can detect discontinuities of length one-half the wavelength. Sensitivity describes the ability to locate small discontinuities. Resolution describes the ability of the system to distinguish discontinuities that are close together within the material or located near the part surface. Sensitivity and resolution generally increase as the frequency increases or, equivalently, with shorter wavelengths. However, coarse grains, such as often found in cast materials, can cause sound scattering at high frequencies and so the sensitivity must be reduced. At higher frequencies the depth at which flaws can be located is reduced due to loss in sound energy due to scattering. The ultrasound signal can be delivered to the material by flexible fibers to permit testing of hard to access locations. A variation on this technique is to project a laser beam on the surface and then measure the amplitude of the ultrasound signal generated as the laser passes a flaw [20]. Such a system allows scanning of the material and can measure smaller defects than the traditional ultrasound techniques.

PROBLEMS

10.1   The surface tension per unit width for glass is 0.0031 lb/in. The glass has elastic constants $E = 9.01 \times 10^6$ psi, $\nu = 0.251$, and the tensile strength is 24,900 psi. Compute the critical stress for a crack of half-length $c = 0.0001$ inches along the longitudinal direction of a thin-walled tube made of this glass which is pressurized from the inside. Assume plane stress.

10.2   The surface tension per unit width for glass is 0.0031 lb/in. The glass has elastic constants $E = 9.01 \times 10^6$ psi, $\nu = 0.251$, and the tensile strength is 24,900 psi. A thin rectangular plate is loaded uniaxially. Compute the critical stress for a crack of length 0.0005 inches that is transverse to the load direction.

10.3   Create a graph of the critical stress versus crack size for a mode I loading on an iron plate in plane stress using Griffith's brittle crack propagation condition. The elastic modulus is 150 GPa and the surface energy is 204×10^{-7} J/cm².

10.4   Show that for the double cantilever beam (Fig. 10.3) subjected to an opening force, P,

$$G = \frac{12P^2c^2}{Ed^3w^2},$$

where w is the beam width, $2d$ is the beam height, and c is the crack length. *Hint*: See equation (10.2). Verify that G increases with c so that the crack propagation behavior is unstable.

10.5   Show that for the double-cantilever beam (Fig. 10.3) subjected to a pure bending moment, M,

$$G = \frac{12M^2}{Ed^3w^2},$$

where w is the beam width, $2d$ is the beam height, and c is the crack length. *Hint*: See equation (10.2). Verify that G is independent of c so that the crack propagation behavior is neutral.

10.6   Which mode crack type would be expected in

(a)   beam bending,

(b)   a rod in axial compression,

(c)   a rod in combined bending and torsion?

10.7   An axially loaded member must carry a stress of 500 MPa. Compute and compare the size of the crack that will induce catastrophic failure if the member is made of a mild steel with $K_{I_c} = 130$ MPa-m$^{1/2}$ or a medium-carbon steel with $K_{I_c} = 40$ MPa-m$^{1/2}$. Assume a shape factor of one.

10.8   Compute the mode I stress intensity factor for the double-cantilever beam subjected to

(a)   a force P and

(b)   a pure bending moment, M.

10.9   Determine the mode I stress intensity factor for the double-cantilever beam test specimen under a constant wedging condition of displacement, $2h$.

10.10   A steel has properties, $\sigma_Y = 185$ MPa, the ASTM grain size number is 3 (the average grain diameter is 0.125 mm), and K_{I_c}, K_{II_c}, and K_{III_c} are 7.0, 22.0, and 29.0 MPa-m$^{1/2}$, respectively. Determine whether yielding or fracture will occur first for a thin-walled cylinder tested in torsion. There is a through-the-thickness crack at 45° to the cylinder longitudinal axis and of length equal to 16 times the average grain diameter. Check both the Tresca and von Mises criteria to investigate yielding.

10.11   Determine the critical stress for an axially loaded plate made of AISI 4340 steel ($\sigma_Y = 345$ MPa) and of width, $2w = 60$ cm, with a crack of length 1.4 cm transverse to the load. Will the plate yield before the crack begins to propagate?

10.12   A polycrystalline material satisfies $K_I = 300$ kgf/mm$^{1.5}$ and Hall-Petch constants $\sigma_0 = 6.0$ kgf/mm² and $k = 2.5$ kgf/mm$^{1.5}$. The average grain diameter is 0.01 mm. Estimate the smallest crack length that is required for brittle fracture to occur just at the point of yielding.

10.13   A thin-walled cylindrical pressure vessel is made of a steel with $K_{I_c} = 165$ MPa-m$^{1/2}$ and $\sigma_Y = 950$ MPa. The wall thickness is 1 cm and the radius is 30 cm.

(a)   Determine the critical circumferential crack size if the internal pressure is 20 MPa.

(b)   Determine the critical size of a crack parallel to the axis of the vessel if the pressure is 20 MPa. Assume a shape factor of one.

10.14   A curved beam with inner radius $a = 6$ cm and outer radius $b = 12$ cm is in pure bending in its plane of curvature due to an applied moment, $M = 10$ kN·m that decreases the radius of curvature of the beam. The thickness is $t = 2$ cm. It is made of an aluminum with $E = 70$ GPa, $\nu = 0.33$, $K_{I_c} = 5$ MPa-m$^{1/2}$, and $\sigma_Y = 80$ MPa.

(a)   Determine its state of stress using the Winkler-Bach theory at $b = 12$ cm.

(b)   Determine the critical crack length at $b = 12$ cm.

10.15   A steel has $K_{I_c} = 30$ ksi-in$^{1/2}$ and $E = 30 \times 10^6$ psi.

(a) Determine G_c for the Griffith theory.

(b) From (a), determine the surface energy γ per unit crack surface area.

10.16 An aluminum has $G_c = 12.5$ kJm^{-2}, $E = 70$ GPa. Compute K_c MPa-m$^{1/2}$. Assume $Y = 1$ and plane stress.

10.17 A steel has surface energy $\gamma_s = 1.2$ Pa.m, $E = 207$ GPa, and $K_{I_c} = 10$ MPa-m$^{1/2}$. Determine the plastic work, γ_p.

10.18 A large thin rectangular plate is loaded axially by an in-plane stress of $\sigma = 30$ ksi. A mid-plate crack of full length 0.06 inches is perpendicular to the load. The plate is made of an aluminum ($E = 10.1 \times 10^6$ psi, $\nu = 0.33$, $\sigma_Y = 64 \times 10^3$ psi and $\sigma_u = 69 \times 10^3$ psi, $K_{I_c} = 24.5$ ksi-in$^{1/2}$). Assume a shape factor of $Y = 1$.

(a) Compute the critical stress for brittle fracture using the Irwin theory.

(b) Estimate the size of the plastic zone for the given stress using the Irwin plane stress estimate. Draw a figure indicating the dimension calculated.

10.19 An aluminum alloy has a yield stress of 50 ksi. A large thin rectangular plate made of this aluminum with a center crack of full length 0.5 inches is observed to fracture under an in-plane axial 30 ksi stress normal to the crack. For an external in-plane axial stress of 28 ksi, determine the stress along the centerline ($\theta = 0$) of the crack at the following distances from the actual crack tip:

(a) 0.05 inches;

(b) 0.15 inches. Assume the shape factor $Y = 1$ and perfectly plastic behavior.

10.20 A thin aluminum plate is axially loaded perpendicularly to its edges by an in-plane stress of 200 MPa. For this aluminum, the plane stress fracture toughness is $K_{I_c} = 30$ MPa-m$^{1/2}$ and $\sigma_Y = 400$ MPa. Assume a shape factor of one.

(a) Compute the critical crack length c for a crack of length $2c$ perpendicular to the load.

(b) Determine the extent of the plastic zone using the Irwin estimate.

(c) Compute the effective crack length.

10.21 Repeat the previous problem using the Dugdale estimate of the extent of the plastic zone.

10.22 A large thin rectangular plate is loaded axially by an in-plane stress of $\sigma = 50$ ksi. A mid-plate crack of full length 0.09 inches is perpendicular to the load. The plate is made of an aluminum ($E = 10.1 \times 10^6$ psi, $\nu = 0.33$, $\sigma_Y = 64 \times 10^3$ psi and $\sigma_u = 69 \times 10^3$ psi, $K_{I_c} = 24.5$ ksi-in$^{1/2}$). Assume a shape factor of $Y = 1$ and plane stress.

(a) Compute the critical stress for brittle fracture using the Irwin theory.

(b) Compute the critical value of the Griffith constant, G_c. Compute the critical stress for brittle fracture using the Griffith theory.

(c) Compute the size of the plastic zone for the applied 50 ksi stress using the Irwin plane stress estimate. Draw a figure indicating the dimension calculated.

(d) Let σ_y be the stress in the direction perpendicular to the crack face. Under the applied 50 ksi stress, what is the stress σ_y at a point along the line extended through the crack from the crack tip into the solid plate but at 0.01 in. from the crack tip? Assume perfectly plastic behavior in the plastic zone.

(e) Compute the required thickness of the plate to transform the problem into one of plane strain. Recall the estimate $t \geq 2.5(K_{I_c}/\sigma_Y)^2$.

10.23 For a crack of length $2c = 2$ mm in a thin plate of width $2w = 80$ mm made of steel with a yield stress of 130 MPa and $E = 200$ GPa compute the crack opening displacement predicted by the Dugdale model. The applied stress is $\sigma = 100$ MPa.

10.24 A steel has the properties $E = 200$ GPa, $K_{I_c} = 69.3$ MPa-m$^{1/2}$, and $\sigma_Y = 300$ MPa. Determine

(a) the critical strain energy release rate, and

(b) the critical crack tip opening displacement. Assume plane stress.

10.25 A large uniaxially loaded, 2014-T4 aluminum plate has a hole of radius $R = 1.5$ cm. A crack of length $c = 0.24$ cm is perpendicular to the stress direction and at the edge of the hole. Estimate at what stress the crack will propagate. Compare your estimate to the exact solution.

10.26 A large rectangular plate of thickness t must carry an axial tensile in-plane distributed load of 1,000 pounds per unit width along two opposite edges. The plate MUST be elastically loaded. The manufacturer of the plate guarantees that the largest crack, due to the rolling process, parallel to the loaded edges of the plate, but not on any plate edge, has full length 0.02 inches. If required, use a shape factor of one.

(a) Determine the required minimum plate thickness if the chosen material is steel ($E = 30 \times 10^6$ psi, $\nu = 0.28$, $\sigma_Y = 34 \times 10^3$ psi and $\sigma_u = 50 \times 10^3$ psi, $K_c = 188$ ksi-in$^{1/2}$, $\gamma = 0.00617$ lb/in).

    (1) Do any brittle fracture analysis using the Griffith theory;

    (2) Do any brittle fracture analysis using the Irwin theory.

(b) Determine the required minimum plate thickness if the chosen material is aluminum ($E = 10 \times 10^6$ psi, $\nu = 0.33$, $\sigma_Y = 64 \times 10^3$ psi and $\sigma_u = 69 \times 10^3$ psi, $K_c = 24.5$ ksi-in$^{1/2}$). Do any brittle fracture analysis using the Irwin theory.

(c) Decide, using a computation, whether to specify steel or aluminum if weight is to be minimized. The specific weight of steel is 0.283 lb/in^3 and of aluminum is 0.10 lb/in^3. Do not use the Griffith result for steel in this computation.

REFERENCES

[1] ANNUAL BOOK OF ASTM STANDARDS, *3*, Part 1, ASTM International, West Conshohocken, PA (2002).

[2] R.W. ARMSTRONG, Cleavage crack propagation within crystals by the Griffith mechanism versus a dislocation mechanism, *Materials Science and Engineering*, *1*, 251–254 (1966).

[3] R.W. ARMSTRONG, Grain size: the fabric of (brittle) fracture of polycrystals, in *Fourth International Conference on Fracture (ICF4)*, D. M. R. Taplin, Ed., *4*, 1977, pp. 83–96.

[4] R.W. ARMSTRONG, The (cleavage) strength of pre-cracked polycrystals. *Engineering Fracture Mechanics*, *28*, 529–538 (1987).

[5] A. G ATKINS AND Y. W. MAI, *Elastic and Plastic Fracture*, Ellis Horwood Ltd., Chichester, UK, 1985.

[6] J. P. BERRY, Brittle behavior of polymeric solids, in *Fracture Processes in Polymeric Solids*, B. Rosen, Ed., Wiley Interscience, New York, 1964, pp. 195–234.

[7] B. A. BILBY, A. H. COTTRELL, AND K. H. SWINDEN, The spread of plastic yield from a notch, *Proceedings Royal Society*, *A272*, 304–314 (1963).

[8] N. E. DOWLING, *Mechanical Behavior of Materials*, Prentice Hall, Englewood Cliffs, NJ, 1993.

[9] M. J. DOYLE, A. MARANCI, E. OROWAN, AND S. T. STORK, The fracture of glassy polymers, *Proceedings of the Royal Society of London A*, *329*, 137–151 (1972).

[10] D. S. DUGDALE, Yielding of steel sheets containing slits, *Journal of the Mechanics and Physics of Solids*, *8*, 100–104 (1960).

[11] A. A. GRIFFITH, The phenomena of rupture and flow in solids. *Philosophical Transactions. Royal Society of London*, *A221*, 163–198 (1920).

[12] A. A. GRIFFITH, The theory of rupture, *Proceedings of the First International Congress for Applied Mechanics, Delft, 1924*, *1925*, 55–63 (1924).

[13] C. E. INGLIS, Stresses in a plate due to the presence of cracks and sharp corners. *Transactions of the Institution of Naval Architects*, *55*, 219–242 (1913).

[14] G. R. IRWIN, Analysis of stresses and strain near the end of a crack traversing a plate. *Journal of Applied Mechanics*, *24*, 361–364 (1957).

[15] G. R. IRWIN, Fracture, in *Encyclopedia of Physics VI*, S. Flugge, Ed., 1958, pp. 551–590.

[16] G. R. IRWIN, Plastic zone near a crack and fracture toughness, *Mechanical and Metallurgical Behavior of Sheet Materials*. 7th Sagamore Ordnance Materials Research Conference (1960), Syracuse University Research Institute, Syracuse, NY, 1961, pp. 63–78.

[17] G. R. IRWIN, Fracture Mechanics, in *Metals Handbook*, H. E. Boyer, T. L. Gall, Eds., *8*, American Society for Metals, Metals Park, OH, 1985, p. 439.

[18] W. G. KNAUSS, The mechanics of polymer fracture, *Applied Mechanics Reviews*, *26*, 1–17 (1973).

[19] J. F. KOTT AND A. R. BOCCACCIN, The fracture mechanics-microstructure linkage, in *Mechanics of Materials* M. A. Meyers, R. W. Armstrong, and H. O. K. Kirchner, Eds., Wiley, NY, 1999, pp. 399–427.

[20] A. K. KROMINE, P. A. FROMITCHOV, S. KRISHNASWAMY, AND J. D. ACHENBACH, Laser ultrasonic detection of surface-breaking defects: Scanning laser source techniques, *Materials Evaluation*, *58*, 173–177 (2000).

[21] D. M. MARSH, Plastic flow and fracture of glass, *Proceedings of the Royal Society of London. Series A, Mathematical and Physical Sciences*, *282*, 33–43 (1964).

[22] F. A. MCCLINTOCK AND A. S. ARGON, *Mechanical Behavior of Materials*, Addison-Wesley, Reading, MA, 1966.

[23] E. OROWAN, The fatigue of glass under stress, *Nature*, *154*, 341–343 (1944).

[24] E. OROWAN, Fracture and strength of solids, *Reports on Progress in Physics*, *12*, 185–232 (1948–1949).

[25] A. R. PAYNE, Hysteresis in rubber vulcanizates, in *Rubber and Rubber Elasticity*, Wiley, New York, 169–196 (1974).

[26] I. I. PEREPECHKO, *An Introduction to Polymer Physics*, Mir, Moscow, 1981.

[27] J. R. RICE, A path-independent integral and the approximate analysis of strain concentration by notches and cracks, *ASME Transactions. Journal of Applied Mechanics*, *33*, 379–385.

[28] P. M. SHALLHORN, Fracture resistance–theory and experiment, *Journal of Pulp and Paper Science*, *20*, J119–J124 (1994).

[29] S. SURESH, *Fatigue of Materials*, 2nd ed., Cambridge University Press, Cambridge, UK, 1998.

[30] J. WEERTMAN, *Dislocation Based Fracture Mechanics*, World Scientific, Singapore, 1996.

[31] J. G. WILLIAMS AND G. P. MARSHALL, Crack stability in PMMA, *Polymer*, *15*, 251–252 (1974).

[32] M. L. WILLIAMS, *Journal of Applied Mechanics*, *20*, 109–114 (1957).

CHAPTER 11

FATIGUE

11.1   INTRODUCTION

A body subjected to repeated loads less than the ultimate failure stress can fracture. The body is said to have become fatigued, in analogy with becoming too "tired" to support the applied loads. The emphasis in Chapter 10 was on fracture due to monotonically increasing loads, rather than variable loads. Failure by fatigue is one of the most common causes of failure that engineers must deal with. Some estimates are that over three-quarters of all failures of dynamically loaded machine parts are due to fatigue. Fatigue failure is especially a problem since it can occur even when the body is always elastically loaded. Therefore designing to keep the maximum stresses below the yield limit is not sufficient to avoid this type of failure.

Engineers have been aware of this problem for centuries. In the 1800s, the primary engineering material was iron. Many engineers believed that the iron structure changed from fibrous to crystal under repeated loading. This is not true; it is now known that the crystal structure of metals is not changed much under cyclic repeated loads. Later it was proposed that embrittlement was the cause of fatigue failure. It was conjectured that in localized areas of high stress, the cyclic load caused the metal to strain harden. This, it was thought, made the metal more brittle until the failure stress was reached. At this point, a crack initiated, and fracture followed as the crack grew. More recent theories place the explanation for fatigue failure in the behavior of dislocations in the metal crystal and tiny voids in the material. In any case, once a crack initiates, the repeated loading causes the crack to open and close. This results in crack growth and ultimately in fracture of the material.

11.2   TYPES OF FATIGUE LOADING

Several types of fatigue loading can be distinguished, depending on their regularity. A cyclic load is one that always varies between the same maximum, σ_{max}, and minimum, σ_{min}, stresses. Cyclic loads are often called "completely reversed" loads. For a cyclic loading (Fig. 11.1), a mean stress, σ_m, and a stress amplitude, σ_a, can be defined by

$$\sigma_m = \frac{1}{2}(\sigma_{max} + \sigma_{min}),$$

and

$$\sigma_a = \frac{1}{2}(\sigma_{max} - \sigma_{min}).$$

The sign convention that tensile normal stresses are positive and compressive normal stresses are negative is maintained. The amplitude, σ_a, is always positive while the mean, σ_m, can be either positive or negative.

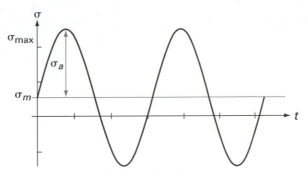

FIGURE 11.1   Cyclic stress loading over time, t.

EXAMPLE 11.1

A stress varies sinusoidally with respect to time between 65 MPa in tension and 25 MPa in compression. The mean stress is

$$\sigma_m = (65 - 25)/2 = 20 \text{MPa},$$

and the amplitude is

$$\sigma_a = [65 - (-25)]/2 = 45 \text{MPa}.$$

Assume that the stress is equal to σ_m at time zero. The stress, then, could be written as a function of time

$$\sigma(t) = 20 + 45 \sin(\omega t),$$

where ω is the frequency of the stress alternation, which was not given. Most empirical models used to predict failure by fatigue assume that the frequency of the stress change does not affect the number of cycles to failure. ∎

In the case that $\sigma_m = 0$, the loading is said to be purely cyclic. Cyclic loads are most common in rotating or vibrating bodies, such as machine parts. These dynamically loaded bodies must be investigated by a combination of dynamic (or vibration) analysis to locate their position over time and of fatigue stress analysis to predict their lifetime.

Other structures are subjected to random fluctuating loads. For example, the deck of a bridge would see a random fatigue loading as various types of vehicles are driven across it. The consequences of random loads are difficult to predict. Statistical techniques may be required. In this text, to lay the foundations, only cyclic fatigue loads will be considered.

11.3 DESIGN AGAINST FATIGUE FAILURE

No techniques have yet been developed to predict, from the lattice structure of a metal, safe stress levels under cyclic loads. Therefore design must be based on experimental results. The most basic set of tests is that presented in what is called an S-N curve. A purely cyclic load, one with zero mean stress, at a stress amplitude, S, is applied to a test specimen and the number of cycles, N, that occur before fracture of the specimen is counted. Several such tests are conducted at one stress amplitude to obtain an average number of cycles to failure. This then gives one point on the S-N curve. Tests are conducted in a similar manner for many different stress amplitudes. The graph of all these is called the S-N curve (Fig. 11.2), or sometimes the Wohler diagram. Clearly it is expensive and time consuming to produce such a curve.

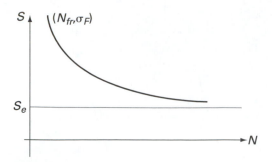

FIGURE 11.2   *S-N* curve with an endurance limit, S_e.

The *S-N* curve can be obtained experimentally either for the material alone, in which case it is called the intrinsic fatigue curve, or for a particular machine part. In the latter case, the curve is called the component fatigue curve. These two curves are related in engineering handbooks by

$$S = S_{\text{int}}C_s C_L \frac{K_s}{K_f},$$

where S_{int} is the intrinsic fatigue strength. K_s is a surface finish factor that also depends on the grain size. K_f is a reduction factor depending on the component shape, the type of loading, and the material. The size factor C_s depends on the magnitude of the stress gradient, and C_L depends on the type of loading. C_s for bending and torsion is about 85% of that in axial loading. The loading factor, C_L, is largest in bending, smaller in axial, and smaller still in torsion loading. These K-factors are not related to the K used in fracture mechanics.

The surface finish on a part can drastically affect its fatigue life. A rough surface contains many flaws at which cracks can initiate since it is marred by many microindentations with stress concentrations. A rough surface offers many more sites for crack initiation than a smooth surface. For the same reasons, environmental actions on the part that cause surface corrosion can be very detrimental to fatigue life. The surface finish factor, K_s, depends on both the surface finish and on the grain size of the material. Usually, it decreases with static tensile strength, which is closely related to grain size. Corroded surfaces have the lowest surface finish factor, between 0.1 and 0.2. The surface factor increases in the following order of finishes: as-forged, hot-rolled, machined, ground, and polished. A polished surface is given a finish factor of 1.

Fatigue testing was first conducted in the late 1800s when steel was the primary engineering material. This historical fact means that many concepts of fatigue analysis are colored by the behavior of steel. For example, the *S-N* curve for steel seems to be asymptotic as the number of cycles increases to a stress, S_e, which was named the "fatigue limit" or "endurance limit" (Fig. 11.2). Early researchers in fatigue then conjectured that, if the stress amplitude was held below the endurance limit, no fatigue failure would occur. A rule of thumb is that for steel with moderate tensile strengths ($< 1,200$ MPa), the endurance limit is about one-half the ultimate strength, $S_e = 0.5S_u$. Unfortunately, when *S-N* curves were produced for other materials, no such endurance limit was visible. FCC metals rarely exhibit an endurance limit. However, some engineers estimate what is essentially an infinite life endurance limit for aluminum alloys and cast iron as $S_e = 0.3S_u$, where S_u is the tensile fracture strength.

Because they believed such an endurance limit existed, engineers in the early days of fatigue analysis tried to design structures for an infinite life by keeping the stress amplitude below the endurance limit. This of course led to overdesign. Today, engineers first decide on the desired lifetime of their design as measured in terms of the number of stress cycles, N, it must survive. They then use the S-N curve to select the allowable stress amplitude corresponding to this N. This stress is called the fatigue strength of the metal for the chosen number of cycles, N. For an individual human, anything surviving over 100 years has for all practical purposes an infinite lifetime. But be cautious; the lifetime in fatigue analysis is measured in numbers of cycles not in time. One of the initial engineering design decisions is how many cycles are expected per unit time.

EXAMPLE 11.2   An aluminum machine part must survive 10 years. It is expected to undergo a million cycles per year. Therefore, the part must survive 10 million cycles. The corresponding stress amplitude on the aluminum S-N curve is taken as the allowable stress, σ_{all}. This stress may be smaller than the yield stress for aluminum. ■

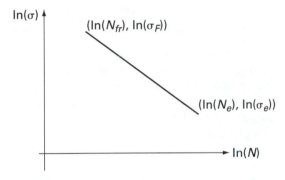

FIGURE 11.3   S-N data on a log-log curve.

A common trick is to assume that, if the S-N data is plotted on a log-log curve, the result is a straight line (Fig. 11.3). In this case, only the (N_{fr}, σ_F) and (N_e, σ_e) data points on the S-N curve need be determined experimentally to draw the curve. Sometimes the endurance limit can be estimated from the ultimate stress. The slope of the line on the log-log curve is

$$b = \frac{\ln(\sigma_F/\sigma_e)}{\ln(N_{fr}/N_e)}.$$

The value of b should be negative since the curve is decreasing. The critical stress, σ_c, is that cyclic stress amplitude that causes fracture in a given number of cycles. If b is known, then the critical stress for any given number of cycles, (N_c, σ_c), the point on the S-N curve is given by taking the slope between (N_c, σ_c) and (N_{fr}, σ_F) as in the previous equation and rewriting the result as

$$N_c = N_{fr} \left(\frac{\sigma_c}{\sigma_F} \right)^{1/b}. \tag{11.1}$$

This formulation was first proposed by O. H. Basquin in 1910. It is given in terms of (N_{fr}, σ_F) since that point is the most easily determined experimentally. N_{fr} is typically 1/2 because fracture occurs on the first half cycle if σ_F is the uniaxial fracture stress.

EXAMPLE 11.3   The ultimate stress for a steel is $\sigma_u = 124$ MPa. Assume that the endurance limit is reached after one million cycles. It is known that for $N = 10^3$ cycles, $\sigma = 0.9\sigma_u = 111.6$ MPa is the fatigue failure stress. Determine the number of cycles to failure corresponding to a stress of 65 MPa.

Solution   Approximate $\sigma_e = 0.5\sigma_u = 62$ MPa. $N_e = 10^6$ is given. Because $N = 10^3$ and $\sigma = 0.9\sigma_u = 111.6$ MPa, the slope can be computed as

$$b = \frac{\ln(111.6/62)}{\ln(10^3/10^6)} = -0.0851.$$

The critical number of cycles to failure is

$$N_c = 10^3 \left(\frac{65}{111.6}\right)^{-1/0.0851} = 5.739 \times 10^5. \qquad \blacksquare$$

The *S-N* curves have been shown experimentally to depend on average grain size, l. The fatigue life improves as the average grain size is reduced. Some experimentalists have reported that the peak stress required to cause failure at a fixed number of cycles varies proportionally to $l^{-1/2}$, a Hall-Petch relation.

11.4   FAILURE CRITERIA FOR UNIAXIAL METAL FATIGUE

The *S-N* curves provide a design technique when the mean applied cyclic stress is zero. Other techniques are required when the mean stress is nonzero. Several empirical criteria have been proposed. Each of these criteria defines a curve in σ_m-σ_a space from which to determine whether or not a combination of mean stress and stress amplitude is safe from fatigue failure. The separatrix curves are defined in terms of the fatigue strength, σ_c, at a given number of cycles from the *S-N* curve for zero mean stresses and in terms of the uniaxial yield stress for the material, σ_Y. The fatigue strength in the case of steel is the endurance limit. Each of these empirical criteria says that any mean stress and stress amplitude combination on the same side of the separatrix curve as the origin is safe.

The Soderberg (1938) criterion separatrix is defined by

$$\frac{\sigma_a}{\sigma_c} + \frac{\sigma_m}{\sigma_Y} = 1. \qquad (11.2)$$

The Soderberg criterion is applied to bodies which are elastic throughout the cyclic loading. The modified Goodman (1899) criterion for nonelastic cyclic loads replaces the yield stress in the Soderberg by the ultimate stress, σ_u.

$$\frac{\sigma_a}{\sigma_c} + \frac{\sigma_m}{\sigma_u} = 1. \qquad (11.3)$$

Both the Soderberg and Goodman criteria have been successful in predicting fatigue failure for steel (Fig. 11.4).

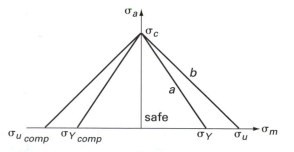

FIGURE 11.4   (a) Safe regions predicted by the Soderberg and (b) by the Goodman criteria.

If the amplitude is zero, then the stress is constant. Yielding or fracture occurs at the respective points on the σ_m-axis where $\sigma_m = \sigma_Y$ or $\sigma_m = \sigma_u$. If the mean stress is zero, the loading is purely cyclic so that fatigue failure occurs at the critical stress from the S-N diagram, $\sigma_a = \sigma_c$. The critical stress, σ_c, is the amplitude at a mean stress of zero for a given lifetime. Each criterion was apparently developed by merely drawing a straight line between these failure points on the axes in σ_m-σ_a space. The criteria are the equations for these straight lines.

Exercise 11.1 Verify that the Soderberg and Goodman criteria are each the equation of a straight line in σ_m-σ_a space. Determine the slope of each line.

In the case of brittle materials, since $\sigma_Y = \sigma_u \equiv \sigma_f$, the Soderberg and Goodman criteria give the same result as the SAE criterion (Society of Automotive Engineers),

$$\frac{\sigma_a}{\sigma_c} + \frac{\sigma_m}{\sigma_f} = 1. \tag{11.4}$$

Another criterion is that of Gerber (1874),

$$\frac{\sigma_a}{\sigma_c} + \left(\frac{\sigma_m}{\sigma_u}\right)^2 = 1. \tag{11.5}$$

The Goodman criterion is more conservative than the Gerber and so is more frequently used.

EXAMPLE 11.4 A steel machine part is uniaxially loaded by a force that fluctuates between a tensile stress of 50 ksi and a compressive stress of 10 ksi. The yield stress for the steel is 120 ksi. An S-N curve for the steel shows that the critical stress for the required number of cycles but with zero mean stress is 70 ksi. Determine whether or not this part will fail due to fatigue and compute a factor of safety.

Solution Since the member is elastically loaded, use the Soderberg criterion. A calculation shows that $\sigma_m = (50 - 10)/2 = 20$ ksi and $\sigma_a = [50 - (-10)]/2 = 30$ ksi. Therefore

$$\sigma_a = \frac{3}{2}\sigma_m.$$

The Soderberg criterion says that

$$\frac{\sigma_a}{70} + \frac{\sigma_m}{120} = 1.$$

Solving these two equation simultaneously gives the critical stresses

$$\sigma_m = 33.6 \text{ ksi,}$$

and

$$\sigma_a = 50.4 \text{ ksi.}$$

The factor of safety is the ratio of this critical value of the mean stress to the actual mean stress.

$$SF = 33.6/20 = 1.68. \qquad\blacksquare$$

11.4.1 Helical Compression Springs

Compression springs are common mechanical components that are placed under a cyclic load. A fatigue life analysis requires relating the axial spring load, F, to the shear stresses induced in the spring wire. The spring wire radius is r, $d = 2r$, and the mean diameter of the spring coils is D (Fig. 11.5).

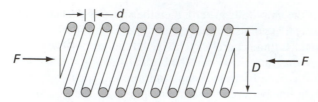

FIGURE 11.5   A helical compression spring.

The maximum shear stress in the wire is due to the superposition of a shear and a torsional stress at the inner radial position of the coil. The shear force in the wire is F and the applied torque is $FD/2$. The area terms are $A = \pi d^2/4$ and $J = \pi d^4/32$. A strength of materials analysis produces, since $r = D/2$,

$$\tau = \frac{Tr}{J} + \frac{F}{A} = \frac{8FD}{\pi d^3} + \frac{4F}{\pi d^2}.$$

The ratio $C = D/d$ is called the spring index. In terms of C,

$$\tau = K_s \frac{8FD}{\pi d^3}, \tag{11.6}$$

where $K_s = 1 + 0.5/C$.

Exercise 11.2   Verify the expression for τ in terms of C.

To also account for the curvature of the wire, many spring designers compute the stress in terms of the Wahl correction factor [12],

$$K = \frac{4C - 1}{4C - 4} + \frac{0.615}{C};$$
$$\tau = K \frac{8FD}{\pi d^3}. \tag{11.7}$$

The spring constant in $F = k\delta$, where δ is the lengthwise spring deflection, can be obtained from the geometry of the spring by computing δ using energy methods. For length of the wire, L, the strength of materials torsional strain energy in the wire is

$$U = \frac{T^2 L}{2GJ} = \frac{4F^2 D^3 n}{d^4 \mu},$$

where n is the number of coils formed from wire length, $L = \pi Dn$, and $\mu = G$ is the material shear modulus. The deflection of the point of application of the force, F, is, by Castigliano's theorem (Chapter 7, Section 7.2),

$$\delta = \frac{dU}{dF} = \frac{8FD^3 n}{d^4 \mu};$$

so that

$$k = \frac{d^4 \mu}{8D^3 n}.$$

The shear modulus, $\mu = 11.5 \times 10^6$ psi for spring steel, chrome vanadium, and chrome silicon. $\mu = 5.5 \times 10^6$ for brass, and $\mu = 10 \times 10^6$ for stainless steel.

In spring design, the tensile yield stress is approximated as 3/4 of the tensile strength, σ_u, and the torsional yield stress is approximated, by the distortion energy failure theory, as 0.577 of the tensile yield stress. Experiment shows that the endurance limit for spring steel is about 45 ksi.

EXAMPLE 11.5 ASTM A 228 cold-drawn music wire (high-carbon steel) has a tensile strength of 230 ksi. The wire has diameter, $d = 0.125$ inches, the outer diameter of the spring is 0.875 inches, the spring length is 2.5 inches, and there are 10 coils. The spring is subject to a cyclic load between 20 and 40 pounds. Use the Soderberg criteria to determine if the spring has infinite life under this load.

Solution The mean diameter is $D = 0.875 - d = 0.75$ inches. The spring index is $C = D/d = 6$, $K_s = 1 + 0.5/C = 1.083$. The mean force, F_m, is 30 pounds and the force amplitude, F_a, is 10 pounds. The mean shear stress and the amplitude shear stress are

$$\tau_m = K_s \frac{8 F_m D}{\pi d^3} = 31{,}770 \text{ psi} \quad \text{and} \quad \tau_a = 10{,}590 \text{ psi.}$$

The torsional yield stress is approximated by $\tau_Y = (0.577)(0.75)230{,}000 = 99{,}533$ psi. The endurance limit is $\tau_e = 45$ ksi. Application of the Soderberg criterion yields,

$$\frac{\tau_a}{\tau_e} + \frac{\tau_m}{\tau_Y} = 0.555.$$

The loading is safe.   ∎

Exercise 11.3 Using $\mu = 11.5 \times 10^6$ psi, verify that the spring constant for the spring of the example is $k = 83.2$ lb/in.

11.5 MULTIAXIAL FAILURE CRITERIA FOR METAL FATIGUE

Much as in the case of yield criteria for plasticity, multiaxial fatigue failure criteria are based on the uniaxial criteria. The strategy is to determine a uniaxial equivalent mean stress and stress amplitude. Then the uniaxial criteria are applied to the uniaxial equivalent stress values. Each of the fatigue failure theories depends on a yield or fracture stress. The equivalent yield stress is determined by the choice of a yield theory such as the von Mises or Tresca criterion.

A common and conservative strategy in classical machine design for applying the von Mises criterion to fatigue is to separately compute the Mises stress corresponding to the mean stress and the Mises stress corresponding to the amplitude. These two values are then used in the Soderberg or Goodman criterion. Because fatigue failure involves crack propagation and fracture, the Goodman criterion is more commonly used. This strategy is most successful if the biaxial loading is proportional. Proportional loading means that the principal stresses remain in the same ratio at all times in the cyclic loading. For example, the principal stresses in a thin-walled pressure vessel in which the pressure varies sinusoidally are in the ratio 2 at all times. One problem is that this method does not seem to accurately account for the effect of a compressive mean normal stress on fatigue life. Also, fatigue cracks are generally initiated on the plane of maximum shear stress. A difficulty in identifying this plane under nonproportional loading is that in a biaxial loading involving cycling both normal and shear stresses, the directions of the principal stresses may change with time.

EXAMPLE 11.6 Suppose $\sigma_x = 0.5\sigma_y + A\sin(\omega t)$, where ω is the frequency and t is time, σ_y is constant, and $\sigma_z = 0$. Compute the mean and amplitude of the Mises equivalent stresses.

Solution The mean stresses are $\sigma_x = 0.5\sigma_y$ and $\sigma_z = 0$. The equivalent mean stress is given by the Mises criterion.

$$\sigma_{mean}^2 = \sigma_x^2 + \sigma_y^2 - \sigma_x\sigma_y = (0.5\sigma_y)^2 + \sigma_y^2 - (0.5\sigma_y)\sigma_y = 0.75\sigma_y^2.$$

So, $\sigma_{mean} = \sqrt{0.75\sigma_y^2}$. The equivalent stress amplitude is the Mises stress of the respective amplitudes

$$\sigma_{amp}^2 = \sigma_x^2 + \sigma_y^2 - \sigma_x\sigma_y = A^2 + 0^2 - (A)0 = A^2.$$

Therefore, $\sigma_{amp} = A$. Note that this loading is not proportional. ∎

EXAMPLE 11.7 A thin-walled shaft is to support a rigid disk of radius $r = 6$ inches which slips over one end and is then fixed to the shaft. The rod is cantilevered at the other end. The disk is at a distance of $L = 10$ inches from the cantilevered end (Fig. 11.6). A load that fluctuates between 1,500 pounds in one direction and 1,500 pounds in the opposite direction is applied tangent to the edge of the disk. The space in which the shaft must operate constrains its outer radius to $R = 1.5$ inches. Determine the wall thickness, t, so that the shaft will not suffer fatigue failure before one million cycles. An S-N diagram indicates that the critical stress for a million cycles is $\sigma_c = 35$ ksi. Use strength of materials to estimate any stresses that are needed.

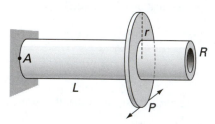

FIGURE 11.6 A disk mounted on a thin-walled shaft.

Solution The loading is purely cyclic. Examine the fatigue behavior at the point, A, at which the risk of static failure is the greatest. Since the disk is approximated as rigid, the load, P, can be moved to standard position at the center of the rod cross section. The rod is subjected to a torsion of $T = Pr$ and a bending load, P. The bending moment at point A is then $M = PL$. The loading is proportional. The polar moment of inertia for a thin-walled circular rod can be approximated by $J = 2\pi R^3 t$, where t is the wall thickness and R is the rod radius. Recall that for a circular cross section, the area moment of inertia about a diameter is one-half the polar moment of inertia, so that $I = \pi R^3 t$.

The state of stress is two dimensional on the surface of the rod. The axial bending stress is $\sigma_x = MR/I = PL/\pi R^2 t$. The shear stress is $\tau_{xy} = TR/J = Pr/2\pi R^2 t$.

The distortion energy theory (Chapter 9, Section 9.6.4) implies that the Mises equivalent axial stress is given by

$$(\sigma_x - \sigma_y)^2 + \sigma_x^2 + \sigma_y^2 + 6\tau_{xy}^2 = 2\sigma_{eq}^2.$$

In this case, since $\sigma_y = 0$,

$$\sigma_{eq}^2 = \sigma_x^2 + 3\tau_{xy}^2 = (PL/\pi R^2 t)^2 + 3(Pr/2\pi R^2 t)^2.$$

The equivalent Mises mean is zero and the equivalent Mises amplitude $2391/t$ is obtained by setting $P = 1,500$ pounds.

In the purely cyclic load, fatigue failure is assumed to occur at $\sigma_{eq} = \sigma_c = 35$ ksi. Substituting these values into either the Soderberg [equation (11.2)] or Goodman criteria [equation (11.3)] and solving for t yields $t = 0.0683$ inches. Notice that $t < R/10$ so that the thin-walled approximations are valid. ∎

EXAMPLE 11.8 A thin-walled cylindrical pressure vessel of radius $R = 6$ inches and wall thickness $t = 0.125$ inches contains a pressure of $p = 100$ psi. The vessel is made of a steel with $\sigma_Y = 50$ ksi in both tension and compression. The vessel is subjected to a bending moment, M, which varies cyclically between $\pm 90,000\pi$ in.-lb. Assume the endurance limit is $\sigma_e = 25,000$ ksi. Determine whether or not the Soderberg criterion predicts an infinite fatigue life for the pressure vessel.

Solution   Take a coordinate system with x along the longitudinal axis of the cylinder and y in the direction perpendicular to the bending axis of the applied moment, M. The highest bending stress magnitudes occur at a point, A, on the outer surface of the vessel at $y = R$. Recall that the area moment of inertia for a thin ring is $I = \pi R^3 t$. Assuming that the structure is elastically loaded, the stresses at point A are, by superposition of the pressure vessel and bending stresses,

$$\sigma_x = \frac{pR}{2t} + \frac{MR}{\pi R^3 t} = 2{,}400 \pm 20{,}000 \text{ psi};$$

$$\sigma_z = \frac{pR}{t} = 4{,}800 \text{ psi}.$$

The principal mean stresses are $\sigma_x = 2{,}400$ psi and $\sigma_z = 4{,}800$ psi. The mean equivalent Mises stress is

$$\sigma_{\text{mean}} = (\sigma_x^2 + \sigma_z^2 - \sigma_x\sigma_z)^{1/2} = [(2{,}400)^2 + (4{,}800)^2 - (2{,}400)(4{,}800)]^{1/2} = 4{,}157 \text{ psi}.$$

The principal amplitudes are $\sigma_x = 20{,}000$ psi and $\sigma_z = 0$ psi. Therefore the equivalent Mises amplitude is 20,000 psi. Under equivalent stress strategy, the Soderberg criterion for an endurance limit of $\sigma_e = 25{,}000$ psi and a yield stress of $\sigma_Y = 50{,}000$ psi is

$$\frac{20{,}000}{25{,}000} + \frac{4{,}157}{50{,}000} < 1.$$

The Soderberg criterion predicts infinite life.

The maximum Mises stress is

$$\sigma_{\text{max}} = [(22{,}400)^2 + (4{,}800)^2 - (22{,}400)(4{,}800)]^{1/2} = 20{,}427 \text{ psi}.$$

Both the maximum and minimum values are less than the absolute value of σ_Y so the structure is elastically loaded as assumed. ∎

11.6 MICROMECHANICS OF FATIGUE

Fatigue behavior is assumed to be rate independent in metals. In other words, the assumption is that fatigue does not depend on the frequency of the reversed loadings, but only on the number of reversals and the magnitude of the amplitude. Fatigue of polymeric materials, like wood, may depend on the frequency of the cycling; the experimental evidence does not lead to a definitive conclusion.

Orowan [7] pointed out in 1948 that fatigue failure can occur even when the maximum stresses are in the elastic range. Yet monotonic loading fracture requires loading into the plastic region. Orowan tried to resolve this apparent contradiction in his theory of mechanical fatigue, first proposed in 1939. He said that fatigue failure is a result of two material facts: the material is not homogeneous and materials strain harden.

He assumed that the plastic response of the metal is time independent; there is no creep. He also assumed that, if the material is perfectly homogeneous, cyclic loading in the elastic range can never cause fatigue failure. On the other hand, if the material is strained into the plastic range, the stress would build up and there would be no safe fatigue amplitude. He concluded that since materials do show fatigue failure under elastic loads and have safe cyclic strain amplitudes, they must not be homogeneous. There must be regions of inhomogeneous material within the body that can become plastic while the remaining bulk portion of the body remains elastic. Such regions are called plastic inclusions. Orowan said that such plastic inclusions may be a region of stress concentration due to the geometry of the member or may be a region in which there is a particularly favorable orientation of slip directions and slip planes. Once the stress in the inclusion reaches a critical state, a crack is initiated. Subsequent fatigue cycles cause the crack to grow and eventually result in failure of the member. Orowan's proof of this hypothesis depends on a three-spring and one-plastic-element model for the behavior of the plastic inclusion within an elastic region. Currently, researchers are more likely to speak in more general terms of damage, often due to local stress concentrations induced by many types of mechanisms, which cause the microcracks leading to fatigue failure.

11.7   FATIGUE FRACTURE

Fatigue failure results from a fracture initiated at microcracks that grow as the load on the body is cycled. The study and control of fatigue depends on understanding how the microcracks grow under a dynamic load even though the stress never exceeds the elastic limit for a static load. After a small portion of the critical number of cycles, slip bands are visible on the metal specimen. These result from dislocation motion along the corresponding plane. But since the load is cyclic, the dislocations must oscillate on the slip plane. Such a motion is unlikely in static loading of any magnitude.

After a number of cycles, the material begins to fatigue harden. This does not result from blockage of dislocation motion by other dislocations as in work hardening. It is conjectured rather that small voids, or intrusions, are created along the slip planes. These voids block dislocation motion. In conjunction with the intrusions are extrusions growing out of the material surface along the slip lines. The voids can act as microcracks at which the fracture is initiated. Even if the microcracks individually grow very slowly, they can meet up with other microcracks and create a crack large enough to initiate fracture. Such a process is called void coalescence.

Fatigue cracks are usually initiated on slip planes at the surface of the body or adjacent to internal voids or inclusions. The growth of the crack along the slip plane is often called a Stage I crack. After the crack is large enough, it may change direction to propagate in a direction normal to the applied tensile load. This growth is often called Stage II. The crack may pass along grain boundaries (intercrystalline growth) or grow through a grain (transcrystalline growth). In the latter case, the fracture may involve either cleavage or localized plasticity, depending on whether the behavior is brittle or ductile.

In the ductile case of localized plastic zones, striations are formed on the crack surface. These striations are a consequence of the creation and relaxation of a plastic zone on each cycle. The striations are similar to tree rings and, in principle, can be counted to determine the number of cycles to failure. The point of crack initiation lies at the center of the striation pattern. Two regions are often visible on the fracture surface of a part which failed due to fatigue (Fig. 11.7). One region is corrugated, but otherwise smooth, with the grooves surrounding the point at which a microcrack initiated the fracture. The fracture proceeded slowly though this fatigue region. The second region is the portion that failed very rapidly once the crack was large enough; it resembles the surface of a brittle fracture as described in Chapter 10.

Substructures in a metal can halt fatigue crack propagation. For example, a twin boundary separates crystalline regions whose lattice structures appear to be mirror images of each other reflected about the twin boundary (Chapter 9, Section 9.3). A shear stress parallel to the twin plane is required to create twinning. Such twins can result

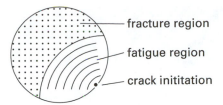

FIGURE 11.7   Regions of a fatigue fracture surface.

from an externally created deformation or by annealing. Twinning can delay fatigue fracture by blocking the growth of cracks (Fig. 11.8 and Fig. 11.9).

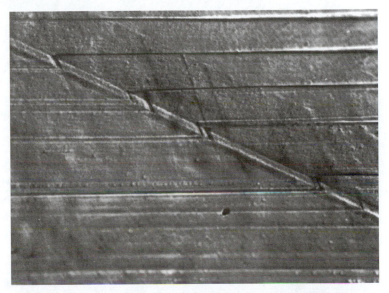

FIGURE 11.8   Basal (0001) slip and (10$\bar{1}$2) twin interactions on the surface of a magnesium crystal after a cyclic fatigue shear test. Note the relatively faint residual slip traces giving evidence of a wider twin width during the earlier stages of deformation, and also evidence of significant blockage of slip at the twin interfaces.

In the aftermath of a catastrophic failure, an engineer or material scientist can inspect a fracture surface and learn at what point the crack was initiated and whether or not it was a fatigue failure by the patterns on the surface. Such information helps the engineer to redesign the structure to avoid the type of loading that caused the failure.

It is difficult to design a component to prevent fatigue failure. Since a flawed surface contributes greatly to fatigue problems and many failures begin at the surface, an engineer might try to control the surface properties by specifying polishing or by hardening the surface of the component.

11.8   FATIGUE FRACTURE MODELS

The traditional endurance tests of smooth laboratory specimens do not distinguish between crack initiation and crack growth in a material under repeated loading. The *S-N* curves ignore any flaws that might exist in the material at the beginning of the test. Therefore, in design, they lead to an overestimate of the life of a component.

FIGURE 11.9 A higher-magnification image of another magnesium sample as for Figure 11.8. The crack openings occur within heavily marked slip bands that were blocked by deformation twins. The remnants of the twins are visible in the background.

Even in a carefully made material, microcracks form after a small number of load cycles. The material is said to be damaged when such microcracks or voids form. An alternative to the endurance method of studying fatigue is to assume that the fatigue life is related to the rate at which the microcracks grow and eventually attain the critical crack length necessary for fracture. Cyclic tensile loading with a maximum stress below the fracture stress can eventually cause fracture.

Beginning in the early 1960s, a method of analyzing this behavior was developed from the ideas initiated by P. C. Paris in his 1962 Ph.D. thesis. The idea was to describe fatigue in terms of the crack growth rate and the fracture mechanics concept of the stress intensity factor. When a specimen is repeatedly loaded, the length of any existing cracks increase over time with the number of cycles. Fatigue is a dynamic process so that the crack growth rate is as significant as the current size of the crack. Two factors influence the crack growth rate, the body configuration, and the load-time history. Crack growth is assumed to be continuous. Some early researchers had suggested that crack growth might periodically cease for loading intervals of from one hundred to millions of cycles, but no evidence has been found for such delay in crack growth. Furthermore, the crack growth rate is assumed not to significantly depend on the loading rate, $d\sigma/dt$. Finally the crack growth rate may be measured as a function of time, da/dt, or as a function of the cycles, da/dN. When the frequency of the load is a constant, v_c Hz, then

$$\frac{da}{dt} = v_c \frac{da}{dN}.$$

To generate experimental information about the rate of crack growth, a mode I specimen with an edge crack of a known length is repeatedly loaded in tension perpendicular to the crack faces. The change in length is measured as a function of the number of load cycles applied to the specimen. The crack growth depends on the material and the stress levels achieved in the load cycle. Paris observed that the process is driven by the change in stress intensity factors between the maximum and minimum stresses in the loading cycle. The stress intensity is $K = Y\sigma\sqrt{\pi a}$, where a is the half crack length and Y is the geometric factor for the specimen and crack orientation as defined in fracture mechanics. Therefore

$$\Delta K = K_{max} - K_{min} = (\sigma_{max} - \sigma_{min})Y\sqrt{\pi a}.$$

Fracture mechanics postulates a critical value of $K = \sigma Y \sqrt{\pi a}$, called K_c, above which the crack will nearly instantly propagate through a brittle material, resulting in fracture. It is assumed that $K_{min} \leq K_{max} < K_c$. Note that in fatigue analysis, the half-crack length is denoted by a.

Experimentally, a simple relationship is obtained if the base 10 logarithm of the crack growth rate per cycle is plotted as a function of the logarithm of ΔK. Experiment shows that the constant crack growth rate da/dN increases with the value of ΔK, which depends on the crack length, associated with the cyclic load. In most brittle materials, the curve has an initial region of slow growth, an intermediate, almost linear region, and finally a region of rapid growth (Fig. 11.10).

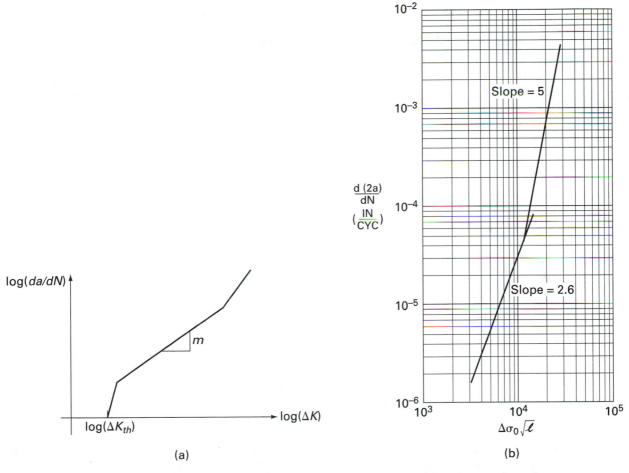

(a)   (b)

FIGURE 11.10 (a) Schematic of the three regions on a crack growth rate versus change in stress intensity curve on a logarithmic coordinate system. (b) Curve for Al 7075-T6. Reprinted from [5] by permission.

Each material appears to have a small threshold value, ΔK_{th}, at which cracks begin significant growth. Notice that it is impossible, in this model, to have $da/dN = 0$ because then $\log_{10}(da/dN)$ approaches negative infinity. Below the threshold, under a cyclic load, the growth must be slow. In ceramics, this is called subcritical crack growth in which no significant crack propagation occurs.

In the intermediate region, the Paris model,

$$\frac{da}{dN} = C(\Delta K)^m, \tag{11.8}$$

is chosen because it reproduces the linear curve observed on the experimental log-log graph.

$$\log_{10}\left(\frac{da}{dN}\right) = \log_{10}\left[C(\Delta K)^m\right] = \log_{10} C + m \log_{10}(\Delta K).$$

The slope of line is m and the vertical axis intercept is $\log_{10} C$.

The third region of rapid growth often gives a curve asymptotic to the vertical line through K_c, the critical fracture strength. In each of the regions, the crack grows but it has not reached the critical state at which extremely rapid, unstoppable growth, and thus fracture, would occur.

Because of the use of ΔK, this is a theory for materials that behave in an essentially brittle manner. If a large portion of the specimen yields plastically during a cycle, the fatigue crack growth rate curve cannot be used in design.

Based on his tests of aluminum 7075-T6 and 2024-T3, Paris suggested that under sinusoidal loading $m = 4$. More recent tests show a range of values of this exponent for other materials. For steels, m lies in a range from about 2.0 to 3.5, and C is on the order of 10^{-10} to 10^{-13} for growth rates measured in m/cycle. ΔK is measured in MPa-m$^{1/2}$ for these values of m and C.

EXAMPLE 11.9 Steel with $m = 2.8$ and $C = 1.2 \times 10^{-11}$ is loaded sinusoidally by $\sigma(t) = 100 \sin(0.2t) + 150$ MPa. Determine the crack growth rate for mode I cracks having a maximum length of 0.001 m. Assume the geometric factor is $Y = 1.12$.

Solution $\Delta K = (\sigma_{max} - \sigma_{min})Y\sqrt{\pi a} = (200)(1.12)(0.03963) = 8.88$ MPa-m$^{1/2}$. The crack growth rate is, because the units for C assume that ΔK is in MPa-m$^{1/2}$,

$$\frac{da}{dN} = 1.2 \times 10^{-11}(8.88)^{2.8} = 5.43 \times 10^{-9} \text{ m/cycle}. \qquad \blacksquare$$

The experimental curves also depend on the ratio of the maximum and minimum stresses in the load cycle, $R = \sigma_{min}/\sigma_{max}$. A given ΔK is more dangerous at a higher minimum stress in the cycle. The prediction of the crack growth rate in design depends on both ΔK and R. The experimental curve is intended to be independent of geometry and depend only on R and the material. Tests are usually done with $0 < R < 0.2$ to compare materials. R has more of an effect in lower growth rates than higher. Therefore ΔK_{th} changes as R changes.

The Walker model describes this relationship by writing ΔK in a form equivalent to that when $R = 0$ ($\sigma_{min} = 0$). Walker assumed that for other ratios, R, $\overline{\Delta K} = K_{max}(1 - R)^\gamma$, where γ depends only on the material. Here $\overline{\Delta K}$ is the stress intensity for $R = 0$ that causes the same crack growth rate as ΔK at $R \neq 0$. But $\Delta K = K_{max} - K_{min} = K_{max}(1 - R)$. Elimination of K_{max} from the expression for $\overline{\Delta K}$ by this relation produces

$$\overline{\Delta K} = \frac{\Delta K}{(1 - R)^{1-\gamma}}.$$

This expression for the crack intensity is substituted in the Paris relation at $R = 0$

$$\frac{da}{dN} = C_0(\overline{\Delta K})^{m_0}$$

to give the Walker relation,

$$\frac{da}{dN} = C_0 \left(\frac{\Delta K}{(1-R)^{1-\gamma}} \right)^{m_0} = C(\Delta K)^{m_0},$$

where C_0 and m_0 are the constants for the case $\sigma_{\min} = 0$. In this model, the exponent, m, is independent of R and the relation of the coefficient C to R is given by

$$C = \frac{C_0}{(1-R)^{m_0(1-\gamma)}}.$$

Another model for the dependence of the crack growth rate on R was proposed by R. G. Forman,

$$\frac{da}{dN} = \frac{C(\Delta K)^m}{(1-R)(K_c - K_{\max})}.$$

This relation predicts extremely rapid growth when $K_{\max} = K_c$. The Forman model is most accurate in the intermediate and high ranges of ΔK. The constants are not identical to those in the Walker model, even for the same experimental data. These theories ignore sequence effects, that is, variations in loading history from cyclic loading at a fixed amplitude.

EXAMPLE 11.10   For aluminum 2024-T3, $m_0 = 3.6$, $\gamma = 0.68$, and $C_0 = 1.42 \times 10^{-11}$. Compare the crack growth rates at $\Delta K = 50$ MPa-m$^{1/2}$ for $R = 0.1$ and $R = 0.2$ as predicted by the Walker relation.

Solution   At $R = 0.1$,

$$\frac{da}{dN} = C_0 \left(\frac{\Delta K}{(1-R)^{1-\gamma}} \right)^{m_0} = 2.10 \times 10^{-5} \text{ mm/cycle},$$

and at $R = 0.2$,

$$\frac{da}{dN} = 2.40 \times 10^{-5} \text{ mm/cycle},$$

a 14.3% difference.   ■

Failure is expected, in these descriptions of fatigue fracture, when the crack is long enough that $K_{\max} = K_c$ for fracture, not when $\Delta K = K_c$. The failure crack half-length, a_{cf}, for a repeated loading between $\sigma_{\min}$ and $\sigma_{\max}$ is obtained by solving $K = \sigma Y \sqrt{\pi a}$ for the crack length corresponding to K_c,

$$a_{cf} = \left(\frac{K_c}{\sqrt{\pi} Y \sigma_{\max}} \right)^2.$$

Brittle fracture can be due to striation and cleavage, called transgranular fracture, or due to cracking along grain boundaries, called intergranular fracture.

Knowledge of the crack growth rate, da/dN, permits calculation of a life estimate for a component. Let a_i be the initial crack length at cycle N_i and a_f be the final half-length at cycle N_f. Assuming that the shape factor, Y, is a constant for all crack lengths, the Paris relation $da/dN = C(\Delta K)^m$ integrates to give a closed-form expression for the relation between crack length and the number of cycles in a constant amplitude repeated loading with stress difference, $\Delta\sigma$. By separation of the variables, N and a,

$$N_f - N_i = \int_{a_i}^{a_f} \frac{da}{C(\Delta K)^m}$$

$$= \int_{a_i}^{a_f} \frac{da}{C(\Delta\sigma Y\sqrt{\pi a})^m} \qquad (11.9)$$

$$= \frac{a_f^{1-m/2} - a_i^{1-m/2}}{C(\Delta\sigma Y\sqrt{\pi})^m(1-m/2)}.$$

Generally, the lifetime is influenced more by the initial crack half-length, a_i, than by the final fracture crack length, a_f. As m tends to 2, the lifetime $N_f - N_i$ increases without limit because the expression is undefined at $m = 2$. When the shape factor is a function of the current crack size, a numerical integration technique must be used to compute the number of cycles required to reach a given final crack length.

EXAMPLE 11.11

A 4340 steel after heat treatment has the properties $C = 1.2 \times 10^{-11}$, $m = 2.8$, $K_c = 185$ MPa-m$^{1/2}$, $\sigma_Y = 1000$ MPa, $\sigma_u = 1050$ MPa and is loaded cyclically between 0 and 290 MPa. The maximum initial crack half-length is estimated to be 0.0001 meter. Assume the geometric factor is $Y = 1$. Determine the number of cycles to fracture.

Solution

First, determine the critical crack half-length for fracture.

$$a_{cf} = \left(\frac{K_c}{\sqrt{\pi}Y\sigma_{\max}}\right)^2 = 0.1295 \text{ meters.}$$

To find the cycles required to grow a crack of half-length 0.0001 to 0.1295 m, apply equation (11.9) with $N_i = 0$. This produces $N_f = 200,723$ cycles.

The Soderberg method predicts the critical stress for $\sigma_a = \sigma_m = 145$ MPa to be $\sigma_c = 169$ which is less than the endurance limit, $\sigma_e \sim 0.5\sigma_u = 525$ MPa. Therefore the method predicts that the loading should be safe forever. The Soderberg method overestimates the fatigue life. This is a typical result for the Soderberg relation. ■

Design against fatigue failure can be performed by the series of steps suggested by Clarke [3]. Experimental crack growth rate data must be obtained to produce the parameters, C and m. The mode I critical fracture intensity, K_{I_c}, must be obtained experimentally. The orientation and size of the largest crack at the beginning of the service life of the member must be estimated. The fracture mechanics geometric factor must be determined from the shape of the body and the orientation of the largest cracks with respect to the load. Finally the nominal stress field expected during the cyclic load must be calculated so that the change in stress intensity factor may be calculated. From this data, the expected fatigue life of the component may be estimated.

EXAMPLE 11.12

A 7079-T6 aluminum has $C = 4 \times 10^{-18}$ and $m = 3$ when $7 \leq \Delta K \leq 40$ ksi-in$^{1/2}$ at room temperature. Also $K_c = 35$ ksi-in$^{1/2}$, $\sigma_Y = 65$ ksi. A rectangular plate is cyclically loaded in-plane by an axial stress ranging between 10 and 30 ksi. The critical defects are determined to be center cracks. Determine the number of cycles to grow an initial maximum crack length of 0.1 inches to fracture.

Solution

The critical crack length is determined from

$$a_{cf} = \left(\frac{K_c}{\sqrt{\pi}Y\sigma_{\max}}\right)^2,$$

where for a plate of width $2w$, half crack length $c = a$ (Chapter 10, Table 10.1),

$$Y = \frac{1 - \frac{c}{2w} + 0.326\frac{c^2}{w^2}}{\sqrt{1 - \frac{c}{w}}}.$$

The number of cycles to failure is given by

$$\frac{da}{dN} = \frac{1}{C(\Delta K)^m} = \frac{\left(\sqrt{1 - \frac{c}{w}}\right)^m}{C\left[\Delta\sigma\left(1 - \frac{c}{2w} + 0.326\frac{c^2}{w^2}\right)\sqrt{\pi a}\right]^m}.$$

In this situation, the geometric factor in the stress intensity depends on the current crack length so that the integral cannot be obtained in closed form. There are many numerical methods by which to solve for the number of cycles to reach a given final crack length. The equation could be solved by a differential equation algorithm such as the Runge-Kutta. Alternatively, the expression could be rewritten as an integral,

$$N_f = \int_{a_i}^{a_f} \frac{da}{C(\Delta K)^m} = \int_{a_i}^{a_f} \frac{da}{C(\Delta\sigma Y\sqrt{\pi a})^m},$$

and then approximation techniques for the integral are used to find the value of N_f. ∎

Ambient environmental factors such as temperature directly affect the material properties and therefore the fatigue behavior. Higher temperatures increase crack propagation. But even environmental factors not corrosive under zero stress, such as moisture present as humidity in the air, can have a significant influence on fatigue behavior (Fig. 11.11). Ambient moisture lowers the fatigue life of some metals through corrosion fatigue which affects both crack initiation and crack propagation. During tests of aluminum in high relative humidity, fatigue damage has resulted in the evolution of hydrogen gas from a reaction between the moisture and the specimen. This emission of hydrogen gas seems to be a sensitive indicator of the onset of fatigue damage or microcracking in aluminum. Most surfaces of aluminum in air have a thin film of oxidation. The cyclic loading aids the oxidation of aluminum which captures the oxygen molecule from water leaving the hydrogen. One hypothesis for the mechanism of increased fatigue damage in high humidity is that a newly formed crack surface is oxidized and so resists reverse slip upon the change in load direction. Therefore the damage cannot be recovered over the load cycle and accumulates. This idea is supported by the fact that coating the aluminum with a thin surface layer of a film-forming liquid reduces fatigue damage. Fatigue in a near vacuum or low air pressure is a significant issue for the space program or for high-altitude aircraft. Fatigue tests performed first in a vacuum and then in air show up to a 15% decrease in fatigue life in the specimen tested in air. The moisture content is the active gas in the air for this phenomenon.

11.9 FATIGUE OF CERAMICS

Despite most ceramic materials being relatively brittle in ambient temperature engineering applications, their fatigue properties have been characterized rather successfully with the same type of fatigue-fracture mechanics relations used for more ductile structural materials. One reason may be that, as shown by indentation hardness tests in silicon crystal specimens without precracks, permanent deformation precedes crack initiation in ceramics. An importance difference between ceramics and metals lies in the ease of crack growth, once nucleated within a crystal grain. Because the ceramic

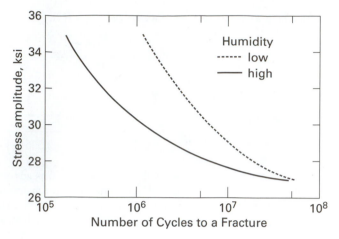

FIGURE 11.11 *S-N* curve for 2024-T4 aluminum in high ($\sim$ 90%) and low ($\sim$ 5%) relative humidity. (Reprinted from [2] by permission.)

crack grows more easily, there has been an important effort to use crystalline ceramics in an ultrafine grain condition.

The qualitative fatigue behavior of ceramics is similar to that of metals, but the quantitative behavior is not. The crack propagation data for ceramics seem to satisfy the Paris model [equation (11.8)],

$$\frac{da}{dN} = C(\Delta K)^m.$$

However, the value of m ranges between 12 and 20, much greater than the value of 2–4 for metals. The crack propagation rate for short cracks, less than 500 μm, is much greater than for long cracks, those on the order of 3 mm.

Stress-life testing of ceramics produces similar results as for metals. The slope of the *S-N* curve representing the logarithm of the lifetime as a function of the applied stress is higher than for metals. Therefore, the lifetime decreases more quickly as the stress amplitude is increased than in the case of metals. A designer must therefore be more cautious when picking the allowable stress for a desired lifetime of a ceramic component. For design, the strategy discussed in Section 11.8 is to integrate the crack propagation rate curve, equation (11.8), to obtain the number of cycles for a crack to grow to a critical value. However, because of the large value of the exponent m for ceramics, a slight error in estimating the applied stresses in use will lead to a large difference in the estimated fatigue life. The initial maximum allowable defect size may therefore be required to be so small as to be undetectable; the calculation is then useless in practice. No fully satisfactory design criteria for fatigue of ceramics currently exists.

Some of the cyclic fatigue propagation curves are not of sigmoidal shape as they are for metals. Therefore Ritchie and Dauskardt [9] recommend that the threshold value of the intensity factor, ΔK_{TH}, be taken as that at a growth rate of 10^{-10} m/cycle. Such a choice usually gives a value about half of K_c.

The fatigue fracture surface of a ceramic shows none of the striations visible for metals. In general, the fracture surface of a ceramic is similar under both monotonic and cyclic loading, although for some materials, such as the composite Al_2O_3-SiC_w with SiC whiskers in a Al_2O_3 matrix, there are slight differences in roughness.

The high values of the exponent in the Paris relation for ceramics seem to be a result of the dependence of the propagation rates on K_{max}, the value of the stress intensity at the maximum stress in the fatigue loading. The Paris law has been rewritten by Ritchie and Dauskardt [9] to include this dependence on K_{max}. For $R = \sigma_{min}/\sigma_{max}$,

$$\frac{da}{dN} = C(1-R)^n (K_{max})^n (\Delta K)^p, \qquad (11.10)$$

where $n + p = m$, where m is the exponent in the Paris law. In this formulation, the values of n and p are on the order of 10 and 5, respectively.

Crack tip shielding refers to physical mechanisms which can lower the propagation rate predicted by the Paris relation. Mechanisms near the crack which help prevent crack tip growth are said to shield the crack. These mechanisms either involve behavior in the zone about the tip called zone shielding or involve processes which either bridge the gap between the crack surfaces or bring them in contact, called contact shielding. For example, in tetragonal Mg-PSZ, the high stress at a crack tip can induce a martensitic transformation from the tetragonal to a monoclinic phase. As a consequence, compressive stresses act locally on the crack surface and shield the crack from the far-field stress. The behavior is called transformation shielding.

A second mechanism of crack shielding is crack closure, which occurs if asperities, or protuberances, in the crack surface make contact with the opposite crack surface. Such contact creates a smaller effective crack length and therefore reduces the crack propagation rate. In the composite Al_2O_3-SiC_w, whiskers may span the crack from surface to surface; this shielding mechanism is called crack bridging. The exact mechanisms that act near the crack tip of a ceramic to propagate the crack are not known. But possible mechanisms are classified as either intrinsic, those in which the fatigue load produces microdamage ahead of the crack tip, and extrinsic, those which are similar to fracture under monotonic loading. In the latter, the unloading portion of the cycle is thought to reduce the effect of crack shielding mechanisms and allow the crack to grow. One possible mechanism is microcracking at the crack tip. Closure can also be induced by debris or corrosion.

11.10 FATIGUE OF POLYMERS

Polymers usually have internal flaws throughout the material due to the manufacturing process. These flaws serve as locations for crack initiation as early as during the first loading cycle. Crack propagation during fatigue fracture of thermoplastics, like crystalline polycarbonate, may produce striations similar to those observed in metals (Fig. 11.12). The uniaxial stress-strain curve for such a material has a long ductile region so that in that sense it is similar to metals.

The nature of the fatigue fracture surface obtained in polymers that deform by crazing seems to be related to the three regions of the S-N curve (Fig. 11.13).

Under the high stress amplitudes of region I, the surface is specular and crazing precedes fracture much as in monotonic load failure. Under region II stress amplitudes, crazing begins at the specimen surface and forms a portion of a ring about the fracture surface. The interior of the ring is featureless. In region III, no crazing occurs and there are often striations produced by the crack growth [10]. In some polymers, those that do not craze, region I may not be present in the S-N curve. The stress at the transition from region I to region II seems to correspond to the stress at which crazing is initiated in a constant strain rate increasing loading, at least for polystyrene. Polycarbonate (PC) has no region I and exhibits little crazing. Polymethylmethacrylate (PMMA), that is, Plexiglas, has a small region I and exhibits a modest amount of crazing. However,

FIGURE 11.12   Striation pattern on the fatigue fracture surface of polycarbonate. Reprinted from [6] by permission.

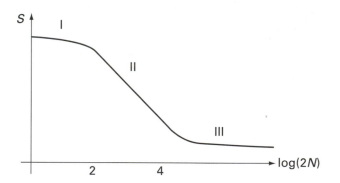

FIGURE 11.13   Schematic *S-N* Curve for polymers.

both PC and PMMA behave like polystyrene (PS) at low temperatures around 77K. Polystyrene is often viewed as a semibrittle material.

Polymers, because they are viscoelastic, may generate significant amounts of heat during cyclic loading. But these materials are not good heat conductors; therefore the specimen temperature may rise enough to affect experimental results. Further, the thicker the specimen the longer it takes to transfer the generated heat to the ambient air. The ambient temperature has a significant influence. For example, after 25,000 cycles at room temperature the temperature rise in nylon 6 has been measured as about 1°C, but at ambient temperatures near the glass transition (30–54°C) the temperature rise after 25,000 cycles can be as much as 50°C. The temperature rise per unit time depends on the cycling frequency; the temperature change is not proportional to the frequency, but does increase with frequency. In polymers, not only is the number of cycles important, but also their frequency because of the viscoelastic nature of polymers. The heating, if it raises the material temperature above the glass transition, reduces the modulus and strength; this behavior is called thermal failure. It may depend on both the specimen geometry and the material. Some of the heat may induce a phase transition.

Thermal failure is not the only fatigue failure mechanism in polymers; crack initiation and growth is also important. Crack initiation may involve various mechanisms, depending on the polymer. Initiation is possible at spherulite boundaries. This is similar to crack initiation at grain boundaries in metals. Cracks may form at craze sites as well as at internal or surface flaws in the material.

The fatigue crack propagation may be represented by the Paris relationship for hard polymers such as PMMA and polycarbonate (PC). The Paris relation is successful for a range of ΔK under 10 MPa-m$^{1/2}$ in amorphous polymers including polyvinyl chloride (PVC), PMMA, PC, and epoxies, as well as in semi-crystalline polymers such as nylon 66, polyamide (PA), and polyvinylidene fluoride (PVDF). The values of m are much larger than those for metals. The Paris relation for polymers depends on the frequency of load cycling. Some polymers such as PMMA, polystyrene, and PVC show a reduced crack growth rate, and thus more fatigue resistance, as the frequency increases at constant ΔK. Other polymers slightly increase the crack growth rate with a frequency increase; these include polycarbonate and polysulfone. Those polymers that show the highest tendency for crazing also show the greatest sensitivity to frequency of load cycling.

11.10.1 Fatigue of Rubber

Growth of a flaw in rubber is partially due to the application of a stress. An empirical relation has been devised to predict this growth, Δc, on each loading cycle.

$$\Delta c = B(2cW)^n, \tag{11.11}$$

where B and n are material constants, and W is the strain energy. The exponent n ranges from 1 to 2 for strain-crystallizing rubbers and from 3 to 6 for amorphous rubbers. This relation applies if $\theta \equiv 2cW$ is above a critical value at which growth begins. A log-log plot of the experimental relation between Δc and θ becomes linear after a short but rapid increase just above the critical value of θ.

A lifetime estimate in terms of an initial crack length, c_o, is obtained by integrating $dc/dN = B(2cW)^n$ over the number of cycles, N.

$$\int_{c_o}^{c_f} \frac{dc}{c^n} = B(2W)^n \int_0^N dN. \tag{11.12}$$

Letting $c_f = \infty$ yields the expression for the number of cycles that the material can survive,

$$\frac{1}{N} = (n-1)c_o^{n-1}B2^nW^n. \tag{11.13}$$

If a crystallizing rubber specimen is subjected to fatigue cycles in which the stress always remains in tension, the crystallization near the crack tip created by the high stresses allows very little crack growth due to tearing.

11.10.2 Fatigue of Paper

An S-N curve has been obtained for kraft paper, the type used to make paper bags. The tests were performed at 73°F and 50% relative humidity under a deformation rate of 0.5 in./min as the load was increased and at 12 in./min as the load was decreased. Tests of three different basis weight (weight per unit surface area) papers of the same type fit a single equation when the load is normalized by dividing by the fracture load for a sheet which has never been loaded. If S is the normalized maximum load, plotting the logarithm of one less than the number of cycles, N, versus the logarithm of the load produces

$$N - 1 = aS^{-b}, \tag{11.14}$$

for constants a and b. The minus 1 term may arise because, on the first cycle, the load stretches out kinks and bends in the wood pulp fibers as much as it stretches the material. Some researchers have debated whether the number of cycles or the rate is more important in predicting failure.

A great deal remains to be learned about fatigue failure for each of the various classes of materials. Because fatigue is a major cause of failure in mechanical components, such research should have a high priority.

11.11   STRAIN ANALYSIS OF FATIGUE

The S-N curves, relating the cycles to failure for a constant stress amplitude cyclic load with zero mean, are most applicable to long-lived materials under stress amplitudes in the elastic range. Materials that have short lives under cyclic loads producing plastic strain can be studied by a cyclic strain amplitude versus cycles to failure curve. The analysis of fatigue using such curves was first proposed by S. S. Manson in 1965 and independently by L. F. Coffin. Each data point is created by determining the cycles to failure, N_f, of the material under a completely reversed constant strain amplitude uniaxial test (Fig. 11.14) in which the strain is controlled.

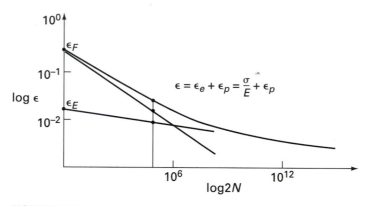

FIGURE 11.14   A typical strain versus cycles to failure curve. The total strain amplitude is ϵ_T.

In each cycle, the strain amplitude has an elastic and a plastic component so that $\epsilon = \epsilon_e + \epsilon_p$ (Fig. 11.15).

The strain amplitude on the chosen final cycle, number N_f, is decomposed similarly into a plastic component, $\log \epsilon_p$, represented by the line asymptotic to the $\log \epsilon$-$\log N$ curve at small $\log N$ and an elastic component, $\log \epsilon_e$, represented by the line asymptotic at large $\log N$, as shown in Figure 11.14. The line asymptotic at large $\log N$ represents the elastic component because, at a large number of cycles, strain hardening causes the plastic component to become very small. For fracture on the first half cycle, the elastic component is very small. The value of the intercept of this curve with the vertical axis is the elastic component $\log \epsilon_E = \log(\sigma_F/E)$ in Figure 11.14, where σ_F

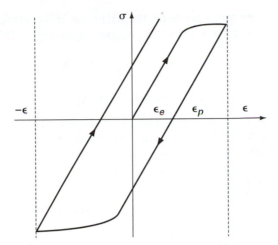

FIGURE 11.15   A constant strain cycle showing the elastic, ϵ_e, and the plastic, ϵ_p, strains in the first half-cycle.

and ϵ_F are the fracture stress and strain in a uniaxial test, respectively, which is taken to be at the one half-cycle.

The Manson-Coffin relation gives an empirical expression for each of the elastic and plastic strains as a function of lifetime, N_f. Apparently, they each arrived at their expression by generalizing the Basquin relation [equation (11.1)] to include the case in which the strain has a plastic as well as an elastic component. In their expression, the number of cycles is taken as $2N$. From the Basquin relation [equation (11.1)],

$$N_c = N_{fr} \left(\frac{\sigma_c}{\sigma_F} \right)^{1/b}.$$

In the Basquin relation, the critical number of cycles is $N_c = 2N_f$ and $N_{fr} = 1$ because the first data point is taken at a strain amplitude that causes failure after a half-cycle. Also the elastic component of the strain is $\sigma_c/E = \epsilon_e$. Therefore

$$\epsilon_e = \frac{\sigma_F}{E}(2N_f)^b,$$

or the elastic strain at a failure at N_c cycles, is

$$\log(\epsilon_e) = \log\left(\frac{\sigma_F}{E}\right) + b\log(2N_f).$$

Then b is the slope of the line corresponding to ϵ_e on a $\ln - \ln$ plot (Fig. 11.14).

The plastic strain satisfies a similar empirical relation,

$$\log(\epsilon_p) - \log\epsilon_F = c\big(\log(2N_f) - \log(1)\big)$$

so that

$$\epsilon_p = \epsilon_F(2N_f)^c,$$

where ϵ_F is the plastic strain component of the strain causing failure in a half-cycle and c is the slope of the asymptotic plastic strain line at ϵ_F in Figure 11.14. The sum of these two strain terms produces the Manson-Coffin relation between the constant cyclic fully reversed strain amplitude, ϵ, and the chosen cycles to failure, N_f;

$$\epsilon = \frac{\sigma_F}{E}(2N_f)^b + \epsilon_F(2N_f)^c. \tag{11.15}$$

The values σ_F, ϵ_F, b, and c are taken to be material constants.

EXAMPLE 11.13 Determine the maximum allowable strain amplitude for a steel 4340 member that must survive 1,000 cycles to failure. Published data, for example, Dowling [4, p. 627], shows that $E = 207$ GPa, $\sigma_F = 1758$ MPa, $\epsilon_F = 2.12$, $b = -0.0977$, and $c = -0.774$.

Solution Use the Manson-Coffin relation,

$$\epsilon = \frac{\sigma_F}{E}(2N_f)^b + \epsilon_F(2N_f)^c = 0.00404 + 0.00591 = 0.00995. \qquad\blacksquare$$

At very long lives, the plastic strain component tends to zero due to hardening so that lifetime estimates can be made from the elastic component alone. The transition from short life to long life for a particular material is often taken as the value of N_f at which the elastic and plastic strain components are equal.

$$\frac{\sigma_F}{E}(2N_f)^b = \epsilon_F(2N_f)^c.$$

Exercise 11.4 Verify that the transition fatigue life is

$$N_f = \frac{1}{2}\left(\frac{\sigma_F}{\epsilon_F E}\right)^{1/(c-b)}.$$

The S-N and ϵ-N curves assume that the amplitude of the load remains the same over the life of the piece. In many applications, the fully reversed load amplitude changes more randomly. As a first approximation to random amplitude reversible loading, examine a loading that involves a set of cycles at one amplitude, another set at a second amplitude, and so on. Let n_i be the number of cycles actually applied to the specimen and let N_i be the number of cycles to failure at the ith amplitude, obtained from an empirical study at that constant amplitude. Each expression, n_i/N_i, is viewed as a fraction of the lifetime of the piece. An empirical rule is that the piece will fail when a hundred percent of its life is used up, $\sum_i(n_i/N_i) = 1$. This elementary expression, which involves no fracture mechanics, was postulated by A. Palmgren in Germany in the 1920s for ball bearing life and rediscovered by M. A. Miner in 1945.

Experiment: To avoid the use of machinery to repeatedly load the specimens, steel paper clips tested in bending will allow a person's hand to apply the fatigue load. The tests will be under strain control in which the strain amplitude is defined by the bend angle. By hand, apply a fully reversed bending load by bending the clip back and forth to the chosen angle until the clip fractures. Count the number of cycles to fracture. A cycle includes one bend in each opposite direction. In this sense, it is similar to a cycle of a sine or cosine function. Test clips at angles of 45, 90, 135, or 180° as the strain amplitude. Determine an average lifetime for each one of the strains.

Finally, choose a sequence of strains (the bend angle) and corresponding number of cycles. Apply this sequence to another paper clip until it fractures. Note the sequence and number of cycles to fracture. Compare the result to the prediction of the Palmgren-Miner rule. Did the Palmgren-Miner rule correctly predict the lifetime of the paper clip bent following a nonuniform sequence of strains and cycles?

Both the stress and strain controlled empirical curves, which depend only the amplitude of loading, assume that the speed at which load is applied, the stress or strain rate, does not affect the lifetime. Repeat the strain-controlled fatigue tests in which one set of paper clips is bent so that it takes 1 second to make the angle (high rate) and another set so that it takes 5 seconds to make the angle (low rate). Does the data support the assumption that the fatigue life is independent of the loading rate in each cycle?

PROBLEMS

11.1 A cantilever beam with length 27 inches and 2 inch by 3 inch rectangular cross section is loaded by an axial load of 6,000 pounds and a vertical load, P, at its free end. An axial test shows that the material of the beam yields at 30 ksi. An S-N curve for the material shows that for a lifetime of 10^5 cycles, the fatigue strength or critical stress for completely reversed stress is 10 ksi.

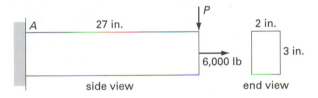

side view   end view

(a) Determine the principal stresses at point A, which is at the wall on top of the beam, in terms of P. Use elementary strength of materials. Neglect shear.

(b) If P is cyclically varied between 0 pounds and 1,000 pounds, determine the mean stress and the amplitude of the resulting cyclic stress at A.

(c) Will the beam undergo a failure after 10^5 cycles due to fatigue at point A? Prove your answer with calculations.

(d) Determine the maximum safe value, P_m, of the load, cyclically varying between 0 and P_m, to avoid fatigue failure after 10^5 cycles at point A.

11.2 Show that the Gerber criterion predicts a longer fatigue life for $\sigma_a = 45$ ksi and $\sigma_m = 5$ ksi than the Goodman criterion in the case that $N_e = 10^6$, $N_f = 10^3$, $\sigma_e = 30$ ksi, and $\sigma_f = 90$ ksi. *Hint*: First compute σ_c.

11.3 Suppose that for a steel alloy, $\sigma_e = 0.5\sigma_u$. Use the Goodman criterion to determine the maximum allowable amplitude, σ_a, of a cyclic stress when the mean stress, σ_m, is equal to the endurance limit, σ_e.

11.4 For a fixed amplitude, σ_a, is the critical stress, σ_c, lower for pure cyclic or cyclic fatigue by the Soderberg criterion at the transition from safe to unsafe loading? How does the critical stress change if the mean stress is increased in tension? How does the critical stress change if the mean stress is increased in compression?

11.5 A uniaxial test of an aluminum material shows an ultimate uniaxial stress $\sigma_u = 200$ MPa.

(a) Compute the fatigue failure stress knowing that failure just occurs under a cyclic load whose amplitude is $\sigma_a = 64$ MPa and mean stress is $\sigma_m = 40$ MPa.

(b) Compute the mean and amplitude of a cyclic stress with a range of 90 MPa required to just cause fatigue failure. Ans. (a) 80 MPa, (b) $\sigma_{max} = \sigma_m + \sigma_a = 132.5$ MPa.

11.6 A rectangular cross section cantilever beam of length 30 cm is loaded by an axial tensile load of $F = 10,000$ N and a load transverse to the width at the free end of $P(t) = 4,500\sin(10t)$ N. The beam is made of steel with $E = 207$ GPa, $\nu = 0.28$, $\sigma_Y = 250$ MPa, and $\sigma_u = 350$ MPa. The S-N diagram has $\sigma_f = \sigma_u$ for $N_f = 1$ and $\sigma_e = 160$ MPa at $N_e = 10^8$ cycles.

The constant rectangular beam cross section must have width $w = 0.05$ m. The beam must remain elastically loaded. Determine the minimum height, h, of the beam cross section so that the beam will survive for 10^5 cycles. Use either the Soderberg or Goodman criteria as appropriate. Ignore the bending shear stress in the beam.

11.7 A helical spring is made of ASTM A 228 cold-drawn music wire (high-carbon steel) which has a torsional yield strength of 80 ksi and ultimate shear stress of 120 ksi. The shear modulus is 11.5×10^6 psi. The wire has diameter, $d = 0.125$ inches, the outer diameter of the spring is 0.875 inches, the spring length is 2.5 inches, and there are 12 coils. The spring is subject to a cyclic force, F, between 10 and 30 pounds.

(a) Compute the shear stress mean and amplitude in the spring wire.

(b) A shear S-N curve for this metal shows that 28 ksi is the critical shear stress for a life of 10^6 cycles. Use the Soderberg or Goodman criterion applied to torsional stresses, whichever is appropriate, to determine if the spring has a life of 10^6 cycles under this load.

(c) Determine the maximum deflection of the spring under the given force.

11.8 Brass spring wire has a tensile strength, σ_u, of 105 ksi. The wire has diameter, $d = 0.065$ inches, the outer diameter of the spring is 0.25 inches, the spring length is 1.5 inches, and there are 12 coils. The spring is subject to a cyclic load between 2 and 8 pounds.

(a) Use the Soderberg criteria to determine if the spring has infinite life under this load. Assume that the maximum amplitude of the shear stress for infinite life is 40 ksi. An estimate for the shear yield stress can be obtained from the approximate relations to the tensile properties for spring wire.

(b) Compute the spring constant.

11.9 A thin-walled shaft is to support a rigid disk of radius $r = 6$ inches which slips over one end and is firmly attached to the shaft. The rod is cantilevered at the other end. The disk is at a distance of $L = 10$ inches from the cantilevered end. A load that fluctuates between zero and 3,000 pounds in tension is applied tangent to the edge of the disk. The space in which the shaft must operate constrains its outer radius to $R = 1.5$ inches. Determine the wall thickness, t, so that the shaft will not suffer fatigue failure before one million cycles. An S-N diagram for the shaft material indicates that the critical stress for a million cycles is $\sigma_c = 35$ ksi. Also $\sigma_Y = 60$ ksi. Use strength of materials to estimate any stresses that are needed.

11.10 Would you expect fatigue fracture to be a design concern in an axially loaded piston rod? Why or why not?

11.11 A thin-walled cylindrical pressure vessel of radius $R = 6$ inches and wall thickness $th = 0.125$ inches contains a pressure that varies sinusoidally with time, t, as $p(t) = 200 + 100 \sin 0.01t$ psi. The vessel is subjected to a constant bending moment, $M = 25,000$ in-lb. The vessel is to be made of a steel. Use the Soderberg criterion to design a required minimum yield stress, σ_Y, for the steel to ensure an infinite fatigue life for the pressure vessel. Assume the endurance limit is $\sigma_e = 0.5\sigma_Y$. Also require a safety factor of 4.

11.12 A cantilevered shaft rotates under a constant torque of $M_t = 6,000$ in-lb and has a mass weighing $W = 200$ pounds attached to its free end. The shaft has length, $L = 12$ inches, and has a circular cross section of radius, $r = 0.5$ inches. Let A be a point on the surface of the shaft at the wall.

(a) Is the point A under a cyclic load? If so describe the state of stress at A. Use elementary strength of materials to estimate the stresses.

(b) If the point A is in a state of biaxial cyclic stress, compute the Mises equivalent mean and amplitude stresses.

(c) The shaft is to be made of a steel. Use the Soderberg criterion to design a required minimum yield stress, σ_Y, for the steel to ensure an infinite fatigue life for the pressure vessel. Assume the endurance limit is $\sigma_e = 0.5\sigma_Y$.

11.13 The material for the 3-inch-long center shaft of a bicycle pedal is to be designed. The shaft is to have maximum radius of 0.25 inches. Assume the shaft is cantilevered into the supporting link; it is actually a threaded joint.

(a) Express the load on the pedal in terms of a trigonometric function as the rider rotates it at a frequency of ω radians per second. The maximum load the rider can exert is assumed to be twice the rider's weight. Choose a design weight for the rider. Is the load fully reversed?

(b) Determine the minimum required ultimate stress for the material so that the pedal has the desired design life using the Goodman criterion. This value gives a criterion for the material choice. Assume the endurance limit is $\sigma_c = 25,000$ ksi for the design lifetime in a uniaxial loading.

11.14 A steel is heat treated so that the average grain size increases by 25%. How will this affect the critical stress according to the Soderberg criterion for fixed σ_m and σ_a? *Hint:* The Hall-Petch model might be useful.

11.15 Steel with Paris constants $m = 2.8$ and $C = 1.2 \times 10^{-11}$ is loaded sinusoidally by $\sigma(t) = 100 \sin(0.2t) + 150$ MPa. Determine the crack growth rate for mode I cracks having a maximum half length of 0.00075 meters. Assume the geometric factor is $Y = 1.12$. Compare the value of ΔK computed for a maximum crack size of 0.001 meter in Example 11.9 of this chapter and the crack growth rates in the two cases.

11.16 Steel with $m = 2.8$ and $C = 1.2 \times 10^{-11}$ is loaded sinusoidally by $\sigma(t) = 100 \sin(0.2t) + 150$ MPa. Determine the crack growth rate for mode I cracks having a maximum half-length of 0.00125 meter. Assume the geometric factor is $Y = 1.12$. Compare the value of ΔK computed for a maximum crack size of 0.001 m in Example 11.9 of this chapter and the crack growth rates in the two cases. Compare these results to those of the previous problem.

11.17 An aluminum has Paris constants $C = 4 \times 10^{-18}$ and $m = 4$. Also $K_c = 35$ ksi-in$^{1/2}$, $\sigma_Y = 65$ ksi. A rectangular cross-sectional beam is loaded by a moment, M, that varies cyclically between 10,000 and 60,000 in-lb. with bending axis parallel to the width of the cross section. There is a through-the-width crack that is initially a_i deep on the top of the beam. The width of the beam is $w = 1$ inch and the depth of the beam is $h = 4$ inches. Brittle fracture occurs after 10^3 cycles. Determine a_i. Assume a shape factor $Y = 1.12$.

11.18 For AISI 4340 steel, $m_0 = 3.24$, $\gamma = 0.42$, and $C_0 = 5.11 \times 10^{-13}$. Compare the crack growth rates at $\Delta K = 50$ MPa-m$^{1/2}$ at $R = 0.2$, $R = 0.1$, and $R = 0$ as predicted by the Walker relation.

11.19 Determine the critical half crack length, a_{cf}, for a cyclic load $\sigma(t) = 400 \sin(t) + 600$ MPa in AISI 4340 steel. The yield stress is 1,200 MPa and the critical fracture intensity is 125 MPa-m$^{1/2}$. Assume that $Y = 1$.

11.20 A 4340 steel has Paris parameters $C = 1.2 \times 10^{-11}$, $m = 2.8$, and $K_c = 185$ MPa-m$^{1/2}$, $\sigma_Y = 1,000$ MPa, $\sigma_u = 1,050$ MPa and is loaded cyclically between 0 and 250 MPa. The initial crack maximum half-length is estimated to be 0.0001 m. Assume the geometric factor is $Y = 1$.

(a) Determine the maximum crack length after 150,000 cycles.

(b) Determine the number of cycles to fracture.

(c) Compare this result to the Soderberg prediction of the critical stress.

11.21 A cantilever beam made of 99% pure ceramic Al_2O_3 is subjected to a fully reversed load so that $R = -1$ at 5 Hz. The Paris parameters are $C = 1.1 \times 10^{-11}$, $m = 14$ for the range $2.7 \leq \Delta K \leq 4.0$ MPa-m$^{1/2}$ at room temperature. $K_{Ic} = 4.0$ MPa-m$^{1/2}$. Suppose that the stress range of the cyclic load is ± 20 MPa. The maximum initial half crack size is 0.000005 meter. In the cantilever beam, mode I cracks are expected to be the most dangerous. Assume a shape factor $Y = 1$. Determine the lifetime of the member.

11.22 The ceramic ZrO_2 partially stabilized with MgO, denoted Mg-PSZ, at room temperature under a fully reversed cyclic load of frequency $v_c = 50$ Hz has Paris parameters $C = 4.9 \times 10^{-22}$, $m = 24$ for the range $3.0 \leq \Delta K \leq 4.2$ MPa-m$^{1/2}$. $K_{Ic} = 10.0$ MPa-m$^{1/2}$. Suppose that the stress range of the cyclic load is ± 20 MPa. The maximum initial crack size is 0.000005 meter. Assume a shape factor $Y = 1$. Determine the lifetime of the member.

11.23 A rod of cross-sectional radius 2.5 cm subject to an axial cyclic stress $\sigma_x = 70 + 60 \sin(\omega t)$ MPa, where x is the longitudinal direction on the rod, ω is the frequency, and t is time, is to be made of a ceramic with Paris parameters $C = 1.1 \times 10^{-11}$, $m = 14$, and $K_{Ic} = 4$ MPa-m$^{1/2}$. The manufacturer guarantees that the largest initial mode I crack is of length $2a = 0.02$ mm. The rod must survive $N = 10^6$ cycles.

(a) Compute the critical crack length for brittle fracture using the Irwin theory. Assume the shape factor is $Y = 1$.

(b) Assuming that all cracks of interest are mode I, should the designer specify this ceramic? Explain your decision using calculations.

11.24 A rod of cross-sectional radius, R, subject to an axial cyclic load $F_x = 80\pi + 60\pi \sin(\omega t)$ kN, where x is the longitudinal direction on the rod, ω is the frequency, and t is time, is to be made of a metal with Paris parameters $C = 1.42 \times 10^{-11}$, $m = 3.6$, and $K_{Ic} = 28.6$ MPa-m$^{1/2}$. Also $E = 70$ GPa and $\sigma_Y = 450$ MPa. (Recall that the sine

takes values between ± 1). The manufacturer guarantees that the largest initial mode I crack is of length $2a = 0.02$ mm. The rod must be elastically loaded.

(a) Compute the maximum and minimum axial stresses σ_x in terms of R. Write $\Delta\sigma$ in terms of R in the units of MPa.

(b) Compute the critical crack length for brittle fracture using the Irwin theory. Assume the shape factor is $Y = 1$.

(c) Assuming that all cracks of interest are mode I, use the integrated Paris relation to determine the minimum allowable radius, R, to have the rod survive the fatigue loading for a life of $N_f = 10^6$ cycles. A numerical solution may be required.

11.25 Determine the maximum allowable strain amplitude for an aluminum 7075-T6 member that must survive 1,000 fully reversed cycles to failure. Data from [4, p. 627] shows that $E = 71$ GPa, $\sigma_F = 1466$ MPa, $\epsilon_F = 2.62$, with Manson-Coffin parameters $b = -0.143$, and $c = -0.619$.

11.26 Determine the maximum allowable strain amplitude for a steel 4340 member that must survive 10,000,000 fully reversed cycles to failure.

11.27 Determine the transition fatigue life for

(a) 4340 steel and

(b) aluminum 7075-T6.

11.28 A cantilever beam of length $L = 30$ cm, and rectangular cross section is made to deform cyclically in bending by a rotating cam placed under the free end. The beam is made of 2024-T351 aluminum.

(a) Determine the maximum allowable strain amplitude at the wall underneath the beam (same side as the cam touches) for a life of 10^4 cycles. The aluminum has Manson-Coffin constants $b = -0.113$, $c = -0.713$, $\epsilon_F = 0.409$, $\sigma_F = 927$ MPa, and $E = 73.1$ GPa.

(b) At how many cycles of lifetime does the allowable plastic component of the strain amplitude equal 0.00001?

REFERENCES

[1] M. BARSOUM, *Fundamentals of Ceramics*, McGraw-Hill, New York, 1997.

[2] J. A. BENNETT, Effect of reactions with the atmosphere during fatigue of metals, in *Fatigue—An Interdisciplinary Approach*, J. J. Burke, N. L. Reed, V. Weiss, Eds., Syracuse University Press, Syracuse, NY, 1964, pp. 209–227.

[3] W. G. CLARKE, JR., Fracture mechanics in fatigue, *Experimental Mechanics* 11(9), 421–428 (1971).

[4] N. E. DOWLING, *Mechanical Behavior of Materials*, Chapters 9–11, 14, Prentice Hall, Englewood Cliffs, NJ, 1993.

[5] H. W. LIU, Discussion, in *Fatigue: An Interdisciplinary Approach*, J. J. Burke, N. L. Reed, and V. Weiss, eds., Syracuse University Press, Syracuse, NY, 1964, p. 129.

[6] A. J. MCEVILY, JR., R. C. BOETTNER, AND T. L. JOHNSTON, On the formation and growth of fatigue cracks in

polymers, in *Fatigue: An Interdisciplinary Approach*, J. J. Burke, N. L. Reed, and V. Weiss, Eds., Syracuse University Press, Syracuse, NY, 1964.

[7] E. OROWAN, Fracture and strength of solids, *Reports on Progress in Physics*, 12, 185–232 (1948–1949).

[8] P. C. PARIS, The fracture mechanics approach to fatigue, in *Fatigue—An Interdisciplinary Approach*, J. J. Reed, V. Weiss, Eds., Syracuse University Press, Syracuse, NY, 1964, pp. 107–132.

[9] R. O. RITCHIE AND R. H. DAUSKARDT, Cyclic fatigue of ceramics: A fracture mechanics approach to subcritical crack growth and life prediction, *Journal of the Ceramic Society of Japan. Int. Ed.*, 99, 1009–1023 (1991).

[10] J. M. SCHULTZ, Fatigue Behavior of Engineering Polymers, in *Treatise on Materials Science and Technology, Vol. 10: Properties of Solid Polymeric Materials, Part B*, Academic Press, New York, 1977, pp. 599–636.

[11] S. SURESH, *Fatigue of Materials*, 2nd ed., Cambridge University Press, Cambridge, UK, 1998.

[12] A. M. WAHL, *Mechanical Springs*, 2nd ed., McGraw-Hill, New York, 1963.

CHAPTER *12*

CREEP

12.1 INTRODUCTION

It might be expected that if a uniaxially loaded member is held at a constant stress, the strain will remain constant indefinitely. This is not the case; the member continues to deform over time even though the stress remains constant. The member is not in equilibrium over time since the strain varies. This phenomenon, a nonequilibrium response of the material to a constant stress, is called creep. Some materials deform more than others under the same constant stress in a fixed time period. The deformation depends on the structure of the material, the temperature, the magnitude of the applied stress, and the relative humidity of the surrounding atmosphere for creep of moisture adsorbing materials like wood or concrete.

Imagine that a plate, perhaps used as a shield, is suspended horizontally by two rods just above a rotating blade in a high-temperature environment. The supporting rods are very likely to creep under the weight of the plate because of the high temperature, and the elongation of the rods may allow the shield to hit the rotating blade. At this point a catastrophe is likely.

A concrete column that supports a roof carries a nearly constant compressive load (ignoring things like snow in winter) over a period of many years. Due to the constant axial force, the column will shorten slightly and perhaps cause the roof or building to settle. The rate of deformation per unit time increases with the amount of constant stress. The designer's job is to keep the stress low enough in the column so that the rate of change in the length is sufficiently small to allow the structure to safely perform its function during its planned lifetime.

Definition   A solid body is said to creep if it deforms continuously over time under the influence of a constant stress.

Possibly this terminology was chosen to suggest a baby's slow creep as it learns to crawl. Every material exhibits some creep at any constant stress; although for small stresses at low or moderate temperatures, the creep may be so small as to be negligible. The creep is measured by strain which changes as a function of time. The change in strain of the material per unit time is called the creep rate (or strain rate). The higher the creep rate, the more the body will deform in a given period of time.

If a piece of steel rod is loaded uniaxially at room temperature, the creep deformation is very hard to measure because it is so small. But if the test is performed on a lead specimen, the change in length is very apparent in a short period of time. The difference in the magnitude of this effect is due to the comparative melting temperatures of steel and lead. In other words, the creep response in metals depends strongly on the temperature. The melting temperature of lead is just a few hundred degrees above room temperature, while that of steel is far above room temperature. Most solid materials have a melting temperature, T_m, at which the material transforms from its solid phase to its liquid phase (some polymers such as thermosets, wood, or paper disintegrate at high temperature). The bonds between various molecules of the material are weaker in the liquid state than in the solid state. As the temperature of the solid approaches the melting temperature of that particular material, the molecular bonds tend to weaken and are more easily broken by an applied stress. The high temperature can also encourage solid diffusion. Several other microstructural mechanisms may also be involved in creep.

Definition   The homologous temperature of a metal is the ratio of its current temperature to the melting temperature, $T_h = T/T_m$, where all temperatures are absolute, that is, on the Kelvin scale.

When the homologous temperature is above 0.5, the creep strain of many materials is quite significant.

Typical melting points of some metals are given in Table 12.1.

TABLE 12.1   Melting Points of Metals [1]

Metal	Temperature (°C)	Temperature (°F)
Aluminum	657.0	1214.6
Copper	1083.0	1981.4
Iron (cast)	1200.0	2,192.0
Lead	327.4	621.3
Molybdenum	2621.0	4,749.8
Silver	960.8	1,761.4
Tin	232.0	449.6

At room temperature, those metals with the lowest melting points should exhibit the highest creep rates since room temperature is closer to their melting points. Experiment shows this is true. Lead, for example, creeps very rapidly at room temperature, compared to iron.

Designers of metal internal combustion engines that operate at high temperatures (for fuel efficiency) must be concerned about the creep of the metal parts. On the other hand, since polymers have much lower melting points than metals, creep is a concern for devices made of polymers even at room temperature.

12.2   CREEP TESTS

A designer must be able to predict the amount of creep deformation in a given period of time to ensure that the structure will function properly and survive for its projected lifetime. Since the creep effects are different for different materials, tests must be performed to determine creep as a function of time. Some structures are to be designed to last for years, but experimental tests can only be conducted economically over shorter periods of time. The experimentalist attempts to devise a short-term test that can be used to predict long-term effects. This is called accelerated testing. A long laboratory test lasts a few thousand hours (less than a year), while the structure might have a required 20-year lifetime. To make predictions of long-term creep from short-term tests, a mathematical model must be constructed from the tests that relates creep and creep rate to the constant applied stress, to the material, and to elapsed time.

The simplest creep test is that in which a constant uniaxial stress is applied to a specimen. Creep is usually investigated by plotting the strain resulting from creep versus time, called a creep curve. Since the stress is constant, the uniaxial stress-strain curve is of no help. If the temperature is constant during the test, the test is said to be isothermal. A typical uniaxial, isothermal, creep curve for a metal is given in Figure 12.1.

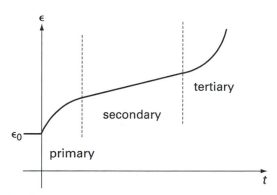

FIGURE 12.1   The regions exhibited by creep curves for metals.

On a creep curve for metal, there is an initial strain, ϵ_0, which occurs very quickly, called the instantaneous strain. As time passes, the metal creep curves typically show three reasonably distinct regions, if the metal is allowed to creep until fracture. In the primary region, the creep strain increases rapidly, and the curve is concave down. The curve is concave down since hardening causes a decrease in the initially high flow rate. The primary region is of relatively short duration at moderate temperatures. The strain in that region is sometimes called transient strain. The strain rate, which is the slope of the creep curve, is nearly constant in the secondary region. This is usually the longest of the three regions in a creep curve for a metal. The lower the constant applied stress and the lower the temperature, the longer the secondary region persists. To use the computed secondary region strain rate in design, the length of time of the secondary region must be known, unless it is assumed that it exceeds the design lifetime. The strain rate increases during the tertiary region so that the curve is concave up. The strain rate increases due to both a reduction in the cross section of the sample and due to damage to the material. This pattern is not exhibited by the creep curves for polymers, which do not show the secondary, almost constant strain rate, region.

The tertiary region is only significant for ductile materials. Brittle materials tend to fracture before leaving the secondary region. During creep of a ductile material, the existence of a tertiary region is partially explained by the necking of the material specimen just before fracture and by the formation of microcracks which can grow into visible cracks.

Because the melting temperatures are not the same for all materials, it is difficult to make a meaningful comparison of their creep rates at the same temperature. For this reason, the homologous temperature was introduced. Creep effects in different materials can be more meaningfully compared when the homologous temperature of both materials is the same.

Polymer creep curves do not have the three-region structure of metal creep curves. In particular, there is no secondary region in which the creep strain rate is a constant. The polymer curves usually resemble the graph of a logarithmic function. Figure 12.2 shows this shape for tests of cellulose acetate and also shows the effect of increasing the creep load.

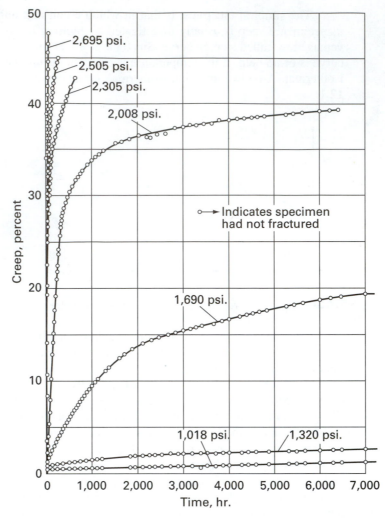

FIGURE 12.2 Tensile creep curves of cellulose acetate under different creep loads in ambient conditions of 77°F and 50% relative humidity. (Reprinted from [9] by permission of ASTM.)

12.3 CREEP MECHANISMS

Creep deformation is due to atomic or molecular movement. Creep of metals, like plastic deformation, is partially due to the multiplication of dislocations and their motion along slip planes, in response to shear stresses induced by the creep load. In addition, slip between grains is possible if the material is polycrystalline. A designer may specify small grains to increase the strength of a metal member, but small grain size can increase the creep response as well because the increased grain boundary increases the slip possibilities. Most structural metals are polycrystalline, but special manufacturing processes have been developed to produce structures made of a single crystal, called monocrystalline materials. A typical application of a monocrystal is in the blades of a high-performance turbine.

Within each grain, the dislocation multiplication phenomenon tends to work harden the material. Conversely, a high homologous temperature activates atomic diffusion which drives the material lattice back to a more ordered state. So, in this sense, there may be competing processes involved in creep. Diffusion occurs much more slowly than dislocation motion, which can be at the speed of sound. For example, diffusion can cause the annihilation of nearby dislocations. The small region of atoms between two neighboring edge dislocations can diffuse to other lattice sites so that the lattice is ordered and the dislocations have disappeared (Fig. 12.3).

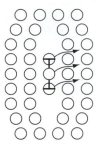

FIGURE 12.3   Diffusion that annihilates two edge dislocations.

Creep mechanisms may or may not involve mass transfer by means of diffusion. Diffusional creep can occur within a grain or along a grain boundary. Diffusion is the motion of atoms or molecules through a solid. It is a nonequilibrium process. The flux, J, is the number of particles that move through an area of the solid per unit time. So J has units of particles/cm^2-s and is assumed to be a function of the concentration gradient, $\partial c / \partial x$, with respect to the position x. Fick made the assumption that this function is the simplest possible, a linear function.

$$J = -D\frac{\partial c}{\partial x}, \tag{12.1}$$

where the diffusion coefficient, D, depends on the temperature, the applied stress, and the activation energy, Q, for particle motion. In the absence of stress, one model for D is the Arrhenius relation,

$$D = D_0 \exp\left(\frac{-Q}{RT}\right). \tag{12.2}$$

The higher the temperature, T, and the smaller the activation energy, Q, the greater the flux J.

To find the change in composition, c, at a position, x, as diffusion occurs over time, Fick proposed

$$\frac{\partial c}{\partial t} = D\frac{\partial^2 c}{\partial x^2}, \tag{12.3}$$

where the diffusion coefficient D is assumed not to vary with position. This expression follows from the nonequilibrium thermodynamic relation on fluxes, $\partial c / \partial t = -\partial J / \partial x$, assumed by Fick and from the Fick relation on J [equation (12.1)]. Using the product rule,

$$\frac{\partial c}{\partial t} = \frac{\partial}{\partial x}(-J) = \frac{\partial D}{\partial x}\frac{\partial c}{\partial x} + D\frac{\partial^2 c}{\partial x^2} = D\frac{\partial^2 c}{\partial x^2}, \tag{12.4}$$

because $\partial D / \partial x = 0$. The partial differential equation (12.3), called the diffusion equation, is difficult to solve in general. Solutions must be obtained numerically in most cases and depend on the boundary conditions.

Diffusional creep within a grain is related to the vacancy concentration, the number of vacancies per unit volume. Recall that vacancies are point defects in a lattice structure. The vacancy concentration, c_v, is usually assumed to obey the Arrhenius relation, $c_v = k_n \exp(-Q_v/RT)$, where Q_v is the activation energy for the formation of vacancies, and T is the absolute temperature. Often k_n is taken as the number of lattice points per volume. For example in copper, $Q_v \sim 20$ kcal/mol, $R = 1.987$ cal/mol/K, and $k_n = n/a^3 = 4/(3.6151 \times 10^{-8})^3 = 8.47 \times 10^{22}$, where n is the number of atoms per cell and a is the lattice parameter. Svante August Arrhenius (1859–1927) was a Swedish chemist who showed experimentally that the equation named after him describes chemical reaction rates.

The activation energy is the energy that must be supplied, in this case, to move the atom from its position in the lattice. The required energy is often called the energy barrier that must be overcome to initiate the process in question. The activation energy is a material property for the process. Here the increase in temperature causes thermal vibration in the lattice and thus increases the chance that an atom will move from its current site. The fact that the creep activation energy and the self diffusion activation energy for most metals and nonmetals are nearly equal supports this viewpoint.

In a homogeneous material, vacancy diffusion involves substitutional atoms leaving a lattice site to file a vacancy. The Arrhenius relation for c_v gives the equilibrium number of vacancies of an unstressed material at the chosen temperature. In general, there are more vacancies in tensile regions than in compressive regions if the body is stressed. Further, it is possible to create a greater vacancy density by heating the material and allowing it to come to an equilibrium number of vacancies. Rapid cooling to prevent diffusion freezes in a number of vacancies much greater than the equilibrium number at the lower temperature.

Nabarro-Herring creep is nearly solely due to atomic diffusion and so dominates creep at low stresses and high temperatures. This mechanism does not depend on line dislocation behavior and so can also be found in amorphous nonmetals. In particular, it is most important in ceramics. The creep strain due to diffusion within a grain of size d is measured by $\Delta d/d$. The secondary region creep rate due to this type of creep under a stress, σ, is modeled by the Nabarro-Herring equation

$$\dot{\epsilon} = \frac{d\epsilon}{dt} = \frac{8\sigma\Omega D}{kTd^2},$$ (12.5)

where Ω is the atomic volume and k is the Boltzmann constant. This strain rate is inversely proportional to the square of the grain size and directly proportional to the stress. Experiment shows creep to be linearly proportional to stress only at low stresses.

Polycrystalline materials with many grain boundaries can undergo Coble creep. This is characterized by atomic diffusion along grain boundaries. The finer-grained the metal or ceramic, the more likely that the Coble creep mechanism due to grain boundary diffusion will dominate. It is most common at moderate temperatures and moderate stresses. The Coble model for this type of creep is

$$\dot{\epsilon} = \alpha \frac{\delta\sigma\Omega D_g}{kTd^3},$$ (12.6)

where α is a parameter, δ is the grain boundary width, and $D_g = Dd/\delta$ is the grain boundary diffusivity coefficient. Coble creep is inversely proportional to the cube of the grain size and directly proportional to the stress. Both Nabarro-Herring and Coble creep assume that the vacancies are initiated and obliterated in the grain boundary. Also the stress and temperature are assumed constant over each grain, and it is assumed that no cavitation, that is, void creation, in the material has occurred.

Viscous creep can occur in a material which has a glassy region at its grain boundaries. Such glass regions occur in ceramics made by liquid phase sintering, in which a portion of the material is liquid at the beginning of sintering. Liquid phase sintering is the most common commercial manufacturing process for ceramics because it produces a more uniform density and is a faster process. One structural ceramic made in this manner is silicon nitride (Si_3N_4). Recall that viscosity is a measure of the effect of shear stress on the rate of deformation, written as the velocity gradient. The usual viscosity, η, is the proportionality constant when the shear and rate of deformation are linearly related, that is, Newtonian. One model for viscous creep assumes that the effective viscosity is η/f^3 where η is the viscosity of the glassy phase and f is the glassy phase volume fraction. The strain rate is then calculated by the Newtonian model

$$\dot\epsilon = \frac{\tau f^3}{\eta},\qquad(12.7)$$

where τ is the local shear stress. This model ignores the fact that grains may cease slipping if the glassy fraction is small enough that it is squeezed from between grains. The model also ignores the fact that cavitation can occur over time in the glass region.

In polycrystalline ceramics, dislocation motion is not a significant contributor to creep. Grain boundary sliding is a more typical cause of ceramic creep. Such sliding can lead to the formation of pores along the grain boundaries, especially at sites where more than two grains meet. The pores then may serve as sites for crack initiation, leading eventually to fracture. The material may then fracture before a significant deformation has occurred. Because grain boundary sliding is a primary cause of creep in ceramics, ceramics with smaller grain sizes, and therefore more grain boundary area, tend to creep more. No ceramic is pure so that those impurities present may help create glass phases along the grain boundary, further increasing the creep strain as the glass region viscously flows.

Dislocation creep tends to occur at higher temperatures and stresses. Two mechanisms are presumed to dominate, dislocation glide and dislocation climb. Here line defects are involved rather than point defects, such as vacancies. An applied stress creates a shear on a slip plane so that an edge dislocation can glide along the plane. Large stress is required. If the motion of the edge dislocation is blocked by an impurity or another dislocation tangle, climb may occur. A sufficient activation energy is required to move a dislocation past a barrier so that there may be some creep strain. Dislocation glide is produced when the extra energy provided is thermal. Dislocation glide, which involves no atomic diffusion, occurs at low to moderate temperatures and high stresses.

Diffusion of atoms can also cause dislocation climb. The end atom on an edge dislocation can interchange with a nearby vacancy in the lattice (Fig. 12.4). The void might have been brought nearby by distortion of the lattice by the creep load. The atoms rearrange themselves into their correct lattice positions on the plane of the former void. In this manner the dislocation seems to have moved one atomic unit in the opposite direction from the void. This is called dislocation climb. Many materials scientists like to think that rather than an atom moving, the void or vacancy has moved from its former lattice position to the end position on the edge dislocation. In dislocation climb, which involves diffusion of an atom or molecule, the dislocation moves perpendicular to the slip plane. On the other hand, stress-induced dislocation motion such as occurs in plastic deformation is parallel to the slip plane. Dislocation creep occurs at high temperatures and moderate stresses.

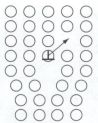

FIGURE 12.4 Dislocation climb of an edge dislocation.

Empirical models for the creep rate associated with dislocation climb generally involve higher than linear powers of the stress. This process does not depend on grain size because the grain boundaries are not involved to a great extent.

Several types of substructures within a grain can develop during creep. There can be subgrains with relatively sharply defined boundary regions. Smaller structures called cells can also be created by the piling up of dislocations in the shape of cell boundaries. Some refer to this as polygonization since the cell boundaries look like polygons (Fig. 12.5). The subgrain and cell boundaries are much harder than the remainder of the material so that the body becomes divided into soft and hard regions. It is no longer homogeneous; the material might be best thought of as two phase. The soft regions have a much lower dislocation density than the hard regions. Subsequent dislocation motion occurs primarily in the soft regions. At higher temperatures the volume fraction of the cell boundaries decreases while that of the interiors increases. At larger creep stresses, the cell boundaries become more diffuse and less well defined [18].

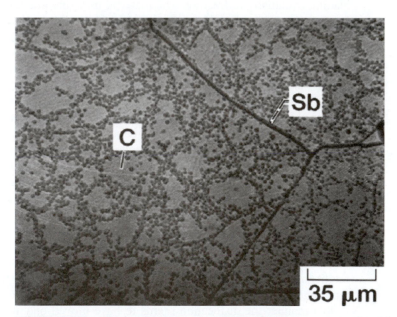

FIGURE 12.5 Creep substructure in sodium chloride true-strained to 0.9 at 873 K, showing cells (C) and subboundaries (Sb), reprinted from [19] with permission from Elsevier.

The logarithmically shaped curve exhibited in the primary region of the creep of metals at low temperature is due to the cross-slip of dislocations past an obstacle (Chapter 9, Section 9.3). The phenomenon slows the strain rate over time to produce a logarithmic curve of the strain versus time [16, Chapter 5].

Fracture due to creep is believed to depend on void growth, much as does fatigue fracture. Three modes of high-temperature creep fracture have been distinguished. The material is said to rupture if the fracture is due to a large reduction in cross-sectional area. This is most common under high stresses. Transgranular creep fracture is a consequence of void nucleation and growth within the metal crystal. This has many similarities to ductile fracture at low temperatures. It too is most often seen under high stress or high strain rate. Intergranular creep fracture results from voids nucleating along grain boundaries. This response is most likely at lower stresses.

12.4 MATHEMATICAL MODELS FOR UNIAXIAL CREEP

Creep in metals is a special case of viscoplasticity, which is characterized by the fact that some state variables, such as temperature and stress, are held constant. In ceramics and polymers, as well as metals, it is a nonequilibrium process. Therefore it is modeled by evolution equations, which are differential equations showing how the remaining state variables vary over time. In the case of creep, the most important is the strain. A design engineer requires an equation relating creep strain to time in order to predict the change in shape of a body due to creep over the lifetime of the structure. A mathematical model for creep must account for the fact that the creep strain at a given time from application of the stress is larger for larger stresses and is larger if the test is conducted at a higher temperature. Many of the models that have been proposed are empirical in the sense that their expression can be forced to fit the experimental data by adjusting a few parameters. They are not derived from basic physical principles. None is able to fit all regions of the typical creep curve for metal.

Creep is thermally activated, and therefore its dependence on temperature is sometimes represented by an Arrhenius function. The Arrhenius equation applied to strain says that the strain rate $\dot{\epsilon}$ depends on the absolute temperature, T, and the creep activation energy, Q_c.

$$\dot{\epsilon} = C \exp\left(\frac{-Q_c}{RT}\right), \tag{12.8}$$

where C is a temperature-independent coefficient and R is the universal gas constant. This equation can be rewritten in terms of the Boltzmann constant, $k = 13.8 \times 10^{-24}$ J/K by setting $q = Q_c/N$ and recalling that $k = R/N$, where N is Avogadro's number, the number of atoms in a number of grams equal to the atomic mass of the element (0.6023×10^{24}). A mole contains N items.

$$\dot{\epsilon} = C \exp\left(\frac{-q}{kT}\right). \tag{12.9}$$

Often C is taken in the form, $C = A\sigma^n$, where σ is the constant creep stress and A and n are material constants. The activation energy for creep, Q_c, has a value close to that for diffusion in the homogeneous material. For example, the creep activation energy for rutile (TiO_2) is on the order of 40 kcal/mol. The Arrhenius equation applies best to creep induced by dislocation climb.

EXAMPLE 12.1   A creep test performed at 900°C produces a strain rate of 0.0006/s. Determine the expected strain rate at 1100°C under the same creep load. The activation energy is known to be $Q_c = 220$ kJ/mol. The dominant creep mechanism is believed to be dislocation climb.

Solution   The Arrhenius equation can be used if the coefficient, C, is determined from the given test. Keep in mind that C also depends on the creep stress. Because the activation energy is in joules, $R = 8.314$ J/mol-K.

$$0.0006 = C \exp[-220,000/(8.314 * 1173)]$$

implies that $C = 3,760,826.17$. Therefore

$$\dot{\epsilon} = (3,760,826.17) \exp[-220,000/(8.314 * 1373)] = 0.01604/s.$$

Notice that knowing the strain rate from two tests permits an engineer to determine both the coefficient, C, and the activation energy, Q_c. From this information the strain rate at any other temperature can be predicted, if the Arrhenius equation is valid. ∎

Exercise 12.1   Examine the strain rate variation with temperature over the range $0.6 \leq T_h \leq 0.8$, while the creep load is fixed, for a copper in which $Q_c = 220$ kJ/mol. Plot $\dot{\epsilon}/C$ versus absolute temperature using the Arrhenius equation.

12.4.1   Creep Life Estimates

The creep life to fracture is an important consideration in design, especially if the designer chooses not to restrict the stress so that only a fixed amount of creep deformation occurs in a given time. The expense and lack of time prevent an engineer from conducting creep tests at all possible constant stresses and temperatures. Equations to estimate creep life have been proposed. For example, the Monkman-Grant model assumes that the product of the life t_f and the creep strain rate is a constant.

$$t_f \dot{\epsilon} = M. \tag{12.10}$$

The constant M for the ceramic silicon nitride, for example, is on the order of $M \sim 10^{-6}$ if time is measured in hours.

Creep data is often extrapolated using the Larson-Miller parameter. The modified Arrhenius model, $\dot{\epsilon} = A\sigma^n \exp(-Q_c/RT)$ implies, by taking the natural log of both sides, that

$$\ln(\dot{\epsilon}) = \ln(A\sigma^n) - \frac{Q_c}{RT}.$$

The strain rate is often inversely proportional to the time to fracture, t_f, so that $\dot{\epsilon} = k_c/t_f$, for some constant, k_c. Therefore,

$$\ln(k_c) - \ln(t_f) = \ln(A\sigma^n) - \frac{Q_c}{RT},$$

or

$$T[\ln(A\sigma^n) - \ln(k_c) + \ln(t_f)] = \frac{Q_c}{R},$$

a material constant.

This suggests that, as proposed by F. R. Larson and J. Miller in 1952 for iron, there is a stress-dependent term, LM, now called the Larson-Miller parameter, such that for T measured in Kelvins,

$$LM = T[\log_{10}(t_f) + C]. \tag{12.11}$$

Recall that the base of a logarithm is changed by $\ln(x) = \log_{10}(x)/\log_{10}(e)$. Usually $C \sim 20$ if t_f is measured in hours.

EXAMPLE 12.2   The Larson-Miller relation can be used to extrapolate the fracture time known at one temperature to that at another temperature, if the stress is the same in both tests. Compute the fracture time at 500°C if that at 400°C is 10,000 hours.

Solution $T_1[\log_{10}(t_{f,1}) + C] = T_2[\log_{10}(t_{f,2}) + C]$ implies

$$673[\log_{10}(10,000) + 20] = 773[\log_{10}(t_f) + 20]$$

so that $t_f = 7.856$ hours.   ∎

EXAMPLE 12.3   Other expressions of the Larson-Miller parameter are given in terms of the natural logarithm, such as

$$LM = \frac{T}{1000}[36 + 0.78\ln(t_f)].$$

Cast iron under a creep stress of 4,000 psi has $LM = 35.9$ and $C = 36$. The Larson-Miller parameter can be used to predict the fracture time at 400°C.

$$36.9 = \frac{673}{1000}[36 + 0.78\ln(t_f)]$$

implies that $t_f = \exp(24.14) = 3.05 \times 10^{10}$ hours.   ∎

EXAMPLE 12.4   An iron support rod must survive a temperature of 600°C for 4 years. The rod must support an axial load of 2,000 pounds. The rod is to have a circular cross section. Design the radius so that the rod will avoid creep fracture. An approximation for the relation of the allowable creep stress to LM for this iron is known to be $\ln\sigma = (0.5\ln 1.5)LM$ psi.

Solution   Four years is 35,040 hours. Then $LM = 873[36 + 0.78\ln(35,040)]/1,000 = 38.6$. The allowable stress is $\sigma = \exp[0.5(\ln 1.5)38.6] = 2,480$ psi. Since $2,000/\sigma = \pi r^2$, the radius must be greater than 0.507 inches.   ∎

The dependence on time in polymer creep fracture is described by the idea of durability. The time from the application of a constant load to fracture is called the durability $\tau(\sigma)$ under that stress. Many models for the durability have been proposed. A simple model for the durability of cross-linked polymers was proposed by the Russian, G. M. Batenev, in the early 1960s,

$$\tau = B\sigma^{-m}\exp\left(\frac{U}{kT}\right),$$

where B and m are material constants, σ is the fracture stress, k is Boltzmann's constant, T is the absolute temperature, and U is the activation energy for fracture. The concept of durability can be used in combination with that of creep deformation as a function of time to compute the fracture strain or percent elongation at fracture.

12.4.2 Secondary Region Strain Rate Models

Several equations have been proposed to relate the constant applied uniaxial stress to the constant strain rate in the secondary region. Many of these equations are found by first plotting the experimental data. The researcher then chooses a function whose graph is of the same form as the experimental graph and adjusts the constants in the function so that its graph and experimental graph coincide as closely as possible. The goal is to pick a functional form which will fit experimental results from many different materials so that it might describe the common features of the phenomenon being tested. No universally accepted theoretical justification for them has been found. Many types of formulas are proposed for the experimental data in the hope that theoreticians can use them as clues in explaining the phenomenon. Until a satisfactory theory is found, the empirical formulas also provide tools to make predictions when designing structures. They describe what happens but not why it happens.

In 1929, Norton proposed a power law model, which is most successful if the applied creep stress, σ, is relatively small.

$$\dot{\epsilon} = \bar{k}\sigma^p, \tag{12.12}$$

where $\bar{k}$ and p are material constants, usually $3 \leq p \leq 7$. This model shows how the slope of the linear secondary region varies with the applied creep stress.

At high constant applied stress, the exponential law is used.

$$\dot{\epsilon} = B'e^{\alpha\sigma}, \tag{12.13}$$

where B' and α are material constants.

Both cases can be included in the hyperbolic sine law suggested by F. Garofalo around 1962,

$$\dot{\epsilon} = B''[\sinh(c\sigma)]^n, \tag{12.14}$$

where B'', c, and n are material constants. By definition, the hyperbolic sine is $\sinh(x) = [e^x - e^{-x}]/2$. The function increases slowly for small values of x and rapidly for large values of x. For small σ, $\sinh(c\sigma) \sim c\sigma$ so that $\dot{\epsilon} \sim B''(c\sigma)^n$, the power law model. For large σ, $\sinh(c\sigma) \sim 0.5\exp(c\sigma)$ so that $\dot{\epsilon} \sim B''(c0.5\exp(c\sigma))^n = B'\exp(a\sigma)$, the exponential model.

To also account for larger creep stresses, Nadai had proposed, in 1937, a simpler hyperbolic sine model for the secondary creep rate in which $n = 1$ in equation (12.14).

$$\dot{\epsilon} = D \sinh\left(\frac{\sigma}{C}\right), \tag{12.15}$$

where C and D are material constants. A Taylor series expansion of the hyperbolic sine shows that for small stresses, $\dot{\epsilon} \sim \sigma/C$, while the relation of creep strain rate to the constant applied stress is non-linear for larger stresses. Both the Norton and Nadai models perform best in tension at temperatures for which the strain rate is small. Notice that neither makes the temperature dependence explicit.

Some typical values of the constants in the hyperbolic sine model (12.14) are given in Table 12.2.

TABLE 12.2 Constants in the Hyperbolic Sine Model [12]

Metal	Temperature (K)	B'' (1/s)	c (1/psi)	n
Copper	673	6.80×10^{-6}	1.11×10^{-4}	3.57
Copper	773	6.10×10^{-6}	1.97×10^{-4}	3.39
Aluminum	477	2.78×10^{-6}	3.98×10^{-4}	5.00
Aluminum	533	1.94×10^{-5}	5.13×10^{-4}	4.55
Steel	1,069	1.67×10^{-8}	1.06×10^{-4}	3.50

EXAMPLE 12.5   Determine the strain rate for steel at 796°C (1,069 K) at a stress of 3,000 psi.

Solution   Apply the hyperbolic sine law to determine $\dot{\epsilon}$.

$$\dot{\epsilon} = (1.67 \times 10^{-8})\{\sinh[(1.06 \times 10^{-4})(3000)]\}^{3.5}$$
$$= 3.21 \times 10^{-10} \text{ microstrain per second.} \tag{12.16}$$

Therefore an axially loaded 100-inch-long steel specimen at 796°C (1069 K) and at a stress of 3,000 psi will creep $(100)\dot{\epsilon} = 3.21 \times 10^{-8}$ in./s. This bar would creep about 1.01 inch in a year or about 1% of its original length. ∎

12.4.3 Strain Models

The previous set of models attempted to predict the secondary region strain rate. Others have given models for the strain as a function of time. These can be compared to the strain rate models by simply taking the derivative with respect to time.

Andrade, around 1910, proposed a model for primary creep. This seems to be the first mathematical model proposed for metal creep.

$$L = L_0(1 + \beta t^{1/3})e^{kt}, \tag{12.17}$$

where L_0 is the original length, L is the length at time t, and the material constants, β and k, are both stress and temperature dependent. Notice that when $t = 0$ the model predicts, as it should, that $L = L_0$.

EXAMPLE 12.6   Andrade originally proposed his equation as a creep model for lead. Determine the constants, β and k, for a lead creep test of a 5-inch specimen in which the deformation was 0.27 inches after 30 seconds and 0.34 inches after 60 seconds.

Solution   Two equations must be solved simultaneously for the constants, β and k.

$$5.27/5 = (1 + \beta 30^{1/3})e^{30k}$$
$$5.34/5 = (1 + \beta 60^{1/3})e^{60k}.$$

To eliminate k, square the first equation and divide it into the second equation to obtain

$$1.068/(1.054)^2 = (1 + \beta 60^{1/3})/(1 + \beta 30^{1/3})^2.$$

This quadratic equation in β,

$$9.2819\beta^2 + 2.0595\beta - 0.03863 = 0,$$

can be solved by the quadratic formula. The positive root is $\beta = 0.017394$. This is then substituted into one of the equations to obtain $k = -1.474 \times 10^{-6}$. ∎

The Andrade model does not make the creep stress and temperature dependence explicit. This model was later generalized by Nutting in 1921 [10, p. 15] to show the stress dependence of creep in the primary region,

$$\epsilon(t) = k\sigma^P t^n, \tag{12.18}$$

where ϵ_c is the creep strain at time t, and k, p, and n are material constants. The value of n is usually between 0 and 0.5 . The Nutting uniaxial equation has been applied to soft materials like rubber. However, the predictions were only good for small times.

The uniaxial creep of steel has been modeled by J. D. Lubahn and R. P. Felgar as $\sigma = C\epsilon^m \dot{\epsilon}^n$. A typical set of values at 1000°F [15, p. 574] is $m = 0.2805, n = 0.1455$, and $C = 10^{6.035}$, where the stress is measured in psi.

Integration of the relation shows that it is equivalent to a power law model.

$$t = \left(\frac{C}{\sigma}\right)^{1/n} \left(\frac{m+n}{n}\right) \epsilon^{(m+n)/n}.$$

This has the form $\epsilon = A\sigma^q t^p$. For the values above $p = n/(m+n) = 0.34155$ and $q = 2.35$.

Exercise 12.2 Use the fact that $t = \int d\epsilon/\dot{\epsilon}$ to solve for t in terms of ϵ and obtain the above result. Then write the result in the form of the Nutting power law model.

EXAMPLE 12.7 A steel rod has a creep stress of 10 ksi applied at 600°C. The creep strain is determined to satisfy $\epsilon = 0.15t^{0.19}$, where time is measured in hours. The creep strain after 5 hours is therefore $\epsilon = 0.2036$. ∎

At low homologous temperatures ($0.05 \le T/T_m \le 0.3$), creep can often be modeled by a logarithmic function of time, t,

$$\epsilon = A \ln(1+t) + C, \tag{12.19}$$

where A and C are constants depending on the material and temperature (Fig. 12.6). At these low temperatures, the creep is in the primary region; it does not reach the secondary region. This model is a good one for graphite, for example.

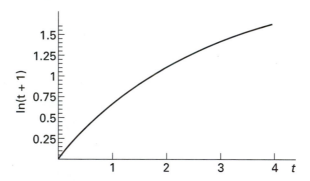

FIGURE 12.6 Logarithmic model for creep strain versus time, with $A = 1$, $C = 0$.

The power and exponential models fit metal creep well at moderate homologous temperatures ($0.2 \le T/T_m \le 0.7$). The power model is

$$\epsilon = \epsilon_0 + Bt^m + At, \tag{12.20}$$

where ϵ_0 is the instantaneous strain and A, B, and m are material constants. The exponential law is

$$\epsilon = \epsilon_0 + B(1 - e^{-rt}) + At, \tag{12.21}$$

where ϵ_0 is the instantaneous strain and A, B, and r are material constants.

Just as for the relation of creep rate and applied constant stress for the secondary region, the logarithmic and exponential models for metals can be combined into a single model,

$$\epsilon = \epsilon_0 + B \ln[1 + C(1 - e^{-rt})] + At. \tag{12.22}$$

A creep model intended for all times is the hyperbolic sine model proposed by Findley, Adams, and Worley [10, p. 14]

$$\epsilon(t) = a \sinh(\sigma/\sigma_o) + bt^n \sinh(\sigma/\sigma_1), \tag{12.23}$$

where $a, b, n, \sigma_o,$ and σ_1 are material constants for a given constant temperature.

12.5 MULTIAXIAL CREEP

The preceding models only predict uniaxial creep strain or strain rate. To analyze multiaxial creep stress behavior, a strategy similar to that taken in the invention of yield criteria for plastic behavior has been used since creep is a viscoplastic, that is, a time-dependent plastic, response. An effective, or equivalent, uniaxial stress, σ_e, is determined from the von Mises distortion energy criterion. The effective stress is the uniaxial stress which would produce the same distortion energy as the given state of multiaxial stress (Chapter 9, Section 9.6.7).

$$\sigma_e^2 = \sigma_x^2 + \sigma_y^2 + \sigma_z^2 - \sigma_x\sigma_y - \sigma_x\sigma_z - \sigma_y\sigma_z + 3(\tau_{xy}^2 + \tau_{xz}^2 + \tau_{yz}^2). \tag{12.24}$$

A uniaxial creep model is applied to the effective stress to predict the behavior. As a first approximation, assume that the stresses remain constant and do not "redistribute" during creep.

EXAMPLE 12.8   A steel, thin-walled pressure vessel of radius 12 inches and wall thickness 1 inch contains a gas at a pressure of 500 psi. The temperature is 500° C. Predict the rate of creep strain to be expected. Assume that at this temperature the strain rate is given by $\dot{\epsilon} = 0.001\sigma^{0.3}/1{,}000$ hours.

Solution   The effective uniaxial stress can be computed from the two-dimensional principal stresses, $\sigma_h = pR/t$ and $\sigma_L = pR/2t$.

$$\sigma_e^2 = \sigma_x^2 + \sigma_y^2 - \sigma_x\sigma_y$$

$$= \sigma_h^2 + \sigma_L^2 - \sigma_h\sigma_L = 0.75\left(\frac{pR}{t}\right)^2.$$

Therefore, the effective stress is $\sigma_e = \sqrt{0.75}\,pR/t$. Applying the uniaxial model to the effective stress implies that the strain rate is $\dot{\epsilon} = 0.001(\sqrt{0.75}\,pR/t)^{0.3} = 0.01302/1{,}000$ hours, a large value. ∎

More contemporary analyses of multiaxial creep for metals in regimes of constant strain rates, analogous to the secondary regime in uniaxial creep, are based on the constant multiaxial creep stress given by the deviatoric stress, S_{ij}, because plastic strain is involved. Again, creep of a metal is a viscoplastic process. The creep response depends on both temperature and stress. Experiment has shown that if the stress magnitude is below a threshold value, C, the strain rate is proportional to a power of the stress. Above the threshold, an exponential relation holds. For FCC metals, C is typically in the range of 5–40 MPa. Further, the temperature dependence of the response in FCC metals changes at a temperature around one-half of the melting temperature, T_m, in Kelvins. The dominate mechanism at high temperatures ($0.5T_m < T < T_m$) and stresses below C is dislocation climb induced by diffusion. The dominant mechanism at low temperature and high stress is dislocation glide, or slip, controlled by the obstacles that the dislocations encounter as they move.

In 1974, F. K. G. Odqvist proposed a multiaxial model for constant strain rate creep that generalizes Norton's power law and uses an Arrhenius function to account for thermally excited dislocation climb. In this model, the creep strain rate is also a tensor, $\dot{\epsilon}_{ij}$. The strain rate is a product of an Arrhenius function and a power function.

$$\dot{\epsilon}_{ij} = \frac{1}{2} A \exp\left(\frac{-Q}{RT}\right) \left(\frac{||\mathbf{S}||}{C}\right)^n \frac{S_{ij}}{||\mathbf{S}||} \qquad \text{for} \qquad ||\mathbf{S}|| < C, \qquad (12.25)$$

where A, C, and n are material constants. The term Q is the activation energy for self-diffusion when dislocation climb is the controlling mechanism. The norm is defined to be $||\mathbf{S}|| = (S_{ij} S_{ij}/2)^{1/2}$ where the index multiplication convention is followed. The material constant n is approximately 4 for the dislocation climb mechanism.

The model generalizes [11] for arbitrary stress magnitudes as suggested by F. Garofalo to

$$\dot{\epsilon}_{ij} = \frac{1}{2} A \exp\left(\frac{-Q}{RT}\right) \sinh^n \left(\frac{||\mathbf{S}||}{C}\right) \frac{S_{ij}}{||\mathbf{S}||}. \qquad (12.26)$$

At lower temperatures $T < 0.5 T_m$, the Arrhenius function is replaced by

$$\exp\left\{\frac{-Q}{RT}\left[\ln\left(\frac{T_m}{2T}\right) + 1\right]\right\}.$$

EXAMPLE 12.9   A copper, thin-walled pressure vessel of radius 25 cm and wall thickness 2 cm contains a gas at pressure, 0.8 MPa. The temperature is 500° C. Predict from the Odqvist model the rate of creep strain to be expected. The activation energy is 180 kJ/mol. The melting temperature 1,356 K. The model parameters are $C = 13$ MPa, $A = 2 \times 10^7$, and $n = 4.5$.

Solution   The deviatoric stress is $S_{11} = pR/2t = 5$ MPa, $S_{22} = 0$, $S_{33} = -pR/2t = -5$ MPa, and $S_{ij} = 0$ for $i \neq j$. It has norm $||\mathbf{S}|| = pR/2t\sqrt{2} = 3.53$ MPa, which is less than C so that the Odqvist model applies.

The temperature is 773 K which is greater than one-half the melting temperature. Therefore, the temperature dependence is $\exp(-Q/RT) = \exp(-180,000/8.31/773) = 6.81 \times 10^{-13}$. The creep rate is

$$\dot{\epsilon}_{11} = \frac{1}{2} A \exp\left(\frac{-Q}{RT}\right) \left(\frac{||\mathbf{S}||}{C}\right)^n \frac{S_{ij}}{||\mathbf{S}||} = 2.75 \times 10^{-8}/\text{s}.$$

Also $\dot{\epsilon}_{22} = 0$ and $\dot{\epsilon}_{33} = -\dot{\epsilon}_{11}$. ∎

12.6   RELAXATION

The creep strain decreases somewhat when the stress is removed from a specimen after creep, but not fracture, has occurred. The phenomenon is called relaxation and occurs because some of the creep strain is elastic and can be recovered when the force is removed (Fig. 12.7). Most polymers exhibit considerably more relaxation than metals. Polymers are built of long curled molecules that stretch under loads without breaking. When the load is removed, the molecule curls again into its lower energy state of equilibrium. This effect is often described by pretending that the polymer is a living being and "remembers" its original state; a polymer is said to be a material with memory. This is a misnomer however; the molecule is merely returning to its equilibrium state.

The relaxation effect in metals shows that part of the creep strain is elastic and part is plastic. The plastic strain results from the permanent deformation of the material. The time required for a material to recover all of its elastic creep strain is called the relaxation time and varies from material to material.

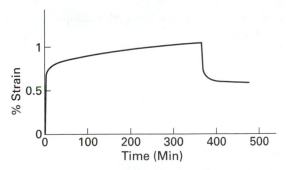

FIGURE 12.7   Creep relaxation resulting from removing the creep load in a test of a paper sheet.

The behavior of a perfectly elastic material only depends on its current state. Other materials seem to respond differently to the same current state if their past histories differ. Such materials are said to have a history-dependent response.

12.7   MECHANICAL MODELS FOR POLYMER CREEP AND RELAXATION

The classical uniaxial, isothermal models for behavior combining elastic and viscous response were based on mechanical models as described in Chapter 3, Section 3.10.4. Recall that the Maxwell model is

$$\dot{\epsilon} = \frac{\dot{\sigma}}{E} + \frac{\sigma}{c}. \tag{12.27}$$

Stress relaxation in which the strain is held constant while the stress changes over time can be expressed by Maxwell's model. Since the strain is constant, $\dot{\epsilon} = 0$ and Maxwell's differential equation becomes

$$\dot{\sigma} = -\frac{E}{c}\sigma, \tag{12.28}$$

which has the solution

$$\sigma(t) = \sigma(0)e^{-Et/c}. \tag{12.29}$$

This implies that as time passes, the stress relaxes to zero. This response is more characteristic of fluids than of solids.

The Maxwell model does not model solid creep well since in creep $\sigma(t) = \sigma_0$ and $\dot{\sigma} = 0$ implies that $\dot{\epsilon} = \sigma_0/c$. The solution of this differential equation is linear $\epsilon(t) = (\sigma_0/c)t + \epsilon_0$, where ϵ_0 is an integration constant. No solid has this creep response. Therefore, the Maxwell model is applied only to fluids.

The Kelvin-Voigt model, a spring and dashpot in parallel, is the differential equation, for constant stress,

$$\sigma = E\epsilon + c\frac{d\epsilon}{dt}. \tag{12.30}$$

The solution to the Kelvin-Voigt model for constant stress σ is obtained by using the integrating factor, $\exp(Et/c)$.

$$\epsilon(t) = \frac{\sigma}{E} + \exp\left(\frac{-Et}{c}\right)\left(\epsilon(0) - \frac{\sigma}{E}\right). \tag{12.31}$$

The strain tends as $t \to \infty$ to the linear elastic equilibrium value $\epsilon = \sigma/E$, assuming that the initial strain is $\epsilon(0)$.

The Kelvin-Voigt equation, written in the integral form of Boltzmann, is if $\epsilon(0) = 0$

$$\epsilon(t) = \int_0^t \frac{1}{c} e^{-E(t-\tau)/c} \sigma(\tau) d\tau. \tag{12.32}$$

The Boltzmann integral form makes it easy to calculate the creep response predicted by the Kelvin-Voigt model. Let the constant creep stress be σ_0. Then the integral is, in agreement with equation (12.31),

$$\epsilon(t) = \int_0^t \frac{1}{c} e^{-E(t-\tau)/c} \sigma_0 d\tau = \frac{\sigma_0}{E} \left(1 - e^{-Et/c}\right). \tag{12.33}$$

The creep strain as time passes approaches the line $\epsilon = \sigma_0/E$ asymptotically (Fig. 12.8). This is certainly a better model for creep of solids. However, for many materials, the creep strain merely increases rather than approaches a horizontal asymptote.

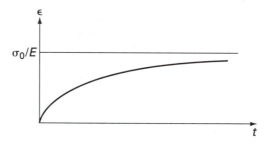

FIGURE 12.8 The creep strain response predicted by the Kelvin-Voigt model.

The standard linear solid is represented by a spring with constant, E_1, in parallel with a Maxwell model having constants, E_2 and c.

$$\dot{\epsilon} + k\epsilon = \frac{1}{E_1 + E_2}\dot{\sigma} + k\frac{\sigma}{E_1}, \tag{12.34}$$

where $k = E_1 E_2/c(E_1 + E_2)$.

The solution is, for constant stress and initial strain $\epsilon(0)$,

$$\epsilon(t) = \frac{\sigma}{E_1} + \exp\left(\frac{-E_1 E_2 t}{c(E_1 + E_2)}\right)\left(\epsilon(0) - \frac{\sigma}{E_1}\right). \tag{12.35}$$

The long-term behavior of the standard linear model is $\sigma = E_1\epsilon$.

The relaxation time $t_r = c/k$ is the time which makes the exponent of e equal to one. In the Maxwell model [equation (12.28)], where $k = E$, the relaxation time is the time required during stress relaxation to reduce the stress by a factor of $1/e$. In the Kelvin-Voigt model [equation (12.30)] applied to creep, the relaxation time, also $t_r = c/E$, is the time required to reduce the difference between the current strain and the asymptotic value by a factor of $1/e$.

12.8   MOISTURE-ACCELERATED CREEP OF HYDROPHILIC MATERIALS

A hydrophilic material is one that bonds with moisture. Nylon, concrete, kevlar, wood, paper, and any other cellulosic material are all hydrophilic. In a changing relative humidity environment, the material continually adsorbs and desorbs moisture so that it is in a nonequilibrium state. A hydrophilic material that, in a cyclic ambient relative humidity, creeps more than it would in a constant relative humidity equal to the highest value in the cycle is said to undergo moisture-accelerated creep.

Both wood and concrete exhibit moisture-accelerated creep. Green wood specimens show accelerated bending creep when dried under a bending load. Dried wood specimens subjected to vapor at saturation pressure followed by drying in a stepwise cyclic pattern show large accelerated creep. In one test of beech beams, the strains increased at 30% relative humidity, a drying phase, and decreased during the wetting phase at 90% relative humidity. The creep strain at failure after 14 complete cycles was about 25 times that of a specimen held in creep at a constant 93% relative humidity. The load in both cases was 3/8 of the stress-strain test failure load at 93% relative humidity. Most of the accelerated creep occurs during the desorption portion of the humidity cycle after the first cycle. If the load is removed in vapor at the saturation pressure, almost all of the strain is recovered. Compressive axial tests on hollow wood tubes made of spruce during changes between 50% and 100% relative humidity also found that, after the first moisture change, the wood crept more during desorption than adsorption.

The first published observation of accelerated creep under variable moisture content in concrete preceded that of wood by 20 years. The bending tests of 2 inch by 2 inch cross section concrete beams 32 inches long with a transverse load at the midsection by Pickett [17] showed that creep increased above that in constant conditions during alternate wetting in water and drying at 50% relative humidity for periods of many days. This is called the Pickett effect by concrete specialists. The accelerated creep of concrete during drying is thought to be a result of both stress-induced shrinkage or swelling and of tensile microcracking in combination with the aging of the concrete. The creep strain when either humidity condition was held constant followed logarithmic shaped creep curves with the creep strain during a constant 50% relative humidity greater at a given time than that in constant wetting in water [7, p. 315]. Paper behaves in the opposite manner with its creep greater in higher relative humidities.

12.8.1   Paper Creep in Variable Relative Humidity

The creep of paper in a constant relative humidity (RH) and temperature follows a logarithmic curve. The mechanical response of paper in a variable relative humidity climate is quite different from that in constant relative humidity. Because water facilitates the bonding of the hydrophilic pulp fibers to form paper, moisture has a strong influence on the mechanical properties of paper. The mechanical response of paper is determined by the interaction of the external forces on the paper structure and the ambient relative humidity, as well as the temperature. A mechanosorptive effect is one arising from the interaction of loads with moisture sorption mechanisms. Moisture-accelerated creep occurs in the special case in which the load on the specimen is held constant while the ambient relative humidity is varied. Variations in the ambient conditions, such as relative humidity and temperature, magnify the mechanical creep strain response in paper compared to that measured at any constant ambient condition and can cause earlier failure. This behavior is observed under both tensile and compressive loads. This mechanosorptive problem has been studied by experimentalists, but there is as yet no universally accepted explanation of the physical mechanisms underlying such behavior.

V. L. Byrd [8] provided the first experiments on the acceleration of tensile creep due to variable ambient relative humidity in paper; although it had been studied earlier in wood by L. D. Armstrong and coworkers. Two constant relative humidity creep tests for a paper handsheet are compared to a creep test during which the ambient relative humidity is cycled in Figure 12.9.

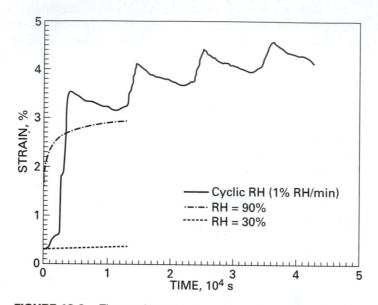

FIGURE 12.9 The accelerated creep response in a cyclic relative humidity varying between 30% and 90% and varying at 1% RH/min compared to that at constant relative humidities of 30% and 90%. Reprinted from [13] by permission of TAPPI.

The specimens were conditioned first at 23° C and 30% relative humidity for 24 hours to bring them to moisture equilibrium with the ambient air. The tensile creep tests of handsheets were conducted at a constant 30% RH, a constant 90% RH, and as the relative humidity linearly cycled from 30% to 90% to 30% at a rate of 1/3% RH/min. through four such cycles. One might have expected the constant 90% test to produce the largest creep strain at any given time. The varying relative humidity, however, causes the tensile specimen to creep more than it would at a constant relative humidity equal to the highest relative humidity reached in the cycle. In this sense, the variation in the relative humidity accelerates the creep. Therefore, attempting worst case testing of paper products at high constant relative humidity is not a safe design strategy. Cyclic variations of ambient temperature also accelerate the creep of paper. However, the strains are an order of magnitude smaller than those observed by varying the relative humidity.

12.9 CREEP FRACTURE

Creep is most significant in polycrystalline metals at high homologous temperatures, say, greater than 0.3. Recall that the homologous temperature is the ratio of the temperature of the metal to its melting temperature. At lower homologous temperatures, grain boundaries help strengthen polycrystalline metals by blocking dislocation motion. The high-deformation regions that tend to be within grains at low temperatures move to the grain boundaries at higher temperatures and may nucleate voids. Cavitation is hypothesized to occur due to excessive strain or stress ahead of a crack tip. Cavities may grow due to diffusive processes. Therefore, some important mechanisms for creep crack growth are the development of voids along grain boundaries, called grain boundary cavitation, the diffusion of lattice vacancies to existing voids, and the action of corrosive materials at the crack tip. The latter may include oxygen from the air in the case of some metals such as the superalloys or some carbides. These processes are often called damage mechanisms.

The creep crack growth rate in metals, especially at higher homologous temperatures at which creep is more significant, has been described using a modification of classical fracture mechanics achieved by writing a strain energy in terms of the strain rate rather than the strain (e.g., see [20]). A constitutive model is assumed in which the stress depends on the rate of strain as opposed to the strain. For example, in the uniaxial case, the Norton power law model for secondary region creep, $\dot{\epsilon} = B\sigma^n$, might be assumed.

The nonlinear elastic application of the J-integral uses the Ramberg-Osgood relation (see Chapter 9, Section 9.4.2) in the form, $\epsilon/\epsilon_Y = c(\sigma/\sigma_Y)^M$, where c is a material constant. The Norton power creep law for secondary creep can be written in the form $\dot{\epsilon}/\dot{\epsilon}_o = c(\sigma/\sigma_o)^N$, where ϵ_o is the strain under a creep stress of σ_o. The similarity of the two relations suggests replacing ϵ in the J-integral by $\dot{\epsilon}$. The strain energy, $W = \int \sigma_{ij} d\epsilon_{ij}$, in the classical derivation of rate-independent fracture mechanics is replaced by $W^* = \int \sigma_{ij} d\dot{\epsilon}_{ij}$.

Assume the deformation field around a crack is two dimensional and at equilibrium. The displacement in the x-direction is $u(x, y)$ and in the y-direction is $v(x, y)$. Also the strain-displacement relations (see Chapter 3) are $\epsilon_x = \partial u/\partial x$ and $\gamma_{xy} = \partial u/\partial y + \partial v/\partial x$. The J-integral (see Chapter 10, Section 10.4.3) is $J = \int_\Gamma [W dy - \sigma_x(\partial u/\partial x)ds - \sigma_y(\partial v/\partial x)ds]$, where Γ is a curve surrounding the crack tip, the stress acts on the outside of the curve Γ, the displacements are along the curve Γ, y is the in-plane direction normal to the crack face, and s is the arclength along the curve, Γ.

The creep fracture analog to the J-integral is called the C^*-integral. For a mode I crack opening loading

$$C^* = \int_\Gamma [W^* dy - \sigma_x(\partial \dot{u}/\partial x)ds - \sigma_y(\partial \dot{v}/\partial x)ds]. \qquad (12.36)$$

The units of C^* are J/m^{2}s. C^* has been experimentally shown to be proportional to the crack growth rate in the creep of some steels, when the value of C^* is large enough. It is not clear whether or not this relationship holds for all metals.

12.10 DESIGN FOR CREEP

A designer examines creep curves at various stress and various temperatures to determine a stress "creep limit." To do this, the designer first determines what plastic strain is permissible during the lifetime of the structure so that it will continue to function safely and properly. The creep limit is the stress at which the creep strain will stay below this design-allowable strain over the planned lifetime and at the design temperature. The allowable strain is divided by the lifetime to determine an allowable creep rate. The secondary regions of creep curves are then examined to determine what constant applied stress produces the allowable creep rate at the design temperature. Alternatively, the mathematical models which predict creep rate as a function of applied stress may be used.

Since creep increases with increasing temperature, the designer may choose the creep (stress) limit at the highest temperature in which the structure is expected to operate. If the temperature is lower in practice, then the structure is all the safer since the creep will be less than projected.

To choose the creep limit from experimental curves alone would require tests at all possible applied stresses, all temperatures, and for times as long as the lifetime of the structure. Experimentalists and creep theoreticians have attempted to produce equations, mathematical models, which describe creep at various stresses and temperatures as a function of time. These models, several of which have been discussed above, tend to be approximate and differ from material to material. Considerable research remains to be done in this area, especially to create models that predict creep due to more complicated loadings and for newer engineering materials such as polymers, ceramics, and composites.

EXAMPLE 12.10   A circular cross-sectional rod of length 10 inches must carry a load of 500 pounds in a 200° C environment. The rod is to made of an aluminum with a yield stress of 20 ksi and with E= 8,000 ksi at 200° C. The maximum allowable stress is 15 ksi. The rod must creep less than 0.1 inches during its 5-year service life. Assume the hyperbolic sine secondary creep law. The weight of the rod is to be minimized. Design the radius, R, of the rod. Ignore primary creep.

Solution   The 5-year life is $t_f = 5(365)24(3600)$ seconds. Neglect the initial deformation when the load is applied. The allowable creep rate determined from the allowable deformation is $\dot{\epsilon} = (0.1)/(10t_f) = 6.342 \times 10^{-11}$/s. Determine the allowable creep stress, σ_c, from the hyperbolic sine model using the appropriate values from Table 12.2 at 477 K.

$$\sigma_c = \frac{1}{c} \sinh^{-1} \left[\left(\frac{\dot{\epsilon}}{B''} \right)^{1/n} \right] = 295.6 \text{ psi.}$$

This stress governs because it is less than the allowable stress of 15 ksi. The axial stress must satisfy $\sigma = 500/\pi R^2 \leq \sigma_c$ so that $R \geq \sqrt{500/\pi \sigma_c} = 0.734$ inches.

Note that the initial deformation when the load is initially applied to this rod is $\delta = \sigma_c L/E = 3.695 \times 10^{-4}$ inches. This initial deformation is small enough to neglect compared to the allowable creep of 0.1 inches.   ∎

A designer can attempt to avoid creep problems by specifying a material which resists creep. The refractory metals, niobium (Nb), molybdenum (Mo), tantalum (Ta), and tungsten (W), have very high melting temperatures, from 2,470 to 3,410° C. Therefore at the operating temperatures, 1,100° C, often found in engines, they would exhibit very little creep since their homologous temperature is under 0.5 in that range. Unfortunately, they are easily oxidized and embrittled.

Another possibility is to design the material so that it includes finely grained precipitates, which would block dislocations and thereby prevent deformation. But most metals form very large grains at high temperatures. Superalloys show small creep strains up to about 1,000° C and have high strength and corrosion resistance as well. The alloys are based on nickel, an iron and nickel combination, or on cobalt.

Two other strategies have been used. The first is to alloy the metal with other particles to prevent large grains from forming at high temperatures to create solid solution strengthening. The second is to control the molding of the part so that it solidifies into a single crystal when cooled.

PROBLEMS

12.1 Determine the concentration of vacancies in copper at 30°C.

12.2 Sketch the strain rate variation, $\dot{\epsilon}/C$, with temperature over the range $0.6 \leq T_h \leq 0.8$ according to the Arrhenius equation for aluminum in which $Q_c = 150$ kJ/mol.

12.3 A creep test performed at 900°C produces a strain rate of 0.0005/s and a test under the same stress at 1,100°C produces a strain rate of 0.01/s. Determine the activation energy and predict the strain rate at 1,000°C. The dominant creep mechanism is believed to be dislocation climb.

12.4 Predict the lifetime of an axially loaded member made of silicon nitride subjected to a strain rate of $\dot{\epsilon} = 10^{-9}$.

12.5 A material has a fracture time at 600°C of 100,000 hours. Estimate its fracture time under the same stress but at a temperature of 650°C.

12.6 Use the hyperbolic sine law to predict the creep rate for copper at 673 K under a constant stress of 4,000 psi. Ans. 4.21×10^{-7}/s.

12.7 Use the hyperbolic sine law to predict the creep rate for aluminum at 533 K under a constant stress of 2,000 psi. Ans. 4.72×10^{-5}/s.

12.8 Use the hyperbolic sine law to determine the constant stress in copper at 773 K that will cause a strain rate of 5×10^{-6}/s in the secondary region of the creep curve. The solution of this problem will require using the inverse hyperbolic sine function on a calculator.

12.9 Plot the creep strain versus time curve for the following data obtained from a lead specimen at a temperature of 85°F. The original length was 5 inches. The bar had cross-sectional dimensions of 0.067 inches by 0.380 inches. The load was 50 pounds.

TABLE 12.3 Creep data for lead

Time (s)	Deformation (in.)
0	0
30	0.27
60	0.34
90	0.36
120	0.38
150	0.42
180	0.46
210	0.49
240	0.53
270	0.61
300	0.82
330	2.01

(a) Determine the approximate times at which the primary region becomes secondary and the secondary becomes tertiary.

(b) Determine the approximate strain rate in the secondary region.

(c) Compute the homologous temperature.

(d) Determine whether or not the model proposed by Andrade fits this data by comparing the graph of the data and of the model.

12.10 A steel, thin walled pressure vessel of radius 12 inches and wall thickness 1 inch contains a gas at pressure, p. The temperature is 500°C. What is the maximum pressure permitted so that the rate of creep strain is no more than 0.2/1000 h? Assume that, at this temperature, the strain rate is given by $\dot{\epsilon} = 0.01\sigma^{0.3}/1,000$ h.

12.11 A nickel, thin-walled pressure vessel of radius 25 cm and wall thickness 2 cm contains a gas at pressure, 1 MPa. The temperature is 700°C. Predict from the Odqvist model the rate of creep strain to be expected. The activation energy is 290 kJ/mol. The melting temperature is 1726 K. The model parameters are $C = 25$ MPa, $A = 1.5 \times 10^9$, and $n = 4.5$.

12.12 A thick aluminum rectangular plate is loaded by an in-plane stress of 4 MPa parallel to one pair of opposite edges and by an in-plane stress of 5 MPa parallel to the other pair of edges. The temperature is 300°C. Predict from the Odqvist model the rate of creep strain to be expected. The activation energy is 140 kJ/mol. The melting temperature is 933 K. The model parameters are $C = 8$ MPa, $A = 5 \times 10^9$, and $n = 4.5$.

12.13 Plot the creep strain response of the standard linear model and of the Kelvin-Voigt model on the same graph under a 20 MPa stress to see which approaches its asymptotic limit more quickly. In the standard model $E_1 = 1.08$ GPa, $E_2 = 2$ GPa, and $c = 0.5$ GPa-s. In the Kelvin-Voigt model $E = 1.08$ GPa and $c = 0.5$ GPa-s. Assume the initial strain is zero.

12.14 Plot the creep strain response of the Kelvin-Voigt model for $c = 0.05$ GPa-s, $c = 0.5$ GPa-s, and $c = 5$ GPa-s on the same graph under a 20 MPa stress to see which approaches its asymptotic limit more quickly. The stiffness is $E = 1.08$ GPa. Assume the initial strain is zero.

12.15 A circular cross-sectional rod of radius 1 cm is made of a Kelvin-Voigt material with $E = 0.98$ GPa and $c = 240$ GPa-s. The rod carries an axial load of 500 N. The initial strain is 800 microstrains.

(a) Determine the strain after 5 minutes.

(b) Determine the asymptotic limit of the creep strain.

12.16 A circular cross-sectional rod of length 10 inches must carry a load of 300 pounds in a 260°C environment. The rod is made of an aluminum with a yield stress of 18 ksi and with $E = 7,500$ ksi at 260°C. The maximum allowable stress is 9 ksi. The rod must creep less than 0.12 inches during its 4-year service life. Assume the hyperbolic sine secondary creep law.

The weight of the rod is to be minimized. Design the radius of the rod. Ignore primary creep.

12.17 A cylindrical thin-walled pressure vessel of radius $r = 10$ cm, length $L = 60$ cm must contain a gas at a pressure of $p = 800$ kPa at $20°$C. Design the wall thickness t to prevent the three failure modes of yielding, fracture, and creep, as follows.

The vessel is made of an aluminum with $E = 70$ GPa, $\nu = 0.33$, $\sigma_Y = 448$ MPa, $\sigma_u = 560$ MPa, and $K_{Ic} = 28.6$ MPa-m$^{1/2}$.

(a)   Write the circumferential ($\sigma_h = pr/t$) and longitudinal ($\sigma_l = pr/2t$) principal stresses in terms of t. The radial normal stress is zero.

(b)   Use the Mises criterion to determine the smallest wall thickness t so that the vessel remains elastically loaded.

(c)   Determine the minimum t to prevent mode I brittle fracture of a longitudinal crack of size 0.0005 m at $20°$C. Assume a shape factor of $Y = 1$.

(d)   To design for misuse, assume that the vessel may be temporarily subjected to an ambient temperature of $100°$C.

(1)   What is the gas pressure at $100°$C if the gas is an ideal gas? Recall that $pV = NRT$ where V is the fixed internal volume of the cylinder (ignore thermal expansion of the cylinder), N is the fixed number of moles of the gas in the cylinder, and the gas constant $R = 8.314$ J/mol-K.

(2)   The creep strain rate in the secondary region must be at most $\dot{\epsilon} = 10^{-7}$/s. Determine the allowable stress in psi from the hyperbolic sine model, assuming for purposes of this calculation that $B'' = 2.78 \times 10^{-6}$, $c = 3.98 \times 10^{-4}$, and $n = 5$. Convert this stress to Pa using 1 psi $\sim$ 7,000 Pa.

(3)   Use the Mises equivalent stress to determine the minimum allowable wall thickness, t.

(e)   To avoid failure by any of these three modes, what is the minimum allowable wall thickness?

REFERENCES

[1] *Mechanical Engineer's Reference Book*, 11th ed., A. Parrish, Ed., Butterworth's, London, 1973, pp. 1–78.

[2] L. D. ARMSTRONG, Deformation of wood in compression during moisture movement, *Wood Science*, 5(2), 81–86 (1972).

[3] L. D. ARMSTRONG AND G. N. CHRISTENSEN, Influence of moisture changes on deformation of wood under stress, *Nature*, *191*, 869–870 (1961).

[4] L. D. ARMSTRONG AND P. U. A. GROSSMAN, The behavior of particleboard and hardwood beams during moisture cycling, *Wood Science and Technology*, *6*, 128–137 (1972).

[5] L. D. ARMSTRONG AND R. S. T. KINGSTON, Effect of moisture changes on creep in wood, *Nature*, *185*, 862 (1960).

[6] Z. P. BAŽANT, Constitutive equation of wood at variable humidity and temperature, *Wood Science and Technology*, *19*, 159–177 (1985).

[7] Z. P. BAŽANT AND KAPLAN, M. F., *Concrete at High Temperatures*, Longman, Harlow, Essex, UK, 1996.

[8] V. L. BYRD, Effect of relative humidity changes during creep on handsheet paper properties, *Tappi*, 55(2), 247–252 (1972).

[9] W. N. FINDLEY, Creep characteristics of Plastics, in *Symposium on Plastics*, ASTM, 1944, pp. 118–134.

[10] W. N. FINDLEY, J. S. LAI, AND K. ONARON, *Creep and Relaxation of Nonlinear Viscoelastic Materials*, Dover, New York, 1989.

[11] A. D. FREED AND K. P. WALKER, Viscoplasticity with creep and plasticity bounds, *International Journal of Plasticity*, *9*, 213–242 (1993).

[12] F. GAROFALO, *Fundamentals of Creep and Creep-rupture in Metals*, MacMillan Co., New York, 1965, p. 52.

[13] H. W. HASLACH, JR., The moisture and rate-dependent mechanical properties of paper: A review, *Mechanics of Time-Dependent Materials*, *4*, 169–210 (2000).

[14] R. F. S. HEARMON AND J. M. PATON, Moisture content changes and creep of wood, *Forest Products Journal*, *14*, 357–359 (1964).

[15] J. D. LUBAHN AND R. P. FELGAR, *Plasticity and Creep of Metals*, Wiley, New York, 1961.

[16] F. A. McCLINTOCK AND A. S. ARGON, *Mechanical Behavior of Materials*, Addison-Wesley, Reading, MA, 1966.

[17] G. PICKETT, The effect of change in moisture content on the creep of concrete under a sustained load, *Journal of the American Concrete Institute* 29/54(10), 857–864 (1942).

[18] S. V. RAJ AND A. D. FREED, Effect of strain and stress on the relative dimensions of the 'hard' and 'soft' regions in crept NaCl single crystals, *Scripta Metallurgica et Materialia*, *27*, 1741–1746 (1992).

[19] S. V. RAJ AND G. M. PHARR, Creep substructure formation in sodium chloride single crystals in the power law and exponential creep regimes, *Materials Science and Engineering*, *A122*, 236 (1989).

[20] H. RIEDEL, *Fracture at High Temperatures*, Springer-Verlag, Berlin, 1987.

CHAPTER 13

STATIC STABILITY

A designed structure might fail in one of several ways. The material may endure stresses larger than the elastic limit. Crack growth can occur at loads below the yield limit if existing cracks are of sufficient length. A dynamically loaded body may eventually fracture from fatigue. At high homologous temperatures, the member deforms due to creep. In addition, design failure might result from excessive buckling deformation.

A statically loaded structure must be designed so that slight perturbations from the design values of either the applied loads or the actual dimensions do not significantly affect its behavior. No structure can be built to zero tolerance nor can small ambient disturbances be permitted to cause failure. A static equilibrium state is said to be stable if, when the system is perturbed slightly from the equilibrium state, it returns to that equilibrium state. It is very important to design statically loaded structures so that they operate at a stable equilibrium state since there are always perturbations in a real system. The stability of an equilibrium state for a conservative, that is, elastic, system is most easily determined from the potential energy function for the system by the principle of stationary potential energy.

A designer often must also decide whether or not stability is lost as a design parameter is varied. If the stability is lost at a particular value of a design parameter, the system can jump from one equilibrium configuration to another, as in the buckling of a column. This can be dangerous. If the graph of the equilibrium states as a function of the design parameter splits at the value at which stability is lost, the graph is said to bifurcate. The critical buckling load of a column defines such a bifurcation point when the behavior of a column is thought of as parametrized by the load on it. A designer should locate and avoid such points for the behavior of the structure to ensure that it can be used safely.

While the calculation of the bifurcation point is often possible using linear methods, the postbuckling displacements of such systems usually cannot be computed from the linearized system equations. Numerical methods, such as the finite element method, are usually inefficient and expensive tools to determine the behavior of a system as a design parameter is changed because many cases must be computed to understand the system response pattern. A qualitative analysis using the potential energy function is often efficient, although the finite element method may be used to approximate the bifurcation point once the completed qualitative analysis can guide the approximation technique. Engineering members including columns, thin rings, and thin tubes are good examples for the application of methods currently used to study the stability of a system.

13.1   STABILITY OF EQUILIBRIUM STATES OF CONSERVATIVE SYSTEMS

A system is conservative if there is no dissipation through friction or other means. The principle of stationary potential energy implies that the behavior of a conservative system is described by its potential energy function. Recall from Chapter 7, Section 7.4:

Definition   An equilibrium state is said to be stable if its potential energy is a minimum. The equilibrium state is said to be unstable if its potential energy is a maximum.

If the potential energy depends only on a single variable and if the second derivative is nonzero, the sign of its second derivative determines stability. If the second derivative is zero, other techniques must be applied to determine the behavior of the potential energy at the equilibrium point.

EXAMPLE 13.1   A ball can spontaneously move along a surface whose height is given by $y = 2x^4 + 2$. Determine the stability of the equilibria.

Solution   The potential energy is the negative of the work done by the weight of the body. In the system with datum at $y = 0$, the body moves from the datum to its current position in a direction opposite that of the force, the weight. The potential energy is $\Pi(x) = mg(2x^4 + 2)$. The only equilibrium occurs when $x = 0$. Its second derivative at the equilibrium $x = 0$ is zero. However, the first derivative changes from negative to positive at $x = 0$. Therefore $x = 0$ is a minimum of the potential energy function and the equilibrium is stable. Notice that the equilibrium at $x = 0$ is stable because the coefficient of x^4 is positive. It would be unstable if the coefficient of x^4 were negative.   ∎

Exercise 13.1   Graph the potential function for the previous example.

If the potential function depends on two variables, $V(x, y)$, then the stability of the equilibria, the simultaneous solutions to $\partial V/\partial x = 0$ and $\partial V/\partial y = 0$, are determined from the following conditions on the second derivatives which are usually derived in calculus. Denote by $V_x = \partial V/\partial x$ and $V_{xx} = \partial^2 V/\partial x^2$, and so on.

$$V_{xx}V_{yy} - V_{xy}^2 > 0, \quad \text{and}$$

$$V_{xx} + V_{yy} > 0 \quad \text{stable} \tag{13.1}$$

$$V_{xx} + V_{yy} < 0 \quad \text{unstable} \tag{13.2}$$

The Hessian is defined to put this criterion in a more general context.

Definition   Suppose that a conservative system with n-degrees of freedom has potential energy, $\Pi(x_1, \ldots, x_n)$, a function of n-state variables, x_i, $i = 1, \ldots, n$. Define $\Pi_{ij} = \partial^2\Pi/\partial x_i \partial x_j$, $i, j = 1, \ldots, n$. Then the Hessian at a point, $(x_1, \ldots, x_n)$, is the n by n matrix of second derivatives evaluated at that point, $H(x_1, \ldots, x_n) = (\Pi_{ij}|_{x_1,\ldots,x_n})$.

Definition   An $n \times n$ matrix is said to be positive definite if each of the minors along the diagonal has positive determinant. The ith minor is the upper left $i \times i$ submatrix.

An equilibrium state is stable if the Hessian at that point is positive definite. In the case of a function of two variables, the Hessian is

$$\begin{pmatrix} V_{xx} & V_{xy} \\ V_{xy} & V_{yy} \end{pmatrix}.$$

It is positive definite if the determinants of the two minors are positive,

$$V_{xx} > 0;$$
$$V_{xx} V_{yy} - V_{xy}^2 > 0. \tag{13.3}$$

An alternative definition of positive definite in two dimensions is that the Hessian at that point has positive determinant and positive trace. This is simply the condition derived in calculus.

$$V_{xx} V_{yy} - V_{xy}^2 > 0;$$
$$V_{xx} + V_{yy} > 0. \tag{13.4}$$

The two conditions, (13.3) and (13.4), are equivalent. A proof by contradiction shows that condition (13.4) implies condition (13.3). Assume that $V_{xx} < 0$. Condition (13.4) implies that $V_{xx} > -V_{yy}$. Multiply both sides of this inequality by the assumed positive $-V_{xx}$ to obtain $-V_{xx}^2 > -V_{xx} V_{yy}$. But by the first line of condition (13.4), $-V_{xx}^2 > -V_{xx} V_{yy} > V_{xy}^2$. This implies that the negative number $-V_{xx}^2$ is positive, which is a contradiction. Therefore $V_{xx} > 0$.

Exercise 13.2 Verify that condition (13.3) implies condition (13.4).

EXAMPLE 13.2 The potential for a system is $V(x, y) = 4x^2 + 9y^2$. Determine the stability all equilibria of this system.

Solution The simultaneous solution of $V_x = 8x = 0$ and $V_y = 18y = 0$ shows that the only equilibrium point is (0,0). The second derivatives are $V_{xx} = 8$, $V_{xy} = 0$, and $V_{yy} = 18$. Therefore the determinant and trace of the Hessian are both positive and the equilibrium, $(0, 0)$, is stable. This can also be easily seen by sketching the graph of $V(x, y)$ (Fig. 13.1). It is an elliptic paraboloid with its single minimum at $(0, 0)$. ∎

13.2 STABILITY OF EQUILIBRIUM STATES OF CONSERVATIVE SYSTEMS SUBJECT TO A VARIABLE DESIGN PARAMETER

A common engineering problem is the study of the behavior of a structure under a load which can vary. For example, if a column is subjected to an axial load that is too large, then the column might buckle. While the actual load on the structure may be fixed in a particular design, the engineering analysis is facilitated by imagining that the load can take arbitrary values in order to decide which load magnitudes are dangerous. Alternatively, a designer may wish to model the behavior of the system as other parameters are varied. It helps to distinguish between the variables that describe the state of the system and other parameters such as the load, the elastic modulus of the material, or the geometric dimensions of the members of the structure.

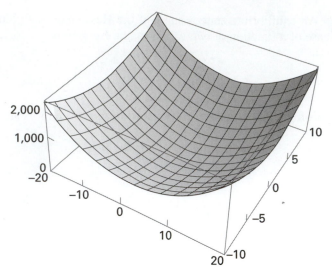

FIGURE 13.1 Graph of $V(x, y) = 4x^2 + 9y^2$.

Definition A state variable is a variable describing the configuration or response of a system. A control variable is one inducing the changes in the system to which the state variables respond.

Knowledge of the values of the equilibrium state variables completely describes the system for a fixed set of control variables. For example, the angle from the vertical which locates the position of a rigid rod supported by a torsional spring and carrying a vertical dead load is a state variable since it describes the configuration of the structure. The load is a control variable, as are the length of the rod and the torsional spring constant.

If a control variable is allowed to change, then one obtains a family of potential energy functions, one for each value of the control variable. The minima and maxima of the family member defined by a particular value of the control variable give the equilibrium states corresponding to that value of the control variable. One, or more, of the functions in this family can have a singularity. This point is important because the behavior of the system undergoes a change, which is often drastic, at the corresponding value of the control variable. For one degree of freedom, the second derivative of the potential function is zero at a singular point. For higher degrees of freedom, the Hessian of the potential function must be singular. This means that its determinant is zero.

Definition A point is a singular point for a function if it is an equilibrium point for the function and if the Hessian of the function is singular at that point.

Repeated application of numerical methods to locate the equilibria for each value of the parameter are not efficient for these problems since the effect of varying the parameter is not easy to decipher from a few trial cases. Instead, if the potential energy is simple enough to analyze, the equilibrium states are often represented by graphs in state variable–control variable space. In a system with a single state variable and a single control parameter, an equilibrium path is the curve in control-state variable space composed of the equilibria corresponding to different values of the control parameter as it is increased.

Definition A bifurcation point is a point on the graph of the equilibrium states as a function of the design control parameters where the graph branches.

The stability changes at a bifurcation point, which must therefore be a singularity. Any equilibrium state at which the stability changes must have a singular Hessian. But a singularity need not be a bifurcation point in a family of functions.

In 1970, J. M. T. Thompson, a British engineer, proved [3, pp. 61–65] that at least two equilibrium paths must intersect at a point where the stability changes in a system with one state variable and one control variable.

THEOREM An initially stable equilibrium path rising monotonically with respect to the control parameter cannot become unstable without intersecting another distinct equilibrium path.

Proof

For a fixed value of the control variable, P, the sign of the first derivative changes from positive to negative as x increases through a minimum, a stable equilibrium, by the first derivative test. This suggests that the equilibrium paths, those where the first derivative is zero, divide the x-P plane into regions in which the first derivative is positive and regions in which it is negative. Since the order of sign change is reversed from negative to positive at an unstable equilibrium, there must be at least one other curve on which the derivative of the potential energy is zero. This is the second equilibrium path. In Figure 13.2, the original equilibrium path is shown as a solid curve and the deduced bifurcating equilibrium path is drawn as a dotted curve.

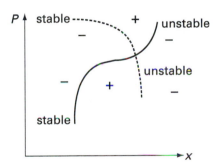

FIGURE 13.2 Exchange of stability at the intersection of two equilibrium paths.

At the bifurcation point, the stability must switch from one path to another. The schematic dividing the x-P plane into regions of positive and negative first derivatives shows that it is impossible for two equilibrium paths to intersect if one is composed completely of stable points and the other is completely composed of unstable points on both sides of the bifurcation point. ∎

EXAMPLE 13.3 Suppose that x is a state variable describing the configuration of a system and that there is one control variable, b. The potential energy is $\Pi(x; b) = x^3 + bx$. Discuss the stability of the equilibria.

Solution The equilibrium states occur at $d\Pi/dx = 3x^2 + b = 0$. If $b > 0$, there are no equilibria. If $b < 0$, there are two equilibria, a maximum at $x = -\sqrt{-b/3}$ and a minimum at $x = \sqrt{-b/3}$. There is a transition in the type of equilibria as b passes through zero. When $b = 0$, there is a singularity at $x = 0$; the second derivative is zero there. So the member of the family of potential functions with a singularity marks the transition in behavior. As b moves from negative values to zero, the equilibria approach each other until they coalesce when $b = 0$ in the position $x = 0$. As b continues to increase, there are no additional equilibria. The graph of the equilibria in x-b space is a parabola, one-half of which is stable equilibria and the half of which is unstable equilibria. This equilibrium path is not monotonic so that Thompson's theorem does not apply. The singular point at $b = x = 0$ is not a bifurcation point.

The singularity of the member of the family corresponding to $b = 0$ is called a limit point. This type of behavior occurs in the buckling of arches. Figure 13.3 shows one member of each type in the family, the members corresponding to $b = 1$, $b = 0$, and $b = -1$. ■

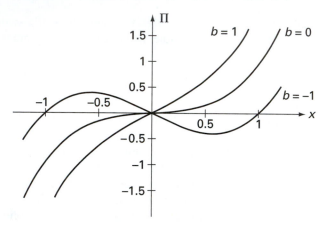

FIGURE 13.3 Typical members of the family of functions, $\Pi(x; b) = x^3 + b\,x$.

EXAMPLE 13.4

A rigid rod of negligible weight and of length, L, is supported at one end by a frictionless pin and a torsional spring, with spring constant, k, and is loaded at the other by a vertical dead weight, P (Fig. 13.4). The angle, θ, between the vertical and the rod is the only degree of freedom since the rod is rigid and can only move in a single plane. The system is called perfect if the rod is initially vertical as the load, P, is increased from zero. When the load is larger than some load, P_c, the rod deflects to one side, the analog to buckling for a rigid member. Determine the equilibria and their stability by (a) energy methods, (b) statics.

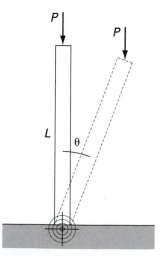

FIGURE 13.4 A load bearing rod supported at its base by a frictionless pin and torsional spring.

Solution   (a) The length of the rod and the spring constant are fixed control variables. The load, P, is a variable control variable. The potential energy for the perfect conservative system is

$$V(\theta) = \frac{1}{2}k\theta^2 - PL(1 - \cos\theta). \tag{13.5}$$

The equilibria satisfy $dV/d\theta = k\theta - PL\sin\theta = 0$. The stability of the equilibria is determined by the second derivative

$$\frac{d^2V}{dx^2} = k - PL\cos\theta.$$

There is a singularity in the vertical position, when $\theta = 0$, if $P = k/L$. If $P < P_c = k/L$, then $k/LP > 1$ and

$$\sin\theta = \frac{k}{LP}\theta > \theta.$$

The only equilibrium solution is at $\theta = 0$, since the fact that the slope of the $\sin\theta$ curve is maximum at the origin and equal to 1 implies that $\sin\theta > \theta$ is impossible. These equilibria are all stable because $d^2V/dx^2 > 0$ when $P < k/L$ and $\theta = 0$.

For a constant $P > P_c$, the energy function has three equilibria. One is $\theta = 0$; these are all unstable because $d^2V/dx^2 < 0$ when $P > k/L$ and $\theta = 0$. Two nonzero solutions exist in the range $-\pi < \theta < \pi$ since the line with slope, $k/LP < 1$, intersects the $\sin\theta$ curve at three points, including the origin. The other two are equidistant from the origin. These two are minima, because $k\theta - PL\sin\theta = 0$ implies that

$$\frac{d^2V}{dx^2} = k - PL\cos\theta = PL\left[\frac{\sin(\theta)}{\theta} - \cos(\theta)\right] > 0,$$

if $\theta \neq 0$.

The graph, in θ-P space (Fig. 13.5), of the equilibrium states has the three-pronged shape of a pitchfork and so is called a pitchfork bifurcation. A bifurcation, or path splitting, as P increases exists at $P_c = k/L$.

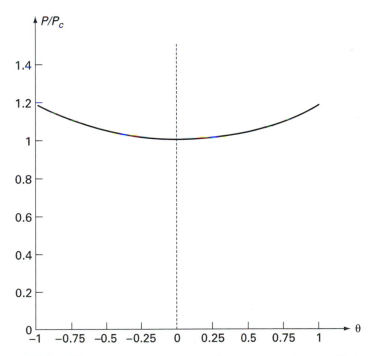

FIGURE 13.5   The pitchfork graph, in θ-P/P_c space, of the equilibrium states of $V(\theta; P) = \frac{1}{2}k\theta^2 - PL(1 - \cos\theta)$.

A somewhat easier approximate analysis is obtained from the Taylor series of the potential to the fourth order,

$$V^{(4)}(\theta) = \left(\frac{PL}{24}\right)\theta^4 + \left(\frac{k}{2} - \frac{PL}{2}\right)\theta^2. \qquad (13.6)$$

This contains all the qualitative information about the shape of the graph of the equilibrium points. The equilibria occur at $(PL/6)\theta^3 + (k - PL)\theta = 0$. The solutions are $\theta = 0$ for all P and

$$\theta = \sqrt{6 - \frac{6k}{PL}},$$

which exist if $k/PL < 1$, that is, $P > k/L$. The singularity occurs at the value of P making the coefficient of θ^2 zero, $P_c = k/L$. The second derivative $d^2V^{(4)}/d\theta^2 = (PL/2)\theta^2 + (k - PL)$. The stability behavior is the same as in the exact solution.

The graph of its equilibria with respect to the load is also a pitchfork and the bifurcation occurs at $P_c = k/L$. The stability at the bifurcation is determined by the sign of θ^4.

(b) The equilibrium states can also be identified for this system by taking moments about the pin to obtain $k\theta - PL\sin\theta = 0$. Viewing the equilibrium load $P = k\theta/L\sin\theta$ as a function of θ and taking the limit as θ tends to zero shows that $P_c = k/L$ is the critical load above which the nonvertical positions are equilibria. The stability cannot be determined from statics techniques. ∎

Exercise 13.3   Use a computer to plot the equilibria for the functions $V(\theta)$ and $V^{(4)}(\theta)$ on the same graph in $\theta - P$ space to see how close the approximation is.

EXAMPLE 13.5   A rigid rod is supported by a frictionless pin at its bottom and at its top by two springs adjusted so that the rod is in the vertical position when the springs are unstretched (Fig. 13.6). The rod has length L and the spring constants are each k. Determine the maximum vertical dead load, P, at which the rod is in a stable equilibrium in the vertical position. Neglect the weight of the rod.

Solve this problem by (a) energy methods; (b) statics.

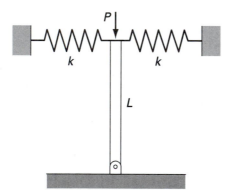

FIGURE 13.6   A load-bearing rigid rod supported by a frictionless pin at its bottom and at its top by two horizontal springs.

Solution   **(a)** The load is thought of as steadily increasing from zero while the rod is initially in the vertical position. The problem can then be viewed as seeking the bifurcation point in the family of energy functions with control parameter, P. The potential energy is

$$\Pi(\theta) = 2\left(\frac{1}{2}kL^2\sin^2\theta\right) - PL(1 - \cos\theta), \qquad (13.7)$$

where θ is the angle between the rod and the vertical. The equilibria, obtained from

$$\frac{d\Pi}{d\theta} = 2kL^2 \sin\theta \cos\theta - PL \sin\theta = 0,$$

are those values of θ satisfying

$$\sin\theta = 0 \quad \text{or} \quad \cos\theta = \frac{P}{2kL}.$$

The first equation shows that the vertical position, $\theta = 0$, is an equilibrium state for all possible loads, P. The behavior of the second derivative,

$$\frac{d^2\Pi}{d\theta^2} = 2kL^2 \cos(2\theta) - PL \cos\theta,$$

is zero if $P = 2kL$ and $\theta = 0$. This is the load at which bifurcation of the graph of the equilibrium positions occurs. In the vertical position, where $\theta = 0$, the second derivative is positive if $P < 2kL$ and so these equilibrium positions are stable. The remaining loads make the rod unstable in the vertical position. A small touch on the rod will cause it to fall permanently away from the vertical position.

Those equilibria defined by $P = 2kL \cos\theta$ have second derivative

$$\begin{aligned}
\frac{d^2\Pi}{d\theta^2} &= 2kL^2 \cos(2\theta) - PL \cos\theta \\
&= 2kL^2(\cos^2\theta - \sin^2\theta) - PL \cos\theta \\
&= 2kL^2(\cos^2\theta - \sin^2\theta) - 2kL^2 \cos^2\theta \\
&= -2kL^2 \sin^2\theta < 0.
\end{aligned}$$

These equilibria are unstable.

(b) The sum of the moments about the pin support must be zero for equilibrium. From the free body diagram,

$$\sum M_O = 2(kL \sin\theta)L \cos\theta - PL \sin\theta = 0.$$

Notice that this is exactly the same equation as obtained from $d\Pi/d\theta = 0$ above. The determination of equilibria is therefore the same. However, statics provides no tool by which to determine whether or not the equilibria are stable. The energy methods analysis produces more information. ∎

Exercise 13.4 On a computer or by hand, sketch the graph of the equilibrium positions in the state-control parameter space, θ-P space. Use this graph to predict how the rod will behave at the equilibria away from the vertical position. Decide whether or not the system always collapses once the load reaches the critical value at the bifurcation.

13.3 LINEAR ELASTIC COLUMNS

The examples in the previous section present structures that are composed of rigid members and that are described by a finite number of state variables. Deformable bodies, on the other hand, require an infinite number of state variables, or degrees of freedom, to describe their change in shape under loads. A column is a simple deformable body that can undergo a qualitative shape change as the load varies. The shape of the deformed column is usually described by the displacement of the centroid of each cross section. The infinite number of state variables are the transverse displacements, $w(x)$, of the centroids of faces indexed by the variable, x, for $0 \le x \le L$, where L is the length of the column.

The contraction of the linear elastic column under the compressive axial load is neglected in simple column analyses, to isolate the consequences of bending. Even though the column can bend, it does not change length due to the compressive effects of the load. Such an idealized column is often called an inextensible column since the length of the neutral surface is unchanged by the load.

The study of column deformations was initiated by Leonhard Euler. Euler (1707–1783) spent his career in two of the major mathematical centers of his century, St. Petersburg in Russia and Berlin in Prussia. After getting his degree at the University of Basel, where the mathematician John Bernoulli taught, he went to St. Petersburg in 1727; the Russian Academy of Science was founded there in 1725 and staffed by the Bernoulli brothers, Daniel and Nicholas, sons of John. He emphasized analytic rather than the fashionable geometric solutions to mechanics problems. He was enticed to the Prussian Academy in Berlin in 1741 by the new king of Prussia, Frederick the Great, but returned to St. Petersburg 25 years later in 1766 at the invitation of Catherine II of Russia. Some of his early work was on elastic curves, called elastica. These can be thought of as columns with zero cross-sectional area, but still having flexural stiffness. To study these, Euler invented the calculus of variations and used Daniel Bernoulli's suggested form for the potential energy of an elastic curve to analyze columns. This work was also based on Jacob Bernoulli's assumption that the bending moment in a beam is proportional to its curvature. In 1744, Euler published a book in which he derived the deflection of the free end of a cantilever beam as $PL^3/3C$ and the critical load for a fixed-free column as $C\pi^2/4L^2$. Euler called the constant, C, the absolute elasticity and his suggestion that his equations be used in an experimental determination of C were followed by many researchers. The constant, C, is the product of the elastic modulus and the area moment of inertia, now called the flexural stiffness. Later, in a 1757 publication, Euler developed the approximate differential equation, $C(d^2w/dx^2) = -Pw$ for the column deflection. The axial load is P and the transverse column displacement is w.

The curvature of a line is used to calculate the exact displacement of a buckled column (Fig. 13.7). (In Chapter 4, Section 4.9, the curvature of a surface was used in the description of the elastic Hertzian contact stresses.) Let s be the arclength along the column and $\psi(s)$ be the angle between the tangent to the column and the vertical at position s. The curvature of the column, viewed as a line (an elastica), is by definition $d\psi/ds$.

FIGURE 13.7   The angle of curvature, $\psi(s)$, of a column (elastica).

By assumption, the internal bending moment is $M = -EI/\rho(s)$, where ρ is the radius of curvature, the reciprocal of the curvature, $1/\rho(s) = d\psi/ds$. Therefore the differential equation describing the deflected column is

$$EI\frac{d\psi}{ds} = M(x).\tag{13.8}$$

To obtain an equation that produces the critical load, the curvature is linearized by assuming that dw/dx is small.

$$\frac{d\psi}{ds} = \frac{\frac{d^2w}{dx^2}}{\sqrt{1+\left(\frac{dw}{dx}\right)^2}} = \frac{d^2w}{dx^2}.$$

The underlying physical assumption is that the buckled displacement from the straight position cannot be very large so that the slope is small. This is not a large displacement theory. Substituting in equation (13.8) produces the linear differential equation,

$$EI\frac{d^2w}{dx^2} = M(x). \tag{13.9}$$

13.3.1 The Critical Buckling Load for Various Boundary Conditions

The column differential equation (13.9) is valid for the undeformed column in which $w(x)$ and $M(x)$ are zero for all x. By assuming that the small $w(x)$ is nonzero the equation can be solved for the critical load at which the column begins to bend. The equation cannot produce the postbuckling displacement. Its solution for the critical load depends on the supports for the column. These supports determine the boundary conditions for the differential equation.

The expression for the internal bending moment at each cross-sectional face in the column depends on the supports. First consider a column supported by frictionless pins at its ends. The internal moment, M, is written in terms of the load and the transverse deflection as $M(x) = -Pw(x)$. Substituting in equation (13.9) produces the linear ordinary differential equation,

$$EI\frac{d^2w}{dx^2} + Pw = 0. \tag{13.10}$$

This homogeneous differential equation has solution,

$$w(x) = A\sin(kx) + B\cos(kx), \tag{13.11}$$

where $k = \sqrt{P/EI}$. This form of the differential equation is identical to that of the linearized harmonic oscillator studied in linear vibrations courses.

A column with pin–pin supports has zero deflection at each end. Applying these two conditions in equation (13.11) determines the unknown coefficients, A and B. If $x = 0$, then $w(0) = 0$ implies that $B = 0$. The coefficient, A, must be nonzero if there is to be any deflection. If $x = L$, then $w(L) = 0$ implies that $0 = A\sin(kL)$. Therefore because $A \neq 0$, $\sin(kL) = 0$ and

$$kL = n\pi, \quad \text{for} \quad n = 1, 2, 3, \ldots . \tag{13.12}$$

The minimum possible load is that when $n = 1$. Using $k = \sqrt{P/EI}$ and squaring

$$P_c = \frac{\pi^2 EI}{L^2}. \tag{13.13}$$

The pin–pin column postbuckling solution for $w(x)$ is therefore

$$w(x) = A\sin(\sqrt{P/EI}x). \tag{13.14}$$

It is not possible to determine the amplitude, A. In fact, the solution (13.14) only satisfies the boundary conditions at P_c; at any other value of $P > 0$, it does not. Equation (13.14) is not the displacement equation for the buckled column. This information is lost when the differential equation, [equation (13.8)], is linearized. Only the critical load can be determined in this manner. As will be shown later, the correct linearized approximation to the postbuckling displacement for a pin–pin supported column is $w(x) = a \sin(\pi x/L)$.

Engineers sometimes wish to write the critical load for columns with other boundary conditions in the same form as that for the pin–pin column.

Definition The effective length, L', of a column with arbitrary supports is the length required to write the critical load in the same form as that for a pin–pin supported column.

The effective length, then, depends on the boundary conditions, that is, the supports, for the column.

EXAMPLE 13.6

A column with fixed-free supports is given coordinates with $x = 0$ at the free end. The boundary conditions are that $w(L) = 0$ and $w'(L) = 0$. Let the displacement at the free end be $w(0) = \delta$. The internal bending moment is then $M(x) = P[\delta - w(x)]$, and from equation (13.9) the governing differential equation is

$$EI\frac{d^2w}{dx^2} + Pw = P\delta.$$

This has solution of the form

$$w(x) = A \sin(kx) + B \cos(kx) + \delta,$$

for $k = \sqrt{P/EI}$. The two boundary conditions imply, after some algebra, that $A = -\delta \sin(kL)$ and $B = -\delta \cos(kL)$. The deflection at the free end is

$$\delta = w(0) = \delta - \delta \cos(kL).$$

This implies that $\cos(kL) = 0$ and therefore that $kL = n\pi/2$ for some integer n. Taking $n = 1$,

$$P_c = \frac{\pi^2 EI}{4L^2}.$$

The effective length, $L' = 2L$. The critical load, written in terms of the effective length, L', has the same form as equation (13.13),

$$P_c = \frac{\pi^2 EI}{(L')^2}. \qquad \blacksquare$$

The linear analysis suggests that a buckled column can take other shapes corresponding to values of k obtained when n is taken larger than 1, often called the higher buckling modes. But the higher modes are in a sense a mathematical fiction for static loads. A higher mode can only occur if the column is somehow constrained by lateral loads or supports to prevent the appearance of all the lower modes.

The critical load, since it depends on the elastic modulus, also depends on the environment in which the column is used. The elastic modulus for metals depends on the temperature of the material. For wood and other cellulosic materials, the elastic modulus in the Euler equation varies with both the material moisture content and temperature. The elastic modulus decreases with increasing temperature and with increasing moisture content. For example, A. Nissan has proposed an exponential model for the isothermal variation of the elastic modulus of paper with moisture content, m.

$$E(m) = E_0 \exp(a - bm), \tag{13.15}$$

where E_0, a, and b are material constants. Structural columns made of paper have been experimented with.

13.3.2 Design of Columns

Design analysis requires knowledge of the state of stress in the members. If the material is to remain elastically loaded, the stress cannot exceed the yield stress, σ_Y. In the design of elastically loaded slender members, the critical buckling stress as well as the yield stress may not be exceeded.

To make the calculation of the critical buckling axial normal stress in the column easier, the critical load is often rewritten in terms of the radius of gyration, r.

Definition   The radius of gyration of an area about an axis is the number, r, such that the area moment of inertia about that axis is $I = Ar^2$.

The radius of gyration depends on the axis about which the area moment of inertia is taken. This radius is similar to the radius of gyration for the mass moment of inertia, used in dynamics. In that case, the mass moment of inertia is $I = mr^2$, where m is the total mass of the body.

EXAMPLE 13.7   The radius of gyration for a circular cross section of radius R about its diameter is obtained from $I = \pi R^4/4 = (\pi R^2)r^2$. Therefore the radius of gyration for the area moment of inertia about a diameter is half the radius, $r = R/2$. ∎

EXAMPLE 13.8   Determine the radius of gyration for a rectangular cross section of sides, b and h, about the centroidal axis parallel to the side of length b.

Solution   For the rectangle with base, b, and height, h,

$$r = \sqrt{\frac{I}{A}} = \sqrt{\frac{\frac{bh^3}{12}}{bh}} = \frac{h}{\sqrt{12}}.$$

The Euler formula in terms of the critical stress, $\sigma_c = P_c/A$, is

$$\sigma_c = \frac{P_c}{A} = \frac{\pi^2 E r^2 A}{AL^2} = \frac{\pi^2 E}{(L/r)^2}. \tag{13.16}$$

∎

The term, L/r, is often called the slenderness ratio and can be considered a design parameter along with the elastic modulus, E. The larger the slenderness ratio, the smaller the stress required to buckle the column.

EXAMPLE 13.9   In the three-member truss shown (Fig. 13.8), determine the minimum cross-sectional dimensions required to prevent both buckling and yielding in member BC, which has a square cross section and is made of aluminum ($E = 70$ GPa, $\sigma_Y = 100$ MPa).

Solution   The compressive member BC in the truss is long and slender and so can be thought of as a column. Statics can be used to estimate the axial load on the column. The applied load is $P =$

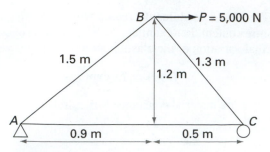

FIGURE 13.8   A three-member truss under a horizontal load, *P*.

5,000 N. An equilibrium analysis of the free-body diagram shows that the force F in member BC is compressive and has magnitude $F = 13P/14 = 4{,}643$ N.

Let a be the unknown length of the edge of the cross section. To prevent yielding,

$$\sigma_Y \geq \frac{P}{a^2}$$

so that the edge length must satisfy

$$a \geq \sqrt{\frac{P}{\sigma_Y}} = 0.00681 \text{ meters,}$$

the cross section must be large enough that the applied load is less than the critical buckling load. The column has pin–pin supports since truss analysis assumes each truss joint is a pin in order that each member be two force. The length of BC is $L = 1.3$ meters.

$$P \leq \frac{\pi^2 EI}{L^2} = \frac{\pi^2 E}{L^2}\frac{a^4}{12}.$$

Solving for a,

$$a \geq \sqrt[4]{\frac{12PL^2}{\pi^2 E}} = 0.01921 \text{ meter.}$$

The cross section must have dimensions greater or equal to the larger of the two computed minimal values of a. Therefore, buckling controls the design, and the cross-sectional dimension must be greater than 0.01921 meter.

The radius of gyration of a square cross section about a centroidal axis parallel to an edge is $a/\sqrt{12}$. Therefore, the slenderness ratio for the smallest allowable edge length is $L/r = 234.4$. Any larger cross-sectional edge reduces the slenderness ratio.   ∎

Whether a column can carry more or less load in the buckled position depends on the stability of the postbuckling states. As will be shown by an energy function analysis of column buckling, if the column is stable in a buckled position, it carries a load greater than the critical load. The designer might actually allow the column to buckle slightly, possibly as a safety factor. If the column postbuckling states are unstable, the column will collapse if the critical load is exceeded. Unfortunately, the linearized buckling equations do not permit an investigation of the design stability. Nonlinear methods are required.

13.3.3   The Critical Buckling Load by the Rayleigh-Ritz Method

This technique is based on the fact that the energy function corresponding to the critical load has a singularity. The strain energy portion, at least, of the potential energy function for a deformable body is expressed as an integral of the unknown equilibrium displacement. By choosing an approximation to the displacement, the energy function can be written in closed form, rather than as an integral. The value of the load producing the required singularity is then easily determined. This process also indicates how the finite element method (Chapter 8), which is based on the Rayleigh-Ritz technique, can be used to determine critical buckling loads.

The column is assumed to be elastic so that it is a conservative mechanical system. It is an infinite degrees of freedom system since the location of each point of the column must be independently specified to describe the shape of the column. Its potential energy function is the sum of a strain energy term due to bending and a term that is the negative of the work of the load. The strain energy for a column in which plane sections remain plane in bending is, as discussed in Chapter 7, Section 7.6, on energy methods,

$$\frac{1}{2} \int_0^L EI \left(\frac{d^2w}{dx^2} \right)^2 dx.$$

The distance through which the axial load, P, moves is the sum of the changes in length of each incremental piece of the column.

$$\int_0^L (ds - dx) = \int_0^L \left(\sqrt{1 + \left(\frac{dw}{dx} \right)^2} - 1 \right) dx = \frac{1}{2} \int_0^L \left(\frac{dw}{dx} \right)^2 dx,$$

where the latter equality is obtained from the Taylor series approximation for small dw/dx,

$$\sqrt{1 + \left(\frac{dw}{dx} \right)^2} = 1 + \frac{1}{2} \left(\frac{dw}{dx} \right)^2 - \dots .$$

The energy function for a buckled column, which depends on the displacement, is then

$$\Pi(w(x)) = \int_0^L \frac{1}{2} EI \left(\frac{d^2w}{dx^2} \right)^2 dx - P \frac{1}{2} \int_0^L \left(\frac{dw}{dx} \right)^2 dx. \tag{13.17}$$

The energy method's attack to determine the behavior of the column seeks an equilibrium configuration, $w(x)$, which is an extremal value for the integral energy function [equation (13.17)]. The Rayleigh-Ritz approximation to this technique requires the choice of a form for the displacement, $w(x)$. The energy function is then written in terms of this $w(x)$ and its undetermined coefficients. The resulting approximate energy function is minimized with respect to the undetermined coefficients. In this way, the original infinite degrees of freedom problem is reduced to a finite degrees of freedom problem which can be handled with elementary calculus.

EXAMPLE 13.10   Use the Rayleigh-Ritz method to estimate the critical buckling load for a pin–pin supported column of length, L, and flexural stiffness, EI.

Solution   The displacement function chosen must satisfy the boundary conditions, $w(0) = w(L) = 0$. One simple possibility is a parabola with an undetermined coefficient, a, so that $w(x) = a[(L/2)^2 - (x - (L/2))^2] = a[Lx - x^2]$. With this displacement, the potential energy function [equation (13.17)] reduces to a function of the single scalar variable, a, after integrating.

$$\Pi(a) = a^2 \left(2EIL - \frac{1}{6} PL^3 \right).$$

This function has an equilibrium state when $d\Pi/da = 0$ so that $a = 0$. The column is undeflected. The singularity occurs at the load, P, which makes the term in the parentheses zero.

$$P_c = \frac{12EI}{L^2}.$$

The second derivative, $d^2\Pi/da^2$, determines the stability of these equilibria. The unbuckled configuration is stable if $P < \frac{12EI}{L^2}$ and unstable if $P > 12EI/L^2$. The transition from stable to unstable as P is increased from zero occurs at the critical load. The energy function was linearized by the choice of the strain energy term and by the use of a small displacement approximation to the distance the load P moves through, and so the buckled equilibrium states cannot be determined.

The exact critical load for a pin–pin supported column is known to be $P = \pi^2 EI/L^2$, which is about 18% less than the critical load predicted by this Rayleigh-Ritz approximation. This choice of displacement cannot be expected to produce a good approximation to the critical load since $d^2w/dx^2 = -2a$ implies that the moments at the ends are not zero.   ∎

The critical load computed by the Rayleigh-Ritz technique is always greater than the exact critical load since the critical load occurs at the minimum of the potential energy. As such, the exact critical load corresponding to the exact displacement field is smaller than any other including the critical load obtained from the Rayleigh-Ritz approximation to the displacement. This upper bound on the critical load is not very useful in design as the critical load estimate since a load lower than it which is also greater than the actual critical load causes buckling.

The Rayleigh-Ritz method can also be used to obtain an estimate of the critical buckling load when the column has a nonconstant cross section.

EXAMPLE 13.11   Estimate the critical buckling load if the cross section is circular with radius, $r(x) = r_0 + r_L x/L$, which varies linearly from r_0 at one end to $r_0 + r_L$ at the other end, so that the cross section is a function of position (Fig. 13.9). The column is pin–pin supported.

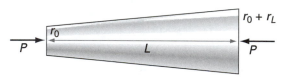

FIGURE 13.9   A long column with linearly varying circular cross section.

Solution   The bending is probably not symmetric about the column midpoint. But for the sake of comparison to the previous example, choose $w(x) = a[Lx - x^2]$. With this displacement, the potential energy function is the same as in the previous example except for the contribution of I in the strain energy.

$$\Pi(a) = a^2 \left[EL\pi \left(\frac{1}{10}r_L^4 + \frac{1}{2}r_L^3 r_0 + r_L^2 r_0^2 + r_L r_0^3 + \frac{1}{2}r_0^4 \right) - \frac{1}{6}PL^3 \right].$$

Again, the critical load occurs at the value of P for which the second derivative of Π equals zero,

$$P_c = \frac{6E\pi}{L^2} \left(\frac{1}{10}r_L^4 + \frac{1}{2}r_L^3 r_0 + r_L^2 r_0^2 + r_L r_0^3 + \frac{1}{2}r_0^4 \right).$$

This reduces to the critical load obtained in the previous example when $r_L = 0$.   ∎

13.3.4 The Exact Solution by Elliptical Integrals

The linearized ordinary differential equation studied in the preceding section only produces the critical load at which buckling is initiated. It suggests that that once buckling begins, the displacement suddenly becomes indeterminate or even infinite. This is contradicted by experience. The linearized equation cannot determine the magnitude of the displacement of the column after the critical load is exceeded. The displacement as a function of load can only be determined by considering the nonlinearized differential equation.

The assumption that plane cross sections remain plane implies that

$$\frac{EI}{\rho(s)} = -M(s), \tag{13.18}$$

where $\rho(s)$ is the column radius of curvature and $M(s)$ is the internal bending moment on the face at position s. By definition, the curvature $1/\rho(s) = d\psi/ds$. Along the column, $\sin\psi = dw/ds$ (Fig. 13.10). The origin is placed at a point where the bending moment is known to be zero. For example, the origin is at one end for a pin–pin supported column or at the free end for a fixed-free column. The internal bending moment is then $M(s) = Pw(s)$.

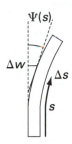

FIGURE 13.10 The displacement, Δw, and the curvature angle, $\psi(s)$, for a bent column.

After substituting the expressions for the curvature and moment, differentiate equation (13.18) with respect to s. Then substitute $\sin\psi$ for dw/ds to obtain the differential equation describing the deflected column.

$$EI\frac{d^2\psi}{ds^2} + P\sin\psi = 0. \tag{13.19}$$

This second-order differential equation is transformed to obtain a first-order differential equation for $d\psi/ds$. The differential equation is integrated once with respect to ψ. Then $d\psi = (d\psi/ds)ds$ is substituted into the left integral to get

$$\int \frac{d^2\psi}{ds^2}\frac{d\psi}{ds}ds = -k^2\int \sin\psi\, d\psi, \tag{13.20}$$

where $k = \sqrt{P/EI}$. But since

$$\int \frac{d^2\psi}{ds^2}\frac{d\psi}{ds}ds = \frac{1}{2}\int \frac{d}{ds}\left(\frac{d\psi}{ds}\right)^2 ds,$$

integration yields

$$\frac{1}{2}\left(\frac{d\psi}{ds}\right)^2 = k^2\cos\psi + C. \tag{13.21}$$

The constant, C, must be determined later from the boundary conditions on the column. Equation (13.21) produces the first-order differential equation for the curvature,

$$\frac{d\psi}{ds} = \pm k\sqrt{2}\sqrt{\cos\psi - C}. \tag{13.22}$$

The plus or minus sign is taken in front of the square root depending on whether $d\psi/ds$ is always positive or negative, that is, whether the angle, ψ, is increasing or decreasing in the coordinates chosen.

Integrate by separating variables to obtain

$$L = \frac{1}{k}\int_{\alpha}^{\beta}\frac{d\psi}{\sqrt{2}\sqrt{\cos\psi - C}}, \tag{13.23}$$

where $\alpha = \psi(0)$ and $\beta = \psi(L)$.

The determination of C and a variable change depending on the boundary conditions often produce a complete elliptic integral of the first kind,

$$L = \frac{1}{k}\int_{0}^{\pi/2}\frac{d\phi}{\sqrt{1 - p^2\sin^2\phi}} = \frac{1}{k}K(p). \tag{13.24}$$

This equation provides the relation between the applied load and the angle ψ, from which the displacement at a point in the column may be computed. The complete elliptic integral on the right side of this equation, $K(p)$, is solved numerically and can be found in many symbolic mathematics computer packages or in handbooks of mathematical functions. The tables are used to compute the value of p and then this is used to obtain the exact deflection.

EXAMPLE 13.12

Compute the exact deflection of the free end for a fixed-free column of length, $L = 0.5$ m, and flexural stiffness, $EI = 2$ N-m^2, when the column is subjected to a load of $P = 1.05P_c$.

Solution

The boundary conditions are that $\psi = \beta$ at the free end, where β is not known, and zero at the fixed end. Since the moment is zero at the free end, $d\psi/ds = 0$ at the free end because $M(s) = EI d\psi/ds$. The equation, assuming that the parameter s is measured from the fixed end,

$$\frac{d\psi}{ds} = k\sqrt{2}\sqrt{\cos\psi - C}$$

applied at the free end implies that $C = \cos\beta$. Therefore

$$L = \int_{0}^{\beta}\frac{d\psi}{k\sqrt{2}\sqrt{\cos\psi - \cos\beta}}$$

$$= \int_{0}^{\beta}\frac{d\psi}{k\sqrt{\sin^2(\beta/2) - \sin^2(\psi/2)}}$$

since $\cos x = [1 - \sin^2(x/2)]/2$. Make the variable change

$$\sin(\psi/2) = \sin(\beta/2)\sin(\phi),$$

and put $p = \sin(\beta/2)$, so that

$$d\psi = \frac{2p\cos\phi d\phi}{\cos(\psi/2)} = \frac{2p\cos\phi d\phi}{\sqrt{1 - p^2\sin^2\phi}}.$$

The value of p is not yet known since the angle β is not yet known. Substituting,

$$L = \frac{1}{k} \int_0^{\pi/2} \frac{d\phi}{\sqrt{1 - p^2 \sin^2 \phi}} = \frac{1}{k} K(p).$$

Since $P_c = \pi^2 EI/4L^2$, the term

$$k = \sqrt{\frac{P}{EI}} = \frac{\pi}{2L} \sqrt{\frac{P}{P_c}}.$$

Therefore

$$\frac{\pi}{2} \sqrt{\frac{P}{P_c}} = kL = K(p).$$

By a table of elliptical integrals of the first kind, the value of p corresponding to $P/P_c = 1.05$ and $K(p) = \pi\sqrt{1.05}/2 = 1.609587$ is $p = 0.09355$. Therefore $\beta = 2\arcsin(p) = 0.1871$ radians or $10.736°$. The value of β appears to be independent of the length only because the load is given as a multiple of the critical load, which depends on the length.

Because $\sin \psi = dw/ds$, the horizontal deflection, w_h, of the free end is

$$w_h = \int_0^L \sin \psi \, ds = \int_0^\beta \frac{\sin \psi \, d\psi}{k\sqrt{2}\sqrt{\cos \psi - \cos \beta}}$$

$$= \int_0^\beta \frac{\sin \psi \, d\psi}{2k\sqrt{\sin^2(\beta/2) - \sin^2(\psi/2)}}.$$

Make the coordinate change $\sin(\psi/2) = \sin(\beta/2)\sin(\phi)$, with $p = \sin(\beta/2)$, and use $k = K(p)/L = 2(1.609587)$,

$$w_h = \frac{2p}{k} \int_0^{\pi/2} \sin \phi \, d\phi = \frac{2p}{k} = \frac{2pL}{K(p)} = \frac{2(0.09355)}{2(1.609587)} = 0.05812 \text{ m}. \qquad (13.25)$$

■

EXAMPLE 13.13   Repeat the previous example for a pin–pin supported column.

Solution   Now the boundary conditions are that the moments are zero at the ends so that $d\psi/ds = 0$ at both ends. By symmetry, the angle at one pinned end is β and the angle at the other end is $-\beta$, where β depends on the axial load,

$$L = \frac{1}{k} \int_{-\beta}^{\beta} \frac{d\psi}{\sqrt{2}\sqrt{\cos \psi - C}} = \frac{2}{k} \int_0^\beta \frac{d\psi}{\sqrt{2}\sqrt{\cos \psi - C}}.$$

Therefore,

$$\frac{\pi}{2} \sqrt{\frac{P}{P_c}} = kL = 2K(p).$$

This example is to be completed as problem 13.

■

13.4   PERTURBATION METHODS

The classical analysis linearizes the governing differential equation to successfully produce the critical value of the load on a column. Linearization loses information; in particular, the linear model cannot determine how far the column will bend under a given load. To compute the deflection, the differential equation was solved exactly in terms of elliptical integrals. Neither of the techniques using the linearized equation or the exact elliptical integral can directly determine the stability of the postbuckling equilibrium states. Using the elliptical integral to calculate the displacement corresponding to several loads higher than the critical load can indicate the stability behavior. To examine stability efficiently, an energy function is written for the system so that one may determine stability of an equilibria by whether it is a minimum or maximum of the energy function. Locating the equilibria of an integral function of a function requires solving the associated Euler-Lagrange differential equation (see Chapter 7, Section 7.8).

If the governing equation for the column is not linearized, the resulting nonlinear Euler-Lagrange differential equation can seldom be solved in closed form. A useful approximation technique is a perturbation analysis in which the solution of the nonlinear partial differential equation is reduced to solving a recursive sequence of relatively simple linear ordinary differential equations. Thompson and Hunt [3, pp. 27–35] showed how to do this for the linear elastic column, and Haslach [1] developed the nonlinear elastic case of a cubic stress-strain curve.

The postbuckling behavior of a column with linear uniaxial constitutive equation, $\sigma = E\epsilon$, can be studied in the pure bending case when the longitudinal compressive stress at the bending neutral axis is neglected, so that the column is inextensible. Since only the initial postbuckling behavior is of concern, the strains are assumed to be small; however, large displacements are still acceptable. Assume that plane sections remain planar and perpendicular to the deformed longitudinal axis and ignore any shear strain. In dealing with elastic structures that buckle by bending, one must be concerned with whether or not the material behaves the same in tension and compression. Here it is assumed that it does.

The distance that the load moves through, used to obtain the potential energy of the load, is the difference of the initial column length, L, and the final column length. Denote by $w(x)$ the displacement transverse to the longitudinal direction of the column. w' is the derivative with respect to the longitudinal coordinate, x. Since the column is inextensible, the column can be parametrized by x rather than s. Therefore, the distance the load, P, moves through is the initial length minus the length of the buckled column projected onto the line of action of P.

$$L - \int_0^L (1 - w'^2)^{1/2}]dx = \int_0^L [1 - (1 - w'^2)^{1/2}]dx.$$

Since plane sections remain plane, the axial strain is a function of the curvature, $1/\rho = d\psi/ds$. Because $dw/dx = \sin\psi$, the curvature is written as

$$\frac{1}{\rho} = \frac{d(\arcsin(w'))}{dx} = w''(1 - w'^2)^{-1/2}.$$

Notice that the curvature is not linearized to w''. The strain-displacement relation is

$$\epsilon_x(x, y, z) = \frac{z}{\rho} = zw''(1 - w'^2)^{-1/2}, \tag{13.26}$$

where z is the coordinate perpendicular to the bending axis and the longitudinal coordinate.

The linear constitutive model for axial strain produces the strain energy density, $U_o = \int \sigma d\epsilon = (1/2)E\epsilon^2$. The strain energy, for cross-sectional area A, is

$$U = \int \left(\frac{1}{2} E\epsilon^2 \right) dx dA.$$

Put $I = \int_A z^2 dA$. It is assumed, when computing values for I that, near the critical buckling load, the bending axis remains at the centroid of the cross section. In this case, the potential energy becomes, after substituting equation (13.26) in the strain energy expression,

$$V = \int_0^L [0.5 EI w''^2 (1 - w'^2)^{-1}] dx - P \int_0^L [1 - (1 - w'^2)^{1/2}] dx. \qquad (13.27)$$

The Euler-Lagrange equation is obtained from V,

$$EI[w''''(1 - w'^2)^{-1} + 4w'w''w'''(1 - w'^2)^{-2} + w''^3(1 + 3w'^2)(1 - w'^2)^{-3}]$$
$$+ Pw''(1 - w'^2)^{-3/2} = 0. \qquad (13.28)$$

Exercise 13.5 Verify that this is the correct Euler equation. Verify that its linearization is $EI w'''' + P w'' = 0$ by assuming small w'. Compare this to the linearized differential equation used previously to compute the critical load for column buckling.

The natural boundary conditions for w in the pin–pin case are $w''(0) = w''(L) = 0$. The forced boundary conditions are $w(0) = w(L) = 0$.

13.4.1 Perturbation Technique

The Euler equations are solved by a perturbation technique that reduces the number of degrees of freedom in the energy function from infinite to one. A perturbation from the unbuckled state is written in terms of a perturbation parameter, b.

$$w(x) = w_1 b + w_2 b^2 + w_3 b^3 + w_4 b^4 + \dots;$$
$$P = P_o + P_1 b + P_2 b^2 + P_3 b^3 + P_4 b^4 + \dots, \qquad (13.29)$$

where the w_n are functions of x. The perturbation parameter b is zero when the column is at the critical buckling load. The additional assumption of the "false" boundary conditions

$$w_1(L/2) = 1;$$
$$w_n(L/2) = 0, \qquad \text{for } n \geq 2 \qquad (13.30)$$

forces the perturbation parameter, b, to be the midcolumn transverse displacement since the column is assumed to be pin–pin supported. Because b must be zero at buckling, P_o is the critical buckling load. Note that in the formulation given here P and P_o are assumed to be positive.

To simplify the calculation, write Taylor series expansions of the following terms about the undeformed configuration, $w' = 0$,

$$(1 - w'^2)^{-1} = 1 + w'^2 + w'^4 + \ldots ;$$
$$(1 - w'^2)^{-2} = 1 + 2w'^2 + 4w'^4 + \ldots ;$$
$$(1 - w'^2)^{-3} = 1 + 3w'^2 + 6w'^4 + \ldots ;$$
$$(1 - w'^2)^{-3/2} = 1 + \frac{3}{2}w'^2 + \frac{15}{8}w'^4 + \ldots .$$

Substitution of equations (13.29), the perturbation expansions of w and P, and the Taylor series expansions above into the Euler equation for w obtained from the energy function equation (13.27) produces a polynomial in b which equals zero. The only way that it can be zero is for the coefficient of each power of b to itself be zero. Each of these is a linear differential equation involving the w_i for $i \leq n$. These equations are solved sequentially beginning with the coefficient of the b, then the coefficient of b^2, etc. to produce formulas for the P_n and w_n.

Exercise 13.6   Carry out the substitution into the Euler equation for w to obtain the polynomial in b.

The coefficient of b is

$$EI w_1'''' + P_0 w_1'' = 0.$$

Therefore $w_1(x)$ has the form

$$w_1(x) = A_1 \sin(\beta x) + B_1 \cos(\beta x) + Cx + F,$$

where $\beta = \sqrt{P_0/EI}$. The natural and forced boundary conditions imply that

$$0 = B_1 + F$$
$$0 = A_1 \sin(\beta L) + B_1 \cos(\beta L) + CL + F$$
$$0 = -B_1 \beta^2$$
$$0 = -A_1 \sin(\beta L) - B_1 \cos(\beta L).$$

Therefore $B_1 = C = F = 0$ and $L\sqrt{P_0/EI} = \pi$. Consequently, $w_1(x) = A_1 \sin(\beta x)$ and $P_0 = \pi^2 EI/L^2$, the critical load.

Determination of the w_n for $n > 1$ requires the solution of an ordinary differential equation of the type

$$w'''' + D^2 w'' = k \sin(Dx) + m \sin(3Dx) + n \sin(5Dx) + \ldots , \qquad (13.31)$$

where $D = \pi/L$. The solution has the form

$$w(x) = A \sin(Dx) + B \cos(Dx) + Cx + F$$
$$+ Kx \cos(Dx) + M \sin(3Dx) + N \sin(5Dx) + \ldots ,$$

where $K = k/2D^3$, $M = m/72D^4$, $N = n/600D^4$.

Exercise 13.7   Verify that this $w(x)$ is the solution of the differential equation (13.31).

Set $K = 0$ to suppress the secular term, that is, the term that tends to infinity as x increases. Such growth in the displacement is physically impossible for the column. The equation $K = 0$ determines P_n in terms of the w_i for $i \leq n$.

The coefficient of b^2 is

$$EIw_2'''' + P_0 w_2'' + P_1 w_1'' = 0.$$

After substituting $w_1 = A_1 \sin(\beta x)$, the coefficient of $\sin(Dx)$ is $k_3 = A_1 P_1$. This secular term is suppressed by setting $k_3 = 0$ and so $P_1 = 0$. The solution to the resulting harmonic equation is either $w_2(x) = 0$ or $w_2 = A_2 \sin(\pi x / L)$. But the false boundary condition requires that $w_2(L/2) = 0$. Therefore $w_2(x) = 0$.

The coefficient of b^3 is

$$EI[w_3'''' + w_1'^2 w_1'''' + 4w_1' w_1'' w_1''' + w_1''^3] + P_0 w_3'' + \frac{3}{2} P_0 w_1'^2 w_1'' + P_2 w_1'' = 0,$$

which becomes after substitution of the known expressions for w_1 and P_0,

$$EI[w_3'''' + D^6 \cos^2(Dx)\sin(Dx) + 4D^6 \cos^2(Dx)\sin(Dx) - D^6 \sin^3(Dx)] + P_0 w_3''$$
$$- \frac{3}{2} P_0 D^4 \cos^2(Dx)\sin(Dx) - P_2 D^2 \sin(Dx) = 0.$$

The trigonometric identity

$$\sin(3x) = 3\sin(x) - 4\sin^3(x)$$

implies that

$$\sin^3(x) = \frac{3}{4}\sin(x) - \frac{1}{4}\sin(3x);$$

$$\cos^2(x)\sin(x) = \frac{1}{4}\sin(x) + \frac{1}{4}\sin(3x).$$

Substituting,

$$EIw_3'''' + P_0 w_3'' = -k_3 \sin(\pi x / L) - m_3 \sin(3\pi x / L),$$

where $k_3 = D^6 EI/8 - D^2 P_2$ and $m_3 = 9D^6 EI/8$. Therefore suppressing the secular term by putting $k_3 = 0$ yields $P_2 = EID^4/8$. Applying the boundary conditions gives

$$w_3(x) = -\frac{D^2 EI}{64}[\sin(\pi x / L) + \sin(3\pi x / L)].$$

By symmetry, the fourth-order terms, w_4 and P_3 are zero. In fact, for this reason, all even power terms in the perturbation expansion for w and all odd power terms in that for P are zero.

Exercise 13.8   Plot the relation between the midcolumn displacement b and the load P, when $P > P_0$, by using the second-order approximation, $P = P_0 + P_2 b^2$.

Exercise 13.9   By plotting their graphs, compare the linearized solution $w(x) = w_1(x)b$ and the third-order nonlinear solution, $w(x) = w_1(x)b + w_3(x)b^3$, for $0 \leq x \leq L$ and $b = 0.5$.

13.4.2 The Energy Function

The single degree of freedom potential energy function, V, of a pin–pin supported column for the inextensible model is obtained by substituting the respective perturbation expressions for $w(x)$ in the powers of b up to the fourth order, w_1 and w_3, into equation (13.27) for V and integrating. Write the result to the fourth order in b,

$$V = V_0 + V_2 b^2 + V_4 b^4. \tag{13.32}$$

The potential energy to the fourth order in b is then

$$V = \left[\frac{7}{128} \pi E I D^5 - \frac{5}{128} P \pi D^3 \right] b^4 + \left[\frac{1}{4} \pi E I D^3 - \frac{1}{4} P \pi D \right] b^2, \tag{13.33}$$

for $D = \pi/L$. The critical load, $P_c = EI(\pi/L)^2$, is recovered by setting the coefficient, V_2, of b^2 equal to zero since the singularity occurs at such a load. At the critical load, the usual second derivative test cannot be applied to determine stability since the second derivative is zero. Higher-order derivatives at $b = 0$ need to be considered. The fourth-order terms are sufficient to determine the stability at the critical and post-buckling loads. At the singularity, V_4 is the first nonzero term. The fourth derivative of $V(b)$ at $b = 0$, which is $24V_4$, is the lowest order nonzero derivative and is positive. Therefore the equilibrium is stable at the critical load, and the buckled equilibria at loads higher than the critical are also stable. This expression of the potential energy therefore permits determination of the stability at the critical load. None of the other methods can do so.

Exercise 13.10 Verify that this is the correct expression for the potential energy.

Exercise 13.11 Verify that the coefficient of b^4 is positive at the critical load, P_0.

13.5 BUCKLING OF COLUMNS THAT ARE NOT LINEAR ELASTIC

The Engesser theory predicts the critical buckling load for columns which are not linear elastic. A generalization of the perturbation method described in the previous section can also be used [1]. The Engesser method may be applied to either nonlinear elastic or plastically loaded columns. Friedrich R. Engesser (1848–1931) was a German engineer who designed many railroad bridges and in 1885 was elected to a professorship at Karlsruhe. He proposed, in an 1889 paper, the tangent modulus theory as an extension of the Euler theory for buckling critical loads. The tangent modulus for a uniaxial stress-strain curve is $E_T = d\sigma/d\epsilon$. On the linear portion of the stress-strain curve, $E_T = E$. About 10 years later, Engesser revised this theory to allow different moduli for the portion of the bent beam in compression and the portion in tension. The original proposal was that the critical load for a column that is not linear elastic is predicted by replacing the elastic modulus in the Euler expression for the critical load [equation (13.13)], by the tangent modulus, computed at the critical buckling strain.

$$\sigma_c = \frac{\pi^2 E_T}{(L/r)^2}.$$

To derive this result, assume that the material behaves the same in tension and compression. The column is also assumed to obey the strength of materials assumption that its planar cross sections remain planar in bending so that the strain is proportional to the curvature, $\epsilon = y/\rho$. Here ρ is the radius of curvature and y is the distance from the centroidal axis parallel to the neutral axis. Recall that the neutral axis only passes through the centroid when the material is linear elastic.

Assume that buckling occurs at stresses that depend nonlinearly on the strain. If the material is plastically loaded rather than nonlinearly elastically loaded at buckling, the load is increased monotonically from zero so that the nonlinear portion of the stress-strain curve is well defined. Recall that the magnitude of the plastic stresses is dependent on the way the material is loaded. The prebuckling stress distribution is constant across a face as the compressive load on the column is increased to the buckling load. After buckling occurs, the compressive stresses on the concave face of the beam are larger than those on the convex side. The concave side is in compression so that the bending adds additional compressive stress, while the convex side is in tension so that the compressive stresses are reduced there. Buckling is assumed to have just initiated at the critical load, P_c, so that the stress, σ_1, on the concave side differs only slightly from that on the convex side, σ_2 (Fig. 13.11). Then the stress difference can be approximated using the slope, E_T, of the tangent to the stress-strain curve at this strain.

$$\Delta\sigma = \sigma_1 - \sigma_2 = E_T \Delta\epsilon.$$

Since plane sections remain plane, $\Delta\epsilon = h/\rho$.

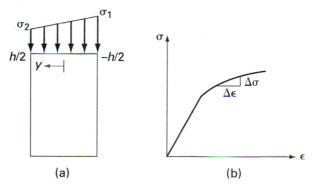

FIGURE 13.11   (a) The stress distribution across the beam face. (b) The stress-strain curve.

A free-body diagram shows that if $w(x)$ is the displacement of the pin–pin supported column at position, x, then $P_c w(x) = M(x)$. The internal bending moment, $M(x)$, can be computed from the stresses on the face. The stress is assumed to vary linearly across the face of the beam since the stress difference across the face is small. The width of the rectangular face is h. The relation is

$$\sigma(y) = \sigma_2 - m\left(y - \frac{h}{2}\right),$$

where the slope of the stress variation is $m = (\sigma_1 - \sigma_2)/h$. Then for beam thickness, t,

$$M(x) = \int_{-h/2}^{h/2} y\sigma(y)t\,dy = -mI = -\frac{\Delta\sigma}{h}I,$$

where $I = h^3t/12$ is the area moment of inertia about the centroidal axis (not the neutral axis). Using the facts that $\Delta\sigma = E_T\Delta\epsilon$, that $\Delta\epsilon = h/\rho$, and that for small slopes, dw/dx, the curvature is $1/\rho = d^2w/dx^2$, the moment is

$$M(x) = -\frac{\Delta\sigma}{h}I = -\frac{E_T\Delta\epsilon}{h}I = -\frac{E_Th}{\rho h}I = -E_T I\frac{d^2w}{dx^2}.$$

This result combined with $P_c w(x) = M(x)$ implies that the displacement of the column must satisfy the differential equation at the critical buckling load,

$$E_T I \frac{d^2 w}{dx^2} + P_c w(x) = 0.$$

This linear ordinary differential equation is exactly the same as that for the linearly elastic column except that E is replaced by E_T. The critical load is computed as before, and again the amplitude of the displacement cannot be determined since the result was obtained by linearizing the various relationships.

Dividing both sides by the cross-sectional area, A, of the beam produces the equation for the critical stress,

$$\sigma_c = \frac{\pi^2 E_T I}{AL^2} = \frac{\pi^2 E_T}{(L/r)^2}. \tag{13.34}$$

This proof for small strains applies to a linear elastic column except that the slope of the linear stress distribution across the beam face is E rather than E_T.

The tangent modulus, E_T, has a hidden dependence on the unknown critical buckling strain, ϵ_c. Since σ is also a nonlinear function of the strain, equation (13.34) is nonlinear in ϵ_c. Even if an analytical expression for the nonlinear portion of the stress curve is known, a numerical technique, such as Newton's method, is usually needed to determine the critical buckling strain, ϵ_c, at which the column starts to buckle.

EXAMPLE 13.14   Determine the critical buckling load for a column with slenderness ratio, $L/r = 150$, if the column is made of a material with stress-strain curve, $\sigma = 10^9 \epsilon^{1/2}$. The column is pin–pin supported.

Solution   The critical buckling load is given by $\sigma_c = \pi^2 E_T/(L/r)^2$, where $E_T = (1/2)10^9 \epsilon_c^{-1/2} = (1/2)10^{18}/\sigma_c$, since from the stress-strain curve, $\epsilon_c^{-1/2} = 10^9/\sigma_c$. Substituting into the Engesser equation for the critical stress produces

$$\sigma_c = \sqrt{\frac{\pi^2 (1/2) 10^{18}}{(L/r)^2}} = 14.81 \text{ MPa.} \qquad \blacksquare$$

EXAMPLE 13.15   Determine the critical buckling stress for a column with slenderness ratio, $L/r = 180$, if the column is made of redwood with stress-strain curve, $\sigma = E_1 \epsilon + E_3 \epsilon^3$, where $E_1 = 8.72$ GPa and $E_3 = -2{,}590$ GPa. The column is pin–pin supported. Note that, in this formulation, the compressive stresses and strains in the column are taken as positive.

Solution   The Engesser relation can be applied if all terms are expressed in terms of the critical strain, ϵ_c. The tangent modulus is $E_T = E_1 + 3E_3 \epsilon^2$. The Engesser equation becomes

$$E_1 \epsilon_c + E_3 \epsilon_c^3 = \frac{\pi^2}{(L/r)^2}(E_1 + 3E_3 \epsilon_c^2).$$

Rearranging gives a cubic polynomial

$$\left(\frac{L}{r}\right)^2 E_3 \epsilon_c^3 - \pi^2 3 E_3 \epsilon_c^2 + \left(\frac{L}{r}\right)^2 E_1 \epsilon_c - \pi^2 E_1 = 0,$$

or when the values of the constants are substituted,

$$\epsilon_c^3 - 9.1739 \times 10^{-4} \epsilon_c^2 - 3.3798 \times 10^{-3} \epsilon_c + 1.02956 \times 10^{-6} = 0.$$

This equation can be solved numerically using Newton's method, making an initial guess of a small value of ϵ_c, to obtain

$$\epsilon_c = 0.0003046.$$

The other two roots of the cubic equation are 0.058445 and -0.057832. Substitution of ϵ_c into the stress-strain relation gives $\sigma_c = 2.656$ MPa in compression. ∎

13.6 BUCKLING OF THIN-WALLED CIRCULAR RINGS AND TUBES

Rings and tubes are important engineering structures. Thin-walled rings and tubes are especially susceptible to buckling. A section of a thin ring can be thought of as a beam. Under the assumption that plane sections remain plane, the change in curvature at a cross section is proportional to the bending moment on the face, as in the derivation of the beam or column bending stress relations.

The thin ring is assumed to be initially circular with a radius of curvature, R. The position on the ring is defined by s, the arclength from a fixed point on the ring. The radius of curvature at position s on the ring after deformation is denoted by $\rho(s)$. The bending moment is positive if it increases the radius of curvature, or equivalently decreases the curvature. The deformation is primarily due to bending. The relation of the radius of curvature to the moment is then for a constant flexural rigidity, EI,

$$EI \left(\frac{1}{\rho(s)} - \frac{1}{R} \right) = -M(s). \tag{13.35}$$

Note that I is the area moment of inertia for the area of the radial face of the ring formed by cutting the ring with a radial plane.

It is easier to work in terms of the parameter, θ, using the fact that the arclength is $s = R\theta$. The radial deflection along the original radial line at position, θ, on the ring is $w(\theta)$, which is positive away from the center of the ring. In polar coordinates, the equation for the deformed ring is $r(\theta) = R - w(\theta)$. The curvature in terms of the parameter, θ, is obtained from the parametric form traditionally given in elementary calculus.

$$\frac{1}{\rho(\theta)} = \frac{r^2 + 2r'^2 - rr''}{(r^2 + r'^2)^{3/2}},$$

where $r' = dr/d\theta$ and $r'' = d^2r/d\theta^2$. The curvature is linearized by assuming that r' is small so that

$$\frac{1}{\rho(\theta)} = \frac{r^2 - rr''}{r^3} = \frac{r - r''}{r^2}.$$

Substitute $r(\theta) = R - w(\theta)$ and drop the term w^2 of second order from the denominator to obtain the classical linear expression for the curvature of a deformed ring.

$$\frac{1}{\rho(\theta)} = \frac{R - w + (d^2w/d\theta^2)}{R^2 - 2wR}.$$

Then get a common denominator and collect terms to write

$$\frac{1}{\rho(\theta)} - \frac{1}{R} = \frac{1}{R^2 - 2wR} \left(w + \frac{d^2w}{d\theta^2} \right).$$

Under the assumption that for small w, $R^2 - 2wR$ is approximately R^2, the differential equation, equation (13.35), governing the deformation of the ring becomes

$$\frac{d^2w}{d\theta^2} + w = -\frac{MR^2}{EI}. \tag{13.36}$$

The deflection, $w(\theta)$, resulting from a particular loading on the ring is determined once the internal bending moments, $M(\theta)$, caused by the loading are known.

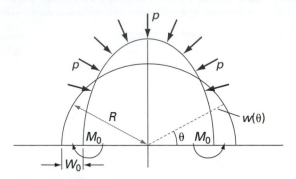

FIGURE 13.12   The deformed ring under constant hydrostatic pressure, p.

Under a large enough hydrostatic pressure P, per unit length of the centerline, the radial displacement, $w(\theta)$, of the ring is no longer the same at each point of the ring (Fig. 13.12). But the nonuniformly deformed ring maintains two orthogonal axes of symmetry. Cut the ring along the axis of symmetry on which the radial displacement is negative, so that the radius of curvature is shorter than originally. The free-body diagram shows that the compressive force, P, at the ends is

$$P = p(R - w_0),$$

where w_0 is the unknown change in radius of curvature at the end. The end moments are also unknown and denoted by M_0. The bending moment at position θ is then

$$M(\theta) = M_0 - pR[w_0 - w(\theta)].$$

Substituting this into the differential equation (13.36) for the displacement yields

$$\frac{d^2w}{d\theta^2} + w = -\frac{R^2}{EI}[M_0 - pR(w_0 - w)].$$

Collect the linear terms in w and let $k^2 = 1 + pR^3/EI$. The sum of the homogeneous and particular solutions produces a displacement of the form

$$w(\theta) = A\sin(k\theta) + B\cos(k\theta) + \frac{-M_0R^2 + pR^3w_0}{EI + pR^3}. \tag{13.37}$$

At the ends of the axis of symmetry, the displacement reaches an extremal value as a function of θ. Therefore $dw/d\theta = 0$ at $\theta = 0$ or $\pi/2$. Consequently, it must be true that $A = 0$ by the first condition. To have nonzero displacements $B \neq 0$ so that from the second condition,

$$\sin\left(\frac{k\pi}{2}\right) = 0.$$

The smallest nonzero value of k satisfying this equation is 2. Therefore, the thin ring has critical buckling compressive pressure, $2 = 1 + p_c R^3 / EI$, or

$$p_c = \frac{3EI}{R^3}. \tag{13.38}$$

This problem was first solved by M. Bresse in 1866. The corresponding value of the radial displacement from equation (13.37) is, using $w(0) = w_0$,

$$w(\theta) = \frac{1}{4}\left(\frac{M_0 R^2}{EI} + w_0\right)\cos(2\theta) - \frac{M_0 R^3}{4EI} + \frac{3w_0}{4}. \tag{13.39}$$

The ring is inextensible if the normal strain in the tangential direction written in polar coordinates is zero.

$$\epsilon(\theta) = \frac{1}{R}\left(\frac{dv}{d\theta} - w\right) = 0 \quad \text{for all } \theta,$$

where v is the tangential displacement in the θ direction. This produces the inextensibility condition, $w = dv/d\theta$. When the ring is assumed to be inextensible, then the tangential displacement is given by

$$v(\theta) = \frac{1}{8}\left(\frac{M_0 R^2}{EI} + w_0\right)\sin(2\theta) + \left(-\frac{M_0 R^3}{4EI} + \frac{3w_0}{4}\right)\theta. \tag{13.40}$$

It remains to determine M_0 and w_0. One can be related to the other by the fact that at the axis of symmetry, $v(0) = v(\pi/2) = 0$ by the symmetry of the deformation. Therefore,

$$M_0 = p_c w_0 R.$$

Again, since the problem was linearized, the displacements can only be found up to an undetermined coefficient, w_0.

$$w(\theta) = w_0 \cos(2\theta);$$
$$v(\theta) = \frac{1}{2}w_0 \sin(2\theta).$$

The so-called higher modes obtained by letting k take larger positive values are physically unrealizable unless additional displacement constraints are applied to the ring (see also [5, Sec. 7.3]).

Exercise 13.12   Sketch a graph of the radial displacement, $w(\theta)$, superimposed on the original circular cross section.

A thin-walled tube under water or inside a pressure vessel could be subjected to a hydrostatic pressure, p, on its external surface which is capable of buckling the tube. The thin ring analysis can be extended to a thin-walled tube. The thin ring is in plane stress. The calculation of the critical pressure for a thin-walled tube is the same except that the thin-walled tube is taken to be in plane strain so that in the ring formula E is replaced by $E/(1 - v^2)$.

The area moment of inertia is that of the radial face of the cylinder. This area has length equal to that of the cylinder and height equal to the thickness, t, of the cylinder wall. The moment of inertia per unit longitudinal length of the cylinder is therefore $I = t^3/12$. The critical compressive pressure is

$$p_c = \frac{3EI}{(1-\nu^2)R^3} = \frac{E}{4(1-\nu^2)}\left(\frac{t}{R}\right)^3. \tag{13.41}$$

Design can be carried out in terms of the thickness to radius ratio.

EXAMPLE 13.16   Design the minimum wall thickness for a pipe made of a steel for which $E = 30 \times 10^6$ psi and $\nu = 0.25$. The radius must be $R = 6$ inches to carry the required amount of fluid. The pipe will operate below the ocean surface where it must support an external hydrostatic pressure of $p = 150$ psi. Assume a factor of safety of 2.

Solution   The factor of safety requires that the design be for a maximum pressure of $p_c = 2p = 300$. Equation (13.41) implies that

$$t^3 = p_c R^3 \frac{4(1-\nu^2)}{E} = 0.0081.$$

Therefore $t = 0.20083$ inches.   ∎

13.7   IMPERFECTIONS

Any real structure has imperfections. For example, the axial load on a column may be slightly offset from the centroid of the cross section, or the column might be slightly bent before it is loaded. The analyses of the perfect system given in the previous sections must be modified to account for imperfections since, in those cases that the critical equilibrium state is unstable, the presence of imperfections can drastically lower the critical load and can lead to an unexpected catastrophic collapse of the structure.

13.7.1   Initially Curved Columns

A column that is slightly curved before the load is applied can be thought of as imperfect. Application of an axial load in this case merely increases the bending of the column. There is no qualitative change of behavior as occurs when a straight column begins to bend at the critical load. If the original shape of the column is $w_0(x)$, then the linearized differential equation governing the column of constant cross section is

$$EI\frac{d^2w}{dx^2} = -M(x) = -P(w_0 + w);$$

so that

$$EI\frac{d^2w}{dx^2} + Pw = -Pw_0. \tag{13.42}$$

The total displacement increment, w, is the sum of the homogeneous and particular solutions. The homogeneous solution is

$$w_h(x) = A\sin\left(\sqrt{\frac{P}{EI}}x\right) + B\cos\left(\sqrt{\frac{P}{EI}}x\right).$$

The coefficients A and B are determined from the boundary conditions on the column.

EXAMPLE 13.17   The initial shape of a pin–pin supported column is known to be

$$w_0(x) = a \sin\left(\frac{\pi x}{L}\right).$$

Determine the displacement under an arbitrary load, $P > 0$.

Solution   In a pin–pin column, $B = 0$ and $A = 0$. The particular solution, $w_p(x)$, depends on $w_0(x)$. Assume that $w_p(x) = C \sin\left(\frac{\pi x}{L}\right)$. Substitution into the differential equation implies that

$$C\left[-EI\left(\frac{\pi}{L}\right)^2 + P\right]\sin\left(\frac{\pi x}{L}\right) = -Pa \sin\left(\frac{\pi x}{L}\right).$$

Solving for C yields

$$w_p(x) = \frac{-Pa}{-EI\left(\frac{\pi}{L}\right)^2 + P}\sin\left(\frac{\pi x}{L}\right).$$

The total deflection is then

$$w(x) = w_0(x) + w_p(x) = \left[a + \frac{-Pa}{-EI\left(\frac{\pi}{L}\right)^2 + P}\right]\sin\left(\frac{\pi x}{L}\right) = \left[\frac{a}{1 - \frac{P}{P_c}}\right]\sin\left(\frac{\pi x}{L}\right),$$

where $P_c = \pi^2 EI/L^2$. As P varies from zero up to P_c, the displacement amplitude increases until it is undefined at a load of $P = P_c$. The material would have yielded long before this load could be reached. This analysis assumes elastic loading. The results for loads significantly greater than zero are not valid, in any case, because the differential equation (13.42) used to describe the deflections is valid only for small dw/dx. ∎

13.7.2 The Eccentrically Loaded Column

A classical treatment of a column imperfection is the case of the axial load offset by a distance, e, called the eccentricity, from the centroid, C, of the cross section (Fig. 13.13).

FIGURE 13.13   Column with an eccentric axial load.

Such a pin–pin column can be treated by assuming there is a moment of eP applied at either end of a perfectly axially loaded column. The differential equation for the displacement, $w(x)$, is

$$EI\frac{d^2w}{dx^2} + Pw = eP. \qquad (13.43)$$

The displacement is the sum of the homogeneous and particular solutions,

$$w(x) = A \sin(kx) + B \cos(kx) + e, \qquad (13.44)$$

where $k = \sqrt{P/EI}$. The boundary condition $w(0) = 0$ implies that $B = -e$, and $w(L) = 0$ implies that $A = e[\cos(kL) - 1]/\sin(kL)$. Again, there is no critical load since the column begins to bend as soon as any compressive axial load is applied. However, the midcolumn deflection, b, can be obtained from the expression for $w(L/2)$

by use of the double-angle trigonometric identities, $\sin(kL) = 2\sin(kL/2)\cos(kL/2)$ and $\cos(kL) = 2\cos^2(kL/2) - 1$.

$$b = w(L/2) = e\left[1 - \sec\left(\sqrt{\frac{P}{EI}}\frac{L}{2}\right)\right]. \tag{13.45}$$

The midcolumn displacement becomes infinite if $P = P_c = \pi^2 EI/L^2$, the critical load for a perfect column. This predicted result is not matched by experiment when the column is still elastically loaded at P_c. This failure is a consequence of the linearization of the curvature when deriving the governing differential equation.

Exercise 13.13   Verify equation (13.45).

EXAMPLE 13.18   Graph the relationship between P and b/e for a pin–pin supported column of length, $L = 0.5$ meter, and flexural stiffness, $EI = 2$ N-m^2.

Solution   The graph is Figure 13.14.   ■

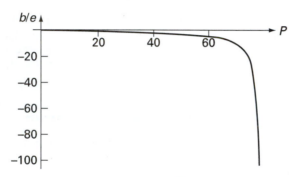

FIGURE 13.14   Graph of the relationship between P and b/e for a pin–pin supported column.

13.7.3   Imperfections Modeled in the Potential Energy

A more realistic model can be built from the potential energy function obtained by the perturbation analysis. The imperfection is modeled by a transverse load, F, applied at midcolumn (Fig. 13.15). This load is independent of the axial load, P.

FIGURE 13.15   Axially loaded column with imperfection load, F.

The potential energy of equation (13.33) is modified by adding the potential of the load, F.

$$V = \left[\frac{7}{128}\pi EID^5 - \frac{5}{128}P\pi D^3\right]b^4 + \left[\frac{1}{4}\pi EID^3 - \frac{1}{4}P\pi D\right]b^2 - Fb. \tag{13.46}$$

The equilibrium states are given by

$$\frac{dV}{db} = 4\left[\frac{7}{128}\pi EID^5 - \frac{5}{128}P\pi D^3\right]b^3 + 2\left[\frac{1}{4}\pi EID^3 - \frac{1}{4}P\pi D\right]b - F = 0.$$
(13.47)

EXAMPLE 13.19   A pin–pin supported column has length, $L = 0.5$ m, and flexural stiffness, $EI = 2$ N-m^2. Graph the equilibrium midcolumn displacements, b, as a function of P with $F = 0.1$.

Solution   The graph is Figure 13.16. The critical load is $P_c = 8\pi^2 \sim 79$. Note that the displacement is not infinite at P_c but that the column carries more load as P is increased.   ∎

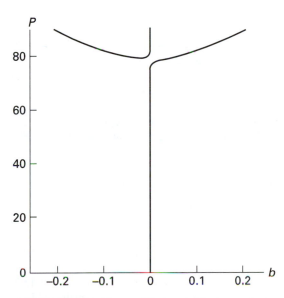

FIGURE 13.16   The equilibrium midcolumn displacements, in b-P-space, for a pin–pin column with $F = 0.1$.

The imperfection analysis is more significant in the case that the equilibrium state at the singularity of the energy function is unstable. The system of a rigid rod supported by linear springs, which was examined in Example 13.5, provides such an example. First consider a general one degree of freedom system for which the Taylor expansion to the fourth order of the potential energy has the form

$$V(x) = ax^4 + bx^2 - Fx,$$
(13.48)

where x is the state variable, the coefficients a and b depend on the load and other system parameters, and F is the imperfection. The coefficient a of x^4 is negative at the singularity. In this case, there are also singularities when $F \neq 0$. The equilibria are the solutions of

$$\frac{dV}{dx} = 4ax^3 + 2bx - F = 0.$$

Those equilibria which are also singularities are found by solving the previous equation simultaneously with

$$\frac{d^2V}{dx^2} = 12ax^2 + 2b = 0.$$

This implies that $x = \sqrt{b/-6a}$, which is defined whenever $b > 0$ and $a < 0$. Substitute in the equilibrium equation to obtain

$$4a\left(\frac{-b}{6a}\right)^{3/2} + b\left(\frac{-b}{6a}\right)^{1/2} - F = 0.$$

This expression can be rewritten to determine the critical value of the load, P, as a function of F since a and b are functions of the load,

$$\frac{8}{27}b^3 = -aF^2. \qquad (13.49)$$

This is the famous 2/3 power rule for the dependence of the critical load on the imperfection.

As the imperfection increases, the critical load decreases. In such structures, the presence of imperfections almost guarantees a disastrous collapse if the design is based on the critical load for the perfection structure in which $F = 0$. This type of investigation is called an imperfection sensitivity analysis.

EXAMPLE 13.20

Determine the postbuckling behavior of a rigid rod of length, $L = 50$ cm, supported by two horizontal springs each having spring constant, $k = 10$ N/cm (Fig. 13.6). The imperfection, F, is an initial moment about the pin.

Solution

The mass of the rod is neglected. The datum for the load, P, is taken at the pin. The horizontal displacement of the rod is $x = L \sin\theta$. The potential energy is $V(\theta; P) = kL^2 \sin^2\theta - PL(1 - \cos\theta) + F\theta$, for an imperfection F. For small angles, θ, the Taylor series about $\theta = 0$ is

$$V(\theta; P) = \left(\frac{PL}{24} - \frac{kL^2}{3}\right)\theta^4 + \left(kL^2 - \frac{1}{2}PL\right)\theta^2 - F\theta. \qquad (13.50)$$

Compare the exact analysis of this problem given in Section 13.2 where the exact form of the potential energy for the perfect system is given by equation (13.7).

In this example, $a = PL/24 - kL^2/3$ and $b = kL^2 - PL/2$. The critical load occurs at $P_c = 2kL$. Also $a < 0$ for $0 < P < 8kL$. By equation (13.49)

$$\frac{8}{27}\left(kL^2 - \frac{1}{2}PL\right)^3 = -\left(\frac{PL}{24} - \frac{kL^2}{3}\right)F^2. \qquad (13.51)$$

The relation between the critical values, P_c, and F is a cusp (Fig. 13.17).

Once the load reaches its critical value and the structure begins to buckle, the load must be reduced in order to maintain equilibrium. If it is held at too large a value, the structure collapses. Such behavior is not seen in those structures whose critical equilibrium state is stable. ∎

13.8 COMPOUND BUCKLING

More complex structures may buckle in several ways. Often, each is called a mode. The critical equilibrium states associated with each mode may occur at different loads for a given set of design parameters. The principle of simultaneous mode design (SMD) proposed by Shanley says that the optimal design is one in which all failure modes occur simultaneously under the action of the loads [2, p. 15]. This design optimization principle requires an engineer to adjust the dimensions or other system parameters so that all possible buckling modes occur at the same critical buckling load. This strategy only produces the optimal design for a perfect structure. Thompson and Hunt [3,

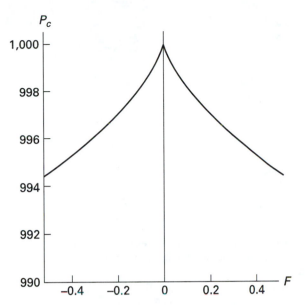

FIGURE 13.17   The critical load for a pin–pin column with imperfection, *F*.

p. 242] point out that this principle almost inevitably leads to an unstable critical equilibrium state. Therefore any imperfections in the structure can result in buckling occurring at a much lower load than the expected critical buckling load. To avoid this design catastrophe, two problems must be examined. First the behavior of coincident buckling modes, including the postbuckling behavior, must be analytically described. This will arise from a singularity analysis of the associated energy function. Second, a design strategy must be developed in terms of the expected imperfections. This should allow the expected load to be as large as possible for the expected range of imperfections.

Multimode buckling examples have been constructed from rigid links and springs. One of the most famous is the example proposed by G. Augusti in 1964 [3] to demonstrate the manner in which two stable critical points can coalesce into a single unstable critical point as a certain design parameter is varied.

The structure consists of a rigid rod supported by a ball and socket and two torsional springs at one end (Fig. 13.18). The rod supports a vertical dead load at its free end. The torsional springs initially lie in perpendicular planes and have torsional stiffnesses, c_1 and c_2. The rod is initially vertical but can rotate in three-dimensional space. The system is said to buckle if the rod rotates about the pin at its base. The buckling in each coordinate plane is controlled by the spring in that plane with critical load either c_1/L or c_2/L. Each spring separately with the rod produces a stable buckling behavior as shown by the two-dimensional Example 13.4 of a rod supported by a single torsional spring and allowed to rotate only in a single plane. Buckling in vertical planes other than the coordinate planes is determined by a combination of spring behavior.

If a designer tries to optimize the load-carrying ability of the perfect system by forcing the buckling loads for the two modes to be equal or coincident, secondary bifurcation points coalesce with the primary critical point. This changes a stable primary critical point into an unstable one. Imperfections are introduced by assuming that the springs are unstretched when the column is not in the vertical position. In this model, the imperfections are independent of one another.

In the Augusti example, the most dangerous case of postbuckling behavior occurs when the imperfections are taken to be equal and the state variables, the angles from

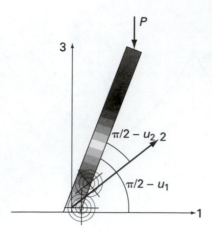

FIGURE 13.18 Augusti model: a load-bearing rod in three-dimensional space supported at one end by a ball and socket joint and two torsional springs, one in the 1-3 plane and one in the 2-3 plane.

the column to the vertical, are also equal. In this case, the buckling load and the initial imperfections obey the two-thirds power rule.

The complement to the angle between the rigid rod and the 1-axis is u_1 and the complement to the angle between the rigid rod and the 2-axis is u_2. The torsional spring constant for the spring initially in plane 1-3 is c_1 and for the spring initially in plane 2-3 is c_2. The imperfection in the system is taken to be such that when the springs are unstretched, the rod initially makes the angle, ϵ, in both coordinate planes. The potential energy is then dependent on the parameter, P.

$$V(u_1, u_2, P) = \frac{1}{2}c_1(u_1 - \epsilon)^2 + \frac{1}{2}c_2(u_2 - \epsilon)^2 + PL\sqrt{1 - \sin^2(u_1) - \sin^2(u_2)}.$$
(13.52)

The Taylor series expansion to the fourth order of the potential energy is

$$\begin{aligned} V(u_1, u_2) = &\frac{1}{2}c_1 u_1^2 + \frac{1}{2}c_2 u_2^2 \\ &+ PL\left[-\frac{1}{2}u_1^2 - \frac{1}{2}u_2^2 + \frac{1}{24}u_1^4 + \frac{1}{24}u_2^4 - \frac{1}{4}u_1^2 u_2^2 \right] \\ &- \epsilon(c_1 u_1 + c_2 u_2). \end{aligned}$$
(13.53)

The equilibria of the perfect system ($\epsilon = 0$) are the simultaneous solutions of

$$\frac{\partial V}{\partial u_1} = c_1 u_1 + PL\left(-u_1 + \frac{1}{6}u_1^3 - \frac{1}{2}u_1 u_2^2 \right) = 0;$$
(13.54)

$$\frac{\partial V}{\partial u_2} = c_2 u_2 + PL\left(-u_2 + \frac{1}{6}u_2^3 - \frac{1}{2}u_2 u_1^2 \right) = 0.$$
(13.55)

As P varies, the equilibrium states change along a one-dimensional curve in u_1-u_2-P space. Such a curve is called an equilibrium path. There are five paths satisfying equation (13.55). The fundamental path is defined by $u_1 = u_2 = 0$ for all positive P. The paths that are similar to the original pitchfork paths when the rod is viewed in a single plane are the uncoupled equilibrium paths defined by

$$u_2 = 0 \quad \text{and} \quad u_1^2 = 6\left(1 - \frac{c_1}{PL}\right); \tag{13.56}$$

$$u_1 = 0 \quad \text{and} \quad u_2^2 = 6\left(1 - \frac{c_1}{PL}\right). \tag{13.57}$$

There is also a pair of coupled paths lying over the 45° lines in the u_1-u_2 plane

$$c_1 + PL\left(-1 + \frac{1}{6}u_1^2 - \frac{1}{2}u_2^2\right) = 0 \tag{13.58}$$

and

$$c_2 + PL\left(-1 + \frac{1}{6}u_2^2 - \frac{1}{2}u_1^2\right) = 0. \tag{13.59}$$

The graphs of these equilibrium curves in u_1-u_2-P-space are shown in Figure 13.19.

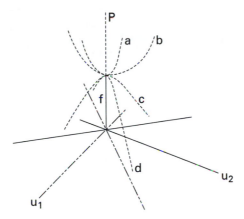

FIGURE 13.19   The equilibrium paths in u_1-u_2-P-space for the Augusti model.

Each equilibrium point on the fundamental path is unstable above the critical load. Those on the rising uncoupled paths are stable, and those on the falling coupled paths are unstable. Often engineers speak of the path itself as being stable or unstable depending on the behavior of the equilibria on it.

The presence of the unstable coupled paths implies that the structure is sensitive to imperfections. For some imperfections, collapse can be expected at loads lower than the critical load for the perfect system. In the case that the two spring constants are equal, the most severe imperfections lie in the 45° planes where $u_1 = u_2$. In this case, denote the equal spring constants by c and the equal angles by u, then the potential energy is

$$V(u; P) = -\frac{1}{6}u^4 + (c - PL)u^2 - 2c\epsilon u. \tag{13.60}$$

This is exactly the form studied in equation (13.48) of the previous section. The imperfection sensitivity satisfies the two-thirds power rule.

A designer may wish to optimize the structure under the assumption of a maximum possible imperfection. Because imperfections will likely be present in real structures, the allowable load cannot be the coincident buckling load for the perfect structure, as proposed in classical simultaneous mode design. The allowable imperfections and load are design parameters. An estimate must first be made of the maximum expected initial imperfection. The design entails calculating the minimum critical load for all allowable imperfections. This value is used as the maximum allowable load on the structure during design.

PROBLEMS

13.1 Determine the stability of the equilibria of a ball of mass m which can move on a surface whose height from a datum is $y = x^3 + 5$. Graph the potential function.

13.2 Determine the stability of the equilibria of a ball of mass m which can move on a surface whose height from a datum is $y = 7 - x^4$. Graph the potential function.

13.3 The potential energy of a system is $\Pi(x) = x^5 + (1 - P)x^2$, for state variable x and load P. Determine the equilibrium states and their stability. Graph the equilibria in x-P space. Locate any bifurcation points.

13.4 The potential energy of a system is $\Pi(x) = x^6 + (1 - P)x^2$, for state variable x and load P. Determine the equilibrium states and their stability. Graph the equilibria in x-P space. Locate any bifurcation points.

13.5 A rigid bar is supported by a frictionless pin at its bottom and at its top by two horizontal springs adjusted so that the bar is in the vertical position when the springs are unstretched (Fig. 13.6). The bar has length 18 inches. Determine the equilibrium position for the spring constants, $k = 20$ lb-in., when the bar supports a load of 100 pounds. Solve this problem by

(a) energy methods;

(b) by statics. Notice that the bar is not required to remain in the vertical position to carry the load.

13.6 A fixed-free linear elastic column has a circular cross section. Determine the minimum slenderness ratio required to prevent both buckling and yielding in the column, which has fixed length L and is made of aluminum ($E = 70$ GPa, $\sigma_Y = 100$ MPa).

13.7 Verify that the critical load for a fixed-pinned linear elastic column is $P_c = \pi^2 EI/(0.699L)^2$. Determine its effective length.

13.8 Verify that the critical load for a fixed-fixed linear elastic column is $P_c = 4\pi^2 EI/L^2$. Determine its effective length.

13.9 Determine the radius of gyration about the centroidal axis parallel to the base for an equilateral triangle.

13.10 Use the Rayleigh-Ritz method to determine the critical load for fixed-free column (i.e., cantilevered) by setting $w(x) = ax^2$. Compare it to the exact solution.

13.11 Verify that, if the coordinate system is chosen with the origin at the fixed end for a fixed-free column, then the buckled displacement function is $w(x) = A[1 - \cos(\sqrt{P/EI}x)]$. *Hint*: Write the bending moment at each point in the column

in terms of the displacement of the free end and the displacement at the point.

13.12 In the three member truss shown, determine the minimum cross-sectional dimensions required to prevent both buckling and yielding in member BC, which is made of aluminum ($E = 70$ GPa, $\sigma_Y = 100$ MPa). Do this for a square cross section and for a circular cross section. Which cross section requires the least amount of material in member BC and therefore gives the lighter structure?

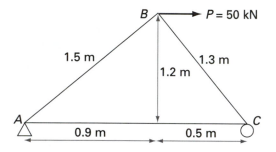

13.13 Complete the calculation of the midcolumn displacement using elliptical integrals for a pin–pin column supporting a load $P/P_c = 1.05$. The column has length $L = 0.5$ m and flexural stiffness $EI = 2$ N-m^2.

13.14 Carry out a perturbation analysis of a fixed-free linear elastic column to show that $w_1(x) = \cos(\pi x/2L)$, $P_2 = EI(\pi/2L)^4/8$, and $w_3(x) = -(\pi/2L)^2[\cos(\pi x/2L) - \cos(3\pi x/2L)]$. The coordinate x is measured from the free end.

13.15 Use the results in the previous problem to make a second-order estimation of the deflection of the free end of a fixed-free column with length $L = 0.5$ m and flexural stiffness $EI = 2$ N-m^2 when the column is subjected to a load of $P = 1.05P_c$. Compare your answer to that obtained in equation (13.25).

13.16 Determine the critical buckling stress for a fixed-free column with slenderness ratio, $L/r = 180$. The column is made of a material with stress-strain curve, $\sigma = E_1\epsilon + E_3\epsilon^3$, where $E_1 = 8.72$ GPa and $E_3 = -2,590$ GPa.

13.17 Determine the critical buckling load for a pin–pin supported column made of a material with a cubic stress-strain curve $\sigma = E\epsilon - E_3\epsilon^3$.

13.18 Use the Engesser relation to determine the critical buckling stress for a pin–pin supported column with slenderness ratio, $L/r = 60$. The column is made of redwood with stress-strain curve, $\sigma = E_1\epsilon + E_3\epsilon^3$, where $E_1 = 8.72$

GPa and $E_3 = -2,590$ GPa. *Hint*: First solve for ϵ_c from the Engesser relation.

13.19 A pipe is made of a steel for which $E = 30 \times 10^6$ psi and $\nu = 0.25$. Its radius is $R = 8$ inches and the wall thickness is $t = 0.5$ inches. The pipe must support an external hydrostatic pressure. Determine the critical value of the external hydrostatic pressure for buckling.

13.20 Design the minimum wall thickness for a pipe made of a steel for which $E = 30 \times 10^6$ psi and $\nu = 0.25$. The radius must be $R = 8$ inches and the pipe must support an external hydrostatic pressure of $p = 200$ psi. Assume a factor of safety of 2.

13.21 A pin–pin linear elastic column is initially imperfect with a shape given by $w_0(x) = 0.05 \sin\left(\frac{\pi x}{L}\right)$ cm. Its length is 3 meters, the circular cross section has radius 1.5 cm, and it is made of a steel with $E = 210$ GPa. Determine the deformed shape under an axial load of 200 N.

13.22 Determine the midcolumn displacement of a pin–pin supported linear elastic column of length, $L = 0.5$ m, and flexural stiffness, $EI = 2$ N-m^2, when an axial load of $P = 40$ N is applied at an eccentricity of $e = 0.001$ m.

13.23 Determine the critical load for a rigid rod of length $L = 50$ cm supported at the bottom by a frictionless pin and supported at the top by two horizontal springs each having spring constant $k = 10$ N/cm. The imperfection, $F = 0.25$ N-m, is an initial moment about the pin. Compare this critical load to that for the perfect system.

13.24 Determine the maximum permissible load on the Augusti structure with equal spring constants and equal imperfections, ϵ in u_1 and u_2 in the case that the maximum allowable imperfection is 0.1 radian. The length of the rod is $L = 10$ inches and the spring constants are both 15 lb.in./radian.

REFERENCES

[1] H. W. HASLACH, JR., Post-buckling behavior of columns with non-linear constitutive equations, *International Journal of Non-linear Mechanics*, 20, 53–67 (1985).

[2] L. SPUNT, *Optimal Structural Design*, Prentice-Hall, Englewood Cliffs, NJ, 1971.

[3] J. M. T. THOMPSON AND G. W. HUNT, *A General Theory of Elastic Stability*, Wiley, New York, 1973.

[4] J. M. T. THOMPSON AND G. W. HUNT, *Elastic Instability Phenomena*, Wiley, New York, 1984.

[5] S. P. TIMOSHENKO AND J. M. GERE, *Theory of Elastic Stability*, 2nd ed., McGraw-Hill, New York, 1961.

CURVATURE

A.1 THE CURVATURE OF A CURVE

The curvature measures the bending of a curve in two- or three-dimensional space. This idea can also be extended to define curvatures for a two-dimensional surface lying in three-space.

A.1.1 Curves Lying in Two-Dimensional Space

Suppose that the curve, $y = f(x)$, is reparametrized by the arclength, s, in a particular direction along the curve. Let $\mathbf{T}$ be the unit tangent vector to the curve in the direction of s. The angle, ϕ, between the x-axis and $\mathbf{T}$ is a measure of the direction of the curve. The curvature, κ, is defined by

$$\kappa = \frac{d\phi}{ds}. \tag{A.1}$$

Using the fact that $\tan \phi = dy/dx$, the curvature can be written in terms of $f(x)$ as

$$\kappa = \frac{d\phi}{ds} = \frac{d\phi/dx}{ds/dx} = \frac{\pm \frac{d^2 y}{dx^2}}{\left[1 + \left(\frac{dy}{dx}\right)^2\right]^{3/2}}. \tag{A.2}$$

The sign of κ is positive if ds/dx is positive. This means that if one visualizes moving along the curve in the direction of positive s, the curvature is positive if the concave side of the curve lies to the left.

The strength of materials beam stress, $\sigma = Mv/I$, where v is the distance from the neutral axis, is obtained by approximating the strain as $\epsilon = \kappa v$. By assuming that the beam neutral axis deflection, $y(x)$, is small so that $dy/dx \sim 0$, the curvature is shown to be approximately,

$$\kappa \sim \frac{d^2 y}{dx^2}.$$

Then because the internal moment, M, is proportional to $d^2 y/dx^2$, so is the longitudinal stress in the beam.

The unit vector, $\mathbf{N}$, is obtained by rotating the unit tangent vector, $\mathbf{T}$, counterclockwise through $90°$. Then

$$\frac{d\mathbf{T}}{d\phi} = \mathbf{N}. \tag{A.3}$$

Therefore by the chain rule,

$$\frac{d\mathbf{T}}{ds} = \kappa\mathbf{N}. \tag{A.4}$$

This vector always points toward the concave side of the curve.

The reciprocal of the curvature is the radius of curvature, $\rho = 1/\kappa$. At a given point, the radius of curvature is the radius of a circle tangent to the curve at the point; the center of the circle of curvature always lies on the concave side of the curve (Fig. A.1).

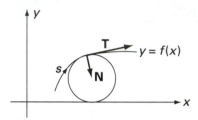

FIGURE A.1   The circle of curvature at a point on the curve, $y = f(x)$.

A.1.2   Curves Lying in Three-Dimensional Space

These ideas can be generalized to the case of a curve lying in three-dimensional Cartesian space,

$$\mathbf{R} = f(s)\mathbf{i} + g(s)\mathbf{j} + h(s)\mathbf{k}. \tag{A.5}$$

The unit tangent vector is given by

$$\frac{d\mathbf{R}}{ds} = \mathbf{T}. \tag{A.6}$$

The curvature is then given as the magnitude of the vector normal to $\mathbf{T}$,

$$\frac{d\mathbf{T}}{ds} = \kappa\mathbf{N}, \tag{A.7}$$

where $\mathbf{N}$ is a unit vector. The binormal vector, $\mathbf{B}$, is defined as the unit vector,

$$\mathbf{B} = \mathbf{T} \times \mathbf{N}. \tag{A.8}$$

The three vectors, $\mathbf{T}$, $\mathbf{N}$, and $\mathbf{B}$ define a moving coordinate system at each point of the curve.

A.1.3   Parametric Form

It is usually quite difficult to calculate the tangent and normal unit vectors and the curvature in terms of the arc length, s. The calculation would be better done in terms of the parameter defining the curve. Assume that the path is a function of a parameter, t, perhaps time, so that

$$\mathbf{R}(t) = x(t)\mathbf{i} + y(t)\mathbf{j} + z(t)\mathbf{k}. \tag{A.9}$$

The velocity vector, which is tangent to the curve, is

$$\mathbf{v}(t) = \dot{x}(t)\mathbf{i} + \dot{y}(t)\mathbf{j} + \dot{z}(t)\mathbf{k}, \tag{A.10}$$

and its magnitude is the speed, $\dot{s} = ds/dt$. Then

$$\dot{s}(t) = \sqrt{\dot{x}(t)^2 + \dot{y}(t)^2 + \dot{z}(t)^2}. \tag{A.11}$$

The unit tangent vector to the curve can then be obtained from the velocity vector as

$$\mathbf{T}(t) = \frac{\mathbf{v}(t)}{\dot{s}(t)}. \tag{A.12}$$

The second derivative gives the unit normal vector and the curvature. The acceleration can be obtained from the velocity written as a multiple of the unit tangent vector, $\mathbf{v}(t) = \dot{s}(t)\mathbf{T}(t)$, so that

$$\mathbf{a}(t) = \ddot{s}(t)\mathbf{T}(t) + \dot{s}(t)\frac{d\mathbf{T}(t)}{ds}. \tag{A.13}$$

Then by equation (A.7),

$$\mathbf{a}(t) = \ddot{s}(t)\mathbf{T}(t) + \dot{s}(t)^2 \kappa \mathbf{N}(t). \tag{A.14}$$

The acceleration vector lies in the plane spanned by the unit tangent and unit normal vectors. This result implies that the curvature can be computed by projecting the acceleration vector onto the $\mathbf{N}$ direction and dividing by the square of the speed.

$$\kappa = \frac{\mathbf{a}(t) \cdot \mathbf{N}(t)}{\dot{s}(t)^2}. \tag{A.15}$$

Exercise A.1   Verify the previous equation.

A.2   SURFACES

A surface is a two-dimensional subspace of three-dimensional space. Portions of the surface can be written in parametric form, for parameters u and v.

$$\mathbf{R}(u, v) = x(u, v)\mathbf{i} + y(u, v)\mathbf{j} + z(u, v)\mathbf{k}. \tag{A.16}$$

A path lying on the surface can be defined by specifying u and v as functions of a parameter, t.

$$\mathbf{R}(t) = x(u(t), v(t))\mathbf{i} + y(u(t), v(t))\mathbf{j} + z(u(t), v(t))\mathbf{k}. \tag{A.17}$$

The velocity, acceleration, unit tangent and normal vectors, and curvature for this curve on the surface are defined exactly as above. However, to aid in computation, these definitions are best given in terms of the derivatives of u and v with respect to t.

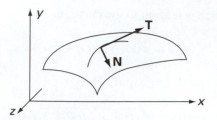

FIGURE A.2   Unit tangent and normal vectors to a curve on a surface in three-dimensional space.

A.2.1   The First Fundamental Form and the Speed Along a Path on a Surface

The velocity of a particle moving on a path along a surface, equation (A.17), is computed by the chain rule.

$$\dot{\mathbf{R}} = \left(\frac{dx}{du}\dot{u} + \frac{dx}{dv}\dot{v}\right)\mathbf{i} + \left(\frac{dy}{du}\dot{u} + \frac{dy}{dv}\dot{v}\right)\mathbf{j} + \left(\frac{dz}{du}\dot{u} + \frac{dz}{dv}\dot{v}\right)\mathbf{k}. \qquad (A.18)$$

This equation can be written in the matrix form, $\dot{\mathbf{R}} = \mathbf{A}(\dot{u} \quad \dot{v})^T$, where the superscript, T, denotes the transpose and

$$\mathbf{A} = \begin{pmatrix} \frac{\partial x}{\partial u} & \frac{\partial x}{\partial v} \\ \frac{\partial y}{\partial u} & \frac{\partial y}{\partial v} \\ \frac{\partial z}{\partial u} & \frac{\partial z}{\partial v} \end{pmatrix}. \qquad (A.19)$$

The speed is then

$$s = \left(\dot{\mathbf{R}}^T \cdot \dot{\mathbf{R}}\right)^{1/2} = \left((\dot{u} \quad \dot{v}) \quad \mathbf{G} \quad (\dot{u} \quad \dot{v})^T\right)^{1/2}, \qquad (A.20)$$

where $\mathbf{G}$ is the first fundamental matrix of the surface:

$$\begin{aligned} \mathbf{G} &= \begin{pmatrix} \left(\frac{\partial x}{\partial u}\right)^2 + \left(\frac{\partial y}{\partial u}\right)^2 + \left(\frac{\partial z}{\partial u}\right)^2 & \frac{\partial x}{\partial u}\frac{\partial x}{\partial v} + \frac{\partial y}{\partial u}\frac{\partial y}{\partial v} + \frac{\partial z}{\partial u}\frac{\partial z}{\partial v} \\ \frac{\partial x}{\partial u}\frac{\partial x}{\partial v} + \frac{\partial y}{\partial u}\frac{\partial y}{\partial v} + \frac{\partial z}{\partial u}\frac{\partial z}{\partial v} & \left(\frac{\partial x}{\partial v}\right)^2 + \left(\frac{\partial y}{\partial v}\right)^2 + \left(\frac{\partial z}{\partial v}\right)^2 \end{pmatrix} \\ &= \begin{pmatrix} \frac{\partial \mathbf{R}}{\partial u} \cdot \frac{\partial \mathbf{R}}{\partial u} & \frac{\partial \mathbf{R}}{\partial u} \cdot \frac{\partial \mathbf{R}}{\partial v} \\ \frac{\partial \mathbf{R}}{\partial v} \cdot \frac{\partial \mathbf{R}}{\partial u} & \frac{\partial \mathbf{R}}{\partial v} \cdot \frac{\partial \mathbf{R}}{\partial v} \end{pmatrix}. \end{aligned} \qquad (A.21)$$

A.2.2   The Curvature of a Surface

The curvature appears in the second derivative of the path lying on the surface [equation (A.17)]. Differentiate the velocity, equation (A.19), written as $\dot{\mathbf{R}} = \dot{s}\mathbf{T}$ to obtain

$\ddot{\mathbf{R}} = \ddot{s}\mathbf{T} + \dot{s}^2\kappa\mathbf{N}$. In terms of the components

$$\ddot{\mathbf{R}} = \left(\frac{\partial^2 x}{\partial u^2}\dot{u}^2 + 2\frac{\partial^2 x}{\partial u \partial v}\dot{u}\dot{v} + \frac{\partial^2 x}{\partial v^2}\dot{v}^2 + \frac{\partial x}{\partial u}\ddot{u} + \frac{\partial x}{\partial v}\ddot{v} \right) \mathbf{i}$$

$$+ \left(\frac{\partial^2 y}{\partial u^2}\dot{u}^2 + 2\frac{\partial^2 y}{\partial u \partial v}\dot{u}\dot{v} + \frac{\partial^2 y}{\partial v^2}\dot{v}^2 + \frac{\partial y}{\partial u}\ddot{u} + \frac{\partial y}{\partial v}\ddot{v} \right) \mathbf{j} \qquad \text{(A.22)}$$

$$+ \left(\frac{\partial^2 z}{\partial u^2}\dot{u}^2 + 2\frac{\partial^2 z}{\partial u \partial v}\dot{u}\dot{v} + \frac{\partial^2 z}{\partial v^2}\dot{v}^2 + \frac{\partial z}{\partial u}\ddot{u} + \frac{\partial z}{\partial v}\ddot{v} \right) \mathbf{k}.$$

Each tangent vector to the surface at a point in the surface determines a curve in the surface defined by the intersection of the surface with the plane spanned by the tangent and the normal, $\mathbf{n}$, to the surface at that point. The tangent vectors to the curves on the surface lying in the coordinate planes perpendicular to the u-v plane at the point are given by the partial derivatives,

$$\frac{\partial \mathbf{R}(u, v)}{\partial u} \quad \text{and} \quad \frac{\partial \mathbf{R}(u, v)}{\partial v}.$$

Therefore,

$$\mathbf{n} = \frac{\frac{\partial \mathbf{R}(u,v)}{\partial u} \times \frac{\partial \mathbf{R}(u,v)}{\partial v}}{\left| \frac{\partial \mathbf{R}(u,v)}{\partial u} \times \frac{\partial \mathbf{R}(u,v)}{\partial v} \right|}. \qquad \text{(A.23)}$$

The curvature of a curve lying in the surface is the magnitude of the component of the acceleration, $\ddot{\mathbf{R}}$, in the direction of the normal, $\mathbf{n}$, to the surface divided by the speed squared, $\dot{s}^2$.

$$\kappa = \frac{(\dot{u} \quad \dot{v}) \; \mathbf{D} \; (\dot{u} \quad \dot{v})^T}{(\dot{u} \quad \dot{v}) \; \mathbf{G} \; (\dot{u} \quad \dot{v})^T}, \qquad \text{(A.24)}$$

where the symmetric matrix, $\mathbf{D}$, is defined by

$$D = \begin{pmatrix} \mathbf{n} \cdot \frac{\partial^2 \mathbf{R}}{\partial u^2} & \mathbf{n} \cdot \frac{\partial^2 \mathbf{R}}{\partial u \partial v} \\ \mathbf{n} \cdot \frac{\partial^2 \mathbf{R}}{\partial v \partial u} & \mathbf{n} \cdot \frac{\partial^2 \mathbf{R}}{\partial v^2} \end{pmatrix}.$$

Definition   The principal curvatures of the surface at a point, P, on the surface are the maximum and minimum of the set of curvatures of all paths lying on the surface at P.

The product of the principal curvatures is called the Gaussian curvature of the surface.

EXAMPLE A.1   The sphere has curves of principal curvature formed by the intersection of the sphere with any plane passing through the center. The radius of curvature is the radius of the sphere.   ■

EXAMPLE A.2   A torus has orthogonal curves of principal curvature at any point (Fig. A.3.).   ■

The curvature corresponding to a given tangent vector at a point on a surface is defined as the curvature of the curve obtained by intersecting the surface with the plane containing the given tangent vector and the normal to the surface. There is one curvature for each possible tangent vector to the surface at the point. The principal curvatures are the maximum and minimum of this set of all possible curvatures. If

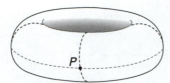

FIGURE A.3   Circles of principal curvature on a torus.

the two principal curvatures are unequal, then their corresponding tangent vectors are perpendicular, that is, the principal paths (those with principal curvature) on the surface are orthogonal.

REFERENCES

[1] I. D. Faux and M. J. Pratt, *Computational Geometry for Design and Manufacture*, Halsted Press, Ellis Horwood, Chichester, UK (Wiley, New York), 1979.

INDEX

A

Addition, 105
Adiabatic, 229
Angle of twist, 173, 174
Airy stress function,
    Cartesian, 125, 126, 127, 136
    polar, 147, 396
Annealing, 331
Arrhenius relation, 80, 110, 377, 461, 462, 465, 466
Atomic bond energy, 60
Atomic bond forces, 61
Atomic bonding, 55
    covalent, 56
    ionic, 55
    metallic, 56
    van der Waals, 56
Axisymmetric 137

B

Backstress, 323, 328, 357, 366, 376
Basis, 8
Basquin equation, 431
Bauschinger effect, 321, 357, 366
Beam differential equation, 194
Beams on an elastic foundation, 193
    infinite beam, 195
    semi-infinite beam, 199
Bifurcation, 481, 485
    cusp, 514
    limit point, 486
    pitchfork, 487
    stability, 505
Biharmonic equation,
    Cartesian, 127, 129
    polar, 148, 170
Biological tissue, 100
Boundary condition 122
    forced, 260
    natural, 260, 265
Bravais, 57
Brittle metal, 67
Buckling 481
    Augusti example, 515
    circular rings and tubes, 507
    columns, 489

    compound, 514
    post-buckling, 481, 494
Bulk modulus of elasticity, 73
Burgers vector, 67, 87, 149

C

Calculus of variations, 253, 258
Castigliano's theorem, 227, 231, 245, 434
Cauchy, 1, 9, 31
Cauchy stress tetrahedron, 10
Cavitation, 477
Ceramics, 73
    ionic, 73
    covalent, 73
Characteristic equation, 19, 351
Clapeyron's theorem, 230
Close-packed, 324, 325
Coefficient of thermal expansion, 74, 105, 135, 143
Collagen, 83
Column,
    eccentrically loaded, 511
    inextensible, 490, 500
    initially curved, 510
    linear elastic, 489
        critical buckling load,
            fixed-free, 492
            pin-pin, 491
        effective length, 492
    non constant cross section, 496
    nonlinear elastic,
        Engesser tangent modulus theory, 504
    post-buckling stability, 500
    radius of gyration, 493
    slenderness ratio, 493
Complementary energy, 228, 230, 231
Complementary energy density, 84, 85, 230, 231
Configuration, 242
Conservative force, 228
Conservative system, 244, 245, 482
Concrete, 90
Condensation, 105
Considère diagram, 53
Constitutive equations, 49, 288
Contact stress, 159
Control variable, 484

Convex set, 340, 357
Crack,
    bridging, 447
    dislocation emission, 413
    front, 383
    initiation, 416
    plastic zone, 406
    propagation, 383
    root, 383
    shielding, 413, 447
    stability, 390
Crack tip opening displacement, 409, 410
Crazing, 421, 447
Creep, 109, 457
    Coble, 462
    diffusional, 461, 462
    dislocation climb, 463, 472
    dislocation glide, 463, 471
    grain boundary sliding, 463
    moisture accelerated, 475
    Nabarro-Herring, 462
    viscous, 463
Creep fracture, 476
    C*-integral, 477
Creep limit, 477
Creep models,
    Andrade, 469
    Arrhenius, 465
    exponential law, 468
    logarithmic law, 470
    Nutting, 469
    hyperbolic sine law, 468, 471
    power law, 468, 477
Creep rate, 457
Critical resolved shear stress, 327
Cumulative plastic strain, 361
Curvature, 160, 266, 267, 490, 500, 507, 521
Curved beams in bending, 183, 234
    elasticity solution, 183
    Winkler-Bach solution, 187
    Castigliano's theorem solution, 234

D

Degrees of freedom, 242
    infinite, 252, 258
Deviatoric stress tensor, 88, 89, 340, 351
Dilatation, 72, 73, 88
Dilatational stress tensor, 88
Direction cosine, 16
Dislocation 64, 65
    edge, 64, 148
    screw, 64, 86
Dislocation pile-up, 328, 416, 417
Dissipation function, 362
Distortional strain energy density, 88, 345

Double cross-slip mechanism, 328
Drucker, D. C., 84
Ductile metal, 67
Durability, 421, 467

E

Eigenvalues, 19
Eigenvectors, 19
Einstein summation convention, 18
Elastic limit, 51
Elastic modulus, 50
Elastic-perfectly plastic, 320, 333, 358
Elastic spring-back, 335, 337
Elliptical coordinates, 391
Elliptical integral, 498
Endurance limit, 430
Energy methods, 227
Energy release rate, 386, 387, 394
Equations of compatibility, 41, 42
Equilibrium,
    stable, 247, 481
    unstable, 247
Euler, L., 490
Euler equation, 141, 151, 184, 272
    nonhomogeneous Euler equation, 146
Euler-Lagrange equation, 253, 258, 260, 264, 500
Eulerian coordinate system, 41
Exact line integral, 228, 244

F

Failure criterion,
    maximum normal stress (Rankine), 342
    maximum distortion energy (von Mises), 345
    maximum octahedral shear stress, 347
    maximum principal normal strain (Saint Venant), 344
    maximum shear stress (Tresca), 343
Failure envelope, 420
Fatigue, 428
    completely reversed load, 428
    Gerber criterion, 433
    Goodman criterion, 432
    purely cyclic load, 429
    random load, 429
    SAE criterion, 433
    Soderberg criterion, 432
Fatigue fracture, 438
Fibered composite, 81
Fick diffusion, 461
Fictitious loads, 236
Finite element method, 283
    area coordinates, 301
    assembly, 291, 293
    beam element, 294
    chapeau function, 286
    compatibility, 285, 295

condition number, 304
constant strain triangle, 296
energy norm, 305
error estimation, 303
Hilbert matrix, 304
h-version, 284
ill-conditioned matrix, 304
invariance, 285
isoparametric element, 302
locking, 304
nodal displacements, 284, 285, 289
nodes, 283
projects, 306
p-version, 284
mesh, 284
stiffness matrix,
    element, 289
    global, 291
strains, 286
truncation error, 305
First law of thermodynamics, 229
First variation, 244, 259
Flaw detection, 424
Four point bending test, 77
Fourier's inequality, 243
Fracture, 383
    brittle, 383, 384
    cleavage, 414
    ductile, 384, 405, 415
    essential work of, 413
    hole-joining, 415
    intercrystalline, 415
    mode I, II, III, 395
    resistance, 390, 412
Frank, F. C., 64
Frank-Read mechanism, 328
Freed-Walker model, 377
Fung model, 100, 251

G
Generalized coordinate, 242, 253
Glass, 79
Glass transition temperature, 81
Gough, J., 94
Grains, 59, 323
    subgrains, 323, 372
Green, G., 83, 84
Griffith, A. A., 384
Griffith criterion, 384, 385

H
Hall-Petch relation, 329, 370, 416, 432
Hardening exponent, 52, 54, 338, 370
Hardness,
    Brinell, 368

Knoop, 370, 374
Meyer, 368
Mohs, 374
Rockwell, 368
scratch, 374
Vickers, 370, 371, 374
Hardness test, 165, 367
Hardening in cyclic loading, 322
Hardening model, 337
Helical compression springs, 433
Hertz, H. R., 159
Hertzian contact stress, 164
Hessian, 482
Homologous temperature, 458, 459, 470, 476
Hookean material, 85
Hooke's law, xii
    Cartesian, 67, 70, 85
    polar, 139
Hydrophilic, 106, 112, 475
Hydrostatic pressure, 73
Hydrostatic stress, 88, 89, 340
Hyperbolic tangent model, 113, 114
Hyperelastic, 84

I
Imperfections, 510
    two-thirds power rule, 514, 517
Incompressible material, 73, 97
Interference fit, 142
Internal state variable, 361, 376
Invariants of a matrix, 19
Irwin, G., 384, 394
Isotropic, 70
Isotropic hardening, 357, 363, 376

J
J-integral, 412, 477
Joule, J. P., 94

K
Kelvin-Voigt model, 109, 110, 111, 112, 473
Kinematic hardening, 323, 357, 366, 376
Kinetic energy, 229
    law of, 229
Kinetic theory for rubber, 95
Kirchhoff, G. R., 265
Kronecker delta, 17

L
Lagrange multiplier, 262
Lagrangian coordinate system, 41
Lamé constants, 72, 125
Lamé equations, 72, 86
Laplace equation, 129, 130, 179

Laplace operator in polar coordinates, 148
Larson-Miller parameter, 466
Lattice friction stress, 326, 416
Lattice structures, 57
    body-centered cubic (BCC), 58
    cesium chloride structure, 75
    diamond cubic, 76
    face-centered cubic (FCC), 58
    hexagonal close-packed (HCP), 58
    sodium chloride structure, 74
    zinc blende structure, 75
Least squares, 256
Lévy-Mises model, 358
Limit point, 486
Linear transformation, 8
Linearly independent, 7
Loss modulus, 111, 112

M

Manson-Coffin relation, 451, 452
Matrix of direction cosines, 17
Maxwell model, 108, 110, 473
Mechanosorptive, 475
Mer, 101
Miller indices, 65, 66
Mises stress, 27, 350
Mises yield criterion, 345, 347, 352, 409, 436, 471
Moisture accelerated creep, 475
Moisture adsorption, 106
Mooney-Rivlin model, 97, 98, 249

N

Navier, 1, 31, 51, 268
Neo-Hookean model, 97, 98, 249, 250
Newton's second law, 229
Nondestructive testing, 424

O

Octahedral plane, 5, 25
Optical fiber, 142
Orthogonal functions, 256
Orthogonal matrix, 17, 28
Orthotropic, 68, 112, 113, 114, 115, 128, 129

P

Paper, 114
Paris, P. C., 440
Paris model, 442, 446
Percent coldworking, 339
Perturbation analysis, 501, 512
Pitchfork bifurcation, 487
Plane strain, 37, 124
Plane stress, 12, 124
Plastic instability, 54

Plasticity, 316
    associated, 362
    deformation theory, 358
    flow theory, 358
    incremental theory, 360
Plastic strain, 52
Plate,
    circular, 271
    flexural rigidity of, 267
    Fourier series solution, 268, 275
    Lévy solution, 271, 276
    orthotropic, 274
    rectangular, 267
    transversely loaded, 265
Plate with a circular hole, 149
Plate with an elliptical hole, 391
Poisson, S. D., 1, 55, 265
Poisson's equation, 176, 222
Poisson's ratio, 54, 68, 69, 79, 81, 82, 83, 91, 94, 113, 116
Polar coordinates, 36, 137
Polymer, 101
Porosity, 78
Potential energy, 229, 244
Prandtl function, 174, 176, 179, 222, 223
Prandtl-Reuss model, 360, 363, 364
Pressurized nonlinear elastic spheres, 248
Principal coordinates, 20
Principal stresses, 18
Principle of,
    complementary energy, 227, 231
        nonlinear elastic, 240
    stationary potential energy, 227, 244, 245, 283, 287, 481
    virtual work, 227, 242, 244, 283
Prism, 173
Product area moment of inertia, 203
    parallel axis theorem, 204
    principal axes, 204
    tensor, 204
Proportional limit, 51
Proportional loading, 354, 364, 435
Positive definite matrix, 482

R

Radius of curvature, 160
    principal, 160
Radius of gyration, 493
Ramberg-Osgood model, 337, 477
Ratchetting, 323
Rate of strain tensor, 359
Rayleigh-Ritz approximation, 253, 284, 288, 494
Relaxation, 472
Relaxation time, 110, 472, 474
Residual stress, 143
Ritz, W., 255
Roots of a cubic equation, 20

Rotating disks and cylinders, 145
Rotation matrix, 17, 204
Rubber, 93
Ruge, A., 42
Rule of mixtures, 82, 83, 129

S

Saccular aneurysm, 250
Saint Venant, B., 174, 318, 344
Schmid's law, 5, 326
Scission, 421
Semi-inverse method, 130
Separation of variables, 130
Shakedown, 322
Shear center, 205, 210
Shear flow, 211
   closed section, 215
Similar matrices, 28
Simmon, Jr., E. E., 42
Simple shear, 34, 353
Singularity, 87, 395, 484
Singular transformation, 19
Slip band, 328
Small angle grain boundary, 59
S-N curve, 429
Softening in cyclic loading, 322
Stacking fault, 325
Standard linear solid, 109, 474
State variable, 484
Statically indeterminate, 238
Storage modulus, 111, 112
Strain,
   engineering strain, 31
   homogeneous strain, 35
   microstrain, 32
   nominal strain, 53
   normal strain, 32, 33, 34
   plane strain, 37
   plastic strain, 52, 320, 332
   shear strain, 33, 34
   strain tensor, 37
   true, 53
Strain concentration factor, 156
Strain control in cyclic loading, 321
Strain-displacement relations,
   Cartesian, 34
   cylindrical, 36
   large displacements, 36
   polar, 138
Strain energy, 84, 228, 230
   of a dislocation, 87, 149
Strain energy density, 84, 85, 230
Strain hardening, 319, 331
Strain hardening exponent, 52, 54, 338
Standard linear solid, 109

Strength, 51
Strengthening,
   age-hardening, 332
   deformation, 331
   dispersion, 332
   solidification, 331
   solid solution, 331
Stress, 2
Stress at a point, 3
Stress compatibility equations, 126, 184
Stress concentration, 152, 153
Stress concentration factor, 152, 156
Stress control in cyclic loading, 321
Stress equilibrium equations,
   Cartesian, 14
   polar, 138, 184
Stress intensity factor, 394, 402, 440
   effective, 407
Stress relaxation, 109
Stress tensor, 7
Stretch, 97
Subgrain, 59, 65, 464
Surface tension, 384
Surface energy, 384, 388, 405
Symmetric matrix, 12, 20

T

Tangent modulus, 111, 504
Taylor, G. I., 64
Tearing of rubber, 419
Temperature,
   homologous, 458, 459, 470, 476
   melting, 458
Tensor, 18
Thick-walled cylinder, 140
   closed ends, 144
Thin-walled beams in bending, 209
   open sections, 209
   closed sections, 215
Three point bending test, 77, 403
Timoshenko beam, 36
Torsion of prismatic rods,
   solid, 173
   thin-walled,
      closed sections, 219
      open sections, 222
Transpose matrix, 12
Transversely isotropic, 69
Tresca, H., 318, 319, 333, 339
Tresca yield condition, 343, 347, 356, 409
True strain, 53
True stress, 53
Twinning, 331, 438
Two-stage slip, 325

U

Ultimate stress, 52
Unsymmetrical bending of beams, 205

V

Vacancy concentration, 462
Variation, 258, 288
Variational principle, 227
Vector space, 7
Virtual displacement, 243, 288
Virtual work, 243, 288
Viscoelasticity, 107, 111, 112, 448
Viscoplasticity, 375, 465, 471
Volume fraction, 82
Volumetric strain energy density, 88
Von Kármán, 266
Vulcanization, 95

W

Walker model, 442
Winkler-Bach curved beam, 187
Work, 228
Work hardening, 328

Y

Yield criterion, 341
Yield strength, 51, 320
Yield surface, 340, 341
Young, T., 51

Z

Zerilli-Armstrong model, 375
Zwicky, F., 62